Universitext

For other titles published in this series, go to
www.springer.com/series/223

Béla Sz.-Nagy • Ciprian Foias
Hari Bercovici • László Kérchy

Harmonic Analysis of Operators on Hilbert Space

Second Edition

Springer

Béla Sz.-Nagy
(Deceased)

Ciprian Foias
Mathematics Department
Texas A & M University
College Station, TX 77843-3368
USA
foias@math.tamu.edu

Hari Bercovici
Mathematics Department
Indiana University
Bloomington, IN 47405
USA
bercovic@indiana.edu

László Kérchy
Bolyai Institute
Szeged University
H-6720 Szeged
Hungary
kerchy@math.u-szeged.hu

ISBN 978-1-4419-6093-1 e-ISBN 978-1-4419-6094-8
DOI 10.1007/978-1-4419-6094-8
Springer New York Dordrecht Heidelberg London

Library of Congress Control Number: 2010934634

Mathematics Subject Classification (2010): 47A45

Printed on acid-free paper

Springer is part of Springer Science+Business Media (www.springer.com)

Foreword

Sz.-Nagy and Foias had been planning for several years to issue an updated edition of their book *Harmonic Analysis of Operators on Hilbert Space* (North-Holland and Akadémiai Kiadó, Amsterdam–Budapest, 1970). This plan was not realized due to Sz.-Nagy's death in 1998. Sz.-Nagy's idea was to include all developments related to dilation theory and commutant lifting. Because there are several other volumes dedicated to some of these developments, we have decided to include in this volume only those subjects that are organically related to the original contents of the book. Thus, the study of $C_{1\cdot}$-contractions and their invariant subspaces in Chap. IX has its origins in Sec. VII.5, while the theory presented in Chap. X completes the study started in Secs. III.4 and IX.4 of the English edition.

The material in the English edition has been reorganized to some extent. The material in the original Chaps. I–VIII was mostly preserved, but the results in the original Chap. IX were dispersed throughout the book. We have added to several chapters a section titled *Further results*, where we discuss some developments related to the material of the corresponding chapter. The selection of topics was dictated by the authors' knowledge, and by space limitations. Many significant results are certainly omitted, and only some of these are listed in the bibliography. We apologize to those authors whose work did not receive proper mention.

Part of the work on this volume was performed during a semester visit by L. Kérchy to Texas A&M University. He wishes to express his gratitude to the Mathematics Department for its hospitality, and to acknowledge additional support from Hungarian research grant OTKA no. K75488. A first version of Chapters I–VIII was expertly typeset by Mrs. Robin Campbell. The authors extend their gratitude to her, as well as the Mathematics Department at Texas A&M University, for their support throughout this project. Jenő Hegedűs kindly translated the foreword to the Russian edition.

Béla Szőkefalvi-Nagy served as a mentor to all three authors. He influenced us through the clarity of his mathematical insight, and through his insistence that published results should answer the highest standards of originality, beauty, and exposition. We dedicate this edition to his memory.

College Station, Bloomington, and Szeged, June, 2009

C. Foias, H. Bercovici, and L. Kérchy

Foreword to the French Edition

In the theory of operators on Hilbert space, definitive results have long been known for self-adjoint, unitary, and normal operators—special types of operators, but types which are especially important in different branches of mathematics and theoretical physics. The theory of nonnormal operators, although also initiated a long time ago, using different methods, has not yet attained any such definitive form. The recent rapid progress in this field was stimulated largely by work of mathematicians in the USSR (M. G. Kreĭn, M. S. Livšic, M. S. Brodskiĭ, etc.) and in the United States (N. Wiener, H. Helson, D. Lowdenslager, P. Masani, etc.). The central concern of the first group was with characteristic functions of operators and the triangular models of operators obtained from them whereas the work of the second group was inspired primarily by prediction theory for stationary stochastic processes. But there is also a third research direction which started from the theorem on unitary dilations of contractions on Hilbert space (Sz.-Nagy, 1953) and was pursued by the authors of the present monograph and others (M. Schreiber, I. Halperin, H. Langer, W. Mlak, etc.). This last research direction has led, for instance, to an effective functional calculus for Hilbert space contractions. It also unifies, in a certain sense, the other two research directions. Thus the characteristic function of a contraction T appears in this study in an altogether natural way, namely by the "harmonic analysis" (or "Fourier analysis") of the unitary dilation of T, and this in turn was inspired by prediction theory.

The purpose of the present monograph is to give a detailed exposition of the information about a contraction that can be obtained from consideration of its unitary dilation.

Chapter I develops the fundamentals of the theory of isometric and unitary dilations, deriving these by several different methods. Most important are the dilations of semigroups with one generator, either discrete ($\{T^n\}, n = 0, 1, \ldots$) or continuous ($\{T(s)\}, 0 \leq s < \infty$). These are used throughout what follows. We also treat dilations of discrete commutative semigroups with several generators; here there are some beautiful and definitive results, but also some difficult unsolved problems. These results (Secs. 6 and 9) are not essential for the reader of the rest of the book.

In Chap. II we establish some geometric and spectral properties of the unitary dilation of a contraction T (or equivalently, of the discrete contraction semigroup $\{T^n\}$). Contractions are classified in terms of the asymptotic behavior of the powers of T and its adjoint T^*. The important notions of quasi-affinity and quasi-similarity are introduced. In Sec. 5 we prove the existence of an abundance of invariant subspaces for certain types of operators, a subject to which we return, with more powerful methods, in Chap. VII.

In Chaps. III and IV we develop a functional calculus for contractions T, based on applying spectral theory to the unitary dilation of T. The relevant functions are analytic on the unit disc, in particular the class of bounded analytic functions. A. Beurling's arithmetic of inner functions plays an essential role in connection with the "minimal functions" of contractions belonging to what we call the class C_0. Outer functions also play a key part in this calculus, especially in extending it to certain classes of analytic functions unbounded on the unit disc. Important applications are to continuous semigroups of contractions (considered as functions of their "cogenerators"), and to functions of accretive and dissipative operators, bounded or not (studied by use of their Cayley transforms). We define and analyze fractional powers of accretive operators, providing an illustration of the methods in a special case that has importance in its own right.

Chapter V, which is independent of the preceding chapters, sets forth the ideas and general theorems of the theory of operator-valued analytic functions. This material (except for Secs. 5 and 8) is used throughout the rest of the book. In particular, we establish the existence and properties of factorizations of these functions. Fundamental in this whole development are two lemmas (Sec. 3) on Fourier representations of Hilbert spaces and certain operators on them, with respect to bilateral or unilateral shifts on the spaces.

The characteristic function of a contraction T makes its appearance in Chap. VI, as the operator-valued analytic function corresponding to a certain orthogonal projection in the space of the unitary dilation of T, when this space is given its Fourier representation according to the lemmas of Chap. V. This yields at once a functional model for T. The functional model affords a tool for analyzing the structure of contractions and the relations among spectrum, minimal function, and characteristic function.

In Chap. VII we establish a one-to-one correspondence between the invariant subspaces of a contraction T and certain factorizations, called the "regular" factorizations, of the characteristic function of T. This correspondence allows us to demonstrate the existence and spectral properties of invariant subspaces for certain types of contractions (class C_{11}), thereby strengthening the results obtained by a more elementary method in Chap. II (Sec. 4).

Chapter VIII deals with contractions T that are "weak", that is, such that the spectrum of T is not the whole unit disc and $I - T^*T$ has finite trace. For these we find a variety of invariant subspaces that furnish a spectral decomposition, in much the same sense as in the theory of normal operators.

Chapter IX contains various further applications of the methods in the book: a criterion for a contraction to be similar to a unitary operator; relations of quasi-similarity for unicellular contractions; criteria for an operator to be unicellular; and finally, extension of these results, by use of a Cayley transformation, to accretive and dissipative operators and to continuous contraction semi-groups.

The Notes at the end of each chapter mention additional results, sketch the history of the subject, and give references to the literature.

The chapters are divided into sections and the sections into subsections. Results are designated as theorems, propositions, lemmas, and corollaries, and these are numbered separately within each section, as are the subsections and the formulas. The form of citations is the following: the second section in a chapter is called Sec. 2. Within that section, the third subsection is denoted by Sec. 2.3; the third formula by (2.3); the third theorem (or proposition, etc.) by 2.3. In references to other chapters, the appropriate roman numeral is prefixed; thus in referring to Chap. I we would write Sec. I.2.3, or (I.2.3), or Theorem I.2.3.

We have presupposed familiarity with the elements of the theory of Hilbert space (in particular with the spectral theory for unitary, self-adjoint, and normal operators). Indeed this monograph may be regarded as a sequel to the book *Leçons d'analyse fonctionnelle*[1] by F. Riesz and B. Sz.-Nagy and to the appendix added to it in 1955 by Sz.-Nagy.

An additional prerequisite is familiarity with the fundamental facts about the Hardy classes of analytic functions on the unit disc or a half-plane; these may be found in Hoffman's book [1]. We should mention also that Chaps. V and VI of our book have points of contact with the recent book by Helson [1]; but the two books overlap only slightly in the material covered.

Our thanks are due to our colleague István Kovács for his remarks offered in the course of reading the manuscript, and to the Publishing House of the Hungarian Academy of Sciences, and the Szeged Printing Shop for the care they showed in the technical preparation of this book.

Szeged and Bucharest, October 1966

Sz.-N.—F.

Foreword to the English Edition

Since this book was written in French three years ago, further progress has been made in several parts of the theory. We have made use of the opportunity of the translation into English to include some of the new results, and we have revised, improved, and completed many parts of the original.

We mention in particular the following changes. It was known (Theorem I.6.4) that every commuting pair of contractions has a (commuting) unitary dilation, but it was an open question whether this holds, without further restrictions, for commuting families of more than two contractions as well. Now we know by an example due to S. Parrott that the answer to this question is negative (Sec. I.6.3). (1) For the interesting subclass of power-bounded operators, consisting of the operators that admit ρ-unitary dilations, it is proved that all of them are similar to contractions (Sec. II.8). (2) A general dilation theorem is proved in Sec. II.2 for the commutants of contractions, and this theorem is applied later to the functional model of contractions of class C_{00} (Sec. VI.3.8). (3) The functional calculus for contractions is slightly extended so as to include certain meromorphic functions on the unit disc also (Sec. IV.1); this generalization is immediate, and proves to be natural and even necessary in the light of some recent research on the contractions of class $C_0(N)$; these are sketched in part 2 of the Notes to Chap. IX. (4) The important norm relation between the inverse of the characteristic function $\Theta_T(\lambda)$ of a contraction T and the resolvent of T, due to Gohberg and Kreĭn, is added as Proposition VI.4.2. (5) Factorizations of a simple example of contractive analytic function are studied in Sec. V.4.5, thus providing useful information in a problem raised by Theorem VII.6.2 (see the last part of the Notes to Chap. VII).

---

[1] References are to the English translation and are indicated by [*Funct. Anal.*].

There are still other places that underwent smaller or greater changes, and we benefited from a number of remarks made by colleagues, in particular by Ju. L. Šmuljan in Odessa, as well as by R. G. Douglas in Ann Arbor and Chandler Davis in Toronto, who kindly revised parts of the manuscript of the present English edition. Our sincere thanks are due to all of them.

Szeged and Bucharest, May 1969

Sz.-N.—F.

Foreword to the Russian Edition

The history of this book, whose Russian translation is recommended to the readers' attention, can be easily traced. In 1953, the famous Hungarian mathematician B. Szőkefalvi-Nagy published in the journal *Acta Scientiarum Mathematicarum (Szeged)* a theorem, now widely known, on the unitary dilation of contractions. This work was soon continued by the author and other researchers. In 1958, the young Romanian mathematician C. Foiaş joined in the elaboration of the theory of contractions. Since then a series of articles by B. Sz.-Nagy and C. Foias, under the common title *On the Contractions of Hilbert Space*, has appeared regularly in *Acta Szeged*.

This research has evolved into a well-developed theory, which plays an important role in modern functional analysis. We are glad to mention that this theory has numerous, sometimes unexpected connections with works of Soviet experts on operator theory. To begin with, B. Sz.-Nagy's original theorem was based on M. A. Naĭmark's result about generalized spectral functions. Later results of the authors of this book yielded explicit connections with the prediction theory of stationary processes, as well as with Beurling's theorem on the invariant subspaces of shifts. At first it seemed that these topics were far from the spectral theory of nonnormal operators developed by Soviet authors, even when they paid special attention to contractions.

The years 1963–1964 were very important in the theory of Hilbert space operators. During that time, B. Sz.-Nagy and C. Foiaş elaborated the functional calculus of contractions, and introduced the basic concept of the minimal function for a certain class of contractive operators. It was very impressive, and in our opinion quite unexpected, when in their work B. Sz.-Nagy and C. Foiaş arrived naturally at the concept of the characteristic function of a contraction, a concept that arose in the research of M. S. Livsič (in connection with operators close to unitaries). The characteristic function played a fundamental role in the research of many Soviet mathematicians for two decades. The authors of this book obtained an essentially new functional model for arbitrary contractive operators, and in this model the characteristic function appeared in a very explicit form. From this point on, the interaction between the research carried out by B. Sz.-Nagy and C. Foiaş, and that of the Soviet school of operator theory in Hilbert space, became clear. This interaction resulted in the solution of a series of hard and important problems in numerous chapters of the theory (operators similar to unitaries; unicellular contractions and dissipative operators; multiplication theorems for characteristic functions; methods connected with minimal functions; and others). For this reason it is not coincidental that the book contains many references to the works of Soviet mathematicians.

Another important event of the period around 1963 is connected with the success achieved by P. Lax and R. Phillips in the scattering theory of acoustic waves. These authors proposed an abstract scheme for scattering problems, and this led to a new interpretation of the S-matrix. Thus this concept, originally introduced in the quantum theory of scattering, has acquired a new life in classical mathematical physics. It turned out that the Lax–Phillips scheme is nothing else than a continuous analogue of the situation considered by B. Sz.-Nagy and C. Foiaş in their study of the special class of C_{00}-contractions. It became clear that the characteristic function of a contraction can also be regarded as the S-matrix of an appropriately formulated scattering problem.

We now witness the creation of a new important branch in the theory of Hilbert space operators. This involves a wide area of research including the theory of characteristic functions of various classes of operators, the calculus of triangular and multiplicative integrals, problems in the similarity theory of linear operators, several chapters of the theory of operators acting on spaces with an indefinite metric, certain aspects of the scattering theory of self-adjoint and non-self-adjoint operators, along with various applications to classical and quantum physics, and to constructive function theory. This research direction can hardly be presented within the framework of a sole monograph. Several books have appeared reflecting different facets of the aforementioned circle of problems. (Cf. for example, L. DE BRANGES [2], M. S. BRODSKIĬ [9], I. C. GOHBERG AND M. G. KREĬN [4], [7], P. D. LAX AND R. S. PHILLIPS [2], M. S. LIVSIČ [4], and H. HELSON [1].) A prominent place is now taken on this list by the monograph of B. Sz.-Nagy and C. Foiaş, summarizing their investigations. We are not sure that the title *Harmonic Analysis of Operators on Hilbert Space* fully reflects the content and the aims of the book, but it is in perfect harmony with the inner beauty of the theory, with its well-proportioned composition, and with its elegant style.

It is worth mentioning that the research topics discussed in the book are supplemented by historical comments and important remarks at the end of each chapter.

The book has been translated into Russian in close collaboration with the authors. Thanks to this cooperation, several small inaccuracies have been corrected, and numerous supplements have been inserted, bringing the contents of the present translation close to that of the English edition.

We have no doubt that the appearance of the Russian translation of this excellent book will be well received by researchers in functional analysis.

M. G. Kreĭn

Contents

Chapter I

Contractions and Their Dilations

1 Unilateral shifts. Wold decomposition

In this book we study linear transformations (or "operators") from a (real or complex) Hilbert space $\mathfrak{H}$ into a Hilbert space $\mathfrak{H}'$; if $\mathfrak{H} = \mathfrak{H}'$ we say that the transformation (or operator) is *on* $\mathfrak{H}$. Note that if T is a bounded linear transformation from $\mathfrak{H}$ into $\mathfrak{H}'$, then its adjoint T^* is the bounded linear transformation from $\mathfrak{H}'$ into $\mathfrak{H}$, defined by the relation

$$(Th, h')_{\mathfrak{H}'} = (h, T^*h')_{\mathfrak{H}} \qquad (h \in \mathfrak{H}, h' \in \mathfrak{H}');$$

we have $\|T\| = \|T^*\|$.

A linear transformation V from $\mathfrak{H}$ into $\mathfrak{H}'$ is said to be *isometric*, or an *isometry*, if

$$(Vh_1, Vh_2)_{\mathfrak{H}'} = (h_1, h_2)_{\mathfrak{H}} \quad \text{for all} \quad h_1, h_2 \in \mathfrak{H},$$

or, equivalently, if

$$V^*V = I_{\mathfrak{H}}$$

(we denote by I the identity transformation on a Hilbert space, indicating this space by a subscript if necessary).

Let V be an isometry on $\mathfrak{H}$. If a subspace $\mathfrak{L}$ of $\mathfrak{H}$ is mapped by V *onto* itself, then $\mathfrak{L}$ *reduces* T. Indeed, $\mathfrak{L} = V\mathfrak{L}$ implies $V^*\mathfrak{L} = V^*V\mathfrak{L} = \mathfrak{L}$; thus $\mathfrak{L}$ is invariant for V as well as for V^*, and hence it reduces V.

The transformation V from $\mathfrak{H}$ into $\mathfrak{H}'$ is said to be *unitary* if V maps $\mathfrak{H}$ isometrically onto $\mathfrak{H}'$, that is, if $V^*V = I_{\mathfrak{H}}$ and $V\mathfrak{H} = \mathfrak{H}'$. The first of these relations implies $(VV^*)V = V(V^*V) = V$, and hence $VV^*h' = h'$ for every element h' of the form $h' = Vh$ $(h \in \mathfrak{H})$. Because $V\mathfrak{H} = \mathfrak{H}'$, we have $VV^* = I_{\mathfrak{H}'}$. Conversely, this relation evidently implies $V\mathfrak{H} = \mathfrak{H}'$. We conclude that the unitary transformations from $\mathfrak{H}$ into $\mathfrak{H}'$ are characterized by the relations

$$V^*V = I_{\mathfrak{H}} \quad \text{and} \quad VV^* = I_{\mathfrak{H}'},$$

that is, by the relation $V^* = V^{-1}$.

B.Sz.-Nagy et al., *Harmonic Analysis of Operators on Hilbert Space*, Universitext,
DOI 10.1007/978-1-4419-6094-8_1, © Springer Science+Business Media, LLC 2010

Let V be an isometry on $\mathfrak{H}$. A subspace $\mathfrak{L}$ of $\mathfrak{H}$ is called a *wandering space* for V if $V^p\mathfrak{L} \perp V^q\mathfrak{L}$ for every pair of integers $p,q \geq 0$, $p \neq q$; because V is an isometry it suffices to suppose that

$$V^n\mathfrak{L} \perp \mathfrak{L} \quad \text{for} \quad n = 1,2,\dots .$$

One can then form the orthogonal sum in $\mathfrak{H}$

$$M_+(\mathfrak{L}) = \bigoplus_0^\infty V^n\mathfrak{L}.$$

Observe that we have

$$VM_+(\mathfrak{L}) = \bigoplus_1^\infty V^n\mathfrak{L} = M_+(\mathfrak{L}) \ominus \mathfrak{L},$$ [1]

and hence

$$\mathfrak{L} = M_+(\mathfrak{L}) \ominus VM_+(\mathfrak{L}). \tag{1.1}$$

An isometry V on $\mathfrak{H}$ is called a *unilateral shift* if there exists in $\mathfrak{H}$ a subspace $\mathfrak{L}$, which is wandering for V and such that $M_+(\mathfrak{L}) = \mathfrak{H}$. This subspace $\mathfrak{L}$, called *generating* for V, is uniquely determined by V: indeed by (1.1) we have $\mathfrak{L} = \mathfrak{H} \ominus V\mathfrak{H}$. The dimension of $\mathfrak{L}$ is called the *multiplicity* of the unilateral shift V. A unilateral shift V is determined up to unitary equivalence by its multiplicity. Indeed, let V and V' be unilateral shifts on $\mathfrak{H}$ and on $\mathfrak{H}'$, respectively, such that $\dim \mathfrak{L} = \dim \mathfrak{L}'$. Then $\mathfrak{L}'$ can be transformed onto $\mathfrak{L}$ by some unitary map φ; this generates a unitary transformation Φ from $\mathfrak{H}'$ onto $\mathfrak{H}$:

$$\Phi \sum_0^\infty V'^n l_n = \sum_0^\infty V^n(\varphi l_n) \quad \left(l_n \in \mathfrak{L}', \quad \sum_0^\infty \|l_n\|^2 < \infty\right),$$

and we have $\Phi V' = V\Phi$ which implies $V' = \Phi^{-1}V\Phi$.

For a unilateral shift V on $\mathfrak{H} = M_+(\mathfrak{L})$ we have $V^*V^n l = V^*VV^{n-1}l = V^{n-1}l$ $(l \in \mathfrak{L}; n \geq 1)$ and $V^*l = 0$ $(l \in \mathfrak{L})$, because $(V^*l,h) = (l,Vh) = 0$ $(l \in \mathfrak{L}, h \in \mathfrak{H})$ owing to the relation $\mathfrak{L} = \mathfrak{H} \ominus V\mathfrak{H} \perp V\mathfrak{H}$. Hence for

$$h = \sum_0^\infty V^n l_n \quad \left(l_n \in \mathfrak{L}, \sum_0^\infty \|l_n\|^2 = \|h\|^2\right) \tag{1.2}$$

we have

$$Vh = \sum_0^\infty V^{n+1} l_n = \sum_1^\infty V^n l_{n-1} \tag{1.3}$$

and

$$V^*h = \sum_1^\infty V^{n-1} l_n = \sum_0^\infty V^n l_{n+1}. \tag{1.3*}$$

---

[1] For a subspace $\mathfrak{B}$ of a Hilbert space $\mathfrak{A}$, we denote by $\mathfrak{A} \ominus \mathfrak{B}$ the orthogonal complement of $\mathfrak{B}$ in $\mathfrak{A}$.

Iterating, (1.3*) yields $V^{*k}h = \sum_0^\infty V^n l_{n+k}$ $(k = 1,2,\ldots)$, and hence

$$\|V^{*k}h\|^2 = \sum_{n=0}^{\infty} \|V^n l_{n+k}\|^2 = \sum_{n=0}^{\infty} \|l_{n+k}\|^2 = \sum_{n=k}^{\infty} \|l_n\|^2 \to 0$$

as $k \to \infty$.

Thus, for a unilateral shift V

$$V^{*k} \to O \quad (k \to \infty). \tag{1.4}$$

The importance of unilateral shifts is shown by the following

Theorem 1.1 (Wold decomposition). *Let V be an arbitrary isometry on the space $\mathfrak{H}$. Then $\mathfrak{H}$ decomposes into an orthogonal sum $\mathfrak{H} = \mathfrak{H}_0 \oplus \mathfrak{H}_1$ such that $\mathfrak{H}_0$ and $\mathfrak{H}_1$ reduce V, the part of V on $\mathfrak{H}_0$ is unitary and the part of V on $\mathfrak{H}_1$ is a unilateral shift. This decomposition is uniquely determined; indeed we have*

$$\mathfrak{H}_0 = \bigcap_{n=0}^{\infty} V^n\mathfrak{H} \quad \textit{and} \quad \mathfrak{H}_1 = M_+(\mathfrak{L}) \quad \textit{where} \quad \mathfrak{L} = \mathfrak{H} \ominus V\mathfrak{H}. \tag{1.5}$$

The space $\mathfrak{H}_0$ or $\mathfrak{H}_1$ may be absent, that is, equal to $\{0\}$.

Proof. The space $\mathfrak{L} = \mathfrak{H} \ominus V\mathfrak{H}$ is wandering for V. Indeed, for $n \geq 1$ we have

$$V^n\mathfrak{L} \subset V^n\mathfrak{H} \subset V\mathfrak{H} \quad \text{and} \quad V\mathfrak{H} \perp \mathfrak{L}.$$

Consider $\mathfrak{H}_1 = M_+(\mathfrak{L})$ and $\mathfrak{H}_0 = \mathfrak{H} \ominus \mathfrak{H}_1$. Observe that h belongs to $\mathfrak{H}_0$ if and only if it is orthogonal to all finite sums $\bigoplus_0^{m-1} V^n\mathfrak{L}$ $(m = 1,2,\ldots)$. Now we have

$$\begin{aligned}\mathfrak{L} \oplus V\mathfrak{L} \oplus \cdots \oplus V^{m-1}\mathfrak{L} &= (\mathfrak{H} \ominus V\mathfrak{H}) \oplus (V\mathfrak{H} \ominus V^2\mathfrak{H}) \oplus \cdots \oplus (V^{m-1}\mathfrak{H} \ominus V^m\mathfrak{H}) \\ &= \mathfrak{H} \ominus V^m\mathfrak{H};\end{aligned}$$

thus $h \in \mathfrak{H}_0$ if and only if $h \in V^m\mathfrak{H}$ for all $m \geq 0$. Hence $\mathfrak{H}_0$ satisfies the first relation (1.5). Because the subspaces $V^m\mathfrak{H}$ $(m = 0,1,2,\ldots)$ form a nonincreasing sequence, we also have $\mathfrak{H}_0 = \bigcap_1^\infty V^n\mathfrak{H}$. It follows that

$$V\mathfrak{H}_0 = V\bigcap_{n=0}^{\infty} V^n\mathfrak{H} = \bigcap_{n=0}^{\infty} V^{n+1}\mathfrak{H} = \bigcap_{m=1}^{\infty} V^m\mathfrak{H} = \mathfrak{H}_0;$$

thus $\mathfrak{H}_0$ reduces V and $V|\mathfrak{H}_0$ is a unitary operator on $\mathfrak{H}_0$. Hence $\mathfrak{H}_1$ also reduces V and the part of V on $\mathfrak{H}_1$ is evidently a unilateral shift. Thus the subspaces given by (1.5) satisfy our conditions.

It remains to prove that if $\mathfrak{H} = \mathfrak{H}_0' \oplus \mathfrak{H}_1'$ is an arbitrary decomposition satisfying these conditions (i.e. if $\mathfrak{H}_1' = M_+(\mathfrak{L}')$, where $\mathfrak{L}'$ is wandering with respect to V, and if $V\mathfrak{H}_0' = \mathfrak{H}_0'$), then $\mathfrak{H}_0' = \mathfrak{H}_0$ and $\mathfrak{H}_1' = \mathfrak{H}_1$. This follows readily from the equations

$$\begin{aligned}\mathfrak{L} = \mathfrak{H} \ominus V\mathfrak{H} &= (\mathfrak{H}_0' \oplus \mathfrak{H}_1') \ominus (V\mathfrak{H}_0' \oplus V\mathfrak{H}_1') = (\mathfrak{H}_0' \oplus \mathfrak{H}_1') \ominus (\mathfrak{H}_0' \oplus V\mathfrak{H}_1') \\ &= \mathfrak{H}_1' \ominus V\mathfrak{H}_1' = \mathfrak{L}'.\end{aligned}$$

2 Bilateral shifts

Let U be a unitary operator on $\mathfrak{H}$ and let $\mathfrak{L}$ be a wandering subspace for U. The operator U^{-1} is also unitary, and hence we have

$$U^p\mathfrak{L} \perp U^q\mathfrak{L}$$

for all integers p,q $(p \neq q)$. Thus we can form the two-way orthogonal sum

$$M(\mathfrak{L}) = \bigoplus_{-\infty}^{\infty} U^n\mathfrak{L};$$

it is obvious that $M(\mathfrak{L})$ reduces U.

Contrary to the case of $M_+(\mathfrak{L})$, the orthogonal sum $M(\mathfrak{L})$ does not determine $\mathfrak{L}$ (e.g., we have $M(\mathfrak{L}) = M(U\mathfrak{L})$). However, the dimension of $\mathfrak{L}$ is determined uniquely by $M(\mathfrak{L})$. This is a corollary of the following proposition:

Proposition 2.1. *If $\mathfrak{L}'$ and $\mathfrak{L}''$ are wandering subspaces for the unitary operator U on $\mathfrak{H}$, such that*

$$M(\mathfrak{L}') \supset M(\mathfrak{L}''), \tag{2.1}$$

then

$$\dim \mathfrak{L}' \geq \dim \mathfrak{L}''. \tag{2.2}$$

If $\dim \mathfrak{L}'$ *is finite, then equality in* (2.2) *implies equality in* (2.1).

Proof. Because $\dim M(\mathfrak{L}') = \aleph_0 \cdot \dim \mathfrak{L}'$ and $\dim M(\mathfrak{L}'') = \aleph_0 \cdot \dim \mathfrak{L}''$, (2.1) implies

$$\aleph_0 \cdot \dim \mathfrak{L}' \geq \aleph_0 \cdot \dim \mathfrak{L}''. \tag{2.3}$$

In the case $\dim \mathfrak{L}' \geq \aleph_0$, the left-hand side of (2.3) equals $\dim \mathfrak{L}'$, and the right-hand side is $\geq \dim \mathfrak{L}''$. Thus, in this case, (2.2) holds. It remains to consider the case when $\dim \mathfrak{L}'$ is a finite number. Choose two orthonormal bases for $\mathfrak{L}'$ and $\mathfrak{L}''$, say

$$\{e'_n\colon\ n \in \Omega'\} \quad \text{and} \quad \{e''_m\colon\ m \in \Omega''\},$$

and observe that

$$\{U^k e'_n\colon\ n \in \Omega';\ k = 0, \pm 1, \ldots\}$$

and

$$\{U^k e''_m\colon\ m \in \Omega''; k = 0, \pm 1, \ldots\}$$

are then orthonormal bases for $M(\mathfrak{L}')$ and $M(\mathfrak{L}'')$, respectively. Applying Bessel's inequality and Parseval's equality we obtain

$$\begin{aligned}\dim \mathfrak{L}' &= \sum_n \|e'_n\|^2 \geq \sum_n \sum_{m,k} |(e'_n, U^k e''_m)|^2 = \sum_m \sum_{n,k} |(U^{-k} e'_n, e''_m)|^2 \\ &= \sum_m \|e''_m\|^2 = \dim \mathfrak{L}''.\end{aligned}$$

Equality holds if and only if e_n' is contained in $M(\mathfrak{L}'')$ for all $n \in \Omega'$. If this is the case, then $\mathfrak{L}' \subset M(\mathfrak{L}'')$, which implies that $M(\mathfrak{L}') \subset M(\mathfrak{L}'')$ and hence, by (2.1), that $M(\mathfrak{L}') = M(\mathfrak{L}'')$.

An operator U on the space $\mathfrak{H}$ is called a *bilateral shift* if U is unitary and if there exists a subspace $\mathfrak{L}$ of $\mathfrak{H}$, such that $\mathfrak{L}$ is wandering for U and $M(\mathfrak{L}) = \mathfrak{H}$. Every such subspace $\mathfrak{L}$ is called a *generating* subspace, and $\dim \mathfrak{L}$ is called the *multiplicity* of the bilateral shift U.

A bilateral shift is determined by its multiplicity up to unitary equivalence. The proof is analogous to that given for unilateral shifts.

Let us note an immediate property of a bilateral shift U, namely that U has no eigenvalue. Indeed, every element of $\mathfrak{H} = M(\mathfrak{L})$ can be written in the form

$$h = \sum_{-\infty}^{\infty} U^n l_n, \quad \text{where} \quad l_n \in \mathfrak{L}, \quad \text{and} \quad \|h\|^2 = \sum_{-\infty}^{\infty} \|U^n l_n\|^2 = \sum_{-\infty}^{\infty} \|l_n\|^2,$$

and hence

$$Uh = \sum_{-\infty}^{\infty} U^{n+1} l_n = \sum_{-\infty}^{\infty} U^n l_{n-1}.$$

Thus if $Uh = \lambda h$, then comparing the components in $U^n \mathfrak{L}$, we get $l_{n-1} = \lambda l_n$ for all n. This contradicts the convergence of the series $\sum_{-\infty}^{\infty} \|l_n\|^2$ unless $l_n = 0$ for all n. Hence we have necessarily $h = 0$.

Proposition 2.2. *Every unilateral shift V on $\mathfrak{H}$ can be extended to a bilateral shift U of the same multiplicity, on some space containing $\mathfrak{H}$ as a subspace.*

Proof. If we set $\mathfrak{L} = \mathfrak{H} \ominus V\mathfrak{H}$, then $\mathfrak{H} = \bigoplus_0^{\infty} V^n \mathfrak{L}$. Form the space $\mathbf{L}$ whose elements are the vectors

$$\mathbf{l} = \{l_n\}_{-\infty}^{\infty}, \quad \text{where} \quad l_n \in \mathfrak{L} \quad \text{and} \quad \|\mathbf{l}\|^2 = \sum_{-\infty}^{\infty} \|l_n\|^2 < \infty.$$

Observe that

$$U\{l_n\} = \{l_{n-1}\}$$

is a bilateral shift on $\mathbf{L}$. One of the generating subspaces for U consists of those vectors $\{l_m\}$ for which $l_n = 0$ if $n \neq 0$ and l_0 is arbitrary (in $\mathfrak{L}$); obviously this subspace has the same dimension as $\mathfrak{L}$.

We embed $\mathfrak{H}$ in $\mathbf{L}$ by identifying the element

$$h = \sum_0^{\infty} V^n l_n \in \mathfrak{H} \quad \left(l_n \in \mathfrak{L}; \sum_0^{\infty} \|l_n\|^2 = \sum_0^{\infty} \|V^n l_n\|^2 = \|h\|^2 \right)$$

with the element

$$\{l_n'\} \in \mathbf{L} \quad \text{for which} \quad l_n' = l_n \ (n \geq 0) \quad \text{and} \quad l_n' = 0 \quad (n < 0).$$

This identification is possible because it preserves the linear and metric structure of $\mathfrak{H}$. Moreover, the element $Vh = \sum_0^\infty V^{n+1} l_n = \sum_1^\infty V^n l_{n-1}$ of $\mathfrak{H}$ will then be identified with the element $\{l'_{n-1}\} = U\{l'_n\}$ of $\mathbf{L}$, which proves that U is an extension of V.

This finishes the proof. Observe that in virtue of the above identifications we have $\mathbf{L} = \bigoplus_{-\infty}^{\infty} U^n \mathfrak{L}$.

Proposition 2.3. *Every isometry V on the space $\mathfrak{H}$ can be extended to a unitary operator U on some space $\mathfrak{K}$ containing $\mathfrak{H}$ as a subspace.*

Proof. By virtue of the Wold decomposition, we have $V = V_0 \oplus V_1$, where V_0 is unitary and V_1 is a unilateral shift. By Proposition 2.2, V_1 can be extended to a bilateral shift U_1; then $U = V_0 \oplus U_1$ is a unitary extension of V.

3 Contractions. Canonical decomposition

1. By a *contraction* from a Hilbert space $\mathfrak{H}$ into a Hilbert space $\mathfrak{H}'$ we mean a linear transformation from $\mathfrak{H}$ into $\mathfrak{H}'$ such that

$$\|Th\|_{\mathfrak{H}'} \leq \|h\|_{\mathfrak{H}} \quad \text{for all} \quad h \in \mathfrak{H}, \tag{3.1}$$

that is, $\|T\| \leq 1$. We always have $\|T\| = \|T^*\|$, therefore T^* will also be a contraction from $\mathfrak{H}'$ into $\mathfrak{H}$. Inequality (3.1) implies $(T^*Th, h) \leq (h, h)$ for all $h \in \mathfrak{H}$, and the analogous inequality for T^* implies $(TT^*h', h') \leq (h', h')$ for all $h' \in \mathfrak{H}'$. Thus, for any contraction T of $\mathfrak{H}$ into $\mathfrak{H}'$ we have $T^*T \leq I_{\mathfrak{H}}$ and $TT^* \leq I_{\mathfrak{H}'}$, and so one can form the operators

$$D_T = (I_{\mathfrak{H}} - T^*T)^{1/2} \quad \text{and} \quad D_{T^*} = (I_{\mathfrak{H}'} - TT^*)^{1/2}, \tag{3.2}$$

which are self-adjoint (D_T on $\mathfrak{H}$ and D_{T^*} on $\mathfrak{H}'$) and bounded by 0 and 1.

We have

$$TD_T^2 = T(I_{\mathfrak{H}} - T^*T) = T - TT^*T = (I_{\mathfrak{H}'} - TT^*)T = D_{T^*}^2 T$$

and hence it follows by iteration that

$$T(D_T^2)^n = (D_{T^*}^2)^n T \quad \text{for} \quad n = 0, 1, 2, \ldots .$$

Consequently

$$Tp(D_T^2) = p(D_{T^*}^2)T \tag{3.3}$$

for every polynomial $p(\lambda) = a_0 + a_1\lambda + \cdots + a_n\lambda^n$. Choose a sequence of polynomials $p_m(\lambda)$ that tends to the function $\lambda^{1/2}$ uniformly on the interval $0 \leq \lambda \leq 1$. The sequence of operators $p_m(A)$ then tends in norm to $A^{1/2}$ for any self-adjoint operator A bounded by 0 and 1. This is a simple consequence of the spectral representation of A. Applying (3.3) to these polynomials we obtain in the limit (as $m \to \infty$)

$$TD_T = D_{T^*}T. \tag{3.4}$$

This relation, and the dual one resulting by taking adjoints,

$$D_T T^* = T^* D_{T^*}, \tag{3.4*}$$

is used repeatedly in the sequel. Let us observe that

$$\|D_T h\|^2 = (D_T^2 h, h) = (h - T^*Th, h) = \|h\|^2 - \|Th\|^2. \tag{3.5}$$

Thus the set $\{h\colon\ h \in \mathfrak{H}, \|Th\| = \|h\|\}$ coincides with the set $\mathfrak{N}_{D_T} = \{h\colon\ h \in \mathfrak{H}, D_T h = 0\}$, $\mathfrak{N}_{D_T}$ is obviously a subspace of $\mathfrak{H}$.

We call D_T and D_{T^*} the *defect operators*,

$$\mathfrak{D}_T = \overline{D_T \mathfrak{H}} = \mathfrak{N}_{D_T}^{\perp} \quad \text{and} \quad \mathfrak{D}_{T^*} = \overline{D_{T^*} \mathfrak{H}'} = \mathfrak{N}_{D_{T^*}}^{\perp}$$

the *defect spaces*, and

$$\mathfrak{d}_T = \dim \mathfrak{D}_T \quad \text{and} \quad \mathfrak{d}_{T^*} = \dim \mathfrak{D}_{T^*}$$

the *defect indices*, of the contraction T.

Observe that $\mathfrak{d}_T = 0$ characterizes the isometric operators, and $\mathfrak{d}_T = \mathfrak{d}_{T^*} = 0$ characterizes the unitary operators. Thus, the defect indices measure, in a sense, the deviation of the contraction T from being unitary. Equations (3.4) and (3.4*) imply

$$T\mathfrak{D}_T \subset \mathfrak{D}_{T^*} \quad \text{and} \quad T^*\mathfrak{D}_{T^*} \subset \mathfrak{D}_T. \tag{3.6}$$

More precisely, the following relations hold.

$$\mathfrak{D}_{T^*} = \overline{T\mathfrak{D}_T} \oplus \mathfrak{N}_{T^*}, \quad \text{where} \quad \mathfrak{N}_{T^*} = \{h'\colon\ h' \in \mathfrak{H}', T^*h' = 0\}, \tag{3.7}$$

and

$$\mathfrak{D}_T = \overline{T^*\mathfrak{D}_{T^*}} \oplus \mathfrak{N}_T, \quad \text{where} \quad \mathfrak{N}_T = \{h\colon\ h \in \mathfrak{H}, Th = 0\}. \tag{3.7*}$$

By reason of symmetry, it suffices to prove (3.7). Let us observe first that for $h' \in \mathfrak{N}_{T^*}$ we have $h' = h' - TT^*h' = D_{T^*}^2 h'$ and hence $\mathfrak{N}_{T^*} \subset D_{T^*}\mathfrak{H}'$. On the other hand, $\mathfrak{N}_{T^*}$ is orthogonal to $T\mathfrak{D}_T$, because

$$(TD_T h, h') = (D_T h, T^*h') = 0 \quad \text{for} \quad h \in \mathfrak{H}, h' \in \mathfrak{N}_{T^*}.$$

Thus (3.7) is proved if we show that any element $g \in \mathfrak{D}_{T^*}$, which is orthogonal to $T\mathfrak{D}_T$, belongs necessarily to $\mathfrak{N}_{T^*}$. Now, indeed, our hypotheses

$$g \in \mathfrak{D}_{T^*}, \quad g \perp T\mathfrak{D}_T$$

imply

$$T^*g \in T^*\mathfrak{D}_{T^*} \subset \mathfrak{D}_T \text{ and } T^*g \perp \mathfrak{D}_T.$$

Thus $T^*g = 0$ and $g \in \mathfrak{N}_{T^*}$.

Observe that the restriction $T|\mathfrak{H}\ominus\mathfrak{D}_T$ of T to $\mathfrak{H}\ominus\mathfrak{D}_T$ is a unitary operator with range $\mathfrak{H}\ominus\mathfrak{D}_{T^*}$ and inverse $T^*|\mathfrak{H}\ominus\mathfrak{D}_{T^*}$.

2. In the sequel we mainly consider the case $\mathfrak{H}=\mathfrak{H}'$, that is, contractions on the space $\mathfrak{H}$. A simple but useful property of these contractions is expressed by the following result.

Proposition 3.1. *A contraction T on $\mathfrak{H}$ and its adjoint T^* have the same invariant vectors, that is, $Th=h$ implies $T^*h=h$, and conversely.*

Proof. If $Th=h$, then $(h,T^*h)=(Th,h)=(h,h)=\|h\|^2$; hence it follows

$$\|h-T^*h\|^2=\|h\|^2-2\,\mathrm{Re}(h,T^*h)+\|T^*h\|^2=\|h\|^2-2\|h\|^2+\|T^*h\|^2\leq 0,$$

because $\|T^*h\|\leq\|h\|$. This proves that $T^*h=h$. The converse assertion follows by symmetry.

3. Two important types of contractions on a Hilbert space are the *unitary* operators and the *completely nonunitary* (c.n.u.) contractions. A contraction T on $\mathfrak{H}$ is said to be c.n.u. if for no nonzero reducing subspace $\mathfrak{L}$ for T is $T|\mathfrak{L}$ a unitary operator. The structure of the unitary operators is well known: for them a spectral theory and an effective functional calculus are available. For these theories, we refer the reader to [*Funct. Anal.*]. As regards c.n.u. contractions, one of the principal aims of the present book is to develop a theory for them that corresponds in some sense to the spectral theory and to the functional calculus for unitary operators. Our theory is based on a simple theorem concerning "unitary dilations of contractions," which we formulate and prove in Sec. 4.

Let us recall that the bilateral shifts are unitary operators. In contrast, the unilateral shifts are c.n.u. In fact, if the unilateral shift V in $\mathfrak{H}$ were reduced by some subspace $\mathfrak{H}_0\neq\{0\}$ to a unitary operator $V_0=V|\mathfrak{H}_0$, then we would have $\|V^{*n}h\|=\|h\|$ for all $h\in\mathfrak{H}_0$. Relation (1.4) implies, however, that $V^{*n}h\to 0$ $(n\to\infty)$ which is a contradiction for $h\neq 0$.

It is an important fact that every contraction can be decomposed into the orthogonal sum of a unitary operator and a c.n.u. contraction. As a consequence, the study of contractions of general type can be reduced to the study of contractions of these two particular types.

Theorem 3.2. *To every contraction T on the space $\mathfrak{H}$ there corresponds a decomposition of $\mathfrak{H}$ into an orthogonal sum of two subspaces reducing T, say $\mathfrak{H}=\mathfrak{H}_0\oplus\mathfrak{H}_1$, such that the part of T on $\mathfrak{H}_0$ is unitary, and the part of T on $\mathfrak{H}_1$ is completely nonunitary; $\mathfrak{H}_0$ or $\mathfrak{H}_1$ may equal the trivial subspace $\{0\}$. This decomposition is uniquely determined. Indeed, $\mathfrak{H}_0$ consists of those elements h of $\mathfrak{H}$ for which*

$$\|T^nh\|=\|h\|=\|T^{*n}h\|\qquad(n=1,2,\ldots).$$

$T_0=T|\mathfrak{H}_0$ and $T_1=T|\mathfrak{H}_1$ are called the unitary part and the completely nonunitary part of T, respectively, and $T=T_0\oplus T_1$ is called the canonical decomposition of T. In particular, for an isometry, the canonical decomposition coincides with the Wold decomposition.

Proof. Let us introduce the notation

$$T(n) = T^n \quad (n \geq 1), \qquad T(0) = I, \qquad T(n) = T^{*|n|} \quad (n \leq -1). \tag{3.8}$$

Because $T(n)$ is a contraction on $\mathfrak{H}$ for every integer n, the set of vectors h for which $\|T(n)h\| = \|h\|$ (n fixed) is equal to the subspace $\mathfrak{N}_{D_{T(n)}}$ formed by the vectors h for which $D_{T(n)}h = 0$. As a consequence the set

$$\mathfrak{H}_0 = \{h\colon \ \|T(n)h\| = \|h\| \qquad (n = 0, \pm 1, \ldots)\}$$

can be expressed as $\mathfrak{H}_0 = \bigcap_{n=-\infty}^{\infty} \mathfrak{N}_{D_{T(n)}}$. It follows that $\mathfrak{H}_0$ is also a subspace of $\mathfrak{H}$. Both T and T^* transform $\mathfrak{H}_0$ into itself. Indeed, for $h \in \mathfrak{H}_0$ we have

$$\|T^n Th\| = \|T^{n+1}h\| = \|h\| = \|Th\| \quad (n = 0, 1, \ldots),$$
$$\|T^{*n}Th\| = \|T^{*n-1}T^*Th\| = \|T^{*n-1}h\| = \|h\| = \|Th\| \quad (n = 1, 2, \ldots);$$

here we have made use of the fact that, for a contraction T, $\|Th\| = \|h\|$ implies $T^*Th = h$. Hence $Th \in \mathfrak{H}_0$. One shows analogously that $T^*h \in \mathfrak{H}_0$. Thus $\mathfrak{H}_0$ *reduces* T. If we set $T_0 = T|\mathfrak{H}_0$, then $T_0^* = T^*|\mathfrak{H}_0$, and

$$T_0^* T_0 = T^*T|\mathfrak{H}_0 = I_{\mathfrak{H}_0}, \quad T_0 T_0^* = TT^*|\mathfrak{H}_0 = I_{\mathfrak{H}_0};$$

thus T_0 is unitary.

The subspace $\mathfrak{H}_1 = \mathfrak{H} \ominus \mathfrak{H}_0$ also reduces T, and $T_1 = T|\mathfrak{H}_1$ is c.n.u. Indeed, suppose $\mathfrak{H}_2$ is a nonzero subspace of $\mathfrak{H}_1$, reducing T, and such that $T|\mathfrak{H}_2$ is unitary. Then for every $h \in \mathfrak{H}_2$ we have $\|T(n)h\| = \|h\|$ and hence $h \in \mathfrak{H}_0$. Therefore $\mathfrak{H}_2 \subset \mathfrak{H}_0$: a contradiction.

It remains to prove the uniqueness of the decomposition. Let $\mathfrak{H} = \mathfrak{H}_0' \oplus \mathfrak{H}_1'$ be an arbitrary decomposition of $\mathfrak{H}$ with the properties in question. Because T is unitary on $\mathfrak{H}_0'$, we have $\|T(n)h\| = \|h\|$ for all $h \in \mathfrak{H}_0'$ and hence $\mathfrak{H}_0' \subset \mathfrak{H}_0$. The spaces $\mathfrak{H}_0$ and $\mathfrak{H}_0'$ reduce T, therefore the same is true for $\mathfrak{H}_0 \ominus \mathfrak{H}_0'$, and $T|\mathfrak{H}_0 \ominus \mathfrak{H}_0'$ is unitary. Because $\mathfrak{H}_0 \ominus \mathfrak{H}_0' \subset \mathfrak{H} \ominus \mathfrak{H}_0' = \mathfrak{H}_1'$ and because T is c.n.u. on $\mathfrak{H}_1'$, we have necessarily $\mathfrak{H}_0 \ominus \mathfrak{H}_0' = \{0\}$; that is, $\mathfrak{H}_0' = \mathfrak{H}_0$.

The last assertion of the theorem follows from the uniqueness of the decomposition.

4 Isometric and unitary dilations

1. For two operators, A on the Hilbert space $\mathfrak{A}$, and B on the Hilbert space $\mathfrak{B}$, we indicate by

$$A = \operatorname{pr} B$$

the relationship defined by the following two requirements.

(i) $\mathfrak{A}$ is a subspace of $\mathfrak{B}$.
(ii) $(Aa, a') = (Ba, a')$ for all $a, a' \in \mathfrak{A}$.

Condition (ii) is obviously equivalent to the condition

(ii′) $Aa = P_{\mathfrak{A}}Ba$ for all $a \in \mathfrak{A}$, where $P_{\mathfrak{A}}$ denotes the orthogonal projection from $\mathfrak{B}$ into $\mathfrak{A}$.

Here are some immediate properties of this relation.

(a) $A \subset B$ (i.e., B is an extension of A) implies $A = \operatorname{pr} B$,
(b) $A = \operatorname{pr} B$, $B = \operatorname{pr} C$ imply $A = \operatorname{pr} C$,
(c) $A = \operatorname{pr} B$ implies $A^* = \operatorname{pr} B^*$,
(d) $A = \operatorname{pr} B$ and $A' = \operatorname{pr} B'$ (A, A' on $\mathfrak{A}$; B, B' on $\mathfrak{B}$) imply $cA + c'A' = \operatorname{pr}(cB + c'B')$ for arbitrary scalar coefficients c, c',
(e) $A = \operatorname{pr} B$, $A' = \operatorname{pr} B'$ imply $A \oplus A' = \operatorname{pr}(B \oplus B')$,
(f) $A_n = \operatorname{pr} B_n$ (A_n or $\mathfrak{A}$, B_n on $\mathfrak{B}$, $n = 1, 2, \ldots$) and

$$B_n \rightharpoonup B, \quad B_n \to B, \quad \text{or} \quad B_n \rightrightarrows B \quad (n \to \infty),$$

imply the convergence of A_n in the same sense (i.e., weakly, strongly, or in norm) to an operator A, and we have $A = \operatorname{pr} B$.

Now we make the following definition.

Definition. Let A and B be two operators, A on the space $\mathfrak{A}$, and B on the space $\mathfrak{B}$. We call B a *dilation* of A if

$$A^n = \operatorname{pr} B^n \quad \text{for} \quad n = 1, 2, \ldots .$$

Two dilations of A, say B on $\mathfrak{B}$ and B' on $\mathfrak{B}'$, are said to be *isomorphic* if there exists a unitary transformation φ from $\mathfrak{B}'$ onto $\mathfrak{B}$, such that

(i) $\varphi a = a$ for all $a \in \mathfrak{A}$,
(ii) $B' = \varphi^{-1} B \varphi$.

2. We can now state our first result on dilations.

Theorem 4.1. *For every contraction T on the Hilbert space $\mathfrak{H}$ there exists an isometric dilation V on some Hilbert space $\mathfrak{K}_+ (\supset \mathfrak{H})$, which is moreover minimal in the sense that*

$$\mathfrak{K}_+ = \bigvee_0^\infty V^n \mathfrak{H}. \tag{4.1}$$

This minimal isometric dilation of T is determined up to isomorphism; thus one can call it "the minimal isometric dilation" of T. The space $\mathfrak{H}$ is invariant for V^* and we have

$$TP_+ = P_+ V \quad \text{and} \quad T^* = V^* | \mathfrak{H}, \tag{4.2}$$

where P_+ denotes the orthogonal projection from $\mathfrak{K}_+$ onto $\mathfrak{H}$.

Proof. Let us form the Hilbert space

$$\mathbf{H}_+ = \bigoplus_0^\infty \mathfrak{H},$$

whose elements are the vectors

$$\mathbf{h} = \{h_0, h_1, \ldots\} \quad \text{with} \quad h_n \in \mathfrak{H}, \quad \|\mathbf{h}\|^2 = \sum_0^\infty \|h_n\|^2 < \infty.$$

We embed $\mathfrak{H}$ in $\mathbf{H}_+$ as a subspace, by identifying the element $h \in \mathfrak{H}$ with the element $\{h, 0, \ldots\} \in \mathbf{H}_+$; this identification is allowed because it obviously preserves the linear and metric structure of $\mathfrak{H}$. Observe that we have then

$$P_{\mathfrak{H}}\{h_0, h_1, \ldots\} = \{h_0, 0, 0, \ldots\} = h_0.$$

We define on $\mathbf{H}_+$ an operator $\mathbf{V}$ by

$$\mathbf{V}\{h_0, h_1, \ldots\} = \{Th_0, D_T h_0, h_1, \ldots\}, \quad \text{where} \quad D_T = (I - T^*T)^{1/2}.$$

From the relation $\|Th\|^2 + \|D_T h\|^2 = \|h\|^2$, which holds for every $h \in \mathfrak{H}$, we deduce

$$\begin{aligned} \|\mathbf{V}\{h_0, h_1, \ldots\}\|^2 &= \|Th_0\|^2 + \|D_T h_0\|^2 + \|h_1\|^2 + \|h_2\|^2 + \cdots \\ &= \|h_0\|^2 + \|h_1\|^2 + \|h_2\|^2 + \cdots = \|\{h_0, h_1, \ldots\}\|^2; \end{aligned}$$

because $\mathbf{V}$ is obviously linear, it is an isometry on $\mathbf{H}_+$. Moreover, we obtain for $n = 1, 2, \ldots$ by induction:

$$\mathbf{V}^n\{h_0, h_1, \ldots\} = \{T^n h_0, D_T T^{n-1} h_0, D_T T^{n-2} h_0, \ldots, D_T h_0, h_1, h_2, \ldots\},$$

and hence it follows for $h \in \mathfrak{H}$ that

$$\begin{aligned} P_{\mathfrak{H}} \mathbf{V}^n h = P_{\mathfrak{H}} \mathbf{V}^n \{h, 0, 0, \ldots\} &= P_{\mathfrak{H}}\{T^n h, D_T T^{n-1} h, \ldots, D_T h, 0, \ldots\} \\ &= \{T^n h, 0, 0, \ldots\} = T^n h. \end{aligned}$$

This proves that $\mathbf{V}$ is an isometric dilation of T. In general, our $\mathbf{V}$ is not minimal. However, it is easy to show that every isometric dilation V_0 of T, say on a space $\mathfrak{K}_0 (\supset \mathfrak{H})$, contains a minimal isometric dilation V in the sense that V is the restriction of V_0 to some subspace $\mathfrak{K}_+$ of $\mathfrak{K}_0$, invariant for V_0. In fact, one has to take

$$\mathfrak{K}_+ = \bigvee_0^\infty V_0^n \mathfrak{H}.$$

Let V be a minimal isometric dilation of T, on the space $\mathfrak{K}_+$. From the dilation property it follows for $h \in \mathfrak{H}$ and $n = 0, 1, \ldots,$

$$TP_+ \cdot V^n h = TT^n h = T^{n+1} h = P_+ V^{n+1} h = P_+ V \cdot V^n h.$$

On account of (4.1) this implies the first relation (4.2) which we now show implies the second. For $h \in \mathfrak{H}$ and $k \in \mathfrak{K}_+$ we have

$$(T^*h,k) = (T^*h,P_+k) = (h,TP_+k) = (h,P_+Vk) = (h,Vk) = (V^*h,k),$$

and thus $T^*h = V^*h$. Hence in particular $V^*\mathfrak{H} \subset \mathfrak{H}$.

Therefore it remains only to show that all minimal isometric dilations of T are isomorphic. To this end let us start by observing that for any isometric dilation V of T and for $h,h' \in \mathfrak{H}$ we have

$$(V^n h, V^m h') = \begin{cases} (V^{n-m}h,h') = (T^{n-m}h,h') & \text{if } n \geq m \geq 0, \\ (h,V^{m-n}h') = (h,T^{m-n}h') & \text{if } m \geq n \geq 0; \end{cases} \tag{4.3}$$

thus $(V^n h, V^m h')$ does not depend upon the particular choice of V. Consequently, the scalar product of two finite sums of the form

$$\sum_{n=0}^{N} V^n h_n, \quad \sum_{m=0}^{N'} V^m h'_m \qquad (h_n, h'_m \in \mathfrak{H})$$

depends only upon the vectors h_n, h'_m, and not upon the particular choice of V. Thus, if V_1 and V_2 are two isometric dilations of T on the spaces $\mathfrak{K}_1$ and $\mathfrak{K}_2$, respectively, then, setting

$$\varphi\left(\sum_0^N V_2^n h_n\right) = \sum_0^N V_1^n h_n \qquad (N = 0,1,\ldots; h_n \in \mathfrak{H}), \tag{4.4}$$

we define an isometric (and consequently, a well-defined and linear) transformation from the linear manifold $\mathfrak{L}_2$ of the elements of the form $\sum_0^N V_2^n h_n$, onto the linear manifold $\mathfrak{L}_1$ of the elements of the form $\sum_0^N V_1^n h_n$.

If the dilations V_1 and V_2 are minimal, that is, if

$$\mathfrak{K}_i = \bigvee_0^\infty V_i^n \mathfrak{H} \qquad (i = 1,2),$$

then $\mathfrak{K}_i = \overline{\mathfrak{L}}_i$ $(i = 1,2)$ and consequently φ can be extended by continuity to a unitary transformation from $\mathfrak{K}_2$ onto $\mathfrak{K}_1$. We have $\varphi h = \varphi(V_2^0 h) = V_1^0 h = h$ for $h \in \mathfrak{H}$. Furthermore, we have $\varphi(V_2 k) = V_1(\varphi k)$ first for $k \in \mathfrak{L}_2$ and hence by continuity for all $k \in \mathfrak{K}_2$. Thus $V_2 = \varphi^{-1} V_1 \varphi$, and this proves that the dilations V_1 and V_2 of T are isomorphic.

This finishes the proof of Theorem 4.1.

As a supplement to Theorem 4.1, we mention that if $\mathbf{V}$ is any isometric operator on some Hilbert space $\mathbf{H}_+$ containing $\mathfrak{H}$ satisfying the condition

$$P_{\mathfrak{H}} \mathbf{V} = T P_{\mathfrak{H}}$$

then the subspace

$$\mathfrak{K}_+ = \bigvee_{n=0}^{\infty} \mathbf{V}^n \mathfrak{H}$$

is *reducing* $\mathbf{V}$.

To prove this we first notice that the above condition is equivalent to

$$\mathbf{V}^*|\mathfrak{H} = T^*$$

and that the set $\mathfrak{K}_+^0$ of all the finite sums

$$k = \sum_{n=0}^{N} \mathbf{V}^n h_n \qquad (N = 0, 1, \ldots; h_n \in \mathfrak{H})$$

is dense in $\mathfrak{K}_+$. For the above k, we obviously have $\mathbf{V}k \in \mathfrak{K}_+^0$ as well as

$$\mathbf{V}^* k = T^* h_0 + \sum_{n=1}^{N} \mathbf{V}^{n-1} h_n \in \mathfrak{K}_+^0,$$

that is, $\mathbf{V}\mathfrak{K}_+^0, \mathbf{V}^*\mathfrak{K}_+^0 \subset \mathfrak{K}_+^0$, and subsequently $\mathfrak{K}_+$ is reducing $\mathbf{V}$.

3. We are now able to prove the following theorem, of fundamental importance for our investigations.

Theorem 4.2. *For every contraction T on the Hilbert space $\mathfrak{H}$ there exists a unitary dilation U on a space $\mathfrak{K}$ containing $\mathfrak{H}$ as a subspace, which is minimal, that is, such that*

$$\mathfrak{K} = \bigvee_{-\infty}^{\infty} U^n \mathfrak{H}. \tag{4.5}$$

This minimal unitary dilation is determined up to isomorphism, and thus can be called "the minimal unitary dilation" of T.

Proof. Let us take an arbitrary isometric dilation of T and extend it to a unitary operator: this is possible by virtue of Proposition 2.3. Thus we obtain a unitary dilation of T, which is not necessarily minimal. However, every unitary dilation U_0 of T contains a minimal one; we have only to take the restriction of U_0 to the subspace

$$\mathfrak{K} = \bigvee_{-\infty}^{\infty} U_0^n \mathfrak{H},$$

which reduces U_0.

Let us observe next that if U is a unitary dilation of T, then we have for h, h' in $\mathfrak{H}$,

$$(U^n h, U^m h') = \begin{cases} (T^{n-m} h, h') & \text{if } n \geq m, \\ (h, T^{m-n} h') & \text{if } m \geq n; \end{cases} \tag{4.6}$$

in this case the integers m, n can be positive, negative, or 0. On the basis of (4.6), one proves that any two minimal unitary dilations of T are isomorphic just as it was done for the isometric dilations, the only difference being that now one has to admit integer exponents m,n of any sign.

This finishes the proof of the theorem.

Let us state some obvious facts. If we denote by U_T the minimal unitary dilation of T then we have

$$U_{T^*} = (U_T)^*, \tag{4.7}$$

$$U_T = U_{T'} \oplus U_{T''} \quad \text{for} \quad T = T' \oplus T'', \tag{4.8}$$

and

$$U_T = T \qquad \text{for unitary } T. \tag{4.9}$$

Consequently, *if $T = T_0 \oplus T_1$ is the canonical decomposition of T into the orthogonal sum of its unitary part T_0 and its c.n.u. part T_1, then*

$$U_T = T_0 \oplus U_{T_1}.$$

4. If T is a contraction on $\mathfrak{H}$ then so is

$$T_a = (T - aI)(I - \bar{a}T)^{-1} \qquad (|a| < 1). \tag{4.10}$$

In fact, for any $h \in \mathfrak{H}$ we have, setting $g = (I - \bar{a}T)^{-1}h$,

$$\|h\|^2 - \|T_a h\|^2 = \|(I - \bar{a}T)g\|^2 - \|(T - aI)g\|^2 = (1 - |a|^2)(\|g\|^2 - \|Tg\|^2) \geq 0.$$

This shows, moreover, that if T is an isometry then so is T_a. We obviously have $(T_a)^* = (T^*)_{\bar{a}}$, therefore we also obtain that if T^* is an isometry then so is $(T_a)^*$. Consequently, if T is unitary then so is T_a.

Proposition 4.3. *Let T be a contraction on $\mathfrak{H}$, and let $|a| < 1$. If V is an isometric dilation of T then V_a is an isometric dilation of T_a; moreover, if V is minimal, then V_a is also minimal. Similarly, if U is a unitary dilation of T then U_a is a unitary dilation of T_a; moreover, if U is minimal, then U_a is also minimal.*

Proof. Consider the Taylor expansions

$$\left(\frac{\lambda - a}{1 - \bar{a}\lambda}\right)^n = \sum_{\nu=0}^{\infty} c_\nu(a;n)\lambda^\nu \qquad (n = 0, 1, 2, \ldots);$$

because their radius of convergence is larger than 1, we have $\sum_\nu |c_\nu(a;n)| < \infty$. This implies that the operator series

$$\sum_{\nu=0}^{\infty} c_\nu(a;n)T^\nu$$

converges in norm and its sum is equal to T_a^n $(n = 0, 1, 2, \ldots)$. (A functional calculus $\sum_0^\infty c_\nu \lambda^\nu \to \sum_0^\infty c_\nu T^\nu$ is studied for the case $\sum_\nu |c_\nu| < \infty$, and for analytic functions of still more general type, in Chaps. III and IV.) Because we also have

$$\sum_{\nu=0}^{\infty} c_\nu(a;n)V^\nu = V_a^n, \tag{4.11}$$

we conclude that

$$T_a^n = \sum_{\nu=0}^{\infty} c_\nu(a;n)T^\nu = \sum_{\nu=0}^{\infty} c_\nu(a;n)\mathrm{pr}\, V^\nu = \mathrm{pr}\, \sum_{\nu=0}^{\infty} c_\nu(a;n)V^\nu = \mathrm{pr}\, V_a^n$$

for $n \geq 0$, and hence V_a is an isometric dilation of T_a. Relation (4.11) implies that

$$\bigvee_{n=0}^{\infty} V_a^n \mathfrak{H} \subset \bigvee_{n=0}^{\infty} V^n \mathfrak{H}. \tag{4.12}$$

Now, $(T_a)_{-a} = T$ and $(V_a)_{-a} = V$, and substituting $V \to V_a$, $a \to -a$ in (4.12) it follows that

$$\bigvee_{n=0}^{\infty} V^n \mathfrak{H} = \bigvee_{n=0}^{\infty} [(V_a)_{-a}]^n \mathfrak{H} \subset \bigvee_{n=0}^{\infty} V_a^n \mathfrak{H},$$

and thus we have

$$\bigvee_{n=0}^{\infty} V_a^n \mathfrak{H} = \bigvee_{n=0}^{\infty} V^n \mathfrak{H}. \tag{4.13}$$

These relations are valid in particular for a unitary dilation U of T, and for the unitary dilation $U^* = U^{-1}$ of T^*. Because $U_a^* = (U^*)_{\bar{a}}$, (4.13) yields, when V and a are replaced by U^* and $\bar{a}$, the relation

$$\bigvee_{n=0}^{\infty} U_a^{*n} \mathfrak{H} = \bigvee_{n=0}^{\infty} [(U^*)_{\bar{a}}]^n \mathfrak{H} = \bigvee_{n=0}^{\infty} U^{*n} \mathfrak{H}.$$

Hence we have in this case

$$\bigvee_{n=0}^{\infty} U_a^n \mathfrak{H} = \bigvee_{n=0}^{\infty} U^n \mathfrak{H}, \quad \bigvee_{n=0}^{\infty} U_a^{-n} \mathfrak{H} = \bigvee_{n=0}^{\infty} U^{-n} \mathfrak{H} \tag{4.14}$$

and consequently

$$\bigvee_{-\infty}^{\infty} U_a^n \mathfrak{H} = \bigvee_{-\infty}^{\infty} U^n \mathfrak{H}. \tag{4.15}$$

Clearly, (4.13) and (4.15) imply our assertions concerning minimality of the corresponding dilations.

5 Matrix construction of the unitary dilation

1. It is possible to construct a unitary dilation of the contraction T on $\mathfrak{H}$ by the following matrix method. Consider the Hilbert space $\mathbf{H} = \bigoplus_{-\infty}^{\infty} \mathfrak{H}$ the elements of which are the vectors

$$\mathbf{h} = \{h_i\}_{-\infty}^{\infty} \quad \text{with} \quad h_i \in \mathfrak{H} \quad \text{and} \quad \|\mathbf{h}\|^2 = \sum_{-\infty}^{\infty} \|h_i\|^2 < \infty.$$

We embed $\mathfrak{H}$ in $\mathbf{H}$ by identifying the element $h \in \mathfrak{H}$ with the vector $\mathbf{h} = \{h_i\}$ for which $h_0 = h$ and $h_i = 0$ $(i \neq 0)$; $\mathfrak{H}$ becomes a subspace of $\mathbf{H}$, and the orthogonal projection from $\mathbf{H}$ into $\mathfrak{H}$ is given by

$$P_{\mathfrak{H}}\{h_i\} = h_0. \tag{5.1}$$

Every (bounded, linear) operator $\mathbf{S}$ on $\mathbf{H}$ can be represented by the matrix (S_{ij}) $(-\infty < i, j < \infty)$, whose entries S_{ij} are the operators on $\mathfrak{H}$ satisfying $(Sh)_i = \sum_{j=-\infty}^{\infty} S_{ij}h_j$; to the sums, products, and adjoints of operators $\mathbf{S}$ there correspond the sums, products, and adjoints of the matrices, where, by definition, we set

$$(S_{ij})^* = (S_{ji}^*).$$

It is important to note that (5.1) implies

$$P_{\mathfrak{H}}\mathbf{S}h = P_{\mathfrak{H}}\{S_{i0}h\} = S_{00}h \quad \text{for} \quad h \in \mathfrak{H}.$$

Consider now the matrix (U_{ij}) with entries

$$U_{00} = T, \quad U_{01} = D_{T^*}, \quad U_{-1,0} = D_T, \quad U_{-1,1} = -T^*, \quad U_{i,i+1} = I_{\mathfrak{H}},$$

for $i \neq 0, 1$, and $U_{ij} = O$ for all other i, j; that is, the matrix

$$\begin{bmatrix} \ddots & & & & & & \\ & I & & & & & \\ & & I & & & & \\ & & & D_T & -T^* & & \\ & & & \boxed{T} & D_{T^*} & & \\ & & & & & I & \\ & & & & & & I \\ & & & & & & & \ddots \end{bmatrix}, \tag{5.2}$$

where (in order to indicate the indices of rows and columns) we have drawn a square around the central entry U_{00}. All the entries not indicated are O, with the exception of the entries just above the diagonal, which are all equal to $I = I_{\mathfrak{H}}$. Setting

$$h_i' = \sum_j U_{ij}h_j \qquad (i = 0, \pm 1, \pm 2, \ldots),$$

that is,

$$h_{-1}' = D_T h_0 - T^* h_1, \quad h_0' = T h_0 + D_{T^*} h_1, \quad h_i' = h_{i+1} \quad (i \neq 0, -1), \tag{5.3}$$

one shows by elementary calculations based on the relations (3.4), (3.4*), and (3.5), that

$$\|D_T h_0 - T^* h_1\|^2 + \|T h_0 + D_{T^*} h_1\|^2 = \|h_0\|^2 + \|h_1\|^2$$

and, consequently, $\sum_{-\infty}^{\infty} \|h'_i\|^2 = \sum_{-\infty}^{\infty} \|h_i\|^2$. Thus the matrix (U_{ij}) defines an isometry $\mathbf{U}$ in $\mathbf{H}$. Moreover, $\mathbf{U}$ is *unitary*, because the system of equations (5.3) has for every given vector $\{h'_i\} \in \mathbf{H}$ the solution $\{h_i\} \in \mathbf{H}$ with

$$h_0 = D_T h'_{-1} + T^* h'_0, \quad h_1 = -T h'_{-1} + D_{T^*} h'_0, \quad h_i = h'_{i-1} \quad (i \neq 0, 1); \tag{5.4}$$

this can be proved easily by means of relations (3.4) and (3.4*).

The matrix (U_{ij}) is triangular; indeed it is superdiagonal (i.e., $U_{ij} = O$ for $i > j$). Now, the product (C_{ij}) of two superdiagonal matrices, say (A_{ij}) and (B_{ij}), is also superdiagonal, and we have $C_{ii} = A_{ii} B_{ii}$. Hence we conclude that the central entry in the matrix of $\mathbf{U}^n$ $(n \geq 1)$ is equal to T^n, that is, $T^n = \operatorname{pr} \mathbf{U}^n$ $(n \geq 1)$: $\mathbf{U}$ is a dilation of T.

2. This unitary dilation need not, however, be minimal. In order to obtain a minimal unitary dilation, we modify the above construction as follows. Instead of the space $\mathbf{H} = \bigoplus_{-\infty}^{\infty} \mathfrak{H}$ we consider its subspace $\mathbf{K}$ consisting of the vectors $\{h_n\} \in \mathbf{H}$ for which

$$h_n \in \mathfrak{D}_T \ (n \leq -1), \quad h_0 \in \mathfrak{H}, \quad h_n \in \mathfrak{D}_{T^*} \ (n \geq 1);$$

obviously $\mathfrak{H} \subset \mathbf{K} \subset \mathbf{H}$. The subspace $\mathbf{K}$ is invariant for $\mathbf{U}$. By virtue of the formulas (5.3) this is established if we prove that

$$h_0 \in \mathfrak{H} \quad \text{and} \quad h_1 \in \mathfrak{D}_{T^*} \quad \text{imply} \quad D_T h_0 - T^* h_1 \in \mathfrak{D}_T.$$

But this follows from the relations $D_T \mathfrak{H} \subset \mathfrak{D}_T$ and $T^* \mathfrak{D}_{T^*} \subset \mathfrak{D}_T$ (cf. (3.6)).

Secondly, $\mathbf{U}$ maps $\mathbf{K}$ onto $\mathbf{K}$. By virtue of the formulas (5.4) one just has to show that

$$h'_0 \in \mathfrak{H} \quad \text{and} \quad h'_{-1} \in \mathfrak{D}_T \quad \text{imply} \quad -T h'_{-1} + D_{T^*} h'_0 \in \mathfrak{D}_{T^*}.$$

But this follows from the relations $D_{T^*} \mathfrak{H} \subset \mathfrak{D}_{T^*}$ and $T \mathfrak{D}_T \subset \mathfrak{D}_{T^*}$ (cf. (3.6)).

It follows that $\mathbf{U}_0 = \mathbf{U}|\mathbf{K}$ is a unitary dilation of T. Moreover, it is a minimal one. To prove this, first we calculate $\mathbf{U}_0^n h$ and $\mathbf{U}_0^{*n} h = \mathbf{U}_0^{-n} h$ for $h \in \mathfrak{H}$ and $n = 1, 2, \ldots$. From formulas (5.3) (for $\mathbf{U}_0$) and (5.4) (for $\mathbf{U}_0^*$) it follows by iteration that

$$\mathbf{U}_0^n h = \{\ldots, 0, D_T h, D_T T h, \ldots, D_T T^{n-1} h, \boxed{T^n h}, 0, \ldots\}$$

and

$$\mathbf{U}_0^{-n} h = \{\ldots, 0, \boxed{T^{*n} h}, D_{T^*} T^{*n-1} h, \ldots, D_{T^*} T^* h, D_{T^*} h, 0, \ldots\}$$

$(n = 1, 2, \ldots)$, where the components are arranged in order of increasing subscripts, the central component (i.e., the one with subscript 0) being indicated by a square. From these formulas we deduce

$$\mathbf{U}_0^n h - \mathbf{U}_0^{n-1} T h = \{\ldots, 0, \overset{-n}{\smile}{D_T h}, 0, \ldots, \boxed{0}, \ldots\}$$

and

$$\mathbf{U}_0^{-n}h - \mathbf{U}_0^{-n+1}T^*h = \{\ldots,\boxed{0},\ldots,0,\underset{}{\overset{n}{\smile}}{D_{T^*}h},0,\ldots\}$$

for $n \geq 1$. It follows that the (closed linear) span of the subspaces $\mathbf{U}_0^n\mathfrak{H}$ $(-\infty < n < \infty)$ contains all the vectors $\{h_n\} \in \mathbf{K}$ whose components are all 0 except the nth one, which is an arbitrary element of $D_T\mathfrak{H}, \mathfrak{H}$, or $D_{T^*}\mathfrak{H}$, according to the sign of n $(n < 0, n = 0$, or $n > 0)$. These vectors obviously span the whole space $\mathbf{K}$ when n varies over all integers, and therefore we have

$$\mathbf{K} = \bigvee_{-\infty}^{\infty} \mathbf{U}_0^n\mathfrak{H},$$

and thus the unitary dilation $\mathbf{U}_0$ of T is a minimal one.

As an elementary example, let us observe that if $T = aI_{\mathfrak{H}}$ with $|a| < 1$, then the minimal unitary dilation of T is given by the matrix

$$\begin{bmatrix} \ddots & & & & & & \\ & I & & & & & \\ & & I & & & & \\ & & & dI & -\bar{a}I & & \\ & & & \boxed{aI} & dI & & \\ & & & & & I & \\ & & & & & & I \\ & & & & & & & \ddots \end{bmatrix},$$

where $d = (1 - |a|^2)^{1/2}$ acting on the space

$$\cdots \oplus \mathfrak{H} \oplus \mathfrak{H} \oplus \mathfrak{H} \oplus \mathfrak{H} \oplus \cdots.$$

3. The unitary dilation $\mathbf{U} = \mathbf{U}_T$ constructed in Subsec. 1 is in general not minimal, but has the advantage that it is defined on a space depending only on the space $\mathfrak{H}$ and not upon the particular choice of the contraction T on $\mathfrak{H}$. The corresponding matrices $(U_{T;i,j})$ are all superdiagonal. It follows that for an arbitrary sequence T_i $(i = 1, \ldots, r)$ of (not necessarily different) contractions on $\mathfrak{H}$ we have $(\mathbf{U}_{T_1} \cdots \mathbf{U}_{T_r})_{00} = (\mathbf{U}_{T_1})_{00} \cdots (\mathbf{U}_{T_r})_{00} = T_1 \cdots T_r$, that is,

$$T_1 \cdots T_r = \operatorname{pr} \mathbf{U}_{T_1} \cdots \mathbf{U}_{T_r},$$

and also

$$T_1^{n_1} \cdots T_r^{n_r} = \operatorname{pr} \mathbf{U}_{T_1}^{n_1} \cdots \mathbf{U}_{T_r}^{n_r} \qquad (n_i \geq 0). \tag{5.5}$$

However, this property of the operators $\mathbf{U}_T$ is of very limited practical value, because, in general, the relation $\mathbf{U}_{T^n} = (\mathbf{U}_T)^n$ is not valid, and $\mathbf{U}_{T_1}$ and $\mathbf{U}_{T_2}$ do not commute even when T_1 and T_2 do. This raises the problem of considering commutative systems of contractions and trying to find a corresponding commutative system of unitary operators so that (5.5) holds. We investigate this problem in Sec. 6 and return to it again in Sec. 9.

6 Commutative systems of contractions

1. Let us start with a generalization of the notion of dilation for systems of operators. Let $\mathscr{A} = \{A_\omega\}_{\omega\in\Omega}$ be a commutative system of bounded operators on the space $\mathfrak{H}$. A system $\mathscr{B} = \{B_\omega\}_{\omega\in\Omega}$ of bounded operators on a space $\mathfrak{K}$ is called a *dilation* of the system $\mathscr{A}$, if (i) $\mathfrak{H}$ is a subspace of $\mathfrak{K}$, (ii) the system $\mathscr{B}$ is commutative, and (iii)

$$A_{\omega_1}^{n_1}\cdots A_{\omega_r}^{n_r} = \operatorname{pr} B_{\omega_1}^{n_1}\cdots B_{\omega_r}^{n_r} \qquad (n_i \geq 0; i = 1,\ldots,r)$$

for every finite set of subscripts $\omega_i \in \Omega$.

The dilation $\mathscr{B}$ is said to be *isometric, unitary*, and so on, when it consists of operators B_ω of the type in question.

Theorems 4.1 and 4.2 raise the question of whether every commutative system of contractions possesses an isometric or unitary dilation. In this section we show that the answer is positive for every system of two commuting contractions, and negative for some commutative systems of more than two contractions. In Sec. 9 we consider commuting systems of more than two contractions, satisfying certain additional conditions, which do admit isometric and unitary dilations.

Theorem 6.1. *For every commuting pair $\mathscr{T} = \{T_1, T_2\}$ of contractions on a Hilbert space $\mathfrak{H}$ there exists an isometric dilation.*

Proof. Let us consider the space $\mathbf{H}_+ = \bigoplus_0^\infty \mathfrak{H}$ as in the proof of Theorem 4.1, $\mathfrak{H}$ being embedded in $\mathbf{H}_+$ as a subspace as indicated there. We define on $\mathbf{H}_+$ the operators $\mathbf{W}_1$ and $\mathbf{W}_2$ by

$$\mathbf{W}_i\{h_0,h_1,h_2,\ldots\} = \{T_ih_0, D_{T_i}h_0, 0, h_1, h_2,\ldots\} \qquad (i = 1,2); \tag{6.1}$$

these operators are isometric because $\|T_ih_0\|^2 + \|D_{T_i}h_0\|^2 = \|h_0\|^2$, but in general they do not commute.

Let us form the space $\mathfrak{G} = \mathfrak{H}\oplus\mathfrak{H}\oplus\mathfrak{H}\oplus\mathfrak{H}$. By the natural identification

$$\{h_0,h_1,h_2,\ldots\} = \{h_0,\{h_1,h_2,h_3,h_4\},\{h_5,h_6,h_7,h_8\},\ldots\}$$

we have

$$\mathbf{H}_+ = \mathfrak{H}\oplus\mathfrak{G}\oplus\mathfrak{G}\oplus\cdots.$$

Let G be a unitary operator on $\mathfrak{G}$, determined later, and define an operator $\mathbf{G}$ on $\mathbf{H}_+$ by

$$\mathbf{G}\{h_0,h_1,\ldots\} = \{h_0, G\{h_1,\ldots,h_4\}, G\{h_5,\ldots,h_8\},\ldots\}. \tag{6.2}$$

Then $\mathbf{G}$ is also unitary and its inverse is given by

$$\mathbf{G}^{-1}\{h_0,h_1,\ldots\} = \{h_0, G^{-1}\{h_1,\ldots,h_4\}, G^{-1}\{h_5,\ldots,h_8\},\ldots\}. \tag{6.3}$$

Set

$$\mathbf{V}_1 = \mathbf{G}\mathbf{W}_1 \quad \text{and} \quad \mathbf{V}_2 = \mathbf{W}_2\mathbf{G}^{-1}. \tag{6.4}$$

These are isometries on $\mathbf{H}_+$. Let us try to find a G such that $\mathbf{V}_1$ and $\mathbf{V}_2$ commute. First we calculate $\mathbf{V}_1\mathbf{V}_2$ and $\mathbf{V}_2\mathbf{V}_1$. By virtue of (6.1)–(6.4) we have

$$\begin{aligned}
&\mathbf{V}_1\mathbf{V}_2\{h_0,h_1,\ldots\}\\
&\quad = \mathbf{G}\mathbf{W}_1\mathbf{W}_2\mathbf{G}^{-1}\{h_0,h_1,\ldots\}\\
&\quad = \mathbf{G}\mathbf{W}_1\mathbf{W}_2\{h_0,G^{-1}\{h_1,\ldots,h_4\},G^{-1}\{h_5,\ldots,h_8\},\ldots\}\\
&\quad = \mathbf{G}\mathbf{W}_1\{T_2h_0,D_{T_2}h_0,0,G^{-1}\{h_1,\ldots,h_4\},G^{-1}\{h_5,\ldots,h_8\},\ldots\}\\
&\quad = \mathbf{G}\{T_1T_2h_0,D_{T_1}T_2h_0,0,D_{T_2}h_0,0,G^{-1}\{h_1,\ldots,h_4\},G^{-1}\{h_5,\ldots,h_8\},\ldots\}\\
&\quad = \{T_1T_2h_0,G\{D_{T_1}T_2h_0,0,D_{T_2}h_0,0\},\{h_1,\ldots,h_4\},\{h_5,\ldots,h_8\},\ldots\}
\end{aligned}$$

and

$$\begin{aligned}
\mathbf{V}_2\mathbf{V}_1\{h_0,h_1,\ldots\} &= \mathbf{W}_2\mathbf{G}^{-1}\mathbf{G}\mathbf{W}_1\{h_0,h_1,\ldots\} = \mathbf{W}_2\mathbf{W}_1\{h_0,h_1,\ldots\}\\
&= \mathbf{W}_2\{T_1h_0,D_{T_1}h_0,0,h_1,h_2,\ldots\}\\
&= \{T_2T_1h_0,D_{T_2}T_1h_0,0,D_{T_1}h_0,0,h_1,h_2,\ldots\}.
\end{aligned}$$

Because $T_1T_2 = T_2T_1, \mathbf{V}_1$ will commute with $\mathbf{V}_2$ if, and only if, G satisfies

$$G\{D_{T_1}T_2h,0,D_{T_2}h,0\} = \{D_{T_2}T_1h,0,D_{T_1}h,0\} \tag{6.5}$$

for every $h \in \mathfrak{H}$. Now a simple calculation yields

$$\begin{aligned}
\|D_{T_1}T_2h\|^2 + \|D_{T_2}h\|^2 &= \|h\|^2 - \|T_1T_2h\|^2\\
&= \|h\|^2 - \|T_2T_1h\|^2 = \|D_{T_2}T_1h\|^2 + \|D_{T_1}h\|^2,
\end{aligned}$$

and hence $\|\{D_{T_1}T_2h,0,D_{T_2}h,0\}\| = \|\{D_{T_2}T_1h,0,D_{T_1}h,0\}\|$ for all $h \in \mathfrak{H}$. This means that (6.5) determines G as an isometric transformation of the linear manifold $\mathfrak{L}_1$ of the vectors of the form $\{D_{T_1}T_2h,0,D_{T_2}h,0\}$, onto the linear manifold $\mathfrak{L}_2$ of the vectors of the form $\{D_{T_2}T_1h,0,D_{T_1}h,0\}$; G extends by continuity to an isometry from $\mathfrak{M}_1 = \overline{\mathfrak{L}}_1$ onto $\mathfrak{M}_2 = \overline{\mathfrak{L}}_2$. It remains to show that G can be extended to an isometry of the whole space $\mathfrak{G}$ onto itself. This is equivalent to the assertion that the subspaces $\mathfrak{M}_1^\perp = \mathfrak{G} \ominus \mathfrak{M}_1$ and $\mathfrak{M}_2^\perp = \mathfrak{G} \ominus \mathfrak{M}_2$ have the same dimension. When $\mathfrak{H}$ and hence also $\mathfrak{G}$ have finite dimension, this is obvious. When $\dim\mathfrak{H}$ is infinite, we have

$$\dim\mathfrak{H} = \dim\mathfrak{G} \geq \dim\mathfrak{M}_i^\perp \geq \dim\mathfrak{H} \qquad (i = 1,2),$$

because both $\mathfrak{M}_1^\perp$ and $\mathfrak{M}_2^\perp$ contain subspaces of the same dimension as $\mathfrak{H}$, for example, the subspace formed by the vectors $\{0,h,0,0\}$ $(h \in \mathfrak{H})$. This proves that $\dim\mathfrak{M}_1^\perp = \dim\mathfrak{M}_2^\perp$.

If the unitary operator G is determined in this way, the operators $\mathbf{V}_1$ and $\mathbf{V}_2$ will be two commuting isometries on $\mathbf{H}_+$. They satisfy

$$\mathbf{V}_i\{h_0, h_1, \ldots\} = \{T_i h_0, \ldots\} \quad (i = 1, 2),$$

and hence

$$\mathbf{V}_1^{n_1}\mathbf{V}_2^{n_2}\{h_0, h_1, \ldots\} = \{T_1^{n_1}T_2^{n_2}h_0, \ldots\} \quad \text{for} \quad n_1, n_2 \geq 0,$$

and consequently,

$$P_{\mathfrak{H}}\mathbf{V}_1^{n_1}\mathbf{V}_2^{n_2}h = T_1^{n_1}T_2^{n_2}h \quad \text{for every} \quad h \in \mathfrak{H} \quad \text{and} \quad n_1, n_2 \geq 0.$$

Thus $\{\mathbf{V}_1, \mathbf{V}_2\}$ is an isometric dilation of $\{T_1, T_2\}$.

Remark. For an arbitrary isometric dilation $\{V_1, V_2\}$ of $\{T_1, T_2\}$ on the space $\mathfrak{K}$, the subspace

$$\mathfrak{K}' = \bigvee_{n_1, n_2 \geq 0} V_1^{n_1}V_2^{n_2}\mathfrak{H}$$

is invariant for V_1 and V_2, and contains $\mathfrak{H}$ as a subspace; thus the restrictions of V_1 and V_2 to $\mathfrak{K}'$ also form an isometric dilation $\{V_1', V_2'\}$ of $\{T_1, T_2\}$, which is, moreover, minimal, that is, such that

$$\mathfrak{K}' = \bigvee_{n_1, n_2 \geq 0} V_1'^{n_1}V_2'^{n_2}\mathfrak{H}. \tag{6.6}$$

However, contrary to the case of a single contraction, one cannot assert that all the minimal isometric dilations are isomorphic.

2. The existence of a unitary dilation follows from the existence of an isometric dilation by virtue of the following result.

Proposition 6.2. *For every commutative system $\{V_\omega\}_{\omega\in\Omega}$ of isometric operators on $\mathfrak{H}$ there exists a commutative system $\{U_\omega\}_{\omega\in\Omega}$ of unitary operators on a space $\mathfrak{K}$ containing $\mathfrak{H}$ as a subspace, such that $U_\omega \supset V_\omega$ for every $\omega \in \Omega$. In brief, every commutative system of isometries can be extended to a commutative system of unitary operators.*

This proposition holds for finite and infinite systems as well. For finite systems one obtains a proof by applying a finite number of times a process that at every stage diminishes the number of the nonunitary operators. (By the way, iteration of this process a transfinite number of times would also yield a proof for infinite systems; however, we are momentarily interested in the finite case only, and in fact, in the case of two isometries. We return to the infinite case in Sec. 9, using another method.)

The process in question is founded on the following result.

Proposition 6.3. *Let $\{V, W_\nu (\nu \in N)\}$ be a commutative system of isometries on a space $\mathfrak{H}$. Then there exists a commutative system $\{\mathbf{V}, \mathbf{W}_\nu (\nu \in N)\}$ of isometries on a space $\mathbf{H}$ containing $\mathfrak{H}$ as a subspace, such that* (i) *$V \subset \mathbf{V}$ and $W_\nu \subset \mathbf{W}_\nu$ $(\nu \in N)$,*

(ii) $\mathbf{V}$ *is unitary, and* (iii) $\mathbf{W}_\nu$ *is unitary* (*on* $\mathbf{H}$) *for every* ν *such that* W_ν *is unitary* (*on* $\mathfrak{H}$).

Proof. First we extend the isometry V to a unitary operator $\mathbf{V}$ on some space $\mathbf{H} \supset \mathfrak{H}$, this being possible by virtue of Proposition 2.3; this extension can be chosen to be minimal in the sense that

$$\mathbf{H} = \bigvee_{-\infty}^{\infty} \mathbf{V}^n \mathfrak{H} \tag{6.7}$$

(this condition is obviously fulfilled by the unitary extension constructed in the proof of Proposition 2.3). Let us observe that for every finite sum

$$\sum \mathbf{V}^n h_n \tag{6.8}$$

($h_n \in \mathfrak{H}$; n runs over a finite set of integers, of arbitrary sign) and for every $\nu \in N$ we have

$$\begin{aligned}\left\|\sum_n \mathbf{V}^n W_\nu h_n\right\|^2 &= \sum_n\sum_m (\mathbf{V}^n W_\nu h_n, \mathbf{V}^m W_\nu h_m)\\ &= \sum_{n\geq m}\sum (\mathbf{V}^{n-m} W_\nu h_n, W_\nu h_m) + \sum_{n<m}\sum (W_\nu h_n, \mathbf{V}^{m-n} W_\nu h_m).\end{aligned}$$

Because $\mathbf{V}$ extends V, because V and W_ν commute, and because W_ν is an isometry the preceding expression is equal to

$$\begin{aligned}\sum_{n\geq m}\sum (V^{n-m} h_n, h_m) + \sum_{n<m}\sum (h_n, V^{m-n} h_m) &= \sum_{n\geq m}\sum (\mathbf{V}^{n-m} h_n, h_m) + \sum_{n<m}\sum (h_n, \mathbf{V}^{m-n} h_m)\\ &= \sum_n\sum_m (\mathbf{V}^n h_n, \mathbf{V}^m h_m) = \left\|\sum_n \mathbf{V}^n h_n\right\|^2 = \|\mathbf{h}\|^2.\end{aligned}$$

Consequently, by setting

$$\mathbf{W}_\nu \sum_n \mathbf{V}^n h_n = \sum_n \mathbf{V}^n W_\nu h_n \tag{6.9}$$

we obtain an isometric transformation $\mathbf{W}_\nu$ of the linear manifold $\mathbf{M}$ of vectors of the form (6.8) into itself. Because $\overline{\mathbf{M}} = \mathbf{H}$ (6.7), $\mathbf{W}_\nu$ extends by continuity to an isometry on $\mathbf{H}$.

If W_ν is unitary on $\mathfrak{H}$, that is, $W_\nu \mathfrak{H} = \mathfrak{H}$, (6.9) implies that $\mathbf{W}_\nu \mathbf{M} = \mathbf{M}$ and hence $\mathbf{W}_\nu \mathbf{H} = \mathbf{H}$. Thus $\mathbf{W}_\nu$ is unitary on $\mathbf{H}$.

Let us show finally that the system $\{\mathbf{V}, \mathbf{W}_\nu\ (\nu \in N)\}$ is commutative. In fact, one has

$$\begin{aligned}\mathbf{V}\mathbf{W}_\nu(\mathbf{V}^n h) &= \mathbf{V}(\mathbf{V}^n W_\nu h) = \mathbf{V}^{n+1} W_\nu h = \mathbf{W}_\nu \mathbf{V}^{n+1} h = \mathbf{W}_\nu \mathbf{V}(\mathbf{V}^n h),\\ \mathbf{W}_{\nu_1}\mathbf{W}_{\nu_2}(\mathbf{V}^n h) &= \mathbf{W}_{\nu_1}\mathbf{V}^n W_{\nu_2} h = \mathbf{V}^n W_{\nu_1} W_{\nu_2} h\end{aligned}$$

and, by the same reasoning,

$$\mathbf{W}_{\nu_2}\mathbf{W}_{\nu_1}(\mathbf{V}^n h) = \mathbf{W}_{\nu_2}\mathbf{V}^n W_{\nu_1} h = \mathbf{V}^n W_{\nu_2} W_{\nu_1} h$$

for $v, v_1, v_2 \in N$, for an arbitrary integer n, and for $h \in \mathfrak{H}$. By virtue of (6.7) these relations imply $\mathbf{V}\mathbf{W}_v = \mathbf{W}_v\mathbf{V}$ and $\mathbf{W}_{v_1}\mathbf{W}_{v_2} = \mathbf{W}_{v_2}\mathbf{W}_{v_1}$; thus the system is indeed commutative.

Starting with Theorem 6.1, and applying Proposition 6.2 for two commuting isometries, we arrive at the following result.

Theorem 6.4. *For every commuting pair of contractions there exists a unitary dilation.*

3. It is a rather striking fact that the above theorem does not hold for more than two commuting contractions. In fact, we construct *a system* $\{T_1, T_2, T_3\}$ *of commuting contractions for which no system* $\{U_1, U_2, U_3\}$ *of commuting unitary operators can be found such that*

$$T_i = \operatorname{pr} U_i \qquad (i = 1, 2, 3). \tag{6.10}$$

To this end we choose unitary operators A_1, A_2, A_3 on a Hilbert space $\mathfrak{A}$, such that

$$A_1 A_2^{-1} A_3 \neq A_3 A_2^{-1} A_1. \tag{6.11}$$

(We choose, e.g., $A_2 = I$, and for A_1 and A_3 any two noncommuting unitary operators on $\mathfrak{A}$.) We consider the space $\mathfrak{H} = \mathfrak{A} \oplus \mathfrak{A}$ of elements $\{a_1, a_2\}$ $(a_1, a_2 \in \mathfrak{A})$, and define the operators T_i $(i = 1, 2, 3)$ on $\mathfrak{H}$ by

$$T_i\{a_1, a_2\} = \{0, A_i a_1\}.$$

Clearly $\|T_i\| = 1$ and $T_i T_j = O = T_j T_i$ for $i, j = 1, 2, 3$. Suppose there exist commuting unitary operators U_1, U_2, U_3 on some Hilbert space $\mathfrak{K}(\supset \mathfrak{H})$, for which the relations (6.10) hold. Then we have

$$P_{\mathfrak{H}} U_i\{a, 0\} = T_i\{a, 0\} = \{0, A_i a\} \qquad (a \in \mathfrak{A}). \tag{6.12}$$

Because $\|U_i\{a, 0\}\| = \|\{a, 0\}\| = \|a\| = \|A_i a\| = \|\{0, A_i a\}\|$ by virtue of the isometry property of U_i and A_i, we infer from (6.12) that

$$U_i\{a, 0\} = \{0, A_i a\} \qquad (a \in \mathfrak{A}).$$

Hence we deduce

$$U_j^{-1} U_i\{a, 0\} = U_j^{-1}\{0, A_i a\} = U_j^{-1}\{0, A_j(A_j^{-1} A_i) a\} = \{A_j^{-1} A_i a, 0\}$$

and

$$U_k U_j^{-1} U_i\{a, 0\} = U_k\{A_j^{-1} A_i a, 0\} = \{0, A_k A_j^{-1} A_i a\}.$$

Because the Us commute, we conclude that $A_k A_j^{-1} A_i = A_i A_j^{-1} A_k$ $(i, j, k = 1, 2, 3)$. This contradicts the assumption (6.11) and thus proves that no commuting Us exist that satisfy (6.10).

7 Positive definite functions on a group

The constructions of the unitary dilation of a contraction, given in Secs. 4 and 5, although rather simple, have the disadvantage of being overly tied to the particular problem in question. By contrast, the method which we follow in Sec. 8 is based on a general theorem on positive definite operator-valued functions on a group.

Definitions. Let G be a group.

(i) A function $T(s)$ on G, whose values are bounded operators on a Hilbert space $\mathfrak{H}$, is said to be *positive definite* if $T(s^{-1}) = T(s)^*$ for every $s \in G$, and

$$\sum_{s\in G}\sum_{t\in G}(T(t^{-1}s)h(s),h(t)) \geq 0 \tag{7.1}$$

for every finitely nonzero function $h(s)$ from G to $\mathfrak{H}$, that is, which has values different from 0 on a finite subset of G only.

(If the space $\mathfrak{H}$ is complex, the condition $T(s^{-1}) = T(s)^*$ $(s \in G)$ is a consequence of (7.1). The proof is elementary and we omit it.)

(ii) By a *unitary representation* of the group G we mean a function $U(s)$ on G, whose values are unitary operators on a Hilbert space $\mathfrak{K}$ and that satisfies the conditions $U(e) = I$ (e being the identity element of G) and $U(s)U(t) = U(st)$ for $s,t \in G$.

There is a connection between the two notions just defined, which we now establish.

Theorem 7.1. (a) *If $U(s)$ is a unitary representation of the group G in the space $\mathfrak{K}$, and if $\mathfrak{H}$ is a subspace of $\mathfrak{K}$, then $T(s) = P_{\mathfrak{H}}U(s)|\mathfrak{H}$ is a positive definite function on G such that $T(e) = I_{\mathfrak{H}}$. If, moreover, G has a topology and $U(s)$ is a continuous function of s (weakly or strongly, which amounts to the same thing because $U(s)$ is unitary), then $T(s)$ is also a continuous function of s.*

(b) *Conversely, for every positive definite function $T(s)$ on G, whose values are operators on $\mathfrak{H}$, with $T(e) = I_{\mathfrak{H}}$, there exists a unitary representation of G on a space $\mathfrak{K}$ containing $\mathfrak{H}$ as a subspace, such that*

$$T(s) = \operatorname{pr} U(s) \qquad (s \in G) \tag{7.2}$$

and

$$\mathfrak{K} = \bigvee_{s\in G} U(s)\mathfrak{H} \quad \text{(minimality condition)}. \tag{7.3}$$

This unitary representation of G is determined by the function $T(s)$ up to isomorphism so that one can call it "the minimal unitary dilation" of the function $T(s)$. If, moreover, the group G has a topology and $T(s)$ is a (weakly) continuous function of s, then $U(s)$ is also a (weakly, hence also strongly) continuous function of s.

Proof. Part (a) of the theorem is easy. In fact, we have $T(e) = P_{\mathfrak{H}}U(e)|\mathfrak{H} = P_{\mathfrak{H}}|\mathfrak{H} = I_{\mathfrak{H}}$,

$$T(s^{-1}) = P_{\mathfrak{H}}U(s^{-1})|\mathfrak{H} = P_{\mathfrak{H}}U(s)^*|\mathfrak{H} = (P_{\mathfrak{H}}U(s)|\mathfrak{H})^* = T(s)^*,$$

and

$$\sum_{s\in G}\sum_{t\in G}(P_{\mathfrak{H}}U(t^{-1}s)h(s),h(t)) = \sum_{s\in G}\sum_{t\in G}(U(t)^*U(s)h(s),h(t))$$
$$= \left\|\sum_{s\in G}U(s)h(s)\right\|^2 \geq 0$$

for every finitely nonzero function $h(s)$ from G to $\mathfrak{H}$. The assertion concerning continuity is obvious.

Part (b). Let us consider the set $\mathbf{H}$, obviously linear, of the finitely nonzero functions $h(s)$ from G to $\mathfrak{H}$, and let us define on $\mathbf{H}$ a bilinear form[2] by

$$\langle \mathbf{h},\mathbf{h}'\rangle = \sum_s\sum_t(T(t^{-1}s)h(s),h'(t)) \qquad [\mathbf{h}=h(s),\mathbf{h}'=h'(s)].$$

By virtue of (7.1) we have $\langle \mathbf{h},\mathbf{h}\rangle \geq 0$ and hence it follows, using Schwarz's inequality $|\langle \mathbf{h},\mathbf{h}'\rangle|^2 \leq \langle \mathbf{h},\mathbf{h}\rangle\cdot\langle \mathbf{h}',\mathbf{h}'\rangle$, that the vectors $\mathbf{h}$ for which $\langle \mathbf{h},\mathbf{h}\rangle = 0$ form a linear manifold $\mathbf{N}$ in $\mathbf{H}$. It also follows that the value of $\langle \mathbf{h},\mathbf{h}'\rangle$ does not change if we replace the functions $\mathbf{h},\mathbf{h}'$ by equivalent ones modulo $\mathbf{N}$. In other words, the form $\langle \mathbf{h},\mathbf{h}\rangle$ defines in the natural way a bilinear form (k,k') on the quotient space $\mathfrak{K}_0 = \mathbf{H}/\mathbf{N}$. The corresponding quadratic form (k,k) is positive definite on $\mathfrak{K}_0$, therefore $\|k\| = (k,k)^{1/2}$ is a norm on $\mathfrak{K}_0$; by completing $\mathfrak{K}_0$ with respect to this norm we obtain a Hilbert space $\mathfrak{K}$.

Now we embed $\mathfrak{H}$ in $\mathfrak{K}$ (and even in $\mathfrak{K}_0$) by identifying the element h of $\mathfrak{H}$ with the function $\mathbf{h} = \delta_e(s)h$ (where $\delta_e(e) = 1$ and $\delta_e(s) = 0$ for $s \neq e$), or, more precisely, with the equivalence class modulo $\mathbf{N}$ determined by this function. This identification is allowed because it preserves the linear and metric structure of $\mathfrak{H}$. Indeed, we have

$$\langle \delta_e h,\delta_e h'\rangle = \sum_s\sum_t(T(t^{-1}s)\delta_e(s)h,\delta_e(t)h')_{\mathfrak{H}} = (T(e)h,h')_{\mathfrak{H}} = (h,h')_{\mathfrak{H}}.$$

Now we set, for $\mathbf{h} = h(s) \in \mathbf{H}$ and $a \in G$,

$$\mathbf{h}_a = h(a^{-1}s).$$

We obviously have $(\mathbf{h}+\mathbf{h}')_a = \mathbf{h}_a + \mathbf{h}'_a$, $(c\mathbf{h})_a = c\mathbf{h}_a$, $\mathbf{h}_e = \mathbf{h}$, $(\mathbf{h}_b)_a = \mathbf{h}_{ab}$, and furthermore,

$$\langle \mathbf{h}_a,\mathbf{h}'_a\rangle = \sum_s\sum_t(T(t^{-1}s)h(a^{-1}s),h'(a^{-1}t))$$
$$= \sum_\sigma\sum_\tau(T(\tau^{-1}\sigma)h(\sigma),h'(\tau)) = \langle \mathbf{h},\mathbf{h}'\rangle.$$

Therefore $\mathbf{h} \in \mathbf{N}$ implies $\mathbf{h}_a \in \mathbf{N}$ and consequently the transformation $\mathbf{h} \to \mathbf{h}_a$ in $\mathbf{H}$ generates a transformation $k \to k_a$ of the equivalence classes modulo $\mathbf{N}$. Setting $U(a)k = k_a$, thus we define for every $a \in G$ a linear transformation of $\mathfrak{K}_0$ onto $\mathfrak{K}_0$,

[2] In the complex case, the bilinearity means linearity in the first variable and conjugate linearity in the second variable.

such that $U(e) = I$, $U(a)U(b) = U(ab)$, and $(U(a)k, U(a)k') = (k, k')$. These transformations on $\mathfrak{K}_0$ extend by continuity to unitary transformations on $\mathfrak{K}$, forming a representation of the group G.

For $h, h' \in \mathfrak{H}$ we obtain (setting $\delta_a(s) = \delta_e(a^{-1}s)$)

$$(U(a)h, h')_{\mathfrak{K}} = \langle \delta_a h, \delta_e h' \rangle = \sum_s \sum_t (T(t^{-1}s)\delta_a(s)h, \delta_e(t)h')_{\mathfrak{H}} = (T(a)h, h')_{\mathfrak{H}},$$

and hence

$$T(a) = \operatorname{pr} U(a) \quad \text{for every} \quad a \in G.$$

Let us observe next that every function $\mathbf{h} = h(s) \in \mathbf{H}$ can be considered as a finite sum of terms of the type $\delta_\sigma(s)h$ (i.e., the type $(\delta_e(s)h)_\sigma$ $(\sigma \in G)$), and hence every element k of $\mathfrak{K}_0$ can be decomposed into a finite sum of terms of the type $U(\sigma)h$ $(\sigma \in G, h \in \mathfrak{H})$. This implies (7.3).

The isomorphism of the unitary representations of G satisfying (7.2) and (7.3) is a consequence of the relation

$$\begin{aligned}(U(s)h, U(t)h') &= (U(t)^*U(s)h, h') = (U(t^{-1})U(s)h, h') \\ &= (U(t^{-1}s)h, h') = (T(t^{-1}s)h, h'),\end{aligned}$$

which shows that the scalar products of the elements of $\mathfrak{K}$ of the form $U(s)h$, $U(t)h'$ $(s, t \in G, h, h' \in \mathfrak{H})$ do not depend upon the particular choice of the unitary representation $U(s)$ satisfying our conditions.

It remains to consider the case when G has a topology and $T(s)$ is a weakly continuous function of s. Let us show that $U(s)$ is then also a weakly continuous function of s, that is, the scalar-valued function $(U(s)k, k')$ is a continuous function of s, for any fixed $k, k' \in \mathfrak{K}$. Because $U(s)$ has a bound independent of s (in fact, $\|U(s)\| = 1$), and because, moreover, the linear combinations of the functions of the form $\delta_\sigma h$ $(\sigma \in G, h \in \mathfrak{H})$ (or, to be more exact, the corresponding equivalence classes modulo $\mathbf{N}$) are dense in $\mathfrak{K}$, one concludes that it suffices to prove that

$$(U(s)\delta_\sigma h, \delta_\tau h')$$

is a continuous function of s for any fixed $h, h' \in \mathfrak{H}$ and $\sigma, \tau \in G$. Now, this scalar product is equal to

$$(U(s)U(\sigma)h, U(\tau)h') = (U(\tau^{-1}s\sigma)h, h') = (T(\tau^{-1}s\sigma)h, h'),$$

and this is a continuous function of s because $T(s)$ was assumed to be a weakly continuous function of s. This finishes the proof of the theorem.

8 Some applications

1. Let us consider a function $T(n)$ on the additive group Z of the integers n, whose values are bounded operators on a Hilbert space $\mathfrak{H}$ and for which $T(0) = I$ and $T(-n) = T(n)^*$. According to the general definition, $T(n)$ is positive definite on Z if

$$\sum_{n=-\infty}^{\infty} \sum_{m=-\infty}^{\infty} (T(n-m)h_n, h_m) \geq 0 \tag{8.1}$$

for every two-way sequence $\{h_n\}_{-\infty}^{\infty}$ of elements of $\mathfrak{H}$, which is finitely nonzero, that is, such that $h_n \neq 0$ for a finite set of subscripts only. For such a sequence we can choose an integer a so that the sequence $\{h'_\nu\}_{-\infty}^{\infty}$ defined by

$$h'_\nu = h_{\nu+a}$$

satisfies $h'_\nu = 0$ for $\nu < 0$. When $\nu = n - a$ and $\mu = m - a$ we have $\nu - \mu = n - m$, and therefore (8.1) holds for the finitely nonzero two-way sequences if and only if it holds for the finitely nonzero one-way sequences $\{h_n\}_0^{\infty}$, that is, if we have

$$\sum_{n=0}^{\infty} \sum_{m=0}^{\infty} (T(n-m)h_n, h_m) \geq 0. \tag{8.1$'$}$$

As a first application let us consider the function $T(n)$ which derives from an operator T in the Hilbert space $\mathfrak{H}$ as follows:

$$T(n) = T^{*|n|} \quad (n \leq -1), \quad T(0) = I, \quad T(n) = T^n \quad (n \geq 1). \tag{8.2}$$

Let us observe that the reciprocal formulas

$$g_n = \sum_{n \leq m < \infty} T^{m-n} h_m, \quad h_n = g_n - T g_{n+1} \quad (n \geq 0) \tag{8.3}$$

define a one-to-one transformation $\{h_n\}_0^{\infty} \to \{g_n\}_0^{\infty}$ of the set of finitely nonzero sequences onto itself.

Thus our function $T(n)$ satisfies (8.1$'$) if and only if it satisfies

$$\sum_{n=0}^{\infty} \sum_{m=0}^{\infty} (T(n-m)(g_n - T g_{n+1}), g_m - T g_{m+1}) \geq 0 \tag{8.4}$$

for every finitely nonzero sequence $\{g_n\}_0^{\infty}$. Now the sum (8.4) can be rearranged into the sum

$$\sum_{n=0}^{\infty} \sum_{m=0}^{\infty} (D(n,m) g_n, g_m), \tag{8.5}$$

where

$$\begin{aligned}
D(0,0) &= T(0) = I, \\
D(1,0) &= T(1) - T(0)T = T - T = O, \\
D(0,1) &= T(-1) - T^*T(0) = T^* - T^* = O, \\
D(n,n) &= T(0) - T(-1)T - T^*T(1) + T^*T(0)T \\
&= I - T^*T - T^*T + T^*T = I - T^*T \quad \text{for } n \geq 1, \\
D(n,m) &= T(k) - T(k-1)T - T^*T(k+1) + T^*T(k)T \\
&= T^k - T^{k-1}T - T^*T^{k+1} + T^*T^kT = O \quad \text{for } n-m = k \geq 1, \\
D(n,m) &= T(-k) - T(-k-1)T - T^*T(-k+1) + T^*T(-k)T \\
&= T^{*k} - T^{*k+1}T - T^*T^{*k-1} + T^*T^{*k}T = O \quad \text{for } n-m = -k \leq -1.
\end{aligned}$$

Hence the sum (8.5) equals

$$(g_0, g_0) + \sum_{n=1}^{\infty} ((I - T^*T)g_n, g_n).$$

In order that this sum be nonnegative for every finitely nonzero sequence $\{g_n\}_0^\infty$, it is necessary and sufficient that we have $I - T^*T \geq O$; that is, $\|T\| \leq 1$. So we have obtained:

The contractions T are characterized by the property that $T(n)$ is a positive definite function on the group Z.

It follows by applying Theorem 7.1 that for every contraction T on $\mathfrak{H}$ there exists a unitary representation $U(n)$ of the group Z on some space $\mathfrak{K} \supset \mathfrak{H}$, such that $T(n) = \text{pr}\, U(n)$ holds for every integer n and that $\mathfrak{K}$ be spanned by the subspaces $U(n)\mathfrak{H}$ $(-\infty < n < \infty)$. Setting $U(1) = U$ we have $U(n) = U^n$ for each n, and hence we have in particular $T^n = \text{pr}\, U^n$ for $n \geq 0$; that is, U is a minimal unitary dilation of T.

So we have obtained *a new proof of the existence of a minimal unitary dilation of a contraction.*

2. Let us consider now a *continuous one-parameter semigroup $\{T(s)\}_{s \geq 0}$ of contractions on $\mathfrak{H}$*. That is, $T(s)$ is, for every value of the real parameter $s \geq 0$, an operator on $\mathfrak{H}$ such that

$$\left.\begin{array}{ll}
T(0) = I; & \\
T(s_1 + s_2) = T(s_1)T(s_2) & \text{for } s_1, s_2 \geq 0; \\
\|T(s)\| \leq 1 & \text{for } s \geq 0; \\
T(s) \rightharpoonup I & \text{as } s \to +0.
\end{array}\right\} \qquad (8.6)$$

These conditions imply the *strong continuity* of $T(s)$. In fact, if $0 \le s_1 < s_2$ and $h \in \mathfrak{H}$, we have for $\sigma = s_2 - s_1$,

$$\begin{aligned}\|T(s_2)h - T(s_1)h\|^2 &= \|T(s_1)[T(\sigma)h - h]\|^2 \\ &\le \|T(\sigma)h - h\|^2 = \|T(\sigma)h\|^2 - 2\,\mathrm{Re}(T(\sigma)h,h) + \|h\|^2 \\ &\le 2\|h\|^2 - 2\,\mathrm{Re}(T(\sigma)h,h) = 2\,\mathrm{Re}(h - T(\sigma)h,h);\end{aligned}$$

the last term tends to 0 as $\sigma \to +0$, and hence the assertion follows.

Relations (8.6) are obviously invariant with respect to taking adjoints: if the one-parameter semigroup $\{T(s)\}_{s\ge 0}$ is continuous, then so is $\{T(s)^*\}_{s\ge 0}$. In particular $T(s)^*$ is also a strongly continuous function of s $(s \ge 0)$.

Let $\{T(s)\}_{s\ge 0}$ be a continuous one-parameter semigroup of contractions. We extend it to a function $T(s)$ on the whole real line R by setting $T(-s) = T(s)^*$. We show that the function thus obtained is positive definite on the additive group R, that is,

$$\sum_s\sum_t (T(s-t)h(s),h(t)) \ge 0 \tag{8.7}$$

for every finitely nonzero function $h(s)$ from R to $\mathfrak{H}$. Suppose $h(s_n) = h_n \neq 0$ for the finite subset $S = \{s_n\}$ of points of R and $h(s) = 0$ for the other points. Then we have to show that the sum

$$\sum_n\sum_m (T(s_n - s_m)h_n,h_m) \tag{8.7'}$$

is nonnegative. In the particular case that the values s_n are commensurable (i.e., of the form $s_n = \nu_n \cdot d$ for some positive real number d and integers ν_n of any sign), the sum (8.7′) can be written in the form

$$\sum_n\sum_m (T_d(\nu_n - \nu_m)h_n,h_m),$$

where $T_d(n)$ is the function associated with the contraction $T_d = T(d)$ in the sense of Subsec. 1. Hence, positivity of the sum (8.7′) follows in this case from the results of Subsec. 1. When the points of the set S are not commensurable, we replace S by a set $S^{(k)} = \{s_\nu^{(k)}\}$ of commensurable (e.g., rational) points converging to S as $k \to \infty$ (i.e., $s_\nu^{(k)} \to s_\nu$ for each ν). For $S^{(k)}$, the sum (8.7′) is nonnegative, and this property is preserved when we pass to the limit S, owing to the weak continuity of $T(s)$.

One can therefore apply Theorem 7.1 to arrive at the following result.

Theorem 8.1. *For every continuous one-parameter semigroup $\{T(s)\}_{s\ge 0}$ of contractions on $\mathfrak{H}$ there exists a continuous one-parameter group $\{U(s)\}_{-\infty}^{\infty}$ of unitary operators on a space $\mathfrak{K} \supset \mathfrak{H}$, such that*

$$T(s) = \mathrm{pr}\, U(s) \qquad (0 \le s < \infty)$$

and

$$\mathfrak{K} = \bigvee_{s\in R} U(s)\mathfrak{H} \qquad \text{(minimality condition).}$$

These conditions determine $U(s)$ up to an isomorphism and we call it "the minimal unitary dilation" of the given semigroup of contractions.

3. Let B_λ be an operator-valued distribution function on the interval $0 < \lambda \leq 2\pi$; thus B_λ is, for every value of λ, a bounded self-adjoint operator on the complex Hilbert space $\mathfrak{H}$, such that $B_\lambda \leq B_\mu$ for $\lambda < \mu$, $B_\lambda = B_{\lambda+0}$, $B_{+0} = O$, and $B_{2\pi} = I$. The integrals

$$T(n) = \int_0^{2\pi} e^{in\lambda}\, dB_\lambda \qquad (n = 0, \pm 1, \ldots)$$

exist in an obvious sense (as limits in operator norm of Riemann-type sums) and define an operator-valued function $T(n)$ on the group Z, such that $T(0) = I$ and $T(-n) = T(n)^*$. Moreover,

$$\begin{aligned}\sum_n \sum_m (T(n-m)h_n, h_m) &= \int_0^{2\pi} \sum_n \sum_m e^{i(n-m)\lambda} d(B_\lambda h_n, h_m) \\ &= \int_0^{2\pi} \sum_n \sum_m e^{i(n-m)\lambda} (B(d\lambda)h_n, h_m) \\ &= \int_0^{2\pi} \left(B(d\lambda) \sum_n e^{in\lambda} h_n, \sum_m e^{im\lambda} h_m \right) \geq 0,\end{aligned}$$

the last integral denoting the limit of the sums

$$\sum_k \left((B(\lambda_{k+1}) - B(\lambda_k)) \sum_n e^{in\lambda_k} h_n, \sum_n e^{in\lambda_k} h_n \right),$$

where $\lambda_0 = 0 < \lambda_1 < \cdots < \lambda_k < \cdots < \lambda_l = 2\pi$ and $\max(\lambda_{k+1} - \lambda_k) \to 0$. Thus we can apply Theorem 7.1; it follows that there exists a unitary operator $U = \int_0^{2\pi} e^{i\lambda} dE_\lambda$ on a complex Hilbert space $\mathfrak{K} \supset \mathfrak{H}$, such that $T(n) = \mathrm{pr}\, U(n)$ $(n = 0, \pm 1, \ldots)$; that is,

$$\int_0^{2\pi} e^{in\lambda} d(B_\lambda h, h') = \int_0^{2\pi} e^{in\lambda} d(E_\lambda h, h') \qquad (h, h' \in \mathfrak{H}) \tag{8.8}$$

for all integers n. If we choose $\{E_\lambda\}$ so that it satisfies the same condition of normalization as $\{B_\lambda\}$ (i.e., $E_\lambda = E_{\lambda+0}$, $E_{+0} = O$, $E_{2\pi} = I_{\mathfrak{K}}$), then (8.8) implies

$$B_\lambda = \mathrm{pr}\, E_\lambda \quad \text{for} \quad 0 < \lambda \leq 2\pi.$$

So we have proved the following result.

Theorem 8.2. *For every operator-valued distribution function B_λ there exists an orthogonal projection-valued distribution function E_λ (i.e. a spectral family) in some larger space such that $B_\lambda = pr\ E_\lambda$.*

Let us note that, in the above proof, the interval of variation of the parameter λ was $(0, 2\pi]$, but the result extends to the case of any finite or infinite interval, by using a continuous monotonic change of variable. Let us also note that the case of real spaces can be reduced to the case of complex spaces by an obvious "complexification."

4. Finally we give an important inequality that can be easily derived from Theorem 4.2. Let T be a contraction on the complex space $\mathfrak{H}$, and let U be a unitary dilation of T on the (complex) space $\mathfrak{K} \supset \mathfrak{H}$. The relations $T^n = \mathrm{pr}\, U^n$ $(n = 0, 1, \ldots)$ imply $p(T) = \mathrm{pr}\, p(U)$ for every polynomial $p(\lambda) = c_0 + c_1\lambda + \cdots + c_n\lambda^n$ with real or complex coefficients, and hence $\|p(T)\| \leq \|p(U)\|$. Now it follows from the spectral representation of unitary operators, that $\|p(U)\|$ is equal to the maximum of $|p(\lambda)|$ on the spectrum of U, and thus $\|p(U)\| \leq \max_{|\lambda|=1} |p(\lambda)|$. Consequently, $\|p(T)\| \leq \max_{|\lambda|=1} |p(\lambda)|$.

So we have proved the following proposition.

Proposition 8.3 (von Neumann inequality). *For every contraction T on the complex space $\mathfrak{H}$ and for every polynomial $p(\lambda)$ we have*

$$\|p(T)\| \leq \max_{|\lambda|\leq 1} |p(\lambda)|. \tag{8.9}$$

9 Regular unitary dilations of commutative systems

Let $\mathscr{T} = \{T_\omega\}_{\omega\in\Omega}$ be a commutative system of contractions on the space $\mathfrak{H}$. Recall that a corresponding system $\mathscr{U} = (U_\omega)_{\omega\in\Omega}$ of operators on a space $\mathfrak{K} \supset \mathfrak{H}$ is a *unitary dilation* of the system $\mathscr{T}$ if $\mathscr{U}$ is also commutative, consists of unitary operators U_ω, and satisfies

$$\prod_1^r T_{\omega_i}^{n_i} = \mathrm{pr} \prod_1^r U_{\omega_i}^{n_i} \tag{9.1}$$

for every finite set of subscripts $\omega_i \in \Omega$ and of corresponding integers $n_i \geq 0$.

This definition can be expressed in a more convenient form if we introduce the class Z^Ω of the "vectors" $\mathbf{n}$ with the components n_ω $(\omega \in \Omega)$, where $\omega \mapsto n_\omega$ is a finitely nonzero function from Ω to the set of integers (of any sign). Z^Ω is an Abelian group with respect to the addition by components, the identity element being the vector $\mathbf{o}$ all of whose components are 0. When we speak in this section of "vectors" $\mathbf{n}$, $\mathbf{m}$, and so on, we always mean vectors in Z^Ω. If $n_\omega \geq 0$ for all ω, we write $\mathbf{n} \geq \mathbf{o}$; $\mathbf{n} \geq \mathbf{m}$ means that $\mathbf{n} - \mathbf{m} \geq \mathbf{o}$. For arbitrary $\mathbf{n}$, $\mathbf{m}$, we set $\mathbf{n} \cup \mathbf{m} = \{\max\{n_\omega, m_\omega\}\}$ and $\mathbf{n} \cap \mathbf{m} = \{\min\{n_\omega, m_\omega\}\}$; finally, we define $\mathbf{n}^+ = \mathbf{n} \cup \mathbf{o}$, $\mathbf{n}^- = -(\mathbf{n} \cap \mathbf{o})$.

Let us set, for a vector $\mathbf{n} \geq \mathbf{o}$,

$$T^{\mathbf{n}} = \prod_{\omega\in\Omega} T_\omega^{n_\omega}; \tag{9.2}$$

this product is well defined because $\mathscr{T}$ is commutative, and because there are only a finite number of factors different from I. Let us also set

$$U^{\mathbf{n}} = \prod_{\omega\in\Omega} U_\omega^{n_\omega}, \tag{9.3}$$

where the restriction $\mathbf{n} \geq \mathbf{o}$ is no longer necessary, because the unitary operators U_ω have (unitary) inverses U_ω^{-1}. Obviously, the operators $U^{\mathbf{n}}$ yield a unitary representation of the group Z^Ω. Conversely, every unitary representation $U(\mathbf{n})$ of the group

Z^Ω can be obtained this way. This follows from the fact that $\mathbf{n}$ can be written in the form

$$\mathbf{n} = \sum_{\rho \in \Omega} n_\rho \mathbf{e}_\rho,$$

where $\mathbf{e}_\rho$ is the vector whose only nonzero component is the ρth one, this being equal to 1; indeed, one has now only to set $U_\rho = U(\mathbf{e}_\rho)$.

Thus the problem of finding the unitary dilations $\mathscr{U}$ of $\mathscr{T}$ is equivalent to finding the unitary representations $U(\mathbf{n})$ of the group Z^Ω for which

$$T^{\mathbf{n}} = \operatorname{pr} U(\mathbf{n}) \quad \text{for} \quad \mathbf{n} \geq \mathbf{o}; \tag{9.1*}$$

that is, $T^{\mathbf{n}} = P_{\mathfrak{H}} U(\mathbf{n})|\mathfrak{H}$. Now if $U(\mathbf{n})$ is a unitary representation of Z^Ω having this property it is possible to extend the function $T^{\mathbf{n}}$ $(\mathbf{n} \geq \mathbf{o})$ to a function $T(\mathbf{n})$ that is defined and positive definite on the whole group Z^Ω: one has only to set

$$T(\mathbf{n}) = P_{\mathfrak{H}} U(\mathbf{n})|\mathfrak{H}, \quad \mathbf{n} \in Z^\Omega. \tag{9.4}$$

Conversely, for every extension of $T^{\mathbf{n}}$ to a positive definite function $T(\mathbf{n})$ on Z^Ω there exists, by Theorem 7.1, a unitary representation $U(\mathbf{n})$ of Z^Ω (even a minimal one) so that (9.4) holds.

Thus the problem of finding the (minimal) unitary dilations of the given system $\mathscr{T}$ of contractions is equivalent to the problem of finding the extensions of the function $T^{\mathbf{n}}$ $(\mathbf{n} \geq \mathbf{o})$ to a positive definite function $T(\mathbf{n})$ on Z^Ω. Clearly, to different extensions there correspond nonisomorphic unitary dilations.

When $\mathscr{T}$ consists of a single contraction T, the function T^n $(n \geq 0)$ has the unique positive definite extension to $Z^1 = Z$ defined by $T(-n) = T(n)^* = T^{*n}$ for $n > 0$; see Subsec. 1 of the preceding section. For $\mathscr{T}$ consisting of two (commuting) contractions the existence of at least one positive definite extension follows from Theorem 6.4. In contrast, it follows from Sec. 6.3 that for commutative systems of more than two contractions there exist in general no positive definite extensions of $T^{\mathbf{n}}$. However, we are able to exhibit certain particular cases in which such extensions do exist.

We proceed as follows. We extend $T^{\mathbf{n}}$ to a function $T(\mathbf{n})$ on Z^Ω by some simple rule which is amenable to calculations, and determine under which additional conditions this extension is positive definite.

One simple rule is to set

$$T(\mathbf{n}) = (T^{\mathbf{n}^-})^* T^{\mathbf{n}^+}; \tag{9.5}$$

let us call it the *regular extension*. It has a *dual* one which results from changing the order of the two factors on the right; however, because this means essentially replacing the system $\{T_\omega\}$ by the system $\{T_\omega^*\}$, the study of this dual extension reduces to the study of the regular one. One sees immediately that the regular extension satisfies the condition $T(-\mathbf{n}) = T(\mathbf{n})^*$. By a reasoning analogous to that given at the beginning of the preceding section (for the case of a single contraction), $T(\mathbf{n})$ is positive definite on Z^{Ω} if (and only if)

$$\sum_{\mathbf{n}\geq\mathbf{o}}\sum_{\mathbf{m}\geq\mathbf{o}}(T(\mathbf{n}-\mathbf{m})h(\mathbf{n}),h(\mathbf{m}))\geq 0 \tag{9.6}$$

for every finitely nonzero function $h(\mathbf{n})$ defined for $\mathbf{n}\geq\mathbf{o}$.

Now we make use of a generalization of the reciprocal formulas (8.3). Let us observe first that for every function $h(\mathbf{n})$ of the above type the function

$$g(\mathbf{n})=\sum_{\mathbf{m}\geq\mathbf{n}}T^{\mathbf{m}-\mathbf{n}}h(\mathbf{m}) \qquad (\mathbf{n}\geq\mathbf{o}) \tag{9.7}$$

is also finitely nonzero. We can retrieve the function $h(\mathbf{n})$ from the function $g(\mathbf{n})$ in the following way. For each finite subset v of Ω let us set

$$\mathbf{e}(v)=\{e_{\omega}(v)\},\quad e_{\omega}(v)=\begin{cases}1 & \text{if } \omega\in v,\\ 0 & \text{if } \omega\in\Omega\backslash v,\end{cases}$$

and let $|v|$ denote the number of the elements of v. With these notations, the reciprocal of the formula (9.7) is

$$h(\mathbf{n})=\sum_{v\subset\Omega}(-1)^{|v|}T^{\mathbf{e}(v)}g(\mathbf{n}+\mathbf{e}(v)) \qquad (\mathbf{n}\geq\mathbf{o}), \tag{9.8}$$

where, as indicated by the notation, v runs over the set of all the finite subsets of Ω.[3] In fact, for any fixed $\mathbf{n}$ ($\geq\mathbf{o}$) we have

$$\begin{aligned}\sum_{v\subset\Omega}(-1)^{|v|}T^{\mathbf{e}(v)}&\sum_{\mathbf{m}\geq\mathbf{n}+\mathbf{e}(v)}T^{\mathbf{m}-\mathbf{n}-\mathbf{e}(v)}h(\mathbf{m})\\ &=\sum_{v\subset\Omega}(-1)^{|v|}\sum_{\mathbf{m}\geq\mathbf{n}+\mathbf{e}(v)}T^{\mathbf{m}-\mathbf{n}}h(\mathbf{m})\\ &=\sum_{\mathbf{m}\geq\mathbf{n}}\left[\sum_{\mathbf{e}(v)\leq\mathbf{m}-\mathbf{n}}(-1)^{|v|}\right]T^{\mathbf{m}-\mathbf{n}}h(\mathbf{m})=h(\mathbf{n});\end{aligned}$$

here we have used the elementary theorem of combinatorics asserting that if v runs through all subsets of a finite set v_0 (the whole set and the empty set also admitted) then

$$\sum_{v\subset v_0}(-1)^{|v|}=\begin{cases}1 & \text{if } v_0 \text{ is empty},\\ 0 & \text{if } v_0 \text{ is not empty}.\end{cases} \tag{9.9}$$

Conversely, if one starts with an arbitrary, finitely nonzero function $g(\mathbf{n})$ ($\mathbf{n}\geq\mathbf{o}$) then the function $h(\mathbf{n})$ which it generates by formula (9.8) is also finitely nonzero,

---

[3] The function $g(\mathbf{n})$ is finitely nonzero, and hence there are only a finite number of nonzero terms of the sum (9.8).

and for every fixed $\mathbf{n}$ ($\geq \mathbf{o}$) we have

$$\begin{aligned}\sum_{\mathbf{m}\geq\mathbf{n}} T^{\mathbf{m}-\mathbf{n}}h(\mathbf{m}) &= \sum_{v\subset\Omega}(-1)^{|v|}\sum_{\mathbf{m}\geq\mathbf{n}} T^{\mathbf{m}-\mathbf{n}+\mathbf{e}(v)}g(\mathbf{m}+\mathbf{e}(v)) \\ &= \sum_{v\subset\Omega}(-1)^{|v|}\sum_{\mathbf{p}\geq\mathbf{n}+\mathbf{e}(v)} T^{\mathbf{p}-\mathbf{n}}g(\mathbf{p}) \\ &= \sum_{\mathbf{p}\geq\mathbf{n}}\left(\sum_{\mathbf{e}(v)\leq\mathbf{p}-\mathbf{n}}(-1)^{|v|}\right)g(\mathbf{p}) = g(\mathbf{n}),\end{aligned}$$

again by virtue of (9.9).

We conclude that formulas (9.7)–(9.8) give a transformation of the set of finitely nonzero functions (defined for $\mathbf{n}\geq\mathbf{o}$) onto itself, and the inverse of this transformation.

Consequently, in order that (9.6) hold for every finitely nonzero function $h(\mathbf{n})$ ($\mathbf{n}\geq\mathbf{o}$) it is necessary and sufficient that the sum

$$\sum_{\mathbf{n}\geq\mathbf{o}}\sum_{\mathbf{m}\geq\mathbf{o}}\left(T(\mathbf{n}-\mathbf{m})\sum_{v\subset\Omega}(-1)^{|v|}T^{\mathbf{e}(v)}g(\mathbf{n}+\mathbf{e}(v)),\ \sum_{w\subset\Omega}(-1)^{|w|}T^{\mathbf{e}(w)}g(\mathbf{m}+\mathbf{e}(w))\right)$$

be ≥ 0 for every finitely nonzero function $g(\mathbf{n})$ ($\mathbf{n}\geq\mathbf{o}$). Now this sum can be written in the form

$$\sum_{\mathbf{p}\geq\mathbf{o}}\sum_{\mathbf{q}\geq\mathbf{o}}(D(\mathbf{p},\mathbf{q})g(\mathbf{p}),g(\mathbf{q})), \tag{9.10}$$

where

$$D(\mathbf{p},\mathbf{q}) = \sum_{v\subset\pi(\mathbf{p})}\sum_{w\subset\pi(\mathbf{q})}(-1)^{|v|+|w|}(T^{\mathbf{e}(w)})^*T(\mathbf{p}-\mathbf{e}(v)-\mathbf{q}+\mathbf{e}(w))T^{\mathbf{e}(v)} \tag{9.11}$$

and $\pi(\mathbf{n})$ is the set defined for every vector $\mathbf{n}$ by

$$\pi(\mathbf{n}) = \{\omega:\ n_\omega > 0\}.$$

We now prove that $D(\mathbf{p},\mathbf{q}) = O$ if $\mathbf{p}\neq\mathbf{q}$. Observe first that if $\mathbf{p}\neq\mathbf{q}$ then the sets $\pi(\mathbf{p}-\mathbf{q})$ and $\pi(\mathbf{q}-\mathbf{p})$ cannot both be empty. By reason of symmetry, it suffices therefore to consider the case when $\pi(\mathbf{p}-\mathbf{q})$ is not empty. The set

$$\delta(w) = \pi(\mathbf{p})\cap\pi(\mathbf{p}-\mathbf{q}+\mathbf{e}(w))$$

is then nonempty for every finite subset w of Ω and, in fact, contains the set $\pi(\mathbf{p}-\mathbf{q})$.

Next we observe that for every fixed w one obtains all subsets v of $\pi(\mathbf{p})$ by taking $v = v'\cup v''$, where v' and v'' satisfy the conditions

$$v'\subset\pi(\mathbf{p})\backslash\delta(w) \quad\text{and}\quad v''\subset\delta(w).$$

We have $|v| = |v'| + |v''|$ and $\mathbf{e}(v) = \mathbf{e}(v') + \mathbf{e}(v'')$; then $D(\mathbf{p},\mathbf{q})$ is equal to

$$\sum_{w\subset\pi(\mathbf{q})} (-1)^{|w|}(T^{\mathbf{e}(w)})^* \sum_{v'\subset\pi(\mathbf{p})\setminus\delta(w)} (-1)^{|v'|}\left[\sum_{v''\subset\delta(w)} (-1)^{|v''|}T(\mathbf{p}-\mathbf{q}+\mathbf{e}(w) - \mathbf{e}(v') - \mathbf{e}(v''))T^{\mathbf{e}(v'')}\right]T^{\mathbf{e}(v')}.$$

Now we show that the sum between the brackets [ ] equals O for every fixed w and v'. In fact, it follows from definition (9.5) of the regular extension that

$$T(\mathbf{p}-\mathbf{q}+\mathbf{e}(w)-\mathbf{e}(v')-\mathbf{e}(v''))T^{\mathbf{e}(v'')} = (T^{\mathbf{a}})^*T^{\mathbf{b}},$$

where

$$\mathbf{a} = [\mathbf{p}-\mathbf{q}+\mathbf{e}(w)-\mathbf{e}(v')-\mathbf{e}(v'')]^-$$

and

$$\mathbf{b} = [\mathbf{p}-\mathbf{q}+\mathbf{e}(w)-\mathbf{e}(v')-\mathbf{e}(v'')]^+ + \mathbf{e}(v'').$$

Now $\mathbf{a}$ and $\mathbf{b}$ do not depend on v''; namely we have

$$\mathbf{a} = [\mathbf{p}-\mathbf{q}+\mathbf{e}(w)-\mathbf{e}(v')]^- \quad \text{and} \quad \mathbf{b} = [\mathbf{p}-\mathbf{q}+\mathbf{e}(w)-\mathbf{e}(v')]^+,$$

which is a consequence of the fact that for $\omega \in v''$ we have $p_\omega - q_\omega + e_\omega(w) \geq 1$ and $e_\omega(v') = 0$. Thus the above sum between the brackets [ ] is equal to $(T^{\mathbf{a}})^*T^{\mathbf{b}}\sum_{v''\subset\delta(w)}(-1)^{|v''|}$, and this is equal to O because the set $\delta(w)$ is not empty. This proves that $D(\mathbf{p},\mathbf{q}) = O$.

The sum (9.10) reduces thus to the form

$$\sum_{\mathbf{p}\geq\mathbf{o}} (D(\mathbf{p},\mathbf{p})g(\mathbf{p}),g(\mathbf{p})).$$

In order that this sum be nonnegative for every finitely nonzero function $g(\mathbf{p})$ ($\mathbf{p} \geq \mathbf{o}$) it is necessary and sufficient that we have $D(\mathbf{p},\mathbf{p}) \geq O$ for every $\mathbf{p} \geq \mathbf{o}$. Now, (9.11) gives

$$D(\mathbf{p},\mathbf{p}) = \sum_{v\subset\pi(\mathbf{p})}\sum_{w\subset\pi(\mathbf{p})} (-1)^{|v|+|w|}(T^{\mathbf{e}(w)+[\mathbf{e}(w)-\mathbf{e}(v)]^-})^*T^{[\mathbf{e}(w)-\mathbf{e}(v)]^+ + \mathbf{e}(v)}.$$

It is easily seen that both exponents of T are equal to $\mathbf{e}(u)$ where $u = v\cup w$. Using (9.9) again we obtain

$$\begin{aligned}\sum_{v\cup w=u}{}^{(v,w)}(-1)^{|v|+|w|} &= \sum_{v\subset u}{}^{(v)}(-1)^{|v|}\sum_{\substack{w=(u\setminus v)\cup v'\\ v'\subset v}}{}^{(w)}(-1)^{|w|}\\ &= \sum_{v\subset u}{}^{(v)}(-1)^{|v|+|u\setminus v|}\sum_{v'\subset v}{}^{(v')}(-1)^{|v'|} = (-1)^{|u|},\end{aligned}$$

and finally

$$D(\mathbf{p},\mathbf{p}) = \sum_{u\subset\pi(\mathbf{p})} (-1)^{|u|}(T^{\mathbf{e}(u)})^* T^{\mathbf{e}(u)}.$$

Thus we have obtained the following result.

Theorem 9.1. *Let $\mathscr{T} = \{T_\omega\}_{\omega\in\Omega}$ be a commutative system of contractions on the space $\mathfrak{H}$. In order that $\mathscr{T}$ have a unitary dilation $\mathscr{U} = \{U_\omega\}_{\omega\in\Omega}$ on a space $\mathfrak{K}\supset\mathfrak{H}$, which is regular, that is, such that*

$$(T^{\mathbf{n}^-})^* T^{\mathbf{n}^+} = \operatorname{pr} U^{\mathbf{n}} \quad \textit{for all} \quad \mathbf{n} = \{n_\omega\}_{\omega\in\Omega} \in Z^\Omega,$$

where

$$U^{\mathbf{n}} = \prod_{\omega\in\Omega} U_\omega^{n_\omega} \quad \textit{for all} \quad \mathbf{n}, \qquad T^{\mathbf{n}} = \prod_{\omega\in\Omega} T_\omega^{n_\omega} \quad \textit{for} \quad \mathbf{n}\geq\mathbf{o},$$

it is necessary and sufficient that we have

$$S(u) = \sum_{v\subset u} (-1)^{|v|}(T^{\mathbf{e}(v)})^* T^{\mathbf{e}(v)} \geq O \tag{9.12}$$

for every finite subset u of Ω. Moreover, one can require that $\mathscr{U}$ be minimal, that is, the subspaces $U^{\mathbf{n}}\mathfrak{H}$ ($\mathbf{n}\in Z^\Omega$) span the space $\mathfrak{K}$. In this case the regular unitary dilation $\mathscr{U}$ is determined up to isomorphism.

Remark 1. Suppose the system $\mathscr{T}$ contains an isometry, say T_{ω_0}. Then $S(u) = O$ for every finite subset u of Ω containing ω_0.

In fact, for any such u we obtain all subsets v of u by taking $v = v_0\cup v_1$, where v_0 and v_1 satisfy $v_0\subset u_0 = \{\omega_0\}$ and $v_1\subset u_1 = u\backslash u_0$. We have

$$T^{\mathbf{e}(v)*}T^{\mathbf{e}(v)} = T^{\mathbf{e}(v_1)*}T^{\mathbf{e}(v_0)*}T^{\mathbf{e}(v_0)}T^{\mathbf{e}(v_1)} = T^{\mathbf{e}(v_1)*}T^{\mathbf{e}(v_1)}$$

because $T^{\mathbf{e}(v_0)}$ equals T_{ω_0} or I accordingly as v_0 is the set u_0 or the empty set (the one-point set u_0 has just these two subsets). Because $|v| = |v_0| + |v_1|$, we have

$$S(u) = \sum_{v_1\subset u_1} (-1)^{|v_1|}\left[\sum_{v_0\subset u_0} (-1)^{|v_0|}\right] T^{\mathbf{e}(v_1)*}T^{\mathbf{e}(v_1)} = O,$$

where we have used (9.9), in this case for the one point set u_0.

Let us introduce the following notion. We say that the operators A and B *doubly commute* if A commutes with B and B^* (and, therefore, B commutes with A and A^*).

Remark 2. Let u_d be a nonempty subset of a finite subset u of Ω such that T_ω and $T_{\omega'}$ doubly commute whenever $\omega\in u_d$ and $\omega'\in u$, $\omega'\neq\omega$. Let $u_c = u\backslash u_d$. If $S(u_c)\geq O$ then $S(u)\geq O$ also.

In fact, one has

$$\begin{aligned} S(u) &= \sum_{v_d \subset u_d} \sum_{v_c \subset u_c} (-1)^{|v_d|+|v_c|} T^{\mathbf{e}(v_d)*} T^{\mathbf{e}(v_c)*} T^{\mathbf{e}(v_c)} T^{\mathbf{e}(v_d)} \\ &= \sum_{v_c \subset u_c} (-1)^{|v_c|} T^{\mathbf{e}(v_c)*} T^{\mathbf{e}(v_c)} \sum_{v_d \subset u_d} (-1)^{|v_d|} \prod_{\omega \in v_d} T_\omega^* T_\omega \\ &= \sum_{v_c \subset u_c} (-1)^{|v_c|} T^{\mathbf{e}(v_c)*} T^{\mathbf{e}(v_c)} \prod_{\omega \in u_d} (I - T_\omega^* T_\omega) \\ &= S(u_c) \cdot \prod_{\omega \in u_d} (I - T_\omega^* T_\omega). \end{aligned}$$

Because the factors $I - T_\omega^* T_\omega$ $(\omega \in u_d)$ are nonnegative and they commute with each other as well as with $S(u_c)$, we see that $S(u_c) \geq O$ implies $S(u) \geq O$.

Remark 3. If $\sum_{\omega \in u} \|T_\omega\|^2 \leq 1$, then $S(u) \geq 0$.

In fact, let $u = \{\omega_1, \ldots, \omega_r\}$ and write, for the sake of brevity, T_i in place of T_{ω_i}. For $0 \leq p \leq r$ and for $h \in \mathfrak{H}$, set

$$a_p(h) = \sum_{\substack{v \subset u \\ |v| = p}} \|T^{\mathbf{e}(v)} h\|^2.$$

Then for $1 \leq p \leq r$ we have

$$\begin{aligned} a_p(h) &= \sum_{1 \leq i_1 < \ldots < i_p \leq r} \|T_{i_p} \ldots T_{i_1} h\|^2 \leq \sum_{1 \leq i_1 < \cdots < i_p \leq r} \|T_{i_p}\|^2 \|T_{i_{p-1}} \ldots T_{i_1} h\|^2 \\ &= \sum_{1 \leq i_1 < \ldots < i_{p-1} \leq r} \|T_{i_{p-1}} \ldots T_{i_1} h\|^2 \sum_{i_{p-1} < i_p \leq r} \|T_{i_p}\|^2 \\ &\leq \sum_{1 \leq i_1 < \ldots < i_{p-1} \leq r} \|T_{i_{p-1}} \ldots T_{i_1} h\|^2 = a_{p-1}(h), \end{aligned}$$

and hence

$$\begin{aligned} (S(u)h, h) &= \sum_{v \subset u} (-1)^{|v|} \|T^{\mathbf{e}(v)} h\|^2 = \sum_{p=0}^{r} (-1)^p a_p(h) \\ &\geq a_0(h) - a_1(h) = \|h\|^2 - \sum_{i=1}^{r} \|T_i h\|^2 \geq \left(1 - \sum_{i=1}^{r} \|T_i\|^2\right) \|h\|^2 \geq 0. \end{aligned}$$

By virtue of the above remarks, Theorem 9.1 implies the following result.

Proposition 9.2. *Let $\mathcal{T}$ be a commutative system of contractions. Delete from $\mathcal{T}$ the isometries, and from the rest, denoted by $\mathcal{T}_1$, delete those operators that doubly commute with every other operator in $\mathcal{T}_1$; let the rest be denoted by $\mathcal{T}_2$. If $\mathcal{T}_2$ has a regular unitary dilation* (in particular, if $\mathcal{T}_2$ is empty) *then so does $\mathcal{T}$.*

In particular $\mathcal{T} = \{T_\omega\}$ has a regular unitary dilation in each of the cases below:

(i) *$\mathcal{T}$ consists of isometries,*
(ii) *$\mathcal{T}$ consists of doubly commuting contractions,*
(iii) *$\mathcal{T}$ is countable and $\sum_\omega \|T_\omega\|^2 \leq 1$.*

When V and U are isometries, the relation $V = \operatorname{pr} U$ implies $V \subset U$ and hence the part (i) of the above proposition implies that *for every commutative system* $\{V_\omega\}$ *of isometries there exists a commutative system* $\{U_\omega\}$ *of unitary operators on some larger space such that* U_ω *is an extension of* V_ω *for every* ω. Thus we have obtained a new proof of Proposition 6.2, valid for finite as well as for infinite systems.

10 Another method to construct isometric dilations

1. We sketch one more method of constructing an isometric dilation of a contraction. Its interest is due mainly to the fact that it carries over, with obvious modifications, to one-parameter semigroups of contractions too.

Let T be a contraction on the Hilbert space $\mathfrak{H}$. For each $h \in \mathfrak{H}$ we have

$$\begin{aligned}\|h\|^2 &= \|D_T h\|^2 + \|Th\|^2 = \|D_T h\|^2 + \|D_T Th\|^2 + \|T^2 h\|^2 = \cdots \\ &= \|D_T h\|^2 + \|D_T Th\|^2 + \cdots + \|D_T T^{n-1} h\|^2 + \|T^n h\|^2.\end{aligned}$$

The inequalities

$$\|h\| \geq \|Th\| \geq \cdots \geq \|T^n h\| \geq \cdots \geq 0 \tag{10.1}$$

imply therefore

$$\|h\|^2 = \sum_{j=0}^{\infty} \|D_T T^j h\|^2 + \lim_{n\to\infty} \|T^n h\|^2. \tag{10.2}$$

From (10.1) it also follows that

$$I \geq T^*T \geq \cdots \geq T^{*n}T^n \geq \cdots \geq O,$$

and hence $S = \lim_{n\to\infty} T^{*n}T^n$ exists in the sense of strong operator convergence. Set $Q = S^{1/2}$. Then

$$T^* Q^2 T = \lim_n T^{*n+1} T^{n+1} = Q^2,$$

so we have $\|QTh\|^2 = \|Qh\|^2$ $(h \in \mathfrak{H})$, and the transformation $Qh \to QTh$ is isometric. It extends by continuity to an isometry W from $\mathfrak{Q} = \overline{Q\mathfrak{H}}$ into $\mathfrak{Q}$. Thus we have

$$QT = WQ. \tag{10.3}$$

Let us now consider the space $\mathbf{H}$ of vectors $\mathbf{h} = \{h_n\}_{-\infty}^{\infty}$ with $h_n \in \mathfrak{D}_T$ and

$$\|\mathbf{h}\|^2 = \sum_{-\infty}^{\infty} \|h_n\|^2 < \infty.$$

We embed $\mathfrak{H}$ in $\mathfrak{K} = \mathbf{H} \oplus \mathfrak{Q}$ by identifying the element h of $\mathfrak{H}$ with the element

$$\{\ldots, 0, 0, \boxed{D_T h}, D_T Th, D_T T^2 h, \ldots\} \oplus Qh$$

of $\mathfrak{K}$, this being justified by the isometry expressed by (10.2). (We have put a square around the 0th component.)

Let V be the bilateral shift on $\mathbf{H}$ defined by

$$V\{h_n\} = \{h'_n\}, \quad \text{where} \quad h'_n = h_{n+1} \qquad (n = 0, \pm 1, \ldots);$$

then $U = V \oplus W$ is an isometry on $\mathfrak{K}$, and by (10.3) it follows for $h \in \mathfrak{H}$ and $m = 1, 2, \ldots$ that

$$U^m h - T^m h = \{h_n^{(m)}\}_{n=-\infty}^{\infty} \oplus 0, \tag{10.4}$$

where

$$h_n^{(m)} = \begin{cases} D_T T^{m+n} h & \text{if } -m \leq n \leq -1, \\ 0 & \text{in the other cases.} \end{cases}$$

Clearly, the vector on the right-hand side of (10.4) is orthogonal to $\mathfrak{H}$ and hence $T^m = \operatorname{pr} U^m$ $(m = 1, 2, \ldots)$. Thus U is an isometric dilation of T.

2. Consider the case of a continuous one-parameter semigroup $\{T(s)\}_{s \geq 0}$ of contractions on $\mathfrak{H}$; see Sec. 8.2. This semigroup has an *infinitesimal generator* A, defined by

$$Ah = \lim_{s \to +0} \frac{1}{s}[T(s) - I]h \tag{10.5}$$

whenever this limit exists (in the strong sense); A is a closed linear operator with domain $\mathfrak{D}(A)$ dense in $\mathfrak{H}$ and we have $(d/ds)T(s)h = AT(s)h = T(s)Ah$ for $h \in \mathfrak{D}(A)$, $s \geq 0$; see [*Func. Anal.*] Sec. 142.

Let us note that $\|T(s)\| \leq 1$ implies

$$\operatorname{Re}((T(s) - I)h, h) = \operatorname{Re}(T(s)h, h) - (h, h) \leq \|T(s)\| \|h\|^2 - \|h\|^2 \leq 0$$

for $h \in \mathfrak{H}$; hence it follows

$$\operatorname{Re}(Ah, h) \leq 0 \quad \text{for} \quad h \in \mathfrak{D}(A). \tag{10.6}$$

We define by

$$[h, k] = -(Ah, k) - (h, Ak) \tag{10.7}$$

a bilinear form on $\mathfrak{D}(A)$; by virtue of (10.6) it is semidefinite, that is,

$$[h, h] = -2 \operatorname{Re}(Ah, h) \geq 0 \qquad (h \in \mathfrak{D}(A)). \tag{10.8}$$

It follows from the definition of A that for $h \in \mathfrak{D}(A)$ and $s \geq 0$,

$$\begin{aligned} \frac{d}{ds}\|T(s)h\|^2 &= 2 \operatorname{Re}(T(s)Ah, T(s)h) \\ &= 2 \operatorname{Re}(AT(s)h, T(s)h) = -[T(s)h, T(s)h] \end{aligned}$$

and consequently

$$\|h\|^2 = \int_0^t [T(s)h, T(s)h] ds + \|T(t)h\|^2.$$

Thus

$$\|h\|^2 = \int_0^\infty [T(s)h, T(s)h]ds + \lim_{t\to\infty} \|T(t)h\|^2, \tag{10.9}$$

where the limit exists because $\|T(t)h\|^2$ is a nonincreasing function of t. It also follows that $\lim_{t\to\infty} T(t)^*T(t)$ exists in the sense of strong operator convergence; let $Q(\geq O)$ be the square root of this limit. Then we construct, in analogy with the discrete case, a continuous semigroup $\{W(s)\}_{s\geq 0}$ of isometries on $\mathfrak{Q} = \overline{Q\mathfrak{H}}$ such that

$$QT(s) = W(s)Q \qquad (s \geq 0). \tag{10.10}$$

Let $\mathfrak{D}$ be the completion, with respect to the metric (10.7)–(10.8), of the pre-Hilbert space $\mathfrak{D}(A)$ modulo the subspace formed by the vectors h for which $[h,h] = 0$. Then we consider the functions $\mathbf{h} = \mathbf{h}(s)$ $(-\infty < s < \infty)$ with values in $\mathfrak{D}$, strongly measurable and such that

$$\|\mathbf{h}\|^2 = \int_{-\infty}^{\infty} \|\mathbf{h}(s)\|^2 ds < \infty.$$

These functions form a Hilbert space $\mathbf{H}$, where as usual we do not distinguish two functions as elements of this space if they coincide almost everywhere. We embed the space $\mathfrak{H}$ in the space

$$\mathfrak{K} = \mathbf{H} \oplus \mathfrak{Q}$$

by identifying the element h of $\mathfrak{H}$ with the element $\mathbf{h} \oplus Qh$ of $\mathfrak{K}$, where

$$\mathbf{h}(s) = \begin{cases} 0 & \text{for } s < 0, \\ T(s)h & \text{for } s \geq 0. \end{cases}$$

Let $\{V(s)\}_{s\geq 0}$ be the continuous semigroup defined by

$$(V(t)\mathbf{h})(s) = \mathbf{h}(s+t) \qquad (-\infty < s < \infty; t \geq 0; \mathbf{h} \in \mathbf{H}).$$

Then the operators $U(s) = V(s) \oplus W(s)$ $(s \geq 0)$ form a continuous semigroup of isometries on $\mathfrak{K}$. For $h \in \mathfrak{H}$ and $t \geq 0$ we have

$$U(t)h - T(t)h = \mathbf{h}^{(t)} \oplus 0, \tag{10.11}$$

where

$$\mathbf{h}^{(t)}(s) = \begin{cases} T(s+t)h & \text{if } -t \leq s < 0, \\ 0 & \text{in the other cases.} \end{cases}$$

The right-hand side of (10.11) is obviously orthogonal to $\mathfrak{H}$, and hence we have $T(t) = \text{pr}\, U(t)$ $(t \geq 0)$. Thus the semigroup $\{U(s)\}_{s\geq 0}$ is a dilation of the semigroup $\{T(s)\}_{s\geq 0}$.

3. For an example we construct an isometric dilation of a concrete continuous semigroup of contractions, generated by a system of differential equations. In the particular case we consider, the use of the infinitesimal generator can be avoided.

The system we consider is of the form

$$\frac{dx_i}{dt} = X_i(x) \qquad (i = 1, \ldots, n), \tag{10.12}$$

where $x = (x_1, \ldots, x_n)$ lies in the n-dimensional real Euclidean space R^n. We suppose that the functions X_i are continuously differentiable and that the divergence of $(X_1, \ldots, X_n)$ is nonpositive, that is,

$$\rho(x) = -\sum_{i=1}^{n} \frac{\partial}{\partial x_i} X_i(x) \geq 0.$$

We also suppose that for every $x \in R^n$ the solution $x(t)$ of (10.12) with $x(0) = x$ exists not only on a small neighborhood of $t = 0$ but on the whole t-axis. In this case

$$\tau_t : \ x \to x(t)$$

is a differentiable transformation of R^n onto itself. The functional determinant

$$\delta_t(x) = \frac{D(\tau_t x)}{D(x)} = \frac{D((\tau_t x)_1, \ldots, (\tau_t x)_n)}{D(x_1, \ldots, x_n)}$$

can be calculated easily: in fact, $\delta_t(x)$ is the Wronskian of the solutions

$$\left\{ \frac{\partial(\tau_t x)_1}{\partial x_j}, \ldots, \frac{\partial(\tau_t x)_n}{\partial x_j} \right\} \qquad (j = 1, \ldots, n)$$

of the system of linear differential equations

$$\frac{du_i}{dt} = \sum_{k=1}^{n} \frac{\partial X_i(\tau_t x)}{\partial x_k} u_k \qquad (i = 1, \ldots, n),$$

associated with the system (10.12), and hence one derives, using Liouville's theorem and the fact that $\delta_0(x) = 1$, the formula

$$\delta_t(x) = \exp\left(-\int_0^t \rho(\tau_s x) ds \right). \tag{10.13}$$

Thus for any Borel-measurable function $\varphi(x)$ integrable on R^n we have

$$\int_{R^n} \varphi(\tau_{-t} x)\, dx = \int_{R^n} \varphi(x) \delta_t(x)\, dx \qquad (dx = dx_1 \ldots dx_n). \tag{10.14}$$

Set, for $f(x) \in L^2(R^n)$ and for $t \geq 0$,

$$(T(t)f)(x) = f(\tau_{-t} x).$$

Choosing $\varphi(x) = |f(x)|^2$ formulas (10.13) and (10.14) imply that $T(t)$ is a contraction of $L^2(R^n)$ for $t \geq 0$. Moreover, the obvious relations $\tau_{t+s} = \tau_t \circ \tau_s$ and $\tau_0 =$ the identity transformation on R^n, imply that $\{T(t)\}_{t\geq 0}$ is a semigroup of contractions on $L^2(R^n)$. For continuous $f(x)$ with compact support it is obvious that $T(t)f \to f$ strongly, as $t \to +0$. These functions being dense in $L^2(R^n)$ we conclude easily that the semigroup $\{T(t)\}$ is continuous.

Let us introduce the measures

$$d\nu(x,s) = \rho(x)dxds \quad (\text{in } R^{n+1}), \quad d\mu(x) = \delta_\infty(x)dx \quad (\text{in } R^n),$$

where

$$\delta_\infty(x) = \exp\left(-\int_0^\infty \rho(\tau_s x)ds\right) = \lim_{t\to\infty} \delta_t(x) \quad (\geq 0).$$

Proposition 10.1. *The continuous one-parameter semigroup* $\{T(t)\}_{t\geq 0}$ *of contractions attached to the system of differential equations* (10.12) *with divergence* $-\rho(x) \leq 0$, *has an isometric dilation that is unitarily equivalent to the semigroup* $\{U(t)\}_{t\geq 0}$ *defined on the space*

$$\mathfrak{K} = L^2(R^{n+1};\nu) \oplus L^2(R^n;\mu)$$

by

$$U(t)[\mathbf{f}(x,s) \oplus f(x)] = \mathbf{f}(x,s+t) \oplus f(\tau_{-t}x).$$

Proof. Let $g(x)$ be continuously differentiable and have compact support in R^n. Then

$$\frac{d}{dt}\|T(t)g\|^2\bigg|_{t=0} = \frac{d}{dt}\int_{R^n} |g(\tau_{-t}x)|^2 dx\bigg|_{t=0} = -\int_{R^n} \sum_{j=1}^n \frac{\partial |g(x)|^2}{\partial x_j} X_j(x)\, dx,$$

and hence by partial integration

$$\frac{d}{dt}\|T(t)g\|^2\bigg|_{t=0} = \int_{R^n} |g|^2 \sum_{j=1}^n \frac{\partial X_j}{\partial x_j}\, dx = -\int_{R^n} |g|^2 \rho\, dx.$$

The function $T(t)g$ is also continuously differentiable and has compact support, thus we obtain from the preceding formula that

$$\begin{aligned}\frac{d}{dt}\|T(t)g\|^2 &= \frac{d}{ds}\|T(s)T(t)g\|^2\bigg|_{s=0} \\ &= -\int_{R^n} |T(t)g|^2 \rho\, dx = -\int_{R^n} |g(\tau_{-t}x)|^2 \rho(x)\, dx,\end{aligned}$$

and consequently, by the obvious relation

$$\|g\|^2 = -\int_0^t \left(\frac{d}{ds}\|T(s)g\|^2\right) ds + \|T(t)g\|^2,$$

we conclude that

$$\|g\|^2 = \int_0^t \int_{R^n} |g(\tau_{-s}x)|^2 \rho(x)\, dx\, ds + \int_{R^n} |g(\tau_{-t}x)|^2 dx.$$

Thus using first (10.14) (with $\varphi = |g|^2$) and then letting $t \to \infty$ we obtain finally

$$\|g\|^2 = \iint\limits_{R^n \times (0,\infty)} |g(\tau_{-s}x)|^2 d\nu(x,s) + \int_{R^n} |g(x)|^2 d\mu(x). \tag{10.15}$$

This formula, valid for every function of the type considered (thus for a set of functions dense in $L^2(R^n)$), is the concrete form of the formula (10.9) for the semigroup under consideration. The rest of the proof proceeds on the basis of the formula (10.15) in the same way as the construction in the preceding section was derived from formula (10.9).

Remark. Finer analysis (which we omit) shows that the semigroup $\{U(t)\}_{t\geq 0}$ obtained consists in fact of *unitary* operators, and that its natural extension to a group $\{U(t)\}_{-\infty}^{\infty}$ yields the *minimal* unitary dilation of $\{T(t)\}$.

11 Unitary ρ-dilations

1. As a generalization of the notion of unitary dilation of an operator we introduce the following concept.

Definition. We call *class* $\mathscr{C}_\rho$ $(\rho > 0)$ the set of operators T on the Hilbert space $\mathfrak{H}$ for which there exists a unitary operator U on some Hilbert space $\mathfrak{K}(\supset \mathfrak{H})$ such that

$$T^n = \rho \cdot \operatorname{pr} U^n \qquad (n = 1, 2, \ldots); \tag{11.1}$$

U is then called a *unitary ρ-dilation* of T.

For $T \in \mathscr{C}_\rho$ we obviously have

$$\|T^n\| \leq \rho \qquad (n = 1, 2, \ldots) \tag{11.2}$$

and hence $\lim \|T^n\|^{1/n} \leq 1$; therefore the spectrum of T is contained in the closed unit disc.

The class $\mathscr{C}_1$ consists precisely of the contractions. The following theorem characterizes each of the classes $\mathscr{C}_\rho$, at least for *complex* spaces.

Theorem 11.1. *Let T be a bounded operator on the complex Hilbert space $\mathfrak{H}$, and let $\rho > 0$. In order that T belong to the class $\mathscr{C}_\rho$ it is necessary and sufficient that the condition*

$$\left(\frac{2}{\rho} - 1\right) \|zTh\|^2 + \left(2 - \frac{2}{\rho}\right) \operatorname{Re}(zTh, h) \leq \|h\|^2, \tag{I_ρ}$$

or equivalently, that the condition

$$(\rho - 2)\|(I - zT)h\|^2 + 2\,\mathrm{Re}((I - zT)h, h) \geq 0 \qquad (\mathrm{I}_\rho^*)$$

be satisfied for all $h \in \mathfrak{H}$, $|z| \leq 1$.

Proof. The equivalence of the two conditions is obvious.

Suppose T has a unitary ρ-dilation U on the space $\mathfrak{K}$. Because U is unitary, the series $I + 2zU + \cdots + 2z^nU^n + \cdots$ converges in the operator norm for $|z| < 1$, its sum being equal to $(I + zU)(I - zU)^{-1}$. From (11.1) it follows that the series

$$I_{\mathfrak{H}} + \frac{2}{\rho} zT + \cdots + \frac{2}{\rho} z^n T^n + \cdots$$

also converges in the operator norm. The sum of this series is then necessarily equal to

$$\left(1 - \frac{2}{\rho}\right) I_{\mathfrak{H}} + \frac{2}{\rho}(I_{\mathfrak{H}} - zT)^{-1},$$

therefore we have

$$\left(1 - \frac{2}{\rho}\right) I_{\mathfrak{H}} + \frac{2}{\rho}(I_{\mathfrak{H}} - zT)^{-1} = \mathrm{pr}(I_{\mathfrak{K}} + zU)(I_{\mathfrak{K}} - zU)^{-1} \quad (|z| < 1). \qquad (11.3)$$

On the other hand, we have

$$\mathrm{Re}((I + zU)k, (I - zU)k) = \|k\|^2 - |z|^2\|Uk\|^2 = (1 - |z|^2)\|k\|^2 \geq 0$$

for $k \in \mathfrak{K}$ and $|z| < 1$, and hence we deduce

$$\mathrm{Re}((I + zU)(I - zU)^{-1}k', k') \geq 0 \qquad (k' \in \mathfrak{K}, |z| < 1).$$

Thus, using (11.3), we obtain

$$\mathrm{Re}\left[\left(1 - \frac{2}{\rho}\right)(l, l) + \frac{2}{\rho}((I - zT)^{-1}l, l)\right] \geq 0 \quad (l \in \mathfrak{H}, |z| < 1). \qquad (11.4)$$

Setting $l = l_z = (I - zT)h, h \in \mathfrak{H}$, and multiplying by ρ we obtain (I_ρ^*) first for $|z| < 1$ and then by continuity for $|z| \leq 1$ also.

We now prove that, conversely, the (equivalent) conditions (I_ρ), (I_ρ^*) imply that T is of class $\mathscr{C}_\rho$. To this end we first show that (I_ρ) implies that the spectrum of T is contained in the closed unit disc. Suppose the contrary. Then there exists a point $1/z_0$ outside the unit circle, belonging to the boundary of the spectrum of T and, consequently, belonging to the approximate point spectrum of T (see HALMOS [4], Problem 63). That is, there exists a sequence $\{h_n\}$ of elements of $\mathfrak{H}$ such that $\|h_n\| = 1$ $(n = 1, 2, \ldots)$, $(I - z_0T)h_n \to 0$ $(n \to \infty)$, and hence

$$z_0(Th_n, h_n) \to 1 \quad \text{and} \quad \|z_0Th_n\| \to 1.$$

Let $0 < r < 1 - |z_0|$. Then $z = z_0 + rz_0$ is also inside the unit circle. So we have, by virtue of (I_ρ^*),

$$(\rho-2)\|(I-z_0T)h_n - rz_0Th_n\|^2 + 2\,\mathrm{Re}[((I-z_0T)h_n,h_n) - rz_0(Th_n,h_n)] \geq 0.$$

Let $n \to \infty$; we obtain in the limit:

$$(\rho-2)r^2 - 2r \geq 0.$$

Dividing by r and letting $r \to 0$ we obtain the contradiction $-2 \geq 0$. This proves the assertion that no point of the spectrum lies outside the unit circle, that is, $(I-zT)^{-1}$ exists as a bounded operator on $\mathfrak{H}$ for every z inside the unit circle, and equals the sum of the series $\sum_0^\infty z^n T^n$, which converges in the operator norm.

Let l be an arbitrary element in $\mathfrak{H}$ and apply (I_ρ^*) to $h = h_z = (I-zT)^{-1}l$. Then

$$(\rho-2)\|l\|^2 + 2\,\mathrm{Re}(l,(I-zT)^{-1}l) \geq 0,$$

and hence

$$(Q(r,\varphi)l,l) \geq 0 \qquad (0 \leq r < 1; 0 \leq \varphi \leq 2\pi; l \in \mathfrak{H}), \tag{11.5}$$

where $Q(r,\varphi)$ is the operator-valued function defined by the series

$$Q(r,\varphi) = I + \frac{1}{\rho}r(e^{i\varphi}T + e^{-i\varphi}T^*) + \cdots + \frac{1}{\rho}r^n(e^{in\varphi}T^n + e^{-in\varphi}T^{*n}) + \cdots,$$

which converges in the operator norm. Let $\{h_n\}_{-\infty}^{\infty}$ be a finitely nonzero sequence of elements of $\mathfrak{H}$ and let

$$h(\varphi) = \sum_{-\infty}^{\infty} h_n e^{-in\varphi}.$$

We deduce from (11.5) that

$$\begin{aligned} 0 &\leq \frac{1}{2\pi}\int_0^{2\pi} (Q(r,\varphi)h(\varphi),h(\varphi))\,d\varphi \\ &= \sum_{-\infty}^{\infty}(h_n,h_n) + \frac{1}{\rho}\sum_{n>m}\sum r^{n-m}(T^{n-m}h_n,h_m) + \frac{1}{\rho}\sum_{m>n}\sum r^{m-n}(T^{*m-n}h_n,h_m) \end{aligned}$$

for every $r, 0 \leq r < 1$. If we let $r \to 1-0$, we obtain

$$\sum_{n=-\infty}^{\infty}\sum_{m=-\infty}^{\infty}(T_\rho(n-m)h_n,h_m) \geq 0,$$

where $T_\rho(n)$ is derived from T by the formulas

$$T_\rho(0) = I, \quad T_\rho(n) = \frac{1}{\rho}T^n \text{ and } T_\rho(-n) = \frac{1}{\rho}T^{*n} \quad (n \geq 1). \tag{11.6}$$

This function $T_\rho(n)$ of n is thus positive definite on the additive group Z of the integers. By virtue of Theorem 7.1 there exists a unitary operator U_ρ on a space $\mathfrak{K}_\rho$

containing $\mathfrak{H}$ as a subspace, such that $T_\rho(n) = \mathrm{pr}\, U_\rho^n$ for all $n \in Z$. By definition (11.6) of $T_\rho(n)$ this means that U_ρ is a unitary ρ-dilation of T.

This concludes the proof of the theorem.

Observe that for $\rho = 1$ condition (I_ρ) reduces to the condition $\|Th\| \leq \|h\|$, that is, $\|T\| \leq 1$. Thus Theorem 11.1 constitutes a generalization of Theorem 4.2. For $\rho = 2$, (I_ρ) reduces to the condition

$$\mathrm{Re}\, z(Th,h) \leq \|h\|^2 \qquad (h \in \mathfrak{H}, |z| \leq 1) \tag{11.7}$$

which is obviously equivalent to

$$\|h\|^2 \geq |(Th,h)| \qquad (h \in \mathfrak{H}). \tag{11.8}$$

So in this particular case our theorem can be formulated as follows.

Proposition 11.2. *In order that the operator T belong to the class $\mathscr{C}_2$ it is necessary and sufficient that the "numerical radius" of T defined by*

$$w(T) = \sup\{|(Th,h)|\colon\ h \in \mathfrak{H}, \|h\| \leq 1\}$$

satisfy $w(T) \leq 1$.

Let us add some further remarks.

Remark 1. *When $0 < \rho < 2$, $\rho \neq 1$ condition* (I_ρ) *reduces to*

$$\|(\mu I - T)h\| \leq \frac{|\mu|}{|\rho - 1|}\|h\| \qquad \left(h \in \mathfrak{H}, \left|\frac{\rho-1}{\rho-2}\right| \leq |\mu| < \infty\right), \tag{I'_ρ}$$

and when $2 < \rho < \infty$ it reduces to

$$\|(\mu I - T)h\| \geq \frac{|\mu|}{\rho - 1}\|h\| \qquad \left(h \in \mathfrak{H}, \frac{\rho-1}{\rho-2} \leq |\mu| < \infty\right). \tag{I''_ρ}$$

In fact, for $0 < |z| \leq 1$ multiply (I_ρ) by the factor

$$\frac{\rho}{2-\rho}\frac{1}{|z|^2},$$

and set

$$\mu = \frac{\rho-1}{\rho-2}\frac{1}{z}.$$

By rearranging we arrive at the above alternative forms of (I_ρ), depending on the sign of the factor, that is the form (I'_ρ) if this factor is positive ($\rho < 2$), and the form (I''_ρ) if it is negative ($\rho > 2$).

Remark 2. *If $1 < \rho < 2$, condition* (I'_ρ) *is equivalent to*

$$\|\mu I - T\| \leq |\mu| + 1 \qquad \left(\frac{\rho-1}{2-\rho} \leq |\mu| < \infty\right). \tag{II'_ρ}$$

In fact, (II'_ρ) implies (I'_ρ) because

$$|\mu|+1 \le \frac{|\mu|}{\rho-1} \quad \text{for} \quad |\mu| \ge \frac{\rho-1}{2-\rho}.$$

Conversely, for such a μ we deduce from (I'_ρ), setting $\varepsilon = \mu/|\mu|$,

$$\|\mu I - T\| \le \left|\mu - \varepsilon\frac{\rho-1}{2-\rho}\right| + \left\|\varepsilon\frac{\rho-1}{2-\rho}I - T\right\| \le |\mu| - \frac{\rho-1}{2-\rho} + \frac{1}{2-\rho} = |\mu|+1;$$

thus condition (II'_ρ) is valid.

Remark 3. *If $2<\rho<\infty$, condition (I''_ρ) is equivalent to the condition that T has its spectrum in the closed unit disc and*

$$\|(\mu I - T)^{-1}\| \le \frac{1}{|\mu|-1} \qquad \left(1<|\mu|<\frac{\rho-1}{\rho-2}\right). \tag{II''_ρ}$$

In fact, (I''_ρ) implies for $1<|\mu|\le r_\rho = (\rho-1)/(\rho-2)$ and $\mu = \varepsilon|\mu|$ that

$$\begin{aligned}\|(\mu I - T)h\| &\ge \|(\varepsilon r_\rho I - T)h\| - \|(\varepsilon r_\rho - \mu)h\| \\ &\ge |\varepsilon r_\rho|\frac{1}{\rho-1}\|h\| - (r_\rho - |\mu|)\|h\| = (|\mu|-1)\|h\|\end{aligned}$$

for all $h \in \mathfrak{H}$. Moreover, because (I''_ρ) implies that $T \in \mathscr{C}_\rho$, the spectrum of T is in the closed unit disc, and we obtain that (II''_ρ) holds.

Conversely, if T satisfies (II''_ρ) and if, moreover, its spectrum is in the closed unit disc, then $(I-zT)^{-1}$ exists and is an analytic function of z inside the unit circle, and we obtain by (II''_ρ) and the maximum principle that

$$\begin{aligned}\max_{|z|\le 1/r_\rho}\|(I-zT)^{-1}\| = \max_{|z|=1/r_\rho}\|(I-zT)^{-1}\| &= r_\rho \cdot \max_{|\zeta|=r_\rho}\|(\zeta I - T)^{-1}\| \\ &\le \frac{r_\rho}{r_\rho-1} = \rho-1.\end{aligned}$$

Thus for $r_\rho \le |\mu| < \infty$ we have

$$\|(\mu I - T)^{-1}\| = \frac{1}{|\mu|}\left\|\left(I - \frac{1}{\mu}T\right)^{-1}\right\| \le \frac{\rho-1}{|\mu|},$$

and consequently (I''_ρ).

Remark 4. *If the spectrum of T lies in the open unit disc, then $T \in \mathscr{C}_\rho$ for ρ large enough.*

Indeed, then inequality (II''_ρ) holds for ρ large enough.

2. From condition (I^*_ρ) it is obvious that the class $\mathscr{C}_\rho$ is a nondecreasing function of ρ. We show that it is in fact an increasing function.

Proposition 11.3. *If* $\dim\mathfrak{H}\geq 2$, *the class* $\mathscr{C}_\rho$ $(0<\rho<\infty)$ *increases with* ρ; *that is,*

$$\mathscr{C}_\rho\subset\mathscr{C}_{\rho'} \quad \text{and} \quad \mathscr{C}_\rho\neq\mathscr{C}_{\rho'} \quad \text{for} \quad 0<\rho<\rho'<\infty.$$

Proof. We construct for every $\rho>0$ an operator T_ρ in $\mathfrak{H}$ such that $T_\rho\in\mathscr{C}_\rho$ and $\|T_\rho\|=\rho$; this operator can belong to none of the classes $\mathscr{C}_\sigma$ with $0<\sigma<\rho$. To this end choose an orthonormal basis $\{\varphi_1,\varphi_2,\psi_\nu(\nu\in\Omega)\}$ in $\mathfrak{H}$ (the set Ω may be empty), and we consider the operator T_ρ defined by

$$T_\rho\varphi_1=\rho\varphi_2,\quad T_\rho\varphi_2=0,\quad T_\rho\psi_\nu=0\quad(\nu\in\Omega).$$

Clearly we have $\|T_\rho\|=\rho$ and $T_\rho^n=O\ (n\geq 2)$. Let $\mathfrak{K}$ be a Hilbert space of dimension $\aleph_0\cdot\dim\mathfrak{H}$ and choose any orthonormal basis in $\mathfrak{K}$; its elements can be arranged in the following way.

$$\{\varphi'_m\ (m=0,\pm1,\pm2,\ldots);\quad \psi'_{\nu m}\ (\nu\in\Omega;m=0,\pm1,\pm2,\ldots)\}.$$

We identify φ_1 with φ'_1,φ_2 with φ'_2, and ψ_ν with $\psi'_{\nu 0}$ $(\nu\in\Omega)$; this defines an isometric embedding of $\mathfrak{H}$ in $\mathfrak{K}$ as a subspace. Next we define a unitary operator U on $\mathfrak{K}$ by setting

$$U\varphi'_m=\varphi'_{m+1},\quad U\psi'_{\nu m}=\psi'_{\nu,m+1}\quad(\nu\in\Omega)$$

for $m=0,\pm1,\pm2,\ldots$. If we denote by P the orthonormal projection of $\mathfrak{K}$ into $\mathfrak{H}$, we obtain

$$\rho PU\varphi_1=\rho P\varphi_2=\rho\varphi_2,\quad \rho PU\varphi_2=\rho P\varphi'_3=0,\quad \rho PU\psi_\nu=\rho P\psi'_{\nu,1}=0,$$

and for $n\geq 2$:

$$\rho PU^n\varphi_i=\rho P\varphi'_{i+n}=0\quad(i=1,2),\quad \rho PU^n\psi_\nu=\rho P\psi'_{\nu,n}=0.$$

Thus $\rho\cdot PU^nh=T_\rho^nh$ for $n\geq 1$ and $h=\varphi_1,\varphi_2,\psi_\nu$. This implies the same relation for arbitrary $h\in\mathfrak{H}$. Thus U is a unitary ρ-dilation of T_ρ.

This concludes the proof.

3. The von Neumann inequality (8.9) can be extended, in an appropriate form, to the classes $\mathscr{C}_\rho$. In fact, (11.1) implies for every polynomial $p(\lambda)$ with complex coefficients:

$$p(T)=\mathrm{pr}[\rho\cdot p(U)+(1-\rho)\cdot p(0)I_{\mathfrak{K}}]. \tag{11.9}$$

The fact that U is unitary yields the following result.

Proposition 11.4. *For* $T\in\mathscr{C}_\rho$ *and for any polynomial* $p(\lambda)$ *of the complex variable* λ *we have*

$$\|p(T)\|\leq\max_{|z|\leq 1}|\rho\cdot p(z)+(1-\rho)\cdot p(0)|. \tag{11.10}$$

It is possible to complete this proposition as follows.

Proposition 11.5. *Let $q(\lambda)$ be a polynomial such that $q(0) = 0$ and $|q(\lambda)| \leq 1$ for $|\lambda| \leq 1$. Then for $T \in \mathscr{C}_\rho$ we also have $q(T) \in \mathscr{C}_\rho$ $(0 < \rho < \infty)$.*

Proof. Let U be a unitary ρ-dilation of T. Applying (11.9) to the polynomials $p(\lambda) = q(\lambda)^n$ $(n = 1, 2, \ldots)$ yields

$$q(T)^n = \rho \cdot \operatorname{pr} q(U)^n \qquad (n = 1, 2, \ldots). \tag{11.11}$$

Because $|q(\lambda)| \leq 1$ for $|\lambda| \leq 1$, it follows from the spectral theory of unitary operators that $\|q(U)\| \leq 1$. Consequently, there exists a unitary operator V in some larger space such that

$$q(U)^n = \operatorname{pr} V^n \qquad (n = 0, 1, \ldots). \tag{11.12}$$

Now (11.11) and (11.12) imply

$$q(T)^n = \rho \cdot \operatorname{pr} V^n \qquad (n = 1, 2, \ldots),$$

and hence $q(T) \in \mathscr{C}_\rho$.

For $\rho = 2$ this result can also be restated, by virtue of Proposition 11.2, in the following form.

Proposition 11.6. *If $w(T) \leq 1$ then $w(q(T)) \leq 1$ for every polynomial $q(\lambda)$ such that $q(0) = 0$ and $|q(\lambda)| \leq 1$ $(|\lambda| \leq 1)$. In particular, $w(T) \leq 1$ implies $w(T^n) \leq 1$ $(n = 1, 2, \ldots)$.*

Let us state again that the results of this section relate to operators on complex Hilbert spaces.

12 Notes

Theorem 1.1 on the decomposition of a space $\mathfrak{H}$ induced by an isometry V on $\mathfrak{H}$ has been formulated in a probabilistic setting by WOLD [1], p. 89. Except for the expression of the subspace $\mathfrak{H}_0$ of the unitary part as the intersection of the ranges of the iterates of V, the theorem already appears in the fundamental paper on abstract Hilbert space of VON NEUMANN [1], p. 96; in its present form the theorem was stated and proved by HALMOS [2], Lemma 1.

Proposition 2.1 on bilateral shifts can also be derived—at least for complex Hilbert space—from the general theory of spectral multiplicity; the direct proof given here, which is valid without restriction on the field of scalars, is due to HALPERIN; see SZ.-N.–F. [V].

Proposition 3.1 on the invariant vectors of a contraction was found by SZ.-NAGY in connection with some ergodic theorems (cf. RIESZ–SZ.-NAGY [1] and [*Func. Anal.*] Sec. 144). Generalizations of this proposition were given in SZ.-N.–F. [1].

Theorem 3.2 on the canonical decomposition of a contraction was proved by LANGER [1] and SZ.-N.–F. [IV].

The notation $A = \operatorname{pr} B$ was introduced by SZ.-NAGY in [P]. For two operators so related, HALMOS [1] says that A is a "compression" of B, and B is a "dilation" of

A, whereas SZ.-NAGY [P] says "projection" instead of "compression". In this book we have preferred to abandon this terminology and preserve for the term "dilation" the meaning given in Sec. 4 (i.e., "power dilation" in the sense of HALMOS [4]). By the way, let us observe that, because $A = \operatorname{pr} B$ if and only if the bilinear form (Bb, b') is an extension of the bilinear form (Aa, a'), we would be justified in calling B a "numerical extension" of A, and A a "numerical restriction" of B (in analogy with "numerical range", "numerical radius", etc.).

The fact that for every contraction T there exists an isometry V such that $T = \operatorname{pr} V$, was observed already by JULIA [1]–[3]. HALMOS [1] has shown that V can be chosen to be unitary, $V = U$.

Theorem 4.2, on the existence of a unitary U such that the relations $T^n = \operatorname{pr} U^n$ hold simultaneously for $n = 1, 2, \ldots$ (i.e., of a unitary dilation of T), was found by SZ.-NAGY [I]. The original proof used the theorem of F. RIESZ on the trigonometric moment problem and the theorem of NAĬMARK [1] (Theorem 8.2) on the existence, for every operator distribution function B_λ, of an orthogonal projection-valued distribution function E_λ (i.e., of a spectral family) such that $B_\lambda = \operatorname{pr} E_\lambda$. The next proof (SZ.-NAGY [I bis], [P], [1]) was based upon the fact that the function $T(n)$, derived from the contraction T by formulas (8.2), is positive definite on the additive group of the integers, so that one can apply the theorem of NAĬMARK [1] on operator-valued positive definite functions on groups (Theorem 7.1). We have reproduced this method in Sec. 8.1, with the only difference that the positive definiteness of the function $T(n)$ is proved here in a simpler way than in the places indicated. It should be remarked that the NAĬMARK theorem (Theorem 7.1) was extended to $*$-semigroups by SZ.-NAGY [P]; from among the various applications of this generalized theorem we should mention the proof of a theorem of HALMOS [1] on subnormal operators.

These first two proofs of Theorem 4.2 were followed by the matrix proof due to SCHÄFFER [1], reproduced in Sec. 5.1. The modification of this construction yielding the minimal unitary dilation, given in Sec. 5.2, is due to SZ.-NAGY [2] (cf. also HALPERIN [1]). The proof in Sec. 4 of this book follows a fourth method: first one finds an isometric dilation, then this is extended to a unitary dilation. Finally, the construction in Sec. 10.1 also yields an isometric dilation, which, moreover, can be shown to be minimal; see DOUGLAS [3].

The problem of finding a unitary (or isometric) dilation of a commutative system of contractions was proposed by SZ.-NAGY; he proved the existence of a unitary dilation under the additional condition that the contractions considered are doubly commuting; see SZ.-NAGY [I bis], [7]. The concept of the regular unitary dilation and the systematic study of the existence problem for such dilations, is due to BREHMER [1]; see also SZ.-NAGY [4] (the terminology used in this book is new). This study was later completed and simplified by HALPERIN [2], [4]; the proof in Sec. 9 is close to that of HALPERIN [2].

Theorems 6.1 and 6.4 establishing the existence, for every commutative pair of contractions, of an isometric dilation and of a unitary dilation, is due to ANDO [1]. The example of a system of three commuting contractions for which no unitary dilation exists (Sec. 6.3), was derived by PARROTT [1].

Theorem 8.1, on the unitary dilation of a continuous one-parameter semigroup of contractions, is due to SZ.-NAGY [I], [I bis], [P]. If $\{V_s\}_{s\geq 0}$ is a semigroup of isometric operators and $\{U_s\}_{s\geq 0}$ is its unitary dilation, then we have necessarily $V_s \subset U_s$ $(s \geq 0)$. So one obtains a new proof of a theorem of COOPER [1] asserting that *every continuous one-parameter semigroup of isometries on the space* $\mathfrak{H}$ *can be extended to a continuous one-parameter semigroup of unitary operators on a space* $\mathfrak{K} \supset \mathfrak{H}$.

The fact that every commutative system of isometries on the space $\mathfrak{H}$ can be extended to a commutative system of unitary operators on a space $\mathfrak{K} \supset \mathfrak{H}$ (Proposition 6.2), was proved by ITÔ [1] and BREHMER [1]. In Sec. 6 we have followed (at least partially) the method of ITÔ, and in Sec. 9, Proposition 9.2 (i), the method of BREHMER. See also DOUGLAS [4].

Inequality (8.9) for contractions can be restated by saying that the closed unit disc $|\lambda| \leq 1$ is a "spectral set" for every contraction on a complex Hilbert space. This theorem was obtained first by VON NEUMANN [4]; his proof used some methods of the theory of analytic functions. The proof was later simplified by HEINZ [1] (cf. [*Func. Anal.*]); in this form the proof is based upon the classical Cauchy–Poisson formula. The proof in Sec. 8, which reduces the problem through the use of unitary dilations to the simpler particular case of unitary operators, was given in SZ.-NAGY [I], [P].

As a natural generalization of von Neumann's inequality one can conjecture the inequality

$$\|p(T_1,\dots,T_n)\| \leq \max_{|\lambda_1|\leq 1,\dots,|\lambda_n|\leq 1} |p(\lambda_1,\dots,\lambda_n)| \tag{*}$$

for any commuting system $\{T_i\}_1^n$ of contractions and any polynomial p of the complex variables $\lambda_1,\dots,\lambda_n$. For $n = 2$, inequality (*) follows from ANDO's theorem (Theorem 6.4) in the same way as inequality (8.9) followed from Theorem 4.2. For $n \geq 3$ this method breaks down, because then the system $\{T_i\}_1^n$ has in general no unitary dilation (cf. Sec. 6.3). Although this does not imply that (*) should fail for $n \geq 3$, it turned out that (*) also fails for $n \geq 3$ (see VAROPOULOS [1]).

Let us also mention that inequality (8.9) characterizes complex Hilbert spaces among complex Banach spaces; indeed, if (8.9) is valid for every contraction T on a complex Banach space X and for every polynomial $p(\lambda)$, then X is necessarily a Hilbert space (cf. FOIAŞ [1]).

For $p(\lambda) = \lambda$, inequality (8.9) reduces to $\|T\| \leq 1$. On the other hand, the contractions on a Hilbert space are characterized by the property of admitting a unitary dilation. Now, in a complex Hilbert space, the unitary operators are those normal operators whose spectrum is situated on the unit circle. Hence, for an operator T on a complex Hilbert space, the validity of von Neumann's inequality is equivalent to the existence of a normal dilation whose spectrum is situated on the unit circle.

This raises the following problem. Let T be a bounded operator on the complex Hilbert space $\mathfrak{H}$ and let S be a compact subset of the plane of complex numbers. *Is the validity of the inequality*

$$\|p(T)\| \leq \max_{z\in S} |p(z)|,$$

for every polynomial $p(z)$, equivalent to the existence of a normal dilation N of T whose spectrum lies on the boundary of S? For bounded Jordan sets S (i.e., whose boundary is a simple closed curve) this equivalence was established by SZ.-N.–F. [III], and for general compact sets S with connected complement by FOIAŞ [4]. (See also LEBOW [1] and BERGER [1].) Later, SARASON [1] showed that the proof of this equivalence (for compact S with connected complement) can be reduced to the case of the disc.

The method used in Sec. 10 to obtain isometric dilations of a contraction or of a continuous one-parameter semigroup of contractions, is not the only one that allows dealing with these two cases in an analogous way. Indeed, the original method of SZ.-NAGY [I] also permitted the two cases to be treated analogously. The method of Sec. 10.2 has been obtained in connection with the problem of finding the unitary dilation of a system of differential equations of negative divergence (cf. Sec. 10.3); this problem was proposed to one of the authors by A. G. KOSTJUČENKO and its solution is contained in Proposition 10.1 and in the remark following it.

The first result on the existence of a unitary ρ-dilation (for $\rho \neq 1$) is due to BERGER [2] (cf. also HALMOS [3]) and concerns the case $\rho = 2$: this is our Proposition 11.2. The concept of unitary ρ-dilations and the first general results in this direction are in SZ.-N.–F. [6]; however, the proof of Theorem 11.1 as given in Sec. 11 is in part different from the original one, and follows the line of the proof given in SZ.-NAGY [10] for the Berger–Halmos theorem (our Proposition 11.2). Moreover, we have omitted in Theorem 11.1 the condition that the spectrum of T be contained in the closed unit disc, because this turns out to be a consequence of condition (I_ρ) not only if $\rho \leq 2$ (as noted originally by the authors) but for any $\rho > 0$. This fact was recently observed by DAVIS [1].

The fact that the class $\mathscr{C}_\rho$ is nondecreasing as a function of ρ $(0 < \rho < \infty)$, and nonconstant for small or for large values of ρ, was observed already in SZ.-N.–F. [6]. That it is strictly increasing (Proposition 11.3), was observed first by DURSZT [1]; he also obtained a criterion for a normal T to belong to $\mathscr{C}_\rho$.

Propositions 11.5 and 11.6 are due to Stampfli (cf. HALMOS [3]). As an obvious consequence we obtain the inequality (conjectured by Halmos):

$$w(T^n) \leq w(T)^n \qquad (n = 1, 2, \ldots),$$

which holds for an arbitrary bounded operator T. See PEARCY [1] for a more elementary proof.

The problem of unitary ρ-dilations can be generalized as follows. Given a self-adjoint operator A on a complex Hilbert space $\mathfrak{H}$, with positive lower and upper bounds, characterize the class $\mathscr{C}_A$ of those operators T on $\mathfrak{H}$ for which there exists a unitary operator U on some space $\mathfrak{K} \supset \mathfrak{H}$ such that

$$QT^nQ = \operatorname{pr} U^n \quad (n = 1, 2, \ldots), \text{ where } Q = A^{-1/2}$$

(the unitary ρ-dilation corresponds to the case $A = \rho I$). Such a characterization is given by the condition:

$$(Ah,h) - 2\mathrm{Re}(z(A-I)Th,h) + |z|^2((A-2I)Th,Th) \geq 0 \quad (h \in \mathfrak{H}, |z| \leq 1).$$

This generalization of Theorem 11.1 was proposed by H. Langer (correspondence). ISTRĂŢESCU [1] observed that $\mathscr{C}_A \subset \mathscr{C}_B$ if $A \leq B$, so that in particular $\mathscr{C}_A \subset \mathscr{C}_\rho$ if $\rho \geq \|A\|$.

From the consequences of the theorem on unitary dilations let us also mention here the result of SZ.-N.–F. [9] that to every contraction T on $\mathfrak{H}$ we can find a unitary operator U on some space $\mathfrak{L}$ such that the operator $T' = T \oplus U$ on $\mathfrak{M} = \mathfrak{H} \oplus \mathfrak{L}$ admits a "continuous scale" of invariant subspaces $\mathfrak{M}_\lambda$ $(0 \leq \lambda \leq 1)$, that is, such that $T'\mathfrak{M}_\lambda \subset \mathfrak{M}_\lambda$, $\mathfrak{M}_0 = \{0\}$, $\mathfrak{M}_1 = \mathfrak{M}$, $\mathfrak{M}_\lambda \subset \mathfrak{M}_\mu$ for $\lambda < \mu$, and finally $\mathfrak{M}_\lambda = \overline{\bigcup_{\varkappa<\lambda} \mathfrak{M}_\varkappa} = \bigcap_{\mu>\lambda} \mathfrak{M}_\mu$.

In connection with the subject treated in this chapter, also see BERBERIAN [1]; BERGER AND STAMPFLI [1], [2]; EGERVÁRY [1]; FURUTA [1], [2]; HOLBROOK [1]; KATO [2]; KENDALL [1], [2]; KORÁNYI [1]; MLAK [5]–[8]; NAKANO [1]; NAĬMARK [2], [3]; SZ.-NAGY [3]; SUCIU [1]; and THORHAUER [1], [2].

13 Further results

1. Many of the results of dilation theory can be put in a more algebraic framework which makes it easier to study their possible extensions. Denote by $\mathscr{B}(\mathfrak{H})$ the algebra of bounded linear operators on the Hilbert space $\mathfrak{H}$. Recall that a Banach algebra $\mathscr{A}$ is called a C^*-algebra if it is isometrically isomorphic with a subalgebra of $\mathscr{B}(\mathfrak{H})$, closed under taking adjoints. Such an algebra has a natural adjoint operation $a \to a^*$ inherited from $\mathscr{B}(\mathfrak{H})$. Let $\mathscr{A}$ be a C^*-algebra with unit, and let $\mathscr{B} \subset \mathscr{A}$ be a subalgebra of $\mathscr{A}$, containing the unit. A *representation* of $\mathscr{B}$ is simply a unital algebra homomorphism $\Phi\colon \mathscr{B} \to \mathscr{B}(\mathfrak{H})$. Such a representation is said to be *contractive* if $\|\Phi(b)\| \leq \|b\|$ for every $b \in \mathscr{B}$. A *dilation* of a representation $\Phi\colon \mathscr{B} \to \mathscr{B}(\mathfrak{H})$ to the algebra $\mathscr{A}$ consists of a Hilbert space $\mathfrak{K} \supset \mathfrak{H}$, and a contractive representation $\Psi\colon \mathscr{A} \to \mathscr{B}(\mathfrak{K})$ such that $\Psi(a^*) = \Psi(a)^*$ for every $a \in \mathscr{A}$, and

$$\Phi(b) = P\Psi(b)|\mathfrak{H} \quad (b \in \mathscr{B}),$$

where $P\colon \mathfrak{K} \to \mathfrak{H}$ denotes the orthogonal projection. An obvious condition for the existence of a dilation is that Φ itself be contractive. If the space $\mathfrak{H}$ is invariant for every $\Psi(b), b \in \mathscr{B}$, and

$$\Phi(b) = \Psi(b)|\mathfrak{H} \quad (b \in \mathscr{B}),$$

we say that Ψ is a *lifting* of Φ. The following result was first observed by SARASON [4].

Lemma 13.1. *Consider a dilation* $\Psi\colon \mathscr{A} \to \mathscr{B}(\mathfrak{K})$ *of a representation* $\Phi\colon \mathscr{B} \to \mathscr{B}(\mathfrak{H})$*. There exist subspaces* $\mathfrak{M} \subset \mathfrak{N} \subset \mathfrak{K}$*, invariant for every* $\Psi(b), b \in \mathscr{B}$*, such that* $\mathfrak{N} \ominus \mathfrak{M} = \mathfrak{H}$.

The spaces $\mathfrak{M}, \mathfrak{N}$ are easily found: $\mathfrak{N}$ is the closed linear span of $\{\Psi(b)h : b \in \mathscr{B}, h \in \mathfrak{H}\}$, and $\mathfrak{M} = \mathfrak{N} \ominus \mathfrak{H}$. This observation makes it possible to study dilations using liftings.

In order to relate this algebraic framework to the material in Chap. I, denote by $\mathscr{A}_C$ the C^*-algebra consisting of all continuous complex functions on the unit circle C, and let $\mathscr{B}_1$ denote the subalgebra of $\mathscr{A}_C$ consisting of all polynomials. A representation $\Phi\colon \mathscr{B}_1 \to \mathscr{B}(\mathfrak{H})$ is entirely determined by the operator $T = \Phi(\lambda)$ because $\Phi(p(\lambda)) = p(T)$ for every polynomial $p \in \mathscr{B}_1$. Theorem 4.2 can now be reformulated as follows.

Theorem 13.2. *Let* T *be a contraction on the Hilbert space* $\mathfrak{H}$*, and define* $\Phi\colon \mathscr{B}_1 \to \mathscr{B}(\mathfrak{H})$ *by* $\Phi(p(\lambda)) = p(T)$*,* $p \in \mathscr{B}_1$*. Then* Φ *has a dilation to* $\mathscr{A}_C$.

Proof. Let $\mathfrak{K}$ and U be provided by Theorem 4.2, and define $\Psi\colon \mathscr{A}_C \to \mathscr{B}(\mathfrak{H})$ by

$$\Psi(f) = f(U) \quad (f \in \mathscr{A}_C),$$

where the functional calculus is defined using the spectral measure of U. It is easy to verify that Ψ is indeed a dilation of Φ.

The von Neumann inequality (Proposition 8.3) is then the statement that Φ is a contractive representation, provided that $\|T\| \leq 1$.

General conditions for the existence of dilations in this context were given by ARVESON [1]. To formulate the result, let us note that a matrix $[T_{ij}]_{i,j=1}^n$ of operators on the Hilbert space $\mathfrak{H}$ can be viewed as an operator on the direct sum $\mathfrak{H} \oplus \mathfrak{H} \oplus \cdots \oplus \mathfrak{H}$ of n copies of $\mathfrak{H}$, and one can therefore speak of the operator norm of such a matrix. Analogously, a matrix $[a_{ij}]_{i,j=1}^n$ with entries in a C^*-algebra $\mathscr{A}$ has a well-defined norm. A representation $\Phi\colon \mathscr{B} \to \mathscr{B}(\mathfrak{H})$ of a subalgebra $\mathscr{B}$ of a C^*-algebra $\mathscr{A}$ is said to be *completely contractive* if

$$\|[\Phi(b_{ij})]_{i,j=1}^n\| \leq \|[b_{ij}]_{i,j=1}^n\|$$

for every positive integer n, and every matrix $[b_{ij}]_{i,j=1}^n$ with $b_{ij} \in \mathscr{B}$. We can now state one of the main results of ARVESON [1].

Theorem 13.3. *Consider a* C^**-algebra* $\mathscr{A}$*, a unital subalgebra* $\mathscr{B} \subset \mathscr{A}$*, and a representation* $\Phi\colon \mathscr{B} \to \mathscr{B}(\mathfrak{H})$*. The representation* Φ *has a dilation to* $\mathscr{A}$ *if and only if it is completely contractive.*

We see in particular that the von Neumann inequality $\|p(T)\| \leq \|p\|$ also extends to matrix polynomials if $\|T\| \leq 1$. More interestingly, this gives some insight into the existence of normal dilations for operators. Thus, consider an operator T on a

Hilbert space $\mathfrak{H}$, and a bounded open set Ω in the complex plane; denote by Γ the boundary of Ω. The closed set $\overline{\Omega}$ is said to be *spectral* for T if the inequality

$$\|p(T)\| \leq \sup_{\lambda\in\Gamma} |p(\lambda)| \tag{13.1}$$

holds for every rational function p with no poles in $\overline{\Omega}$. A normal operator N on a Hilbert space $\mathfrak{K} \supset \mathfrak{H}$ is called a *normal boundary dilation* of T if $\sigma(N) \subset \Gamma$, and

$$p(T) = Pp(N)|\mathfrak{H}$$

for any rational function p with no poles in $\overline{\Omega}$. As before P is the orthogonal projection onto $\mathfrak{H}$. The appropriate algebraic context here is given by the C^*-algebra $\mathscr{A}_\Gamma$ of continuous complex functions on Γ, and the subalgebra $\mathscr{B}_\Omega$ consisting of rational functions with no poles in $\overline{\Omega}$. The existence of a normal boundary dilation is then clearly equivalent to the existence of a dilation of the representation $p \to p(T)$ $(p \in \mathscr{B}_\Omega)$, and Theorem 13.3 implies that such dilations exist if and only if the inequality (13.1) is true for all matrix-valued rational functions. In other words, the requirement is that $\overline{\Omega}$ be a *complete spectral set* for T. Theorem 13.2 implies that the closed unit disk is a spectral set for T if and only if it is a complete spectral set. Extensions of this theorem depend therefore on finding other sets in the plane for which this implication is true. In the positive direction, AGLER [2] proved the following result for an annulus $\Omega = \{\lambda : |\lambda| \in (\alpha, \beta)\}$, where $0 < \alpha < \beta$.

Theorem 13.4. *Let T be an operator on the Hilbert space $\mathfrak{H}$ such that the annulus $\overline{\Omega}$ is a spectral set for T. Then T has a normal boundary dilation.*

Assume on the other hand that Ω has higher connectivity, for instance,

$$\Omega = \Omega(\alpha_1, \alpha_2; \rho_1, \rho_2) = \{\lambda : |\lambda| < 1, |\lambda - \alpha_1| > \rho_1, |\lambda - \alpha_2| > \rho_2\},$$

where $\alpha_1, \alpha_2 \in D$, $|\alpha_1| + \rho_1 < 1$, $|\alpha_2| + \rho_2 < 1$, and $\rho_1 + \rho_2 < |\alpha_1 - \alpha_2|$. Then DRITSCHEL AND MCCULLOUGH [1] proved the following result.

Theorem 13.5. *With the above notation, there exists an operator T on some Hilbert space such that $\overline{\Omega(\alpha_1, \alpha_2; \rho_1, \rho_2)}$ is a spectral set for T, but T does not have a normal boundary dilation.*

An explicit operator for a specific set $\Omega(\alpha_1, \alpha_2; \rho_1, \rho_2)$ had been found numerically by AGLER, HARLAND, AND RAPHAEL [1]. In this example, T is a finite matrix.

2. Fix now a positive integer n, and consider the algebra $\mathscr{A}_{C^n}$ of continuous complex functions on the product of n copies of C. Denote by $\mathscr{B}_n$ the subalgebra of $\mathscr{A}_{C^n}$ consisting of polynomials. A representation Φ of $\mathscr{B}_n$ is entirely determined by the operators $T_1 = \Phi(\lambda_1), \ldots, T_n = \Phi(\lambda_n)$, where λ_j denote the coordinate functions on C^n. These operators must commute with each other. A dilation of Φ to $\mathscr{A}_{C^n}$ is the same thing as a commuting unitary dilation of the n-tuple $(T_1, T_2, \ldots, T_n)$. Theorem 6.1 can thus be reformulated as follows.

Theorem 13.6. *Let T_1, T_2 be two commuting contractions on a Hilbert space $\mathfrak{H}$, and denote by $\Phi: \mathscr{B}_2 \to \mathscr{B}(\mathfrak{H})$ the representation defined by $\Phi(p) = p(T_1, T_2)$ for $p \in \mathscr{B}_2$. Then Φ is completely contractive.*

For the special example in Sec. 6.3 the corresponding representation Φ of $\mathscr{B}_3$ can in fact be shown to be contractive. The operators T_1, T_2, T_3 do not have commuting unitary dilations, thus it follows that Φ is not completely contractive. Examples of commuting contractions T_1, T_2, T_3 for which Φ is not even contractive were given by CRABB AND DAVIE [1] and VAROPOULOS [1].

3. Consider again a unital subalgebra $\mathscr{B}$ of a C^*-algebra $\mathscr{A}$, and a representation $\Phi: \mathscr{B} \to \mathscr{B}(\mathfrak{H})$. When Φ does not have a dilation to $\mathscr{A}$, one can ask whether the representation $b \to X^{-1}\Phi(b)X$ does have such a dilation for some invertible operator X. An obvious necessary condition for the existence of such an operator X is the existence of a constant k such that

$$\|[\Phi(b_{ij})]_{i,j=1}^n\| \leq k\|[b_{ij}]_{i,j=1}^n\|$$

for every positive integer n, and every matrix $[b_{ij}]_{i,j=1}^n$ with $b_{ij} \in \mathscr{B}$. When Φ satisfies this condition, we say that it is *completely bounded*. PAULSEN [1] proved that this necessary condition is sufficient as well.

Theorem 13.7. *Let Φ be a representation of a unital subalgebra $\mathscr{B}$ of a C^*-algebra $\mathscr{A}$. There exists an invertible operator X, such that the representation $b \to X^{-1}\Phi(b)X$ has a dilation to $\mathscr{A}$, if and only if Φ is completely bounded.*

In particular, this gives a criterion for a single operator T to be similar to a contraction: the representation $p \to p(T)$, $p \in \mathscr{B}_1$, must be completely bounded. SZ.-NAGY [11] asked whether a power-bounded operator must be similar to a contraction. After the negative answer provided by FOGUEL [1], HALMOS [5],[6] asked whether T must be similar to a contraction if the representation $p \to p(T)$, $p \in \mathscr{B}_1$, is bounded. The question is therefore whether this boundedness condition implies complete boundedness. PISIER [1] settled the problem by showing that the map may be bounded, but not completely bounded.

4. Let us return now to the case of an operator T with a spectral set $\overline{\Omega}$, where Ω is a finitely connected subset of the plane, with smooth boundary Γ. The following result was proved by DOUGLAS AND PAULSEN [1].

Theorem 13.8. *Assume that $\overline{\Omega}$ is a spectral set for T. Then the representation $p \to p(T)$, $p \in \mathscr{B}_\Omega$, is completely bounded. In particular, T is similar to an operator that has a normal boundary dilation.*

For further developments related to this material, see PAULSEN [2] and DOUGLAS AND PAULSEN [2].

5. In order to study lifting theory for more general operators, AGLER [3] proposed the study of *families* of representations. Fix an algebra $\mathscr{B}$ with unit, and a collection $\mathscr{F}$ of unital representations $\Phi: \mathscr{B} \to \mathscr{B}(\mathfrak{H})$. Then $\mathscr{F}$ is called a *family* if the following conditions are satisfied.

1. For each $b \in \mathscr{B}$ we have $\sup\{\|\Phi(b)\| : \Phi \in \mathscr{F}\} < \infty$,
2. For any set $\{\Phi_i\}_{i\in I} \subset \mathscr{F}$, the representation $b \to \bigoplus_{i\in I} \Phi_i(b)$ belongs to $\mathscr{F}$,
3. If $\mathfrak{M} \subset \mathfrak{H}$ is invariant for $\Phi(b), b \in \mathscr{B}$, then the representation $b \longrightarrow \Phi(b)|\mathfrak{M}$ belongs to $\mathscr{F}$,
4. If Ψ is a *-representation of the C^*-algebra generated by $\Phi(\mathscr{B})$, then $b \to \Psi(\Phi(b))$ belongs to $\mathscr{F}$.

Observe that the algebra $\mathscr{B}$ is not assumed to sit inside a fixed C^*-algebra. As before, Φ is called a lifting of a representation of the form $b \longrightarrow \Phi(b)|\mathfrak{M}$. Such a lifting is said to be *trivial* if $\mathfrak{M}$ is reducing for $\Phi(b), b \in \mathscr{B}$. Finally, an element $\Phi \in \mathscr{F}$ is said to be *extremal* if all its liftings are trivial. The following result is from AGLER [3].

Theorem 13.9. *Every element Φ of a family $\mathscr{F}$ has an extremal lifting.*

The simplest example of a family $\mathscr{F}$ consists of the contractive representations of the algebra $\mathscr{B}_1$ considered earlier, and which correspond to contractions T on a Hilbert space. The extremal elements correspond to operators T such that T^* is isometric.

Another interesting family is described as follows. Denote by $\mathscr{B}$ the algebra of all polynomials with complex coefficients in the (commuting) variables $X_1, X_2, \dots, X_n$, and let $\mathscr{F}$ consist of those representations Φ satisfying

$$\sum_{j=1}^{n} \Phi(X_j)^* \Phi(X_j) \leq I.$$

The result of DRURY [1] can be interpreted as a characterization of the extremal elements of $\mathscr{F}$: they are the ones for which all the operators $\Phi(X_j)^*$ are isometric. For further information about this family see ARVESON [3].

VASILESCU [1] and CURTO AND VASILESCU [1,2] study another interesting example related to the polydisc.

Other examples of interest can be (and have been) studied by defining appropriate families, though determining the extremal elements can be difficult. For instance, the extremals are not known for the family of contractive representations of the algebra $\mathscr{B}_3$ (of polynomials in three variables).

6. Isometric dilations for noncommuting operators have also been studied. The following result was proved by FRAZHO [1] for $n = 2$, and DURSZT AND SZ.-NAGY [1], and BUNCE [1] for arbitrary (even infinite) n. See also FRAZHO [2] and POPESCU [1],[2] for a thorough analysis of this situation, including an appropriate analogue of the Wold decomposition.

Theorem 13.10. *Let $T_1, T_2, \dots, T_n$ be operators on a Hilbert space $\mathfrak{H}$ satisfying the inequality*

$$T_1T_1^* + T_2T_2^* + \cdots + T_nT_n^* \leq I_{\mathfrak{H}}.$$

There exist isometric operators with pairwise orthogonal ranges $V_1, V_2, \dots, V_n$ on a Hilbert space $\mathfrak{K} \supset \mathfrak{H}$ such that $V_j^*\mathfrak{H} \subset \mathfrak{H}$ and $V_j^*|\mathfrak{H} = T_j$ for $j = 1, 2, \dots, n$.

Under appropriate hypotheses, the operators V_j can be chosen to be creation operators on a full Fock space.

See also MUHLY AND SOLEL [1] for a dilation theory in the context of correspondences over a von Neumann algebra.

7. Theorem 8.1 was extended to bounded representations of amenable groups; see DIXMIER [2].

8. The decomposition of a contraction into unitary and completely nonunitary parts was extended in DURSZT [3] to arbitrary operators on Hilbert space.

9. The norms of invertible operators realizing the similarity of an operator of class C_ρ to a contraction are estimated in OKUBO AND ANDO [1]. These can also be estimated using complete boundedness; see PAULSEN [2].

Chapter II

Geometrical and Spectral Properties of Dilations

1 Structure of the minimal unitary dilations

In the sequel we consider a contraction T on the real or complex Hilbert space $\mathfrak{H}$, and its minimal unitary dilation U on the Hilbert space $\mathfrak{K}$, real or complex, respectively ($\mathfrak{K} \supset \mathfrak{H}$). The linear manifolds

$$\mathfrak{L}_0 = (U - T)\mathfrak{H} \quad \text{and} \quad \mathfrak{L}_0^* = (U^* - T^*)\mathfrak{H} \quad (\subset \mathfrak{K}) \tag{1.1}$$

and their closures

$$\mathfrak{L} = \overline{(U - T)\mathfrak{H}}, \quad \mathfrak{L}^* = \overline{(U^* - T^*)\mathfrak{H}} \tag{1.2}$$

play an important role in our investigations.

Theorem 1.1. (i) *The subspaces $\mathfrak{L}$ and $\mathfrak{L}^*$ are wandering subspaces for U, their dimensions being equal to the defect indices of T:*

$$\dim \mathfrak{L} = \mathfrak{d}_T, \quad \dim \mathfrak{L}^* = \mathfrak{d}_{T^*}. \tag{1.3}$$

(ii) *The space $\mathfrak{K}$ can be decomposed into the orthogonal sum*

$$\mathfrak{K} = \cdots \oplus U^{*2}\mathfrak{L}^* \oplus U^*\mathfrak{L}^* \oplus \mathfrak{L}^* \oplus \mathfrak{H} \oplus \mathfrak{L} \oplus U\mathfrak{L} \oplus U^2\mathfrak{L} \oplus \cdots. \tag{1.4}$$

Proof. It would be easy to obtain these properties from the matrix form of U constructed in Sec. I.5.2, but we prefer to give a direct proof, independent of the particular realization of U.

Part (i): To prove that $\mathfrak{L}$ and $\mathfrak{L}^*$ are wandering subspaces, it suffices to show that $U^n\mathfrak{L}_0 \perp \mathfrak{L}_0$ and $U^{*n}\mathfrak{L}_0^* \perp \mathfrak{L}_0^*$ for $n = 1, 2, \ldots$; by reason of symmetry it even suffices to consider one of these cases, say that of $\mathfrak{L}_0$. Now for $h, h' \in \mathfrak{H}$ and $n = 1, 2, \ldots$ we

B.Sz.-Nagy et al., *Harmonic Analysis of Operators on Hilbert Space*, Universitext,
DOI 10.1007/978-1-4419-6094-8_2, © Springer Science+Business Media, LLC 2010

have

$$\begin{aligned}
&(U^n(U-T)h,(U-T)h') \\
&\quad = (U^n h,h') - (U^{n-1}Th,h') - (U^{n+1}h,Th') + (U^n Th,Th') \\
&\quad = (T^n h,h') - (T^{n-1}Th,h') - (T^{n+1}h,Th') + (T^n Th,Th') = 0.
\end{aligned}$$

In order to prove that $\dim \mathfrak{L} = \mathfrak{d}_T$, let us observe that for $h \in \mathfrak{H}$

$$\begin{aligned}
\|(U-T)h\|^2 &= \|Uh\|^2 - 2\,\mathrm{Re}(Uh,Th) + \|Th\|^2 \\
&= \|h\|^2 - 2\,\mathrm{Re}(Th,Th) + \|Th\|^2 = \|h\|^2 - \|Th\|^2 \\
&= \|D_T h\|^2.
\end{aligned} \tag{1.5}$$

By virtue of this relation, the transformation φ defined by

$$\varphi(U-T)h = D_T h \tag{1.6}$$

maps $\mathfrak{L}_0$ isometrically onto $D_T\mathfrak{H}$, and consequently φ extends by continuity to a unitary transformation of $\mathfrak{L}$ onto the defect space $\mathfrak{D}_T$. This proves that $\dim \mathfrak{L} = \dim \mathfrak{D}_T = \mathfrak{d}_T$. The equality $\dim \mathfrak{L}^* = \mathfrak{d}_{T^*}$ can be proved analogously.

Part (ii): Let us show first that the terms of the right-hand side of (1.4) are mutually orthogonal. We have already proved that $\mathfrak{L}$ and $\mathfrak{L}^*$ are wandering subspaces, therefore it remains only to show that

$$U^n\mathfrak{L} \perp U^{*m}\mathfrak{L}^*, \quad U^n\mathfrak{L} \perp \mathfrak{H} \text{ and } U^{*m}\mathfrak{L}^* \perp \mathfrak{H} \quad \text{for } m,n \geq 0;$$

it even suffices to establish these relations for $\mathfrak{L}_0$ and $\mathfrak{L}_0^*$ instead of $\mathfrak{L}$ and $\mathfrak{L}^*$. Now we have for $h,h' \in \mathfrak{H}$,

$$\begin{aligned}
&(U^n(U-T)h,U^{*m}(U^*-T^*)h') \\
&\quad = (U^{n+m+2}h,h') - (U^{n+m+1}h,T^*h') - (U^{n+m+1}Th,h') + (U^{n+m}Th,T^*h') \\
&\quad = (T^{n+m+2}h,h') - (T^{n+m+1}h,T^*h') - (T^{n+m+1}Th,h') + (T^{n+m}Th,T^*h') \\
&\quad = 0,
\end{aligned}$$

$$(U^n(U-T)h,h') = (U^{n+1}h,h') - (U^nTh,h') = (T^{n+1}h,h') - (T^nTh,h') = 0,$$

and

$$\begin{aligned}
(U^{*m}(U^*-T^*)h,h') &= (U^{*m+1}h,h') - (U^{*m}T^*h,h') \\
&= (T^{*m+1}h,h') - (T^{*m}T^*h,h') = 0,
\end{aligned}$$

so the orthogonality relations are established. Let us denote the orthogonal sum on the right-hand side of (1.4) by $\mathfrak{K}'$. Applying U term by term we obtain

$$U\mathfrak{K}' = \cdots \oplus U^*\mathfrak{L}^* \oplus \mathfrak{L}^* \oplus U\mathfrak{L}^* \oplus U\mathfrak{H} \oplus U\mathfrak{L} \oplus U^2\mathfrak{L} \oplus \cdots.$$

Now $U\mathfrak{K}'$ equals $\mathfrak{K}'$, because, as we show in a moment, we have

$$U\mathfrak{L}^* \oplus U\mathfrak{H} = \mathfrak{H} \oplus \mathfrak{L}. \tag{1.7}$$

Therefore $\mathfrak{K}'$ is a subspace of $\mathfrak{K}$ reducing U and containing $\mathfrak{H}$, and this implies by the minimality of U that

$$\mathfrak{K}' = \mathfrak{K}.$$

In order to establish (1.7) it suffices to show that $U\mathfrak{L}_0^* \oplus U\mathfrak{H} = \mathfrak{H} \oplus \mathfrak{L}_0$, that is,

$$U(U^* - T^*)\mathfrak{H} \oplus U\mathfrak{H} = \mathfrak{H} \oplus (U-T)\mathfrak{H}. \tag{1.8}$$

Now this follows from the fact that, for an element $u \in \mathfrak{K}$, the possibility of a representation of the form

$$u = h' + (U-T)h'' \qquad (h', h'' \in \mathfrak{H})$$

is equivalent to the possibility of a representation of the form

$$u = Uh_1 + U(U^* - T^*)h_2 \qquad (h_1, h_2 \in \mathfrak{H}).$$

Indeed, we have only to set

$$h_1 = T^*h' + (I - T^*T)h'', \quad h_2 = h' - Th''$$

and, conversely,

$$h' = Th_1 + (I - TT^*)h_2, \quad h'' = h_1 - T^*h_2.$$

This completes the proof of Theorem 1.1.

Theorem 1.2. *In order that*

$$\text{(a) } M(\mathfrak{L}) = \mathfrak{K} \ \ \textit{or} \ \ \text{(a}^*\text{) } M(\mathfrak{L}^*) = \mathfrak{K},$$

it is necessary and sufficient that the condition

$$\text{(b) } T^n \to O \ (n \to \infty) \ \ \textit{or} \ \ \text{(b}^*\text{) } T^{*n} \to O \ (n \to \infty)$$

be satisfied, respectively. Thus each of conditions (b) *and* (b*) *implies that U is a bilateral shift, of multiplicity $\mathfrak{d}_T$ or $\mathfrak{d}_{T^*}$, respectively.*

Proof. For $h \in \mathfrak{H}$ and for $n = 1, 2, \ldots$ we have

$$M(\mathfrak{L}) \ni \sum_{k=0}^{n-1} U^{-k-1}(U-T)T^k h = \sum_{k=0}^{n-1} (U^{-k}T^k - U^{-k-1}T^{k+1})h = h - U^{-n}T^n h,$$

and if condition (b) holds, then

$$h = \lim_{n\to\infty}(h - U^{-n}T^n h) \in M(\mathfrak{L}).$$

Thus (b) implies

$$\mathfrak{H} \subset M(\mathfrak{L}), \quad U^n\mathfrak{H} \subset U^n M(\mathfrak{L}) = M(\mathfrak{L}) \quad (n = 0, \pm 1, \ldots)$$

and consequently $\mathfrak{K} = M(\mathfrak{L})$. The implication (b*) $\Rightarrow$ (a*) can be proved similarly. We now turn to the proof of the converse implications. Condition (a) implies in particular that every element $h \in \mathfrak{H}$ can be represented as the sum of an orthogonal series of the form

$$h = \sum_{-\infty}^{\infty} U^k l_k, \quad \text{where} \quad l_k \in \mathfrak{L}, \sum_{-\infty}^{\infty} \|l_k\|^2 = \|h\|^2;$$

hence we deduce for $n = 1, 2, \ldots$,

$$T^n h = P_{\mathfrak{H}} U^n h = P_{\mathfrak{H}} \sum_{k=-\infty}^{\infty} U^{n+k} l_k.$$

By virtue of (1.4) $\mathfrak{H}$ is orthogonal to $U^m\mathfrak{L}$ for $m \geq 0$, so we have

$$T^n h = P_{\mathfrak{H}} \sum_{k=-\infty}^{-n-1} U^{n+k} l_k,$$

$$\|T^n h\|^2 \leq \left\| \sum_{k=-\infty}^{-n-1} U^{n+k} l_k \right\|^2 = \sum_{k=-\infty}^{-n-1} \|l_k\|^2,$$

and hence

$$T^n h \to 0 \quad \text{for} \quad n \to \infty.$$

The implication (a*) $\Rightarrow$ (b*) can be proved similarly.

Theorem 1.2 has in certain particular cases a converse.

Proposition 1.3. *If the defect index $\mathfrak{d}_T$ is finite and if the minimal unitary dilation U of T is a bilateral shift of multiplicity equal to $\mathfrak{d}_T$, then $T^n \to O$ as $n \to \infty$. Similarly, if $\mathfrak{d}_{T^*}$ is finite and U is a bilateral shift of multiplicity $\mathfrak{d}_{T^*}$, then $T^{*n} \to O$ as $n \to \infty$.*

Proof. It suffices to consider the first of the two assertions. The hypothesis that U is a bilateral shift of multiplicity $\mathfrak{d}_T$ means that there exists a wandering subspace $\mathfrak{L}'$ for U such that $\mathfrak{K} = M(\mathfrak{L}')$, $\dim \mathfrak{L}' = \mathfrak{d}_T$. We also have $M(\mathfrak{L}) \subset \mathfrak{K}$, $\dim \mathfrak{L} = \mathfrak{d}_T$, and $\mathfrak{d}_T$ is finite, thus it follows from Proposition I.2.1 that $M(\mathfrak{L}) = M(\mathfrak{L}') = \mathfrak{K}$, and hence $T^n \to O$, by the preceding proposition.

Proposition 1.4. *For every contraction T on $\mathfrak{H}$ and for its minimal unitary dilation U on $\mathfrak{K}$, we have*

$$M(\mathfrak{L}) \vee M(\mathfrak{L}^*) = \mathfrak{K} \ominus \mathfrak{H}_0, \tag{1.9}$$

where $\mathfrak{H}_0$ denotes the maximal subspace of $\mathfrak{H}$ on which T is unitary (cf. Theorem I.3.2*). In particular, if T is completely nonunitary, then*

$$M(\mathfrak{L}) \vee M(\mathfrak{L}^*) = \mathfrak{K}. \tag{1.10}$$

Proof. Let f be an element of $\mathfrak{K}$, orthogonal to $M(\mathfrak{L})$ and to $M(\mathfrak{L}^*)$. Because f is orthogonal, in particular, to $U^n\mathfrak{L}$ and to $U^{*n}\mathfrak{L}^*$ for $n = 0, 1, \ldots$, by (1.4) it is necessarily contained in $\mathfrak{H}$. The vector f is also orthogonal to $U^{*n}\mathfrak{L}$ $(n \geq 1)$, thus we have for any $h \in \mathfrak{H}$

$$0 = (f, U^{*n}(U-T)h) = (U^{n-1}f, h) - (U^n f, Th) = (T^{n-1}f, h) - (T^n f, Th);$$

choosing $h = T^{n-1}f$ we obtain

$$\|T^{n-1}f\|^2 - \|T^n f\|^2 = 0 \qquad (n = 1, 2, \ldots),$$

and thus

$$\|f\| = \|Tf\| = \|T^2 f\| = \cdots .$$

Similarly, from the orthogonality of f to $U^n\mathfrak{L}^*$ $(n \geq 1)$ we obtain that

$$\|f\| = \|T^* f\| = \|T^{*2} f\| = \cdots .$$

We conclude that $f \in \mathfrak{H}_0$.

Conversely, for every $f \in \mathfrak{H}_0$ we have $U^n f = T^n f$ for $n \geq 0$ and $U^n f = T^{*|n|} f$ for $n \leq 0$, and thus $U^n f \in \mathfrak{H}$ for every integer n; this implies that $U^n f \perp \mathfrak{L}$, $f \perp U^{-n}\mathfrak{L}$, and consequently $f \perp M(\mathfrak{L})$. By similar reasoning, we have $f \perp M(\mathfrak{L}^*)$. Hence $f \perp M(\mathfrak{L}) \vee M(\mathfrak{L}^*)$.

This concludes the proof of (1.9); (1.10) follows because for a c.n.u. T we have $\mathfrak{H}_0 = \{0\}$.

2 Isometric dilations. Dilation of commutants

1. Let us begin by observing that the subspaces $M(\mathfrak{L})$ and $M(\mathfrak{L}^*)$ reduce the operator U, and hence the same is true for the subspaces

$$\mathfrak{R} = \mathfrak{K} \ominus M(\mathfrak{L}^*) \quad \text{and} \quad \mathfrak{R}_* = \mathfrak{K} \ominus M(\mathfrak{L}). \tag{2.1}$$

We call

$$R = U|\mathfrak{R} \quad \text{and} \quad R_* = U|\mathfrak{R}_* \tag{2.2}$$

the residual part and the *dual residual* (or $*$*-residual) part of* U*, respectively*. These are unitary operators on $\mathfrak{R}$ and $\mathfrak{R}_*$. We investigate them more closely in the next section.

Let us consider now the subspace

$$\mathfrak{K}_+ = \bigvee_0^\infty U^n\mathfrak{H} \qquad (\subset \mathfrak{K}). \tag{2.3}$$

This is invariant for U and contains $\mathfrak{H}$ as a subspace; hence

$$U_+ = U|\mathfrak{K}_+ \tag{2.4}$$

is a minimal isometric dilation of T (cf. Theorem I.4.1). When speaking in the sequel of the minimal isometric dilation of T we always consider it as "embedded" in this way in the minimal unitary dilation.

By virtue of the obvious relation

$$U^n h = T^n h + (U-T)T^{n-1}h + U(U-T)T^{n-2}h + \cdots + U^{n-1}(U-T)h$$

for $h \in \mathfrak{H}$ and $n \geq 0$, the space $\mathfrak{K}_+$ is contained in $\mathfrak{H} \oplus M_+(\mathfrak{L})$. On the other hand, $\mathfrak{H} \subset \mathfrak{K}_+$, thus

$$U^n(U-T)\mathfrak{H} \subset U^{n+1}\mathfrak{H} \vee U^n\mathfrak{H} \subset \mathfrak{K}_+ \qquad (n \geq 0),$$

and hence $\mathfrak{H} \oplus M_+(\mathfrak{L})$ is contained in $\mathfrak{K}_+$. This proves the relation

$$\mathfrak{K}_+ = \mathfrak{H} \oplus M_+(\mathfrak{L}). \tag{2.5}$$

Because $\mathfrak{L} \subset \mathfrak{K}_+$, there is no difference in defining $M_+(\mathfrak{L})$ with respect to U or to U_+.

Comparing (2.5) with (1.4) and (2.1) we obtain

$$\mathfrak{K}_+ = \mathfrak{K} \ominus \left[\bigoplus_0^\infty U^{*n}\mathfrak{L}^*\right] = [\mathfrak{R} \oplus M(\mathfrak{L}^*)] \ominus \left[\bigoplus_0^\infty U^{-n}\mathfrak{L}^*\right] = \mathfrak{R} \oplus \left[\bigoplus_1^\infty U^n\mathfrak{L}^*\right];$$

setting

$$\mathfrak{L}_* = U\mathfrak{L}^* = \overline{U(U^* - T^*)\mathfrak{H}} = \overline{(I - UT^*)\mathfrak{H}} \tag{2.6}$$

we can also write this as

$$\mathfrak{K}_+ = \mathfrak{R} \oplus M_+(\mathfrak{L}_*). \tag{2.7}$$

We have $\mathfrak{L}_* \subset \mathfrak{H} \vee U\mathfrak{H} \subset \mathfrak{K}_+$, thus there is again no difference in defining $M_+(\mathfrak{L}_*)$ with respect to U or to U_+.

By virtue of (1.7) we have

$$\mathfrak{L}_* \oplus U\mathfrak{H} = \mathfrak{H} \oplus \mathfrak{L}. \tag{2.8}$$

We deduce hence that

$$\mathfrak{L} \cap \mathfrak{L}_* = \{0\}. \tag{2.9}$$

In fact, taking the orthogonal complements of $\mathfrak{L}$ and of $\mathfrak{L}_*$ in $\mathfrak{H} \oplus \mathfrak{L}$, which by virtue of (2.8) are equal to $\mathfrak{H}$ and to $U\mathfrak{H}$, we see that (2.9) is equivalent to the relation

$$\mathfrak{H} \vee U\mathfrak{H} = \mathfrak{H} \oplus \mathfrak{L}. \tag{2.10}$$

This relation obviously follows from

$$U\mathfrak{H} = [(U-T)+T]\mathfrak{H} \subset \mathfrak{L} \oplus \mathfrak{H} \quad \text{and} \quad \mathfrak{L} = \overline{(U-T)\mathfrak{H}} \subset U\mathfrak{H} \vee \mathfrak{H}.$$

Let us recall that $\mathfrak{R}$ reduces U to a unitary operator R, the residual part of U. Because $\mathfrak{R} \subset \mathfrak{K}_+$, we also have $R = U_+|\mathfrak{R}$. Thus the decomposition (2.7) necessarily coincides with the Wold decomposition of $\mathfrak{K}_+$ with respect to the isometric operator U_+. Consequently,

$$\mathfrak{L}_* = \mathfrak{K}_+ \ominus U_+\mathfrak{K}_+, \quad \mathfrak{R} = \bigcap_{n\geq 0} U_+^n \mathfrak{K}_+ = \bigcap_{n\geq 0} \bigvee_{k\geq n} U^k \mathfrak{H}. \tag{2.11}$$

By virtue of Theorem 1.2 and the definition (2.1) of $\mathfrak{R}$, we have $\mathfrak{R} = \{0\}$ if and only if $T^{*n} \to O$ $(n \to \infty)$.

Decomposition (1.4) of $\mathfrak{K}$ shows that $U^n\mathfrak{L} \perp U^{-m}\mathfrak{L}^* = U^{-(m+1)}\mathfrak{L}_*$ for $n, m \geq 0$. Denoting by

$$P^{\mathfrak{L}_*}$$

the orthogonal projection of $\mathfrak{K}$ onto $M(\mathfrak{L}_*) (= M(\mathfrak{L}^*))$ we have thus

$$P^{\mathfrak{L}_*} M_+(\mathfrak{L}) \subset M_+(\mathfrak{L}_*). \tag{2.12}$$

If T is c.n.u., then it follows from (1.10) that

$$(I - P^{\mathfrak{L}_*})\mathfrak{K} = \overline{(I - P^{\mathfrak{L}_*})M(\mathfrak{L})}.$$

With regard to (2.1) this implies

$$\mathfrak{R} = \overline{(I - P^{\mathfrak{L}_*})M(\mathfrak{L})}. \tag{2.13}$$

Summing up we have the following result.

Theorem 2.1. *Let T be a contraction on $\mathfrak{H}$, U its minimal unitary dilation on $\mathfrak{K}$, and U_+ its minimal isometric dilation on $\mathfrak{K}_+$ ($\subset \mathfrak{K}$). Then we have*

$$\mathfrak{K} = M(\mathfrak{L}_*) \oplus \mathfrak{R} \qquad (cf.\ (2.1)),$$
$$\mathfrak{K}_+ = M_+(\mathfrak{L}_*) \oplus \mathfrak{R} = \mathfrak{H} \oplus M_+(\mathfrak{L}) \qquad (cf.\ (2.5)\ and\ (2.7)),$$

where

$$\mathfrak{L} = \overline{(U-T)\mathfrak{H}} \quad and \quad \mathfrak{L}_* = \overline{(I - UT^*)\mathfrak{H}}$$

are subspaces of $\mathfrak{K}_+$, wandering for U_+ (and hence for U), and $\mathfrak{R}$ is the subspace of $\mathfrak{K}_+$ that reduces U_+ (and U) to the unitary part R of U_+. Moreover, we have

$$\mathfrak{L} \cap \mathfrak{L}_* = \{0\} \qquad (cf.\ (2.9)),$$
$$P^{\mathfrak{L}_*} M_+(\mathfrak{L}) \subset M_+(\mathfrak{L}_*) \qquad (cf.\ (2.12)),$$

and, if T is completely nonunitary,

$$\mathfrak{R} = \overline{(I - P^{\mathfrak{L}_*})M(\mathfrak{L})} \qquad (cf.\ (2.13)).$$

The subspace $\mathfrak{R}$ reduces to $\{0\}$ if and only if $T^{*n} \to O$ as $n \to \infty$.

2. Let T be as above, and let W be an isometry on a space $\mathfrak{G}$. As a first application of Theorem 2.1, more particularly of decomposition (2.5), we consider the connections between the solutions X and Y of the operator equations

(a) $$TX = XW$$

and

(b) $$U_+Y = YW,$$

where X is a bounded operator from $\mathfrak{G}$ to $\mathfrak{H}$, and Y is a bounded operator from $\mathfrak{G}$ to $\mathfrak{K}_+$.

Firstly, let us observe that every solution Y of (b) gives rise to a solution X of (a) by setting

(c) $$X = P_+Y,$$

where P_+ denotes the orthogonal projection of $\mathfrak{K}_+$ into $\mathfrak{H}$. Indeed, this follows immediately from the relation $TP_+ = P_+U_+$; see (I.4.2).

We show that every X can be obtained this way, that is, for every X there exists a Y such that (c) holds. Actually, there can exist more than just one such Y, and as a consequence of (c) all of them obviously satisfy the inequality $\|X\| \leq \|Y\|$. We find a Y for which this inequality is, in fact, an equality, and for this Y it is obviously sufficient to prove that $\|Y\| \leq \|X\|$. By reason of homogeneity it is possible to restrict our study to the case $\|X\| = 1$ (the case $X = O$ is trivial: set $Y = O$). Then we have to find Y so that

(d) $$\|Y\| \leq 1.$$

Observe first that by virtue of decomposition (2.5) the general form of an operator Y from $\mathfrak{G}$ to $\mathfrak{K}_+$ satisfying (c) is:

$$Y = X + B_0 + U_+B_1 + U_+^2B_2 + \cdots, \tag{2.14}$$

where each B_n is an operator from $\mathfrak{G}$ to $\mathfrak{L}$. The additional condition (d) means that

$$(\|Yg\|^2 =)\|Xg\|^2 + \sum_0^\infty \|B_ng\|^2 \leq \|g\|^2 \quad \text{for} \quad g \in \mathfrak{G}. \tag{2.15}$$

From (2.14) we deduce

$$
\begin{aligned}
U_+Y - YW &= \left(U_+X + \sum_0^\infty U_+^{n+1}B_n\right) - \left(XW + \sum_0^\infty U_+^n B_n W\right) \\
&= \sum_0^\infty U_+^n (B_{n-1} - B_n W)
\end{aligned}
$$

with $B_{-1} = U_+X - XW$. Because of (a) we have $B_{-1} = (U_+ - T)X$, and thus B_{-1} is an operator from $\mathfrak{G}$ to $\mathfrak{L}$.

In order that Y satisfy (b), it is therefore necessary and sufficient that the following equations hold.

$$
B_n W = B_{n-1} \ (n = 0, 1 \ldots), \quad B_{-1} = (U_+ - T)X. \tag{2.16}
$$

To sum up: the form of an operator Y from $\mathfrak{G}$ to $\mathfrak{K}_+$ satisfying (b), (c), and (d) is given by (2.14), the operators B_n (from $\mathfrak{G}$ to $\mathfrak{L}$) being subject to conditions (2.15) and (2.16).

We construct such a sequence of operators B_n by recurrence. Suppose that, for an $N \geq 0$, the operators B_n $(n < N)$ are already determined so that they satisfy the conditions

$(\mathrm{e})_N$ $$s_N(g) \equiv \|Xg\|^2 + \sum_{0 \leq n < N} \|B_n g\|^2 \leq \|g\|^2 \quad (g \in \mathfrak{G})$$

and

$(\mathrm{f})_N$ $$B_n W = B_{n-1} \ (0 \leq n < N), \quad B_{-1} = (U_+ - T)X.$$

(For $N = 0$ these conditions are satisfied: $s_0(g) \equiv \|Xg\|^2 \leq \|g\|^2$ because $\|X\| = 1$, and condition $(\mathrm{f})_0$ reduces to the equation defining B_{-1}.)

Let us note that $(\mathrm{e})_N$ can be written in the equivalent form

$$
I_{\mathfrak{G}} - X^*X - \sum_{0 \leq n < N} B_n^* B_n \geq O;
$$

let us denote the positive square root of this positive operator by D_N.

In order to determine the next operator in the sequence (i.e., B_N), observe first that

$$
\|B_{-1}g\|^2 = \|(U_+ - T)Xg\|^2 = \|Xg\|^2 - \|TXg\|^2 = \|Xg\|^2 - \|XWg\|^2, \tag{2.17}
$$

as a consequence of (a). In the case $N \geq 1$ we obtain using $(\mathrm{f})_N$ that

$$
\begin{aligned}
s_N(Wg) &= \|XWg\|^2 + \sum_{0 \leq n < N} \|B_n W g\|^2 \\
&= \|XWg\|^2 + \|B_{-1}g\|^2 + \sum_{0 \leq m < N-1} \|B_m g\|^2;
\end{aligned}
$$

hence it follows by (2.17) and $(\mathrm{e})_N$ that

$$s_N(Wg) = s_N(g) - \|B_{n-1}g\|^2 \leq \|g\|^2 - \|B_{N-1}g\|^2.$$

The equality $\|g\| = \|Wg\|$ implies

$$\begin{aligned}\|B_{N-1}g\|^2 &\leq \|Wg\|^2 - s_N(Wg)\\ &= \|Wg\|^2 - \|XWg\|^2 - \sum_{0\leq n<N} \|B_nWg\|^2 = \|D_NWg\|^2.\end{aligned}$$

On the other hand, $\|Xg\| \leq \|g\| = \|Wg\|$, therefore we obtain from (2.17) that

$$\|B_{-1}g\|^2 \leq \|Wg\|^2 - \|XWg\|^2 = \|D_0Wg\|^2.$$

Thus we conclude that the inequality

$$\|B_{N-1}g\|^2 \leq \|D_NWg\|^2 \qquad (g \in \mathfrak{G}) \tag{2.18}$$

holds in all cases ($N \geq 0$). Hence we infer that there exists a contraction C_N of $D_NW\mathfrak{G}$ into $\mathfrak{L}$ such that

$$B_{N-1} = C_ND_NW. \tag{2.19}$$

The contraction C_N extends by continuity to the closure $\mathfrak{G}_1$ of $D_NW\mathfrak{G}$; it even extends to a contraction of the whole space $\mathfrak{G}$ into $\mathfrak{L}$ if one defines it, for example, to be O on the orthogonal complement $\mathfrak{G} \ominus \mathfrak{G}_1$. With the contraction C_N of $\mathfrak{G}$ into $\mathfrak{L}$ thus obtained, we define

$$B_N = C_ND_N. \tag{2.20}$$

The relation

$$B_NW = B_{N-1}$$

is then obvious. Moreover, we have

$$\|B_Ng\|^2 = \|C_ND_Ng\|^2 \leq \|D_Ng\|^2 = \|g\|^2 - \|Xg\|^2 - \sum_{0\leq n<N} \|B_ng\|^2,$$

and hence $(2.15)_{N+1}$ holds. The operators B_n, defined in such a way by recurrence for every $n \geq 0$, will satisfy (2.15) and (2.16), limit cases of $(\mathrm{e})_N$ and $(\mathrm{f})_N$. Thus we have proved the following result.

Proposition 2.2. *For a contraction T on $\mathfrak{H}$ and an isometry W on $\mathfrak{G}$, the general solution X of* (a) *is obtained by* (c) *from the general solution Y of* (b); *moreover, for given X one can choose Y so that $\|X\| = \|Y\|$.*

3. It is now easy to deduce the following more general theorem.

Theorem 2.3. *Let T and T' be contractions on the Hilbert spaces $\mathfrak{H}$ and $\mathfrak{H}'$, and let U_+ and U'_+ be their minimal isometric dilations on the spaces $\mathfrak{K}_+$ and $\mathfrak{K}'_+$, respectively. For every bounded operator X from $\mathfrak{H}'$ into $\mathfrak{H}$ satisfying*

(a) $TX = XT'$

there exists a bounded operator Y from $\mathfrak{K}'_+$ into $\mathfrak{K}_+$ satisfying the conditions

(b) $U_+Y = YU'_+$,
(c) $X = P_+Y|\mathfrak{H}'$,
(c′) $Y(\mathfrak{K}'_+ \ominus \mathfrak{H}') \subset \mathfrak{K}_+ \ominus \mathfrak{H}$,
(d) $\|X\| = \|Y\|$.

Conversely, every bounded solution Y of (b) *satisfying* (c′) *gives rise by* (c) *to a solution X of* (a).

Conditions (c) *and* (c′) *together are equivalent to the condition*

(c″) $X^* = Y^*|\mathfrak{H}$.

Proof. Consider first a bounded operator Y from $\mathfrak{K}'_+$ into $\mathfrak{K}_+$ satisfying (b) and (c′). We show that the operator X to which it gives rise by (c) satisfies (a). Indeed, using relation (I.4.2) we get

$$TX = TP_+Y|\mathfrak{H}' = P_+U_+Y|\mathfrak{H}' = P_+YU'_+|\mathfrak{H}'$$
$$= P_+YP'_+U'_+|\mathfrak{H}' + P_+Y(I-P'_+)U'_+|\mathfrak{H}'.$$

Because $P'_+U'_+|\mathfrak{H}' = T'$ by the dilation property, and because $P_+Y(I-P'_+) = O$ by virtue of (c′), we obtain (a).

Consider next an arbitrary bounded solution X of (a). Multiplying in (a) by P'_+ from the right and using the relation $T'P'_+ = P'_+U'_+$ (cf. (I.4.2)), we obtain

$$TX_0 = X_0U'_+ \quad \text{with} \quad X_0 = XP'_+;$$

X_0 is an operator from $\mathfrak{K}'_+$ into $\mathfrak{H}$. If we now apply Proposition 2.2 with $W = U'_+$, we obtain that there exists an operator Y from $\mathfrak{K}'_+$ into $\mathfrak{K}_+$ such that

$$U_+Y = YU'_+,\ P_+Y = X_0, \quad \text{and} \quad \|X_0\| = \|Y\|.$$

We obviously have $X_0|\mathfrak{H}' = X$ and $\|X_0\| = \|X\|$, thus it follows that Y satisfies (c) and (d). It also satisfies (c′) because

$$P_+Y(I-P'_+) = X_0(I-P'_+) = XP'_+(I-P'_+) = O.$$

Finally, the equivalence of conditions (c) and (c′) to (c″) is straightforward. So the proof is complete.

4. Theorem 2.3 can be extended by replacing the conditions of minimality on U_+ and U'_+ with the weaker conditions

$$U^*_+|\mathfrak{H} = T^* \quad \text{and} \quad U'^*_+|\mathfrak{H}' = T'^*, \tag{2.21}$$

respectively. Indeed, if these conditions hold then the spaces

$$\mathfrak{K}_{+0} = \bigvee_{n\geq 0} U^n_+\mathfrak{H}, \quad \mathfrak{K}'_{+0} = \bigvee_{n\geq 0} U'^n_+\mathfrak{H}'$$

are reducing U_+ and U'_+ respectively, moreover

$$U_{+0} = U_+|\mathfrak{K}_{+0} \quad \text{and} \quad U'_{+0} = U'_+|\mathfrak{K}'_{+0}$$

are the minimal isometric dilations of T and T' (see the remark at the end of Sec. I.2). Therefore for a given operator X satisfying the property (a) above, we can apply Theorem 2.3 to obtain an operator Y_0 from $\mathfrak{K}'_{+0}$ into $\mathfrak{K}_{+0}$ satisfying

$$U_{+0}Y_0 = Y_0U'_{+0}, \quad Y_0(\mathfrak{K}'_{+0} \ominus \mathfrak{H}') \subset \mathfrak{K}_{+0} \ominus \mathfrak{H}$$

and

$$X = P_{\mathfrak{H}}Y_0|\mathfrak{H}, \quad \|Y_0\| = \|X\|.$$

Defining $Y = Y_0P_{\mathfrak{K}'_{+0}}$ we obtain an operator from $\mathfrak{K}'_+$ into $\mathfrak{K}_+$ satisfying all the properties (b), (c), (c$'$), (c$''$), and (d) in Theorem 2.3. The proof of the fact that in the present more general case, the relations (b), (c), (c$'$) imply (a) is identical to that given in the proof of Theorem 2.3.

3 The residual parts and quasi-similarities

1. We resume the study of the residual and $*$-residual parts of the minimal unitary dilation U of T, defined by (2.1) and (2.2). Let

$$P_{\mathfrak{H}}, \; P_{\mathfrak{R}}, \; P_{\mathfrak{R}_*}$$

denote the orthogonal projections of the space $\mathfrak{K}$ onto the subspaces $\mathfrak{H}, \mathfrak{R}$, and $\mathfrak{R}_*$, respectively.

Proposition 3.1. *For every $h \in \mathfrak{H}$ we have*

$$P_{\mathfrak{R}}h = \lim_{n\to\infty} U^nT^{*n}h, \quad P_{\mathfrak{R}_*}h = \lim_{n\to\infty} U^{-n}T^nh \tag{3.1}$$

and consequently

$$\|P_{\mathfrak{R}}h\| = \lim_{n\to\infty} \|T^{*n}h\|, \quad \|P_{\mathfrak{R}_*}h\| = \lim_{n\to\infty} \|T^nh\|, \tag{3.2}$$

$$P_{\mathfrak{H}}P_{\mathfrak{R}}h = \lim_{n\to\infty} T^nT^{*n}h, \quad P_{\mathfrak{H}}P_{\mathfrak{R}_*}h = \lim_{n\to\infty} T^{*n}T^nh. \tag{3.3}$$

Proof. Relations (3.2) and (3.3) follow immediately from (3.1) because U is a unitary dilation of T. So it suffices to prove relations (3.1), or, by reason of symmetry, one of them, say the one concerning $\mathfrak{R}_*$.

The inequalities $\|T^{n+1}h\| \leq \|T\| \cdot \|T^nh\| \leq \|T^nh\|$ $(n \geq 0)$, show that the sequence $\{\|T^nh\|\}$ is nonincreasing, and hence is convergent. For $0 \leq m \leq n$ we have

$$\begin{aligned}(U^{-n}T^nh, U^{-m}T^mh) &= (U^{m-n}T^nh, T^mh) = (T^{*n-m}T^nh, T^mh) \\ &= (T^{*m}T^{*n-m}T^nh, h) = (T^{*n}T^nh, h) = \|T^nh\|^2,\end{aligned}$$

and hence

$$\begin{aligned}&\|U^{-n}T^nh-U^{-m}T^mh\|^2\\ &=\|U^{-n}T^nh\|^2+\|U^{-m}T^mh\|^2-2\operatorname{Re}(U^{-n}T^nh,U^{-m}T^mh)\\ &=\|T^nh\|^2+\|T^mh\|^2-2\|T^nh\|^2=\|T^mh\|^2-\|T^nh\|^2.\end{aligned}$$

Thus the convergence of the numerical sequence $\{\|T^nh\|^2\}$ implies the convergence of the sequence $\{U^{-n}T^nh\}$ in $\mathfrak{K}$ $(n\to\infty)$. Setting

$$k=\lim_{n\to\infty}U^{-n}T^nh,$$

let us show that $k=P_{\mathfrak{R}_*}h$; that is,

$$\text{(a)}\quad k\perp M(\mathfrak{L})\quad\text{and}\quad\text{(b)}\quad h-k\in M(\mathfrak{L}).$$

Now (a) means that k is orthogonal to $U^m\mathfrak{L}$ for every integer m, and this is indeed true because

$$U^{-n}T^nh\perp U^m\mathfrak{L}\quad\text{for}\quad n\geq -m,$$

as a consequence of the relation $\mathfrak{H}\perp U^{m+n}\mathfrak{L}$ $(m+n\geq 0)$ which follows from (1.4). As to (b), we have only to observe that

$$\begin{aligned}&h-U^{-n}T^nh\\ &=U^{-1}(U-T)h+U^{-2}(U-T)Th+\cdots+U^{-n}(U-T)T^{n-1}h\in M(\mathfrak{L})\end{aligned}$$

and therefore

$$h-k=\lim_{n\to\infty}(h-U^{-n}T^nh)\in M(\mathfrak{L}).$$

This concludes the proof of Proposition 3.1.

Proposition 3.2. *If at least one point in the interior of the unit circle is not an eigenvalue of T, then*

$$\overline{P_{\mathfrak{R}}\mathfrak{H}}=\mathfrak{R}. \tag{3.4}$$

The analogous fact holds for the dual case.

Proof. Assume to the contrary that (3.4) is false, so that there exists a nonzero $k\in\mathfrak{R}$, such that $k\perp P_{\mathfrak{R}}\mathfrak{H}$, or equivalently,

$$k\perp M(\mathfrak{L}^*)\quad\text{and}\quad k\perp\mathfrak{H}.$$

Then by (1.4) we have $k\in M_+(\mathfrak{L})$ and hence k has an orthogonal expansion

$$k=\sum_{n=0}^{\infty}U^nl_n,\quad\text{where}\quad l_n\in\mathfrak{L},\ \sum_{n=0}^{\infty}\|l_n\|^2=\|k\|^2.$$

Because $k \neq 0$, at least one coefficient l_n is nonzero; let l_ν be the first of these nonzero coefficients. Then we have

$$U^{-\nu-1}k = U^{-1}l_\nu + \sum_{\mu=0}^{\infty} U^\mu l_{\nu+\mu+1}. \tag{3.5}$$

Because k belongs to $\mathfrak{R}$, $U^{-\nu-1}k = R^{-\nu-1}k$ also belongs to $\mathfrak{R} = \mathfrak{K} \ominus M(\mathfrak{L}^*)$ and hence, in particular, $U^{-\nu-1}k \perp \mathfrak{L}^*$. On the other hand (1.4) implies $U^\mu \mathfrak{L} \perp \mathfrak{L}^*$ for $\mu \geq 0$, therefore we deduce from (3.5) that $U^{-1}l_\nu \perp \mathfrak{L}^*$ and $l_\nu \perp U\mathfrak{L}^*$. Now, we obviously have $l_\nu \in \mathfrak{L} \subset \mathfrak{H} \oplus \mathfrak{L}$ and, by (1.7), $\mathfrak{H} \oplus \mathfrak{L} = U\mathfrak{L}^* \oplus U\mathfrak{H}$, so we conclude that $l_\nu \in U\mathfrak{H}$. Thus there exists an $h \in \mathfrak{H}$ such that $l_\nu = Uh$; consequently $P_\mathfrak{H} l_\nu = P_\mathfrak{H} Uh = Th$. Because $\mathfrak{L} \perp \mathfrak{H}$, we have $P_\mathfrak{H} l_\nu = 0$; hence $Th = 0$. But $l_\nu \neq 0$ implies $h \neq 0$, and this means that 0 is an eigenvalue of T.

Let us now consider the contraction $T_a = (T - aI)(I - \bar{a}T)^{-1}$ with $|a| < 1$, and the analogous transform $(U_+)_a$ of U_+; $(U_+)_a$ is the minimal isometric dilation of T_a (cf. Proposition I.4.3). An operator S is unitary if and only if its transform S_a is unitary, thus we conclude that the maximal subspace of $\mathfrak{K}_+$ in which $(U_+)_a$ is unitary, coincides with the maximal subspace of $\mathfrak{K}_+$ in which U_+ is unitary, that is with $\mathfrak{R}$. Thus, applying the results already obtained to T_a instead of T, we conclude that if (3.4) does not hold then 0 is an eigenvalue of T_a and hence a is an eigenvalue of T.

So if (3.4) does not hold, then every point in the interior of the unit circle is an eigenvalue of T. This concludes the proof.

2. It is convenient to introduce some further notions. (In what follows, $\mathfrak{H}_1, \mathfrak{H}_2, \ldots$ are assumed to be Hilbert spaces, but most of the notions introduced make sense for Banach spaces as well.)

Definition 1. By an *affinity from* $\mathfrak{H}_1$ *to* $\mathfrak{H}_2$ we mean a linear, one-to-one, and bicontinuous transformation X from $\mathfrak{H}_1$ onto $\mathfrak{H}_2$. Two bounded operators, say S_1 on $\mathfrak{H}_1$ and S_2 on $\mathfrak{H}_2$, are said to be *similar* if there exists an affinity X from $\mathfrak{H}_1$ to $\mathfrak{H}_2$ such that $XS_1 = S_2X$ (and consequently $X^{-1}S_2 = S_1X^{-1}$).

Definition 2. By a *quasi-affinity from* $\mathfrak{H}_1$ *to* $\mathfrak{H}_2$ we mean a linear, one-to-one, and continuous transformation X from $\mathfrak{H}_1$ onto a dense linear manifold in $\mathfrak{H}_2$. (Thus X^{-1} exists on this dense domain, but is not necessarily continuous.[1]) If S_1 and S_2 are bounded operators, S_1 on $\mathfrak{H}_1$ and S_2 on $\mathfrak{H}_2$, we say that S_1 is a *quasi-affine transform* of S_2 if there exists a quasi-affinity X from $\mathfrak{H}_1$ to $\mathfrak{H}_2$ such that $XS_1 = S_2X$. The operators S_1 and S_2 are called *quasi-similar* if they are quasi-affine transforms of one another.

Remark. Similarity is a rather strong relation, which preserves for example, the spectrum. Quasi-similarity does not have such strong implications. Nevertheless, it

[1] Every one-to-one linear transformation A from $\mathfrak{H}_1$ into $\mathfrak{H}_2$ is *invertible*, the inverse transformation A^{-1} being a linear transformation with domain equal to the range of A. When A^{-1} is defined on the whole space $\mathfrak{H}_2$ and is continuous (i.e., bounded), we say that A is *boundedly invertible* (or that it has a *bounded inverse*).

will be clear from the results obtained in this book that quasi-affinity and quasi-similarity are very natural and useful concepts.

Some of the first consequences of Definition 2 are listed in the following two propositions.

Proposition 3.3. (1) *If X is a quasi-affinity from $\mathfrak{H}_1$ to $\mathfrak{H}_2$, and Y is a quasi-affinity from $\mathfrak{H}_2$ to $\mathfrak{H}_3$, then YX is a quasi-affinity from $\mathfrak{H}_1$ to $\mathfrak{H}_3$.*

(2) *If X is a quasi-affinity from $\mathfrak{H}_1$ to $\mathfrak{H}_2$, then X^* is a quasi-affinity from $\mathfrak{H}_2$ to $\mathfrak{H}_1$.*

(3) *If X is a quasi-affinity from $\mathfrak{H}_1$ to $\mathfrak{H}_2$, then $|X| = (X^*X)^{1/2}$ is a quasi-affinity on $\mathfrak{H}_1$ (i.e., from $\mathfrak{H}_1$ to $\mathfrak{H}_1$). Moreover, $X \cdot |X|^{-1}$ extends by continuity to a unitary transformation V_X from $\mathfrak{H}_1$ to $\mathfrak{H}_2$.*

Proof. A simple exercise left to the reader.

Proposition 3.4. (1) *If S_1 is a quasi-affine transform of S_2 and S_2 is a quasi-affine transform of S_3, then S_1 is a quasi-affine transform of S_3.*

(2) *If S_1 is a quasi-affine transform of S_2, then S_2^* is a quasi-affine transform of S_1^*.*

(3) *If a unitary operator S_1 on $\mathfrak{H}_1$ is the quasi-affine transform of a unitary operator S_2 on $\mathfrak{H}_2$, then S_1 and S_2 are unitary equivalent.*

Proof. (1) and (2) follow immediately from the corresponding parts of the preceding proposition. As to (3), let us observe first that (a) $XS_1 = S_2X$ (where X is a quasi-affinity from $\mathfrak{H}_1$ to $\mathfrak{H}_2$) implies, because S_1 and S_2 are unitary, that $S_2^*X = S_2^{-1}X = XS_1^{-1} = XS_1^*$, and hence (b) $X^*S_2 = S_1X^*$. From (a) and (b) one obtains $|X|^2S_1 = X^*XS_1 = X^*S_2X = S_1X^*X = S_1|X|^2$ and, by iteration, $|X|^{2n}S_1 = S_1|X|^{2n}$ $(n = 0, 1, \ldots)$; hence $p(|X|^2)S_1 = S_1p(|X|^2)$ for every polynomial $p(x)$. Let $\{p_n(x)\}$ be a sequence of polynomials tending to $|x|^{1/2}$ uniformly on the interval $0 \le x \le \|X\|^2$. Then $p_n(|X|^2)$ converges (in the operator norm) to $|X|$ so that we obtain as a limit the relation (c) $|X|S_1 = S_1|X|$. From (a) and (c) it follows that

$$S_2V_X|X| = S_2X = XS_1 = V_X|X|S_1 = V_XS_1|X|;$$

because $|X|\mathfrak{H}_1$ is dense in $\mathfrak{H}_1$ it results that $S_2V_X = V_XS_1$. By virtue of part (3) of the preceding proposition, V_X is unitary, and thus S_1 and S_2 are unitarily equivalent.

3. Let us return to the subject of Subsec. 1 and deduce the following consequences of relations (3.1):

$$U^*P_{\mathfrak{R}}h = P_{\mathfrak{R}}T^*h \quad \text{and} \quad UP_{\mathfrak{R}_*}h = P_{\mathfrak{R}_*}Th \quad (h \in \mathfrak{H}). \tag{3.6}$$

Indeed, we have only to notice that

$$U^*(\lim U^nT^{*n}h) = \lim U^{n-1}T^{*n}h = (\lim U^{n-1}T^{*n-1})T^*h$$

and

$$U(\lim U^{-n}T^nh) = \lim U^{-(n-1)}T^nh = (\lim U^{-(n-1)}T^{n-1})Th.$$

Proposition 3.5. (i) *If Th does not equal 0 and $T^{*n}h$ does not converge to 0 as $n \to \infty$, for any nonzero $h \in \mathfrak{H}$, then*

$$X = P_{\mathfrak{R}}|\mathfrak{H}$$

is a quasi-affinity from $\mathfrak{H}$ to $\mathfrak{R}$ and we have

$$R^*X = XT^*, \quad X^*R = TX^*, \tag{3.7}$$

so that R is a quasi-affine transform of T. (ii) *If T^*h does not equal 0 and $T^n h$ does not converge to 0 as $n \to \infty$, for any nonzero $h \in \mathfrak{H}$, then $Y = P_{\mathfrak{R}_*}|\mathfrak{H}$ is a quasi-affinity from $\mathfrak{H}$ to $\mathfrak{R}_*$ and we have*

$$R_*Y = YT, \tag{3.8}$$

so that T is a quasi-affine transform of R_*. (iii) *If for no nonzero $h \in \mathfrak{H}$ does either $T^n h$ or $T^{*n}h$ converge to 0 as $n \to \infty$, then T is quasi-similar to R as well as to R_*, which are, in this case, unitarily equivalent.*

Proof. Under the conditions of (i), the first of the relations (3.2) implies that $Xh \neq 0$ for $h \neq 0$, and Proposition 3.2 implies $\overline{X\mathfrak{H}} = \mathfrak{R}$. Thus X is a quasi-affinity from $\mathfrak{H}$ to $\mathfrak{R}$ and (3.7) follows from the first of the relations (3.6). Case (ii) is the dual of (i). In case (iii), the conditions of (i) and (ii) are simultaneously satisfied, and thus R is a quasi-affine transform of T and T is a quasi-affine transform of R_*. By Proposition 3.4, R is then a quasi-affine transform of R_*, and hence R_* is unitarily equivalent to R. Hence we conclude that T is a quasi-affine transform of R too. This concludes the proof.

4. Part (iii) of Proposition 3.5 easily provides a characterization of those contractions T that are quasi-similar to a unitary operator. If quasi-similarity is replaced by the stronger relation of similarity we have the following results.

Proposition 3.6. *If T is a c.n.u. contraction, then the following conditions are equivalent.*

(a) *T is similar to a unitary operator V; that is, $T = S^{-1}VS$ holds with an affinity S.*
(b) *The transformation $X = P_{\mathfrak{R}}|\mathfrak{H}$ from $\mathfrak{H}$ to $\mathfrak{R}$ is an affinity and so T is similar to R.*
(c) *The transformation $Q = P^{\mathfrak{L}_*}|M_+(\mathfrak{L})$ is onto $M_+(\mathfrak{L}_*)$ and boundedly invertible.*

Furthermore, $\|X^{-1}\| = \|Q^{-1}\| \leq \|S\|\|S^{-1}\|$ holds.

In the proof of Proposition 3.6 we need the following lemma.

Lemma 3.7. *Let*

$$\mathfrak{M} = \mathfrak{A} \oplus \mathfrak{B} \quad \text{and} \quad \mathfrak{M} = \mathfrak{X} \oplus \mathfrak{Y}$$

be two orthogonal decompositions of the Hilbert space $\mathfrak{M}$, and let $P_{\mathfrak{A}}$ and $P_{\mathfrak{B}}$ denote the orthogonal projections from $\mathfrak{M}$ onto $\mathfrak{A}$ and $\mathfrak{B}$, respectively. If

$$P_{\mathfrak{A}}\mathfrak{X} = \mathfrak{A} \quad \text{and} \quad \|P_{\mathfrak{A}}x\| \geq c\|x\| \quad (x \in \mathfrak{X}) \tag{3.9}$$

with some positive constant c, then we also have

$$P_{\mathfrak{B}}\mathfrak{Y} = \mathfrak{B} \quad \textit{and} \quad \|P_{\mathfrak{B}}y\| \geq c\|y\| \quad (y \in \mathfrak{Y}). \tag{3.10}$$

Proof. The inequality

$$\|P_{\mathfrak{A}}x\|^2 \geq c^2\|x\|^2 = c^2[\|P_{\mathfrak{A}}x\|^2 + \|P_{\mathfrak{B}}x\|^2]$$

implies

$$C^2\|P_{\mathfrak{A}}x\|^2 \geq \|P_{\mathfrak{B}}x\|^2 \quad \text{with} \quad C = \sqrt{1-c^2}/c.$$

Thus the hypotheses (3.9) imply that the formula

$$A(P_{\mathfrak{A}}x) = P_{\mathfrak{B}}x \quad (x \in \mathfrak{X})$$

defines an operator A from $\mathfrak{A}$ into $\mathfrak{B}$, bounded by the constant C. The graph $\{a \oplus Aa\colon\ a \in \mathfrak{A}\}$ of A in $\mathfrak{M} = \mathfrak{A} \oplus \mathfrak{B}$ equals $\{P_{\mathfrak{A}}x \oplus P_{\mathfrak{B}}x\colon\ x \in \mathfrak{X}\}$, and consequently $\mathfrak{X}$, its orthogonal complement $\{-A^*b \oplus b\colon\ b \in \mathfrak{B}\}$ will be equal to $\mathfrak{Y}$. This implies that $P_{\mathfrak{B}}\mathfrak{Y} = \mathfrak{B}$ and $P_{\mathfrak{A}}y = -A^*P_{\mathfrak{B}}y$ for $y \in \mathfrak{Y}$. It follows that

$$\|P_{\mathfrak{A}}y\| \leq C\|P_{\mathfrak{B}}y\|, \quad \|y\|^2 = \|P_{\mathfrak{A}}y\|^2 + \|P_{\mathfrak{B}}y\|^2 \leq (1+C^2)\|P_{\mathfrak{B}}y\|^2 = \frac{1}{c^2}\|P_{\mathfrak{B}}y\|^2,$$

and this concludes the proof of the relations (3.10).

Proof (Proof of Proposition 3.6). Suppose T is a c.n.u. contraction on the space $\mathfrak{H}$, similar to a unitary operator V, so that $T = S^{-1}VS$ for some affinity S on $\mathfrak{H}$. Then T and T^* are boundedly invertible and we have for every integer n:

$$T^{-n} = S^{-1}V^{-n}S, \quad T^{*-n} = S^*V^nS^{*-1},$$

and hence

$$\|T^{-n}\| \leq k, \quad \|T^{*-n}\| \leq k, \quad \text{with } k = \|S\|\ \|S^{-1}\| = \|S^*\|\ \|S^{*-1}\|;$$

consequently, setting $c = 1/k$ we obtain

$$\|T^nh\| \geq c\|h\|, \quad \|T^{*n}h\| \geq c\|h\| \quad (h \in \mathfrak{H}). \tag{3.11}$$

From the second inequality we deduce by (3.2) that

$$\|P_{\mathfrak{R}}h\| \geq c\|h\| \quad (h \in \mathfrak{H}), \tag{3.12}$$

where $P_{\mathfrak{R}}$ denotes the orthogonal projection from the space $\mathfrak{K}$ (of the minimal unitary dilation U of T) onto the subspace $\mathfrak{R}$ (of the residual part R of U). Because 0 is not an eigenvalue of T, Proposition 3.2 asserts that $\overline{P_{\mathfrak{R}}\mathfrak{H}} = \mathfrak{R}$. The inequality (3.12) implies that $P_{\mathfrak{R}}\mathfrak{H}$ is closed, and we conclude that

$$P_{\mathfrak{R}}\mathfrak{H} = \mathfrak{R}. \tag{3.13}$$

On account of (3.12) and (3.13),

$$X = P_{\mathfrak{R}}|\mathfrak{H} \tag{3.14}$$

is an affinity from $\mathfrak{H}$ to $\mathfrak{R}$, and hence X^* is an affinity from $\mathfrak{R}$ to $\mathfrak{H}$. Using (3.14) we deduce from Proposition 3.5 that $X^*R = TX^*$; thus T is similar to R. Therefore, if a contraction is similar to a unitary operator, then it is similar in particular to the residual part of its minimal unitary dilation.

Now the space $\mathfrak{K}_+$ of the minimal isometric dilation of T admits the decompositions

$$\mathfrak{K}_+ = M_+(\mathfrak{L}_*) \oplus \mathfrak{R}, \quad \mathfrak{K}_+ = M_+(\mathfrak{L}) \oplus \mathfrak{H}. \tag{3.15}$$

Thus relations (3.12) and (3.13) imply, by virtue of Lemma 3.7, that

$$P^{\mathfrak{L}_*}M_+(\mathfrak{L}) = M_+(\mathfrak{L}_*) \quad \text{and} \quad \|P^{\mathfrak{L}_*}l\| \geq c\|l\| \quad (l \in M_+(\mathfrak{L})). \tag{3.16}$$

It follows that

$$Q = P^{\mathfrak{L}_*}|M_+(\mathfrak{L})$$

is a contraction from $M_+(\mathfrak{L})$ onto $M_+(\mathfrak{L}_*)$, which is boundedly invertible with $\|Q^{-1}\| \leq 1/c$.

Conversely, if Q is invertible with $\|Q^{-1}\| \leq 1/c$, then we infer by Lemma 3.7 that $X = P_{\mathfrak{R}}|\mathfrak{H}$ is invertible with $\|X^{-1}\| \leq 1/c$. Thus the proof is complete.

4 A classification of contractions

The preceding results motivate, to some extent, the introduction of the following classes $C_{..}$ of contractions.

$$T \in C_{0\cdot} \text{ if } T^n h \to 0 \text{ for all } h;\ T \in C_{1\cdot} \text{ if } T^n h \not\to 0 \text{ for all } h \neq 0;$$
$$T \in C_{\cdot 0} \text{ if } T^{*n} h \to 0 \text{ for all } h;\ T \in C_{\cdot 1} \text{ if } T^{*n} h \not\to 0 \text{ for all } h \neq 0.$$

Furthermore, set

$$C_{\alpha\beta} = C_{\alpha\cdot} \cap C_{\cdot\beta} \qquad (\alpha, \beta = 0, 1).$$

These special classes play an important role in the study of general contractions. To begin with, let us recall that, given a decomposition

$$\mathfrak{H} = \mathfrak{H}_1 \oplus \mathfrak{H}_2 \oplus \cdots \oplus \mathfrak{H}_p \tag{4.1}$$

of the Hilbert space $\mathfrak{H}$ into the orthogonal sum of subspaces $\mathfrak{H}_i$ $(i = 1, \ldots, p)$, to every bounded operator T in $\mathfrak{H}$ there corresponds a matrix $[T_{ij}]$ $(i, j = 1, \ldots, p)$, whose entries T_{ij} are the bounded operators from $\mathfrak{H}_j$ to $\mathfrak{H}_i$ defined by

$$T_{ij} = P_i T|\mathfrak{H}_j,$$

where P_i denotes the orthogonal projection from $\mathfrak{H}$ into $\mathfrak{H}_i$. It is obvious that if T is a contraction on $\mathfrak{H}$ then T_{ij} is a contraction from $\mathfrak{H}_j$ into $\mathfrak{H}_i$. *If, moreover, T is* *c.n.u., then all the diagonal entries T_{ii} are also c.n.u.* Indeed, if there exists in $\mathfrak{H}_i$ a subspace $\mathfrak{M}$ such that $T_{ii}|\mathfrak{M}$ is unitary, then for $f \in \mathfrak{M}$ we have $T_{ii}f = P_iTf$ and $\|f\| = \|T_{ii}f\| = \|P_iTf\| \le \|Tf\| \le \|f\|$. Hence $T_{ii}f = Tf$ so that T coincides on $\mathfrak{M}$ with T_{ii}; thus T is unitary on $\mathfrak{M}$. This contradicts the assumption that T is c.n.u. unless $\mathfrak{M} = \{0\}$. Thus T_{ii} is c.n.u.

We say that the decomposition (4.1) generates a *triangulation* (or, rather, a super-diagonalization) of T if $T_{ij} = O$ for $i > j$. This is equivalent to the set of conditions

$$T\mathfrak{H}_j \subset \mathfrak{H}_1 \oplus \cdots \oplus \mathfrak{H}_j \quad \text{for} \quad j = 1,2,\ldots,p. \tag{4.2}$$

In this case the subspace $\mathfrak{H}_1$ is therefore invariant for T and the matrices of the operators T^n $(n = 1,2,\ldots)$ are of the same type, with $(T^n)_{ii} = (T_{ii})^n$.

Theorem 4.1. *Every contraction T on the space $\mathfrak{H}$ has triangulations of the following types,*

$$\text{(a)} \quad \begin{bmatrix} C_{0\cdot} & * \\ O & C_{1\cdot} \end{bmatrix} \quad \textit{and} \quad \text{(a}^*\text{)} \quad \begin{bmatrix} C_{\cdot 1} & * \\ O & C_{\cdot 0} \end{bmatrix},$$

where, in the diagonal, one has indicated the class of the respective operator only (with the operator O on the space $\{0\}$ belonging to all these classes); the type of the entries denoted by $*$ is not specified. These triangulations are uniquely determined and are called the "canonical triangulations" of T. There also exists a triangulation of type

$$\text{(b)} \quad \begin{bmatrix} C_{01} & * & * & * & * \\ O & C_{00} & * & * & * \\ O & O & C_{11} & * & * \\ O & O & O & C_{00} & * \\ O & O & O & O & C_{10} \end{bmatrix}.$$

Proof. Let us set

$$\mathfrak{H}_1 = \{h\colon\ h \in \mathfrak{H}, T^n h \to 0\}; \tag{4.3}$$

$\mathfrak{H}_1$ is obviously a subspace of $\mathfrak{H}$, invariant for T. Set $\mathfrak{H}_2 = \mathfrak{H} \ominus \mathfrak{H}_1$. The decomposition $\mathfrak{H} = \mathfrak{H}_1 \oplus \mathfrak{H}_2$ then yields a triangulation of T:

$$T = \begin{bmatrix} T_1 & * \\ O & T_2 \end{bmatrix}, \quad \text{where} \quad T_1 = T|\mathfrak{H}_1, \quad T_2 = P_2T|\mathfrak{H}_2.$$

Because $T_1^n = T^n|\mathfrak{H}_1$, we have $T_1 \in C_{0\cdot}$ by the definition of $\mathfrak{H}_1$. We show that $T_2 \in C_{1\cdot}$. To this end we use the second one of the relations (3.1), which asserts that

$$Qh = \lim_{n\to\infty} U^{-n}T^n h \quad (h \in \mathfrak{H}), \tag{4.4}$$

where we have written for simplification Q instead of $P_{\mathfrak{R}_*}$. It then follows for an arbitrary integer $m \geq 0$ that

$$\begin{aligned} Qh &= \lim_{n\to\infty} U^{-(n+m)} T^{n+m} h \\ &= \lim_{n\to\infty} (U^{-m} U^{-n} T^n P_2 T^m h + U^{-(n+m)} T^n P_1 T^m h) \\ &= U^{-m} \cdot \lim_{n\to\infty} U^{-n} T^n \cdot P_2 T^m h = U^{-m} Q P_2 T^m h, \end{aligned}$$

because $\lim T^n h_1 = 0$ for $h_1 = P_1 T^m h \in \mathfrak{H}_1$. From this result we derive the inequalities

$$\|Qh\| \leq \|P_2 T^m h\| \quad (h \in \mathfrak{H}; m = 0, 1, \ldots).$$

It then follows that if $\lim_{m\to\infty} P_2 T^m h = 0$ for an $h \in \mathfrak{H}$, then $Qh = 0$, and hence, by (3.2), $\lim_{n\to\infty} T^n h = 0$ (i.e., $h \in \mathfrak{H}_1$). Because $T_2^m = P_2 T^m | \mathfrak{H}_2$, we conclude that there exists no nonzero $h \in \mathfrak{H}_2$ for which $T_2^m h$ would converge to 0 as $m \to \infty$; that is, $T_2 \in C_{1\cdot}$. Thus the triangulation under consideration is of type (a).

Now we show that if

$$T = \begin{bmatrix} T_1' & * \\ O & T_2' \end{bmatrix}, \quad \mathfrak{H} = \mathfrak{H}_1' \oplus \mathfrak{H}_2', \quad T_1' = T|\mathfrak{H}_1', \quad T_2' = P_2' T | \mathfrak{H}_2',$$

is a triangulation of T of type (a), then it necessarily coincides with the triangulation just obtained. For this it suffices to show that

$$\mathfrak{H}_1 = \mathfrak{H}_1'. \tag{4.5}$$

Now for $h \in \mathfrak{H}_1'$ we have $T^n h = T_1'^n h \to 0$ because $T_1' \in C_{0\cdot}$; hence it follows by the definition (4.3) of $\mathfrak{H}_1$ that $h \in \mathfrak{H}_1$. Thus we have

$$\mathfrak{H}_1' \subset \mathfrak{H}_1.$$

Let now $h \in \mathfrak{H}_1 \ominus \mathfrak{H}_1'$. Then we have $T^n h \to 0$ because $h \in \mathfrak{H}_1$, and $P_2' T^n h = T_2'^n h$ because $h \in \mathfrak{H}_2'$ (P_2' denotes, of course, the orthogonal projection of $\mathfrak{H}$ into $\mathfrak{H}_2'$); thus $T_2'^n h \to 0$. As $T_2' \in C_{1\cdot}$, this implies $h = 0$. Hence $\mathfrak{H}_1 \ominus \mathfrak{H}_1' = \{0\}$, which proves (4.5).

So we have proved that every contraction T on $\mathfrak{H}$ has a unique triangulation of type (a).

If we take the triangulation of type (a) for T^* and then interchange the order of the corresponding subspaces $\mathfrak{H}_1$ and $\mathfrak{H}_2$, we obtain for T^* a matrix of type

$$\begin{bmatrix} C_{1\cdot} & O \\ * & C_{0\cdot} \end{bmatrix};$$

taking adjoints we obtain thus for T a matrix of type (a*). Therefore the existence and uniqueness of the triangulation of type (a*) follow from the same facts as for the triangulation of type (a).

Let us observe now that if $T \in C_{0\cdot}$ and if $[T_{ij}]$ is an arbitrary triangulation of T (i.e., $T_{ij} = O$ for $i > j$), then $T_{ii} \in C_{0\cdot}$ for every i; in fact $T_{ii}^n = P_i T^n | \mathfrak{H}_i \to O$ as $n \to \infty$.

In particular, if

$$T = \begin{bmatrix} T_1 & * \\ O & T_2 \end{bmatrix}$$

is the triangulation of T of type (a*), then

$$T_1 \in C_{0\cdot} \cap C_{\cdot 1} = C_{01} \text{ and } T_2 \in C_{0\cdot} \cap C_{\cdot 0} = C_{00}.$$

On the other hand, if $T \in C_{1\cdot}$ and if $[T_{ij}]$ is an arbitrary triangulation of T, then $T_{11} \in C_{1\cdot}$, because $T_{11}^n h = T^n h$ for $h \in \mathfrak{H}$. In particular, if

$$T = \begin{bmatrix} T_1 & * \\ O & T_2 \end{bmatrix}$$

is the triangulation of type (a*), we have $T_1 \in C_{1\cdot} \cap C_{\cdot 1} = C_{11}$ and, of course, $T_2 \in C_{\cdot 0}$.

These results can be expressed by the self-explanatory formulas

$$C_{0\cdot} = \begin{bmatrix} C_{01} & * \\ O & C_{00} \end{bmatrix}, \quad C_{1\cdot} = \begin{bmatrix} C_{11} & * \\ O & C_{\cdot 0} \end{bmatrix}. \tag{4.6}$$

One can derive from these the formulas

$$C_{\cdot 0} = \begin{bmatrix} C_{00} & * \\ O & C_{10} \end{bmatrix}, \quad C_{\cdot 1} = \begin{bmatrix} C_{0\cdot} & * \\ O & C_{11} \end{bmatrix} \tag{4.7}$$

by interchanging the order of the subspaces of decomposition and by taking adjoints.

Now, starting with the triangulation of type (a) and then applying formulas (4.6) and (4.7), one obtains

$$\begin{bmatrix} C_{0\cdot} & * \\ O & C_{1\cdot} \end{bmatrix} = \begin{bmatrix} \begin{bmatrix} C_{01} & * \\ O & C_{00} \end{bmatrix} & * \\ O & \begin{bmatrix} C_{11} & * \\ O & C_{\cdot 0} \end{bmatrix} \end{bmatrix} = \begin{bmatrix} \begin{bmatrix} C_{01} & * \\ O & C_{00} \end{bmatrix} & * \\ O & \begin{bmatrix} C_{11} & * \\ O & \begin{bmatrix} C_{00} & * \\ O & C_{10} \end{bmatrix} \end{bmatrix} \end{bmatrix},$$

that is, a triangulation of type (b).

Let us recall that *every contraction T of class C_{11} is quasi-similar to a unitary operator, indeed to the residual part R of the minimal isometric dilation of T*; see Proposition 3.5(iii).

Let

$$T = \begin{bmatrix} T_1 & * \\ O & T_2 \end{bmatrix}$$

be a triangulation of type

$$T = \begin{bmatrix} C_{\cdot 1} & * \\ O & C_{\cdot 0} \end{bmatrix}$$

corresponding to the decomposition $\mathfrak{H} = \mathfrak{H}_1 \oplus \mathfrak{H}_2$. It is obvious that $\mathfrak{H}_2$ is the largest invariant subspace for T^* where the compression of T is of class $C_{\cdot 0}$.

Proposition 4.2. *With the above notation, $\mathfrak{H}_1$ is the largest invariant subspace for T where the restriction of T is of class $C_{\cdot 1}$.*

Proof. Let $\mathfrak{H}' \subset \mathfrak{H}$ be invariant for T, and assume that $T' = T|\mathfrak{H}'$ is of class $C_{\cdot 1}$. There is then a decomposition $\mathfrak{H} = \mathfrak{H}' \oplus \mathfrak{H}'_1 \oplus \mathfrak{H}'_2$ relative to which we have

$$T = \begin{bmatrix} T' & * & * \\ O & T'_1 & * \\ O & O & T'_2 \end{bmatrix}$$

with $T'_1 \in C_{\cdot 1}$ and $T'_2 \in C_{\cdot 0}$. We show that the restriction

$$T'' = \begin{bmatrix} T' & * \\ O & T'_1 \end{bmatrix}$$

of T to $\mathfrak{H}' \oplus \mathfrak{H}'_1$ is of class $C_{\cdot 1}$. This implies that $\mathfrak{H}_1 = \mathfrak{H}' \oplus \mathfrak{H}'_1$, and therefore $\mathfrak{H}' \subset \mathfrak{H}_1$, thus establishing the maximality of $\mathfrak{H}_1$. Assume that we have $\lim_{n\to\infty} \|T''^{*n}(h' \oplus h'_1)\| = 0$ for some $h' \in \mathfrak{H}'$ and $h'_1 \in \mathfrak{H}'_1$. Writing

$$T''^{*n}(h' \oplus h'_1) = k'_n \oplus k'_{1n} \quad (n \geq 1)$$

with $k'_n \in \mathfrak{H}'$ and $k'_{1n} \in \mathfrak{H}'_1$, we observe first that $k'_n = T'^{*n}h'$, and therefore $h' = 0$ because T' is of class $C_{\cdot 1}$. It follows that $k'_{1n} = T_1'^{*n}h'_1$, and therefore $h'_1 = 0$ because T'_1 is also of class $C_{\cdot 1}$.

5 Invariant subspaces and quasi-similarity

1. To begin with, let us recall the definition of the notion of invariant subspace and add the definition of some further related notions.

Definition. Let T be a bounded operator on $\mathfrak{H}$, and let $\mathfrak{L}$ be a subspace of $\mathfrak{H}$.

(a) $\mathfrak{L}$ is said to be *invariant* for T if $T\mathfrak{L} \subset \mathfrak{L}$.
(b) $\mathfrak{L}$ is said to be *regular* for T if $\overline{T\mathfrak{L}} = \mathfrak{L}$.
(c) $\mathfrak{L}$ is said to be *hyperinvariant* for T if it is invariant for every bounded operator which commutes with T.

From the definition it follows that if $\{\mathfrak{L}_\alpha\}$ is a system of invariant (hyperinvariant) subspaces for T, then

$$\bigcap_\alpha \mathfrak{L}_\alpha \quad \text{and} \quad \bigvee_\alpha \mathfrak{L}_\alpha$$

are also invariant (hyperinvariant) subspaces for T.

It is also obvious that if U is a *unitary* operator on $\mathfrak{H}$, then $\mathfrak{L}$ is regular for U if, and only if, it reduces U. In other words, $\mathfrak{L}$ is regular for U if, and only if, $P_{\mathfrak{L}}$ (the orthogonal projection of $\mathfrak{H}$ onto $\mathfrak{L}$) commutes with U.

Let us prove, similarly, that $\mathfrak{L}$ is hyperinvariant for the unitary U if, and only if, $P_{\mathfrak{L}}$ commutes with all the bounded operators that commute with U.

In fact, if A commutes with $P_{\mathfrak{L}}$, then $A\mathfrak{L} = AP_{\mathfrak{L}}\mathfrak{H} = P_{\mathfrak{L}}A\mathfrak{H} \subset P_{\mathfrak{L}}\mathfrak{H} = \mathfrak{L}$; thus if $P_{\mathfrak{L}}$ commutes with all these A then $\mathfrak{L}$ is hyperinvariant for U. Conversely, if $\mathfrak{L}$ is hyperinvariant for U, then $AP_{\mathfrak{L}} = P_{\mathfrak{L}}AP_{\mathfrak{L}}$ for every A commuting with U. Then A^* also commutes with U (in fact, $AU = UA$ implies $U^*A = U^*AUU^* = U^*UAU^* = AU^*$, $A^*U = UA^*$), so we also have $A^*P_{\mathfrak{L}} = P_{\mathfrak{L}}A^*P_{\mathfrak{L}}$ and hence

$$AP_{\mathfrak{L}} = P_{\mathfrak{L}}AP_{\mathfrak{L}} = (P_{\mathfrak{L}}A^*P_{\mathfrak{L}})^* = (A^*P_{\mathfrak{L}})^* = P_{\mathfrak{L}}A.$$

By virtue of these remarks, if $\mathfrak{L}$ is hyperinvariant for U then its orthogonal complement is also hyperinvariant for U.

After these preliminaries we prove the following result.

Proposition 5.1. *Let T be a bounded operator on the space $\mathfrak{H}$, quasi-similar to a unitary operator U on the space $\mathfrak{K}$. With every subspace $\mathfrak{L}$ of $\mathfrak{K}$, which is hyperinvariant for U, we can associate a subspace $q(\mathfrak{L})$ of $\mathfrak{H}$, regular and hyperinvariant for T, so that we have:*

(a) $q(\{0\}) = \{0\}$, (b) $q(\mathfrak{K}) = \mathfrak{H}$,

(c) $q(\mathfrak{L}) \subset q(\mathfrak{L}')$ *if* $\mathfrak{L} \subset \mathfrak{L}'$, (d) $q(\mathfrak{L}) \neq q(\mathfrak{L}')$ *if* $\mathfrak{L} \neq \mathfrak{L}'$,

(e) $\bigcap_\alpha q(\mathfrak{L}_\alpha) = \{0\}$ *if* $\bigcap_\alpha \mathfrak{L}_\alpha = \{0\}$, (f) $\bigvee_\alpha q(\mathfrak{L}_\alpha) = q(\mathfrak{L})$ *if* $\bigvee_\alpha \mathfrak{L}_\alpha = \mathfrak{L}$.

Proof. By our hypothesis there exists a quasi-affinity X from $\mathfrak{K}$ to $\mathfrak{H}$ and a quasi-affinity Y from $\mathfrak{H}$ to $\mathfrak{K}$, such that

$$TX = XU \quad \text{and} \quad UY = YT. \tag{5.1}$$

Let $\mathfrak{L}$ be a subspace of $\mathfrak{K}$, hyperinvariant for U. We set

$$a(\mathfrak{L}) = \overline{X\mathfrak{L}} \quad \text{and} \quad b(\mathfrak{L}) = \{h\colon\ h \in \mathfrak{H}, Yh \in \mathfrak{L}\}; \tag{5.2}$$

clearly, $a(\mathfrak{L})$ and $b(\mathfrak{L})$ are subspaces of $\mathfrak{H}$. Let A be a bounded operator on $\mathfrak{H}$, commuting with T. Using relations (5.1) we see that

$$\begin{aligned} U(YAX) = (UY)(AX) &= (YT)(AX) = Y(TA)X = Y(AT)X \\ &= (YA)(TX) = (YA)(XU) = (YAX)U; \end{aligned} \tag{5.3}$$

$\mathfrak{L}$ being hyperinvariant for U, (5.3) implies

$$(YAX)\mathfrak{L} \subset \mathfrak{L}. \tag{5.4}$$

By virtue of definitions (5.2), we deduce from (5.4) that

$$Aa(\mathfrak{L}) \subset b(\mathfrak{L}). \tag{5.5}$$

Now we set

$$q(\mathfrak{L}) = \bigvee_A Aa(\mathfrak{L}), \tag{5.6}$$

where A runs over the set of bounded operators on $\mathfrak{H}$ that commute with T. In particular, the identity operator on $\mathfrak{H}$ belongs to this set, thus we have $a(\mathfrak{L}) \subset q(\mathfrak{L})$; on the other hand (5.5) implies that $q(\mathfrak{L}) \subset b(\mathfrak{L})$. We conclude that

$$a(\mathfrak{L}) \subset q(\mathfrak{L}) \subset b(\mathfrak{L}). \tag{5.7}$$

By its definition (5.6), $q(\mathfrak{L})$ is obviously hyperinvariant for T. Moreover it is regular for T. In fact, the equality $U\mathfrak{L} = \mathfrak{L}$ implies

$$\overline{Ta(\mathfrak{L})} = \overline{T\overline{X\mathfrak{L}}} = \overline{TX\mathfrak{L}} = \overline{XU\mathfrak{L}} = \overline{X\mathfrak{L}} = a(\mathfrak{L})$$ [2]

and, for A commuting with T,

$$\overline{TAa(\mathfrak{L})} = \overline{ATa(\mathfrak{L})} = \overline{A\overline{Ta(\mathfrak{L})}} = \overline{Aa(\mathfrak{L})};$$

hence:

$$\overline{Tq(\mathfrak{L})} = \bigvee_A \overline{TAa(\mathfrak{L})} = \bigvee_A \overline{Aa(\mathfrak{L})} = \bigvee_A Aa(\mathfrak{L}) = q(\mathfrak{L}).$$

From $q(\mathfrak{L})$ it is possible to retrieve $\mathfrak{L}$ by the formula

$$\overline{Yq(\mathfrak{L})} = \mathfrak{L}. \tag{5.8}$$

In fact, (5.2) and (5.7) imply

$$\overline{YX\mathfrak{L}} = \overline{Y\overline{X\mathfrak{L}}} = \overline{Ya(\mathfrak{L})} \subset \overline{Yq(\mathfrak{L})} \subset \overline{Yb(\mathfrak{L})} \subset \mathfrak{L}; \tag{5.9}$$

on the other hand, because YX commutes with U (apply (5.3) with $A = I$), YX also commutes with $P_{\mathfrak{L}}$, and hence

$$\overline{YX\mathfrak{L}} = \overline{YXP_{\mathfrak{L}}\mathfrak{K}} = \overline{P_{\mathfrak{L}}YX\mathfrak{K}} = \overline{P_{\mathfrak{L}}\overline{YX\mathfrak{K}}} = P_{\mathfrak{L}}\mathfrak{K} = \mathfrak{L}. \tag{5.10}$$

Thus from (5.9) and (5.10) we deduce (5.8).

It remains to establish properties (a)–(f).

(a) For $\mathfrak{L} = \{0\}$ we have $b(\mathfrak{L}) = \{0\}$, because $Yh = 0$ implies $h = 0$. In view of (5.7) it follows that $q(\{0\}) = \{0\}$.

(b) For $\mathfrak{L} = \mathfrak{K}$ we have $a(\mathfrak{K}) = \overline{X\mathfrak{K}} = \mathfrak{H}$; in view of (5.7) this implies $q(\mathfrak{K}) = \mathfrak{H}$.

[2] Here and in the sequel we make use of the following obvious fact: *If $\mathfrak{A}$ is an arbitrary set of points in a topological space $\mathfrak{M}$, and S is a continuous transformation of $\mathfrak{M}$ into a topological space $\mathfrak{M}$, then $\overline{S\overline{\mathfrak{A}}} = \overline{S\mathfrak{A}}$.*

(c) If $\mathfrak{L} \subset \mathfrak{L}'$, then $a(\mathfrak{L}) \subset a(\mathfrak{L}')$, and hence by (5.6) also $q(\mathfrak{L}) \subset q(\mathfrak{L}')$.
(d) This follows immediately from (5.8).
(e) $\bigcap_\alpha q(\mathfrak{L}_\alpha) \subset \bigcap_\alpha b(\mathfrak{L}_\alpha) \subset b(\bigcap_\alpha \mathfrak{L}_\alpha) = b(\{0\}) = \{0\}$ if $\bigcap_\alpha \mathfrak{L}_\alpha = \{0\}$.
(f) Observe first that if each $\mathfrak{L}_\alpha$ is hyperinvariant for U then so is $\mathfrak{L} = \bigvee_\alpha \mathfrak{L}_\alpha$.

Now by (5.6) and (5.2),

$$q(\mathfrak{L}) = \bigvee_A A\, a(\mathfrak{L}) = \bigvee_A A\, \overline{X \bigvee_\alpha \mathfrak{L}_\alpha} = \bigvee_{A,\alpha} AX\mathfrak{L}_\alpha = \bigvee_\alpha \bigvee_A A\, \overline{X\mathfrak{L}_\alpha} = \bigvee_\alpha q(\mathfrak{L}_\alpha).$$

This finishes the proof of Proposition 5.1.

It is to be remarked that if $E(\sigma)$ is the spectral measure corresponding to the unitary operator U (defined for the Borel subsets σ of the unit circle) then $E(\sigma)$ commutes with U and with all the bounded operators commuting with U (cf. [*Func. Anal.*] Sec. 109); hence the subspaces

$$\mathfrak{L}(\sigma) = E(\sigma)\mathfrak{K}$$

are all hyperinvariant for U. Thus the preceding proposition has the following corollary.

Corollary 5.2. *If a bounded operator T on the complex Hilbert space $\mathfrak{H}$ is quasi-similar to a unitary operator U, then there exist at least as many nontrivial subspaces of $\mathfrak{H}$, regular and hyperinvariant for T, as there are values different from O and I of the spectral measure $E(\sigma)$ corresponding to U.*

2. The above results apply in particular to contractions of class C_{11}, because as remarked at the end of Sec. 4 these are quasi-similar to unitary operators. Actually, we can prove this property for a larger class of operators:

Proposition 5.3. *Let T be a power-bounded operator on $\mathfrak{H}$, that is, such that $\|T^n\| \le M$ $(n = 1, 2, \ldots)$. Suppose that for every nonzero h $(\in \mathfrak{H})$ neither $T^n h$ nor $T^{*n} h$ converges to 0 as $n \to \infty$. Then T is quasi-similar to a unitary operator.*

Proof. Observe first that, under the conditions stated,

$$\inf_{n \ge 0} \|T^n h\| = \mu(h) > 0 \quad \text{for all} \quad h \ne 0.$$

In fact, $\mu(h) = 0$ means that for every $\varepsilon > 0$ there exists an integer $n_0 = n_0(h, \varepsilon)$ such that $\|T^{n_0} h\| < \varepsilon / M$; hence we obtain

$$\|T^n h\| = \|T^{n-n_0} T^{n_0} h\| \le M \|T^{n_0} h\| < \varepsilon \quad \text{for} \quad n \ge n_0.$$

Thus $T^n h \to 0$ as $n \to \infty$, and for $h \ne 0$ this contradicts one of our hypotheses.

We shall now use a Banach generalized limit. This is a linear functional defined on the space of bounded numerical sequences $\{c_n\}_{n\geq 0}$, say $L\{c_n\}$, with the following properties:

$$L\{c_n\} \geq 0 \text{ if } c_n \geq 0 \quad (n=0,1,\ldots), \quad L\{1\}=1, \quad L\{c_{n+1}\}=L\{c_n\}.$$

For any $h,k \in \mathfrak{H}$ let us set

$$\langle h,k\rangle = L\{(T^n h, T^n k)\};$$

this is a bilinear form on $\mathfrak{H}$ such that

$$\langle h,h\rangle = L\{\|T^n h\|^2\} \begin{cases} \geq L\{\mu^2(h)\} = \mu^2(h)\cdot L\{1\} = \mu^2(h), \\ \leq L\{M^2\|h\|^2\} = M^2\|h\|^2\cdot L\{1\} = M^2\|h\|^2, \end{cases}$$

and

$$\langle Th,Tk\rangle = L\{(T^{n+1}h, T^{n+1}k)\} = L\{(T^n h, T^n k)\} = \langle h,k\rangle.$$

It follows that there exists a self-adjoint operator A on $\mathfrak{H}$ such that

$$\langle h,k\rangle = (Ah,k) \quad (h,k\in\mathfrak{H}), \quad 0<(Ah,h)\leq M^2\|h\|^2 \quad (h\in\mathfrak{H}; h\neq 0), \tag{5.11}$$

and

$$(ATh,Tk) = (Ah,k) \qquad (h,k\in\mathfrak{H}). \tag{5.12}$$

Now (5.11) implies $Ah\neq 0$ for $h\neq 0$, and the same is then true for the positive self-adjoint square root X of A. Hence X is a quasi-affinity on $\mathfrak{H}$. By (5.12) we have for every $h\in\mathfrak{H}$

$$\|XTh\|^2 = (X^2Th,Th) = (ATh,Th) = (Ah,h) = \|Xh\|^2.$$

In particular $\|XTX^{-1}k\| = \|k\|$ for the elements k in the domain of X^{-1}. This domain being dense in $\mathfrak{H}$, XTX^{-1} extends by continuity to an isometry U on $\mathfrak{H}$ such that

$$XT = UX. \tag{5.13}$$

Now our hypotheses assure in particular that $T^*h\neq 0$ for $h\neq 0$, and hence $\overline{T\mathfrak{H}}=\mathfrak{H}$. So we derive from (5.13):

$$U\mathfrak{H} = U\overline{X\mathfrak{H}} = \overline{UX\mathfrak{H}} = \overline{XT\mathfrak{H}} = \overline{X\overline{T\mathfrak{H}}} = \overline{X\mathfrak{H}} = \mathfrak{H};$$

thus the isometry U is actually unitary on $\mathfrak{H}$.

The hypotheses of our proposition being symmetrical in T and T^*, the results already obtained are also valid for T replaced by T^*. So we obtain that there exist a quasi-affinity Y on $\mathfrak{H}$ and a unitary operator V on $\mathfrak{H}$, such that $YT^*=VY$, and hence

$$TZ = ZW, \tag{5.14}$$

where $Z = Y^*$ is also a quasi-affinity and $W = V^*$ is also unitary on $\mathfrak{H}$.

From this point on, the proof proceeds similarly to that of part (iii) of Proposition 3.5. In fact, (5.13) and (5.14) imply that W is a quasi-affine transform of U, so by Proposition 3.4, U and W are unitarily equivalent. Consequently, by virtue of (5.13) and (5.14), T and U are quasi-similar.

Due to Proposition 5.3, the results concerning hyperinvariant subspaces, obtained in the preceding subsection, apply in particular to the operators T just considered. Moreover, we can prove the following theorem:

Theorem 5.4. *Let T be a power-bounded operator on the complex Hilbert space $\mathfrak{H}$ with $\dim\mathfrak{H} > 1$, such that neither T^n nor T^{*n} converge (strongly) to O as $n \to \infty$. Then either $T = cI$ with $|c| = 1$, or there exists a nontrivial subspace of $\mathfrak{H}$, hyperinvariant for T.*

Proof. We distinguish three cases.

Case 1: There exists a nonzero h_0 in $\mathfrak{H}$ such that $T^n h_0 \to 0$. It is easy to see that

$$\mathfrak{L} = \{h:\ h \in \mathfrak{H},\ T^n h \to 0\}$$

is a subspace of $\mathfrak{H}$, hyperinvariant for T. $\mathfrak{L}$ contains h_0, and hence $\mathfrak{L} \neq \{0\}$. On the other hand, $\mathfrak{L} \neq \mathfrak{H}$, for the contrary would imply $T^n \to O$. Thus $\mathfrak{L}$ is a nontrivial subspace of $\mathfrak{H}$.

Case 2: There exists a nonzero h_0 in $\mathfrak{H}$ such that $T^{*n} h_0 \to 0$. If we set

$$\mathfrak{L} = \mathfrak{H} \ominus \{h:\ h \in \mathfrak{H},\quad T^{*n} h \to 0\},$$

then $\mathfrak{L}$ is again a nontrivial subspace of $\mathfrak{H}$, hyperinvariant for T.

Case 3: There is no nonzero $h \in \mathfrak{H}$ for which $T^n h$ or $T^{*n} h$ converges to 0 as $n \to \infty$. By virtue of Proposition 5.3, T is then quasi-similar to a unitary operator U on $\mathfrak{H}$. As $\dim\mathfrak{H} > 1$, the spectral measure of U has values different from O and I unless U has a one-point spectrum $\{c\}$, $|c| = 1$. In the latter case $U = cI$, which implies $T = cI$. Thus if T is not of this form then by Corollary 5.2 there exists a nontrivial subspace of $\mathfrak{H}$, hyperinvariant for T. The proof is complete.

6 Spectral relations

1. To begin with let us recall that a bilateral shift has no nonzero invariant vector; see Sec. I.2.

Let us consider a contraction T on the space $\mathfrak{H}$ and its minimal unitary dilation U on the space $\mathfrak{K}$; let $\mathfrak{L}$ and $\mathfrak{L}^*$ be the corresponding wandering subspaces defined by (1.2). Let f be an element of $\mathfrak{K}$ invariant for U. Because $M(\mathfrak{L})$ and $M(\mathfrak{L}^*)$ reduce U, the orthogonal projections of f to these subspaces, say f' and f'', are also invariant for U. But these subspaces reduce U to bilateral shifts, so we must have $f' = 0$, $f'' = 0$. Thus f is orthogonal to $M(\mathfrak{L})$ and to $M(\mathfrak{L}^*)$. Consequently, by (1.4), f belongs to $\mathfrak{H}$ and so we have $Tf = P_{\mathfrak{H}} U f = f$.

Conversely, if $Th = h$ for an $h \in \mathfrak{H}$, then we also have $Uh = h$ because of the relation

$$\begin{aligned}\|Uh-h\|^2 &= \|Uh\|^2+\|h\|^2-2\mathrm{Re}(Uh,h) \\ &= 2\|h\|^2-2\mathrm{Re}(Th,h) \\ &= 2\mathrm{Re}(h-Th,h) \leq 2\|h-Th\|\cdot\|h\|,\end{aligned} \tag{6.1}$$

valid for every element h of $\mathfrak{H}$ and for every isometric operator U such that $T = \mathrm{pr}\, U$.

So we have proved that T and U have the same invariant vectors. If we apply this result to cT instead of T, with $|c| = 1$, then, as the minimal unitary dilation of cT equals cU, we obtain the following result.

Proposition 6.1. *Let T be a contraction and let U be its minimal unitary dilation. Every eigenvalue of T of modulus 1 is also an eigenvalue of U, and conversely. The corresponding eigenvectors are the same for T and for U.*

Let us consider now any c such that $T - cI$ is boundedly invertible. Let k be an element of the subspace $\mathfrak{R}$ that reduces U to its residual part R (cf. Sec. 2). Owing to the relation

$$\mathfrak{R} = \bigcap_{n\geq 0} U^n\mathfrak{K}_+ = \bigcap_{n\geq 0} U^n(\mathfrak{H}\oplus M_+(\mathfrak{L})) = \bigcap_{n\geq 0}\left(U^n\mathfrak{H}\oplus\left(\bigoplus_{m\geq n} U^m\mathfrak{L}\right)\right)$$

(cf. (2.11) and (2.5)), k can be expanded for every $n \geq 0$ into a series of the form

$$k = U^n h_n + \sum_{m=n}^{\infty} U^m l_m,$$

where $U^n h_n$ is the orthogonal projection of k into $U^h\mathfrak{H}$, and $U^m l_m$ is the orthogonal projection of k into $U^m\mathfrak{L}$. We have

$$\|k\|^2 \geq \sum_0^{\infty}\|U^m l_m\|^2,$$

and hence

$$\left\|\sum_{m=n}^{\infty} U^m l_m\right\|^2 = \sum_{m=n}^{\infty}\|U^m l_m\|^2 \to 0 \qquad (n\to\infty);$$

consequently

$$k = \lim_{n\to\infty} U^n h_n.$$

This implies

$$\begin{aligned}\|(R-cI_{\mathfrak{R}})k\| &= \|(U-cI)k\| = \lim\|(U-cI)U^n h_n\| = \lim\|U^n(U-cI)h_n\| = \\ &= \lim\|(U-cI)h_n\| \geq \underline{\lim}\|P_{\mathfrak{H}}(U-cI)h_n\| = \underline{\lim}\|(T-cI)h_n\| \geq \\ &\geq C\,\underline{\lim}\|h_n\| = C\,\underline{\lim}\|U^n h_n\| = C\|k\|,\end{aligned}$$

where C stands for $\|(T-cI)^{-1}\|^{-1}$. Because R is unitary, one concludes that $R-cI_{\mathfrak{R}}$ is boundedly invertible, and

$$\|(R-cI_{\mathfrak{R}})^{-1}\| \leq 1/C.$$

So we have proved the following result.

Proposition 6.2. *For each value c such that $T-cI$ is boundedly invertible $R-cI_{\mathfrak{R}}$ is also boundedly invertible and*

$$\|(R-cI_{\mathfrak{R}})^{-1}\| \leq \|(T-cI)^{-1}\|;$$

R denotes here the residual part of the minimal unitary dilation of the contraction T.

If T is not itself unitary then at least one of the wandering subspaces $\mathfrak{L}$, $\mathfrak{L}^*$ is different from $\{0\}$, and hence there exists a nonzero subspace of $\mathfrak{K}$ that reduces U to a bilateral shift, or, in other words, *U contains a bilateral shift.*

Instead of the whole space $\mathfrak{K}$ let us now consider the subspaces

$$M(h) = \bigvee_{n=-\infty}^{\infty} U^n h, \quad M_+(h) = \bigvee_{n=0}^{\infty} U^n h, \quad M_-(h) = \bigvee_{n=0}^{\infty} U^{-n} h \tag{6.2}$$

generated by a nonzero element h of $\mathfrak{H}$. Suppose that $M_+(h)$ and $M_-(h)$ reduce U; then $M_+(h) \cap M_-(h)$ also reduces U. Because $M_+(h)$ is contained in $\mathfrak{K}_+ = \mathfrak{H} \oplus \mathfrak{L} \oplus U\mathfrak{L} \oplus \cdots$ (cf. (2.5)) and, similarly, $M_-(h)$ is contained in $\mathfrak{K}_- = \mathfrak{H} \oplus \mathfrak{L}^* \oplus U^{-1}\mathfrak{L}^* \oplus \cdots$, the space $M_+(h) \cap M_-(h)$ is contained in $\mathfrak{K}_+ \cap \mathfrak{K}_- = \mathfrak{H}$. On the other hand, $M_+(h) \cap M_-(h)$ contains in particular the element h. Thus, $\mathfrak{H}$ contains a subspace reducing U and containing h; as $T = \operatorname{pr} U$, T coincides on this subspace with the unitary operator U. We conclude that h belongs to the subspace $\mathfrak{H}_0$ of $\mathfrak{H}$ on which T is unitary.

If T is c.n.u., then $\mathfrak{H}_0 = \{0\}$ so it cannot contain the nonzero h. Hence, in this case, $M_+(h)$ and $M_-(h)$ cannot both reduce U so that at least one of the subspaces $M_+(h) \ominus UM_+(h)$ and $M_-(h) \ominus U^{-1}M_-(h)$ is different from $\{0\}$. But this implies that there exists in $M(h)$ a nonzero subspace that is wandering for U. In other words, the part of U in $M(h)$ contains a bilateral shift.

Let us sum up.

Proposition 6.3. *For a nonunitary contraction T, the minimal unitary dilation U always contains a bilateral shift. If T is completely nonunitary, then for any nonzero element h of $\mathfrak{H}$ the restriction of U to $M(h)$ also contains a bilateral shift.*

2. All the above considerations relate to real as well as to complex Hilbert space. For complex Hilbert space we can complete these results in the following manner.

In a complex space, every unitary operator U has a spectral representation $U = \int_0^{2\pi} e^{it}\,dE_t$ by means of a spectral family $\{E_t\}$ $(0 \leq t \leq 2\pi)$. Let $E(\sigma)$ be the corresponding spectral measure defined for the Borel subsets σ of the unit circle $C = \{z\colon\ |z| = 1\}$; thus in particular:

$$E((e^{it_1}, e^{it_2}]) = E_{t_2} - E_{t_1} \qquad (0 \leq t_1 < t_2 \leq 2\pi).$$

Let us observe that if $\mathfrak{A}$ is a wandering subspace for U, then we have for $a \in \mathfrak{A}$

$$\int_0^{2\pi} e^{int}\, d(E_t a, a) = (U^n a, a) = \begin{cases} 0 & (n \neq 0), \\ \|a\|^2 & (n = 0); \end{cases}$$

by the uniqueness theorem for the Fourier–Stieltjes series this implies $(E_t a, a) = (t/2\pi)\|a\|^2$ and therefore

$$\|E(\sigma)a\|^2 = (E(\sigma)a, a) = m(\sigma) \cdot \|a\|^2, \tag{6.3}$$

where $m(\sigma)$ denotes the normalized Lebesgue measure on C. So, in particular, $m(\sigma) = 0$ implies $E(\sigma)a = 0$ and, because $E(\sigma)$ commutes with U, also $E(\sigma)U^n a = 0$ for all n. Thus $m(\sigma) = 0$ implies $E(\sigma)f = 0$ for all $f \in M(\mathfrak{A})$.

Now let us consider a c.n.u. contraction and let U be its minimal unitary dilation, with the spectral measure $E(\sigma)$. From what precedes we obtain that $m(\sigma) = 0$ implies $E(\sigma)f = 0$ for all $f \in M(\mathfrak{L})$ as well as for all $f \in M(\mathfrak{L}^*)$, hence also for all

$$f \in M(\mathfrak{L}) \vee M(\mathfrak{L}^*) = \mathfrak{K}.$$

So $m(\sigma) = 0$ implies $E(\sigma) = O$.

Conversely, $E(\sigma) = O$ implies $m(\sigma) = 0$. It even suffices to suppose that $E(\sigma)h = 0$ for a nonzero h in $\mathfrak{H}$. In fact, $E(\sigma)h = 0$ implies $E(\sigma)f = 0$ for all f belonging to $M(h) = \bigvee_{-\infty}^{\infty} U^n h$, in particular for a nonzero wandering vector a for U, the existence of which has been stated in Proposition 6.3. Thus, (6.3) implies $m(\sigma) = 0$.

So we have proved the following result.

Theorem 6.4. *For a completely nonunitary contraction T on the space $\mathfrak{H}$, the spectral measure $E(\sigma)$ of U is equivalent to the normalized Lebesgue measure $m(\sigma)$ on C, that is, if σ is a Borel set on C for which one of these measures is zero then so is the other. Moreover, the scalar-valued measures*

$$\mu_h(\sigma) = (E(\sigma)h, h) \qquad (h \in \mathfrak{H}, h \neq 0) \tag{6.4}$$

are also equivalent to Lebesgue measure.

The fact that $\mu_h(\sigma)$ is absolutely continuous with respect to Lebesgue measure implies that the nondecreasing function $(E_t h, h)$ is the integral of its derivative

$$f_h(t) = \frac{d}{dt}(E_t h, h),$$

which exists almost everywhere. This result can be completed as follows:

Proposition 6.5. *Under the conditions of the preceding proposition, we have* $\log f_h(t) \in L(0,2\pi)$ *for every nonzero h in* $\mathfrak{H}$.

Proof. We make use of the following theorem of Szegő (cf. HOFFMAN [1], p. 49): *Let* $f(t)$ *be a nonnegative, real valued, Lebesgue-integral function on* $(0,2\pi)$. *Set*

$$d(f) = \inf_p \frac{1}{2\pi} \int_0^{2\pi} |1-p(e^{it})|^2 f(t)dt,$$

where p runs over the class A_0 *of polynomials* $p(z)$ *such that* $p(0)=0$. *Then*

$$d(f) = \begin{cases} \exp \frac{1}{2\pi} \int_0^{2\pi} \log f(t)dt & \text{if } \log f(t) \in L(0,2\pi), \\ 0 & \text{if } \log f(t) \notin L(0,2\pi). \end{cases}$$

By virtue of this theorem our proposition means that $d(f_h) > 0$ for all nonzero h in $\mathfrak{H}$.

Let us suppose the contrary, thus $d(f_h) = 0$ for some $h \neq 0$. By virtue of the relation

$$\int_0^{2\pi} |1-p(e^{it})|^2 f_h(t)\,dt = \int_0^{2\pi} |1-p(e^{it})|^2 d(E_t h, h) = \|h - p(U)h\|^2,$$

h can then be approximated, as closely as we wish, by elements of the form $p(U)h$ with $p \in A_0$, that is, h is contained in $\bigvee_{n=1}^{\infty} U^n h = UM_+(h)$. From this it follows that $UM_+(h) = M_+(h)$, and hence $M_+(h)$ reduces U. Furthermore,

$$\int_0^{2\pi} |1-p(e^{it})|^2 d(E_t h, h) = \int_0^{2\pi} |1-p\tilde{}(e^{-it})|^2 d(E_t h, h) = \|h - p\tilde{}(U^{-1}h)\|^2,$$

where

$$p\tilde{}(z) = \overline{p(\bar{z})} \in A_0.$$

Thus $d(f_h) = 0$ implies that $h \in \bigvee_{n=1}^{\infty} U^{-n}h = U^{-1}M_-(h)$. Thus $U^{-1}M_-(h) = M_-(h)$, $M_-(h) = UM_-(h)$, and consequently $M_-(h)$ also reduces U. But we have shown in the proof of Proposition 6.3 that if T is c.n.u. and if $h \neq 0$, then $M_+(h)$ and $M_-(h)$ cannot both reduce U. Thus the assumption $d(f_h) = 0$ leads to a contradiction.

It follows that $d(f_h) > 0$, thus $\log f_h(t) \in L(0,2\pi)$.

Corollary 6.6. *For every nonunitary contraction T the spectrum of the minimal unitary dilation U of T is the whole unit circle C.*

Proof. The operator T has a nontrivial c.n.u. part $T^{(0)}$. It follows from Theorem 6.4 that the spectrum of the minimal unitary dilation $U^{(0)}$ of $T^{(0)}$ coincides with C. Because $\sigma(U) \supset \sigma(U^{(0)})$, one also has $\sigma(U) = C$.

To conclude this section let us prove the following proposition.

Proposition 6.7. *Let T be a completely nonunitary contraction, and suppose that the intersection of the spectrum of T with the unit circle C has Lebesgue measure 0. Then $\mathfrak{K} = M(\mathfrak{L}) = M(\mathfrak{L}^*)$ and so T is of class C_{00}.*

Proof. Let us consider the decomposition $\mathfrak{K} = M(\mathfrak{L}^*) \oplus \mathfrak{R}$ of the dilation space (cf. Sec. 2), and the corresponding decompositions of U and of the spectral measure $E(\sigma)$ of U:

$$U = U' \oplus R, \quad E^U(\sigma) = E^{U'}(\sigma) \oplus E^R(\sigma).$$

By virtue of Proposition 6.2 the spectrum σ_R of R is included in the spectrum of T, and because σ_R is situated on C it follows from our hypothesis that $m(\sigma_R) = 0$. Now, the spectral measure of U is absolutely continuous with respect to Lebesgue measure (cf. Theorem 6.4), thus we have $E^U(\sigma_R) = O$ and $E^R(\sigma_R) = E^U(\sigma_R)|\mathfrak{R} = O$. On the other hand, $E^R(\sigma_R)$ is the identity operator on $\mathfrak{R}$, thus we must have that $\mathfrak{R} = \{0\}$ and $\mathfrak{K} = M(\mathfrak{L}^*)$. The same reasoning, when applied to T^* instead of T, yields $\mathfrak{K} = M(\mathfrak{L})$. By Theorem 1.2 one then obtains that $T \in C_{00}$. This concludes the proof.

7 Spectral multiplicity

1. For any unitary operator U on the (real or complex) Hilbert space $\mathfrak{K}$, and for any nonempty subset $\mathfrak{S}$ of $\mathfrak{K}$, let us denote:

$$M(\mathfrak{S}) = \bigvee_{n=-\infty}^{\infty} U^n \mathfrak{S};$$

$M(\mathfrak{S})$ is a subspace reducing U. Thus the orthogonal projection of $\mathfrak{K}$ onto $M(\mathfrak{S})$, which we denote by $P^{\mathfrak{S}}$, commutes with U. This notation generalizes that already used in the particular case that $\mathfrak{S}$ is a wandering subspace for U, or a single vector k; in the latter case we write $M(k)$ and P^k instead of $M(\{k\})$ and $P^{\{k\}}$.

Because $P^{\mathfrak{S}}$ commutes with U, we have

$$(I - P^{\mathfrak{S}})M(\mathfrak{S}') = M((I - P^{\mathfrak{S}})\mathfrak{S}')$$

for arbitrary $\mathfrak{S}, \mathfrak{S}'$.

Let $\mathfrak{S} = \mathfrak{S}_1 \cup \mathfrak{S}_2$. We have then $M(\mathfrak{S}) = M(\mathfrak{S}_1) \vee M(\mathfrak{S}_2)$; consequently

$$M(\mathfrak{S}) = M(\mathfrak{S}_1) \oplus (I - P^{\mathfrak{S}_1})M(\mathfrak{S}) = M(\mathfrak{S}_1) \oplus M((I - P^{\mathfrak{S}_1})\mathfrak{S}).$$

Hence

$$M(\mathfrak{S}_1 \cup \mathfrak{S}_2) = M(\mathfrak{S}_1) \oplus M(\mathfrak{S}_2') \quad \text{where} \quad \mathfrak{S}_2' = (I - P^{\mathfrak{S}_1})\mathfrak{S}_2; \tag{7.1}$$

here we have used the fact that $\mathfrak{S}_1 \subset M(\mathfrak{S}_1)$ and hence $(I - P^{\mathfrak{S}_1})\mathfrak{S}_1 = \{0\}$.

For any given $\mathfrak{G} \neq \{0\}$ there exists a maximal system Σ of nonzero elements of $M(\mathfrak{S})$, such that $M(k) \perp M(k')$ if $k, k' \in \Sigma$, $k \neq k'$. This follows by Zorn's lemma.

We have then

$$M(\mathfrak{S}) = \bigoplus_{k\in\Sigma} M(k). \tag{7.2}$$

Indeed, otherwise one could increase the system Σ by at least one more nonzero element.

Lemma 7.1. *Let $\mathfrak{S}$ be a subspace of $\mathfrak{K}$, of dimension $d \geq 1$. Then it is possible to choose the decomposition* (7.2) *so that the number of terms does not exceed d.*

Proof. If d is infinite (countable or not), this condition is satisfied automatically. In fact, we then have $d \leq \dim M(\mathfrak{S}) \leq \aleph_0 \cdot \dim \mathfrak{S} = \aleph_0 \cdot d = d$, and hence $d = \dim M(\mathfrak{S})$; on the other hand, because $\dim M(k) \geq 1$ for every $k \in \mathfrak{S}$, the number of terms in (7.2) cannot exceed $\dim M(\mathfrak{S})$. This completes the proof for an infinite d.

When d is finite we proceed by recurrence. For $d = 1$ our assertion is obvious. We suppose that it holds for d smaller than some integer $N(\geq 2)$, and we want to show that it also holds for $d = N$. Let $\mathfrak{S}$ be any subspace of $\mathfrak{K}$ of dimension N. Choose in $\mathfrak{S}$ a nonzero element k_0 and denote by $\mathfrak{S}_0$ the subspace of $\mathfrak{S}$ formed by the elements orthogonal to k_0. By virtue of (7.1) we have

$$M(\mathfrak{S}) = M(\{k_0\} \cup \mathfrak{S}_0) = M(k_0) \oplus M(\mathfrak{S}_0') \tag{7.3}$$

with $\mathfrak{S}_0' = (I - P^{k_0})\mathfrak{S}_0$. We have $\dim \mathfrak{S}_0 = N - 1$, thus $\mathfrak{S}_0'$ is a subspace of dimension $\leq N - 1$. By hypothesis there exists therefore a decomposition

$$M(\mathfrak{S}_0') = \bigoplus_{i=1}^{r} M(k_i), \tag{7.4}$$

where $r \leq N - 1$. By virtue of (7.3) and (7.4) we have

$$M(\mathfrak{S}) = \bigoplus_{i=0}^{r} M(k_i),$$

the number of terms being $r + 1$, and hence $\leq N$.

This concludes the proof of the lemma.

2. Let us suppose now that U is the minimal unitary dilation of a c.n.u. contraction T. According to (1.10) we have

$$\mathfrak{K} = M(\mathfrak{L}) \vee M(\mathfrak{L}^*) = M(\mathfrak{L} \cup \mathfrak{L}^*), \tag{7.5}$$

where $\mathfrak{L}$ and $\mathfrak{L}^*$ are the wandering subspaces for U defined by (1.2). By virtue of (7.1) and (7.5) we have the decomposition

$$\mathfrak{K} = M(\mathfrak{L}) \oplus M(\mathfrak{S}), \quad \text{where} \quad \mathfrak{S} = \overline{(I - P^{\mathfrak{L}})\mathfrak{L}^*};$$

note that $\dim \mathfrak{L} = \mathfrak{d}_T$ and $\dim \mathfrak{S} \leq \dim \mathfrak{L}^* = \mathfrak{d}_{T^*}$.

If we choose in $\mathfrak{L}$ a complete orthonormal system $\Sigma' = \{l\}$; then

$$M(\mathfrak{L}) = \bigoplus_{l \in \Sigma'} M(l),$$

where the number of terms is equal to $\mathfrak{d}_T$. Let us observe that the vectors $l \in \mathfrak{L}$ are wandering for U (i.e., $U^n l \perp U^m l$ if $n \neq m$). On the other hand, there exists, according to the lemma, a decomposition

$$M(\mathfrak{S}) = \bigoplus_{k \in \Sigma''} M(k),$$

where the number of terms does not exceed $\mathfrak{d}_{T^*}$. Thus we obtain the decomposition

$$\mathfrak{K} = \bigoplus_{a \in \Sigma} M(a) \qquad (\Sigma = \Sigma' \cup \Sigma''),$$

where the number of terms does not exceed $\mathfrak{d}_T + \mathfrak{d}_{T^*}$, and for at least $\mathfrak{d}_T$ terms the corresponding vector a is wandering for U.

As (7.5) is symmetrical in $\mathfrak{L}$ and $\mathfrak{L}^*$, we can repeat this reasoning with the roles of $\mathfrak{L}$ and $\mathfrak{L}^*$ interchanged. Thus, after setting

$$\mathfrak{d}_{\max} = \max\{\mathfrak{d}_T, \mathfrak{d}_{T^*}\},$$

we arrive at the following result.

Proposition 7.2. *If U is the minimal unitary dilation, on the space $\mathfrak{K}$, of a completely nonunitary contraction T, then there exists a decomposition of $\mathfrak{K}$ of the form*

$$\mathfrak{K} = \bigoplus_{\alpha} M(a_\alpha) \qquad (a_\alpha \in \mathfrak{K},\ a_\alpha \neq 0), \tag{7.6}$$

with at most $\mathfrak{d}_T + \mathfrak{d}_{T^*}$ terms, and where at least $\mathfrak{d}_{\max}$ of the vectors a_α are wandering for U.

3. In the rest of this section we suppose that the space $\mathfrak{H}$ of the c.n.u. contraction T is complex; then so is the space $\mathfrak{K}$ of the minimal unitary dilation U of T.

Let $\{E_t\}_{0 \le t \le 2\pi}$ be the spectral family associated with U; by Theorem 6.4, E_t is an absolutely continuous function of t. For an arbitrary $a \in \mathfrak{K}$ and for all integers m, n one then has

$$(U^m a, U^n a) = \int_0^{2\pi} e^{i(m-n)t}\, d(E_t a, a) = \int_0^{2\pi} e^{i(m-n)t} p(t)\, dt,$$

where

$$p(t) = \frac{d}{dt}(E_t a, a).$$

It follows that for any finite linear combination of elements of the form $U^n a$ ($n = 0, \pm 1, \pm 2, \ldots$):

$$\left\| \sum_n c_n U^n a \right\|^2 = \int_0^{2\pi} \left| \sum_n c_n e^{int} \right|^2 p(t)\, dt.$$

Thus

$$\sum_n c_n U^n a \to \sum_n c_n e^{int}[2\pi p(t)]^{1/2}$$

is an isometric transformation of a linear manifold, dense in $M(a)$, onto a linear manifold $\mathfrak{M}$ of the space $L^2(\Omega)$, where

$$\Omega = \{t\colon\ t \in (0,2\pi), p(t) > 0\},$$

and the measure considered is always the normalized Lebesgue measure $m(\sigma)$. Now it is easy to prove that $\mathfrak{M}$ is dense in $L^2(\Omega)$. Consequently, the above isometric transformation extends by continuity to a unitary one, from $M(a)$ onto $L^2(\Omega)$; let us denote it by Φ.

Let us observe that if a is a nonzero wandering vector for U, then, by virtue of (6.3), $p(t) = (1/2\pi)\|a\|^2$ so that in this case $\Omega = (0,2\pi)$.

Let us also observe that the restriction of U to $M(a)$ is transformed by Φ into the operator $U^\times(\Omega)$ on $L^2(\Omega)$ defined by

$$(U^\times(\Omega)u)(t) = e^{it}u(t),$$

that is, the operator of multiplication by e^{it} on $L^2(\Omega)$.

Using the sign $\sim$ to indicate *unitary equivalence* of operators, we deduce from Proposition 7.2 the following

Proposition 7.3. *Let T be a completely nonunitary contraction on the complex Hilbert space $\mathfrak{H}$. For its minimal unitary dilation U we have*

$$U \sim \bigoplus_\alpha U^\times(\Omega_\alpha), \tag{7.7}$$

where the sets $\Omega_\alpha \subset (0,2\pi)$ are measurable, the number of the terms does not exceed $\mathfrak{d}_T + \mathfrak{d}_{T^*}$, and for at least $\mathfrak{d}_{\max}$ terms the set Ω_α coincides with the whole interval $(0,2\pi)$.

4. Here are some more or less immediate consequences of the preceding propositions.

Theorem 7.4. *Let T be a completely nonunitary contraction on the complex space $\mathfrak{H}$, and let U be its minimal unitary dilation on $\mathfrak{K}$.*

(a) *If $\mathfrak{d}_{\max}$ is infinite, then U is a bilateral shift of multiplicity $\mathfrak{d}_{\max}$. The same is true if $\dim\mathfrak{H}$ is finite. If $\dim\mathfrak{H} > \aleph_0$, then $\mathfrak{d}_{\max}$ is always infinite, and indeed, we have then $\mathfrak{d}_{\max} = \dim\mathfrak{H}$.*

(b) *There always exists a (not necessarily minimal) unitary dilation U of T that is a bilateral shift of multiplicity not exceeding $\mathfrak{d}_T + \mathfrak{d}_{T^*}$.*

Proof. The case when $\dim\mathfrak{H}$ is finite is simple. As T is c.n.u., its spectrum lies entirely in the interior of the unit circle, and hence it follows that $T^n \to O$ and $T^{*n} \to O$ as $n \to \infty$. In view of Theorem 1.2, U is then a bilateral shift of multiplicity equal to $\mathfrak{d}_T$ and to $\mathfrak{d}_{T^*}$ (hence $\mathfrak{d}_{\max} = \mathfrak{d}_T = \mathfrak{d}_{T^*}$).

In general we have

$$\begin{aligned}\mathfrak{d}_{\max} &\le \dim \mathfrak{H} \le \dim \mathfrak{K} \le \dim M(\mathfrak{L}) + \dim M(\mathfrak{L}^*) \\ &= \aleph_0 \cdot \mathfrak{d}_T + \aleph_0 \cdot \mathfrak{d}_{T^*} = \aleph_0 \cdot \mathfrak{d}_{\max};\end{aligned}$$

if $\dim \mathfrak{H} > \aleph_0$, then $\aleph_0 \cdot \mathfrak{d}_{\max} > \aleph_0$, and hence $\mathfrak{d}_{\max} > \aleph_0$. Consequently, $\mathfrak{d}_{\max} = \aleph_0 \cdot \mathfrak{d}_{\max}$ and thus $\mathfrak{d}_{\max} = \dim \mathfrak{H}$.

To conclude the proof of (a) it remains to show that if $\mathfrak{d}_{\max}$ is infinite, then U is a bilateral shift of multiplicity $\mathfrak{d}_{\max}$.

To this end we begin with relation (7.7). Let A denote the set of subscripts α for which Ω_α coincides with the whole interval $(0, 2\pi)$, and let B be the set of the remaining subscripts α. For the cardinal numbers $|A|$, $|B|$ of A and B, respectively, we have

$$\mathfrak{d}_{\max} \le |A| \le \mathfrak{d}_T + \mathfrak{d}_{T^*} \le 2\mathfrak{d}_{\max}, \quad |B| \le \mathfrak{d}_T + \mathfrak{d}_{T^*} \le 2\mathfrak{d}_{\max}.$$

Because $\mathfrak{d}_{\max}$ is assumed infinite, we have $\mathfrak{d}_{\max} = 2 \cdot \mathfrak{d}_{\max} = \aleph_0 \cdot \mathfrak{d}_{\max}$, and consequently

$$|A| = \mathfrak{d}_{\max} = \aleph_0 \cdot \mathfrak{d}_{\max} \ge \aleph_0 \cdot |B|, \quad \text{which implies} \quad |A| = \aleph_0 \cdot |B| + r, \tag{7.8}$$

where r is a cardinal number ≥ 0.

As a consequence of this relation between the cardinal numbers $|A|$ and $|B|$, we can rearrange the sum (7.7) so that each term with subscript $\beta \in B$ is accompanied by $\aleph_0$ terms with subscripts $\alpha \in A$. Thus we obtain

$$U \sim \bigoplus_{\beta \in B} [U^\times(\Omega_\beta) \oplus U^\times(0, 2\pi) \oplus U^\times(0, 2\pi) \oplus \cdots] \oplus \left[\bigoplus_{r \text{ terms}} U^\times(0, 2\pi) \right]. \tag{7.9}$$

We now make use of the obvious relation

$$U^\times(0, 2\pi) \sim U^\times(\Omega) \oplus U^\times(\Omega'), \tag{7.10}$$

valid for any measurable subset Ω of $(0, 2\pi)$, and its complement Ω' with respect to this interval.

In view of this relation we deduce

$$\begin{aligned} &U^\times(\Omega_\beta) \oplus U^\times(0, 2\pi) \oplus U^\times(0, 2\pi) \oplus \cdots \\ &\quad \sim U^\times(\Omega_\beta) \oplus [U^\times(\Omega'_\beta) \oplus U^\times(\Omega_\beta)] \oplus [U^\times(\Omega'_\beta) \oplus U^\times(\Omega_\beta)] \oplus \cdots \\ &\quad \sim [U^\times(\Omega_\beta) \oplus U^\times(\Omega'_\beta)] \oplus [U^\times(\Omega_\beta) \oplus U^\times(\Omega'_\beta)] \oplus \cdots \\ &\quad \sim U^\times(0, 2\pi) \oplus U^\times(0, 2\pi) \oplus \cdots, \end{aligned}$$

and hence it follows from (7.9) that U is unitarily equivalent to the orthogonal sum of $\aleph_0 \cdot |B| + r = |A| = \mathfrak{d}_{\max}$ replicas (cf. (7.8)) of $U^\times(0, 2\pi)$. Now $U^\times(0, 2\pi)$ is obviously a bilateral shift of multiplicity 1; a generating subspace consists of the constant functions in $L^2(0,2\pi)$. Consequently, the orthogonal sum under consideration will be a bilateral shift of multiplicity $\mathfrak{d}_{\max}$. Operators unitarily equivalent to bilateral shifts are also bilateral shifts of the same multiplicity, thus it follows that U is a bilateral shift of multiplicity $\mathfrak{d}_{\max}$.

Part (a) of the theorem is herewith established. Part (b) follows immediately from (7.7) and (7.10). In fact, we have only to set

$$\widetilde{U} = U \oplus U', \quad \text{where} \quad U' = \bigoplus_\alpha U^\times(\Omega'_\alpha),$$

for then we have

$$\widetilde{U} \sim \left[\bigoplus_\alpha U^\times(\Omega_\alpha)\right] \oplus \left[\bigoplus_\alpha U^\times(\Omega'_\alpha)\right] \sim \bigoplus_\alpha [U^\times(\Omega_\alpha) \oplus U^\times(\Omega'_\alpha)] \sim \bigoplus_\alpha U^\times(0,2\pi);$$

the last sum is a bilateral shift of multiplicity equal to the number of terms of this sum and hence not exceeding $\mathfrak{d}_T + \mathfrak{d}_{T^*}$.

Corollary 7.5. *If T is a contraction on the complex Hilbert space $\mathfrak{H}$ such that $\|Th\| < \|h\|$ for every nonzero h in $\mathfrak{H}$, then the minimal unitary dilation of T is a bilateral shift of multiplicity equal to* $\dim \mathfrak{H}$.

Proof. The operator T is obviously c.n.u., and $(I - T^*T)\ h \neq 0$ for $h \neq 0$. Thus $\overline{D_T \mathfrak{H}} = \mathfrak{H}$, $\mathfrak{d}_{\max} = \mathfrak{d}_T = \dim \mathfrak{H}$, and therefore we can apply part a) of Theorem 7.4.

8 Similarity of operators in $\mathscr{C}_\rho$ to contractions

In Sec I.11 we introduced the classes $\mathscr{C}_\rho$ $(\rho > 0)$ of operators: the operator T on the space $\mathfrak{H}$ belongs to the class $\mathscr{C}_\rho$ if there exists a unitary operator U_ρ on some space $\mathfrak{K}_\rho$ containing $\mathfrak{H}$ as a subspace, such that

$$T^n = \rho \cdot \mathrm{pr}\, U_\rho^n \qquad (n = 1, 2, \ldots). \tag{8.1}$$

The class $\mathscr{C}_1$ consists of the contractions. The aim of this section is to prove the following result, showing that the operators belonging to any of these classes are not very far from being contractions.

Theorem 8.1. *Every operator belonging to a class $\mathscr{C}_\rho$ is similar to a contraction.*

Proof. We essentially use Proposition I.11.3, showing that the class $\mathscr{C}_\rho$ is an increasing function of the parameter ρ (indeed, it suffices to know that it is a non-decreasing function of ρ). A further essential point in the proof is an appropriate modification of the method of Sec. 3 to the case of unitary ρ-dilations.

Let us suppose, then, that T is an operator on $\mathfrak{H}$, of class $\mathscr{C}_r$ with $r > 1$. Then T belongs to every class $\mathscr{C}_\rho$ with $\rho \geq r$, and therefore it has, for each $\rho \geq r$, a unitary ρ-dilation U_ρ in some space $\mathfrak{K}_\rho$. We set for $\rho \geq r$

$$\mathfrak{M}_\rho = \bigvee_{n \geq 0} U_\rho^{*n}(U_\rho^* - T^*)\mathfrak{H} \qquad (\subset \mathfrak{K}_\rho) \tag{8.2}$$

and denote by $P_{\mathfrak{M}_\rho}$ the orthogonal projection of $\mathfrak{K}_\rho$ into $\mathfrak{M}_\rho$. We also set

$$t_\rho = \|P_{\mathfrak{M}_\rho}|\mathfrak{H}\| \qquad (\rho \geq r); \tag{8.3}$$

clearly t_ρ is the smallest value such that the inequality

$$|(h, m_\rho)| \leq t_\rho \cdot \|h\| \cdot \|m_\rho\| \tag{8.4}$$

holds for every $h \in \mathfrak{H}$ and every $m_\rho \in \mathfrak{M}_\rho$; it suffices to consider elements of the form

$$m_\rho = \sum_{n \geq 0} U_\rho^{*n}(U_\rho^* - T^*)h_n \tag{8.5}$$

with $h_0, h_1, \ldots \in \mathfrak{H}$ and $h_n = 0$ for n large enough, because these are dense in $\mathfrak{M}_\rho$.

Relation (8.1) implies the analogous relation for the adjoints, so that we have, setting $\delta = 1/\rho$,

$$\operatorname{pr} U_\rho^{*n}(U_\rho^* - T^*) = \begin{cases} \delta T^{*n+1} - \delta T^{*n}T^* = O & \text{if } n \geq 1, \\ \delta T^* - T^* = (\delta - 1)T^* & \text{if } n = 0. \end{cases}$$

Thus we obtain for $h \in \mathfrak{H}$ and for m_ρ defined by (8.5) that

$$(h, m_\rho) = \sum_{n \geq 0} (h, U_\rho^{*n}(U_\rho^* - T^*)h_n) = (h, (\delta - 1)T^*h_0),$$

and hence (8.4) is equivalent to

$$|(h, (\delta - 1)T^*h_0)| \leq t_\rho \cdot \|h\| \cdot \|m_\rho\|,$$

and this in turn is equivalent to

$$(\delta - 1)^2 \|T^*h_0\|^2 \leq t_\rho^2 \|m_\rho\|^2. \tag{8.6}$$

Using (8.1), we obtain by an easy calculation that

$$\begin{aligned} &(U_\rho^{*j}(U_\rho^* - T^*)h_j, U_\rho^{*k}(U_\rho^* - T^*)h_k) \\ &= \begin{cases} \|h_j\|^2 + (1 - 2\delta)\|T^*h_j\|^2 & \text{if } j = k, \\ (\delta - 1)(Th_j, h_k) & \text{if } k - j = 1, \\ (\delta - 1)(h_j, Th_k) & \text{if } k - j = -1, \\ 0 & \text{in all the other cases.} \end{cases} \end{aligned}$$

So we obtain for the element m_ρ defined by (8.5) that

$$\|m_\rho\|^2 = \sum_{j\geq 0} [\|h_j\|^2 + (1-2\delta)\|T^*h_j\|^2 + 2(\delta-1)\operatorname{Re}(h_j, T^*h_{j+1})],$$

and hence

$$\rho\|m_\rho\|^2 = \rho[\|T^*h_0\|^2 + \sum_{j\geq 0}\|h_j - T^*h_{j+1}\|^2] - 2\sum_{j\geq 0}[\|T^*h_j\|^2 - \operatorname{Re}(h_j, T^*h_{j+1})].$$

It follows that if $m_{\rho'}$ $(r \leq \rho' \leq \rho)$ corresponds to the same sequence $\{h_n\}$ of vectors, then

$$\rho\|m_\rho\|^2 - \rho'\|m_{\rho'}\|^2 = (\rho-\rho')[\|T^*h_0\|^2 + \sum_{j\geq 0}\|h_j - T^*h_{j+1}\|^2] \geq (\rho-\rho')\|T^*h_0\|^2;$$

in particular

$$\rho\|m_\rho\|^2 - r\|m_r\|^2 \geq (\rho - r)\|T^*h_0\|^2. \tag{8.7}$$

Obviously $t_r \leq 1$, thus it follows from (8.6) (for $\rho = r$) that

$$\left(\frac{1}{r}-1\right)^2 \|T^*h_0\|^2 \leq \|m_r\|^2. \tag{8.8}$$

From (8.7) and (8.8) we deduce

$$\rho\|m_\rho\|^2 \geq (\rho - r)\|T^*h_0\|^2 + r\left(\frac{1}{r}-1\right)^2 \|T^*h_0\|^2$$

and hence

$$\left(\frac{1}{\rho}-1\right)^2 \|T^*h_0\|^2 \leq \frac{\rho - 2 + 1/\rho}{\rho - 2 + 1/r}\|m_\rho\|^2 \qquad (\rho \geq r). \tag{8.9}$$

Recalling that t_ρ is the smallest nonnegative value for which the inequality (8.6) holds, we conclude from (8.9) that

$$t_\rho^2 \leq \frac{\rho - 2 + 1/\rho}{\rho - 2 + 1/r};$$

hence $t_\rho < 1$ if $\rho > r$.

In the rest of the proof ρ is a fixed number larger than r; thus it is not necessary to indicate this value ρ by subscripts, so we write U, $\mathfrak{K}$, $\mathfrak{M}$, t instead of U_ρ, $\mathfrak{K}_\rho$, $\mathfrak{M}_\rho$, t_ρ.

Set $\mathfrak{N} = \mathfrak{K} \ominus \mathfrak{M}$. The elements k of $\mathfrak{N}$ are characterized by the equations

$$TP_{\mathfrak{H}}U^n k = P_{\mathfrak{H}}U^{n+1}k \qquad (n = 0, 1, \ldots). \tag{8.10}$$

Indeed, $k \perp \mathfrak{M}$ means that

$$((U^{*n+1} - U^{*n}T^*)h,k) = 0 \quad \text{for all} \quad h \in \mathfrak{H} \quad \text{and} \quad n \geq 0;$$

this can be reduced easily to (8.10). These equations, characterizing $\mathfrak{N}$, show in particular that $\mathfrak{N}$ is invariant for U.

For $h \in \mathfrak{H}$ we have by (8.3) that

$$\|h\|^2 = \|P_{\mathfrak{M}}h\|^2 + \|P_{\mathfrak{N}}h\|^2 \leq t^2\|h\|^2 + \|P_{\mathfrak{N}}h\|^2,$$

and hence

$$\|P_{\mathfrak{N}}h\|^2 \geq (1-t^2)\|h\|^2. \tag{8.11}$$

Because $t(=t_\rho) < 1$, it follows from (8.11) that the linear manifold

$$\mathfrak{N}' = P_{\mathfrak{N}}\mathfrak{H} \tag{8.12}$$

is closed and hence a subspace of $\mathfrak{N}$ and, moreover, that the operator

$$Y = P_{\mathfrak{N}}|\mathfrak{H}$$

is an affinity from $\mathfrak{H}$ to $\mathfrak{N}'$. Consequently, $X = Y^*$ is an affinity from $\mathfrak{N}'$ to $\mathfrak{H}$. By virtue of the obvious relation

$$(P_{\mathfrak{N}}h,k) = (h,k) = (h,P_{\mathfrak{H}}k), \tag{8.13}$$

valid for $h \in \mathfrak{H}$ and $k \in \mathfrak{N}$ (and hence in particular for $k \in \mathfrak{N}'$), we obtain that

$$X = P_{\mathfrak{H}}|\mathfrak{N}'. \tag{8.14}$$

Let us observe that (8.13) also implies $P_{\mathfrak{H}}k = 0$ for every $k \in \mathfrak{N}$ which is orthogonal to $P_{\mathfrak{N}}\mathfrak{H}$ (i.e., to $\mathfrak{N}'$). Thus we have $P_{\mathfrak{H}}(I - P_{\mathfrak{N}'})g = 0$ and hence

$$P_{\mathfrak{H}}g = P_{\mathfrak{H}}P_{\mathfrak{N}'}g \quad \text{for every} \quad g \in \mathfrak{N}. \tag{8.15}$$

If $f \in \mathfrak{N}'$, then we have in particular $Uf \in U\mathfrak{N} \subset \mathfrak{N}$, and hence, setting $g = Uf$, we obtain from (8.14), (8.15), and the case $n = 0$ of (8.10), that

$$TXf = TP_{\mathfrak{H}}f = P_{\mathfrak{H}}Uf = P_{\mathfrak{H}}P_{\mathfrak{N}'}Uf = XVf$$

with

$$V = P_{\mathfrak{N}'}U|\mathfrak{N}'.$$

Clearly, V is a contraction on the space $\mathfrak{N}'$, and $T = XVX^{-1}$. Therefore T is similar to the contraction V.

This concludes the proof.

On account of Proposition I.11.2 we can formulate the following corollary.

Corollary 8.2. *Every operator T with numerical radius $w(T) \leq 1$ is similar to a contraction.*

9 Notes

Theorem 1.1 on the structure of the minimal unitary dilation U of a contraction T on $\mathfrak{H}$, appears implicitly in SZ.-NAGY [2] and explicitly in HALPERIN [1] and SZ.-N.–F. [V].

If $T \in C_{\cdot 0}$ then by virtue of Theorem 2.1 we have $\mathfrak{R}_+ = M_+(\mathfrak{L}_*)$ and hence U_+ is a unilateral shift. Using relations (I.4.2) we obtain therefore that *every contraction of class $C_{0\cdot}$ is the restriction of the adjoint of a unilateral shift.* For this explicit statement see FOIAŞ [5] and DE BRANGES AND ROVNYAK [1].

The first general result concerning the spectral type of U was that of SCHREIBER [1] to the effect that if $\|T\| < 1$ then U is a bilateral shift, of multiplicity equal to $\dim \mathfrak{H}$; for an alternative proof, see SZ.-NAGY [II]. (In these proofs, the space $\mathfrak{H}$ is assumed to be complex.) The fact that the condition $T^n \to O$ implies that U is a bilateral shift of multiplicity equal to the defect index $\mathfrak{d}_T$, and the dual of this fact with T^* instead of T, were proved first by DE BRUIJN [1] by a matrix method valid for real and complex spaces. The result of DE BRUIJN was subsequently generalized by HALPERIN [1] to combine the cases $T^n \to O$ and $T^{*n} \to O$ in the following manner. *Let us suppose that there exist in $\mathfrak{H}$ orthogonal projections, say Q_1 and Q_2, such that*

$$Q_2Q_1 = 0, \quad Q_2TQ_1 = O, \quad (I-Q_2)T(I-Q_1) = O,$$
$$(TQ_1)^n \to 0 \text{ and } (T^*Q_2)^n \to O \qquad (n \to \infty).$$

Then U is a bilateral shift, with the generating subspace

$$\overline{(U-T)Q_1\mathfrak{H}} \oplus (I-Q_1-Q_2)\mathfrak{H} \oplus \overline{(U^*-T^*)Q_2\mathfrak{H}}.$$

The problem whether the condition $\|Th\| < \|h\|$ (for all nonzero $h \in \mathfrak{H}$) implies that U is a bilateral shift of multiplicity equal to $\dim \mathfrak{H}$, was raised by DE BRUIJN [1]. The affirmative answer to this problem was given in SZ.-N.–F. [V] (Corollary 7.5 above) at least for complex spaces.

The role of the residual and $*$-residual parts of U was indicated in SZ.-N.–F. [V], [VII]. In [3] and [VII], the authors introduced the notions of quasi-affinity and quasi-similarity, and the classes $C_{\alpha\beta}$ of contractions (however, only for c.n.u. ones). In particular, they proved that every contraction of class C_{11} is quasi-similar both to the residual part R and the $*$-residual part R_* of the minimal unitary dilation of U (Proposition 3.5(iii) above). This theorem has been used, first in connection with the theory of characteristic functions, and then directly, to obtain information on the invariant subspaces for T (cf. SZ.-N.–F. [IX] and [5]). In Sec. 5 above, the results of the latter paper are presented in a more developed form, giving more detailed information on the invariant (regular and hyperinvariant) subspaces, and moreover these results are extended from contractions to general power-bounded operators (Theorem 5.4). The same theorem (with simple invariance instead of hyperinvariance) was obtained independently by S. Parrott (correspondence). The geometric characterization of the similarity of a contraction to a unitary operator given in Proposition 3.6 is an essential step in the proof of the main result in SZ.-N.–F. [X] (presented later as Theorem VI.4.5).

Let us observe that quasi-similarity preserves the commutativity of the commutant $\{T\}'$, as well as the existence of nontrivial hyperinvariant subspaces for T. This is easily verified using the methods of Sec. 5.

The important fact that if T is a c.n.u. contraction then its minimal unitary dilation U has absolutely continuous spectral measure, was proved first by the authors in [IV], where they used rather deep theorems on analytic functions. The more "geometric" proof given in Sec. 6 was found later (cf. HALPERIN [1] and the authors [V]).

The other results of Sec. 6 and Sec. 7 were obtained by the authors in [III] and [V], with the exception of Propositions 6.3 and 6.5 which are due to MLAK [1]–[3], [5].

In connection with Secs. 1 and 2.1 cf. also see HALPERIN [3] and [5], where the structure of the spaces of minimal regular isometric and unitary dilations are studied for systems of contractions; conditions are also given for the minimal regular unitary dilation to behave in a certain sense as bilateral shifts.

Proposition 2.2 and Theorem 2.3 on the dilation of commutants are due to the present authors; see SZ.-N.–F. [12] (the proof given in this book differs slightly from the original one). Some special cases of Theorem 2.3 were obtained earlier, using entirely different and more involved methods, by SARASON [3] ($T \in C_{00}$, $\mathfrak{d}_T = \mathfrak{d}_{T^*} = 1$) and by SZ.-N.–F. [11] ($T \in C_{00}$). It was Sarason's paper that inspired these investigations of the authors. An alternative proof of Theorem 2.3 was found by DOUGLAS, MUHLY, AND PEARCY [1]. Let us add that Theorem 2.3 implies, and is implied by, Ando's theorem (Theorem I.6.1), as remarked independently by PARROTT [1] also.

Section 8 reproduces the paper SZ.-N.–F. [10] (with some simplification in the part following formula (8.11), suggested by Ju. L. Šmul'jan). The interest in Theorem 8.1 is due mainly to the fact that not all power-bounded operators are similar to contractions (cf. FOGUEL [1], HALMOS [5]; the compact ones *are*, cf. SZ.-NAGY [11]). However, every power-bounded operator T can be approximated in the operator norm, as closely as we wish, by operators T' belonging to some class $\mathscr{C}_\rho$ (cf. HOLBROOK [3]). Indeed, this follows easily from Remark 4 to Proposition I.11.2, by choosing $T' = cT$ with $1 > c \to 1$.

Part of the structural relations treated in Sec. 1 as well as some of the spectral relations treated in Sec. 6 can be extended to unitary ρ-dilations; see DURSZT [2]. In particular, if an operator of class $\mathscr{C}_\rho$ is c.n.u., then its minimal unitary ρ-dilation has absolutely continuous spectral measure; see DURSZT [2], RÁCZ [1], [2], and MLAK [10].

10 Further results

1. The earliest form of the commutant lifting theorem appeared in SARASON [3], and was motivated by the study of interpolation problems for bounded analytic functions in the unit disk. The general form of the theorem has also been used in a variety of interpolation problems. The solutions of the classical interpolation problems can be parametrized by fractional linear transformations applied to an arbitrary analytic function (or Schur parameter) from the unit disk to itself. Extensions of this parametrization to the solutions of commutant lifting problems have been found as well.

An interesting connection was made in HELTON [1], who showed that commutant lifting can be used in the study of control problems. Other applications arose in the study of layered media and systems theory. In this context, a new class of (nonstationary) interpolation problems has been pursued by I. Gohberg and his collaborators. In fact, I. Gohberg and W. Helton initiated an entirely new direction of research in operator theory, inspired by engineering problems. It turned out that these interpolation problems can also be connected to commutant lifting.

We refer to FOIAS AND FRAZHO [1] for an exposition of the parametrization of the solutions to the commutant lifting problem and its applications, including the study of multi-layered media. The connections between commutant lifting and interpolation are explored in FOIAS, FRAZHO, GOHBERG, AND KAASHOEK [1]. For applications to control theory see FOIAS, ÖZBAY, AND TANNENBAUM [1].

A new kind of interpolation problem, which is not covered in this book or the monographs just mentioned, is as follows. Given a contraction T with minimal isometric dilation U_+ and an operator X commuting with T, what is the smallest possible spectral radius of a dilation Y of X in the commutant of U_+? This problem originates in control theory, and it was first considered in BERCOVICI, FOIAS, AND TANNENBAUM [2], where the optimal value is found. The results of this paper do not lead to an effective algorithm to estimate that optimum, or to calculate the optimal solution. In connection with this topic, see also AGLER AND YOUNG [1]–[3] and BERCOVICI, FOIAS, AND TANNENBAUM [3].

Other variants of the commutant lifting theorem were also considered; see, for instance TREIL AND VOLBERG [1] and FOIAS AND TANNENBAUM [1]. For generalizations of the commutant lifting theorem, see also BISWAS, FOIAS, AND FRAZHO [1] and FOIAS, FRAZHO, AND KAASHOEK [3], and for related topics, see KAFTAL, LARSON, AND WEISS [1] and FOIAS, FRAZHO, AND LI [1].

2. The commutant lifting theorem has been extended to the context of dilations for noncommuting contractive n-tuples of operators by POPESCU [2]. In the commutative case, see POPESCU [6] and BHATTACHARYYA, ESCHMEIER, AND SARKAR [1]. In the context of the dilations studied in DRURY [1] and ARVESON [3], the corresponding commutant lifting theorem was proved in BALL, TRENT, AND VINNIKOV [1]. See also BALL, LI, TIMOTIN, AND TRENT [1] for a commutant lifting result in the case of a system of commuting contractions.

3. A theorem about similarity to a contraction, inspired by systems theory, and implying Theorem 8.1, was given in HOLBROOK [3]. For a different approach to similarity problems see VAN CASTEREN [1,2].

4. The invariance of various properties of an operator under quasi-similarity has been studied by many authors; see, for instance, FIALKOW [1],[2]; HERRERO [2],[3]; CLARY [1]; L. M. YANG [1]; AGLER, FRANKS AND HERRERO [1]; MCCARTHY [1]; CHEN, HERRERO, AND WU [1]; MÜLLER AND TOMILOV [1]; and TAKAHASHI [5].

Chapter III

Functional Calculus

1 Hardy classes. Inner and outer functions

1. In the rest of this book we consider only complex Hilbert spaces. For the contractions T on these spaces we construct, in this chapter and in the next one, a functional calculus with the aid of the minimal unitary dilation of T.

Let us begin by introducing some classes of functions, holomorphic on the open unit disk

$$D = \{\lambda : \ |\lambda| < 1\}.$$

First a notation: for any function u on D we define its "adjoint" $u\tilde{}$ by

$$u\tilde{}(\lambda) = \overline{u(\bar{\lambda})}; \tag{1.1}$$

the transformation $u \to u\tilde{}$ is obviously involutive. If u is holomorphic on D then so is $u\tilde{}$, and for the corresponding power series we have

$$u(\lambda) = \sum_0^\infty c_n \lambda^n, \qquad u\tilde{}(\lambda) = \sum_0^\infty \overline{c_n} \lambda^n.$$

Let H^p $(0 < p \leq \infty)$ be the (Hardy) class of functions u, holomorphic on D and such that the corresponding norm

$$\|u\|_p = \begin{cases} \sup_{0<r<1} \left[\frac{1}{2\pi}\int_0^{2\pi} |u(re^{it})|^p \, dt\right]^{1/p} & (0 < p < \infty), \\ \sup_{\lambda \in D} |u(\lambda)| & (p = \infty) \end{cases}$$

is finite. We have $H^p \supset H^{p'} \supset H^\infty$ for $0 < p < p' < \infty$.

Each of the classes H^p is linear and invariant for the involution $u \to u\tilde{}$; H^∞ is even an algebra.

Let us recall some fundamental theorems on the Hardy classes, due to Fatou, Riesz, and Szegő. For the proofs we refer to the monograph of PRIVALOV [1] or to that of HOFFMAN [1], and to the original papers cited there.

B.Sz.-Nagy et al., *Harmonic Analysis of Operators on Hilbert Space*, Universitext,
DOI 10.1007/978-1-4419-6094-8_3,

For $u \in H^p$ the *radial limit*

$$u(e^{it}) = \lim_{r \to 1-0} u(re^{it}) \tag{1.2}$$

exists almost everywhere (a.e.) on the unit circle C, and if $u \not\equiv 0$ we have[1]

$$\log |u(e^{it})| \in L^1 \tag{1.3}$$

and consequently $u(e^{it}) \neq 0$ a.e. Moreover, $u(e^{it})$ exists a.e. even as the *nontangential limit* of $u(\lambda)$, that is, for λ converging to e^{it} in the angle formed by two chords of C issuing from the point e^{it}. The limit function $u(e^{it})$ belongs to the Lebesgue space L^p and, if $0 < p < \infty$, it is also the strong limit of $u(re^{it})$:

$$\int_0^{2\pi} |u(e^{it}) - u(re^{it})|^p \, dt \to 0 \qquad (r \to 1-0); \tag{1.4}$$

if $p \geq 1$ this implies weak convergence, that is,

$$\int_0^{2\pi} f(t) u(re^{it}) \, dt \to \int_0^{2\pi} f(t) u(e^{it}) \, dt \qquad (r \to 1-0) \tag{1.5}$$

for an arbitrary $f \in L^q$ $((1/p) + (1/q) = 1)$. In particular, the Cauchy and Poisson formulas hold for the limit functions; thus for $u \in H^p$ $(p \geq 1)$ we have

$$\frac{1}{2\pi} \int_0^{2\pi} e^{int} u(e^{it}) \, dt = \begin{cases} u(0) & (n = 0), \\ 0 & (n = 1, 2, \ldots) \end{cases} \tag{1.6}$$

and

$$\frac{1}{2\pi} \int_0^{2\pi} P(\rho, \tau - t) u(e^{it}) \, dt = u(\lambda) \qquad (\lambda = \rho e^{i\tau}, \ 0 \leq \rho < 1), \tag{1.7}$$

with

$$P(\rho, \tau) = \frac{1 - \rho^2}{1 - 2\rho \cos \tau + \rho^2}.$$

Conversely, every function $f \in L^p$ $(1 \leq p \leq \infty)$ whose Fourier series is of the type

$$f(t) \sim \sum_0^\infty c_n e^{int}$$

generates the function

$$u(\lambda) = \sum_0^\infty c_n \lambda^n = \frac{1}{2\pi} \int_0^{2\pi} P(\rho, \tau - t) f(t) \, dt \qquad (\lambda = \rho e^{i\tau}) \tag{1.8}$$

[1] For the interval $0 \leq t \leq 2\pi$, the Lebesgue spaces L^p are defined with respect to the normalized measure $dt/2\pi$.

belonging to H^p and such that

$$u(e^{it}) = f(t) \quad \text{a.e.} \tag{1.9}$$

These functions f form a subspace of L^p denoted by L^p_+.

Formulas (1.8) and (1.9) establish therefore, for a fixed p $(1 \le p \le \infty)$, a one-to-one correspondence between the elements of H^p and L^p_+. This correspondence is obviously linear, and it even preserves the metric structure, because

$$\|u\|_p = \begin{cases} \left[\frac{1}{2\pi}\int_0^{2\pi}|u(e^{it})|^p\,dt\right]^{1/p} & (1 \le p < \infty), \\ \operatorname{ess\,sup}|u(e^{it})| & (p = \infty). \end{cases}$$

Consequently, one can identify H^p with L^p_+. In particular, H^2 is identified with L^2_+ so that we have

$$(u_1, u_2)_{H^2} = \frac{1}{2\pi}\int_0^{2\pi} u_1(e^{it})\overline{u_2(e^{it})}\,dt.$$

We call an *inner function* every function $u \in H^\infty$ such that

$$|u(e^{it})| = 1 \quad \text{a.e.} \quad \text{on} \quad C; \tag{1.10}$$

from the integral formula (1.7) and from (1.10) it follows that $|u(\lambda)| \le 1$ for every $\lambda \in D$ also. The general form of the inner functions is

$$u(\lambda) = \varkappa B(\lambda)S(\lambda), \tag{1.11}$$

where $\varkappa$ is a constant factor of modulus 1, $B(\lambda)$ is a Blaschke product:

$$B(\lambda) = \prod \frac{\bar{a}_k}{|a_k|}\frac{a_k - \lambda}{1 - \bar{a}_k\lambda} \quad (|a_k| < 1, \quad \Sigma(1 - |a_k|) < \infty), \tag{1.12}$$

and

$$S(\lambda) = \exp\left[-\int_0^{2\pi}\frac{e^{it}+\lambda}{e^{it}-\lambda}\,d\mu_t\right], \tag{1.13}$$

where μ is a finite nonnegative measure, singular with respect to Lebesgue measure and uniquely determined by the function $u(\lambda)$. If some a_k equals 0, then the corresponding factor in the Blaschke product has to be taken equal to λ. One of the factors $B(\lambda)$ and $S(\lambda)$, or both, may be absent, that is, reduce to the constant function 1.

We call an *outer function* every function on D that can be represented as

$$u(\lambda) = \varkappa \exp\left[\frac{1}{2\pi}\int_0^{2\pi}\frac{e^{it}+\lambda}{e^{it}-\lambda}\log k(t)\,dt\right] \qquad (\lambda \in D), \tag{1.14}$$

where

$$k(t) \ge 0, \quad \log k(t) \in L^1, \tag{1.15}$$

and $\varkappa$ is a complex number of modulus 1. This function u belongs to H^p $(0 < p \leq \infty)$ if and only if $k \in L^p$; in this case

$$|u(e^{it})| = k(t) \quad \text{a.e.} \tag{1.16}$$

It is obvious that the only functions which are at the same time inner and outer, are the constant functions of modulus 1.

The class of the outer functions belonging to H^p is denoted by E^p.

Every function $u \in H^p$ $(0 < p \leq \infty)$ such that $u \not\equiv 0$, has a "canonical" factorization

$$u = u_i u_e$$

into the product of an inner function u_i and an outer function u_e, which are determined up to constant factors of modulus 1. The function u_e belongs to the class E^p and is given by the formula

$$u_e(\lambda) = \varkappa \exp\left[\frac{1}{2\pi}\int_0^{2\pi} \frac{e^{it}+\lambda}{e^{it}-\lambda} \log|u(e^{it})|\, dt\right] \qquad (\lambda \in D), \tag{1.17}$$

where $|\varkappa| = 1$ (cf. (1.3)); u_i and u_e are called the *inner factor* and the *outer factor* of u, respectively.

From (1.17) it follows easily that if u, v, and uv belong to Hardy classes and do not vanish identically, then

$$(uv)_e = u_e v_e \text{ and } (uv)_i = u_i v_i;$$

this holds in particular if $u \in H^\infty$ and $v \in H^p$, because then $uv \in H^p$.

2. We need some characteristic properties of outer functions. It is convenient to introduce first the folowing notation:

$$L^p_{+0} \qquad (1 \leq p \leq \infty)$$

stands for the subspace of L^p consisting of those functions f whose Fourier series is of the form

$$f(t) \sim \sum_1^\infty c_n e^{int}.$$

Let us observe that the only real-valued function belonging to L^p_{+0} is the function $f(t) = 0$ (a.e.).

Proposition 1.1. (a) *For an outer function $u \in H^p$ $(0 < p \leq \infty, u \not\equiv 0)$ the following implications are valid.*

$$f(t) \in L^1 \quad \textit{and} \quad u(e^{it})f(t) \in L^1_{+0} \quad \textit{imply that} \quad f(t) \in L^1_{+0}, \tag{1.18}$$

$$v(\lambda) \in H^1 \quad \textit{and} \quad u(e^{it})\overline{v(e^{it})} \in L^1_{+0} \quad \textit{imply that} \quad v(\lambda) \equiv 0. \tag{1.18'}$$

(b) *In order that a function $u \in H^1$ be outer, it suffices that the following implication be valid.*

$$v(\lambda) \in H^\infty \quad \text{and} \quad u(e^{it})\overline{v(e^{it})} \in L^1_{+0} \quad \text{imply} \quad v(\lambda) \equiv 0. \tag{1.19}$$

Proof. Part (a). The case $f(t) = 0$ a.e. is trivial, so we suppose $f(t) \neq 0$ on a set of positive measure. Because $u(e^{it}) \neq 0$ a.e., the function

$$g(t) = u(e^{it}) f(t) \tag{1.20}$$

is also $\neq 0$ on a set of positive measure. Moreover, $g \in L^1_{+0}$ by the hypothesis in (1.18), thus there exists a function $G \in H^1$ such that $G(0) = 0$, $G \not\equiv 0$, and

$$G(e^{it}) = g(t) \quad \text{a.e.} \tag{1.21}$$

By virtue of (1.3) the functions

$$\log|G(e^{it})| \quad \text{and} \quad \log|u(e^{it})|$$

are integrable, and hence the function

$$\log|f(t)| = \log|g(t)| - \log|u(e^{it})| \tag{1.22}$$

is also integrable. Let us consider the outer function

$$F(\lambda) = \exp\left[\frac{1}{2\pi}\int_0^{2\pi} \frac{e^{it}+\lambda}{e^{it}-\lambda} \log|f(t)|\, dt\right]; \tag{1.23}$$

because $f \in L^1$, we have $F \in H^1$.

Using the fact that u is outer, we deduce from (1.21)–(1.23) that

$$F(\lambda) = G_e(\lambda)/u(\lambda), \tag{1.24}$$

where G_e is the outer factor of G; see (1.17). We have

$$|G(\lambda)| = |G_i(\lambda) G_e(\lambda)| \leq |G_e(\lambda)| \qquad (\lambda \in D),$$

where G_i is the inner factor of G, thus (1.24) implies

$$\left|\lambda^n \frac{G(\lambda)}{u(\lambda)}\right| \leq |F(\lambda)| \qquad (\lambda \in D;\ n = 0, 1, \ldots).$$

Because $F \in H^1$, this inequality shows that the functions $\lambda^n G(\lambda)/u(\lambda)$ ($n = 0, 1, \ldots$) also belong to H^1. Consequently,

$$\int_0^{2\pi} e^{int} \frac{G(e^{it})}{u(e^{it})}\, dt = 2\pi \left[\lambda^n \frac{G(\lambda)}{u(\lambda)}\right]_{\lambda=0} = 0 \qquad (n = 0, 1, 2, \ldots). \tag{1.25}$$

This means that the function $f(t) = g(t)/u(e^{it})$, which by hypothesis belongs to L^1, is actually contained in L^1_{+0}.

This proves that the implication (1.18) is valid. As to (1.18′), we have only to take $f(t) = \overline{v(e^{it})}$ and to observe that the conditions $\overline{v(e^{it})} \in L^1_{+0}$ and $v(e^{it}) \in L^1_+$ (the first of which follows from (1.18) and the second from the fact that $v \in H^1$) imply $v(e^{it}) = 0$ a.e. and thus $v(\lambda) = 0$ $(\lambda \in D)$.

Part (b). The implication (1.19) is impossible for $u \equiv 0$, thus we have $u \not\equiv 0$, and hence u has a canonical factorization $u = u_i u_e$. Let us set

$$v(\lambda) = 1 - \overline{u_i(0)}u_i(\lambda);$$

clearly $v \in H^\infty$. Now $u \in H^1$ and $v \in H^\infty$ imply $u(e^{it}) \in L^1$ and $v(e^{it}) \in L^\infty$, and hence $u(e^{it})\overline{v(e^{it})} \in L^1$. Because $|u_i(e^{it})|^2 = 1$ a.e., we also have

$$\begin{aligned}\int_0^{2\pi} e^{int}u(e^{it})\overline{v(e^{it})}\,dt &= \int_0^{2\pi} e^{int}u(e^{it})\,dt - u_i(0)\int_0^{2\pi} e^{int}u_e(e^{it})\,dt \\ &= 2\pi[\lambda^n u(\lambda) - u_i(0)\cdot\lambda^n u_e(\lambda)]_{\lambda=0} = 0\end{aligned}$$

for $n = 0, 1, \ldots$; thus $u(e^{it})\overline{v(e^{it})} \in L^1_{+0}$. According to (1.19) this implies $v \equiv 0$, and hence $|u_i(0)|^2 = 1 - v(0) = 1$. Using the maximum principle one concludes that $u_i(\lambda) \equiv \varkappa$ (constant, of modulus 1). Thus u is indeed an outer function.

This concludes the proof of Proposition 1.1.

Proposition 1.2 (Beurling's Theorem). *Let $u \in H^2$. In order that u be an outer function it is necessary and sufficient that the functions $\lambda^n u(\lambda)$ $(n = 0, 1, \ldots)$, as elements of the Hilbert space H^2, span H^2.*

Proof. Let v be an element of H^2, orthogonal to $\lambda^n u$ $(n = 0, 1, \ldots)$, thus

$$\int_0^{2\pi} e^{int}u(e^{it})\overline{v(e^{it})}\,dt = 0 \qquad (n = 0, 1, \ldots).$$

Hence $v \in H^2 \subset H^1$ and $u(e^{it})\overline{v(e^{it})} \in L^1_{+0}$. If u is an outer function, we have then, by virtue of Proposition 1.1(a), $v \equiv 0$.

Conversely, if one supposes that the functions $\lambda^n u$ $(n = 0, 1, \ldots)$ span H^2, then the implication

$$[u(e^{it})\overline{v(e^{it})} \in L^1_{+0}] \Rightarrow [v(\lambda) \equiv 0]$$

is valid for the functions $v \in H^2$, and hence *a fortiori* for the functions $v \in H^\infty$. By Proposition 1.1(b), u is therefore outer.

3. It follows immediately from the definition (1.10) of inner functions that the product of two inner functions, and the "adjoint" $u^\sim$ of an inner function u, are also inner functions.

For outer functions, it follows from the definition by (1.14)–(1.17) that the product and the quotient of two outer functions, and the adjoint of an outer function, are also outer functions. The class H^∞ is closed with respect to multiplication and adjunction of its elements, therefore we conclude that the class E^∞ of the bounded outer functions is also closed with respect to these operations. We exhibit an important subclass of E^∞ that also possesses these properties.

Definition. We denote by $E^{\rm reg}$ the class of functions u in H^∞ that have no zeros in D and for which there exists a constant $M = M(u)$ such that

$$\left|\frac{u(\lambda)}{u(r\lambda)}\right| \leq M \quad \text{for} \quad \lambda \in D, \qquad 0 < r < 1.$$

Proposition 1.3. *The class $E^{\rm reg}$ is contained in E^∞. It is closed with respect to multiplication and adjunction of its elements, and contains in particular*

(i) *The functions in H^∞ that are continuous and different from 0 in the closed unit disk $\overline{D}$.*
(ii) *The functions of the form $(1-\alpha\lambda)^\nu$ where $|\alpha| \leq 1$, $\nu \geq 0$.* (We choose the branch with value 1 at $\lambda = 0$.)

Proof. Let $u \in E^{\rm reg}$. We show that for every $v \in H^\infty$ such that

$$u(e^{it})\overline{v(e^{it})} \in L^1_{+0}, \tag{1.26}$$

we have $v \equiv 0$; by virtue of Proposition 1.1(b) this proves that u is outer, that is $u \in E^\infty$.

Now for any fixed r, $0 < r < 1$, the function $1/u(r\lambda)$ has a Taylor series expansion, uniformly convergent on $\overline{D}$:

$$\frac{1}{u(r\lambda)} = \sum_{m=0}^{\infty} a_m^{(r)} \lambda^m.$$

This expansion is uniformly convergent in particular on the unit circle and remains so when multiplied by the bounded function $e^{int}u(e^{it})\overline{v(e^{it})}$. Term-by-term integration yields, in view of (1.26),

$$\int_0^{2\pi} e^{int}\frac{u(e^{it})}{u(re^{it})}\overline{v(e^{it})}\,dt = \sum_{m=0}^{\infty} a_m^{(r)} \int_0^{2\pi} e^{i(n+m)t}u(e^{it})\overline{v(e^{it})}\,dt = 0 \tag{1.27}$$

for $n = 0, 1, \ldots$. The function $u(e^{it})/u(re^{it})$ is bounded in absolute value by the constant M, and as $r \to 1-0$ it tends a.e. to 1. By virtue of Lebesgue's dominated convergence theorem, (1.27) yields in the limit:

$$\int_0^{2\pi} e^{int}\overline{v(e^{it})}\,dt = 0 \qquad (n = 0, 1, \ldots),$$

whence it follows that $v \equiv 0$.

So we have proved that $E^{\rm reg} \subset E^\infty$. The fact that $E^{\rm reg}$ is closed with respect to multiplication and adjunction of its elements is obvious.

For a function u satisfying the hypotheses in (i),

$$u(\lambda)/u(r\lambda)$$

is a continuous function of the variables λ, r, on the compact set $|\lambda| \leq 1, 0 \leq r \leq 1$. Hence this function is bounded along with the function $u(\lambda)$. As to (ii), one proves without difficulty that if $|\alpha| \leq 1$ and $\nu \geq 0$, then

$$\left|\frac{(1-\alpha\lambda)^\nu}{(1-\alpha r\lambda)^\nu}\right| < 2^\nu \qquad (\lambda \in D, 0 < r < 1).$$

This concludes the proof of Proposition 1.3.

4. Given two functions in H^∞, say u and v, we call v a *divisor* of u, and u a *multiple* of v, if $w = u/v$ also belongs to H^∞. The divisor v of u is *nontrivial* if neither v nor w are constant functions. The corresponding factorization $u = vw$ is then also said to be nontrivial.

As an immediate consequence of the maximum principle we see that an inner function w is constant (of modulus 1) if and only if $|w(0)| = 1$. It follows that w and $1/w$ cannot both be inner functions unless w is constant (of modulus 1). Consequently, two inner functions, which are divisors of each other, necessarily "coincide" in the sense that they are equal up to a constant factor of modulus 1. It is convenient not to distinguish two inner functions that coincide in this sense.

Let us consider the parametric representation (1.11)–(1.13) of an inner function as the product of a Blaschke product $B(\lambda)$ and of a *singular* inner function $S(\lambda)$. By virtue of this representation, every inner function u is determined by a sequence $\{a_1, a_2, \ldots\}$ (finite, infinite, or empty) of complex numbers, such that $|a_k| < 1$ and $\Sigma(1 - |a_k|) < \infty$, and by a nonnegative, bounded, singular measure μ on the Borel subsets of the unit circle C (possibly $\mu \equiv 0$); the sequence and the measure are otherwise arbitrary. In order that the function corresponding to the sequence $\{a'_1, a'_2, \ldots\}$ and the measure μ' be a divisor of the function corresponding to the sequence $\{a_1, a_2, \ldots\}$ and the measure μ, it is necessary and sufficient that, taking account of the multiplicities, $\{a'_k\}$ be a subset of $\{a_k\}$ and μ' be a minorant of μ.

Let $\{u_\alpha\}$ be a finite or infinite system of inner functions, u_α corresponding to the sequence $\{a_{\alpha k}\}$ and the measure μ_α These functions have, in an obvious sense, a *largest common inner divisor* $u_\wedge$; $u_\wedge$ corresponds to the intersection of the sets $A_\alpha = \{a_{\alpha k}\}$ (multiplicities taken into account) and to the largest minorant $\mu_\wedge$ of the measures μ_α. Similarly, if the functions u_α have a common inner multiple v, corresponding to the sequence $\{b_k\}$ and to the measure ν, then they also have a *least common inner multiple* $u_\vee$; $u_\vee$ corresponds to the union of the sets A_α (multiplicities taken into account) and to the least common majorant $\mu_\vee$ of the measures μ_α. Let us note that $\bigcup_\alpha A_\alpha \subset \{b_k\}$ and that ν is a majorant of $\mu_\vee$, hence also of $\mu_\wedge$; because ν is singular, $\mu_\vee$ and $\mu_\wedge$ are then singular too.[2] Let us note that if the system $\{u_\alpha\}$ is finite, a common inner multiple v is furnished by the product $\prod_\alpha u_\alpha$.

---

[2] For a construction of $\mu_\wedge$ and $\mu_\vee$ see, for example, DUNFORD AND SCHWARTZ [1], pp. 162–163.

Inner functions without a nonconstant common inner divisor are said to be *relatively prime*.

From these considerations it follows that the only nonconstant inner functions without nontrivial factorizations are the functions

$$\varkappa \frac{\lambda - a}{1 - \bar{a}\lambda}, \quad \text{where} \quad |a| < 1 \quad \text{and} \quad |\varkappa| = 1. \tag{1.28}$$

5. Here are two more propositions on H^∞ that we need later.

Proposition 1.4. *Let $\{u_n\}$ be a uniformly bounded sequence of functions in H^∞, converging to 0 on D. Then*

$$\int_0^{2\pi} u_n(e^{it}) f(t)\, dt \to 0 \qquad (n \to \infty) \tag{1.29}$$

for every function $f \in L^1$.[3]

Proof. The functions $\lambda^{\nu-1} u_n(\lambda)$ $(n = 1, 2, \ldots;\ \nu = 0, \pm 1, \pm 2, \ldots)$ are holomorphic on the domain $0 < |\lambda| < 1$, so we have

$$\begin{aligned}\int_0^{2\pi} u_n(e^{it}) e^{i\nu t}\, dt &= \lim_{r \to 1} \int_0^{2\pi} u_n(re^{it}) r^\nu e^{i\nu t}\, dt \\ &= \lim_{r \to 1} \frac{1}{i} \int_{|\lambda| = r} u_n(\lambda) \lambda^{\nu - 1}\, d\lambda = \frac{1}{i} \int_{|\lambda| = 1/2} u_n(\lambda) \lambda^{\nu-1}\, d\lambda;\end{aligned}$$

the last integral tends to 0 as $n \to \infty$ because $u_n(\lambda)$ tends boundedly to 0 on the circle $|\lambda| = \frac{1}{2}$. Thus (1.29) is satisfied if $f(t) = e^{i\nu t}$ and consequently also if $f(t)$ is an arbitrary trigonometric polynomial. Every function $f \in L^1$ can be approximated in the metric of L^1, as closely as we wish, by trigonometric polynomials, therefore we conclude that (1.29) is satisfied for every $f \in L^1$.

Proposition 1.5. *Let $\{u_\alpha\}$ be a finite or infinite system of inner functions and let v be their largest common inner divisor. For every function $f \in L^1$ such that*

$$u_\alpha(e^{it}) f(t) \in L^1_{+0} \quad \textit{for all} \quad \alpha, \tag{1.30}$$

we also have

$$v(e^{it}) f(t) \in L^1_{+0}. \tag{1.31}$$

Proof. By virtue of (1.30), there exist functions $F_\alpha \in H^1$ such that

$$F_\alpha(e^{it}) = u_\alpha(e^{it}) f(t) e^{-it} \quad \text{a.e.}$$

Let us fix one of the subscripts α, say the subscript 1. The functions

$$d_\alpha(\lambda) = F_1(\lambda) u_\alpha(\lambda) - F_\alpha(\lambda) u_1(\lambda)$$

---

[3] Due to the uniform boundedness principle (cf. e.g., DUNFORD AND SCHWARTZ [1], p. 66) and the formula (1.7), the converse statement also holds.

also belong to H^1 and their radial limits $d_\alpha(e^{it})$ are equal a.e. to 0. This implies $d_\alpha \equiv 0$, $F_1 u_\alpha = F_\alpha u_1$, or setting $u_\alpha / v = w_\alpha$,

$$F_1 w_\alpha = F_\alpha w_1. \tag{1.32}$$

Omitting the trivial case that $f(t) = 0$ a.e., none of the functions $F_\alpha(\lambda)$ is identically 0. Let $F_\alpha = F_{\alpha i} F_{\alpha e}$ be the canonical factorization of F_α. From (1.32) we obtain, by taking the inner factors,

$$F_{1i} w_\alpha = F_{\alpha i} w_1. \tag{1.33}$$

Thus w_1 is a common inner divisor of the functions $F_{1i} w_\alpha$. Now, v is the largest common inner divisor of the functions u_α, thus the functions $w_\alpha = u_\alpha / v$ have no nonconstant common inner divisor. Therefore w_1 must be a divisor of F_{1i}. Consequently, we have

$$G_1 = F_1 / w_1 = F_{1e} \cdot F_{1i} / w_1 \in H^1. \tag{1.34}$$

Because

$$G_1(e^{it}) = F_1(e^{it}) / w_1(e^{it}) = u_1(e^{it}) f(t) e^{-it} / w_1(e^{it}) = v(e^{it}) f(t) e^{-it},$$

(1.34) implies (1.31).

2 Functional calculus: The classes H^∞ and H_T^∞

1. Consider first the functions

$$a(\lambda) = \sum_{k=0}^{\infty} c_k \lambda^k \quad \text{with} \quad \sum_{k=0}^{\infty} |c_k| < \infty; \tag{2.1}$$

they are holomorphic on D and continuous on $\overline{D}$. The class of these functions is denoted by A: this is obviously an algebra with respect to the usual addition and multiplication of functions, and with the involution $a \to \tilde{a}$.

If T is a contraction of the Hilbert space $\mathfrak{H}$, we associate to the function (2.1) the operator

$$a(T) = \sum_{k=0}^{\infty} c_k T^k, \tag{2.2}$$

this operator series converging *in the operator norm*. For fixed T one obtains in this way a mapping

$$a \to a(T)$$

of the algebra A into the algebra $B(\mathfrak{H})$ of the bounded operators on $\mathfrak{H}$; this mapping is an algebra homomorphism and we also have

$$a(T)^* = \tilde{a}(T^*). \tag{2.3}$$

When T is a normal operator with the spectral representation

$$T^n = \int_{\sigma(T)} \lambda^n \, dK_\lambda \qquad (n = 0, 1, \ldots), \tag{2.4}$$

definition (2.2) of $a(T)$ is equivalent to the usual definition

$$a(T) = \int_{\sigma(T)} a(\lambda) \, dK_\lambda, \tag{2.5}$$

a consequence of the fact that the series (2.1) converges uniformly on $\overline{D}$ and that $\sigma(T) \subset \overline{D}$.

Let U be the minimal unitary dilation of the contraction T. The relations $T^n = \operatorname{pr} U^n$ $(n = 0, 1, \ldots)$ imply

$$a(T) = \operatorname{pr} a(U) \qquad (a \in A), \tag{2.6}$$

from which the inequality of von Neumann

$$\|a(T)\| \leq \sup |a(\lambda)| \qquad (\lambda \in D) \tag{2.7}$$

follows in just the same way as for polynomials; see Sec. I.8.

Let us observe then that for every function $\varphi(\lambda)$, holomorphic on D, the functions

$$\varphi_r(\lambda) = \varphi(r\lambda) \qquad (0 < r < 1)$$

belong to the class A. For $u \in H^\infty$ the functions u_r are uniformly bounded,

$$|u_r(\lambda)| \leq \|u\|_\infty \qquad (0 < r < 1, \lambda \in D). \tag{2.8}$$

The operators $u_r(T)$ make sense for $0 < r < 1$, so we can introduce the following definition.

Definition. For a contraction T on the space $\mathfrak{H}$ we denote by H_T^∞ the set of those functions $u \in H^\infty$ for which $u_r(T)$ has a limit in the *strong* sense as $r \to 1 - 0$, and for $u \in H_T^\infty$ we define

$$u(T) = \lim_{r \to 1-0} u_r(T). \tag{2.9}$$

For $a \in A$ this is consistent with the earlier definition. In fact, $a_r(T)$ converges then to $a(T)$ even in the operator norm:

$$\left\| \sum_0^\infty c_k T^k - \sum_0^\infty c_k r^k T^k \right\| \leq \sum_0^\infty |c_k|(1 - r^k) \to 0 \quad \text{as} \quad r \to 1 - 0.$$

From the obvious relations

$$(cu)_r = cu_r, \quad (u + v)_r = u_r + v_r, \quad (uv)_r = u_r v_r \quad (u, v \in H^\infty)$$

it follows that the class H_T^∞ is a subalgebra of H^∞ and the mapping

$$u \to u(T)$$

defined above is an *algebra homomorphism* of H_T^∞ into $B(\mathfrak{H})$.

If $T = T_0 \oplus T_1$ then obviously

$$a(T) = a(T_0) \oplus a(T_1)$$

for $a \in A$; hence it follows readily that

$$H_T^\infty = H_{T_0}^\infty \cap H_{T_1}^\infty \tag{2.10}$$

and, for $u \in H_T^\infty$,

$$u(T) = u(T_0) \oplus u(T_1). \tag{2.11}$$

By taking in particular the canonical decomposition of T (cf. Theorem I.3.2) we can thus reduce the study of the mapping $u \to u(T)$ to the two opposite cases: the unitary operators and the c.n.u. contractions.

2. Let us first consider the case of a completely nonunitary contraction T on $\mathfrak{H}$. Let U be the minimal unitary dilation of T on $\mathfrak{K}$ ($\supset \mathfrak{H}$). We show that in this case

$$H_T^\infty = H_U^\infty = H^\infty \tag{2.12}$$

and, for $u \in H^\infty$,

$$u(T) = \text{pr}\, u(U). \tag{2.13}$$

Relation (2.6) implies $u_r(T) = \text{pr}\, u_r(U)$ $(0 < r < 1)$, thus all we have to show is that if $u \in H^\infty$, then $u_r(U)$ converges strongly as $r \to 1-0$. Let us recall to this end that the spectral measure E, induced on C by the spectral family $\{E_t\}_0^{2\pi}$ of U, is absolutely continuous with respect to Lebesgue measure (cf. Theorem II.6.4). The limit function $u(e^{it})$, existing a.e. on C, is thus also integrable with respect to E: the integral

$$u^{\mathbf{s}}(U) = \int_0^{2\pi} u(e^{it})\, dE_t$$

exists (here we indicate by $\mathbf{s}$ functions of the unitary operator U, defined via the spectral integral). On the other hand, (2.5) implies that $u_r(U) = u_r^{\mathbf{s}}(U)$. Now for every $f \in \mathfrak{K}$,

$$\|[u^{\mathbf{s}}(U) - u_r^{\mathbf{s}}(U)]f\|^2 = \int_0^{2\pi} |u(e^{it}) - u_r(e^{it})|^2\, d(E_t f, f) \to 0 \qquad (r \to 1-0),$$

because the integrand converges boundedly to 0 as $t \to 1-0$, a.e. with respect to Lebesgue measure and hence also a.e. with respect to E. Thus we have proved the existence of $u(U)$ and that

$$u(U) = u^{\mathbf{s}}(U). \tag{2.14}$$

Because $\|u^s(U)\| \leq \text{sup ess}|u(e^{it})| = \|u\|_\infty$, (2.13) and (2.14) imply

$$\|u(T)\| \leq \|u\|_\infty \quad \text{for every} \quad u \in H^\infty. \tag{2.15}$$

Let us apply relation (2.3) to the function $a = u_r$ with $0 < r < 1$. By the obvious relation $(u_r)^{\tilde{}} = (u^{\tilde{}})_r$ we obtain

$$u_r(T)^* = (u^{\tilde{}})_r(T^*). \tag{2.16}$$

Now T^* being c.n.u. along with T and $u^{\tilde{}}$ belonging to H^∞ along with u, we conclude that the right-hand side of (2.16) converges strongly to $u^{\tilde{}}(T^*)$ as $r \to 1-0$. Thus $u_r(T)^*$ converges to $u(T)^*$ not only weakly (this follows immediately from the convergence $u_r(T) \to (T)$), but also strongly, and we have

$$u(T)^* = u^{\tilde{}}(T^*).$$

Let us turn now to the problems regarding the continuity of the mapping $u \to u(T)$.

Inequality (2.15) implies at once that if $u_n(\lambda)$ tends to $u(\lambda)$ uniformly on D, then $u_n(T)$ tends to $u(T)$ *in operator norm:* $u_n(T) \rightrightarrows u(T)$.

We show that for the *strong* convergence $u_n(T) \to u(T)$ it suffices that the functions $u_n(e^{it})$ tend boundedly to $u(e^{it})$ a.e. on C:

$$\|u_n\|_\infty \leq K \quad (n = 1, 2, \ldots), \qquad u_n(e^{it}) \to u(e^{it}) \quad \text{a.e. on} \quad C.$$

In fact, setting $v_n = u_n - u$ it follows from (2.13) and (2.14) that for all $h \in \mathfrak{H}$

$$\|v_n(T)h\|^2 \leq \|v_n(U)h\|^2 = \int_0^{2\pi} |v_n(e^{it})|^2 \, d(E_t h, h), \tag{2.17}$$

and this integral tends to 0 as $n \to \infty$, by Lebesgue's theorem. Here we have again used the fact that the spectral measure E of U is absolutely continuous and therefore

$$d(E_t h, h) = \varphi_h(t) dt, \quad \varphi_h \in L^1.$$

Finally, for the *weak* convergence $u_n(T) \rightharpoonup u(T)$ it suffices that the functions u_n converge boundedly to u on D. For a proof one starts again with (2.13) and (2.14) for $v_n = u_n - u$; hence it results

$$(v_n(T)h, h') = (v_n(U)h, h') = \int_0^{2\pi} v_n(e^{it}) \, d(E_t h, h') \quad (h, h' \in \mathfrak{H}).$$

This integral tends to 0 as $n \to \infty$ on account of Proposition 1.4, because $d(E_t h, h') = \varphi_{h,h'}(t) dt$ and $\varphi_{h,h'} \in L^1$.

Let us note that bounded convergence a.e. on C implies bounded convergence on D by virtue of the Poisson formula (1.7); the converse is not true: consider, for example, the functions λ^n $(n = 1, 2, \ldots)$.

For a *normal* c.n.u. contraction T, our definition of $u(T)$ for $u \in H^\infty$ is compatible with the usual definition via the spectral integral. In fact, this being true for the functions u_r $(\in A)$, we have only to show that

$$\int_{\sigma(T)} u_r(\lambda)\, dK_\lambda \to \int_{\sigma(T)} u(\lambda)\, dK_\lambda \qquad (r \to 1-0).$$

Now this follows again from Lebesgue's convergence theorem, because $u_r(\lambda) \to u(\lambda)$ boundedly on D, and $\sigma(T)\backslash D$ has O measure with respect to the spectral measure K (otherwise T could not be c.n.u.).

Let us return to the case of an arbitrary c.n.u. contraction T. Let $v \in H^\infty$ be such that $|v(\lambda)| < 1$ on D. By virtue of (2.15),

$$T' = v(T) \tag{2.18}$$

is then also a contraction. Let us show that T' is also c.n.u.

To this end let us consider the function

$$w(\lambda) = \frac{v(\lambda) - v(0)}{1 - \overline{v(0)}v(\lambda)};$$

we have $w \in H^\infty$ and $|w(\lambda)| < 1$ on D. Consequently, $w(T)$ is also a contraction and, obviously,

$$w(T) = [v(T) - v(0)I][I - \overline{v(0)}v(T)]^{-1}.$$

We recall that if V is a unitary operator on a Hilbert space, then so is $(V - aI)(I - \bar{a}V)^{-1}$ for $|a| < 1$, and conversely; see Sec. I.4.3. Hence it follows that the space $\mathfrak{H}_0'$ of the unitary part of $T' = v(T)$ is the same as that of the unitary part of $w(T)$. Because $|w(\lambda)| < 1$ on D and $w(0) = 0$, it follows from Schwarz's lemma that $w(\lambda) = \lambda \cdot z(\lambda)$ with $|z(\lambda)| \leq 1$ on D; $z(T)$ is therefore a contraction and we have

$$w(T) = T \cdot z(T) = z(T) \cdot T.$$

For $h \in \mathfrak{H}_0'$ we have

$$\|h\| = \|w(T)^n h\| = \|z(T)^n T^n h\| \leq \|T^n h\|$$
$$\|h\| = \|w(T)^{*n} h\| = \|z(T)^{*n} T^{*n} h\| \leq \|T^{*n} h\|$$

for $n = 1, 2, \ldots$; thus $\|T^n h\| = \|h\| = \|T^{*n} h\|$ because T is a contraction, and therefore $h = 0$ because T is c.n.u. We conclude that $\mathfrak{H}_0' = \{0\}$ and hence $T' = v(T)$ is c.n.u.

So $u(T') = u(v(T))$ makes sense for all $u \in H^\infty$. From (2.18) it follows at once that $p(T') = (p \circ v)(T)$ for all polynomials p; here we used the following notation for the composite function,

$$(f \circ g)(\lambda) = f(g(\lambda)).$$

Applying this relation to the partial sums p_n of the power series of a function $a \in A$, and letting $n \to \infty$, we see that $a(T') = (a \circ v)(T)$ because $p_n \to a$ and $p_n \circ v \to a \circ v$ uniformly on D. Hence we have in particular $u_r(T') = (u_r \circ v)(T)$ for all $u \in H^\infty$ and for $0 < r < 1$. If $r \to 1-0$, $u_r(T')$ tends to $u(T')$ by definition and $(u_r \circ v)(T)$ tends to $(u \circ v)(T)$ weakly, because $u_r \circ v$ tends to $u \circ v$ boundedly on D. Thus we have

$$u(v(T)) = (u \circ v)(T)$$

for all $u \in H^\infty$.

Summing up, we have the following theorem.

Theorem 2.1. *For a completely nonunitary contraction T on $\mathfrak{H}$ and for its minimal unitary dilation U we have $H_T^\infty = H_U^\infty = H^\infty$. The mapping $u \to u(T)$ of H^∞ into the algebra $B(\mathfrak{H})$, defined by*

$$u(T) = \lim_{r \to 1-0} \sum_{k=0}^{\infty} r^k c_k T^k \quad \textit{for} \quad u(\lambda) = \sum_{k=0}^{\infty} c_k \lambda^k \in H^\infty,$$

is an algebra homomorphism of H^∞ into $B(\mathfrak{H})$, with the following properties:

(a) $u(T) = \begin{cases} I & \textit{if } u(\lambda) = 1, \\ T & \textit{if } u(\lambda) = \lambda, \end{cases}$

(b) $\|u(T)\| \leq \|u\|_\infty$,

(c) $u_n(T) \Rightarrow u(T)$ *if u_n tends to u uniformly on D,*

(c′) $u_n(T) \to u(T)$ *if u_n tends boundedly to u almost everywhere on C,*

(c″) $u_n(T) \rightharpoonup u(T)$ *if u_n tends boundedly to u on D,*

(d) $u(T)^* = u^\sim(T^*)$,

(e) *For $v \in H^\infty$ such that $|v(\lambda)| < 1$ on D, $T' = v(T)$ is a completely nonunitary contraction, and we have $u(T') = (u \circ v)(T)$ for every $u \in H^\infty$;*

(f) *If T is normal, $u(T)$ is equal to the integral $u^{\mathfrak{s}}(T)$ of $u(\lambda)$ with respect to the spectral measure corresponding to T;*

(g) $u(T) = \mathrm{pr}\, u(U)$.

Remark 1. The mapping $u \to u(T)$ defined above is the only algebra homomorphism of H^∞ into $B(\mathfrak{H})$ having properties (a) and (c′).

Indeed, (a) implies the uniqueness for polynomials. For the classes A and H^∞, uniqueness follows successively, on account of the fact that the partial sums p_n of the power series of a function $a \in A$, as well as the functions u_r derived from a function $u \in H^\infty$, converge boundedly a.e. on C ($p_n \to a, u_r \to u$).

Remark 2. By virtue of the one-to-one correspondence between the elements of H^∞ and L_+^∞, discussed in Sec. 1.1, the algebraic homomorphism $u \to u(T)$ of H^∞ into $B(\mathfrak{H})$ can also be considered as an algebra homomorphism of L_+^∞ into $B(\mathfrak{H})$. This homomorphism $f \to f(T)$ has in particular the following properties.

(a′) $f(T) = \begin{cases} I \text{ if } f(t) = 1 \text{ a.e.,} \\ T \text{ if } f(t) = e^{it} \text{ a.e.} \end{cases}$

(b′) $\|f(T)\| \leq \|f\|_\infty$.

Proposition 2.2. *Let T be a c.n.u. contraction on $\mathfrak{H}$ ($\neq \{0\}$). Let M be a subalgebra of the algebra L^∞ containing L^∞_+ and let us suppose that there exists an algebra homomorphism $f \to f(T)$ of M into $B(\mathfrak{H})$ with properties* (a′) *and* (b′). *Then $M = L^\infty_+$.*

Proof. On account of (b′) we can extend the homomorphism under consideration to the closure $\overline{M}$ of M in L^∞. Now it is known that every closed subalgebra of L^∞, containing L^∞_+ as a proper subalgebra, contains the function

$$f_1(t) = e^{-it}$$

(cf. HOFFMAN [1], Chap. 10. Maximality). Hence if $M \neq L^\infty_+$, then $f_1(T)$ makes sense, and in view of the relations $e^{it} f_1(t) = f_1(t) e^{it} = 1$, (a′) gives

$$T f_1(T) = f_1(T) T = I;$$

that is, $f_1(T) = T^{-1}$. On the other hand, (b′) implies

$$\|T^{-1}\| = \|f_1(T)\| \leq \|f_1\|_\infty = 1.$$

Now $\|T\| \leq 1$ and $\|T^{-1}\| \leq 1$ obviously imply that T is unitary, and this contradicts the hypothesis that T is c.n.u. on $\mathfrak{H} \neq \{0\}$. This concludes the proof of Proposition 2.2.

By virtue of the preceding two remarks and of Proposition 2.2, our functional calculus for the c.n.u. contractions is unique and maximal.

3. For $u \in H^\infty$ we denote by $u(e^{it})$ the *nontangential limit* of $u(\lambda)$ at the point $z = e^{it}$ of C, if it exists. Let us note that this nontangential limit can exist without being the limit of $u(\lambda)$ at the point z in the sense that $u(\lambda)$ should tend to the value $u(z)$ as λ converges in D to z arbitrarily.

We introduce the following sets of points z on C, associated with a function $u \in H^\infty$.

$C_u = \{z:\ u(\lambda)$ does not have a nontangential limit at $z\}$.
$\mathbf{C}_u = \{z:\ u(\lambda)$ does not have a limit at $z\}$.
$C^0_u = \{z:\ u(\lambda)$ does not have a nonzero nontangential limit at $z\}$[4].
$\mathbf{C}^0_u = \{z:\ u(\lambda)$ does not have a nonzero limit at $z\}$.

All these sets are Borel sets, C_u and (if $u(\lambda) \not\equiv 0$) C^0_u are of Lebesgue measure 0 (cf. Sec. 1), $C_u \subset \mathbf{C}_u$ and $C^0_u \subset \mathbf{C}^0_u$.

Let us consider then a unitary operator V on a Hilbert space $\mathfrak{H}$, and let E_V be the corresponding spectral measure on C. It follows from (2.5) that

$$a(V) = a^{\mathrm{s}}(V) = \int_0^{2\pi} a(e^{it})\, dE_{V,t}$$

[4] In other words, the nontangential limit either does not exist, or it equals zero.

for every $a \in A$, and in particular for the functions u_r $(0 < r < 1)$, where $u \in H^\infty$. Let us note that $|u_r(e^{it})| \leq \|u\|_\infty$ $(0 < r < 1)$ and that $u_r(e^{it}) \to u(e^{it})$ $(r \to 1-0)$, with the exception of the points $e^{it} \in C_u$. If

$$E_V(C_u) = O \tag{2.19}$$

then $u_r^{\mathrm{s}}(V)$ therefore converges strongly to $u^{\mathrm{s}}(V)$ as $r \to 1-0$; indeed, by Lebesgue's dominated convergence theorem we have then

$$\|u_r^{\mathrm{s}}(V)h - u^{\mathrm{s}}(V)h\|^2 = \int_0^{2\pi} |u_r(e^{it}) - u(e^{it})|^2 \, d(E_{V,t}h,h) \to 0 \qquad (h \in \mathfrak{H}).$$

Hence (2.19) ensures that $u \in H_V^\infty$ and $u(V) = u^{\mathrm{s}}(V)$.

Let us examine the validity of the properties of our functional calculus, established in Theorem 2.1 for c.n.u. contractions T, in the actual case $T = V$. We note first that in this case $U = V$, so property (g) is trivial; property (a) is obvious, and property (f) has just been established, at least under the condition (2.19).

Property (b) follows from the inequality (2.7), valid for the class A, if one applies it to the function u_r and if one observes that $\|u_r\|_\infty \leq \|u\|_\infty$ and $u_r(V) \to u(V)$ (strongly) for all $u \in H_V^\infty$. As to the convergence criteria, (c) follows from (b); (c′) subsists if the term "almost everywhere on C" is meant with respect to the spectral measure E_V (and if it is understood that $u_n(e^{it})$ and $u(e^{it})$ exist a.e. with respect to E_V (i.e. $E_V(C_{u_n}) = E_V(C_u) = O$)), and (c″) is no longer valid: a counterexample is furnished by $V = I$ and $u_n(\lambda) = \lambda^n$.

Property (d) remains valid, with the addition that $H_{V^*}^\infty$ consists exactly of the adjoints $u^\sim$ of the functions $u \in H_V^\infty$; that is,

$$H_{V^*}^\infty = (H_V^\infty)^\sim.$$

In fact, V being unitary its polynomials and their strong limits are normal operators; thus in particular $u_r(V)$ is a normal operator for $u \in H_V^\infty$ and for $0 < r < 1$. We just have then to recall relation (2.3) and the fact that, for a sequence of commuting normal operators N_n, $N_n \to N$ strongly if and only if $N_n^* \to N^*$ strongly. Let us also note that for this reason $u(V)$ is a normal operator for all $u \in H_V^\infty$.

As to property (e), it follows from (b) that if $v \in H_V^\infty$ and if $|v(\lambda)| < 1$ on D, then $N = v(V)$ is a contraction. The relation $p(N) = (p \circ v)(V)$ for polynomials p follows because the functional calculus is an algebra homomorphism, the relation $a(N) = (a \circ v)(V)$ for $a \in A$ follows hence by the use of (c); in fact, the partial sums p_n of the power series of a yield a sequence of polynomials such that $p_n \to a$ and $p_n \circ v \to a \circ v$ uniformly on D. In particular,

$$u_r(N) = (u_r \circ v)(V) \tag{2.20}$$

for $u \in H^\infty, 0 < r < 1$. Let us make the additional hypotheses that the sets

$$C_v \quad \text{and} \quad v^{-1}(\mathbf{C}_u) = \{e^{it} : \; v(e^{it}) \in \mathbf{C}_u\} \tag{2.21}$$

are of O measure with respect to E_V. These hypotheses imply that the limits

$$v(e^{it}) = \lim_{r\to 1} v(re^{it}) \quad \text{and} \quad (u\circ v)(e^{it}) = \lim_{r\to 1} u(v(re^{it}))$$

exist and

$$(u\circ v)(e^{it}) = \lim_{r\to 1} u(rv(e^{it}))$$

holds E_V-almost everywhere on C. This implies that $u\circ v \in H_V^\infty$ and, by the convergence criterion (c′),

$$(u_r\circ v)(V) \to (u\circ v)(V) \qquad (r\to 1-0).$$

In view of (2.20) one concludes that $u(N) = \lim_{r\to 1-0} u_r(N)$ exists and equals $(u\circ v)(V)$.

Hence all the properties (a)–(g) of the functional calculus, valid by Theorem 2.1 for c.n.u. contractions, can be extended with some precaution to the case of a unitary $T = V$, except property (c″).

It is easy now to treat the general case. Let T be an arbitrary contraction, with the unitary part T_0 and the c.n.u. part T_1. Then

$$T = T_0 \oplus T_1 \quad \text{and} \quad U_T = U_{T_0} \oplus U_{T_1}, \quad \text{where} \quad U_{T_0} = T_0.$$

Recalling (2.11) and combining Theorem 2.1 with the results just obtained, we arrive at the following theorem.

Theorem 2.3. *Let T be a contraction on $\mathfrak{H}$ and let U be its minimal unitary dilation. Then we have $H_T^\infty = H_U^\infty$, and H_T^∞ contains in particular the functions $u \in H^\infty$ for which the set C_u has O measure with respect to the spectral measure E_T corresponding to the unitary part of T.*

The mapping $u \to u(T)$ of H_T^∞ into $B(\mathfrak{H})$ defined by

$$u(T) = \lim_{r\to 1-0} \sum_{k=0}^{\infty} r^k c_k T^k \quad \textit{for} \quad u(\lambda) = \sum_{k=0}^{\infty} c_k \lambda^k \in H_T^\infty$$

is an algebra homomorphism with the following properties.

(a) $u(T) = \begin{cases} I & \textit{if}\ u(\lambda) = 1, \\ T & \textit{if}\ u(\lambda) = \lambda. \end{cases}$

(b) $\|u(T)\| \le \|u\|_\infty$.

(c) $u_n(T) \Rightarrow u(T)$ *if u_n tends to u uniformly on D.*

(c′) $u_n(T) \to u(T)$ *if u_n tends boundedly to u almost everywhere as well as E_T-almost everywhere on C.*

(d) $H_{T^*}^\infty = (H_T^\infty)^\sim$*, and* $u(T)^* = u^\sim(T^*)$ *for* $u \in H_T^\infty$.

(e) *If u, v are such that $|v(\lambda)| < 1$ on D, and the sets C_v and $v^{-1}(\mathbf{C}_u)$ have measure O with respect to E_T, then we have $v \in H_T^\infty$, $u\circ v \in H_T^\infty$, moreover $T' = v(T)$ is a contraction, u belongs to $H_{T'}^\infty$, and*

$$u(T') = (u\circ v)(T).$$

(f) *For a normal contraction T and for $u \in H^\infty$ such that $E_T(C_u) = O$, $u(T)$ exists and equals the integral $u^{\mathfrak{s}}(T)$ of the function $u(\lambda)$ with respect to the spectral measure corresponding to T.*

(g) $u(T) = \operatorname{pr} u(U)$ *for every* $u \in H^\infty_T$.

3 The role of outer functions

1. The first step towards an extension of our functional calculus to unbounded functions is to study the conditions under which the operator $u(T)$ has a (not necessarily bounded) inverse. The detailed study of this extension is found in the next chapter.

We consider the case of c.n.u. contractions.

Proposition 3.1. *For every completely nonunitary contraction T on $\mathfrak{H}$, and for every outer function $u \in H^\infty$ (i.e., $u \in E^\infty$), the operator $u(T)$ has an inverse with dense domain in $\mathfrak{H}$ (i.e., $u(T)$ is a quasi-affinity on $\mathfrak{H}$).*

Proof. Let $u \in E^\infty$ and $h \in \mathfrak{H}$ be such that $u(T)h = 0$. If U is the minimal unitary dilation of T and if E is the spectral measure corresponding to U, then

$$0 = (T^n u(T)h, h) = (U^n u(U)h, h) = \int_0^{2\pi} e^{int} u(e^{it}) \varphi_h(t)\, dt \qquad (n = 0, 1, \ldots),$$

where $\varphi_h(t) = d(E_t h, h)/dt \in L^1$ and $\varphi_h(t) \geq 0$ a.e. Hence $u(e^{it})\varphi_h(t) \in L^1_{+0}$ and by virtue of Proposition 1.1 we have $\varphi_h(t) \in L^1_{+0}$. Because $\varphi_h(t)$ is real-valued, this implies that $\varphi_h(t) = 0$ a.e. Consequently,

$$\|h\|^2 = \int_0^{2\pi} d(E_t h, h) = \int_0^{2\pi} \varphi_h(t)\, dt = 0.$$

Therefore $u(T)h = 0$ implies that $h = 0$, and hence $u(T)$ is invertible.

Now the operator T^* is also a c.n.u. contraction and $u^\sim$ also belongs to E^∞. Thus $u^\sim(T^*)$ is invertible and because $u^\sim(T^*) = u(T)^*$, this implies that the range of $u(T)$, that is, the domain of $u(T)^{-1}$, is dense in $\mathfrak{H}$. This concludes the proof.

The condition on the function u to be outer turns out to be also necessary in the following sense.

Proposition 3.2. *For every nonouter function $u \in H^\infty$ there exists a c.n.u. contraction T on a space $\mathfrak{H} \neq \{0\}$ such that $u(T) = O$.*

Proof. We can assume that $u(\lambda) \not\equiv 0$. Let $u = u_i u_e$ be the canonical factorization of u. By hypothesis, the inner factor u_i is nontrivial. If we find a c.n.u. contraction T for which $u_i(T) = O$, then we also have $u(T) = u_i(T)u_e(T) = O$.

Let $u_i H^2$ denote the set of the functions $u_i v$ where v runs over H^2. (These products belong to H^2 because $u_i \in H^\infty$.) Because $|u_i(e^{it})| = 1$ a.e., the mapping $v \to u_i v$ is linear and isometric in the Hilbert space metric of H^2. Consequently, $u_i H^2$ is a subspace of H^2. Because u_i is not outer, it follows from Proposition 1.2 (Beurling's theorem) that u_iH^2 does not coincide with H^2. Hence, setting

$$\mathfrak{H} = H^2 \ominus u_iH^2,$$

we have $\mathfrak{H} \neq \{0\}$. Let V denote multiplication on H^2 by the variable λ; V is a unilateral shift on H^2, and consequently $V^{*n} \to O$ as $n \to \infty$. The explicit form of V^* is

$$V^*v = \frac{1}{\lambda}[v(\lambda) - v(0)] \quad \text{for} \quad v \in H^2.$$

Let us note that the subspace u_iH^2 is invariant for V, and hence its orthogonal complement in $\mathfrak{H}$ is invariant for V^*. Setting

$$T = (V^*|\mathfrak{H})^* \tag{3.1}$$

we thus obtain a contraction T on $\mathfrak{H}$ such that

$$T^{*n} = V^{*n}|\mathfrak{H} \to O \qquad (n \to \infty). \tag{3.2}$$

Consequently, V and T are c.n.u. Moreover, (3.2) implies

$$T^n = PV^n|\mathfrak{H} \qquad (n = 0, 1, \ldots), \tag{3.3}$$

where P denotes the orthogonal projection of H^2 onto $\mathfrak{H}$. It follows that

$$w(T) = Pw(V)|\mathfrak{H}$$

for any $w \in H^\infty$, $w(V)$ being the operator of multiplication by $w(\lambda)$ on H^2.

In particular, we have

$$u_i(T)h = Pu_i(V)h = Pu_i(\lambda)h(\lambda) \quad \text{for all} \quad h \in \mathfrak{H}.$$

Because $u_ih \in u_iH^2$ and hence $u_ih \perp \mathfrak{H}$ for every $h \in H^2$, we obtain that $u_i(T) = O$. Thus the operator T has all the desired properties: it is a c.n.u. contraction on a space $\mathfrak{H} \neq \{0\}$, such that $u_i(T) = O$.

It should be added that every other nonzero function $w \in H^\infty$ for which $w(T) = O$, is a multiple of u_i in H^∞. To show this, let us first observe that the function $h_0(\lambda) = 1 - \overline{u_i(0)}u_i(\lambda)$ belongs to $\mathfrak{H}$; indeed we have for every $v \in H^2$:

$$\begin{aligned}(u_iv, h_0)_{\mathfrak{H}} &= \frac{1}{2\pi}\int_0^{2\pi} u_i(e^{it})v(e^{it})[1 - u_i(0)\overline{u_i(e^{it})}]\,dt \\ &= \frac{1}{2\pi}\int_0^{2\pi} v(e^{it})[u_i(e^{it}) - u_i(0)]\,dt = 0.\end{aligned}$$

Thus $w(T) = O$ implies in particular that

$$0 = w(T)h_0 = Pwh_0$$

and hence $wh_0 \in u_i H^2$. It follows that $wh_0 = u_i v$ for some $v \in H^2$ and therefore we have that $w = u_i v_1$ with $v_1 = v + \overline{u_i(0)} w \in H^2$. This implies

$$|v_1(e^{it})| = |u_i(e^{it})||v_1(e^{it})| = |w(e^{it})| \leq \|w\|_\infty$$

a.e. on C. By virtue of Poisson's formula connecting the values of the function v_1 on D with its values (i.e., radial limits) on C, we have $|v_1(\lambda)| \leq \|w\|_\infty$ for every $\lambda \in D$. Hence $v_1 \in H^\infty$, and thus w is a multiple of u_i in H^∞.

From Propositions 3.1 and 3.2 we obtain the following result immediately.

Proposition 3.3. *Let $u \in H^\infty$. In order that $u(T)$ be invertible for every completely nonunitary contraction T on $\mathfrak{H}$, it is necessary and sufficient that u be an outer function, that is, $u \in E^\infty$. For such a function, $u(T)^{-1}$ exists and has dense domain in $\mathfrak{H}$.*

Corollary. *Let $v \in H^\infty$ be such that $|v(\lambda)| < 1$ for $\lambda \in D$. Then $u \in E^\infty$ implies $u \circ v \in E^\infty$.*

Proof. By virtue of Theorem 2.1, if T is any c.n.u. contraction then so is $T' = v(T)$ and we have

$$u(T') = (u \circ v)(T).$$

Now, for $u \in E^\infty, u(T')$ is invertible and so is $(u \circ v)(T)$, and hence $u \circ v \in E^\infty$.

2. Now we make the following definition.

Definition. For an arbitrary contraction T let K_T^∞ denote the class of functions $u \in H_T^\infty$ for which $u(T)^{-1}$ *exists* and has *dense domain*, that is, $u(T)$ is a quasi-affinity.

For $u \in K_T^\infty$ we have that

$$u\tilde{}(T^*) = u(T)^* \tag{3.4}$$

is a quasi-affinity, hence $u\tilde{} \in K_{T^*}^\infty$. By reason of symmetry we have therefore

$$K_{T^*}^\infty = (K_T^\infty)\tilde{}. \tag{3.5}$$

Similarly, if u, v are in K_T^∞ then so is uv, because $(uv)(T) = u(T)v(T)$. Thus the class K_T^∞ is *multiplicative.*

By virtue of Proposition 3.3 we have for a c.n.u. T:

$$E^\infty \subset K_T^\infty.$$

Let T be unitary, $T = V$, and let $u \in H^\infty$. We know that if $u(e^{it})$ exists E_V-a.e. (condition (2.19)) then $u \in H_V^\infty$ and $u(V) = u^s(V)$. In order that $u(V)$ be invertible, it is necessary and sufficient that $u(e^{it})$ exist and be different from 0 E_V-a.e., that is,

$$E_V(C_u^0) = O. \tag{3.6}$$

Under this condition, according to the theory of spectral integrals, $u^{\mathbf{s}}(V)^{-1} = (1/u)^{\mathbf{s}}(V)$, and hence

$$u(V)^{-1} = \int_0^{2\pi} [1/u(e^{it})]\, dE_{V,t},$$

and this operator has dense domain. Condition (3.6) thus ensures that $u \in K_V^\infty$ and that $u(V)^{-1} = (1/u)^{\mathbf{s}}(V)$.

Let $u, v \in H^\infty$ such that $|v(\lambda)| < 1$ on D and that the sets

$$C_v \quad \text{and} \quad v^{-1}(\mathbf{C}_u^0)$$

are of E_V-measure O. These conditions ensure that $(u \circ v)(e^{it})$ exists and is not zero E_V-a.e., and this implies by what precedes that $u \circ v \in K_V^\infty$.

General contractions T can be dealt with by using the canonical decomposition $T = T_0 \oplus T_1$ and applying Proposition 3.3, its corollary, and the remarks which we have just made in the case of a unitary operator. So we arrive at the following result.

Theorem 3.4. (i) *For a contraction T of general kind, the class K_T^∞ is multiplicative and we have $K_{T^*}^\infty = (K_T^\infty)^\sim$.*

(ii) *K_T^∞ contains in particular the functions $u \in E^\infty$ for which the set C_u^0 is of O measure with respect to the spectral measure E_T corresponding to the unitary part of T.*
(iii) *For functions $u \in E^\infty, v \in H^\infty$ such that $|v(\lambda)| < 1$ on D and both C_v, $v^{-1}(\mathbf{C}_u^0)$ are of E_T-measure O, we have $u \circ v \in K_T^\infty$.*

In the present chapter we apply our functional calculus to study a new class of contractions, called class C_0, and the continuous one-parameter semigroups of contractions. In these investigations we only use a part of our results, in particular those established in Theorems 2.1 and 2.3. We further exploit these results in the next chapter when we extend our functional calculus to some classes of unbounded analytic functions.

4 Contractions of class C_0

1. Let T be a contraction on the space $\mathfrak{H}$ and let U be its minimal unitary dilation. We know that the limit

$$Lh = \lim_{n\to\infty} U^{-n} T^n h \tag{4.1}$$

exists for every $h \in \mathfrak{H}$ (cf. Proposition II.3.1, where we have also proved that Lh is the orthogonal projection of h into the subspace $\mathfrak{R}_*$ of the dilation space). From (4.1) we obtain

$$U^{-m} L T^m h = \lim_{n\to\infty} U^{-m-n} T^{m+n} h = Lh,$$

and hence

$$LT^m h = U^m Lh \quad \text{for} \quad m = 0, 1, \ldots . \tag{4.2}$$

If we assume that T is c.n.u. then (4.2) implies that

$$Lu(T)h = u(U)Lh \tag{4.3}$$

for all $u \in H^\infty$ and all $h \in \mathfrak{H}$.

Choose u and h so that $u \not\equiv 0$ and

$$u(T)h = 0; \tag{4.4}$$

by virtue of (4.3) we have

$$u(U)Lh = 0. \tag{4.5}$$

Now $u(e^{it})$ exists and is nonzero a.e. with respect to Lebesgue measure, and hence also with respect to the spectral measure corresponding to U (recall that this spectral measure is absolutely continuous). This implies that $u(U)$ is invertible: $u(U)^{-1} = (1/u)^{\mathfrak{s}}(U)$. So we deduce from (4.5) that $Lh = 0$, and hence

$$\lim_{n\to\infty} \|T^n h\| = \lim_{n\to\infty} \|U^{-n}T^n h\| = \|Lh\| = 0.$$

We can formulate our result as follows.

Proposition 4.1. *Let T be a completely nonunitary contraction on the space $\mathfrak{H}$, and let u be a nonzero function in H^∞. For every $h \in \mathfrak{H}$ such that $u(T)h = 0$ we have then $T^n h \to 0$ as $n \to \infty$.*

If $u(T)h = 0$ for *all* $h \in \mathfrak{H}$ (i.e., if $u(T) = O$), then $u^{\sim}(T^*) = u(T)^* = O$, so we can apply the preceding result to $u^{\sim}$ and T^*, and thus obtain the following proposition.

Proposition 4.2. *Let T be a completely nonunitary contraction such that $u(T) = O$ for some nonzero function $u \in H^\infty$. Then we have $T^n \to O$ and $T^{*n} \to O$, i.e. $T \in C_{00}$.*

The class of these contractions merits further investigation. First we make a definition.

Definition. We call C_0 the *class* of those completely nonunitary contractions T for which there exists a nonzero function $u \in H^\infty$ such that $u(T) = O$.

Proposition 4.2 can then be expressed by the formula

$$C_0 \subset C_{00}.$$

Let us observe that for $T \in C_0$, the function u in the definition can always be taken to be *inner*. Indeed, if $u = u_e u_i$ is the canonical factorization of u, then $u_e(T)u_i(T) = u(T) = O$, and hence $u_i(T) = O$, because $u_e(T)$ is invertible by Proposition 3.1.

It is obvious that if $u(T) = O$ for some $u \in H^\infty$ ($u \not\equiv 0$) then also $v(T) = O$ for the multiples v of u in H^∞. There arises the question of whether for every contraction $T \in C_0$ there exists an inner function u with $u(T) = O$, such that every other function $v \in H^\infty$ with $v(T) = O$ is a multiple of u. Such a function, if it exists, is called a *minimal function* for T and denoted by m_T; this function is then determined up to a constant factor of modulus 1. This follows from the fact that if two inner functions are divisors of each other, then they necessarily coincide; see Sec. 1.

An important example for these notions was constructed in the proof of Proposition 3.2 (see also the discussion following that proof). Let us formulate it as Part (a) of the following proposition.

Proposition 4.3. (a) *For every nonconstant inner function m, the space*

$$\mathfrak{H} = H^2 \ominus mH^2$$

is different from $\{0\}$*; the operator T defined on* $\mathfrak{H}$ *by*

$$Tv = P_{\mathfrak{H}}[\lambda v(\lambda)], \quad T^*v = \frac{1}{\lambda}[v(\lambda) - v(0)] \quad (v \in \mathfrak{H})$$

belongs to the class C_0 *and its minimal function is m.*

(b) *A subspace* $\mathfrak{H}_1$ *of* $\mathfrak{H}$ *is invariant for T if and only if it has the form*

$$\mathfrak{H}_1 = m_2(H^2 \ominus m_1 H^2),$$

where $m = m_1 m_2$ *is a factorization of m into a product of inner factors* m_1 *and* m_2.

Proof. It is enough to prove (b). Let $m = m_1 m_2$ be a factorization of m where m_1 and m_2 are inner functions. If m_1 or m_2 is constant then $\mathfrak{H}_1 = \{0\}$ or $\mathfrak{H}_1 = \mathfrak{H}$, respectively, is trivially invariant. Thus we can assume that neither m_1 nor m_2 is constant. We have

$$\mathfrak{H} = (H^2 \ominus m_2 H^2) \oplus m_2(H^2 \ominus m_1 H^2),$$

so that $\mathfrak{H}_2 = H^2 \ominus m_2 H^2 \subset \mathfrak{H}$ and the restriction of T^* to $\mathfrak{H}_2$ is precisely the analogue of the operator T^* for the case when m is replaced by m_2. Therefore, by virtue of Part (a), $T^*\mathfrak{H}_2 \subset \mathfrak{H}_2$ and consequently $T\mathfrak{H}_1 \subset \mathfrak{H}_1$.

Conversely, let $\mathfrak{H}_1$ be an invariant subspace of T. The cases $\mathfrak{H}_1 = \{0\}$ or $\mathfrak{H}_1 = \mathfrak{H}$ correspond to the trivial factorizations $m = 1 \cdot m$ or $m = m \cdot 1$, respectively, therefore we assume that $\{0\} \neq \mathfrak{H}_1 \neq \mathfrak{H}$. The space $\mathfrak{H}_2 = \mathfrak{H} \ominus \mathfrak{H}_1$ is invariant for T^* and (with the notation as in the proof of Proposition 3.2) also for V^* because $T^* = V^*|\mathfrak{H}$. For $T_2 = (T^*|\mathfrak{H}_2)^*$ we also have $T_2^* = V^*|\mathfrak{H}_2$. This implies that $P_{\mathfrak{H}_2}V = T_2 P_{\mathfrak{H}_2}$ which in turn yields that V is an isometric dilation of T_2. If V were nonminimal, then V would be reducible, by virtue of the remark following the proof of Theorem I.4.1. This contradicts the fact that V is obviously of multiplicity 1. For the same reason $(V - T_2)\mathfrak{H}_2 \neq \{0\}$, because in the opposite case $\mathscr{H}_2$ is reducing V. But by virtue of Theorem II.2.1, $\mathfrak{L} = \overline{(V - T_2)\mathfrak{H}_2}$ is a wandering space of V. From Proposition I.2.1 we can now infer that $\dim \mathfrak{L} = 1$, hence $\mathscr{L}$ is formed by the scalar multiples of some $m_2 \in H^2$, $\|m_2\| = 1$. But $\mathscr{L}$ is a wandering space for V, thus we have

$$0 = (V^n m_2, m_2) = \frac{1}{2\pi} \int_0^{2\pi} e^{int} |m_2(e^{it})|^2 \, dt \qquad (n = 1, 2, \ldots)$$

whence $|m_2(e^{it})|^2 = 1$ a.e.; that is, $m_2(\lambda)$ is an inner function. Now from Theorem II.2.1 it follows that $H^2 \ominus \mathfrak{H}_2 = M_+(\mathfrak{L}) = m_2 H^2$. But $H^2 \ominus \mathfrak{H}_2 = \mathfrak{H}_1 \oplus mH^2$, so $mH^2 \subset m_2 H^2$. Consequently $m = m_2 m_1$ where $m_1 \in H^2$. Because m and m_2 are inner so must be m_1. We have thus obtained that

$$\mathfrak{H}_1 = m_2 H^2 \ominus mH^2 = m_2(H^2 \ominus m_1 H^2),$$

concluding the proof.

Further important examples of contractions of class C_0 are studied later, particularly in Chaps. VIII and X.

2. We now state an important result.

Proposition 4.4. *For every contraction T of class C_0 there exists a minimal function m_T.*

Proof. By hypothesis, the class $\mathscr{J}$ of inner functions u with $u(T) = O$ is not empty. From the lemma below it follows as a special case that $v(T) = O$ for the largest common inner divisor v of the functions u in $\mathscr{J}$. Obviously, $m_T = v$.

The lemma in question reads as follows.

Lemma 4.5. *Let u_α $(\alpha \in A)$ be inner functions, and let v be their largest common inner divisor. Suppose that the equation $u_\alpha(T)h = 0$ holds for a completely nonunitary contraction T on the space $\mathfrak{H}$, for an element h of $\mathfrak{H}$, and for every $\alpha \in A$. Then we have also $v(T)h = 0$.*

Proof. Let U be the minimal unitary dilation of T and let $\{E_t\}$ $(0 \leq t \leq 2\pi)$ be the spectral family of U; E_t is an absolutely continuous function of t. For an arbitrary element g of the space $\mathfrak{H}$ we have

$$\varphi_{h,g}(t) = d(E_t h, g)/dt \in L^1.$$

As $u_\alpha(T)h = 0$ for every $\alpha \in A$, we have for $\nu = 0, 1, \ldots,$

$$0 = (T^\nu u_\alpha(T)h, g) = (U^\nu u_\alpha(U)h, g) = \int_0^{2\pi} e^{i\nu t} u_\alpha(e^{it}) \varphi_{h,g}(t)\, dt;$$

that is,

$$u_\alpha(e^{it}) \varphi_{h,g}(t) \in L^1_{+0} \qquad (\alpha \in A).$$

By virtue of Proposition 1.5 this implies

$$v(e^{it}) \varphi_{h,g}(t) \in L^1_{+0},$$

and hence

$$(v(T)h, g) = \int_0^{2\pi} v(e^{it}) \varphi_{h,g}(t)\, dt = 0.$$

Because g is arbitrary, this implies $v(T)h = 0$.

3. The minimal function of a contraction of class C_0 plays a role analogous in many respects to the well-known role of the minimal polynomials of finite matrices in linear algebra. We know, for instance, that similar matrices have the same minimal polynomial. We prove an analogous fact for the minimal functions of contractions, even under a less stringent condition.

Proposition 4.6. *Let T_1, T_2 be two completely nonunitary contractions on the spaces $\mathfrak{H}_1, \mathfrak{H}_2$, respectively, and suppose that T_2 is a quasi-affine transform of T_1* (cf. Sec. II.3). *If one of these contractions is of class C_0 then so is the other and their minimal functions coincide.*

Proof. Let X be a quasi-affinity from $\mathfrak{H}_2$ to $\mathfrak{H}_1$ such that $XT_2 = T_1X$. Then $XT_2^n = T_1^nX$ $(n = 0,1,\ldots)$, hence also $Xu(T_2) = u(T_1)X$ for every function $u \in H^\infty$. If $T_1 \in C_0$ we have thus $Xm_{T_1}(T_2) = m_{T_1}(T_1)X = O$. We conclude that $m_{T_1}(T_2) = O$ because X is invertible, thus $T_2 \in C_0$ and m_{T_2} is a divisor of m_{T_1}. Conversely, $T_2 \in C_0$ implies $m_{T_2}(T_1)X = Xm_{T_2}(T_2) = O$; because $\overline{X\mathfrak{H}_2} = \mathfrak{H}_1$, it follows that $m_{T_2}(T_1) = O$, thus $T_1 \in C_0$ and m_{T_1} is a divisor of m_{T_2}. Thus we see that if one of the contractions belongs to C_0 then so does the other, and their minimal functions are divisors of each other, and hence they coincide.

This result shows in particular that *quasi-similar contractions of class C_0 have the same minimal function.*

4. We add the following result.

Proposition 4.7. (a) *If $T \in C_0$ then $T^* \in C_0$ also, and $m_{T^*} = m_{\tilde{T}}$.*

(b) *Let $T \in C_0$ and let u be an inner function. We have*[5] *$u \in K_T^\infty$ if and only if u and m_T have no nonconstant common inner divisor.*

Proof. Part (a) is obvious from the relation $u\tilde{\ }(T^*) = u(T)^*$, holding for every $u \in H_T^\infty$.

To prove part (b), set $p = u \wedge m_T$ (the sign $\wedge$ indicating the largest common inner divisor). Because $m_T(T) = O$, we deduce from Lemma 4.5 that $u(T)h = 0$ implies $p(T)h = 0$; the converse implication holds trivially. Thus $u(T)$ is invertible if and only if $p(T)$ is invertible. Now we have $m_T = pq$ with some inner function q and hence $p(T)q(T) = m_T(T) = O$. If $p(T)$ is invertible this implies $q(T) = O$ and hence, by virtue of the minimality property of m_T, we have $q = m_T$ and $p = 1$. As $1(T) = I$ is invertible, we conclude that $u(T)$ is invertible if and only if $p = 1$.

If $p = 1$ then $u\tilde{\ } \wedge m_{\tilde{T}} = p\tilde{\ } = 1$ and hence, using (a) also, we conclude from the above result (applied to $u\tilde{\ }$ instead of u) that $u(T)^*$ $(= u\tilde{\ }(T^*))$ is invertible, which implies that $u(T)^{-1}$ has dense domain; thus in this case we have $u \in K_T^\infty$.

This concludes the proof.

---

[5] The definition of K_T^∞ was given in Sec. 3.2.

5 Minimal function and spectrum

A well-known fact in linear algebra is that the zeros of the minimal polynomial of a matrix are exactly the characteristic values of the matrix. We prove an analogous fact for the class C_0.

Theorem 5.1. *Let m_T be the minimal function and $\sigma(T)$ the spectrum of the contraction T of class C_0. Let s_T be the set consisting of the zeros of m_T in the open unit disk D and of the complement, in the unit circle C, of the union of the arcs of C on which m_T is analytic (i.e., through which it can be continued analytically). Then*

$$\sigma(T) = s_T.$$

Proof. Let α be a point of the closed disk $\overline{D}$ not belonging to s_T. Then $m_T(\alpha) \neq 0$; indeed, if $\alpha \in D$ this follows immediately from the definition of s_T, and if $\alpha \in C$ from the fact that m_T being an inner function we have $|m_T(e^{it})| = 1$ a.e. and thus in particular at the points of C where m_T is analytic. Let

$$u(\lambda) = \frac{1}{\alpha - \lambda}[m_T(\alpha) - m_T(\lambda)].$$

It is obvious that $u \in H^\infty$; by our functional calculus for contractions we have therefore

$$(\alpha I - T)u(T) = u(T)(\alpha I - T) = m_T(\alpha)I - m_T(T) = m_T(\alpha)I.$$

This shows that $\alpha I - T$ is boundedly invertible, with

$$(\alpha I - T)^{-1} = \frac{1}{m_T(\alpha)}u(T);$$

hence α does not belong to $\sigma(T)$. So we have proved the inclusion

$$\sigma(T) \subset s_T. \tag{5.1}$$

Let now $\alpha \in D$ be such that $m_T(\alpha) = 0$. Then we have

$$m_T(\lambda) = \frac{\lambda - \alpha}{1 - \bar{\alpha}\lambda} n_\alpha(\lambda)$$

with some inner function n_α. By virtue of the functional calculus this implies

$$(T - \alpha I)n_\alpha(T) = (I - \bar{\alpha}T)m_T(T) = O;$$

because n_α is not a multiple of m_T, so we have $n_\alpha(T) \neq O$, and therefore $T - \alpha I$ is not invertible. Hence $\alpha \in \sigma(T)$. Thus we have

$$s_T \cap D \subset \sigma(T). \tag{5.2}$$

By virtue of (5.1) and (5.2) it only remains to prove

$$s_T \cap C \subset \sigma(T), \tag{5.3}$$

or, equivalently,

$$C \backslash s_T \supset \rho(T) \cap C, \tag{5.4}$$

$\rho(T)$ denoting the resolvent set of T.

We start to this end with the factorization

$$m_T(\lambda) = \varkappa B(\lambda)S(\lambda) \qquad (|\varkappa| = 1), \tag{5.5}$$

where $B(\lambda)$ is a Blaschke product and $S(\lambda)$ is of the form

$$S(\lambda) = \exp\left(-\int_0^{2\pi} \frac{e^{it}+\lambda}{e^{it}-\lambda}\, d\mu_t\right),$$

μ being a nonnegative, finite singular measure (with respect to Lebesgue measure); see (1.11)–(1.13).

In order to prove (5.4) we have to show that if an open arc ω of C is contained in $\rho(T)$, then the function m_T is analytic on ω. Every point of ω is at a positive distance from $\sigma(T)$, therefore it follows from (5.2) that the zeros of m_T cannot accumulate to a point of ω. This ensures that the function $B(\lambda)$ is analytic on ω. Hence it remains to consider $S(\lambda)$. This function is certainly analytic on ω if $\mu(\omega) = 0$, so it suffices to prove this equation.

Now if $\mu(\omega) > 0$ then there exists a closed subarc ω_1 of ω, with $\mu(\omega_1) > 0$. The function[6]

$$m_1(\lambda) = \exp\left(-\int_{(\omega_1)} \frac{e^{it}+\lambda}{e^{it}-\lambda}\, d\mu_t\right)$$

is then an inner divisor of $m_T(\lambda)$; as $|m_1(0)| = \exp[-\mu(\omega_1)] < 1$, the function $m_1(\lambda)$ is not constant. The function $m_2 = m_T/m_1$ is also inner, and we have $m_1 m_2 = m_T$; hence

$$m_1(T)m_2(T) = m_T(T) = O. \tag{5.6}$$

If the operator $m_1(T)$ were invertible, (5.6) would imply $m_2(T) = O$, and hence m_T would be a divisor of m_2, which is impossible. Thus $m_1(T)$ is not invertible: the subspace

$$\mathfrak{H}_1 = \{h:\ h \in \mathfrak{H}, m_1(T)h = 0\}$$

does not reduce to $\{0\}$. $\mathfrak{H}_1$ is obviously invariant (even hyperinvariant) for T. Let $T_1 = T|\mathfrak{H}_1$. Because $m_1(T_1) = m_1(T)|\mathfrak{H}_1$, the minimal function m_{T_1} of T_1 is a divisor of m_1, so it must be of the form

$$m_{T_1}(\lambda) = \exp\left(-\int_0^{2\pi} \frac{e^{it}+\lambda}{e^{it}-\lambda}\, d\mu_{1t}\right)$$

[6] For any subset β of C we denote by (β) the set of points t in $[0, 2\pi)$ for which $e^{it} \in \beta$.

with a nonnegative, singular measure μ_1, having the majorant μ on ω_1 and vanishing elsewhere. This shows that the set s_{T_1} corresponding to the function m_{T_1} is included in the arc ω_1. Relation (5.1), when applied to T_1, then yields $\sigma(T_1) \subset s_{T_1} \subset \omega_1$.

Let $|\nu| > 1$. By virtue of the reciprocal relations

$$g = (\nu I - T)h, \quad h = \sum_{n=0}^{\infty} \nu^{-n-1} T^n g,$$

any invariant subspace for T is transformed by $\nu I - T$ onto itself. Hence it follows that

$$(\nu I_1 - T_1)^{-1} = (\nu I - T)^{-1} | \mathfrak{H}_1.$$

Thus if ν tends to a point ξ of C belonging to $\rho(T)$, $(\nu I_1 - T_1)^{-1}$ remains bounded, and this shows that ξ also belongs to $\rho(T_1)$. In particular, we have $\omega_1 \subset \rho(T_1)$. When combined with the preceding result $\sigma(T_1) \subset \omega_1$ this gives $\sigma(T_1) \subset \rho(T_1)$, an obvious absurdity. This contradiction followed from the hypothesis that $\mu(\omega) > 0$. Thus we must have $\mu(\omega) = 0$, and so the proof of the theorem is complete.

Corollary 5.2. *Let T be a contraction of class C_0 and let $\lambda_1, \lambda_2, \ldots$ be its different eigenvalues in D. Then $\sum(1 - |\lambda_n|) < \infty$.*

In fact, the values λ_n being the zeros of $m_T(\lambda)$, the convergence of the series in question follows from a theorem of Blaschke on the zeros of a function in H^∞ (cf. HOFFMAN [1], p. 64).

Corollary 5.3. *There exists a $T \in C_0$ such that $\sigma(T) = C$.*

In fact, let μ be a nonnegative, finite, singular measure whose support is the entire unit circle C (e.g., choose a countable dense subset of points of C and assign to each of these points a positive mass so that their sum is finite). The corresponding function

$$m(\lambda) = \exp\left(-\int_0^{2\pi} \frac{e^{it} + \lambda}{e^{it} - \lambda}\, d\mu_t\right),$$

being inner, generates by virtue of Proposition 4.3 a contraction $T \in C_0$ such that $m_T = m$. By Theorem 5.1 we have $\sigma(T) = s_T$, and in this example s_T equals C.

6 Minimal function and invariant subspaces

1. The problem of constructing invariant subspaces for an arbitrary (not necessarily normal or compact) operator, and in this way reducing the study of the operator to the study of its "parts" in these subspaces, has been resolved only in certain cases; one such case has been considered in Sec. II.5. We deal with this problem in this book repeatedly and from various aspects; in the present section we show that for contractions T of class C_0 an approach to this problem is offered by the factorizations of the minimal function m_T.

Proposition 6.1. *Let T be a contraction of class C_0 on $\mathfrak{H}$ $(\neq \{0\})$, and let $\mathfrak{H}_1$ be a nontrivial subspace of $\mathfrak{H}$, invariant for T. Let*

$$T = \begin{bmatrix} T_1 & * \\ O & T_2 \end{bmatrix}$$

be the triangulation of T corresponding to the decomposition $\mathfrak{H} = \mathfrak{H}_1 \oplus \mathfrak{H}_2$. Then T_1 and T_2 are also of class C_0, their minimal functions m_{T_1} and m_{T_2} are divisors of m_T, and m_T is a divisor of $m_{T_1} m_{T_2}$.

Proof. We have obviously

$$T_1^n = T^n|\mathfrak{H}_1 \quad \text{and} \quad T_2^n = P_2 T^n|\mathfrak{H}_2 \qquad (n = 0, 1, \ldots), \tag{6.1}$$

where P_2 denotes the orthogonal projection of $\mathfrak{H}$ into $\mathfrak{H}_2$. Because T is c.n.u. (indeed, $T \in C_0$), so are T_1 and T_2, and hence $u(T_1)$ and $u(T_2)$ exist for every $u \in H^\infty$; from (6.1) we obtain easily that

$$u(T_1) = u(T)|\mathfrak{H}_1, \quad u(T_2) = P_2 u(T)|\mathfrak{H}_2. \tag{6.2}$$

Choosing $u = m_T$, this yields $m_T(T_1) = O$ and $m_T(T_2) = O$; hence T_1 and T_2 belong to C_0 and their minimal functions m_{T_1}, m_{T_2} are divisors of m_T.

From (6.2) it also follows that

$$(m_{T_1} m_{T_2})(T)h_1 = (m_{T_1} m_{T_2})(T_1)h_1 = m_{T_1}(T_1) m_{T_2}(T_1)h_1 = 0 \quad \text{for} \quad h_1 \in \mathfrak{H}_1$$

and

$$P_2 m_{T_2}(T)h_2 = m_{T_2}(T_2)h_2 = 0 \quad \text{for} \quad h_2 \in \mathfrak{H}_2.$$

By the second relation, $h = m_{T_2}(T)h_2$ is orthogonal to $\mathfrak{H}_2$ and so it belongs to $\mathfrak{H}_1$. Hence:

$$(m_{T_1} m_{T_2})(T)h_2 = m_{T_1}(T) m_{T_2}(T)h_2 = m_{T_1}(T)h = m_{T_1}(T_1)h = 0.$$

Thus we have $(m_{T_1} m_{T_2})(T)h = 0$ for $h \in \mathfrak{H}_1$ as well as for $h \in \mathfrak{H}_2$, and hence for every $h \in \mathfrak{H}$ (i.e., $(m_{T_1} m_{T_2})(T) = O$). Consequently, m_T is a divisor of $m_{T_1} m_{T_2}$. This concludes the proof.

Proposition 6.2. *Let T be a contraction of class C_0 on $\mathfrak{H}$ $(\neq \{0\})$, and let $\mathfrak{H}_\alpha$ $(\alpha \in A)$ be invariant subspaces for T. $\mathfrak{H}_\vee = \bigvee_{\alpha \in A} \mathfrak{H}_\alpha$ is then also invariant for T, and the minimal function of $T_\vee = T|\mathfrak{H}_\vee$ is the least common inner multiple of the minimal functions of the contractions $T_\alpha = T|\mathfrak{H}_\alpha$ $(\alpha \in A)$.*

Proof. The invariance of $\mathfrak{H}_\vee$ is obvious; the fact that T_α and $T_\vee$ are of class C_0 follows from the preceding proposition. Because T_α can also be considered as the restriction of $T_\vee$ to $\mathfrak{H}_\alpha$, $m_{T_\vee}$ is divisible by m_{T_α} for each $\alpha \in A$, and hence $m_{T_\vee}$ is also divisible by the least common inner multiple $m_\vee$ of the functions m_{T_α} $(\alpha \in A)$. On the other hand we have

$$m_\vee(T_\vee)h_\alpha = m_\vee(T)h_\alpha = m_\vee(T_\alpha)h_\alpha = 0 \quad \text{for} \quad h_\alpha \in \mathfrak{H}_\alpha$$

by (6.2) and because $m_\vee$ is a multiple of m_{T_α}. The equation $m_\vee(T_\vee)h = 0$ holds then for the elements h of the span of the subspaces $\mathfrak{H}_\alpha$, that is, for $h \in \mathfrak{H}_\vee$; hence $m_\vee(T_\vee) = O$. Consequently, $m_{T_\vee}$ is a divisor of $m_\vee$. Being divisors of each other, the (inner) functions $m_{T_\vee}$ and $m_\vee$ *coincide.*

2. It is convenient also to admit the operator $T = O$ on the trivial space $\{0\}$ as belonging to the class C_0; its minimal function is then obviously the constant function 1. It is obvious that except for this trivial case no minimal function is constant.

The following theorem establishes a mutual correspondence between the inner divisors of the minimal function of T and some of the invariant subspaces for T.

Theorem 6.3. *Let T be a contraction on $\mathfrak{H}$, of class C_0, with minimal function m_T. To each inner divisor m of m_T we let correspond the subspace*

$$\mathfrak{H}_m = \{h\colon\ h \in \mathfrak{H}, m(T)h = 0\}; \tag{6.3}$$

$\mathfrak{H}_m$ is hyperinvariant for T. Set $\mathfrak{H}'_m = \mathfrak{H} \ominus \mathfrak{H}_m$. The contractions T_m $(= T|\mathfrak{H}_m)$ and T'_m appearing in the triangulation

$$T = \begin{bmatrix} T_m & X_m \\ O & T'_m \end{bmatrix}, \qquad \mathfrak{H} = \mathfrak{H}_m \oplus \mathfrak{H}'_m$$

have their minimal functions equal to m and $m' = m_T/m$, respectively. Moreover, we have

(i) $\mathfrak{H}_m = \begin{cases} \{0\} & \text{if } m = 1, \\ \mathfrak{H} & \text{if } m = m_T. \end{cases}$

(ii) $\mathfrak{H}_{m_1} \subset \mathfrak{H}_{m_2}$ *if, and only if, m_1 is a divisor of m_2.*

(iii) *If $\{m_\alpha\}$ $(\alpha \in A)$ is a finite or infinite system of inner divisors of m_T, with the largest common inner divisor $m_\wedge$ and the least common inner multiple $m_\vee$, then*

$$\bigcap_\alpha \mathfrak{H}_{m_\alpha} = \mathfrak{H}_{m_\wedge} \quad \textit{and} \quad \bigvee_\alpha \mathfrak{H}_{m_\alpha} = \mathfrak{H}_{m_\vee}.$$

Proof. The hyperinvariance of $\mathfrak{H}_m$ for T follows immediately from the fact that if a bounded operator S commutes with T then it also commutes with $m(T)$.

By virtue of (6.2) and of the definition (6.3) of $\mathfrak{H}_m$ we have $m(T_m) = m(T)|\mathfrak{H}_m = O$. Consequently, m_{T_m} is a divisor of m, and hence

$$m = m_{T_m} p$$

for some inner function p.

On the other hand, for every $h \in \mathfrak{H}$ we have $m(T)\ m'(T)h = m_T(T)h = 0$, and hence $h_1 = m'(T)h$ belongs to $\mathfrak{H}_m$. Applying (6.2) one obtains for $h \in \mathfrak{H}'_m$

$$m'(T'_m)h = P'm'(T)h = P'h_1 = 0,$$

where P' is the orthogonal projection of $\mathfrak{H}$ onto $\mathfrak{H}'_m$. So we have $m'(T'_m) = O$, and hence $m_{T'_m}$ is a divisor of m'; that is,

$$m' = m_{T'_m} q$$

for some inner function q. These results imply

$$m_T = mm' = m_{T_m} p \cdot m_{T'_m} q = m_{T_m} m_{T'_m} \cdot pq.$$

On the other hand it follows from Proposition 6.1 that m_T is a divisor of $m_{T_m} m_{T'_m}$. Consequently, $pq = 1$, and hence $p = 1$, $q = 1$, $m = m_{T_m}$, $m' = m_{T'_m}$.

Assertion (i) is obvious. As to (ii), it follows readily from the definition (6.3) that if m_1 is a divisor of m_2, then $\mathfrak{H}_{m_1} \subset \mathfrak{H}_{m_2}$. Conversely, $\mathfrak{H}_{m_1} \subset \mathfrak{H}_{m_2}$ implies $T_{m_1} = T_{m_2}|\mathfrak{H}_{m_1}$ and hence

$$m_2(T_{m_1}) = m_2(T_{m_2})|\mathfrak{H}_{m_1} = O,$$

because $m_2(T_2) = O$ by the definition of T_{m_2}. Hence we conclude that the minimal function of T_{m_1}, that is, m_1, is a divisor of m_2.

Finally, as to (iii), let us observe first that because $m_\wedge$ is a divisor of m_α we have $\mathfrak{H}_{m_\wedge} \subset \mathfrak{H}_{m_\alpha}$ $(\alpha \in A)$; consequently

$$\mathfrak{H}_{m_\wedge} \subset \bigcap_\alpha \mathfrak{H}_{m_\alpha}. \tag{6.4}$$

Let h be an element of this intersection, that is, such that $m_\alpha(T)h = 0$ for all $\alpha \in A$. By virtue of Lemma 4.5 we have then $m_\wedge(T)h = 0$, thus $h \in \mathfrak{H}_{m_\wedge}$. Thus the opposite of the inclusion (6.4) is also valid, and so there is equality in (6.4).

Let us turn now to $m_\vee$. Because $m_\vee$ is a multiple of m_α, we have $\mathfrak{H}_{m_\alpha} \subset \mathfrak{H}_{m_\vee}$ for all $\alpha \in A$; hence also

$$\bigvee_\alpha \mathfrak{H}_{m_\alpha} \subset \mathfrak{H}_{m_\vee}. \tag{6.5}$$

Let $\mathfrak{H}^+$ denote the left-hand side of (6.5): this is an invariant subspace for T. Let $\mathfrak{H}^0 = \mathfrak{H}_{m_\vee} \ominus \mathfrak{H}^+$, and consider the triangulation

$$T_{m_\vee} = \begin{bmatrix} T^+ & X \\ O & T^0 \end{bmatrix}$$

corresponding to the decomposition $\mathfrak{H}_{m_\vee} = \mathfrak{H}^+ \oplus \mathfrak{H}^0$. Setting

$$m'_\alpha = m_\vee / m_\alpha \qquad (\alpha \in A) \tag{6.6}$$

we have

$$m_\vee(T) = m_\alpha(T) m'_\alpha(T);$$

hence

$$m'_\alpha(T)\mathfrak{H}_{m_\vee} \subset \mathfrak{H}_{m\alpha}$$

and consequently

$$m'_\alpha(T^0)\mathfrak{H}^0 = P^0 m'_\alpha(T_{m_\vee})\mathfrak{H}^0 \subset P^0 m'_\alpha(T)\mathfrak{H}_{m_\vee} \subset P^0\mathfrak{H}_{m\alpha} \subset P^0\mathfrak{H}^+ = \{0\};$$

that is, $m'_\alpha(T^0) = O$. As this is true for every $\alpha \in A$, we have by Lemma 4.5 that $m'_\wedge(T^0) = O$, where $m'_\wedge$ denotes the largest common inner divisor of the functions m'_α $(\alpha \in A)$. Now by definition (6.6) the functions m'_α are relatively prime (i.e., $m'_\wedge = 1$), and this implies that $\mathfrak{H}^0 = \{0\}$, $\mathfrak{H}_{m_\vee} = \mathfrak{H}^+$; thus there is equality in (6.5). This completes the proof of Theorem 6.3.

3. One of the consequences of this theorem is that if m_1 and m_2 are relatively prime inner divisors of m_T, then

$$\mathfrak{H}_{m_1} \cap \mathfrak{H}_{m_2} = \{0\} \quad \text{and} \quad \mathfrak{H}_{m_1} \vee \mathfrak{H}_{m_2} = \mathfrak{H}_{m_1 m_2}. \tag{6.7}$$

This is the case in particular if there exist $u_1, u_2 \in H^\infty$ such that

$$m_1 u_1 + m_2 u_2 = 1. \tag{6.8}$$

Then we have even

$$\mathfrak{H}_{m_1} \dot{+} \mathfrak{H}_{m_2} = \mathfrak{H}_{m_1 m_2}, \tag{6.9}$$

the sign $\dot{+}$ denoting direct (not necessarily orthogonal) sum. Indeed, (6.8) implies

$$m_1(T)u_1(T)h + m_2(T)u_2(T)h = h$$

for every $h \in \mathfrak{H}$; now $h_1 = m_2(T)u_2(T)h \in \mathfrak{H}_{m_1}$ for $h \in \mathfrak{H}_{m_1 m_2}$, because

$$m_1(T)h_1 = u_2(T)m_1(T)m_2(T)h = u_2(T)\cdot(m_1 m_2)(T)h = 0,$$

and by analogous reason $h_2 = m_1(T)u_1(T)h \in \mathfrak{H}_{m_2}$.

Thus we have proved the following result.

Proposition 6.4. *Let m_1, m_2 be inner divisors of m_T such that there exist $u_1, u_2 \in H^\infty$ satisfying the equation $m_1 u_1 + m_2 u_2 = 1$. Then we have $\mathfrak{H}_{m_1 m_2} = \mathfrak{H}_{m_1} \dot{+} \mathfrak{H}_{m_2}$.*

Another consequence of Theorem 6.3 is that if m is a nontrivial inner divisor of m_T then the corresponding subspace $\mathfrak{H}_m$ of $\mathfrak{H}$ is also nontrivial. In fact, none of the subspaces $\mathfrak{H}_m$ and $\mathfrak{H}'_m = \mathfrak{H} \ominus \mathfrak{H}_m$ equals $\{0\}$, because the operators T_m and T'_m appearing in the corresponding triangulation of T have the nonconstant minimal functions m and $m' = m_T/m$.

Now we know that every nonconstant inner function has a nontrivial inner divisor, except the functions

$$\varkappa \frac{\lambda - a}{1 - \bar{a}\lambda} \qquad (|\varkappa| = 1, |a| < 1);$$

see Sec. 1.4. If m_T is of this exceptional form, then $\lambda - a = \bar{\varkappa}(1 - \bar{a}\lambda)m_T(\lambda)$, and hence

$$T - aI = \bar{\varkappa}(I - \bar{a}T)m_T(T) = O, \quad T = aI;$$

in this case every subspace of $\mathfrak{H}$ is invariant for T, thus if $\dim \mathfrak{H} > 1$ there are nontrivial invariant subspaces.

We conclude that *if* $\dim \mathfrak{H} > 1$, *every contraction T of class C_0 on $\mathfrak{H}$ has a nontrivial invariant subspace.*

This result can be generalized as follows. Let $\mathfrak{H}'$ and $\mathfrak{H}''$ be two subspaces of $\mathfrak{H}$, invariant for T, and such that

$$\mathfrak{H}' \supset \mathfrak{H}'' \quad \text{and} \quad \dim(\mathfrak{H}' \ominus \mathfrak{H}'') > 1.$$

By virtue of Proposition 6.1 we have $T' = T|\mathfrak{H}' \in C_0$. Let

$$T' = \begin{bmatrix} T'' & * \\ O & T''' \end{bmatrix}$$

be the triangulation of T' corresponding to the decomposition $\mathfrak{H}' = \mathfrak{H}'' \oplus \mathfrak{H}'''$; then $T''' \in C_0$, and as $\dim \mathfrak{H}''' > 1$, there exists in $\mathfrak{H}'''$ a nontrivial subspace $\mathfrak{H}_1'''$, invariant for T'''. $\mathfrak{H}'' \oplus \mathfrak{H}_1'''$ is then a subspace of $\mathfrak{H}'$, invariant for T and properly contained between $\mathfrak{H}''$ and $\mathfrak{H}'$.

Let us state this result.

Theorem 6.5. *Let $\mathfrak{H}'$ and $\mathfrak{H}''$ be invariant subspaces for the contraction $T \in C_0$, such that $\mathfrak{H}' \supset \mathfrak{H}''$ and* $\dim(\mathfrak{H}' \ominus \mathfrak{H}'') > 1$. *Then there exists a subspace, invariant for T, and properly contained between $\mathfrak{H}'$ and $\mathfrak{H}''$.*

7 Characteristic vectors and unicellularity

1. Let $T \in C_0$ and let a be a point of the spectrum of T in D, that is, a zero of the minimal function m_T (cf. Theorem 5.1). Setting

$$b_a(\lambda) = \frac{\lambda - a}{1 - \bar{a}\lambda} \tag{7.1}$$

we have then

$$m_T(\lambda) = b_a^k(\lambda) \cdot m_a(\lambda), \tag{7.2}$$

where $k \geq 1$ and m_a is an inner function such that $m_a(a) \neq 0$. The factors in (7.2) are relatively prime. We show in fact that there exist $u, v \in H^\infty$ such that

$$b_a^k \cdot u + m_a \cdot v = 1.$$

By a homography we can reduce this assertion to the case $a = 0$. Then $m_0(0) \neq 0$ so that $1/m_0(\lambda)$ has a Taylor series expansion around the point 0; let $v(\lambda)$ be the

$(k-1)$th partial sum of this series. Then we have

$$1 = m_0(\lambda)\frac{1}{m_0(\lambda)} = m_0(\lambda)v(\lambda) + \lambda^k u(\lambda)$$

with a function $u(\lambda)$, which is holomorphic in some neighborhood of 0 and whose definition extends by means of the relation $\lambda^k u(\lambda) = 1 - m_0(\lambda)v(\lambda)$ to the whole of D. The function $u(\lambda)$ thus obtained belongs obviously to H^∞. So we have

$$\lambda^k u(\lambda) + m_0(\lambda)v(\lambda) = 1 \qquad (\lambda \in D)$$

with $u, v \in H^\infty$.

Let $\mathfrak{H}_a, \mathfrak{H}'_a$ be the hyperinvariant subspaces corresponding to the factors in (7.2), that is,

$$\mathfrak{H}_a = \{h:\ h \in \mathfrak{H},\ b_a^k(T)h = 0\}, \quad \mathfrak{H}'_a = \{h:\ h \in \mathfrak{H},\ m_a(T)h = 0\}.$$

By virtue of Proposition 6.4 we have

$$\mathfrak{H} = \mathfrak{H}_a \dotplus \mathfrak{H}'_a.$$

Then, by Theorem 6.3, the minimal functions of $T_a = T|\mathfrak{H}_a$ and $T'_a = T|\mathfrak{H}'_a$ are equal to b_a^k and m_a, respectively. Now the conditions

$$b_a^n(T)h = 0 \quad \text{and} \quad (T - aI)^n h = 0$$

for an $h \in \mathfrak{H}$ are obviously equivalent. Hence a is a characteristic value of T of index k, and $\mathfrak{H}_a$ consists precisely of the characteristic vectors of T associated with the characteristic value a (and of 0).[7] As to T'_a, $m_a(a) \neq 0$ implies by Theorem 5.1 that a does not belong to the spectrum of T'_a.

So we have proved the following proposition.

Proposition 7.1. *For a contraction T of class C_0 on $\mathfrak{H}$, the points a of the spectrum in the interior of the unit circle are eigenvalues of T. As a characteristic value of T, a has finite index, equal to its multiplicity as zero of the minimal function $m_T(\lambda)$. Furthermore, for every such a, $\mathfrak{H}$ decomposes into the (not necessarily orthogonal) direct sum of two subspaces, hyperinvariant for T, say $\mathfrak{H}_a$ and $\mathfrak{H}'_{a}$, so that $\mathfrak{H}_a$ consists of the characteristic vectors of T associated with the value a (and of the vector 0), whereas $T'_a = T|\mathfrak{H}'_a$ has a in its resolvent set.*

There is some interest in the following result.

---

[7] A complex number a is called a *characteristic value* of T if there exists a vector $h \neq 0$ such that $(T-aI)^n h = 0$ for n large enough; h is called a *characteristic vector* associated with the value a. If $(T-aI)^n h = 0$, but $h_0 = (T-aI)^{n-1}h \neq 0$, then h_0 is obviously an eigenvector of T corresponding to the eigenvalue a; thus every characteristic value is also an eigenvalue. The characteristic value a is said to have a (finite) *index* k if $(T-aI)^k h = 0$ for every characteristic vector h associated with a, but $(T-aI)^{k-1}h_0 \neq 0$ for at least one characteristic vector h_0 associated with a.

Proposition 7.2. *Let $T \in C_0$. In order that the characteristic vectors of T associated with the points of the spectrum of T in D span the whole space $\mathfrak{H}$, it is necessary and sufficient that the minimal function m_T be a Blaschke product.*

Proof. Let us suppose that $\mathfrak{H} = \bigvee_a \mathfrak{H}_a$, a running over the points of the spectrum of T in D, and $\mathfrak{H}_a$ being the subspace formed by the characteristic vectors of T associated with the value a (and by the vector 0). Let B denote the Blaschke factor in the factorization of type (1.11) of the inner function m_T. If $k = k(a)$ is the index of a as a characteristic value of T, then B is divisible by b_a^k so that for $h \in \mathfrak{H}_a$ we have

$$B(T)h = (B/b_a^k)(T) \cdot b_a^k(T)h = 0;$$

this implies $B(T) = O$ because $\mathfrak{H} = \bigvee_a \mathfrak{H}_a$. Then m_T must be a divisor of B; hence $m_T = B$.

Conversely, if we suppose that m_T is a Blaschke product, then m_T is obviously equal to the least common inner multiple of its factors b_a^k, so that by Theorem 6.3 we have

$$\bigvee \mathfrak{H}_a = \mathfrak{H}_{m_T} = \mathfrak{H}.$$

This concludes the proof of Proposition 7.2.

2. A bounded operator T on $\mathfrak{H}$ is said to be *unicellular* if its invariant subspaces are totally ordered by inclusion, that is, if for any two of these subspaces, say $\mathfrak{M}$ and $\mathfrak{N}$, we have $\mathfrak{M} \subset \mathfrak{N}$ or $\mathfrak{M} \supset \mathfrak{N}$.

Observe that *if T is unicellular then so is T^**, a consequence of the fact that if $\{\mathfrak{M}\}$ is the collection of invariant subspaces for T then $\{\mathfrak{M}^{\perp}\}$ is the collection of invariant subspaces for T^*.

We consider the unicellular contractions T of class C_0.

Proposition 7.3. *Let T be a contraction on $\mathfrak{H}$ of class C_0 and unicellular. Then $m_T(\lambda)$ is of the form*

$$\left(\frac{\lambda - \alpha}{1 - \bar{\alpha}\lambda}\right)^n \qquad (|\alpha| < 1; n \text{ a positive integer}), \tag{7.3}$$

or of the form

$$\exp\left(s\frac{\lambda + \alpha}{\lambda - \alpha}\right) \qquad (|\alpha| = 1; s \text{ a positive real number}), \tag{7.4}$$

accordingly as $\dim \mathfrak{H} = n$ *or* $\dim \mathfrak{H} = \infty$.

Proof. Let us observe first that the minimal function m_T cannot have relatively prime nontrivial inner divisors, for in that case there would exist nontrivial invariant subspace $\mathfrak{H}_1$, $\mathfrak{H}_2$ for T such that $\mathfrak{H}_1 \cap \mathfrak{H}_2 = \{0\}$ (cf. Theorem 6.3 (iii)), which is impossible on account of the unicellularity of T. From the general representation (1.11)–(1.13) of the inner functions we deduce then that m_T is either a Blaschke product all factors of which correspond to the same zero α in D, or an inner function S generated by a nonnegative, finite, singular measure on C, whose support is a one-point set $\{\alpha\}$ $(\alpha \in C)$. Thus the only functions to be considered are those given by (7.3) and (7.4). It remains to show that the case (7.3) occurs precisely if $\dim \mathfrak{H} = n$.

As a unicellular operator T can have at most one eigenvector (up to a scalar factor), it follows in the case $\dim \mathfrak{H} = n$ that T has, with respect to an appropriately chosen basis in $\mathfrak{H}$, a matrix

$$\left.\begin{bmatrix} \alpha & 1 & & & \\ & \alpha & 1 & & \\ & & \ddots & \ddots & \\ & & & \alpha & 1 \\ & & & & \alpha \end{bmatrix}\right\} n$$

consisting of just one Jordan cell (whence the term "unicellular"). If $T \in C_0$, then $T^n \to O$, and hence $|\alpha| < 1$. From this representation of T we obtain that the subspaces

$$\mathfrak{H}_\nu = \{h\colon\ h \in \mathfrak{H}, (T - \alpha I)^\nu h = 0\} \qquad (\nu = 0, \ldots, n) \tag{7.5}$$

satisfy the relations

$$\{0\} = \mathfrak{H}_0 \subset \ldots \subset \mathfrak{H}_n = \mathfrak{H} \quad \text{and} \quad \dim \mathfrak{H}_\nu = \nu, \tag{7.6}$$

and hence in particular $\mathfrak{H}_{n-1} \neq \mathfrak{H}_n$. This is obviously equivalent to the fact that $b_\alpha^n(T) = O$ and $b_\alpha^{n-1}(T) \neq O$, where $b_\alpha(\lambda) = (\lambda - \alpha)/(1 - \bar{\alpha}\lambda)$. Because the spectrum of T is the one-point set $\{\alpha\}$, $m_T(\lambda)$ is equal to a power of $b_\alpha(\lambda)$, so we have necessarily $m_T(\lambda) = b_\alpha^n(\lambda)$.

Let us now show that, conversely, if a unicellular contraction T $(\in C_0)$ on a space $\mathfrak{H}$ is such that $m_T = b_\alpha^n$ with $|\alpha| < 1$ and a natural number n, then $\dim \mathfrak{H} < \infty$. In fact, $b_\alpha^n(T) = O$ implies that $(T - \alpha I)^n = O$ and hence

$$T^n = -\sum_{k=0}^{n-1} \binom{n}{k} (-\alpha)^{n-k} T^k.$$

It follows that for every $h \in \mathfrak{H}$ the subspace $M(h)$ spanned by $h, Th, \ldots, T^{n-1}h$ is invariant for T; clearly $\dim M(h) \leq n$. Let h_0 be chosen so that $M(h_0)$ have maximal dimension; we show that then $M(h_0) = \mathfrak{H}$. In the contrary case there would exist an $h_1 \in \mathfrak{H}$ not belonging to $M(h_0)$. As T is unicellular, $M(h_0)$ is necessarily included in $M(h_1)$, and hence $\dim M(h_1) > \dim M(h_0)$, which is impossible. This proves that $\dim \mathfrak{H} \leq n$. (Actually we have here an equality: this follows from what was proved above for the case of spaces of finite dimension.)

From these results we conclude that the case (7.4) occurs precisely if $\dim \mathfrak{H} = \infty$. This concludes the proof.

Corollary. *For a unicellular contraction of class C_0 on $\mathfrak{H}$, the spectrum $\sigma(T)$ consists of a single point α. We have $|\alpha| < 1$ or $|\alpha| = 1$, accordingly as* $\dim \mathfrak{H}$ *is finite or infinite.*

We add a further remark.

Proposition 7.4. *For a unicellular contraction T of class C_0, on a space $\mathfrak{H}$ with* $\dim \mathfrak{H} = n < \infty$, *the subspaces $\mathfrak{H}_\nu$ defined by* (7.5) *(where $\{\alpha\} = \sigma(T)$) are the only invariant subspaces for T.*

Proof. The invariance (even hyperinvariance) of $\mathfrak{H}_\nu$ for T is obvious. Let $\mathfrak{M}$ be an arbitrary invariant subspace. From the unicellularity of T and from the inclusions (7.6) it follows that $\mathfrak{H}_\mu \subset \mathfrak{M} \subset \mathfrak{H}_{\mu+1}$ for some μ. Because $\dim \mathfrak{H}_\nu = \nu$ for every ν, we have $\mu \leq \dim \mathfrak{M} \leq \mu + 1$ and hence $\mathfrak{M}$ coincides with $\mathfrak{H}_\mu$ or $\mathfrak{H}_{\mu+1}$. This concludes the proof.

Let us remark that in the definition of $\mathfrak{H}_\nu$ the condition $(T - \alpha I)^\nu h = 0$ can be replaced by the equivalent one $b_\alpha^\nu(T)h = 0$, and that the functions b_α^ν $(\nu = 0, \ldots, n)$ are the only inner divisors of $m_T (= b_\alpha^n)$.

Analogous facts can be established in the case of infinite-dimensional spaces. Indeed, consider a contraction T of class C_0 with $\sigma(T) = \{\alpha\}$, $|\alpha| = 1$. Replacing T by $\bar{\alpha}T$ we can reduce our study to the case $\sigma(T) = \{1\}$. The minimal function of T then has the form (7.4) with $\alpha = 1$. These functions play an important role in the sequel, therefore we introduce the shorter notation

$$e_s(\lambda) \equiv \exp\left(s\frac{\lambda+1}{\lambda-1}\right) \qquad (s \geq 0). \tag{7.7}$$

These functions are inner, and the only inner divisors of e_a are the functions e_s with $0 \leq s \leq a$.

Proposition 7.5. *Let T be a contraction of class C_0 on $\mathfrak{H}$, such that*

$$m_T(\lambda) = e_a(\lambda) \quad \textit{with} \quad a = a_T > 0.$$

The subspaces

$$\mathfrak{H}_s = \{h\colon\ h \in \mathfrak{H}, e_s(T)h = 0\} \qquad (0 \leq s \leq a) \tag{7.8}$$

are hyperinvariant for T, and we have

$$\mathfrak{H}_0 = \{0\};\ \mathfrak{H}_a = \mathfrak{H};\ \mathfrak{H}_{s_1} \subsetneqq \mathfrak{H}_{s_2} \quad \textit{for} \quad 0 \leq s_1 < s_2 \leq a; \tag{7.9}$$

$$\mathfrak{H}_s = \bigcap_{x>s} \mathfrak{H}_x \qquad \textit{for} \quad 0 \leq s < a; \tag{7.10}$$

$$\mathfrak{H}_s = \overline{\bigcup_{x<s} \mathfrak{H}_x} \qquad \textit{for} \quad 0 < s \leq a. \tag{7.11}$$

Thus the corresponding orthogonal projections E_s form a strictly increasing, continuous spectral family in the interval $[0, a]$.

If, moreover, T is unicellular then the subspaces $\mathfrak{H}_s$ are the only invariant subspaces for T.

Proof. Let us set $T_s = T|\mathfrak{H}_s$ $(0 \le s \le a)$. By Theorem 6.3 we have $m_{T_s} = e_s$; hence it follows in particular that if $s_1 \ne s_2$ then $T_{s_1} \ne T_{s_2}$ and consequently $\mathfrak{H}_{s_1} \ne \mathfrak{H}_{s_2}$. The other assertions in (7.9) are obvious.

Relations (7.10) and (7.11) follow from Theorem 6.3(iii) by the obvious fact that the function e_s is the largest common inner divisor of the functions e_x $(x > s)$, and the least common inner multiple of the functions e_x $(x < s)$.

It remains only to consider the case of a unicellular T. Let $\mathfrak{M}$ be a nontrivial invariant subspace for T and set

$$s = \sup\{x\colon\ \mathfrak{H}_x \subset \mathfrak{M}\}.$$

By (7.9) and (7.11) we also have then $\mathfrak{H}_s \subset \mathfrak{M}$; consequently $s < a$. If $s < x < a$, $\mathfrak{H}_x$ is not included in $\mathfrak{M}$, therefore $\mathfrak{M}$ must be included in $\mathfrak{H}_x$. On account of (7.10) we then have

$$\mathfrak{M} \subset \bigcap_{x>s} \mathfrak{H}_x = \mathfrak{H}_s$$

also. So we have simultaneously $\mathfrak{H}_s \subset \mathfrak{M}$ and $\mathfrak{M} \subset \mathfrak{H}_s$, and hence $\mathfrak{M} = \mathfrak{H}_s$.

This completes the proof of Proposition 7.5.

An example of an operator T of the type considered in Proposition 7.5 is the operator defined in Proposition 4.3(a) with $m = e_a$. This operator is unicellular. Indeed, by Proposition 4.3(b) any invariant subspace $\mathfrak{H}_1$ of this operator T is given by the formula

$$\mathfrak{H}_1 = m_2(H^2 \ominus m_1 H^2) = m_2 H^2 \ominus e_a H^2,$$

where $e_a = m_1 m_2$ and m_1, m_2 are inner functions. Therefore in the definition of $\mathfrak{H}_1$ we can take $m_1 = e_s$, $m_2 = e_{a-s}$ for some $0 \le s \le a$. So the invariant subspaces of T are of the form

$$e_{a-s} H^2 \ominus e_a H^2$$

which clearly increase with s. This concludes the proof of the unicellularity of T.

3. We conclude this section by establishing two properties of arbitrary unicellular operators on a (complex) Hilbert or Banach space $\mathfrak{H}$. The second property concerns "cyclic" vectors, whose definition is as follow.

Definition. For an operator T on $\mathfrak{H}$ a vector h is called *cyclic* if the vectors $T^n h$ $(n = 0, 1, \ldots)$ span $\mathfrak{H}$.

Proposition 7.6. *For every bounded unicellular operator T, the invariant subspaces are hyperinvariant.*

Proof. Let $\mathfrak{L}$ be an invariant subspace for T and let X be a bounded operator commuting with T. Then for any complex λ the subspace $\mathfrak{L}_\lambda = \overline{(\lambda I - X)\mathfrak{L}}$ is also invariant for T, and hence we have either $\mathfrak{L}_\lambda \subset \mathfrak{L}$ or $\mathfrak{L}_\lambda \supset \mathfrak{L}$, depending on λ. If the first case occurs for at least one λ, then $X\mathfrak{L} \subset \mathfrak{L}$. If the second case occurs for every

λ, then we have in particular $\mathfrak{L} \supset (\lambda I - X)^{-1}\mathfrak{L}$ for $|\lambda| > \|X\|$ and hence, by virtue of the relation

$$X = \frac{1}{2\pi i} \oint_{|\lambda|=\rho} \lambda(\lambda I - X)^{-1}\, d\lambda \qquad (\rho > \|X\|),^{8}$$

we also have $\mathfrak{L} \supset X\mathfrak{L}$. Thus $X\mathfrak{L} \subset \mathfrak{L}$ holds in every case. This proves that $\mathfrak{L}$ is hyperinvariant for T.

Proposition 7.7. *If T is a unicellular operator on a separable space $\mathfrak{H}$*[9] *then the set $\mathfrak{S}$ of noncyclic vectors for T is a union of at most countably many proper*[10] *invariant subspaces of $\mathfrak{H}$.*

Proof. If $\{\mathfrak{M}_\alpha\}$ is the family of all proper invariant subspaces of T then clearly $\mathfrak{S} = \cup\mathfrak{M}_\alpha$. Let $\{x_i\}$ be a countable dense subset of $\mathfrak{S}$. For each i choose an $\mathfrak{M}_{\alpha_i}$ such that x_i is in $\mathfrak{M}_{\alpha_i}$, and set $\mathfrak{S}' = \cup\mathfrak{M}_{\alpha_i}$. If $\mathfrak{S}' = \mathfrak{S}$ then $\mathfrak{S}$ is exhibited as a countable union of subspaces of $\mathfrak{H}$. If $\mathfrak{S}'$ is a proper subset of $\mathfrak{S}$ then there exists an $\mathfrak{M}_\beta$ that is contained in none of the $\mathfrak{M}_{\alpha_i}$ and hence contains all the $\mathfrak{M}_{\alpha_i}$. But then $\mathfrak{M}_\beta$ contains $\{x_i\}$ and therefore $\mathfrak{S}$; thus we have $\mathfrak{S} = \mathfrak{M}_\beta$. The proof is now complete.

Because a proper subspace is nowhere dense in $\mathfrak{H}$ we can apply the Baire category theorem (cf., e.g., DUNFORD AND SCHWARTZ [1] p. 20) to obtain from Proposition 7.7 the following corollary.

Corollary 7.8. *For any countable set $\{T_i\}$ of unicellular operators on a separable space $\mathfrak{H}$ there exist vectors which are cyclic for each T_i. In particular, if T is unicellular on a Hilbert space then there exist vectors which are cyclic both for T and T^*.*

8 One parameter semigroups

1. We now indicate how our functional calculus can be applied to the study of continuous one-parameter semigroups $\{T(s)\}_{s\geq 0}$ of contractions on the space $\mathfrak{H}$; see Secs. I.8.2 and I.10.2.

Let us recall that for such a semigroup the *(infinitesimal) generator* A, defined by

$$Ah = \lim_{s\to+0} \frac{1}{s}[T(s) - I]h \tag{8.1}$$

whenever this limit exists, is a closed operator with domain $\mathfrak{D}(A)$ dense in $\mathfrak{H}$, $A - I$ is boundedly invertible and, moreover, A determines the semigroup $\{T(s)\}$ uniquely. (Theorem of Hille and Yosida, cf. HILLE [1] p. 238, or [*Func. Anal.*] Secs. 142 and 143.)

---

[8] For $|\lambda| > \|X\|$ we have $\lambda(\lambda I - X)^{-1} = I + \lambda^{-1}X + \lambda^{-2}X^2 + \cdots$.

[9] It is easy to see that if $\mathfrak{H}$ is a nonseparable Hilbert space then no operator on $\mathfrak{H}$ can be unicellular.

[10] That is, different from $\mathfrak{H}$.

Let us note that $\|T(s)\| \leq 1$ implies

$$\mathrm{Re}(Ah,h) \leq 0 \tag{8.2}$$

for $h \in \mathfrak{D}(A)$; see (I.10.6). This implies further

$$\|(A+I)h\|^2 - \|(A-I)h\|^2 = 4\,\mathrm{Re}(Ah,h) \leq 0 \quad \text{for} \quad h \in \mathfrak{D}(A);$$

hence we obtain that the operator T defined by

$$T = (A+I)(A-I)^{-1} \tag{8.3}$$

is a contraction (its domain of definition is the whole of $\mathfrak{H}$, for so is the domain of $(A-I)^{-1}$).

We call this operator T the *cogenerator* of the semigroup $\{T(s)\}$; this is justified by the fact that T determines A, and hence also $\{T(s)\}$, uniquely. In fact, (8.3) implies that

$$T - I = (A+I)(A-I)^{-1} - I = 2(A-I)^{-1}; \tag{8.4}$$

hence we see that $(T-I)^{-1}$ exists and equals $\frac{1}{2}(A-I)$, and therefore

$$A = (T+I)(T-I)^{-1}. \tag{8.5}$$

The existence of the (generally unbounded) operator $(T-I)^{-1}$ simply means that 1 is not an eigenvalue of T.

By virtue of what we said above, the cogenerator T can replace the generator A in the study of the semigroup $\{T(s)\}$; the advantage of using T instead of A is obvious: T is a bounded operator (indeed, a contraction), whereas A is not bounded in general. We show that several properties of the semigroup are reflected in a striking way by analogous properties of the cogenerator.

The following problems should be addressed first: (1) to characterize the contractions T that are cogenerators of some continuous one-parameter semigroups $\{T(s)\}$ of contractions; and (2) to make explicit the relations between T and $\{T(s)\}$ without referring to the infinitesimal generator A.

Before formulating the solution of these problems, let us remark that if 1 is not an eigenvalue of a contraction T, then it is not an eigenvalue of the unitary part of T, either. Thus, if E_T is the spectral measure on C corresponding to the unitary part of T, then $E_T(\{1\}) = O$. Consequently, every function $u \in H^\infty$ that is defined and continuous on $\overline{D}\backslash\{1\}$, belongs to the class H_T^∞. This is in particular the case for the functions e_s $(s \geq 0)$ (cf. (7.7)), which are holomorphic on the whole complex plane except the point 1, and satisfy

$$|e_s(\lambda)| \leq 1 \text{ on } D \quad \text{and} \quad |e_s(\lambda)| = 1 \text{ on } C\backslash\{1\}.$$

Theorem 8.1. *Let T be a contraction on $\mathfrak{H}$. In order that there exist a continuous semigroup $\{T(s)\}_{s\geq 0}$ of contractions whose cogenerator equals T, it is necessary and sufficient that* 1 *not be an eigenvalue of T. If this is the case, then T and $\{T(s)\}$*

determine each other by the relations

$$T(s) = e_s(T) \qquad (s \geq 0) \tag{8.6}$$

and

$$T = \lim_{s \to +0} \varphi_s(T(s)), \tag{8.7}$$

where

$$\varphi_s(\lambda) = \frac{\lambda - 1 + s}{\lambda - 1 - s} = \frac{1-s}{1+s} - \frac{2s}{1+s}\sum_{n=1}^{\infty} \frac{\lambda^n}{(1+s)^n} \in A. \tag{8.8}$$

Proof. We have already observed that cogenerators are contractions for which 1 is not an eigenvalue. Let T be an arbitrary contraction not having the eigenvalue 1. Applying Theorem 2.3 we deduce from the obvious relations

$$e_0(\lambda) = 1, \quad e_s(\lambda)e_t(\lambda) = e_{s+t}(\lambda); \quad |e_s(\lambda)| \leq 1 \quad \text{for} \quad \lambda \in D$$

and

$$\lim_{s \to +0} e_s(\lambda) = 1 \quad \text{for} \quad \lambda \in \overline{D} \backslash \{1\},$$

that the operators $e_s(T)$ $(s \geq 0)$ form a continuous semigroup of contractions. Let us denote its generator by A' and its cogenerator by T'. We show that $T' = T$. To this end we consider the functions

$$u_s(\lambda) = (\varphi_s \circ e_s)(\lambda) = [e_s(\lambda) - 1 + s][e_s(\lambda) - 1 - s]^{-1} \qquad (s > 0). \tag{8.9}$$

It is easy to show that they are holomorphic and bounded by 1 on D, continuous on $\overline{D} \backslash \{1\}$, and such that

$$\lim_{s \to +0} u_s(\lambda) = \lambda \quad \text{for} \quad \lambda \neq 1.$$

Applying again Theorem 2.3 we obtain that $u_s \in H_T^\infty$, and that

$$\|u_s(T)\| \leq 1 \ (s \geq 0) \quad \text{and} \quad \lim_{s \to +0} u_s(T) = T. \tag{8.10}$$

On the other hand, (8.9) implies

$$u_s(T) = (\varphi_s \circ e_s)(T) = \varphi_s[e_s(T)]. \tag{8.11}$$

By virtue of (8.9) we also have

$$u_s(T)\left[\frac{1}{s}(e_s(T) - I) - I\right] = \frac{1}{s}(e_s(T) - I) + I;$$

applying both sides to an arbitrary element h of the domain of A' and letting $s \to +0$ we obtain, in view of (8.10), that

$$T(A' - I)h = (A' + I)h;$$

thus $T(A'-I) = A'+I$ and $T = (A'+I)(A'-I)^{-1}$; that is, $T = T'$.

Equation (8.6) follows from the fact that a semigroup is determined uniquely by its cogenerator.

Remark. It follows from Proposition I.3.1 that if 1 is not an eigenvalue of the contraction T then it is not an eigenvalue of T^*, either. Moreover, as $e_{\tilde{s}} = e_s$ and hence $e_s(T)^* = e_s(T^*)$, we conclude that *if T is the cogenerator of the semigroup $\{T(s)\}_{s\geq 0}$ then T^* is the cogenerator of the adjoint semigroup $\{T(s)^*\}_{s\geq 0}$.*

Consequently, the strong convergence (8.7) remains valid with T replaced by T^* and $T(s)$ replaced by $T(s)^*$. We use this fact in the following proof.

Proposition 8.2. *A continuous semigroup of contractions $\{T(s)\}_{s\geq 0}$ consists of normal, self-adjoint, or unitary operators, if and only if its cogenerator T is normal, self-adjoint, or unitary, respectively.*

Proof. If $T(s)$ is normal then so is $\varphi_s[T(s)]$. Now

$$\varphi_s[T(s)] \to T \quad \text{and} \quad \varphi_s[T(s)^*] \to T^* \quad \text{for} \quad s \to +0$$

by virtue of (8.7) and the last remark. As $\varphi_s(T(s)^*) = \varphi_s(T(s))^*$ (because $\varphi_{\tilde{s}} = \varphi_s$), it follows that T is also normal.

If $T(s)$ is self-adjoint then so is $\varphi_s(T(s))$ as well as $T = \lim_{s\to+0} \varphi_s(T(s))$.

If $T(s)$ is unitary then $\varphi_s[T(s)]$ is the integral of the function $\varphi_s(\lambda)$ on the unit circle with respect to the spectral measure corresponding to the unitary operator $T(s)$. Now one shows (e.g., by using the Apollonius circles) that

$$1 \geq |\varphi_s(\lambda)| \geq \frac{2-s}{2+s} \quad \text{for} \quad |\lambda| = 1 \quad \text{and} \quad 0 < s \leq 1,$$

which implies

$$\|h\| \geq \|\varphi_s[T(s)]h\| \geq \frac{2-s}{2+s}\|h\| \quad \text{for} \quad h \in \mathfrak{H}, \quad 0 < s \leq 1.$$

Letting $s \to +0$ this yields

$$\|Th\| = \lim_{s\to+0} \|\varphi_s[T(s)]h\| = \|h\|.$$

Thus T is isometric, and because it is also normal, it is unitary.

We turn now to the converse implications. Let us suppose that T is normal, with the spectral representation

$$T = \int \lambda \, dK_\lambda.$$

From Theorem 2.3 (f) we derive that

$$T(s) = e_s(T) = \int e_s(\lambda)\, dK_\lambda, \tag{8.12}$$

and hence $T(s)$ is also normal. As it suffices to restrict the domain of integration to the spectrum of T, and as $e_s(\lambda)$ is real-valued on the real axis and of modulus 1 on the unit circle, it results from (8.12) that if T is self-adjoint, or unitary, then so is $T(s)$ for every $s \geq 0$, respectively. This concludes the proof.

2. It is of interest to note the following particular consequences of (8.12).

If $T(s)$ (and T) are unitary, (8.12) takes the form

$$T(s) = \int_{|\lambda|=1} \exp\left(s\frac{\lambda+1}{\lambda-1}\right) dK_\lambda$$

and this representation reduces to Stone's theorem

$$T(s) = \int_{-\infty}^{\infty} e^{isx} \, dE_x \tag{8.13}$$

by means of the mapping

$$\lambda \to x = -i\frac{\lambda+1}{\lambda-1} \qquad (|\lambda| = 1, \lambda \neq 1)$$

of the unit circle, except the point $\lambda = 1$, onto the real axis.[11]

Similarly, if $T(s)$ (and T) are self-adjoint, we deduce from (8.12) the spectral representation

$$T(s) = \int_0^{\infty} e^{-sy} \, dE_y \tag{8.14}$$

by means of the mapping

$$\lambda \to y = \frac{1+\lambda}{1-\lambda}$$

of the interval $-1 \leq \lambda < 1$ onto the semiaxis $0 \leq y < \infty$.[12]

3. Let T be the cogenerator of the semigroup $\{T(s)\}$ on $\mathfrak{H}$, and let $\mathfrak{H} = \mathfrak{H}_0 \oplus \mathfrak{H}_1$ be the decomposition of the space corresponding to the unitary part $T_0 = T|\mathfrak{H}_0$ and the c.n.u. part $T_1 = T|\mathfrak{H}_1$ of T:

$$T = T_0 \oplus T_1. \tag{8.15}$$

Then we have $u(T) = u(T_0) \oplus u(T_1)$ for every $u \in H_T^\infty$ and in particular for the functions e_s; so we obtain

$$T(s) = T_0(s) \oplus T_1(s) \qquad (s \geq 0) \tag{8.16}$$

[11] This way of obtaining Stone's theorem is close to the one followed by VON NEUMANN [2].

[12] This is a particular case of a theorem of Sz.-Nagy and Hille (cf. [*Func. Anal.*] Sec. 141). The general case can be reduced to this one, because it can be proved directly that if $\{N(s)\}_{s\geq 0}$ is a strongly continuous semigroup of normal operators then there exists a real number α such that $T(s) = e^{-s\alpha}N(s)$ is a contraction for every $s \geq 0$.

for the corresponding semigroups, that is, for

$$T_0(s) = e_s(T_0) \quad \text{and} \quad T_1(s) = e_s(T_1). \tag{8.17}$$

As T_0 is unitary, $T_0(s)$ is also unitary for every $s \geq 0$; see Proposition 8.2. On the other hand, there is no subspace $\mathfrak{H}_1'$ of $\mathfrak{H}_1$ reducing all the operators $T_1(s)$ to unitary ones except the trivial one $\mathfrak{H}_1' = \{0\}$. Indeed, in the contrary case the operators $T_1'(s) = T_1(s)|\mathfrak{H}_1'$ would form a unitary semigroup on $\mathfrak{H}_1'$, so the corresponding cogenerator $T_1' = T_1|H_1'$ would be unitary too (cf. Proposition 8.2), which in the case $\mathfrak{H}_1' \neq \{0\}$ would contradict the fact that T_1 is c.n.u.

It is natural to call a semigroup of contractions $\{T(s)\}_{s\geq 0}$ on $\mathfrak{H}$ *completely nonunitary*, if none of the subspaces $\mathfrak{H}' \neq \{0\}$ of $\mathfrak{H}$ reduces all the operators $T(s)$ to unitary ones.

We can then formulate our result as follows.

Proposition 8.3. *For every continuous one-parameter semigroup of contractions, the canonical decomposition* (8.15) *of the cogenerator induces by means of* (8.16) *and* (8.17) *a decomposition of the semigroup into the orthogonal sum of a unitary semigroup and of a completely nonunitary semigroup.*

The uniqueness of this decomposition follows from the uniqueness of the canonical decomposition of the cogenerator.

9 Unitary dilation of semigroups

1. Let $\{T(s)\}_{s\geq 0}$ be a continuous one-parameter semigroup of contractions on $\mathfrak{H}$, let T be its cogenerator, and let U be the minimal unitary dilation of T acting on the space $\mathfrak{K}$,

$$\mathfrak{K} = \bigvee_{-\infty<n<\infty} U^n \mathfrak{H}. \tag{9.1}$$

As 1 is not an eigenvalue of T, it is not an eigenvalue of U either (*cf.* Proposition II.6.1). Consequently, U is the cogenerator of some continuous semigroup $\{U(s)\}_{s\geq 0}$ of unitary operators, or what amounts to the same thing by setting $U(-s) = U(s)^{-1}$, of a continuous group of unitary operator $\{U(s)\}_{-\infty<s<\infty}$. By virtue of Theorem 8.1 we have

$$T(s) = e_s(t), \quad U(s) = e_s(U) \quad (s \geq 0) \tag{9.2}$$

and

$$T = \lim \varphi_s[T(s)], \quad U = \lim \varphi_s[U(s)] \quad (s \to +0). \tag{9.3}$$

Relations (9.2) imply by Theorem 2.3 (g) that

$$T(s) = \operatorname{pr} U(s) \quad (s \geq 0). \tag{9.4}$$

So we have obtained another proof of Theorem I.8.1 on the existence of a unitary dilation $\{U(s)\}$ of a continuous semigroup of contractions $\{T(s)\}$; this proof derives the existence of the unitary dilation of the semigroup from the existence of the unitary dilation of the cogenerator.

Moreover, the fact that U is the minimal unitary dilation of T implies that $\{U(s)\}$ is the minimal unitary dilation of $\{T(s)\}$, that is,

$$\mathfrak{K}' = \bigvee_{-\infty<s<\infty} U(s)\mathfrak{H}$$

equals $\mathfrak{K}$. Indeed, $\mathfrak{K}'$ reduces each operator $U(t)$, hence also $\varphi_t[U(t)]$ as well as its limit as $t \to +0$; that is, $\mathfrak{K}'$ reduces U. Because $\mathfrak{H}$ is contained in $\mathfrak{K}'$, $U^n\mathfrak{H}$ is also contained in $\mathfrak{K}'$ for $n = 0, \pm 1, \ldots$. On account of (9.1) this implies $\mathfrak{K} \subset \mathfrak{K}'$. Thus $\mathfrak{K}' = \mathfrak{K}$.

We also have

$$\bigvee_{n\geq 0} U^n\mathfrak{H} = \bigvee_{s\geq 0} U(s)\mathfrak{H} \tag{9.5}$$

and, generally,

$$\bigvee_{n\geq 0} U^n\mathfrak{K}_1 = \bigvee_{s\geq 0} U(s)\mathfrak{K}_1 \quad \text{for an arbitrary subset} \quad \mathfrak{K}_1 \text{ of } \mathfrak{K}. \tag{9.6}$$

In fact, if we use the Taylor series expansions of the functions $e_s(\lambda) \in H^\infty$ and $[\varphi_s(\lambda)]^n \in A$ $(s > 0, n > 0)$, say

$$e_s(\lambda) = \sum_{k=0}^{\infty} c_k(s)\lambda^k, \quad [\varphi_s(\lambda)]^n = \sum_{k=0}^{\infty} d_k(s,n)\lambda^k, \tag{9.7}$$

as well as relations (8.6) and (8.7), we obtain

$$U(s) = e_s(U) = \lim_{r\to 1-0} \sum_{k=0}^{\infty} r^k c_k(s)U^k \quad (s > 0) \tag{9.8}$$

and

$$U^n = \lim_{s\to+0} \varphi_s^n[U(s)] = \lim_{s\to+0} \sum_{k=0}^{\infty} d_k(s,n)U(ks) \quad (n > 0); \tag{9.9}$$

hence (9.6) follows in an obvious manner. The dual relation

$$\bigvee_{n\geq 0} U^{*n}\mathfrak{K}_1 = \bigvee_{s\geq 0} U(s)^*\mathfrak{K}_1 \tag{9.10}$$

is derived in the same way, considering the adjoint semigroup $\{U(s)^*\}$.

The following relations can be proved similarly.

$$\bigvee_{n\geq 0} (U^n - T^n)\mathfrak{H}_1 = \bigvee_{s\geq 0} (U(s) - T(s))\mathfrak{H}_1 \tag{9.11}$$

and

$$\bigvee_{n\geq 0} (U^{*n} - T^{*n})\mathfrak{H}_1 = \bigvee_{s\geq 0} (U(s)^* - T(s)^*)\mathfrak{H}_1, \tag{9.12}$$

where $\mathfrak{H}_1$ is an arbitrary subset of $\mathfrak{H}$. In fact, (9.11) results from

$$U(s)-T(s)=e_s(U)-e_s(T)=\lim_{r\to 1-0}\sum_{k=0}^{\infty} r^k c_k(s)(U^k-T^k) \quad (s>0)$$

and

$$U^n-T^n=\lim_{s\to+0}\varphi_s^n[U(s)]-\lim_{s\to+0}\varphi_s^n[T(s)]=\lim_{s\to+0}\sum_{k=0}^{\infty} d_k(n,s)(U(ks)-T(ks)),$$

and (9.12) follows by considering the adjoint semigroups. As a combination of the relations (9.10) and (9.11), and of the relations (9.6) and (9.12), we obtain

$$\bigvee_{m,n\geq 0} U^{*m}(U^n-T^n)\mathfrak{H}_1=\bigvee_{t,s\geq 0} U(t)^*(U(s)-T(s))\mathfrak{H}_1 \tag{9.13}$$

and

$$\bigvee_{m,n\geq 0} U^{m}(U^{*n}-T^{*n})\mathfrak{H}_1=\bigvee_{t,s\geq 0} U(t)(U(s)^*-T(s)^*)\mathfrak{H}_1 \tag{9.14}$$

for an arbitrary subset $\mathfrak{H}_1$ of $\mathfrak{H}$.

These relations are interesting mainly because they indicate once again the symmetry between semigroups and their cogenerators. But they also have an interest all their own. In fact, the space (9.5) is the space of the minimal isometric dilation of T (cf. Sec. I.4), and the spaces (9.11)–(9.14) are, for $\mathfrak{H}_1=\mathfrak{H}$, equal in this order to the spaces

$$\bigoplus_0^{\infty} U^n\mathfrak{L},\quad \bigoplus_0^{\infty} U^{*n}\mathfrak{L}^*,\quad M(\mathfrak{L})=\bigoplus_{-\infty}^{\infty} U^n\mathfrak{L},\quad\text{and}\quad M(\mathfrak{L}^*)=\bigoplus_{-\infty}^{\infty} U^n\mathfrak{L}^*, \tag{9.15}$$

where

$$\mathfrak{L}=\overline{(U-T)\mathfrak{H}}\quad\text{and}\quad \mathfrak{L}^*=\overline{(U^*-T^*)\mathfrak{H}};$$

see Sec. II.1. To this end we just have to observe that

$$U^{n+1}-T^{n+1}=\sum_{k=0}^{n} U^k(U-T)T^{n-k} \qquad (n=0,1,\ldots),$$

and conversely,

$$U^n(U-T)=(U^{n+1}-T^{n+1})-(U^n-T^n)T \qquad (n=0,1,\ldots).$$

Owing to the importance of the spaces (9.15) in the study of a contraction (cf. Chaps. II and VI), the interest of the above relations is apparent.

The following application is connected with Sec. II.3, and deserves particular attention. There we proved by a simple calculation that

$$\|U^{*n}T^n h-U^{*m}T^m h\|^2=\|T^m h\|^2-\|T^n h\|^2 \quad (0\leq m<n)$$

for $h \in \mathfrak{H}$, and we have derived from this that the limit

$$a_h = \lim_{n\to\infty} U^{*n}T^n h \tag{9.16}$$

exists. Now we obtain by an analogous calculation

$$\|U(s)^*T(s)h - U(t)^*T(t)h\|^2 = \|T(t)h\|^2 - \|T(s)h\|^2 \qquad (0 \le t < s)$$

from which follows the existence of the limit

$$a'_h = \lim_{s\to\infty} U(s)^*T(s)h. \tag{9.16$'$}$$

If $\mathfrak{H}_1$ consists of a single element h, let us denote the spaces appearing on the left-hand side and the right-hand side of (9.13) by $\mathfrak{A}_h$ and $\mathfrak{A}'_h$, respectively. From (9.16) and (9.16$'$) we deduce

$$h - a_h = \lim_{n\to\infty} U^{*n}(U^n - T^n)h \in \mathfrak{A}_h, \quad h - a'_h = \lim_{s\to\infty} U(s)^*(U(s) - T(s))h \in \mathfrak{A}'_h.$$

Furthermore we have for any nonnegative integers p,q

$$\begin{aligned}(U^{*n}T^n h, U^{*p}(U^q - T^q)h) &= (T^n h, U^{n-p}(U^q - T^q)h)\\ &= (T^n h, T^{n-p+q}h - T^{n-p}T^q h) = 0\end{aligned}$$

whenever $n \ge p$, and hence $a_h \perp \mathfrak{A}_h$. An analogous reasoning yields $a'_h \perp \mathfrak{A}'_h$. We conclude that $h - a_h$ and $h - a'_h$ are the orthogonal projections of h into $\mathfrak{A}_h$ and $\mathfrak{A}'_h$, respectively. By virtue of (9.13), $\mathfrak{A}_h = \mathfrak{A}'_h$, therefore $a_h = a'_h$.

This proves one-part of the following proposition; the other part can be proved similarly.

Proposition 9.1. *Let $\{T(s)\}_{s\ge 0}$ be a continuous one-parameter semigroup of contractions and let T be its cogenerator. Let $\{U(s)\}$ and U be the corresponding minimal unitary dilations: $U(s) = e_s(U)$. Then we have*

$$\lim_{n\to\infty} U^{*n}T^n h = \lim_{s\to\infty} U(s)^*T(s)h, \quad \lim_{n\to\infty} U^n T^{*n} h = \lim_{s\to\infty} U(s)T(s)^*h \tag{9.17}$$

and consequently

$$\lim_{n\to\infty} \|T^n h\| = \lim_{s\to\infty} \|T(s)h\|, \quad \lim_{n\to\infty} \|T^{*n}h\| = \lim_{s\to\infty} \|T(s)^*h\|. \tag{9.18}$$

2. Let us introduce, for continuous semigroups of contractions $\{T(s)\}_{s\ge 0}$, the classes $C_{0\cdot}$, $C_{1\cdot}$, and so on, in analogy to the corresponding classes for a single contraction T; see Sec. II.3. For example, the semigroup is said to be of class $C_{0\cdot}$ if, for every $h \in \mathfrak{H}$, $T(s)h$ tends to 0 as $s \to +\infty$.

Relations (9.18) imply the following result.

Corollary. *In order that $\{T(s)\}_{s\geq 0}$ belong to one of the classes $C_{0\cdot}$, $C_{1\cdot}$, and so on, of semigroups, it is necessary and sufficient that its cogenerator T belong to the corresponding class of single contractions.*

Let us recall Theorem II.4.1, which asserts the existence of certain triangulations of a contraction, in connection with the classes considered. When applied to the cogenerator T of a semigroup of contractions $\{T(s)\}$, *these triangulations generate triangulations of analogous types of the semigroup.* To this end, we simply observe that if an arbitrary contraction T has a triangulation

$$T = \begin{bmatrix} T_1 & & * \\ & \ddots & \\ O & & T_r \end{bmatrix},$$

then each function $u \in H_T^\infty$ also belongs to $H_{T_i}^\infty$ $(i = 1, \ldots, r)$, and we have

$$u(T) = \begin{bmatrix} u(T_1) & & * \\ & \ddots & \\ O & & u(T_r) \end{bmatrix}.$$

3. Let us return to the first relation (9.18). As $\|T^n h\|$ and $\|T(s)h\|$ are nonincreasing functions of n and s, respectively $(n = 0, 1, \ldots; 0 \leq s < \infty)$, their common limit cannot equal $\|h\|$ unless $\|T^n h\| = \|h\| = \|T(s)h\|$ for every $n \geq 0$ and $s \geq 0$. Hence we obtain the following complement of Proposition 8.2.

Proposition 9.2. *A continuous one-parameter semigroup of contractions consists of isometries if and only if its cogenerator is an isometry.*

Let $\{V(s)\}_{s\geq 0}$ be a continuous semigroup of isometries on the space $\mathfrak{H}$ and let V be its cogenerator. Because V is isometric, it induces a Wold decomposition (cf. Sec. I.1)

$$\mathfrak{H} = \mathfrak{H}_0 \oplus \mathfrak{H}_1 \tag{9.19}$$

with

$$\mathfrak{H}_0 = \bigcap_{n\geq 0} V^n \mathfrak{H} \quad \text{and} \quad \mathfrak{H}_1 = \bigoplus_0^\infty V^n \mathfrak{A}, \quad \mathfrak{A} = \mathfrak{H} \ominus V\mathfrak{H}; \tag{9.20}$$

$\mathfrak{H}_0$ reduces V to a unitary operator V_0, and $\mathfrak{H}_1$ reduces V to a unilateral shift V_1. (One of the subspaces $\mathfrak{H}_0, \mathfrak{H}_1$ may be equal to the trivial subspace $\{0\}$.) In the corresponding decomposition

$$V(s) = V_0(s) \oplus V_1(s), \quad \text{where} \quad V_0(s) = e_s(V_0) \quad \text{and} \quad V_1(s) = e_s(V_1),$$

$\{V_0(s)\}$ is a unitary semigroup (because V_0 is unitary), and $\{V_1(s)\}$ is a c.n.u. semigroup (because V_1 is c.n.u.). Now set

$$\mathfrak{I}_1 = \bigcap_{s\geq a} V_1(s)\mathfrak{H}_1 \quad (a \geq 0). \tag{9.21}$$

The subspace $V_1(s)\mathfrak{H}_1$ is a nonincreasing function of s, that is,

$$V_1(t)\mathfrak{H}_1 = V_1(s)V_1(t-s)\mathfrak{H}_1 \subset V_1(s)\mathfrak{H}_1 \quad \text{for} \quad 0 \leq s < t,$$

thus definition (9.21) does not depend on a. So we have for $t \geq 0$:

$$V_1(t)\mathfrak{I}_1 = V_1(t) \bigcap_{s\geq 0} V_1(s)\mathfrak{H}_1 = \bigcap_{s\geq 0} V_1(t+s)\mathfrak{H}_1 = \bigcap_{s\geq t} V_1(s)\mathfrak{H}_1 = \mathfrak{I}_1;$$

hence it follows that the subspace $\mathfrak{I}_1$ reduces the isometries $V_1(t)$ $(t \geq 0)$ to unitary operators (cf. Sec. I.1) and consequently $V_1|\mathfrak{I}_1$ is also unitary. As V_1 is c.n.u. this implies $\mathfrak{I}_1 = \{0\}$. On the other hand, as we have obviously $V_0(s)\mathfrak{H}_0 = \mathfrak{H}_0$ for every $s \geq 0$, it follows that

$$\bigcap_{s\geq 0} V(s)\mathfrak{H} = \bigcap_{s\geq 0} V_0(s)\mathfrak{H}_0 \oplus \bigcap_{s\geq 0} V_1(s)\mathfrak{H}_1 = \mathfrak{H}_0.$$

Thus for $\mathfrak{H}_0$ the following two representations are valid,

$$\mathfrak{H}_0 = \bigcap_{n\geq 0} V^n\mathfrak{H} = \bigcap_{s\geq 0} V(s)\mathfrak{H}, \tag{9.22}$$

one more relation where a semigroup and its cogenerator play a symmetric role.

Let us consider the case $\mathfrak{H}_1 \neq \{0\}$, that is,

$$\mathfrak{d} = \dim \mathfrak{A} \geq 1;$$

$\mathfrak{d}$ is the multiplicity of the unilateral shift V_1 and we also call it *the multiplicity of the c.n.u. semigroup of isometries* $\{V_1(s)\}$. The cardinal number $\mathfrak{d}$ determines V_1—hence also $\{V_1(s)\}$—up to unitary equivalence.

An example of a completely nonunitary continuous semigroup of isometries is the *continuous unilateral shift* $\{v(s)\}_{s\geq 0}$ on the space $L^2(0,\infty;\mathfrak{N})$ of functions $f(x)$ with values in a Hilbert space $\mathfrak{N}$, defined by

$$v(s)f(x) = f(x-s) \quad \text{for} \quad s \geq x, \quad \text{and } f(x) = 0 \quad \text{for} \quad 0 \leq x < s.$$

We show that the multiplicity of this semigroup of isometries equals $\dim \mathfrak{N}$.

The fact that $\{v(s)\}_{s\geq 0}$ is a continuous semigroup of isometries, is obvious. Moreover, we have

$$v(s)^* f(x) = f(x+s) \quad (s \geq 0).$$

Hence it follows that

$$\|v(s)^* f\|^2 = \int_s^\infty \|f(\xi)\|^2_{\mathfrak{N}}\, d\xi \to 0 \quad \text{as} \quad s \to \infty;$$

this shows that the semigroup $\{v(s)\}$ is c.n.u. Direct calculation gives that the generator a of $\{v(s)\}$ is defined by

$$af(x) = -(d/dx)f(x)$$

for the functions $f \in L^2(0,\infty;\mathfrak{N})$ which are absolutely continuous and such that $f(0) = 0$ and $(d/dx)f(x) \in L^2(0,\infty;\mathfrak{N})$. It then follows that the cogenerator $v = (a+I)(a-I)^{-1}$ and its adjoint v^* are given explicitly by the formulas

$$(vf)(x) = f(x) - 2e^{-x}\int_0^x f(\xi)e^{\xi}\,d\xi, \quad (v^*f)(x) = f(x) - 2e^{x}\int_x^\infty f(\xi)e^{-\xi}\,d\xi.$$

Because

$$L^2(0,\infty;\mathfrak{N}) \ominus vL^2(0,\infty;\mathfrak{N}) = \{f\colon\ f \in L^2(0,\infty;\mathfrak{N}), v^*f = 0\}$$

and equation $v^*f = 0$ obviously has the only solutions

$$f(x) = e^{-x}h \quad (h \in \mathfrak{N}),$$

we conclude that, indeed, the multiplicity of the semigroup $\{v(s)\}$ equals $\dim \mathfrak{N}$.

We have proved the following result.

Theorem 9.3. *Every continuous semigroup $\{V(s)\}_{s\geq 0}$ of isometries on the space $\mathfrak{H}$ is the orthogonal sum of a continuous semigroup of unitary operators and of a completely nonunitary semigroup of isometries. The latter one is unitarily equivalent to the continuous unilateral shift on $L^2(0,\infty;\mathfrak{N})$, where $\mathfrak{N}$ is any Hilbert space whose dimension equals that of $\mathfrak{H} \ominus V\mathfrak{H}$, V denoting the cogenerator of the semigroup $\{V(s)\}$.*

It is understood that one of the components may be absent: the corresponding subspace may reduce to $\{0\}$.

This theorem has an easy consequence for continuous groups $\{U(s)\}_{-\infty<s<\infty}$ of unitary operators on a space $\mathfrak{H}$, for which there exists an "outgoing" subspace, that is, a subspace $\mathfrak{H}_0$ such that

(i) $U(s)\mathfrak{H}_0 \subset \mathfrak{H}_0$ for all $s > 0$,
(ii) $\bigcap_{s>0} U(s)\mathfrak{H}_0 = \{0\}$,
(iii) $\bigvee_{s<0} U(s)\mathfrak{H}_0 = \mathfrak{H}$.

The prototype of a group with these properties is the *continuous bilateral shift* $\{u(s)\}_{-\infty<x<\infty}$ on the space $L^2(-\infty,\infty;\mathfrak{N})$ of functions $f(x)$ with values in a Hilbert space $\mathfrak{N}$, defined by

$$u(s)f(x) = f(x-s) \qquad (-\infty < s < \infty).$$

Here we have the outgoing space $L^2(0,\infty;\mathfrak{N})$ (embedded in the natural way as a subspace in $L^2(-\infty,\infty;\mathfrak{N})$).

Proposition 9.4. *Every continuous group $\{U(s)\}_{-\infty<s<\infty}$ of unitary operators on $\mathfrak{H}$, for which there is an outgoing subspace $\mathfrak{H}_0$, is unitarily equivalent to the continuous bilateral shift $\{u(s)\}_{-\infty<s<\infty}$ on $L^2(-\infty,\infty;\mathfrak{N})$, where $\mathfrak{N}$ is any Hilbert space whose dimension equals that of $\mathfrak{H}_0 \ominus V\mathfrak{H}_0$, V denoting the cogenerator of the semigroup* *formed by the restrictions* $V(s) = U(s)|\mathfrak{H}_0$ $(s \geq 0)$. *Moreover, the unitary operator effectuating this unitary equivalence can be chosen so that it maps* $\mathfrak{H}_0$ *onto the subspace* $L^2(0,\infty;\mathfrak{N})$ *of* $L^2(-\infty,\infty;\mathfrak{N})$.

Proof. By virtue of (i) and (ii), $\{V(s)\}_{s\geq 0}$ is a c.n.u. semigroup of isometries on $\mathfrak{H}_0$; thus by Theorem 9.3 it is unitarily equivalent to the continuous unilateral shift $\{v(s)\}s \geq 0$ on $L^2(0,\infty;\mathfrak{N})$. Let τ be the map of $\mathfrak{H}_0$ onto $L^2(0,\infty;\mathfrak{N})$, which effectuates this unitary equivalence. For any $f \in \mathfrak{H}$ such that $f \in U(-s)\mathfrak{H}_0$ for some $s > 0$, let us define

$$\tau f = u(-s) \cdot \tau(U(s)f);$$

by the unitarity of $U(s)$ and $u(s)$ (for all real s) it follows that τ is thereby extended in a unique way to an isometric map of a dense subset of $\mathfrak{H}$ (use property (iii)) onto a dense subset of $L^2(-\infty,\infty;\mathfrak{N})$. By taking closures, we arrive at an extension of τ (denoted by the same letter) that maps $\mathfrak{H}$ unitarily on $L^2(-\infty,\infty;\mathfrak{N})$ and satisfies the condition $\tau \cdot U(s) = u(s) \cdot \tau$ for all real s.

10 Notes

The results on the Hardy classes of scalar valued functions dealt with in Sec. 1 are mostly classical, (see, e.g., HOFFMAN [1]). Proposition 1.1, generalizing the theorem of Beurling, is new; also see SZ.-N.–F. [VI], Théorème 2. Propositions 1.3–1.5 appear in SZ.-N.–F. [VI] and [VII].

It has been apparent since the paper of SZ.-NAGY [I] that the existence of a unitary dilation for every contraction T provided a possibility for an extended functional calculus on the basis of the well-known functional calculus for the unitary operators. A detailed study of such a calculus was given first by SCHREIBER [2] and then by SZ.-N.–F. [III]. It is in the last-mentioned paper that the functional relations between a continuous one-parameter semigroup of contractions and its cogenerator first appear; these relations form the subject of Sec. 8 of the present chapter.

At this stage the theory had a limited range, due to the fact that one did not know general criteria in order that the minimal unitary dilation U of T has an absolutely continuous spectrum, except the rather strong condition $\|T\| < 1$ found by SCHREIBER [1]. This limitation was removed by the discovery (cf. SZ.-N.–F. [IV]) of the fact that U has an absolutely continuous spectrum for every completely nonunitary T. This made it possible to construct the functional calculus for contractions such as presented in Secs. 2–3 above; see SZ.-N.–F. [VI].

It should be mentioned that in the particular case of an isometry (or, what amounts to the same by using Cayley transforms, in the case of a maximal symmetric operator), a functional calculus somewhat related to ours was proposed earlier by PLESSNER [1]–[3].

The essential difference between the Riesz–Dunford functional calculus (cf. [*Func. Anal.*] Chap. XI) and our functional calculus for contractions T is that the analytic functions which we admit may not be regular at some points of the boundary of the spectrum of T, for example at a finite number of points of C not belonging to the point spectrum of T. A functional calculus of similar type was proposed by

FOIAŞ [2]; this is founded upon the notion of spectral sets (introduced by VON NEUMANN, cf. [*Func. Anal.*] Chap. XI), and allows analytic functions that may have a finite number of singular points at the boundary of the spectrum, but not belonging to the point spectrum. An application of this functional calculus to the theory of contraction semigroups has also been given in FOIAŞ [2], [3].

Proposition 2.2 on the maximality of our functional calculus is new; it generalizes a former result of FOIAŞ [6].

For our functional calculus the "spectral mapping theorem" is also valid, at least if T is a c.n.u. contraction and $u(\lambda)$ is a function belonging to H^∞ and also continuous at the points of $\sigma(T) \cap C$. Then we have the relation $\sigma(u(T)) = u(\sigma(T))$; see FOIAŞ AND MLAK [1].

The contractions of class C_0 and their minimal functions were introduced in SZ.-N.–F. [VII], where most of the results of Secs. 4–7 also appeared. Proposition 7.5 was proved first in SZ.-N.–F. [XI], whereas the results concerning the continuous semigroups of contractions appeared in SZ.-N.–F. [III], and those concerning continuous semigroups of isometries (Theorem 9.3) in SZ.-NAGY [8]. Propositions 7.6 and 7.7 are due to ROSENTHAL [1]; the existence of cyclic vectors for unicellular operators on Hilbert space was also proved in GOHBERG AND KREĬN [7], pp. 52–53.

Proposition 9.4 was obtained by SINAĬ [1] as a consequence of the following theorem of VON NEUMANN [3].

Theorem. *Let $\{U(s)\}$ and $\{V(s)\}$ be two one-parameter continuous groups of unitary operators on $\mathfrak{H}$, satisfying the Weyl commutation relation*

$$(*) \qquad U(s)V(t) = e^{-ist}V(t)U(s) \qquad (-\infty < s,t < \infty).$$

Then there exist a Hilbert space $\mathfrak{N}$ and a unitary operator from $\mathfrak{H}$ to $L^2(-\infty,\infty;\mathfrak{N})$ transforming $\{U(s),V(s)\}$ to $\{u(s),v(s)\}$ defined by

$$(u(s)f)(x) = f(x-s), \quad (v(s)f)(x) = e^{isx}f(x),$$

the (auxiliary) Hilbert space $\mathfrak{N}$ being determined up to unitary equivalence.

Conversely, this theorem can be obtained from Proposition 9.4 (cf. LAX AND PHILLIPS [2]). Indeed, if $V(s) = \int_{-\infty}^{\infty} e^{i\lambda s}\, dE_\lambda$ is the Stone representation of the group $\{V(s)\}$, then it follows from $(*)$ readily that

$$U(s)E_\lambda = E_{\lambda+s}U(s) \qquad (-\infty < s,\lambda < \infty).$$

Setting $\mathfrak{H}_0 = (I-E_0)\mathfrak{H}$ we have therefore

$$\binom{*}{*} \qquad U(s)\mathfrak{H}_0 = (I-E_s)\mathfrak{H} \qquad (\subset \mathfrak{H}_0 \text{ for } s \geq 0),$$

and this makes it apparent that $\mathfrak{H}_0$ is an outgoing subspace for the group $\{U(s)\}$. So we have, up to a unitary equivalence,

$$\mathfrak{H} = L^2(-\infty,\infty;\mathfrak{N}), \quad \mathfrak{H}_0 = L^2(0,\infty;\mathfrak{N}), \quad (U(s)f)(x) = f(x-s),$$

and therefore $U(s)\mathfrak{H}_0 = L^2(s,\infty;\mathfrak{N})$. Now the projection from $L^2(-\infty,\infty;\mathfrak{N})$ onto its subspace $L^2(s,\infty;\mathfrak{N})$ is the multiplication by $1-\chi_s(x)$, where $\chi_s(x)$ is the characteristic function of the interval $-\infty < x < s$. So we get from $\binom{*}{*}$ that $(E_s f)(x) = \chi_s(x)f(x)$, and hence we conclude that

$$(V(s)f)(x) = \int_{-\infty}^{\infty} e^{i\lambda s} d\chi_\lambda(x) \cdot f(x) = e^{isx} f(x).$$

In connection with this chapter also see COOPER [1]; FOIAŞ [1]; FOIAŞ, GEHÉR, AND SZ.-NAGY [1]; HELSON [1]; LAX AND PHILLIPS [1], [2]; MASANI [3], [4]; MLAK [5], [7]; PHILLIPS [3]; SCHREIBER [4]; and SZ.-NAGY [3], [5]–[7], [13].

11 Further results

1. There has been much progress in the study of the class C_0, and Chap. X is dedicated to some of this material. Here we only mention that analogues of this class have been found with multiply connected regions in the plane in place of the disk D; see BALL [3]. Much of the theory developed for the disk was transferred to this context in ZUCCHI [1]; see also PATA AND ZUCCHI [1]. Some results in the context of the bidisk D^2 are found in R. YANG [3], based on the theory started in DOUGLAS AND YANG [1] and R. YANG [1,2].

Operators of class C_0 with finite defect indices also have been studied from the point of view of their polar decompositions in WU [7] and their singular unitary dilations in WU AND TAKAHASHI [2].

Some of the techniques developed for the study of the class C_0 were extended to operators of class $C_{\cdot 0}$, at least when one of the defect indices is finite; see SZ.-N.–F. [24] for operators with both defect indices finite, and SZ.-NAGY [15] for operators with one finite defect index.

2. As we have seen, the spectral mapping theorem $\sigma(u(T)) = u(\sigma(T))$ is valid if T is a completely nonunitary contraction and $u \in H^\infty$ extends continuously to $\sigma(T) \cap C$. If this continuity hypothesis is relaxed, spectral mapping can fail spectacularly, as shown, for instance, in FOIAS AND PEARCY [1] and BERCOVICI, FOIAS, AND PEARCY [2]. These results are based on the characterization given in SZ.-N.–F. [22] of invertible operators of the form $u(T)$.

3. A remarkable connection between the functional calculus developed in this chapter and the existence of invariant subspaces has been discovered after the seminal work in BROWN [1]. For this discussion, we fix a contraction T on $\mathfrak{H}$, whose unitary part has absolutely continuous spectral measure relative to arclength on C. We have therefore $H_T^\infty = H^\infty$. Assume for the moment that, for some $\lambda \in D$, there exist vectors $h, k \in \mathfrak{H}$ satisfying the identity

$$(u(T)h,k) = u(\lambda) \tag{11.1}$$

for every polynomial u (and, by continuity of the functional calculus, for every $u \in H^\infty$). Denote by $\mathfrak{H}_h$ and $\mathfrak{H}_{Th}$ the invariant subspaces for T generated by h and

Th, respectively. Clearly then $k \perp \mathfrak{H}_{Th}$, and $(h,k) = 1$. It follows that one of these invariant subspaces is proper. Note that (11.1) can be rewritten as

$$(u(T)h,k) = \frac{1}{2\pi}\int_0^{2\pi} u(e^{it})f(e^{it})\,dt, \quad u \in H^\infty, \tag{11.2}$$

where the function $f \in L^1$ is defined by $f(e^{it}) = 1/(1-\lambda e^{-it})$. The existence of invariant subspaces can thus be deduced from the existence, for an arbitrary $f \in L^1$, of vectors $h,k \in \mathfrak{H}$ satisfying (11.2). This method is used in BROWN, CHEVREAU, AND PEARCY [1],[2] and BROWN [2] to prove the following remarkable result.

Theorem 11.1. *Assume that T is a contraction such that $\sigma(T) \supset C$. Then T has nontrivial invariant subspaces.*

A complete exposition of the proof appears in BERCOVICI [5]. The proof proceeds by a reduction to the case when (11.2) can indeed be solved for every $f \in L^1$. It is easy to see that (11.1) implies the inequality

$$\|u(T)\| \geq |u(\lambda)|, \quad u \in H^\infty.$$

Thus, if (11.2) can be solved for every $f \in L^1$, we must have

$$\|u(T)\| = \|u\|_\infty, \quad u \in H^\infty,$$

so the functional calculus must be an isometry. The following converse was proved in BERCOVICI [4] and CHEVREAU [1].

Theorem 11.2. *Assume that T is a contraction for which the functional calculus is isometric on H^∞. Then the equation* (11.2) *can be solved for every $f \in L^1$.*

Theorem 11.1 has been extended by AMBROZIE AND MÜLLER [1] to the case of polynomially bounded operators on a Banach space. This is an improvement even in the Hilbert space situation, as seen from the examples in PISIER [1].

The techniques used in these papers generally yield a wealth of information about the structure of the invariant subspace lattice of T. The following result is in BROWN AND CHEVREAU [1].

Theorem 11.3. *If the functional calculus for T is isometric on H^∞, then T is a reflexive operator.*

See Sec. IX.3 for a discussion of reflexivity, as well as a special case of this result. The monograph by BERCOVICI, FOIAS, AND PEARCY [1] contains further information on the rich dilation theory arising from these methods.

4. For contractions T such that $XT = SX$, where S is the unilateral shift and X has dense range, it was shown in TAKAHASHI [3] that $\{T\}'' = \{u(T) : u \in H^\infty\}$. Here $\{T\}''$ denotes the double commutant of T. See also Theorem X.4.2 (3) for a related result in the case of operators of class C_0; this result is from SZ.-N.–F. [26]. In SZ.-N.–F. [28] it is shown that $\{T\}' = \{u(T) : u \in H^\infty\}$ provided that T is of class C_{10}, it has a cyclic vector, and $I - T^*T$ has finite trace.

Chapter IV

Extended Functional Calculus

1 Calculation rules

1. We extend our functional calculus for a contraction T on $\mathfrak{H}$ so that certain unbounded functions are also allowed. Let us recall the definitions of the classes H_T^∞ and K_T^∞ as given in Secs. 2 and 3 of the preceding chapter: H_T^∞ consists of the functions $u \in H^\infty$ for which the strong operator limit $u(T) = \lim_{r \to 1-0} u_r(T)$ exists, and K_T^∞ consists of those functions $u \in H_T^\infty$ for which $u(T)^{-1}$ exists and is densely defined in $\mathfrak{H}$. The class H_T^∞ is an algebra, and the class K_T^∞ is multiplicative.

Definition. Let N_T be the class of the functions φ (meromorphic on the open unit disc) that admit a representation of the form

$$\varphi = \frac{u}{v} \quad \text{with} \quad u \in H_T^\infty \quad \text{and} \quad v \in K_T^\infty \tag{1.1}$$

(note that $v \not\equiv 0$). For such a function φ we define

$$\varphi(T) = v(T)^{-1} u(T). \tag{1.2}$$

As H_T^∞ is an algebra and K_T^∞ is multiplicative, it follows from the relations

$$\alpha \frac{u}{v} = \frac{\alpha u}{v}, \quad \frac{u_1}{v_1} + \frac{u_2}{v_2} = \frac{u_1 v_2 + u_2 v_1}{v_1 v_2}, \quad \frac{u_1}{v_1} \cdot \frac{u_2}{v_2} = \frac{u_1 u_2}{v_1 v_2} \tag{1.3}$$

that the class N_T is also an algebra.

Theorem 1.1. (i) *The definition* (1.2) *of* $\varphi(T)$ *does not depend upon the particular choice of the functions* u, v *in the representation* (1.1). *The operator* $\varphi(T)$ *is closed, with dense domain in* $\mathfrak{H}$, *and commutes with* T *as well as with every bounded operator* S *commuting with* T, *that is,*

$$TS = ST \quad \textit{implies} \quad \varphi(T)S \supset S\varphi(T).$$

B.Sz.-Nagy et al., *Harmonic Analysis of Operators on Hilbert Space*, Universitext,
DOI 10.1007/978-1-4419-6094-8_4,

In particular we have

$$v(T)^{-1}u(T) \supset u(T)v(T)^{-1} \quad \textit{for} \quad u \in H_T^\infty, v \in K_T^\infty. \tag{1.4}$$

Furthermore, we have

(ii) $(c\varphi)(T) = c\varphi(T)$ *for all constants c.*

(iii) $(\varphi_1 + \varphi_2)(T) \supset \varphi_1(T) + \varphi_2(T)$, *with equality, for example, if*

$$\mathfrak{D}[\varphi_2(T)] \supset \mathfrak{D}[(\varphi_1 + \varphi_2)(T)],^1$$

and hence in particular if $\varphi_2 \in H_T^\infty$;

(iv) $(\varphi_1\varphi_2)(T) \supset \varphi_1(T)\varphi_2(T)$, *with equality, for example, if*

$$\mathfrak{D}[\varphi_2(T)] \supset \mathfrak{D}[(\varphi_1\varphi_2)(T)],$$

and hence in particular if $\varphi_2 \in H_T^\infty$. *We have always*

$$\varphi_1(T)\varphi_2(T) = (\varphi_1\varphi_2)(T)\Big|[\mathfrak{D}[\varphi_2(T)] \cap \mathfrak{D}[(\varphi_1\varphi_2)(T)]]; \tag{1.5}$$

in particular we have $(1/\varphi)(T) = \varphi(T)^{-1}$ *if* $\varphi, 1/\varphi \in N_T$.

(v) *If* $\varphi \in N_T$ *then* $\varphi\tilde{} \in N_{T^*}$ *and* $\varphi\tilde{}(T^*) \supset \varphi(T)^*$.

(vi) *If* $\varphi \in N_T$ *and* $\varphi \in H^\infty$ *then* $\varphi \in H_T^\infty$ *and therefore* $\|\varphi(T)\| \leq \|\varphi\|_\infty$.

(vii) *If* $\varphi \in N_T$ *and* $1/\varphi \in H^\infty$ *then* $1/\varphi \in N_T$.

(viii) *Let* $\varphi = u/v$ *with* $u \in H^\infty$, $v \in E^\infty$, *and let* $w \in H^\infty$, $|w(\lambda)| < 1$ *on D. Let us suppose, furthermore, that the sets*

$$C_w, \quad w^{-1}(\mathbf{C}_u), \quad w^{-1}(\mathbf{C}_v^0) \qquad \text{(cf. Secs. III.2 and 3)} \tag{1.6}$$

have O measure with respect to the spectral measure E_T *corresponding to the unitary part of T. Then* $w \in H_T^\infty$, *and* $T' = w(T)$ *is a contraction; furthermore we have* $\varphi \circ w \in N_T$, $\varphi \in N_{T'}$, *and* $\varphi(T') = (\varphi \circ w)(T)$.

(ix) *If the contraction T is normal, with the corresponding spectral measure* $K = K_T$, *and if the function* $\varphi = u/v$ *is such that*

$$u \in H^\infty, \quad v \in E^\infty, \quad E_T(C_u) = O, \quad E_T(C_v^0) = O,^2$$

then $\varphi \in N_T$ *and* $\varphi(T)$ *equals the integral* $\varphi^{\mathbf{s}}(T)$ *of* φ *with respect to K.*

Proof. Part (i): Let $\varphi = u/v$ and $\varphi = u'/v'$ be two representations of type (1.1) of the same function φ. The relation $uv' = vu'$ implies

$$u(T)v'(T) = v(T)u'(T),$$

---

[1] If L is a linear transformation, $\mathfrak{D}(L)$ denotes its domain of definition.

[2] If T is c.n.u., $E_T(\omega)$ is equal for every ω to the operator O on the trivial space $\{0\}$.

and hence

$$v(T)^{-1}u(T) = v(T)^{-1}v'(T)^{-1}v'(T)u(T) = v'(T)^{-1}v(T)^{-1}u(T)v'(T) =$$
$$= v'(T)^{-1}v(T)^{-1}v(T)u'(T) = v'(T)^{-1}u'(T).$$

Thus by formula (1.2), $\varphi(T)$ is uniquely determined by φ.

Let us set $u(T) = U, v(T) = V$. The operator

$$\varphi(T) = V^{-1}U$$

is closed, for the hypotheses

$$h_n \in \mathfrak{D}[\varphi(T)], \quad h_n \to h, \quad \varphi(T)h_n = V^{-1}Uh_n \to g \qquad (n \to \infty)$$

imply $Uh_n \to Uh, Uh_n = V\varphi(T)h_n \to Vg$, and thus

$$Uh = Vg,$$

which shows that

$$h \in \mathfrak{D}[\varphi(T)] \quad \text{and} \quad \varphi(T)h = V^{-1}Uh = g.$$

Relation (1.4) follows from the fact that U and V are bounded and commuting. Indeed, we have

$$V^{-1}U \supset V^{-1}UVV^{-1} = V^{-1}VUV^{-1} = UV^{-1}.$$

By virtue of the definition of the class K_T^∞, V^{-1} has domain dense in $\mathfrak{H}$, and hence so does $\varphi(T)$.

If S is a bounded operator commuting with T, then S also commutes with the functions of T of class H_T^∞, for these are limits of polynomials of T. Consequently, we have

$$\varphi(T)S = V^{-1}US = V^{-1}SU \supset V^{-1}SVV^{-1}U = V^{-1}VSV^{-1}U =$$
$$= SV^{-1}U = S\varphi(T),$$

that is, S commutes with $\varphi(T)$.

Parts (ii)–(iv): (ii) is obvious. To obtain (iii) and (iv), let us consider the functions

$$\varphi_k = u_k/v_k \quad (k = 1,2), \quad \text{with} \quad u_k \in H_T^\infty, \quad v_k \in K_T^\infty,$$

and set $U_k = u_k(T)$ and $V_k = v_k(T)$. On account of (1.3) we have, by definition,

$$(\varphi_1 + \varphi_2)(T) = V_1^{-1}V_2^{-1}(U_1V_2 + U_2V_1), \quad (\varphi_1\varphi_2)(T) = V_1^{-1}V_2^{-1}U_1U_2.$$

Using (1.4), we obtain

$$\begin{aligned}(\varphi_1+\varphi_2)(T) &\supset V_1^{-1}V_2^{-1}U_1V_2+V_2^{-1}V_1^{-1}U_2V_1\\ &\supset V_1^{-1}U_1V_2^{-1}V_2+V_2^{-1}U_2V_1^{-1}V_1=\\ &=\varphi_1(T)+\varphi_2(T).\end{aligned}$$

For the same reason,

$$\varphi_1(T)+\varphi_2(T)=[(\varphi_1+\varphi_2)-\varphi_2](T)+\varphi_2(T)\supset[(\varphi_1+\varphi_2)(T)-\varphi_2(T)]+\varphi_2(T).$$

If $\mathfrak{D}[\varphi_2(T)]\supset\mathfrak{D}[(\varphi_1+\varphi_2)(T)]$ this implies

$$\varphi_1(T)+\varphi_2(T)\supset(\varphi_1+\varphi_2)(T);$$

as the opposite inclusion is always valid we conclude that in this case

$$\varphi_1(T)+\varphi_2(T)=(\varphi_1+\varphi_2)(T).$$

From (1.4) it also follows that

$$(\varphi_1\varphi_2)(T)=V_1^{-1}V_2^{-1}U_1U_2\supset V_1^{-1}U_1V_2^{-1}U_2=\varphi_1(T)\varphi_2(T).$$

This clearly implies

$$\mathfrak{D}[\varphi_1(T)\varphi_2(T)]\subset\mathfrak{D}[(\varphi_1\varphi_2)(T)]\cap\mathfrak{D}[\varphi_2(T)].$$

In order to prove (1.5) it only remains to show that the opposite inclusion also holds; that is,

$$h\in\mathfrak{D}[(\varphi_1\varphi_2)(T)]\cap\mathfrak{D}[\varphi_2(T)]\quad\text{implies}\quad h\in\mathfrak{D}[\varphi_1(T)\varphi_2(T)].$$

Now for such h we have

$$\begin{aligned}V_1(\varphi_1\varphi_2)(T)h&=V_1V_1^{-1}V_2^{-1}U_1U_2h=V_2^{-1}U_1U_2h\\ &=V_2^{-1}U_1V_2\varphi_2(T)h=V_2^{-1}V_2U_1\varphi_2(T)h=U_1\varphi_2(T)h,\end{aligned}$$

and this relation implies

$$\varphi_2(T)h\in\mathfrak{D}[\varphi_1(T)]\quad\text{and}\quad\varphi_1(T)\varphi_2(T)h=(\varphi_1\varphi_2)(T)h.$$

The last assertion in (iv) follows by applying (1.5) to the functions φ, $1/\varphi$ in both orders, and by observing that

$$(\varphi\cdot 1/\varphi)(T)=(1/\varphi\cdot\varphi)(T)=1(T)=I.$$

Part (v): If $\varphi = u/v$ with $u \in H_T^\infty$ and $v \in K_T^\infty$, then $\varphi\tilde{} = u\tilde{}/v\tilde{}$ with $u\tilde{} \in H_{T^*}^\infty$ and $v\tilde{} \in K_{T^*}^\infty$, and

$$\begin{aligned}\varphi\tilde{}(T^*) &= v\tilde{}(T^*)^{-1}u\tilde{}(T^*) = [v(T)^*]^{-1}u(T)^* \\ &= [v(T)^{-1}]^*u(T)^* = [u(T)v(T)^{-1}]^* \supset [v(T)^{-1}u(T)]^* = \varphi(T)^*,\end{aligned}$$

where we have used (1.4) and the fact that $A \supset B$ implies $B^* \supset A^*$.

Part (vi): If $\varphi = u/v$ $(u \in H_T^\infty, v \in K_T^\infty)$ and $|\varphi(\lambda)| \leq M$ on D, then the functions u_r, v_r, φ_r $(0 \leq r < 1)$ belong to the class A, and satisfy $u_r = \varphi_r \cdot v_r$ and $|\varphi_r(\lambda)| \leq M$ on D. According to Sec. III.2.1 we have $u_r(T) = \varphi_r(T) \cdot v_r(T)$ and $\|\varphi_r(T)\| \leq M$. This implies that $\varphi_r(T)v(T) = u_r(T) - \varphi_r(T)[v_r(T) - v(T)] \to u(T)$ as $r \to 1-0$. The range of $v(T)$ is dense in $\mathfrak{H}$, thus we conclude that $\varphi_r(T)h$ converges for every $h \in \mathfrak{H}$ and therefore we have $\varphi \in H_T^\infty$. The inequality $\|\varphi(T)\| \leq \|\varphi\|_\infty$ follows by Theorem III.2.1(b).

Part (vii): If $\varphi = u/v$ $(u \in H_T^\infty, v \in K_T^\infty)$ and $|1/\varphi(\lambda)| \leq M$, then it follows by a reasoning similar to the above, that $\|v(T)h\| \leq M\|u(T)h\|$ for every $h \in \mathfrak{H}$. As $v(T)$ is invertible this inequality implies that $u(T)$ is also invertible. Applying this result to the function $\varphi\tilde{} = u\tilde{}/v\tilde{}$ yields that $u\tilde{}(T^*)$ is also invertible. Because $u\tilde{}(T^*) = u(T)^*$, this means that $u(T)$ has dense range in $\mathfrak{H}$. Thus $u \in K_T^\infty$ and therefore $1/\varphi \in N_T$.

Part (viii): By virtue of Theorem III.2.3, $T' = w(T)$ is a contraction and we have that

$$u(T') = (u \circ w)(T), \quad v(T') = (v \circ w)(T)$$

(the operators indicated exist) and $v \in H_{T'}$. Now the third condition (1.6) implies that $v(T')$ is a quasi-affinity, therefore $v \in K_{T'}^\infty$ and $\varphi \in N_{T'}$. Moreover, by Theorem III. 3.4 we have $v \circ w \in K_T^\infty$. Since $\varphi = u/v$ and $\varphi \circ w = (u \circ w)/(v \circ w)$, it follows that $\varphi \circ w \in N_T$, and

$$\varphi(T') = v(T')^{-1}u(T') = [(v \circ w)(T)]^{-1}(u \circ w)(T) = (\varphi \circ w)(T).$$

Part (ix): Under the conditions stated for u and v we deduce from Theorems III.2.3 and III.3.4, by the usual rules of calculation with spectral integrals, that

$$\varphi(T) = v^{\mathfrak{s}}(T)^{-1}u^{\mathfrak{s}}(T) = (u/v)^{\mathfrak{s}}(T) = \varphi^{\mathfrak{s}}(T).$$

This concludes the proof of Theorem 1.1.

The next proposition follows readily from properties (vi) and (vii).

Proposition 1.2. *If $\varphi \in N_T$, the spectrum of $\varphi(T)$ is contained in the closure of the set of the values of $\varphi(\lambda)$ on D:*

$$\sigma[\varphi(T)] \subset \overline{\varphi(D)}.$$

Proof. One just has to observe that if ζ is at distance $d > 0$ from the set $\varphi(D)$, then $|[\zeta - \varphi(\lambda)]^{-1}| \leq 1/d$ on D, and consequently $\zeta I - \varphi(T)$ is boundedly invertible, with $\|\zeta I - \varphi(T)\| \leq 1/d$.

2. The above reasoning makes it clear that it is of considerable interest to know under which conditions there is equality in (1.4). Here is an answer to this question.

Proposition 1.3. (i) *Suppose* u, v *satisfy the following condition.*

$$(\mathrm{Q}) \qquad \begin{cases} u, v \text{ are continuous on } \overline{D}, \text{ holomorphic on } D, \text{ and} \\ \text{have no common zeros in } \overline{D}. \end{cases}$$

Moreover, let $v \in K_T^\infty$, *where* T *is a contraction on* $\mathfrak{H}$.[3] *Then*

$$v(T)^{-1}u(T) = u(T)v(T)^{-1} \tag{1.7}$$

and, for $\varphi = u/v$,

$$\varphi(T)^* = \varphi^{\sim}(T^*). \tag{1.8}$$

(ii) *Let* $\varphi_k = u_k/v_k$, *where* $u_k \in H_T^\infty, v_k \in K_T^\infty$ $(k = 1, 2)$, *and suppose that the pair of functions* u_1 *and* v_2 *satisfies condition* (Q). *Then*

$$(\varphi_1, \varphi_2)(T) = \varphi_1(T)\varphi_2(T).$$

Proof. We know (cf. HOFFMAN [1] p. 88) that if u, v satisfy condition (Q), then there exist functions a, b, continuous on $\overline{D}$ and holomorphic on D, such that

$$au + bv = 1 \quad \text{on} \quad \overline{D}.$$

Let us set $a(T) = A, b(T) = B, u(T) = U$, and $v(T) = V$. Then we have

$$AU + BV = I.$$

Hence from the relation

$$AV^{-1} \subset V^{-1}A,$$

valid by virtue of (1.4), we obtain

$$\begin{aligned} V^{-1}U &= (AU + BV)V^{-1}U = UAV^{-1}U + BVV^{-1}U \\ &\subset UV^{-1}AU + BU = U(V^{-1}AU + B) = UV^{-1}(AU + VB) = UV^{-1}; \end{aligned}$$

combined with the relation

$$UV^{-1} \subset V^{-1}U \qquad \text{(cf. (1.4))}$$

this yields

$$V^{-1}U = UV^{-1},$$

that is, (1.7).

To obtain (1.8) we just have to repeat the argument in the proof of assertion (v) of Theorem 1.1, using now (1.7) instead of (1.4).

---

[3] Because u, v are continuous on $\overline{D}$ and holomorphic on D, we have $u, v \in H_T^\infty$ for every contraction T; see Theorem III.2.3.

This proves (i). To obtain (ii) we observe that, on account of (1.7),

$$v_2(T)^{-1}u_1(T) = u_1(T)v_2(T)^{-1};$$

it follows that

$$(\varphi_1\varphi_2)(T) = v_1(T)^{-1}v_2(T)^{-1}u_1(T)u_2(T) = v_1(T)^{-1}u_1(T)v_2(T)^{-1}u_2(T) \\ = \varphi_1(T)\varphi_2(T).$$

3. If for some function $\varphi \in N_T$ the operator $\varphi(T)$ happens to be bounded, then we can ask whether there exists a function $w \in H_T^\infty$ such that $\varphi(T) = w(T)$. The following example shows that this is not always the case.

Let T be an operator with $\|T\| < 1$ and with nondiscrete spectrum. Then T is c.n.u., and as a consequence of Theorem III.5.1, T does not belong to the class C_0. Choose a number a such that $\|T\| < a < 1$, and set $\varphi(\lambda) = 1/(\lambda - a)$. Clearly $\varphi \in N_T$, and $\varphi(T) = (T - aI)^{-1}$ is a bounded operator. Suppose there exists $w \in H^\infty$ such that $\varphi(T) = w(T)$. Then we have $(T - aI)\ w(T) = I$, and consequently $u(T) = O$ for $u(\lambda) = (\lambda - a)\ w(\lambda) - 1$. Because T does not belong to C_0 this implies that $u(\lambda) \equiv 0$, which contradicts the equation $u(a) = -1$. Thus there exists no $w \in H_T^\infty$ with $w(T) = \varphi(T)$.

2 Representation of $\varphi(T)$ as a limit of $\varphi_r(T)$

In Sec. III.2 we defined the functions of the contraction T of class H_T^∞ as limits of functions of T of class A, that is, by

$$u(T) = \lim_{r \to 1-0} u_r(T).$$

Now for an arbitrary holomorphic function φ on D, the functions φ_r $(0 \le r < 1)$ belong to the class A, so $\varphi_r(T)$ makes sense and one can ask whether $\varphi_r(T)$ tends to $\varphi(T)$ as $r \to 1$, if φ belongs to the class N_T or at least to an appropriate subclass of N_T.

Theorem 2.1. (i) *Let T be a contraction on $\mathfrak{H}$, and let φ be a function belonging to the class N_T and holomorphic on D. Then every vector $h \in \mathfrak{H}$ such that*

$$\sup_{0<r<1} \|\varphi_r(T)h\| < \infty, \tag{2.1}$$

belongs to the domain of definition of $\varphi(T)$ and the weak convergence relation holds:

$$\varphi_r(T)h \rightharpoonup \varphi(T)h \qquad (r \to 1-0). \tag{2.2}$$

(ii) *Suppose the functions u, v satisfy condition* (Q) *of Proposition* 1.3, *v belongs to the class E^{reg} (cf.* Sec. III.1.3*) and does not vanish on C except at the points of a set of E_T-measure O. Then $\varphi = u/v$ belongs to the class N_T and is holomorphic on*

D, condition (2.1) *characterizes the vectors h for which* $\varphi(T)h$ *is defined, and for every such vector h the strong convergence relation holds:*

$$\varphi_r(T)h \to \varphi(T)h \qquad (r \to 1-0). \tag{2.3}$$

Proof. Part (i): Let $\varphi = u/v$ ($u \in H_T^\infty, v \in K_T^\infty$) and let us consider a sequence $\{r_n\}$ tending to $1-0$. From (2.1) it follows that there exists a subsequence $\{\rho_n\}$ for which $\varphi_{\rho_n}(T)h$ converges weakly, say to g:

$$\varphi_{\rho_n}(T)h \rightharpoonup g. \tag{2.4}$$

Because $v_r(T)^*f \to v(T)^*f$ (as $r \to 1-0$) for every $f \in \mathfrak{H}$, we deduce from (2.4) that

$$\begin{aligned}(u_{\rho_n}(T)h, f) &= (v_{\rho_n}(T)\varphi_{\rho_n}(T)h, f)\\ &= (\varphi_{\rho_n}(T)h, v_{\rho_n}(T^*)f) \to (g, v(T)^*f) = (v(T)g, f).\end{aligned}$$

On the other hand, we have $u_r(T)h \to u(T)h$ $(r \to 1-0)$, and hence in particular

$$(u_{\rho_n}(T)h, f) \to (u(T)h, f).$$

Comparing these results we obtain

$$(v(T)g, f) = (u(T)h, f)$$

for every $f \in \mathfrak{H}$, and hence $v(T)g = u(T)h$. This shows that $\varphi(T)h$ exists and equals g.

So we have proved that (2.1) implies that $\varphi(T)h$ exists and that every sequence $r_n \to 1-0$ contains a subsequence $\rho_n \to 1-0$ for which

$$\varphi_{\rho_n}(T)h \rightharpoonup \varphi(T)h \quad (n \to \infty). \tag{2.5}$$

From this we infer that

$$\varphi_r(T)h \rightharpoonup \varphi(T)h \qquad (r \to 1-0).$$

Indeed, in the contrary case there would exist an $f \in \mathfrak{H}$ and a sequence $r_n \to 1-0$ such that

$$|(\varphi_{r_n}(T)h - \varphi(T)h, f)| \geq \varepsilon > 0 \qquad (n = 1, 2, \ldots),$$

and this contradicts the existence of a subsequence satisfying (2.5).

This concludes the proof of (i).

Part (ii): According to our hypotheses the function v is continuous on $\overline{D}$, holomorphic on D, and vanishes at most at the points of a subset of E_T-measure O of C; furthermore the function

$$w(r;\lambda) = \frac{v(\lambda)}{v(r\lambda)}$$

is bounded on $\overline{D}$ by a constant M independent of r $(0 \le r < 1)$. These conditions imply that $v \in K_T^\infty$ (cf. Proposition III.1.3 and Theorem III.3.4),

$$|w(r;e^{it})| \le M,$$

and

$$\lim_{r\to 1-0} w(r;e^{it}) = 1 \quad \text{a.e. and} \quad E_T\text{- a.e. on} \quad C.$$

From Theorem III.2.3 it follows that

$$v_r(T)^{-1}v(T) = w(r;T) \to I \qquad (r \to 1-0).$$

On the other hand, in view of the condition (Q) for u and v, we have by Proposition 1.3

$$\varphi(T) = u(T)v(T)^{-1},$$

and hence every element h of the domain of $\varphi(T)$ is of the form $h = v(T)g$. Consequently, we have

$$\begin{aligned}\varphi_r(T)h &= v_r(T)^{-1}u_r(T)h = v_r(T)^{-1}u_r(T)v(T)g = v_r(T)^{-1}v(T)u_r(T)g \\ &= w(r;T)u_r(T)g \to u(T)g = u(T)v(T)^{-1}h = \varphi(T)h\end{aligned}$$

as $r \to 1-0$. This concludes the proof.

3 Functions limited by a sector

In view of some later applications we consider in this section functions φ whose values lie in some sector of the plane of complex numbers, whose angle does not exceed π.

We begin with functions belonging to the simple class A.

Proposition 3.1. *Let $\varphi \in A$ be such that*

$$|\arg\varphi(\lambda)| \le \alpha\frac{\pi}{2} \quad \textit{for} \quad \lambda \in D \tag{3.1}$$

with $0 \le \alpha \le 1$. Then we have for every contraction T on $\mathfrak{H}$ and for every $h \in \mathfrak{H}$:

$$|\arg(\varphi(T)h,h)| \le \alpha\frac{\pi}{2}. \tag{3.2}$$

If (3.1) *holds with $0 \le \alpha < \frac{1}{2}$ then the following inequalities are valid.*

(i) $\mathrm{Re}\,(\varphi(T)h, \varphi(T)^*h) \ge \cos\alpha\pi \cdot \max\{\|\varphi(T)h\|^2, \|\varphi(T)^*h\|^2\}$.
(ii) $\|\varphi(T)h\| \ge \cos\alpha\pi \cdot \|\varphi(T)^*h\|, \quad \|\varphi(T)^*h\| \ge \cos\alpha\pi \cdot \|\varphi(T)h\|$.
(iii) $\mathrm{Re}(\varphi(T)h, [\mathrm{Re}\varphi(T)]h) \ge \cos^2(\alpha\pi/2) \cdot \|\varphi(T)h\|^2$,
(iv) $(\cos^2(\alpha\pi/2)/\cos\alpha\pi) \cdot \|\varphi(T)h\| \ge \|[\mathrm{Re}\varphi(T)]h\| \ge \cos^2(\alpha\pi/2) \cdot \|\varphi(T)h\|$,
(v) $\|\mathrm{Im}\varphi(T)h\| \le \tan(\alpha\pi/2) \cdot \|\mathrm{Re}\varphi(T)h\|$, *where* $\mathrm{Re}\varphi(T) = [\varphi(T) + \varphi(T)^*]/2$, *and* $\mathrm{Im}\varphi(T) = [\varphi(T) - \varphi(T)^*]/2i$.

Proof. By virtue of (III.2.6) we have $\varphi(T) = \mathrm{pr}\ \varphi(U), U$ being the minimal unitary dilation of T. If E is the spectral measure corresponding to the unitary operator U, then we have for every $h \in \mathfrak{H}$,

$$|\arg(\varphi(T)h,h)| = |\arg(\varphi(U)h,h)| = \left|\arg \int_\varphi^{2\pi} \varphi(e^{it})\, d(E_t h,h)\right|$$
$$\leq \max_t |\arg(\varphi e^{it})| \leq \alpha\frac{\pi}{2},$$

that is, (3.2). If (3.1) holds with $0 \leq \alpha < \frac{1}{2}$ then

$$|\arg[\varphi(\lambda)]^2| \leq \alpha\pi < \frac{\pi}{2}.$$

Hence

$$\mathrm{Re}[\varphi(\lambda)]^2 \geq \cos\alpha\pi \cdot |\varphi(\lambda)|^2$$

and consequently

$$\begin{aligned}\mathrm{Re}(\varphi(T)h,\varphi(T)^*h) &= \mathrm{Re}(\varphi(T)^2h,h) = \mathrm{Re}(\varphi(U)^2h,h)\\ &= \int_0^{2\pi} \mathrm{Re}[\varphi(e^{it})]^2\, d(E_t h,h)\\ &\geq \cos\alpha\pi \cdot \int_0^{2\pi} |\varphi(e^{it})|^2\, d(E_t h,h)\\ &= \begin{cases}\cos\alpha\pi \cdot \|\varphi(U)h\|^2 \geq \cos\alpha\pi \cdot \|\varphi(T)h\|^2,\\ \cos\alpha\pi \cdot \|\varphi(U)^*h\|^2 \geq \cos\alpha\pi \cdot \|\varphi(T)^*h\|^2;\end{cases}\end{aligned}$$

here we have also used the relations

$$\varphi(T)^2 = \mathrm{pr}\ \varphi(U)^2, \quad \varphi(T)^* = \mathrm{pr}\ \varphi(U)^*;$$

see Sec. III.2.1. So we have proved (i). The constant $\cos\alpha\pi$ is the best one, as it can be seen by the example of the constant function $\varphi(\lambda) \equiv \exp(i\alpha\pi/2)$. The other inequalities derive from (i) as follows. Part (ii) is implied by the Schwarz inequality:

$$\mathrm{Re}(\varphi(T)h,\varphi(T)^*h) \leq \|\varphi(T)h\| \cdot \|\varphi(T)^*h\|.$$

Inequality (iii) follows immediately:

$$\begin{aligned}\mathrm{Re}(\varphi(T)h,[\mathrm{Re}\ \varphi(T)]h) &= \frac{1}{2}\|\varphi(T)h\|^2 + \frac{1}{2}\mathrm{Re}(\varphi(T)h,\varphi(T)^*h)\\ &\geq \frac{1}{2}(1+\cos\alpha\pi)\cdot\|\varphi(T)h\|^2 = \cos^2\frac{\alpha\pi}{2}\cdot\|\varphi(T)h\|^2.\end{aligned}$$

The second of the inequalities (iv) is a consequence of (iii), by the Schwarz inequality. The first inequality (iv) follows from the first inequality (ii):

$$\begin{aligned}\|[\mathrm{Re}\varphi(T)]h\| &\le \frac{1}{2}[\|\varphi(T)h\| + \|\varphi(T)^*h\|]\\ &\le \frac{1}{2}\left(1+\frac{1}{\cos\alpha\pi}\right)\cdot\|\varphi(T)h\| = \frac{\cos^2(\alpha\pi/2)}{\cos\alpha\pi}\cdot\|\varphi(T)h\|.\end{aligned}$$

Finally, (v) is derived from (i) as follows. By (i) we have

$$2\,\mathrm{Re}(\varphi(T)h,\varphi(T)^*h) \ge \cos\alpha\pi\cdot[\|\varphi(T)h\|^2+\|\varphi(T)^*h\|^2].$$

Because

$$\cos\alpha\pi = \frac{1-\tan^2(\alpha\pi/2)}{1+\tan^2(\alpha\pi/2)},$$

we obtain

$$\begin{aligned}&\|\varphi(T)h\|^2 - 2\,\mathrm{Re}(\varphi(T)h,\varphi(T)^*h) + \|\varphi(T)^*h\|^2\\ \le&\tan^2\frac{\alpha\pi}{2}\cdot\left[\|\varphi(T)h\|^2 + 2\,\mathrm{Re}(\varphi(T)h,\varphi(T)^*h)+\|\varphi(T)^*h\|^2\right],\end{aligned}$$

and this yields (v).

Theorem 3.2. *Let $\varphi = u/v$ be as in Theorem* 2.1(ii)*; that is, u and v satisfy condition* (Q)*, v belongs to E^{reg}, and does not vanish on C except at the points of a set of E_T-measure O. Let us also suppose that, for $\lambda \in D$,*

$$|\arg\varphi(\lambda)| \le \frac{\alpha\pi}{2} \quad (0\le\alpha\le 1).$$

Inequality (3.2) *is then valid for every h in the domain of $\varphi(T)$. If, moreover, $0 \le \alpha < \frac{1}{2}$ then the operators $\varphi(T)$, $\varphi(T)^*$, and hence also*

$$\mathrm{Re}\ \varphi(T) = \frac{1}{2}[\varphi(T)+\varphi(T)^*] \quad \textit{and} \quad \mathrm{Im}\ \varphi(T) = \frac{1}{2i}[\varphi(T)-\varphi(T)^*]$$

have the same domain $\mathfrak{D}$, and the inequalities (i)–(v) *of Proposition* 3.1 *hold for every $h \in \mathfrak{D}$. Also,* Re $\varphi(T)$ *is then a positive self-adjoint operator.*

Proof. Because $\varphi_r \in A$ for $0 \le r < 1$, we can apply Proposition 3.1 to φ_r. By virtue of Theorem 2.1 we have

$$\varphi(T) = \lim_{r\to 1}\varphi_r(T).$$

and hence inequality (3.2) is valid for $\varphi(T)$ and for every $h \in \mathfrak{D}[\varphi(T)]$. By the same Theorem 2.1, $\mathfrak{D}[\varphi(T)]$ consists of the elements h for which

$$\sup_r \|\varphi_r(T)h\| < \infty. \tag{3.3}$$

Similarly, $\varphi(T)^* = \varphi^\sim(T^*)$ and $\varphi_r(T)^* = (\varphi^\sim)_r(T^*)$, therefore $\mathfrak{D}[\varphi(T)^*]$ consists of the elements h for which

$$\sup_r \|\varphi_r(T)^* h\| < \infty. \tag{3.4}$$

Now if $0 \le \alpha < \frac{1}{2}$, conditions (3.3) and (3.4) are equivalent on account of the inequalities

$$\|\varphi_r(T)h\| \ge \cos\alpha\pi \cdot \|\varphi_r(T)^* h\| \quad \text{and} \quad \|\varphi_r(T)^* h\| \ge \cos\alpha\pi \cdot \|\varphi_r(T)h\|.$$

Thus $\mathfrak{D}[\varphi(T)] = \mathfrak{D}[\varphi(T)^*]$, and because for an element h of this common domain $\mathfrak{D}$ we have $\varphi_r(T)h \to \varphi(T)h$ and

$$\varphi_r(T)^* h = \varphi_r^\sim(T^*)h \to \varphi^\sim(T^*)h = \varphi(T)^* h \quad \text{as} \quad r \to 1-0,$$

inequalities (i)–(v) of Proposition 3.1 are valid in the limit case $r = 1$ also.

From (3.2) it follows that

$$([\mathrm{Re}\, \varphi(T)]h, h) = \mathrm{Re}(\varphi(T)h, h) \ge 0 \quad \text{for} \quad h \in \mathfrak{D},$$

and thus $\mathrm{Re}\, \varphi(T)$ is a positive symmetric operator. It is actually self-adjoint. To this end it suffices to show that

$$(I + \mathrm{Re}\, \varphi(T))\mathfrak{D} = \mathfrak{H}.^4$$

From inequality (v) and from the positivity of $\mathrm{Re}\, \varphi(T)$ it follows that

$$\|\mathrm{Im}\, \varphi(T)h\| \le c\|(I + \mathrm{Re}\, \varphi(T))h\| \quad \text{with} \quad c = \tan\frac{\alpha\pi}{2} < 1;$$

as a consequence we deduce that there exists an operator B, which can even be assumed to be defined everywhere on $\mathfrak{H}$, such that

$$\mathrm{Im}\, \varphi(T) = B[I + \mathrm{Re}\, \varphi(T)] \quad \text{and} \quad \|B\| \le c.$$

Hence

$$I + \varphi(T) = I + \mathrm{Re}\, \varphi(T) + i\mathrm{Im}\, \varphi(T) = (I + iB)(I + \mathrm{Re}\, \varphi(T)).$$

Now $[I + \varphi(T)]\mathfrak{D} = \mathfrak{H}$ because -1 does not belong to the spectrum of $\varphi(T)$ (cf. Proposition 1.2), and $I + iB$ is boundedly invertible because $\|iB\| < 1$. As a result we have

$$(I + \mathrm{Re}\, \varphi(T))\mathfrak{D} = (I + iB)^{-1}(I + \varphi(T))\mathfrak{D} = (I + iB)^{-1}\mathfrak{H} = \mathfrak{H},$$

and this concludes the proof.

---

[4] See Sec. 125 in [*Func. Anal*].

4 Accretive and dissipative operators

1. Let A_0 be an operator, not necessarily bounded, but with domain $\mathfrak{D}(A_0)$ dense in $\mathfrak{H}$. A_0 is said to be *accretive* if

$$\operatorname{Re}(A_0 f, f) \geq 0 \quad \text{for every} \quad f \in \mathfrak{D}(A_0),$$

and *dissipative* if

$$\operatorname{Im}(A_0 f, f) \geq 0 \quad \text{for every} \quad f \in \mathfrak{D}(A_0).$$

As dissipative operators can be derived from accretive ones by multiplication by i, all we say about accretive operators carries over immediately to dissipative operators.

For an accretive A_0 we have

$$\|A_0 f + f\|^2 \geq \|A_0 f\|^2 + \|f\|^2 \geq \|A_0 f - f\|^2, \tag{4.1}$$

and hence we see that $(A_0 + I)f = 0$ implies $f = 0$ Thus $A_0 + I$ is invertible, and furthermore it follows that the operator T_0 defined by

$$T_0(A_0 + I)f = (A_0 - I)f \qquad (f \in \mathfrak{D}(A_0)) \tag{4.2}$$

is a *contraction* of $(A_0 + I)\mathfrak{D}(A_0)$ onto $(A_0 - I)\mathfrak{D}(A_0)$; we have

$$T_0 = (A_0 - I)(A_0 + I)^{-1}. \tag{4.3}$$

Hence $T_0 = I - 2(A_0 + I)^{-1}$ and

$$I - T_0 = 2(A_0 + I)^{-1}; \tag{4.4}$$

thus $I - T_0$ is invertible and

$$A_0 = (I + T_0)(I - T_0)^{-1}. \tag{4.5}$$

We call T_0 the *Cayley transform* of the accretive operator A_0. From the reciprocal relations (4.3), (4.5) it follows that an accretive operator A_1 is a proper extension of A_0 if and only if its Cayley transform T_1 is a proper extension of T_0.

An accretive operator that has no proper accretive extension is said to be *maximal accretive.*

An obvious sufficient condition for the accretive operator A_0 to be maximal is that its Cayley transform T_0 be defined on the whole space $\mathfrak{H}$: that is $\mathfrak{D}(T_0) = (A_0 + I)\mathfrak{D}(A_0) = \mathfrak{D}$. This condition turns out also to be necessary. Moreover, we prove that every accretive operator has a maximal accretive extension. To this end let us suppose that $\mathfrak{D}(T_0) \neq \mathfrak{H}$. Let T be an extension of T_0 to a contraction defined everywhere on $\mathfrak{H}$. (Such a T exists because T_0 extends to the subspace $\overline{\mathfrak{D}(T_0)}$ by continuity and then to $\mathfrak{H}$ by linearity, setting, e.g., $Th = 0$ for $h \perp \mathfrak{D}(T_0)$.) There is no nonzero invariant vector for T. In fact, if h is invariant for T then it is invariant for T^* too (cf. Proposition I.3.1), and we have for every $f \in \mathfrak{D}(A_0)$:

$$\begin{aligned}(h,(A_0+I)f) &= (T^*h,(A_0+I)f) = (h,T(A_0+I)f) \\ &= (h,T_0(A_0+I)f) = (h,(A_0-I)f);\end{aligned}$$

hence $(h,f)=0$, and as $\mathfrak{D}(A_0)$ is dense in $\mathfrak{H}$ this implies that $h=0$.

Now every contraction T on $\mathfrak{H}$ that has no nonzero invariant vector is the Cayley transform of an accretive operator A, indeed of

$$A=(I+T)(I-T)^{-1}. \tag{4.6}$$

As a matter of fact, because $\mathfrak{D}(A)$ consists of the vectors of the form $f=(I-T)g$, we have

$$\operatorname{Re}(Af,f)=\operatorname{Re}((I+T)g,(I-T)g)=\|g\|^2-\|Tg\|^2\geq 0,$$

and these vectors f are dense in $\mathfrak{H}$ because T^* has no nonzero invariant vector. Thus A is accretive, and it is obvious that its Cayley transform equals T. As $\mathfrak{D}(T)=\mathfrak{H}$, we have that A is *maximal accretive*.

Let us also observe that whenever T is a contraction on $\mathfrak{H}$ not having the eigenvalue 1, then so is T^* (cf. Proposition I.3.1). Hence $I-T^*$ is invertible, $(I-T^*)^{-1}=[(I-T)^{-1}]^*$, and

$$\begin{aligned}(I+T^*)(I-T^*)^{-1} &= -I+2(I-T^*)^{-1} \\ &= [-I+2(I-T)^{-1}]^* = [(I+T)(I-T)^{-1}]^*.\end{aligned}$$

Thus the maximal accretive operators corresponding to T and T^* are adjoints of one another. Taking account of Theorem III.8.1 and of the relations (III.8.3) and (III.8.4) between generators and cogenerators of semigroups of contractions, we can state the following result.

Theorem 4.1. *Every accretive operator in $\mathfrak{H}$ has a maximal accretive extension. For an operator A in $\mathfrak{H}$ the following conditions are equivalent.*

(a) *A is maximal accretive.*
(b) *A is accretive and $(A+I)\mathfrak{D}(A)=\mathfrak{H}$.*
(c) *$A=(I+T)(I-T)^{-1}$ with a contraction T not having the eigenvalue* 1.
(d) *$-A$ is the generator of a continuous semigroup of contractions $\{T_s\}_{s\geq 0}$.*

If A is maximal accretive then so is A^*, and the corresponding Cayley transforms are adjoints of one another.

Let us note that if A is maximal accretive, then the semigroup of contractions generated by $-A$ has its cogenerator equal to the Cayley transform of A (compare the definitions (III.8.3) and (4.5)). Let us also observe that a normal operator A (bounded or not) is maximal accretive if and only if its spectrum is contained in the half-plane $\operatorname{Re} s\geq 0$ of the plane of the complex numbers s.

2. The following result is not used in the remainder of this book.

Proposition 4.2. *Let A_0, B_0 be two accretive operators in $\mathfrak{H}$, such that*

$$(A_0 f, g) = (f, B_0 g) \quad \textit{for} \quad f \in \mathfrak{D}(A_0), \quad g \in \mathfrak{D}(B_0). \tag{4.7}$$

Then we can extend A_0 and B_0 to maximal accretive operators A and B so that we still have

$$(Af, g) = (f, Bg) \quad \textit{for} \quad f \in \mathfrak{D}(A), \quad g \in \mathfrak{D}(B). \tag{4.8}$$

Proof. Equation (4.7) implies for $f \in \mathfrak{D}(A_0)$ and $g \in \mathfrak{D}(B_0)$ that

$$\begin{aligned} ((A_0 + I)f, (B_0 - I)g) &= (A_0 f, B_0 g) - (A_0 f, g) + (f, B_0 g) - (f, g) \\ &= (A_0 f, B_0 g) - (f, B_0 g) + (A_0 f, g) - (f, g) \\ &= ((A_0 - I)f, (B_0 + I)g). \end{aligned}$$

Hence it follows that for the Cayley transforms T_0 of A_0 and S_0 of B_0 we have

$$(T_0 \varphi, \psi) = (\varphi, S_0 \psi) \qquad (\varphi \in \mathfrak{D}(T_0), \psi \in \mathfrak{D}(S_0)). \tag{4.9}$$

If we construct extensions of T_0 and S_0 to contractions T and S defined on the whole space $\mathfrak{H}$, such that

$$(Th, k) = (h, Sk) \quad \text{for} \quad h, k \in \mathfrak{H}, \tag{4.10}$$

then the corresponding maximal accretive operators A and B will satisfy (4.8). This is in fact a consequence of the following relation which follows from (4.10):

$$\begin{aligned} ((I + T)h, (I - S)k) &= (h, k) + (Th, k) - (h, Sk) - (Th, Sk) \\ &= (h, k) + (h, Sk) - (Th, k) - (Th, Sk) \\ &= ((I - T)h, (I + S)k) \quad (h, k \in \mathfrak{H}). \end{aligned}$$

To obtain T and S we first extend T_0 and S_0 by continuity to the closures of their domains; these extensions are also denoted by T_0 and S_0; they satisfy (4.9). For the sake of brevity, we write $\mathfrak{L}$ for $\mathfrak{D}(S_0)$ (this is now a subspace of $\mathfrak{H}$), and we set $\mathfrak{M} = \mathfrak{H} \ominus \mathfrak{L}$.

For any fixed h in $\mathfrak{H}$, $(h, S_0 \psi)$ defines a conjugate-linear form on $\mathfrak{L}$ such that $|(h, S_0 \psi)| \leq \|h\| \cdot \|\psi\|$. Thus there exists a (unique) $h^* \in \mathfrak{L}$ such that $(h, S_0 \psi) = (h^*, \psi)$ and $\|h^*\| \leq \|h\|$. Setting $h^* = L_0 h$ we have defined a contraction L_0 of $\mathfrak{H}$ into $\mathfrak{L}$ such that

$$(L_0 h, \psi) = (h, S_0 \psi) \qquad (h \in \mathfrak{H}, \psi \in \mathfrak{L}). \tag{4.11}$$

In particular, if $h = \varphi \in \mathfrak{D}(T_0)$ then $(L_0 \varphi, \psi) = (\varphi, S_0 \psi) = (T_0 \varphi, \psi)$; hence it follows that $L_0 \varphi$ is the orthogonal projection of $T_0 \varphi$ into $\mathfrak{L}$:

$$L_0 \varphi = P_{\mathfrak{L}} T_0 \varphi.$$

Thus

$$\|P_{\mathfrak{M}}T_0\varphi\|^2 = \|T_0\varphi\|^2 - \|L_0\varphi\|^2 \le \|\varphi\|^2 - \|L_0\varphi\|^2 \quad \text{for} \quad \varphi \in \mathfrak{D}(T_0).$$

We conclude, by an argument due to M. G. Kreĭn, that there exists an extension of $P_{\mathfrak{M}}T_0$ to an operator L_1 defined on the whole space $\mathfrak{H}$, with values in $\mathfrak{M}$, and such that

$$\|L_1 h\|^2 \le \|h\|^2 - \|L_0 h\|^2 \quad \text{for every} \quad h \in \mathfrak{H};$$

see [*Func. Anal.*] Sec. 125. The operator T defined by

$$Th = L_0 h + L_1 h \quad \text{for} \quad h \in \mathfrak{H}$$

is thus a contraction and

$$T\varphi = L_0\varphi + L_1\varphi = P_{\mathfrak{L}}T_0\varphi + P_{\mathfrak{M}}T_0\varphi = T_0\varphi \quad \text{for} \quad \varphi \in \mathfrak{D}(T_0);$$

that is, $T \supset T_0$. Moreover,

$$(Th, \psi) = (L_0 h + L_1 h, \psi) = (L_0 h, \psi) = (h, S_0\psi) \quad \text{for} \quad h \in \mathfrak{H}, \psi \in \mathfrak{D}(S_0).$$

This shows that $T^* \supset S_0$; setting $S = T^*$ the pair $\{T, S\}$ satisfies(4.10), $T \supset T_0$, and $S \supset S_0$, so the proof is complete.

Corollary. *For every accretive A_0 with $\mathfrak{D}(A_0) \subset \mathfrak{D}(A_0^*)$, there exists a maximal accretive A so that*

$$A_0 \subset A \subset B_0^* \quad \textit{with} \quad B_0 = A_0^* | \mathfrak{D}(A_0).$$

In fact, we have $(A_0 f, g) = (f, A_0^* g) = (f, B_0 g)$ for $f, g \in \mathfrak{D}(A_0) = \mathfrak{D}(B_0)$, and in particular $\mathrm{Re}(f, B_0 f) = \mathrm{Re}(A_0 f, f) \ge 0$. Thus B_0 is accretive and there exists a maximal accretive extension A or A_0 such that $(Af, g) = (f, B_0 g)$ for all $f \in \mathfrak{D}(A)$ and $g \in \mathfrak{D}(B_0)$; hence $A \subset B_0^*$.

3. Let A be maximal accretive in $\mathfrak{H}$ and let T be its Cayley transform; T is a contraction defined on the whole space $\mathfrak{H}$, and

$$T = (A - I)(A + I)^{-1}, \quad A = (I + T)(I - T)^{-1}. \tag{4.12}$$

By virtue of these relations, a subspace $\mathfrak{K}$ of $\mathfrak{H}$ reduces[5] A if and only if it reduces T; it also follows that T is unitary if and only if A has the form iH with self-adjoint H.

The operators $A = iH$ are characterized by the property

$$A^* = -A,$$

[5] That is, $PA \subset AP$, P being the orthogonal projection of $\mathfrak{H}$ onto $\mathfrak{K}$.

so we call them *antiadjoint* operators. An operator is completely nonantiadjoint if no nonzero subspace reduces it to an antiadjoint operator. A maximal accretive, completely nonantiadjoint operator is said to be *purely maximal accretive.*

The canonical decomposition of the Cayley transform T of a maximal accretive A generates a decomposition of A.

Proposition 4.3. *For every maximal accretive operator A on $\mathfrak{H}$ there exists a decomposition $\mathfrak{H} = \mathfrak{H}_0 \oplus \mathfrak{H}_1$, reducing A, and such that the part of A in $\mathfrak{H}_0$ is antiadjoint and the part of A in $\mathfrak{H}_1$ is purely maximal accretive. This decomposition is unique; $\mathfrak{H}_0$ or $\mathfrak{H}_1$ may equal $\{0\}$.*

It follows from Proposition I.3.1 that if T is a contraction on $\mathfrak{H}$, then every subspace $\mathfrak{H}_a = \{h\colon\ h \in \mathfrak{H}, Th = ah\}$ with $|a| = 1$ reduces T. This implies that if A is maximal accretive in $\mathfrak{H}$, then every subspace $\{f\colon\ f \in \mathfrak{H}, AF = bf\}$, with purely imaginary number b, reduces A. In particular, the subspace

$$\mathfrak{N}_A = \{f\colon\ Af = 0\} \tag{4.13}$$

reduces A, or what amounts to the same, $Af = 0$ implies $A^*f = 0$, and conversely.

If A is purely maximal accretive, then necessarily $\mathfrak{N}_A = \{0\}$. We conclude that in this case A^{-1} exists and its domain is dense in $\mathfrak{H}$.

4. The relations (4.12) between a maximal accretive operator A and its Cayley transform T yield a method for constructing a functional calculus for A based on the functional calculus for T.

To this end let us consider the homography

$$\lambda \to \delta = \frac{1+\lambda}{1-\lambda} \equiv \omega(\lambda) \tag{4.14}$$

that maps the unit disc D onto the right half-plane

$$\Delta = \{\delta\colon\ \mathrm{Re}\ \delta > 0\};$$

the inverse map is

$$\delta \to \lambda = \frac{\delta - 1}{\delta + 1}. \tag{4.15}$$

We then define the functions of A by the formula

$$f(A) = (f \circ \omega)(T) \tag{4.16}$$

whenever $f \circ \omega \in N_T$.

In this way our functional calculus for contractions gives rise to a functional calculus for maximal accretive operators.

As 1 is not an eigenvalue of T, the one-point set $\{1\}$ is of measure O with respect to the spectral measure E_T corresponding to the unitary part of T. Therefore the class H_T^∞ contains the functions $v \in H^\infty$ that are continuous on $\overline{D}\backslash\{1\}$. Because $\overline{D}\backslash\{1\}$ is mapped by the homography (4.14) onto the closed half-plane $\bar{\Delta} = \{\delta\colon\ \mathrm{Re}\ \delta \geq 0\}$

the point $\delta = \infty$ not included, we see that the class of functions $f(\delta)$ admitted contains in particular the functions that are continuous and bounded on $\overline{\Delta}$, and holomorphic on Δ. One such function is $e^{-t\delta} = \exp(t(\lambda+1)/(\lambda-1)) = e_t(\lambda)$ for $t \geq 0$. We conclude that e^{-tA} makes sense for $t \geq 0$ and equals $e_t(T)$. For fixed z ($\operatorname{Re} z < 0$) and M ($0 < M < \infty$) the integral $f_M(z;\delta) = \int_0^M e^{tz}e^{-t\delta}\,dt$ is the limit, by bounded convergence on $\overline{\Delta}$, of the corresponding Riemann sums; hence it follows that $f_M(z;A) = \int_0^M e^{tz}e^{-tA}\,dt$. As $M \to \infty$, $f_M(z;\delta)$ tends to $(\delta - z)^{-1}$ uniformly on $\overline{\Delta}$; thus on account of Theorem III.2.3(c′) and Theorem 1.1(iv) we have $f_M(z;A) \to (A - zI)^{-1}$. So we arrive within the framework of our functional calculus at the relation

$$(A - zI)^{-1} = \int_0^\infty e^{tz}e^{-tA}\,dt, \tag{4.17}$$

valid for any maximal accretive operator A and for any complex number z with $\operatorname{Re} z < 0$.

5. The results of this section also apply in an obvious way to dissipative operators A': one just has to consider the accretive operator $A = -iA'$. Thus the Cayley transform of the dissipative operator A' will be, by definition, equal to the Cayley transform of the accretive operator $A = -iA'$; A' and T are therefore connected by the relations

$$T = (A' - iI)(A' + iI)^{-1}, \quad A' = i(I + T)(I - T)^{-1}. \tag{4.18}$$

In the case of a *maximal dissipative* A', the canonical decomposition of T generates a decomposition of A' into the orthogonal sum of a self-adjoint operator and of a *purely maximal dissipative* operator.

5 Fractional powers

1. Let A be a maximal accretive operator in the space $\mathfrak{H}$, and let T be its Cayley transform. In this section we define and study the powers A^α of A, where $0 < \alpha < 1$. For this purpose let us consider the functions

$$f_\alpha(\delta) = \delta^\alpha \qquad (\alpha \geq 0)$$

on Δ (for $z = re^{i\varphi}$ ($r > 0, |\varphi| \leq \pi/2$) we define z^α by $r^\alpha e^{i\alpha\varphi}$). Then we have

$$(f_\alpha \circ \omega)(\lambda) = \omega^\alpha(\lambda) = \frac{u^\alpha(\lambda)}{v^\alpha(\lambda)}$$

with

$$u^\alpha(\lambda) = (1+\lambda)^\alpha, \quad v^\alpha(\lambda) = (1-\lambda)^\alpha \qquad (\lambda \in D).$$

The functions u^α and v^α are continuous on $\overline{D}$, holomorphic on D, and belong to the class E^∞ (cf. Proposition III.1.3). Moreover, u^α has only a zero at -1 and v^α has only a zero at $+1$. Because $E_T(\{1\}) = O$, it follows that $\omega^\alpha \in N_T$ and that

$$f_\alpha(A) = \omega^\alpha(T) = v^\alpha(T)^{-1}u^\alpha(T). \tag{5.1}$$

Furthermore, by virtue of Proposition 1.3 we have for any $\alpha, \beta \geq 0$,

$$v^\beta(T)^{-1}u^\alpha(T) = u^\alpha(T)v^\beta(T)^{-1}. \tag{5.2}$$

We use the shorter notation A^α for $f_\alpha(A)$. This is justified if we show that the following relations hold.

$$A^0 = I, \quad A^1 = A, \text{ and } A^\alpha A^\beta = A^{\alpha+\beta} \quad \text{for} \quad \alpha, \beta \geq 0.$$

Now $A^0 = I$, because $\omega^0(\lambda) = 1$. Then, on account of (5.2),

$$A^1 = v^1(T)^{-1}u^1(T) = u^1(T)v^1(T)^{-1} = (I+T)(I-T)^{-1} = A$$

and

$$A^{\alpha+\beta} = \omega^{\alpha+\beta}(T) = (\omega^\alpha \cdot \omega^\beta)(T) = \omega^\alpha(T)\omega^\beta(T) = A^\alpha A^\beta$$

(cf. Proposition 1.3). Because $(\omega^\alpha)^{\breve{}} = \omega^\alpha$, the same proposition also implies

$$(A^\alpha)^* = [\omega^\alpha(T)]^* = \omega^\alpha(T^*) = (A^*)^\alpha,$$

because T^* is the Cayley transform of A^* (cf. Theorem 4.1).

The inequality

$$|\arg \omega^\alpha(\lambda)| \leq \frac{\alpha\pi}{2} \quad \text{on} \quad D,$$

for $0 \leq \alpha \leq 1$, allows us to apply Theorem 3.2 to $A^\alpha = \omega^\alpha(T)$ and to $(A^\alpha)^*$. We consider, then, the functions

$$u^\beta(\lambda) = (1+\lambda)^\beta, \quad v^\beta(\lambda) = (1-\lambda)^\beta, \quad w_\alpha(\lambda) = \frac{u^\alpha(\lambda) - v^\alpha(\lambda)}{u^\alpha(\lambda) + v^\alpha(\lambda)}, \tag{5.3}$$

where $0 \leq \beta$ and $0 \leq \alpha \leq 1$, and show that they satisfy the hypotheses of Theorem 1.1(viii).

In fact, $u^\alpha(\lambda)$ and $v^\alpha(\lambda)$ have no common zero and their values lie in the sector

$$\left\{z\colon\ |\arg z| \leq \frac{\alpha\pi}{2}\right\},$$

thus it follows that $u^\alpha(\lambda) + v^\alpha(\lambda)$ has no zero in $\overline{D}$ for $0 \leq \alpha < 1$. Therefore the function $w_\alpha(\lambda)$ is continuous on $\overline{D}$ and holomorphic on D; moreover, it is easy to see that

$$|w_\alpha(\lambda)| < 1$$

on $\overline{D}$ with the exception of the points $\lambda = \pm 1$, where we have $w_\alpha(1) = 1$ and $w_\alpha(-1) = -1$; otherwise if $\alpha = 1$ then $w_1(\lambda) \equiv \lambda$. So we have for $0 \leq \alpha \leq 1$

and for $\beta \geq 0$

$$C_{w_\alpha} = \varnothing, \quad w_\alpha^{-1}(\mathbf{C}_{u\beta}) = \varnothing \quad \text{and} \quad w_\alpha^{-1}(\mathbf{C}^0_{v\beta}) = w_\alpha^{-1}(\{1\}) = \{1\},$$

and all these sets are of measure O with respect to E_T.

Applying Theorem 1.1(viii) it follows that

$$T_\alpha = w_\alpha(T) \tag{5.4}$$

is a contraction, $\omega^\beta(T_\alpha)$ and $(\omega^\beta \circ w_\alpha)(T)$ exist, and

$$\omega^\beta(T_\alpha) = (\omega^\beta \circ w_\alpha)(T). \tag{5.5}$$

Now, because

$$(\omega^\beta \circ w_\alpha)(\lambda) = \left(\frac{1 + w_\alpha(\lambda)}{1 - w_\alpha(\lambda)}\right)^\beta = \left(\frac{u^\alpha(\lambda)}{v^\alpha(\lambda)}\right)^\beta = \omega^{\alpha\beta}(\lambda),$$

relation (5.5) means that

$$\omega^\beta(T_\alpha) = A^{\alpha\beta}. \tag{5.6}$$

For $\beta = 1$ this relation becomes

$$A^\alpha = \omega^1(T_\alpha) = v^1(T_\alpha)^{-1} \cdot u^1(T_\alpha) = (I - T_\alpha)^{-1}(I + T_\alpha); \tag{5.7}$$

this shows that A^α is maximal accretive, with the Cayley transform T_α. Hence we have $f(A^\alpha) = f \circ \omega(T_\alpha)$ whenever $f \circ \omega(T_\alpha)$ makes sense, so in particular $(A^\alpha)^\beta = \omega^\beta(T_\alpha)$. Thus (5.6) becomes

$$(A^\alpha)^\beta = A^{\alpha\beta} \qquad (0 \leq \alpha \leq 1,\ \beta \geq 0).$$[6]

2. Because A^α is maximal accretive if $0 \leq \alpha \leq 1$, there exists a continuous semigroup $\{T_\alpha(s)\}_{s\geq 0}$ of contractions whose generator is $-A^\alpha$. The cogenerator of this semigroup is then equal to the Cayley transform of A^α, that is, to T_α. By virtue of Theorem III.8.1 we have therefore

$$T_\alpha(s) = e_s(T_\alpha) = e_s(w_\alpha(T)) \tag{5.8}$$

with $e_s(\lambda) = \exp(s(\lambda + 1)(\lambda - 1))$. This function is holomorphic and bounded on D, and continuous on $\overline{D}\backslash\{1\}$, therefore we can apply Theorem 1.1 (viii) to obtain that

$$e_s(w_\alpha(T)) = (e_s \circ w_\alpha)(T). \tag{5.9}$$

---

[6] The restriction $\alpha \leq 1$ is motivated by the fact that, for $\alpha > 1$, A^α is not necessarily accretive (this is the case, e.g., if A is antiadjoint and $A \neq O$).

Now we have

$$(e_s \circ w_\alpha)(\lambda) = \exp\left(s\frac{w_\alpha(\lambda)+1}{w_\alpha(\lambda)-1}\right) = \exp(-s\omega^\alpha(\lambda));$$

hence we see that if $0 \le \alpha < 1$ this function belongs to the class H_T^∞ and is bounded in absolute value by 1 on D, for all the real or complex values of the parameter s lying in the sector

$$\overline{\Delta}_{1-\alpha} = \left\{s\colon\ |\arg s| \le (1-\alpha)\frac{\pi}{2}\right\}.$$

The identities

$$(e_{s_1} \circ w_\alpha)(e_{s_2} \circ w_\alpha) = e_{s_1+s_2} \circ w_\alpha \quad \text{for} \quad s_1, s_2 \in \overline{\Delta}_{1-\alpha},$$

imply that the operators

$$T_\alpha(s) = (e_s \circ w_\alpha)(T) \quad (s \in \overline{\Delta}_{1-\alpha})$$

form a semigroup:

$$T_\alpha(0) = I, \quad T_\alpha(s_1)T_\alpha(s_2) = T_\alpha(s_1+s_2) \quad (s_1, s_2 \in \overline{\Delta}_{1-\alpha}).$$

Moreover, this semigroup is holomorphic in the interior $\Delta_{1-\alpha}$ of $\overline{\Delta}_{1-\alpha}$. To prove this, let us fix a point $s_0 > 0$ and a point s lying in the largest open disc contained in $\Delta_{1-\alpha}$, with center s_0, that is, such that

$$|s - s_0| < \rho_0 = s_0 \cos\frac{\alpha\pi}{2}.$$

Let us consider then the expansion

$$e^{-s\delta^\alpha} = e^{-(s-s_0)\delta^\alpha} \cdot e^{-s_0\delta^\alpha} = \sum_{n=0}^{\infty}\left(\frac{s-s_0}{\rho_0}\right)^n b_n(\delta) \quad (\delta \in \Delta), \tag{5.10}$$

where

$$b_n(\delta) = \frac{(-1)^n}{n!}(\rho_0\delta^\alpha)^n e^{-s_0\delta^\alpha};$$

then we have

$$|b_n(\delta)| \le \sum_{m=0}^{\infty}\frac{1}{m!}|\rho_0\delta^\alpha|^m \cdot \exp(-s_0 \operatorname{Re}\delta^\alpha) = \exp[\rho_0|\delta^\alpha| - s_0 \operatorname{Re}\delta^\alpha] \le 1,$$

because

$$\operatorname{Re}\delta^\alpha \ge |\delta^\alpha|\cos\frac{\alpha\pi}{2} \quad \text{for} \quad \delta \in \overline{\Delta}.$$

The expansion (5.10) is therefore uniformly convergent with respect to δ on $\overline{\Delta}$. Hence it follows for the corresponding operators:

$$T_\alpha(s) = \sum_{n=0}^{\infty} \left(\frac{s - s_0}{\rho_0} \right)^n B_n,$$

where $B_n = b_n(A) = (b_n \circ \omega)(T)$, $\|B_n\| \leq 1$. This concludes the proof of the fact that $T_\alpha(s)$ is a holomorphic function of s on $\Delta_{1-\alpha}$.

Finally we note that Theorem 3.2 can be applied to the functions $\omega^\alpha(\lambda)$, $0 \leq \alpha \leq 1$.

Summing up, we have proved the following result.

Theorem 5.1. *Let A be a maximal accretive operator in $\mathfrak{H}$ and let A^α $(\alpha \geq 0)$ be the operator that corresponds to the function δ^α in the sense of the functional calculus $f \to f(A)$ defined in Sec.* 4.4. *A^α is then a closed operator with domain dense in $\mathfrak{H}$, and we have*

$$A^0 = I; \quad A^1 = A; \quad A^{\alpha+\beta} = A^\alpha A^\beta \ (\alpha, \beta \geq 0); \quad (A^\alpha)^* = (A^*)^\alpha \quad (\alpha \geq 0).$$

For $0 \leq \alpha \leq 1$, A^α is maximal accretive and we have $(A^\alpha)^\beta = A^{\alpha\beta}$ for $\beta \geq 0$; moreover,

$$|\arg(A^\alpha h, h)| \leq \frac{\alpha\pi}{2} \quad \textit{for} \quad h \in \mathfrak{D}(A^\alpha).$$

Let $\{T_\alpha(s)\}_{s \geq 0}$ be the semigroup of contractions whose generator is equal to $-A^\alpha$. If $0 \leq \alpha < 1$ then this semigroup can be extended to the complex values of the parameter s lying in the sector $|\arg s| \leq (1-\alpha)\,\pi/2$ so that the semigroup property is preserved and $T_\alpha(s)$ is a contraction valued function of s, holomorphic in this sector.

If $0 \leq \alpha < \frac{1}{2}$, the operators A^α, $A^{\alpha *}$, and hence also Re A^α *and* Im A^α *have the same domain; moreover,* Re A^α *is positive self-adjoint and the following inequalities hold.*

(i) $\operatorname{Re}(A^\alpha h, A^{\alpha *} h) \geq \cos\alpha\pi \cdot \max\{\|A^\alpha h\|^2, \|A^{\alpha *} h\|^2\}$.

(ii) $\|A^\alpha h\| \geq \cos\alpha\pi \cdot \|A^{\alpha *} h\|, \quad \|A^{\alpha *} h\| \geq \cos\alpha\pi \cdot \|A^\alpha h\|$.

(iii) $\operatorname{Re}(A^\alpha h, [\operatorname{Re} A^\alpha] h) \geq \cos^2 \frac{\alpha\pi}{2} \cdot \|A^\alpha h\|^2$.

(iv) $\left[\cos^2 \frac{\alpha\pi}{2} \Big/ \cos\alpha\pi\right] \cdot \|A^\alpha h\| \geq \|[\operatorname{Re} A^\alpha] h\| \geq \cos^2 \frac{\alpha\pi}{2} \cdot \|A^\alpha h\|$.

(v) $\|[\operatorname{Im} A^\alpha] h\| \leq \tan \frac{\alpha\pi}{2} \cdot \|[\operatorname{Re} A^\alpha] h\|$.

3. This theorem implies in particular that every maximal accretive operator A possesses, for $n \geq 2$, an nth root $B = A^{1/n}$ that is also maximal accretive and such that

$$|\arg(Bh, h)| \leq \frac{\pi}{2n}. \tag{5.11}$$

We prove that there is only one nth root with these properties. To this end we need some supplementary lemmas.

Lemma 5.2. (a) *Let $P \geq O$ and $Q = Q^*$ satisfy*

$$|(Qh,h)| \leq \|Ph\|^2 \quad \textit{for all} \quad h \in \mathfrak{H}.$$

Then the operator $|Q| = (Q^2)^{1/2}$ satisfies

$$\||Q|^{1/2}h\|^2 = (|Q|h,h) \leq 2^{1/2}\|h\|\|Ph\|\|P\| \quad \textit{for all} \quad h \in \mathfrak{H}.$$

(b) *Let T be a contraction and let $R = (T+T^*)/2$, $Q = (T-T^*)/2i$. If Q satisfies for some $0 \leq \theta < \infty$,*

$$|(Qh,h)| \leq \theta\|D_T h\|^2 \quad \textit{for all} \quad h \in \mathfrak{H}$$

then

$$\|D_R h\|^2 \leq (1+(2^{1/2}\theta)^{1/2})^2\|h\|^{3/2}\|D_T h\|^{1/2} \quad \textit{for all} \quad h \in \mathfrak{H}.$$

(c) *With the notation above, if $D_R h_0 = 0$ for some $h_0 \in \mathfrak{H}$, then*

$$h = h_{+1} + h_{-1}, \quad \textit{where} \quad Th_{\pm 1} = Rh_{\pm 1} = \pm h_{\pm 1}.$$

Proof. *Part* (a). Let $Q = \int_{-\|Q\|}^{\|Q\|} \lambda \; dE_\lambda$ be the spectral representation of Q and let $E_+ = E_{\|Q\|} - E_{0-}$, $E_- = E_{0-}$. For $h_\pm = E_\pm h_\pm$ and any complex scalars $a_\pm$ we have

$$\big||a_+|^2(Qh_+,h_+) + |a_-|^2(Qh_-,h_-)\big| \leq \|P(a_+h_+ + a_-h_-)\|^2.$$

Note that $q_\pm = \pm(Qh_\pm,h_\pm) \geq 0$, $q_\pm \leq \|Ph_\pm\|^2$, and

$$\big||a_+|^2 q_+ - |a_-|^2 q_-\big| \leq |a_+|^2\|Ph_+\|^2 + 2\,\mathrm{Re}[a_+\bar{a}_-(Ph_+Ph_-)] + |a_-|^2\|Ph\|^2.$$

Because a_+, a_- are arbitrary, we infer that

$$|\lambda^2 q_+ - q_-| \leq \lambda^2\|Ph_+\|^2 - 2\lambda|(Ph_+,Ph_-)| + \|Ph_-\|^2$$

for all real values of λ. In other words,

$$\lambda^2(\|Ph_+\|^2 \mp q_+) - 2\lambda|(Ph_+,Ph_-)| + (\|Ph_-\|^2 \pm q_-) \geq 0$$

for all λ, and taking discriminants we obtain

$$|(Ph_+,Ph_-)|^2 \leq (\|Ph_+\|^2 \mp q_+)(\|Ph_-\|^2 \pm q_-),$$

and consequently

$$|(Ph_+,Ph_-)| \leq (\|Ph_+\|^4 - q_+^2)^{1/4}(\|Ph_-\|^4 - q_-^2)^{1/4}.$$

Setting $\rho_\pm = q_\pm/\|Ph_\pm\|^2$, we apply this relation to $h_\pm = E_\pm h$ ($h \in \mathfrak{H}$) to obtain

$$\begin{aligned}\|Ph\|^2 &= \|Ph_+ + Ph_-\|^2 = \|Ph_+\|^2 + 2\,\mathrm{Re}(Ph_+, Ph_-) + \|Ph_-\|^2 \\ &\geq \|Ph_+\|^2 - 2\|Ph_+\|\|Ph_-\|(1-\rho_+^2)^{1/4}(1-\rho_-^2)^{1/4} + \|Ph_-\|^2 \\ &\geq \|Ph_+\|^2(1-(1-\rho_+^2)^{1/2}) + \|Ph_-\|^2(1-(1-\rho_-^2)^{1/4}) \\ &\geq (\|Ph_+\|^2\rho_+^2 + \|Ph_-\|^2\rho_-^2)/2 \\ &= (q_+^2/\|Ph_+\|^2 + q_-^2/\|Ph_-\|^2)/2 \\ &\geq (q_+^2/\|h_+\|^2 + q_-^2/\|h_-\|^2)/(2\|P\|^2) \\ &\geq (q_+ + q_-)^2/(2\|h\|^2\|P\|^2) \\ &= (|Q|h,h)^2/(2\|h\|^2\|P\|^2),\end{aligned}$$

where in the last inequality we used the fact that for $a,b,c > 0$, $0 < x < c$, $(a^2/x) + (b^2/(c-x)) \geq (a+b)^2/c$.

Part (b). Let $h \in \mathfrak{H}$. Then

$$\|D_T h\|^2 = \|h\|^2 - \|Th\|^2 = \|D_R h\|^2 - i(Rh, Qh) + i(Qh, Rh) - \|Qh\|^2,$$

hence, by applying (a) and taking into account that Q, R, and D_T are contractions, we have

$$\begin{aligned}\|D_R h\|^2 &\leq \|D_T h\|\|h\| + 2\||Q|^{1/2}h\|\|h\| + \||Q|^{1/2}h\|^2 \\ &\leq \|D_T h\|\|h\| + 2(2^{1/2}\theta)^{1/2}\|h\|^{3/2}\|D_T h\|^{1/2} + 2^{1/2}\theta\|h\|\|D_T h\| \\ &\leq (1 + 2(2^{1/2}\theta)^{1/2} + 2^{1/2}\theta)\|h\|^{3/2}\|D_T h\|^{1/2}.\end{aligned}$$

Part (c). If $\|D_R h_0\|^2 = 0$, then (because R is a self-adjoint contraction) it is obvious that $h_0 = h_{+1} + h_{-1}$ where $Rh_{\pm 1} = \pm h_{\pm 1}$. But in

$$\begin{aligned}\|h_{\pm 1}\|^2 = \pm(Rh_{\pm 1}, h_{\pm 1}) = \pm(Th_{\pm 1}, h_{\pm 1}) = |(Th_{\pm 1}, h_{\pm 1})| \leq \\ \leq \|Th_{\pm 1}\|\|h_{\pm 1}\| \leq \|h_{\pm 1}\|^2\end{aligned}$$

all inequalities turn out to be equalities, thus we must have $Th_{\pm 1} = \pm h_{\pm 1} = Rh_{\pm 1}$ and this concludes the proof of the lemma.

Lemma 5.3. *Let A be a maximal accretive operator in $\mathfrak{H}$ satisfying the condition*

$$|\arg(Af, f)| \leq \alpha\pi/2 \quad \textit{for all} \quad f \in \mathfrak{D}(A) \tag{5.12}$$

with some $\alpha \in [0,1)$. Then $(Af_0, f_0) = 0$ implies $Af_0 = 0$.

Proof. Let T be the Cayley transform of A, that is, the contraction T on $\mathfrak{H}$ defined by

$$Th = (A-I)f \quad \text{for} \quad h = (A+I)f, \qquad f \in \mathfrak{D}(A).$$

Then (with the notation of Lemma 5.2(b)

$$4(Af,f) = ((I+T)h,(I-T)h) = \|D_T h\|^2 + 2i(Qh,h),$$

hence T satisfies the conditions of Lemma 5.2(b) with $\theta = \tan(\pi\alpha/2)/2$.

If $(Af_0, f_0) = 0$, then $D_T h_0 = 0$ and by virtue of Lemma 5.2(b) and (c) we have $h_0 = h_{+1} + h_{-1}$ where $Th_{\pm 1} = \pm h_{\pm 1}$. But 1 is not an eigenvalue of T, hence $h_0 = h_{-1}$ and $Th_0 = -h_0$. It follows that

$$2Af_0 = (T+I)h_0 = 0,$$

which completes the proof.

We also use the following commutativity property.

Lemma 5.4. *Let A, A' be two maximal accretive operators such that*

$$\mathfrak{D}(A) \subset \mathfrak{D}(A') \quad \text{and} \quad A'A \subset AA'. \tag{5.13}$$

Then their Cayley transforms T, T' commute.

Proof. The operators $A+I$ and $A'+I$ are boundedly invertible, thus for every $f \in \mathfrak{H}$ there exists $g \in \mathfrak{H}$ such that

$$f = (A'+I)(A+I)g. \tag{5.14}$$

Because $g \in \mathfrak{D}(A)$, from (5.13) we have also $g \in \mathfrak{D}(A')$; hence $Ag = (A+I)g - g \in \mathfrak{D}(A')$. So we can write (5.14) in the form $f = (A'A + A' + A + I)g$; from the second relation (5.13) we have then

$$f = (AA' + A' + A + I)g = (A+I)(A'+I)g. \tag{5.15}$$

The relations (5.14) and (5.15) imply

$$(A+I)^{-1}(A'+I)^{-1}f = (A'+I)^{-1}(A+I)^{-1}f \qquad (\text{for all} \quad f \in \mathfrak{H}).$$

The relations $T = I - 2(A+I)^{-1}$ and $T' = I - 2(A'+I)^{-1}$ show now that T and T' commute.

After these preliminaries we can turn to the proof of the following result.

Proposition 5.5. *For a maximal accretive A and for every integer $n \geq 2$ there exists a maximal accretive B, and only one, satisfying the conditions:*

$$B^n = A, \tag{5.16}$$

$$|\arg(Bf,f)| \leq \frac{\pi}{2n} \quad (\textit{for all } f \in \mathfrak{D}(B)). \tag{5.17}$$

Proof. Let $\{A^\alpha\}$ be the system of the fractional powers of A constructed in the proof of Theorem 5.1; $B = A^{1/n}$ is then a solution of our problem. We just have to show that no other solution exists. Let B' be any solution, and let S and S' be the Cayley transforms of B and B', respectively; by virtue of (5.4) we have

$$S = w_{1/n}(T),$$

where T denotes the Cayley transform of A and $w_{1/n}$ is the function defined by (5.3).

Let us observe that

$$B'A = B'B'^n = B'^{n+1} = B'^n B' = AB'$$

and that $\mathfrak{D}(A) = \mathfrak{D}(B'^n) \subset \mathfrak{D}(B')$; applying Lemma 5.4 we obtain then that S' commutes with T and consequently with every function of T, in particular S' commutes with S.

Let $B' = B'_0 \oplus B'_1$ be the canonical decomposition of the maximal accretive operator B' into the orthogonal sum of its antiadjoint part B'_0 and its purely maximal accretive part B'_1. Let f be such that $B'^n f = 0$. If $f = f_0 + f_1$ is the corresponding decomposition of f, then we have $B'^n_0 f_0 = 0$ and $B'^n_1 f_1 = 0$. Now B'_1 is invertible (cf. the end of Sec. 4.3), and hence $f_1 = 0$. On the other hand, because $iB'_0 = H$ with a self-adjoint H, it follows from the spectral theory for H that $B'^n_0 f_0 = 0$ implies $B'_0 f_0 = 0$. Thus $B'^n f = 0$ implies $B' f = 0$. As $B'^n = A$, we conclude that the null spaces of A and B' are the same, and as all this applies to B too (instead of B'), it follows that the maximal accretive operators A, B, B' all have the same null space $\mathfrak{N}$. By virtue of the remark following Proposition 4.3, $\mathfrak{N}$ reduces A, B, B', and hence their Cayley transforms too, and for $f \in \mathfrak{N}$ we have $Bf = B'f (= 0)$. Thus to prove $B = B'$ it suffices to consider in the sequel the parts of these operators in $\mathfrak{M} = \mathfrak{H} \ominus \mathfrak{N}$.

Let us set

$$V = (I + S)(I - S') \quad \text{and} \quad W = (I + S')(I - S).$$

As V and W commute we have

$$V^n - W^n = \prod_{k=0}^{n-1} (V - \varepsilon^k W), \quad \varepsilon = e^{2\pi i/n}. \tag{5.18}$$

Because $I + S = B(I - S)$, $I + S' = B'(I - S')$, and $B^n = A = B'^n$, we have

$$V^n = B^n (I - S)^n (I - S')^n = B'^n (I - S')^n (I - S)^n = W^n,$$

and thus

$$\prod_{k=0}^{n-1} (V - \varepsilon^k W) = O. \tag{5.19}$$

Let $g \in \mathfrak{M}$ be such that for some k

$$(V - \varepsilon^k W) g = 0. \tag{5.20}$$

Setting $f=(I-S)(I-S')g$ we have then

$$Bf=(I+S)(I-S')g=Vg=\varepsilon^k Wg=\varepsilon^k(I+S')(I-S)g=\varepsilon^k B'f, \quad (5.21)$$

$$(Bf,f)=\varepsilon^k(B'f,f). \quad (5.22)$$

If $1\leq k\leq n-1$ this implies $(Bf,f)=(B'f,f)=0$, for otherwise the relations

$$|\arg(Bf,f)|\leq\frac{\pi}{2n},\quad |\arg(B'f,f)|\leq\frac{\pi}{2n},\quad \arg\varepsilon^k=\frac{2\pi}{n}k \quad (1\leq k\leq n-1)$$

would contradict (5.22). In view of Lemma 5.2, relation (5.20) implies therefore in this case that $Bf=0$; because B is invertible on $\mathfrak{M}$ it follows that $f=0$, and hence $g=0$.

Thus (5.19) implies $V-W=O$; that is,

$$(I-S)(I+S')=(I+S)(I-S'),$$

hence $S=S'$ and therefore $B=B'$. (All this for the parts of the operators on $\mathfrak{M}$.)

This concludes the proof of the proposition.

6 Notes

The functional calculus of contractions, for not necessarily bounded functions, was developed in SZ.-N.–F. [VI]. There one studied the corresponding calculus for maximal accretive operators also; a slightly more restricted calculus was proposed independently (for maximal dissipative operators) by LANGER [2]. In earlier work[7] one considered only holomorphic functions; meromorphic functions are admitted in the present work. This natural and almost immediate extension is motivated by some results of the authors on the commutant and bicommutant of contractions of class C_0 with finite defect indices; see SZ.-N.–F. [14], [15].

The method of extending an accretive (or dissipative) operator to a maximal one via Cayley transforms, modeled on von Neumann's theory of symmetric operators, is due to PHILLIPS [2]. The characterization of the infinitesimal generators of continuous semigroups of contractions by their Cayley transforms (i.e. the equivalence of conditions (c) and (d) in Theorem 4.1) was given in SZ.-NAGY [II]; also see FOIAŞ [2]. Proposition 4.2 is new.

In a slightly more restricted form, Proposition 4.3 (on the decomposition of maximal accretive or dissipative operators) was given by LANGER [1].

Fractional powers of operators A in Hilbert space, or even in Banach spaces, such that $-A$ is the infinitesimal generator of a continuous one-parameter semigroup of contractions, have been constructed by different authors using different methods.

[7] Including the original French edition of this book.

The definition proposed by BOCHNER [1] and PHILLIPS [1] introduces A^α $(0 < \alpha < 1)$ by means of the continuous semigroup

$$T_{\alpha,t} = e^{-tA^\alpha} \quad (t \geq 0).^8$$

That is, one defines

$$T_{\alpha,t} = \int_0^\infty T_s \, dm_{\alpha,t}(s),$$

where $T_s = e^{-sA}$ and the measure is determined by the Laplace integral

$$e^{-t\rho^\alpha} \int_0^\infty e^{-s\rho} \, dm_{\alpha,t}(s) \quad (t \geq 0;\ \operatorname{Re} \rho \geq 0;\ 0 < \alpha < 1).$$

Further formulas, due to BALAKRISHNAN [1], are the following:

$$A^\alpha h = \frac{\sin \alpha\pi}{\pi} \int_0^\infty \lambda^{\alpha-1}(\lambda I + A)^{-1} Ah \, d\lambda,$$
$$A^\alpha h = \Gamma(-\alpha)^{-1} \int_0^\infty \lambda^{-\alpha-1}(e^{-\lambda A} - I)h \, d\lambda$$

$(h \in \mathfrak{D}(A))$; see also YOSIDA [1], [2] p. 259. In the Hilbert space situation, these formulas may be obtained easily by our functional calculus, in a similar way as (4.17).

The uniqueness theorem (our Proposition 5.5) is due (in its form on dissipative operators) to MACAEV AND PALANT [1] (in the case of bounded operators) and to LANGER [2] (in the general case); our proof is a (slightly simplified) variant of that of Langer. Later, NOLLAU [1] extended the uniqueness theorem to operators in arbitrary Banach spaces.

As to Theorem 5.1, the fact that the semigroup is holomorphic in a sector of the complex plane, was first proved by YOSIDA [1]; the fact that for $0 \leq \alpha < 1/2$ the operators A^α and $A^{\alpha *}$ have the same domain of definition and satisfy the inequalities (i)–(v), was first established (partially with other constant factors) by KATO [1]. His methods are different from ours. Lemma 5.2, which corrects and completes the argument presented in the earlier editions of this book, may be useful in other applications.

In connection with this chapter see also DOLPH [1]; DOLPH AND PENZLIN [1]; LANGER AND NOLLAU [1]; and NOLLAU [2].

7 Further results

If T is a unilateral shift of multiplicity one, then Lemma V.3.2 implies that every operator X commuting with T is of the form $u(T)$ for some $u \in H^\infty$. SARASON [5],[6] proved that this statement extends to unbounded operators. Thus, if X is a closed, densely defined operator satisfying $TX \subset XT$, then there exists a function

[8] This notation only indicates that $T_{\alpha,t}$ is the semigroup whose generator is $-A^\alpha$.

$u \in N_T$ such that $X = u(T)$. Both of these statements remain true if T is replaced by an operator of class C_0 with $\mathfrak{d}_T = 1$. This is proved in SARASON [3] and [6] for the bounded and unbounded case, respectively. Similar results can be proved for operators in the double commutant of an arbitrary operator of class C_0; see Theorem X.4.2.

In connection with this material, see also MARTIN [1]; BERCOVICI, DOUGLAS, FOIAS, AND PEARCY [1]; and BERCOVICI [6].

Chapter V

Operator-Valued Analytic Functions

1 The spaces $L^2(\mathfrak{A})$ and $H^2(\mathfrak{A})$

1. For any separable Hilbert space $\mathfrak{A}$ we denote by $L^2(\mathfrak{A})$ the class of functions $v(t)$ $(0 \leq t \leq 2\pi)$ with values in $\mathfrak{A}$, measurable[1] (strongly or weakly, which are equivalent due to the separability of $\mathfrak{A}$) and such that

$$\|v\|^2 = \frac{1}{2\pi} \int_0^{2\pi} \|v(t)\|_{\mathfrak{A}}^2 \, dt < \infty. \tag{1.1}$$

With this definition of the norm $\|v\|$, $L^2(\mathfrak{A})$ becomes a (separable) Hilbert space; it is understood that two functions in $L^2(\mathfrak{A})$ are considered identical if they coincide almost everywhere (with respect to Lebesgue measure). If $\dim \mathfrak{A} = 1$ (i.e., if $L^2(\mathfrak{A})$ consists of scalar-valued functions), we write L^2 instead of $L^2(\mathfrak{A})$.

Let $\{v_n(t)\}$ $(n = 1, 2, \ldots)$ be a sequence converging to $v(t)$ in $L^2(\mathfrak{A})$, that is, in the mean:

$$\|v_n - v\|^2 = \frac{1}{2\pi} \int_0^{2\pi} \|v_n(t) - v(t)\|_{\mathfrak{A}}^2 \, dt \to 0 \quad (n \to \infty).$$

Then we can choose a subsequence $\{v_{n_k}(t)\}$ $(k = 1, 2, \ldots)$ such that

$$\sum_k \int_0^{2\pi} \|v_{n_k}(t) - v(t)\|_{\mathfrak{A}}^2 \, dt < \infty.$$

By virtue of the theorem of Beppo Levi we have then a.e.

$$\sum_n \|v_{n_k}(t) - v(t)\|_{\mathfrak{A}}^2 < \infty$$

[1] For the theory of integration of vector-valued functions as well as for that of analytic vector-valued function see HILLE [1], Chap. III.

B.Sz.-Nagy et al., *Harmonic Analysis of Operators on Hilbert Space*, Universitext,
DOI 10.1007/978-1-4419-6094-8_5, © Springer Science+Business Media, LLC 2010

and consequently a.e.

$$\|v_{n_k}(t) - v(t)\|_{\mathfrak{A}} \to 0 \quad (k \to \infty).$$

Thus from every sequence converging in $L^2(\mathfrak{A})$ we can choose a subsequence that converges pointwise a.e.

For any integer k let us denote by $\mathfrak{E}_k$ the subspace of $L^2(\mathfrak{A})$ consisting of the functions of the form $e^{ikt}a$ ($a \in \mathfrak{A}$). Clearly, $\mathfrak{E}_k \perp \mathfrak{E}_j$ for $k \neq j$ and, moreover, we have

$$L^2(\mathfrak{A}) = \bigoplus_{-\infty}^{\infty} \mathfrak{E}_k. \tag{1.2}$$

Indeed, let $v \in L^2(\mathfrak{A})$ be orthogonal to all $\mathfrak{E}_k$:

$$\frac{1}{2\pi}\int_0^{2\pi} e^{-ikt}(v(t),a)_{\mathfrak{A}}\, dt = 0 \qquad (a \in \mathfrak{A}; k = 0, \pm 1, \pm 2, \ldots).$$

Then we have $(v(t),a)_{\mathfrak{A}} = 0$ everywhere, with the possible exception of the points t of a set E_a depending on a and of zero measure. Letting a run over a countable dense subset of $\mathfrak{A}$, and taking the union of the corresponding sets E_a one obtains a set E of zero measure; then $v(t) = 0$ for every point t not belonging to E, and consequently $v = 0$ as an element of $L^2(\mathfrak{A})$. This proves (1.2). Let us also observe that

$$\|e^{ikt}a\|_{L^2(\mathfrak{A})} = \|a\|_{\mathfrak{A}} \quad (a \in \mathfrak{A}). \tag{1.3}$$

From (1.2) and (1.3) it follows that there exists a one-to-one correspondence between the elements v of $L^2(\mathfrak{A})$ and the sequences $\{a_k\}_{-\infty}^{\infty}$ ($a_k \in \mathfrak{A}$) with $\sum \|a_k\|_{\mathfrak{A}}^2 < \infty$, such that for corresponding v and $\{a_k\}$ we have

$$v(t) = \sum_{-\infty}^{\infty} e^{ikt} a_k \tag{1.4}$$

and

$$\|v\|^2 = \sum_{-\infty}^{\infty} \|a_k\|_{\mathfrak{A}}^2, \tag{1.5}$$

where (1.4) is understood in the sense of convergence in the mean:

$$\int_0^{2\pi} \left\| v(t) - \sum_{-m}^{n} e^{ikt} a_k \right\|_{\mathfrak{A}}^2 dt \to 0 \quad (m, n \to \infty);$$

by virtue of (1.4) we have

$$a_k = \frac{1}{2\pi}\int_0^{2\pi} e^{-ikt} v(t)\, dt \qquad (k = 0, \pm 1, \ldots);$$

that is, (1.4) is the *Fourier series* of v.

An important subspace of $L^2(\mathfrak{A})$, which we denote by $L^2_+(\mathfrak{A})$, consists of those functions for which $a_k = 0$ $(k < 0)$. With any function

$$v(t) = \sum_0^\infty e^{ikt} a_k \in L^2_+(\mathfrak{A})$$

we associate the function

$$u(\lambda) = \sum_0^\infty \lambda^k a_k$$

of the complex variable λ; $u(\lambda)$ is defined and holomorphic on the unit disc D, because

$$\left\| \sum_m^n \lambda^k a_k \right\|_{\mathfrak{A}} \leq \sum_m^n |\lambda|^k \|a_k\|_{\mathfrak{A}} \leq (1-|\lambda|^2)^{-1/2} \left[\sum_m^n \|a_k\|^2_{\mathfrak{A}} \right]^{1/2} \to 0$$

for $n > m \to \infty$ and for $|\lambda| < 1$, uniformly for $|\lambda| \leq r_0 < 1$. One can recover $v(t)$ from $u(\lambda)$ as a *radial limit in the mean*, in fact,

$$\begin{aligned} \frac{1}{2\pi}\int_0^{2\pi} \|v(t) - u(re^{it})\|^2_{\mathfrak{A}}\, dt &= \frac{1}{2\pi}\int_0^{2\pi} \left\| \sum_0^\infty (1-r^k) e^{ikt} a_k \right\|^2_{\mathfrak{A}} dt \\ &= \sum_0^\infty (1-r^k)^2 \|a_k\|^2_{\mathfrak{A}} \to 0 \end{aligned}$$

as $r \to 1-0$; moreover, we have

$$\frac{1}{2\pi}\int_0^{2\pi} \|u(re^{it})\|^2_{\mathfrak{A}}\, dt = \sum_0^\infty r^{2k} \|a_k\|^2_{\mathfrak{A}} \leq \sum_0^\infty \|a_k\|^2_{\mathfrak{A}} < \infty \qquad (0 \leq r < 1).$$

Let us denote by $H^2(\mathfrak{A})$ the class of functions

$$u(\lambda) = \sum_0^\infty \lambda^k a_k$$

with values in $\mathfrak{A}$, holomorphic on D, and such that

$$\frac{1}{2\pi}\int_0^{2\pi} \|u(re^{it})\|^2_{\mathfrak{A}}\, dt \qquad (0 \leq r < 1)$$

has a bound independent of r. This integral being equal to $\sum_0^\infty r^{2k}\|a_k\|^2_{\mathfrak{A}}$, the last condition is equivalent to the condition $\sum_0^\infty \|a_k\|^2_{\mathfrak{A}} < \infty$. Hence we see that every function $u(\lambda) \in H^2(\mathfrak{A})$ arises from a function $v(t) \in L^2_+(\mathfrak{A})$, indeed from $v(t) = \sum_0^\infty e^{ikt} a_k$. Because $u(\lambda)$ and $v(t)$ determine each other, we can identify the classes $H^2(\mathfrak{A})$ and $L^2_+(\mathfrak{A})$, thus providing $H^2(\mathfrak{A})$ with the Hilbert space structure of $L^2_+(\mathfrak{A})$ and embedding it in $L^2(\mathfrak{A})$ as a subspace.

2. The functions $u(\lambda)$ and $v(t)$ thus associated are also connected by Poisson's formula

$$u(re^{it}) = \frac{1}{2\pi}\int_0^{2\pi} P_r(t-s)v(s)\,ds \qquad (0 \leq r < 1), \tag{1.6}$$

where

$$P_r(t) = \frac{1-r^2}{1-2r\cos t + r^2}. \tag{1.7}$$

This is a simple consequence of the elementary relation

$$(re^{it})^k = \frac{1}{2\pi}\int_0^{2\pi} P_r(t-s)e^{iks}\,ds \qquad (k = 0, 1, \ldots)$$

and of the fact that $v(t)$ is the limit in the mean of its Fourier series.

Making use of the formula (1.6) we can prove that $v(t)$ is the radial limit of $u(re^{it})$ not only in the mean but also pointwise, almost everywhere. More precisely, $u(\lambda)$ tends to $v(t)$ strongly (in $\mathfrak{A}$) as λ tends to e^{it} nontangentially with respect to the unit circle at every point t such that

$$\frac{1}{2s}\int_{t-s}^{t+s} v(\tau)\,d\tau \to v(t) \quad \text{strongly} \quad (s \to 0), \tag{1.8}$$

thus a.e. (generalized Fatou theorem). The proof is the same as in the scalar-valued case; see HOFFMAN [1].

By virtue of this theorem we may write $u(e^{it})$ instead of $v(t)$ when considering functions in $L^2_+(\mathfrak{A})$.

2 Inner and outer functions

1. Consider a function $\Theta(\lambda)$ whose values are bounded operators from a Hilbert space $\mathfrak{A}$ to a Hilbert space $\mathfrak{A}_*$, both separable, and which has a power series expansion

$$\Theta(\lambda) = \sum_0^\infty \lambda^k \Theta_k \tag{2.1}$$

whose coefficients are bounded operators from $\mathfrak{A}$ to $\mathfrak{A}_*$; the series is supposed to be convergent in the open unit disc D (weakly, strongly, or in the norm, which amounts to the same for power series). Let us suppose, moreover, that

$$\|\Theta(\lambda)\| \leq M \quad \text{(bounded independent of } \lambda \text{ in } D). \tag{2.2}$$

Such a function $\{\mathfrak{A}, \mathfrak{A}_*, \Theta(\lambda)\}$ is called a *bounded analytic function* (on D).

Condition (2.2) implies

$$\frac{1}{2\pi}\int_0^{2\pi} \|\Theta(re^{it})a\|^2_{\mathfrak{A}_*}\,dt \leq M^2\|a\|^2_{\mathfrak{A}} \quad (0 \leq r < 1)$$

and consequently (by Sec. 1)

$$\sum_0^\infty \|\Theta_k a\|^2_{\mathfrak{A}_*} \leq M^2 \|a\|^2_{\mathfrak{A}} \tag{2.3}$$

for all $a \in \mathfrak{A}$. It follows by virtue of Sec. 1 that the limit

$$\lim \Theta(\lambda)a \quad (\lambda \to e^{it} \text{ nontangentially}) \tag{2.4}$$

exists in the strong sense in $\mathfrak{A}_*$, everywhere, with the possible exception of the points t of a set E_a of zero measure. Letting a run over a countable dense subset of $\mathfrak{A}$ and taking the union of the corresponding sets E_a we obtain a set E of zero measure, such that, on account of (2.2), the limit (2.4) will exist for every t not belonging to E and for every $a \in \mathfrak{A}$.

Hence

$$\Theta(e^{it}) = \lim \Theta(\lambda) \quad (\lambda \to e^{it} \text{ nontangentially}) \tag{2.5}$$

exists a.e. as a strong limit of operators. In particular, we have a.e.

$$\Theta(e^{it}) = \lim_{r \to 1-0} \Theta(re^{it}) \quad \text{(strongly)}. \tag{2.6}$$

Moreover, also by virtue of Sec. 1, $\Theta(re^{it})a$ converges in the mean (i.e., in $L^2(\mathfrak{A}_*)$) to $\Theta(e^{it})a$ as $r \to 1-0$, and this limit has the Fourier expansion

$$\Theta(e^{it})a = \sum_0^\infty e^{ikt}\Theta_k a, \tag{2.7}$$

which converges in $L^2(\mathfrak{A}_*)$.

With the bounded analytic function $\{\mathfrak{A}, \mathfrak{A}_*, \Theta(\lambda)\}$ we associate its "adjoint" $\{\mathfrak{A}_*, \mathfrak{A}, \Theta^\sim(\lambda)\}$ defined by

$$\Theta^\sim(\lambda) = \Theta(\bar{\lambda})^* \qquad (\lambda \in D) \tag{2.8}$$

which is also analytic and bounded by the same bound; in fact we have

$$\Theta^\sim(\lambda) = \sum_0^\infty \lambda^k \Theta_k^*$$

and $\|\Theta^\sim(\lambda)\| = \|\Theta(\bar{\lambda})^*\| = \|\Theta(\bar{\lambda})\| \leq M$ for $\lambda \in D$. Consequently

$$\Theta^\sim(e^{it}) = \lim \Theta^\sim(\lambda) \quad (\lambda \to e^{it} \text{ nontangentially})$$

exists a.e. as a strong limit of operators. This implies that $\Theta^\sim(e^{-it})$ is a.e. the strong limit of $\Theta(\lambda)^*$ as $\lambda \to e^{it}$ nontangentially. On the other hand, $\Theta(\lambda)$ tends to $\Theta(e^{it})$ strongly, hence $\Theta(\lambda)^*$ tends to $\Theta(e^{it})^*$ weakly, as $\lambda \to e^{it}$ nontangentially, for almost every t. Thus for almost every t, $\Theta^\sim(e^{-it}) = \Theta(e^{it})^*$ and

$$\Theta(e^{it})^* = \lim \Theta(\lambda)^* \quad (\text{as } \lambda \to e^{it} \text{ nontangentially}) \tag{2.9}$$

in the sense of the strong convergence of operators.

In particular, we have

$$\Theta(e^{it})^* = \lim_{r\to 1-0} \Theta(re^{it})^* \quad \text{strongly, a.e.} \tag{2.10}$$

2. In the sequel we deal mostly with analytic functions such that $\|\Theta(\lambda)\| \leq 1$ on D: we call them *contractive* analytic functions. Such a function $\{\mathfrak{A}, \mathfrak{A}_*, \Theta(\lambda)\}$ is said to be *purely contractive* if

$$\|\Theta(0)a\| < \|a\| \quad \text{for all} \quad a \in \mathfrak{A}, a \neq 0.$$

At the other extreme, if $\Theta(0)$ is a unitary operator from $\mathfrak{A}$ onto $\mathfrak{A}_*$, then the maximum principle implies (see the proof of the following result) that $\Theta(\lambda) = \Theta(0)$ for every $\lambda \in D$. Such a function is called a *unitary constant.*

Proposition 2.1. *For every contractive analytic function* $\{\mathfrak{A}, \mathfrak{A}_*, \Theta(\lambda)\}$ *there exist uniquely determined decompositions* $\mathfrak{A} = \mathfrak{A}^0 \oplus \mathfrak{A}'$, $\mathfrak{A}_* = \mathfrak{A}_*^0 \oplus \mathfrak{A}_*'$ *so that, for every fixed* λ, $\Theta^0(\lambda) = \Theta(\lambda)|\mathfrak{A}^0$ *has its range in* $\mathfrak{A}_*^0$ *and* $\Theta'(\lambda) = \Theta(\lambda)|\mathfrak{A}'$ *has its range in* $\mathfrak{A}_*'$, *and that* $\{\mathfrak{A}^0, \mathfrak{A}_*^0, \Theta^0(\lambda)\}$ *is a purely contractive analytic function, and* $\{\mathfrak{A}', \mathfrak{A}_*', \Theta'(\lambda)\}$ *is a unitary constant. The function* $\Theta^0(\lambda)$ *is called the "purely contractive part" of* $\Theta(\lambda)$.

Proof. Let us set

$$\mathfrak{A}' = \{a\colon\ a \in \mathfrak{A}, a = \Theta(0)^*\Theta(0)a\}, \quad \mathfrak{A}_* = \{a_*\colon\ a_* \in \mathfrak{A}_*, a_* = \Theta(0)\Theta(0)^*a_*\}.$$

For $a \in \mathfrak{A}'$ we have $\Theta(0)a = \Theta(0)\Theta(0)^*\Theta(0)a$, and hence $\Theta(0)a \in \mathfrak{A}_*'$, that is, $\Theta(0)\mathfrak{A}' \subset \mathfrak{A}_*'$. By similar reasoning, $\Theta(0)^*\mathfrak{A}_*' \subset \mathfrak{A}'$; by virtue of the relation $a_* = \Theta(0)\Theta(0)^*a_*$ $(a_* \in \mathfrak{A}_*')$ this implies $\mathfrak{A}_*' \subset \Theta(0)\mathfrak{A}'$. Thus $\Theta(0)$ maps $\mathfrak{A}'$ onto $\mathfrak{A}_*'$. This is a *unitary* operator, because for any $a \in \mathfrak{A}'$ we have

$$\|\Theta(0)a\|^2 = (\Theta(0)^*\Theta(0)a, a) = (a, a) = \|a\|^2.$$

For $a \in \mathfrak{A}$ we have $f_a(\lambda) = \Theta(\lambda)a \in H^2(\mathfrak{A}_*)$. If in particular $a \in \mathfrak{A}'$, then

$$\begin{aligned}
(f_a, \Theta(0)a)_{H^2(\mathfrak{A}_*)} &= \frac{1}{2\pi}\int_0^{2\pi} (\Theta(e^{it})a, \Theta(0)a)\, dt \\
&= (\Theta(0)a, \Theta(0)a) = \|a\|^2 = \|a\|\,\|\Theta(0)a\| \\
&\geq \left[\frac{1}{2\pi}\int_0^{2\pi} \|\Theta(e^{it})a\|^2\, dt\right]^{1/2} \|\Theta(0)a\| \\
&= \|f_a\|_{H^2(\mathfrak{A}_*)} \|\Theta(0)a\|_{H^2(\mathfrak{A}_*)}.
\end{aligned}$$

By virtue of the Schwarz inequality we have therefore that $f_a(e^{it}) = \alpha\Theta(0)a$ for almost every t, and hence $f_a(\lambda) = \alpha\Theta(0)a$ for every $\lambda \in D$ and some numerical constant α. Obviously $\alpha = 1$ if $a \neq 0$ and hence $\Theta(\lambda) = \Theta(0)a$ for every $a \in \mathfrak{A}'$. Thus $\Theta'(\lambda) = \Theta(\lambda)|\mathfrak{A}'$ is a constant unitary operator from $\mathfrak{A}'$ onto $\mathfrak{A}_*'$ for $\lambda \in D$.

If we replace in this reasoning $\Theta(\lambda)$ by $\Theta^{\tilde{}}(\lambda)$, the spaces $\mathfrak{A}'$ and $\mathfrak{A}'_*$ interchange their roles and we obtain that the function $\Theta^{\tilde{}}(\lambda)|\mathfrak{A}'_*$ is on D a constant unitary operator from $\mathfrak{A}'_*$ onto $\mathfrak{A}'$.

Consequently, we have for all $a \in \mathfrak{A}^0 = \mathfrak{A} \ominus \mathfrak{A}'$ and for all $a_* \in \mathfrak{A}'_*$:

$$(\Theta(\lambda)a, a_*) = (a, \Theta(\lambda)^* a_*) = (a, \Theta^{\tilde{}}(\bar{\lambda})a_*) = 0 \qquad (\lambda \in D);$$

hence $\Theta(\lambda)a \in \mathfrak{A}^0_* = \mathfrak{A}_* \ominus \mathfrak{A}'_*$; that is, $\Theta^0(\lambda) = \Theta(\lambda)|\mathfrak{A}^0$ maps $\mathfrak{A}^0$ into $\mathfrak{A}^0_*$.

The analytic function $\{\mathfrak{A}^0, \mathfrak{A}^0_*, \Theta^0(\lambda)\}$ thus obtained is *purely contractive.* In fact, if $\|\Theta^0(0)a\| = \|a\|$ for an $a \in \mathfrak{A}^0$, then

$$((I - \Theta(0)^*\Theta(0))a, a) = \|a\|^2 - \|\Theta(0)a\|^2 = \|a\|^2 - \|\Theta^0(0)a\|^2 = 0,$$

and hence $(I - \Theta(0)^*\Theta(0))a = 0$, that is, $a \in \mathfrak{A}'$. As $\mathfrak{A}' \perp \mathfrak{A}^0$, this implies $a = 0$.

The decompositions $\mathfrak{A} = \mathfrak{A}^0 \oplus \mathfrak{A}'$, $\mathfrak{A}_* = \mathfrak{A}^0_* \oplus \mathfrak{A}'_*$ constructed above satisfy therefore the condition stated in the Proposition. It remains to prove uniqueness. To this effect, consider any decompositions $\mathfrak{A} = \mathfrak{B}^0 \oplus \mathfrak{B}'$, $\mathfrak{A}_* = \mathfrak{B}^0_* \oplus \mathfrak{B}'_*$ satisfying the same conditions. Because $\Theta(0)$ maps $\mathfrak{B}'$ unitarily onto $\mathfrak{B}'_*$, we have $\|a\| = \|\Theta(0)a\|$ for $a \in \mathfrak{B}'$, and hence $(I - \Theta(0)^*\Theta(0))a = 0$ (i.e.,$a \in \mathfrak{A}'$); thus $\mathfrak{B}' \subset \mathfrak{A}'$. If there existed in $\mathfrak{A}'$ an element $a \neq 0$ orthogonal to $\mathfrak{B}'$, we should have $\|\Theta(0)a\| = \|a\|$ because $a \in \mathfrak{A}'$, and $\|\Theta(0)a\| < \|a\|$ because $a \in \mathfrak{B}^0$; a contradiction. Thus $\mathfrak{B}' = \mathfrak{A}'$, and hence $\mathfrak{B}'_* = \Theta(\lambda)\mathfrak{B}' = \Theta(\lambda)\mathfrak{A}' = \mathfrak{A}'_*$, $\mathfrak{B}^0 = \mathfrak{A} \ominus \mathfrak{B}' = \mathfrak{A} \ominus \mathfrak{A}' = \mathfrak{A}^0$, $\mathfrak{B}^0_* = \mathfrak{A}_* \ominus \mathfrak{B}'_* = \mathfrak{A}_* \ominus \mathfrak{A}'_* = \mathfrak{A}^0_*$. This completes the proof.

3. With every bounded analytic function $\{\mathfrak{A}, \mathfrak{A}_*, \Theta(\lambda)\}$ we associate the operator Θ from $L^2(\mathfrak{A})$ into $L^2(\mathfrak{A}_*)$ defined by

$$(\Theta v)(t) = \Theta(e^{it})v(t) \quad \text{for} \quad v \in L^2(\mathfrak{A}),$$

and the operator Θ_+ from $H^2(\mathfrak{A})$ into $H^2(\mathfrak{A}_*)$ defined by

$$(\Theta_+ u)(\lambda) = \Theta(\lambda)u(\lambda) \quad \text{for} \quad u \in H^2(\mathfrak{A}).$$

By virtue of the possible identification of $H^2(\mathfrak{A})$ with the subspace $L^2_+(\mathfrak{A})$ of $L^2(\mathfrak{A})$ (and the same for $\mathfrak{A}_*$) we can consider Θ_+ as the restriction of Θ to the subspace $H^2(\mathfrak{A})$ of $L^2(\mathfrak{A})$. Obviously, we have

$$\|\Theta\| = \operatorname*{ess\,sup}_{0 \leq t \leq 2\pi} \|\Theta(e^{it})\| = \sup_{|\lambda < 1} \|\Theta(\lambda)\|$$

so that if the function $\Theta(\lambda)$ is contractive then the operators Θ and Θ_+ are contractions. Let us also observe that for the adjoint of the operator Θ we have

$$(\Theta^* v)(t) = \Theta(e^{it})^* v(t).$$

Proposition 2.2. *In order that the operator Θ_+ from $H^2(\mathfrak{A})$ into $H^2(\mathfrak{A}_*)$ be an isometry it is necessary and sufficient that $\Theta(e^{it})$ be an isometry from $\mathfrak{A}$ into $\mathfrak{A}_*$ for almost every value of t.*

Proof. The sufficiency of the condition is obvious. To prove its necessity suppose that Θ_+ is an isometry, that is, $\|\Theta_+ u\| = \|u\|$ for all $u \in H^2(\mathfrak{A})$. From this we infer that Θ also is an isometry. This follows from the obvious relations

$$\|\Theta e^{-int}u\| = \|e^{-int}\Theta u\| = \|\Theta_+ u\| = \|u\| = \|e^{-int}u\|$$

for $u \in H^2(\mathfrak{A})$ and $n \geq 1$, and from the fact that the functions $e^{-int}u(e^{it})$ ($u \in H^2(\mathfrak{A})$; $n \geq 1$) are dense in $L^2(\mathfrak{A})$. Thus the relation

$$\int_0^{2\pi} \|\Theta(e^{it})v(t)\|^2_{\mathfrak{A}_*}\, dt = \int_0^{2\pi} \|v(t)\|^2_{\mathfrak{A}}\, dt$$

holds true for every function $v \in L^2(\mathfrak{A})$.

Let us choose in particular $v(t) = \varepsilon(\tau,\delta;t)a$, where $a \in \mathfrak{A}$ and $\varepsilon(\tau,\delta;t)$ is the characteristic function of the interval $(\tau,\tau+\delta)$. Thus we obtain, dividing by δ,

$$\frac{1}{\delta}\int_\tau^{\tau+\delta} \|\Theta(e^{it})a\|^2_{\mathfrak{A}_*}\, dt = \|a\|^2_{\mathfrak{A}},$$

and this implies

$$\|\Theta(e^{it})a\|_{\mathfrak{A}_*} = \|a\|_{\mathfrak{A}} \tag{2.11}$$

almost everywhere, the set E_a of the exceptional points t depending on a. Letting a run over a countable set $\{a_n\}$ dense in $\mathfrak{A}$, and taking the union of the corresponding sets E_a, we obtain a set E of zero measure; if $t \notin E$, equation (2.11) holds simultaneously for every a_n, and hence for every $a \in \mathfrak{A}$; thus $\Theta(e^{it})$ is an isometry.

Now we make the following definition.

Definitions. The contractive analytic function $\{\mathfrak{A},\mathfrak{A}_*,\Theta(\lambda)\}$ is said to be

(i) *Inner* if $\Theta(e^{it})$ is an isometry from $\mathfrak{A}$ into $\mathfrak{A}_*$ for almost every t, or equivalently (see Proposition 2.2), if Θ_+ is an isometry from $H^2(\mathfrak{A})$ into $H^2(\mathfrak{A}_*)$.
(ii) *Outer* if $\overline{\Theta_+ H^2(\mathfrak{A})} = H^2(\mathfrak{A}_*)$, the closure being taken in $L^2(\mathfrak{A}_*)$.
(iii) $*$-*inner* if the function $\{\mathfrak{A}_*,\mathfrak{A},\Theta^{\tilde{}}(\lambda)\}$ is inner.
(iv) $*$-*outer* if the function $\{\mathfrak{A}_*,\mathfrak{A},\Theta^{\tilde{}}(\lambda)\}$ is outer.
(v) *Inner from both sides* if it is both inner and $*$-inner, that is, if $\Theta(e^{it})$ is unitary a.e.
(vi) *Outer from both sides* if it is both outer and $*$-outer.

In the scalar case (i.e., if both $\mathfrak{A}$ and $\mathfrak{A}_*$ are of dimension 1) definitions (i) and (ii) reduce to those given in Sec. III.1.1. Indeed, for a scalar-valued analytic function $u(\lambda)$ such that $|u(\lambda)| \leq 1$ on D, (i) means that $|u(e^{it})| = 1$ a.e., and (ii) means that the functions $\{\lambda^k u(\lambda)\}_{k=0}^\infty$ span the space H^2, and this property characterizes outer functions by Beurling's theorem; see Sec. III.1.2. Moreover, it is obvious that in the scalar case every inner function is also $*$-inner, and every outer function is also $*$-outer.

Proposition 2.3. *The only contractive analytic functions* $\{\mathfrak{A},\mathfrak{A}_*,\Theta(\lambda)\}$ *that are simultaneously inner and outer, are the constant unitary functions, that is, for which* $\Theta(\lambda)\equiv\Theta_0$, *where* Θ_0 *is a unitary operator from* $\mathfrak{A}$ *to* $\mathfrak{A}_*$.

Proof. The fact that $\{\mathfrak{A},\mathfrak{A}_*,\Theta(\lambda)\}$ is simultaneously inner and outer means that Θ_+ is a unitary operator from $H^2(\mathfrak{A})$ to $H^2(\mathfrak{A}_*)$. As Θ_+ commutes with multiplication by the variable λ, we have

$$\Theta_+[H^2(\mathfrak{A})\ominus\lambda\cdot H^2(\mathfrak{A})]=H^2(\mathfrak{A}_*)\ominus\lambda\cdot H^2(\mathfrak{A}_*).$$

Now $H^2(\mathfrak{A})\ominus\lambda\cdot H^2(\mathfrak{A})$ and $H^2(\mathfrak{A}_*)\ominus\lambda\cdot H^2(\mathfrak{A}_*)$ consist of the constant functions with values in $\mathfrak{A}$ and in $\mathfrak{A}_*$, respectively. For every $a\in\mathfrak{A}$ we have therefore $\Theta(\lambda)a\equiv a_*$ with some $a_*\in\mathfrak{A}_*$, and a_* runs over $\mathfrak{A}_*$ when a runs over $\mathfrak{A}$. Thus $\Theta(\lambda)$ is constant, its value being an operator from $\mathfrak{A}$ onto $\mathfrak{A}_*$. The limit values of $\Theta(\lambda)$ on the unit circle have to be isometries, therefore we conclude that the constant value of $\Theta(\lambda)$ is a unitary operator. This proves one implication in the proposition. The other one is obvious.

Proposition 2.4. *For every outer function* $\{\mathfrak{A},\mathfrak{A}_*,\Theta(\lambda)\}$ *we have*

(a) $\overline{\Theta(\lambda)\mathfrak{A}}=\mathfrak{A}_*$ *for all* $\lambda\in D$
(b) $\overline{\Theta(e^{it})\mathfrak{A}}=\mathfrak{A}_*$ *for almost all* $t\in(0,2\pi)$

the closure being taken in $\mathfrak{A}_*$.

Proof. Part (a): By virtue of the Cauchy integral formula we have

$$\begin{aligned}(\Theta u,(1-\bar{\lambda}_0\lambda)^{-1}a_*)_{L^2(\mathfrak{A}_*)}&=\frac{1}{2\pi}\int_0^{2\pi}\frac{(\Theta(e^{it})u(e^{it}),a_*)_{\mathfrak{A}_*}}{1-\bar{\lambda}_0e^{-it}}\,dt\\&=(\Theta(\lambda_0)u(\lambda_0),a_*)_{\mathfrak{A}_*}\end{aligned}$$

for $u\in H^2(\mathfrak{A})$, $a_*\in\mathfrak{A}_*$, and $|\lambda_0|<1$. Hence, if a_* is orthogonal to $\Theta(\lambda_0)\mathfrak{A}$ and $a_*\neq0$, then $(1-\bar{\lambda}_0\lambda)^{-1}a_*$ is a nonzero element of $H^2(\mathfrak{A}_*)$, orthogonal to $\Theta H^2(\mathfrak{A})$.

Part (b): Because $\Theta H^2(\mathfrak{A})$ is dense in $H^2(\mathfrak{A}_*)$, it follows in particular that for every constant function $a_*(\lambda)\equiv a_*\in\mathfrak{A}_*$ there exists a sequence of elements $u_n\in H^2(\mathfrak{A})$ such that Θu_n converges to a_* in the mean, and also pointwise a.e. So we have

$$\Theta(e^{it})u_n(e^{it})\to a_*\quad(n\to\infty)$$

at every point t with the possible exception of the points t of a set E_{a_*} of zero measure. Consequently, for t not in E_{a_*}, the closure of $\Theta(e^{it})\mathfrak{A}$ contains the vector a_*. Let a_* run over a countable set $\{a_{*n}\}$ dense in $\mathfrak{A}_*$, and let E denote the union of the sets $E_{a_{*n}}$; E is also of zero measure. For t not in E the closure of $\Theta(e^{it})\mathfrak{A}$ will contain all the vectors a_{*n}, hence it will coincide with the whole space $\mathfrak{A}_*$.

As an immediate corollary of Propositions 2.2 and 2.4 we state that *if the function* $\{\mathfrak{A},\mathfrak{A}_*,\Theta(\lambda)\}$ *is inner, then* $\dim\mathfrak{A}\leq\dim\mathfrak{A}_*$, whereas *if it is outer, then* $\dim\mathfrak{A}\geq\dim\mathfrak{A}_*$.

4. The following definition extends a notion already used for inner functions in H^∞.

Definition. We say that the functions $\{\mathfrak{A}, \mathfrak{A}_*, \Theta(\lambda)\}$ and $\{\mathfrak{A}', \mathfrak{A}'_*, \Theta'(\lambda)\}$ $(\lambda \in D)$ *coincide* if there exist a unitary operator τ from $\mathfrak{A}$ to $\mathfrak{A}'$ and a unitary operator τ_* from $\mathfrak{A}_*$ to $\mathfrak{A}'_*$ such that $\Theta'(\lambda) = \tau_* \Theta(\lambda) \tau^{-1}$ $(|\lambda| < 1)$.

3 Lemmas on Fourier representations

1. Let U be a bilateral shift on a (complex, separable) Hilbert space $\mathfrak{R}$ and let $\mathfrak{A}$ be a generating subspace for U (i.e., such that $\mathfrak{R} = M(\mathfrak{A})$), where

$$M(\mathfrak{A}) = \bigoplus_{-\infty}^{\infty} U^k \mathfrak{A}.$$

We denote by $\Phi^{\mathfrak{A}}$ the unitary transformation from $M(\mathfrak{A})$ to $L^2(\mathfrak{A})$ defined by

$$\left[\Phi^{\mathfrak{A}} \sum_{-\infty}^{\infty} U^k a_k\right](t) = \sum_{-\infty}^{\infty} e^{ikt} a_k \quad \left(a_k \in \mathfrak{A}; \sum_{-\infty}^{\infty} \|a_k\|^2 < \infty\right), \tag{3.1}$$

the series on the right-hand side being convergent in the mean. If we denote by $U^\times$ the operator of multiplication by e^{it} on the space $L^2(\mathfrak{A})$, then we have

$$\Phi^{\mathfrak{A}} U = U^\times \Phi^{\mathfrak{A}}; \tag{3.2}$$

that is, $\Phi^{\mathfrak{A}}$ transforms U to $U^\times$. We call $\Phi^{\mathfrak{A}}$ the *Fourier representation* of $M(\mathfrak{A})$. Similarly, if U_+ is a unilateral shift on a (complex, separable) Hilbert space $\mathfrak{R}_+$ with the generating subspace $\mathfrak{A}$, that is, $\mathfrak{R}_+ = M_+(\mathfrak{A})$, where

$$M_+(\mathfrak{A}) = \bigoplus_{0}^{\infty} U_+^n \mathfrak{A},$$

then the *Fourier representation* of $M_+(\mathfrak{A})$, denoted by $\Phi_+^{\mathfrak{A}}$, is the unitary transformation from $M_+(\mathfrak{A})$ to $H^2(\mathfrak{A})$ defined by

$$\left[\Phi_+^{\mathfrak{A}} \sum_{0}^{\infty} U_+^k a_k\right](\lambda) = \sum_{0}^{\infty} \lambda^k a_k \quad \left(a_k \in \mathfrak{A}; \sum_{0}^{\infty} \|a_k\|^2 < \infty; |\lambda| < 1\right). \tag{3.3}$$

If $U_+^\times$ denotes multiplication by λ on $H^2(\mathfrak{A})$, we have in analogy to (3.2):

$$\Phi_+^{\mathfrak{A}} U_+ = U_+^\times \Phi_+^{\mathfrak{A}}. \tag{3.4}$$

2. The following lemmas play an important role in the sequel.

Lemma 3.1. *Let U and U' be bilateral shifts on the (complex, separable) Hilbert spaces $\mathfrak{R}$ and $\mathfrak{R}'$, with the generating subspaces $\mathfrak{A}$ and $\mathfrak{A}'$, respectively. Let Q be a* *contraction of* $\mathfrak{R}$ *into* $\mathfrak{R}'$ *such that*

$$QU = U'Q \tag{3.5}$$

and

$$QM_+(\mathfrak{A}) \subset M_+(\mathfrak{A}'). \tag{3.6}$$

Then there exists a unique contractive analytic function $\{\mathfrak{A}, \mathfrak{A}', \Theta(\lambda)\}$ *such that*

$$\Phi^{\mathfrak{A}'} Q = \Theta \Phi^{\mathfrak{A}}. \tag{3.7}$$

Proof. By virtue of (3.6) we have in particular for $a \in \mathfrak{A}$:

$$Qa = \sum_0^\infty U'^k a'_k \quad \text{with} \quad a'_k \in \mathfrak{A}', \quad \sum_0^\infty \|a'_k\|^2 = \|Qa\|^2 \leq \|a\|^2.$$

Setting $a'_k = \Theta_k a$ we define a sequence of bounded operators Θ_k $(k = 0, 1, \ldots)$, indeed contractions of $\mathfrak{A}$ into $\mathfrak{A}'$. So we have

$$\Phi^{\mathfrak{A}'} Qa = v_a, \quad \text{where} \quad v_a(t) = \sum_0^\infty e^{ikt}(\Theta_k a) \quad (\text{convergence in } L^2(\mathfrak{A}')).$$

On account of (3.5) and (3.2) this implies

$$\Phi^{\mathfrak{A}'} Q\varphi(U)a = \Phi^{\mathfrak{A}'} \varphi(U')Qa = \varphi(U^\times)\Phi^{\mathfrak{A}'} Qa = \varphi v_a$$

for every trigonometric polynomial $\varphi(e^{it})$ with scalar coefficients, and hence

$$\begin{aligned} \|\varphi(U)a\|^2_{\mathfrak{R}} \geq \|Q\varphi(U)a\|^2_{\mathfrak{R}'} &= \|\Phi^{\mathfrak{A}'} Q\varphi(U)a\|^2_{L^2(\mathfrak{A}')} \\ &= \frac{1}{2\pi} \int_0^{2\pi} |\varphi(e^{it})|^2 \|v_a(t)\|^2_{\mathfrak{A}'} \, dt. \end{aligned}$$

On the other hand we have

$$\begin{aligned} \|\varphi(U)a\|^2_{\mathfrak{R}} = \|\Phi^{\mathfrak{A}} \varphi(U)a\|^2_{L^2(\mathfrak{A})} &= \|\varphi(U^\times)\Phi^{\mathfrak{A}} a\|^2_{L^2(\mathfrak{A})} = \|\varphi a\|^2_{L^2(\mathfrak{A})} \\ &= \frac{1}{2\pi} \int_0^{2\pi} |\varphi(e^{it})|^2 \, dt \cdot \|a\|^2_{\mathfrak{A}}. \end{aligned}$$

These two results together imply

$$\int_0^{2\pi} |\varphi(e^{it})|^2 \|v_a(t)\|^2_{\mathfrak{A}'} \, dt \leq \int_0^{2\pi} |\varphi(e^{it})|^2 \, dt \cdot \|a\|^2_{\mathfrak{A}}.$$

This inequality extends by virtue of the Weierstrass approximation theorem to all continuous functions $\rho(t) \geq 0$ of period 2π, thus

$$\int_0^{2\pi} \rho(t) \|v_a(t)\|^2_{\mathfrak{A}'} \, dt \leq \int_0^{2\pi} \rho(t) \, dt \cdot \|a\|^2_{\mathfrak{A}}.$$

Next, using Lebesgue's dominated convergence theorem, we extend this inequality to all functions $\rho(t) \geq 0$, measurable and bounded on $(0, 2\pi)$. Choosing for $\rho(t)$ in particular the characteristic function, divided by δ, of the interval $(\tau, \tau + \delta)$, and letting $\delta \to 0$ we obtain that

$$\|v_a(t)\|_{\mathfrak{A}'} \leq \|a\|_{\mathfrak{A}} \quad \text{a.e.} \tag{3.8}$$

By virtue of the Poisson formula (1.6) connecting the function $v_a(t)$ with the associated function

$$u_a(\lambda) = \sum_0^\infty \lambda^k \Theta_k a \quad (\lambda \in D) \tag{3.9}$$

of class $H^2(\mathfrak{A}')$, and by well-known properties of the kernel $P_r(t)$, inequality (3.8) implies

$$\|u_a(\lambda)\|_{\mathfrak{A}'} \leq \|a\|_{\mathfrak{A}} \quad (\lambda \in D). \tag{3.10}$$

As the series (3.9) converges strongly in $\mathfrak{A}'$ for each $a \in \mathfrak{A}$, the operator series

$$\Theta(\lambda) = \sum_0^\infty \lambda^k \Theta_k \quad (\lambda \in D)$$

also converges and we have $\Theta(\lambda)a = u_a(\lambda)$; on account of (3.10), $\{\mathfrak{A}, \mathfrak{A}', \Theta(\lambda)\}$ is thus a contractive analytic function. Therefore the nontangential strong operator limit $\Theta(e^{it})$ exists a.e., and we have a.e.:

$$\Theta(e^{it})a = \lim \Theta(\lambda)a = \lim u_a(\lambda) = v_a(t) \quad (\lambda \to e^{it} \text{ nontangentially});$$

thus

$$\Phi^{\mathfrak{A}'} Qa = \Theta(e^{it})a \quad (a \in \mathfrak{A}). \tag{3.11}$$

From (3.5), (3.2) and (3.11) we deduce

$$\begin{aligned} \Phi^{\mathfrak{A}'} Q \sum_k U^k a_k &= \sum_k \Phi^{\mathfrak{A}'} U'^k Q a_k = \sum_k e^{ikt} \Phi^{\mathfrak{A}'} Q a_k \\ &= \sum_k e^{ikt} \Theta(e^{it}) a_k = \Theta(e^{it}) \sum_k e^{ikt} a_k = \Theta(e^{it}) \Phi^{\mathfrak{A}} \sum_k U^k a_k. \end{aligned}$$

Hence it follows

$$\Phi^{\mathfrak{A}'} Qh = \Theta \Phi^{\mathfrak{A}} h$$

first for finite sums $h = \sum_k U^k a_k$ ($a_k \in \mathfrak{A}$), and then by continuity for all $h \in M(\mathfrak{A})$. This completes the proof of Lemma 3.1.

We deduce now an analogous lemma for unilateral shifts along with some results on possible properties of the function $\Theta(\lambda)$.

Lemma 3.2. *Let U_+ and U'_+ be unilateral shifts on the (complex, separable) Hilbert spaces $\mathfrak{R}_+$ and $\mathfrak{R}'_+$, and let $\mathfrak{A}$ and $\mathfrak{A}'$ be the corresponding generating subspaces. Let Q be a contraction of $\mathfrak{R}_+$ into $\mathfrak{R}'_+$ such that*

$$QU_+ = U'_+ Q. \tag{3.12}$$

Then there exists a unique contractive analytic function $\{\mathfrak{A}, \mathfrak{A}', \Theta(\lambda)\}$ *such that*

$$\Phi_+^{\mathfrak{A}'} Q = \Theta_+ \Phi_+^{\mathfrak{A}}. \tag{3.13}$$

In order that this function be

(a) *Purely contractive*
(b) *Inner*
(c) *Outer*
(d) *A unitary constant*
(e) *Boundedly invertible, with a uniform bound k, that is, with*

$$\|\Theta(\lambda)^{-1}\| \leq k \quad \textit{for} \quad \lambda \in D$$

(f) *A constant operator*

it is necessary and sufficient that the following conditions hold, respectively.

(a) $\|P_{\mathfrak{A}'} Qa\| < \|a\|$ *for all nonzero* $a \in \mathfrak{A}$, $P_{\mathfrak{A}'}$ *denoting the orthogonal projection of* $\mathfrak{R}'_+$ *into* $\mathfrak{A}'$.
(b) Q *is an isometry from* $\mathfrak{R}_+$ *into* $\mathfrak{R}'_+$.
(c) $\overline{Q\mathfrak{R}_+} = \mathfrak{R}'_+$.
(d) Q *is unitary from* $\mathfrak{R}_+$ *to* $\mathfrak{R}'_+$.
(e) Q *is boundedly invertible with* $\|Q^{-1}\| \leq k$.
(f) $Q^* U'_+ = U_+ Q^*$.

Proof. By virtue of Proposition I.2.2 the unilateral shifts U_+, U'_+ can be extended to bilateral shifts U, U' on the spaces $\mathfrak{R}, \mathfrak{R}'$, respectively, with multiplicities preserved: we have

$$\mathfrak{R} = \bigoplus_{-\infty}^{\infty} U^n \mathfrak{A}, \quad \mathfrak{R}' = \bigoplus_{-\infty}^{\infty} U'^n \mathfrak{A}'.$$

If $-\infty < p < q < \infty$ and $a_n \in \mathfrak{A}$, then

$$\sum_p^q U^n a_n = U^p \sum_p^q U^{n-p} a_n = U^p \sum_p^q U_+^{n-p} a_n$$

and

$$\sum_p^q U'^n Q a_n = U'^p \sum_p^q U'^{n-p} Q a_n = U'^p \sum_p^q U'^{n-p}_+ Q a_n = U'^p Q \sum_p^q U_+^{n-p} a_n;$$

hence

$$\left\| \sum_p^q U'^n Q a_n \right\| = \left\| Q \sum_p^q U_+^{n-p} a_n \right\| \leq \left\| \sum_p^q U_+^{n-p} a_n \right\| = \left\| \sum_p^q U^n a_n \right\|.$$

This shows that setting

$$\widehat{Q} \sum_p^q U^n a_n = \sum_p^q U'^n Q a_n$$

we obtain a contraction $\widehat{Q}$ of the linear manifold of all vectors of the form $\sum_p^q U^n a_n$, into $\mathfrak{R}'$. Now $\widehat{Q}$ extends by continuity to a contraction of $\mathfrak{R}$ into $\mathfrak{R}'$, which we denote by the same letter $\widehat{Q}$. We have

$$\widehat{Q}\sum_0^q U_+^n a_n = \sum_0^q U'^n_+ Q a_n = Q\sum_0^q U_+^n a_n \quad (q \geq 0),$$

and hence it follows that $\widehat{Q} \supset Q$. Moreover, we have

$$\widehat{Q}U\sum_p^q U^n a_n = \widehat{Q}\sum_p^q U^{n+1} a_n = \sum_p^q U'^{n+1} Q a_n = U'\widehat{Q}\sum_p^q U^n a_n,$$

and thus $\widehat{Q}U = U'\widehat{Q}$. Because we also have

$$\widehat{Q}M_+(\mathfrak{A}) = Q\mathfrak{R}_+ \subset \mathfrak{R}'_+ = M_+(\mathfrak{A}'),$$

we can apply Lemma 3.1 to $\mathfrak{R}$, $\mathfrak{R}'$, U, U', and $\widehat{Q}$. Thus there exists a contractive analytic function $\{\mathfrak{A}, \mathfrak{A}', \Theta(\lambda)\}$ such that

$$\Phi^{\mathfrak{A}'}\widehat{Q} = \Theta\Phi^{\mathfrak{A}}$$

and, taking the restriction to $\mathfrak{R}_+$,

$$\Phi_+^{\mathfrak{A}'} Q = \Theta_+ \Phi_+^{\mathfrak{A}}.$$

Let us now consider the conditions for the properties (a)–(e). For $a \in \mathfrak{A}$ we have

$$\Theta(\lambda)a = \Theta(\lambda)(\Phi_+^{\mathfrak{A}} a)(\lambda) = (\Phi_+^{\mathfrak{A}'} Q a)(\lambda) = \sum_0^\infty \lambda^n a'_n \quad (|\lambda| < 1), \tag{3.14}$$

where the coefficients a'_n are determined by the expansion

$$Qa = \sum_0^\infty U'^n a'_n \quad (a'_n \in \mathfrak{A}'); \tag{3.15}$$

in particular we have

$$a'_0 = P_{\mathfrak{A}'} Q a. \tag{3.16}$$

From (3.14), $\Theta(0)a = a'_0$. Thus, in order that $\Theta(\lambda)$ be purely contractive, it is necessary and sufficient that the inequality $\|a'_0\| < \|a\|$ hold for every nonzero $a \in \mathfrak{A}$. By virtue of (3.16) this establishes the case (a). The cases (b) and (c) are immediate consequences of the fact that the operator Θ_+ is unitarily equivalent to the operator Q; see (3.13). By virtue of Proposition 2.3, the case (d) follows from (b) and (c).

Case (e): Suppose Q is boundedly invertible, with $\|Q^{-1}\| \leq k$. Then $Q_1 = (1/k)Q^{-1}$ is a contraction of $\mathfrak{R}'_+$ into $\mathfrak{R}_+$; and (3.12) implies that $Q_1 U'_+ = U_+ Q_1$. Thus there exists, by the first assertion (already proved) of our lemma, a contractive analytic function $\{\mathfrak{A}', \mathfrak{A}, \Omega_1(\lambda)\}$ such that

$$\Phi_+^{\mathfrak{A}} Q_1 = \Omega_{1+} \Phi_+^{\mathfrak{A}'}.$$

Setting $\Omega(\lambda) = k\Omega_1(\lambda)$ we obtain a bounded analytic function $\{\mathfrak{A}', \mathfrak{A}, \Omega(\lambda)\}$, $\|\Omega(\lambda)\| \le k$, such that

$$\Phi_+^{\mathfrak{A}} Q^{-1} = \Omega_+ \Phi_+^{\mathfrak{A}'}. \tag{3.17}$$

From (3.17) and (3.13) we deduce that the operators Θ_+ and Ω_+ are the inverses of each other:

$$\Omega_+ \Theta_+ = I_{H^2(\mathfrak{A})}, \quad \Theta_+ \Omega_+ = I_{H^2(\mathfrak{A}')}.$$

This implies

$$\Omega(\lambda)\Theta(\lambda) = I_{\mathfrak{A}}, \quad \Theta(\lambda)\Omega(\lambda) = I_{\mathfrak{A}'} \quad (\lambda \in D),$$

and hence $\Omega(\lambda) = \Theta(\lambda)^{-1}$ and $\|\Theta(\lambda)^{-1}\| \le k$ for $\lambda \in D$. Conversely, if $\Theta(\lambda)$ has a uniformly bounded inverse $\Omega(\lambda) = \Theta(\lambda)^{-1}$ on D, say $\|\Omega(\lambda)\| \le k$, then this is a holomorphic function on D, because its derivative exists (in the operator norm topology):

$$\frac{d}{d\lambda}\Omega(\lambda) = -\Theta(\lambda)^{-1} \cdot \frac{d}{d\lambda}\Theta(\lambda) \cdot \Theta(\lambda)^{-1}.$$

The bounded analytic function $\{\mathfrak{A}', \mathfrak{A}, \Omega(\lambda)\}$ induces a bounded operator Ω_+ from $H^2(\mathfrak{A}')$ into $H^2(\mathfrak{A})$, $\|\Omega_+\| \le k$. Because

$$(\Omega_+\Theta_+ v)(\lambda) = \Omega(\lambda)(\Theta_+ v)(\lambda) = \Omega(\lambda)\Theta(\lambda)v(\lambda) = v(\lambda) \quad (v \in H^2(\mathfrak{A}))$$

and

$$(\Theta_+\Omega_+ u)(\lambda) = \Theta(\lambda)(\Omega_+ u)(\lambda) = \Theta(\lambda)\Omega(\lambda)u(\lambda) = u(\lambda) \quad (u \in H^2(\mathfrak{A}')),$$

Ω_+ is the inverse of Θ_+. The operator $(\Phi_+^{\mathfrak{A}})^{-1}\Omega_+\Phi_+^{\mathfrak{A}'}$, which is also bounded by k, is then the inverse of the operator

$$(\Phi_+^{\mathfrak{A}'})^{-1}\Theta_+\Phi_+^{\mathfrak{A}} = Q.$$

Finally, concerning the case (f), if $\Theta(\lambda) = \Theta_0$ is constant, the adjoint of the operator Θ_+ is the multiplication by Θ_0^* from $H^2(\mathfrak{A}')$ into $H^2(\mathfrak{A})$ and hence it obviously satisfies $\Theta_+^* U_+'^{\times} = U_+^{\times}\Theta_+^*$, where $U_+^{\times}$ and $U_+'^{\times}$ are the multiplications by λ on $H^2(\mathfrak{A})$ and $H^2(\mathfrak{A}')$, respectively. Due to the relation (3.2) for U_+ and the similar relation for U_+', we also have $Q^*U_+' = U_+Q^*$. Conversely the latter equation implies that Q takes the kernel of U_+^* into that of $U_+'^*$. Consequently, Θ_+ takes the kernel $\mathfrak{A}(\subset H^2(\mathfrak{A}))$ of $U_+^{\times *}$ into the kernel $\mathfrak{A}'(\subset H^2(\mathfrak{A}'))$ of $U_+'^{\times *}$. This means that $\Theta(\lambda)$ is a constant function (in λ). This concludes the proof of Lemma 3.2.

3. As a first application of Lemma 3.2 we determine the invariant subspaces for unilateral shifts of countable multiplicity, on a complex Hilbert space. By virtue of the Fourier representation of such operators (cf. Subsect. 1), our problem reduces to finding the invariant subspaces for the operator $U_+^\times$ on the space $H^2(\mathfrak{E})$, $\mathfrak{E}$ being a complex, separable Hilbert space.

Theorem 3.3. *For the operator $U_+^\times$ of multiplication by λ on the space $H^2(\mathfrak{E})$, the invariant subspaces are precisely those of the form*

$$\mathfrak{H} = \Theta_+ H^2(\mathfrak{F}) \tag{3.18}$$

where $\{\mathfrak{F}, \mathfrak{E}, \Theta(\lambda)\}$, is an arbitrary inner function.

Proof. If $\{\mathfrak{F}, \mathfrak{E}, \Theta(\lambda)\}$ is an inner function, the corresponding operator Θ_+ is isometric, and hence $\Theta_+ H^2(\mathfrak{F})$ is closed, that is, a *subspace* of $H^2(\mathfrak{E})$. Its invariance for $U_+^\times$ is obvious.

We shall show that every invariant subspace $\mathfrak{H}$ for $U_+^\times$ has the form (3.18). To this end we first embed $\mathfrak{E}$ in $H^2(\mathfrak{E})$ as a subspace by identifying the element $e \in \mathfrak{E}$ with the constant function $e(\lambda) \equiv e$; $\mathfrak{E}$ is then wandering for $U_+^\times$ and

$$H^2(\mathfrak{E}) = \bigoplus_0^\infty U_+^{\times n} \mathfrak{E} \quad (= M_+(\mathfrak{E})); \qquad \text{see (1.2).} \tag{3.19}$$

Let V denote the restriction of $U_+^\times$ to the invariant subspace $\mathfrak{H}$; this is an isometry on $\mathfrak{H}$. We have

$$\bigcap_{n=0}^\infty V^n \mathfrak{H} \subset \bigcap_{n=0}^\infty U_+^{\times n} H^2(\mathfrak{E}) = \{0\},$$

and thus V has no unitary part so that the corresponding Wold decomposition is of the form

$$\mathfrak{H} = \bigoplus_0^\infty V^n \mathfrak{F}, \quad \text{where} \quad \mathfrak{F} = \mathfrak{H} \ominus V\mathfrak{H}. \tag{3.20}$$

Let us now apply Lemma 3.2 to

$$\mathfrak{R}_+ = \mathfrak{H}, \quad U_+ = V, \quad \mathfrak{A} = \mathfrak{F}; \quad \mathfrak{R}'_+ = H^2(\mathfrak{E}), \quad U'_+ = U_+^\times, \quad \mathfrak{A}' = \mathfrak{E}$$

and

$$Q = \text{ the identity transformation of } \mathfrak{H} \text{ into } H^2(\mathfrak{E});$$

(3.12) is satisfied obviously. Thus there exists an *inner* function $\{\mathfrak{F}, \mathfrak{E}, \Theta(\lambda)\}$ such that

$$\Phi_+^{\mathfrak{E}} Q = \Theta_+ \Phi_+^{\mathfrak{F}} \quad (\text{on } \mathfrak{H}). \tag{3.21}$$

Because $\mathfrak{E}$ consists of the constant functions in $H^2(\mathfrak{E})$, the Fourier representation of $H^2(\mathfrak{E})$ with respect to $U_+^\times$ is the identity transformation. On the other hand, we have $Qh = h$ for $h \in \mathfrak{H}$. Thus (3.21) reduces to the relation

$$h = \Theta_+ \Phi_+^{\mathfrak{F}} h \quad (h \in \mathfrak{H})$$

and hence we have

$$\mathfrak{H} = \Theta_+ \Phi_+^{\mathfrak{F}} \mathfrak{H} = \Theta_+ H^2(\mathfrak{F}),$$

which completes the proof.

4. Let us state—without proof—that *the subspaces $\mathfrak{H} \neq \{0\}$ of $H^2(\mathfrak{E})$, which are hyperinvariant for the unilateral shift $U_+^\times$, are precisely those that can be represented in the form*

$$\mathfrak{H} = u \cdot H^2(\mathfrak{E})$$

with a scalar inner function u (i.e., in this case, $\mathfrak{F} = \mathfrak{E}$ and $\Theta(\lambda) = u(\lambda)I_{\mathfrak{E}}$).

4 Factorizations

1. We begin with the following result.

Proposition 4.1. (a) *Let $\{\mathfrak{E},\mathfrak{F},\Theta(\lambda)\}$ and $\{\mathfrak{E},\mathfrak{F}_1,\Theta_1(\lambda)\}$ be two contractive analytic functions, the second one being outer, and suppose that*

$$\Theta(e^{it})^*\Theta(e^{it}) \leq \Theta_1(e^{it})^*\Theta_1(e^{it}) \quad \text{a.e.} \tag{4.1}$$

Then there exists a contractive analytic function $\{\mathfrak{F}_1,\mathfrak{F},\Theta_2(\lambda)\}$ such that

$$\Theta(\lambda) = \Theta_2(\lambda)\Theta_1(\lambda) \qquad (\lambda \in D). \tag{4.2}$$

(b) *If in* (4.1) *the equality sign holds a.e., then $\Theta_2(\lambda)$ is an inner function. If, moreover, $\Theta(\lambda)$ is outer, then $\Theta_2(\lambda)$ is a unitary constant.*

Proof. Part (a): Inequality (4.1) implies that the operator X defined by

$$X(\Theta_1 u) = \Theta u \qquad (u \in H^2(\mathfrak{E}))^2 \tag{4.3}$$

is a contraction of $\Theta_1 H^2(\mathfrak{E})$ into $H^2(\mathfrak{F})$; this extends by continuity to a contraction (also denoted by X) of $H^2(\mathfrak{F}_1) = \overline{\Theta_1 H^2(\mathfrak{E})}$ into $H^2(\mathfrak{F})$. We obviously have $\Theta \cdot \lambda u = \lambda \cdot \Theta u$ and $\Theta_1 \cdot \lambda u = \lambda \cdot \Theta_1 u$, therefore it follows from the definition (4.3) that

$$XU_{1+}^\times = U_+^\times X,$$

where $U_+^\times$ and $U_{1+}^\times$ denote the operators of multiplication by the variable λ on $H^2(\mathfrak{F})$ and $H^2(\mathfrak{F}_1)$, respectively.

Let us apply Lemma 3.2 to the case

$$\mathfrak{K}_+ = H^2(\mathfrak{F}_1), \quad U_+ = U_{1+}^\times; \quad \mathfrak{K}'_+ = H^2(\mathfrak{F}), \quad U'_+ = U_+^\times; \quad Q = X$$

($\mathfrak{F}$ and $\mathfrak{F}_1$ are supposed to be embedded in $H^2(\mathfrak{F})$ and $H^2(\mathfrak{F}_1)$ in the usual way, as the subspaces of the constant functions). It follows that there exists a contractive analytic function $\{\mathfrak{F}_1,\mathfrak{F},\Theta_2(\lambda)\}$ such that

$$Xv = \Theta_2 v \quad \text{for} \quad v \in H^2(\mathfrak{F}_1) \tag{4.4}$$

---

[2] As the operator Θ_+ can be considered as a restriction of the operator Θ, we are allowed to write Θu instead of $\Theta_+ u$ if $u \in H^2(\mathfrak{E})$, and we do so often in the sequel. But of course we have to distinguish between $\Theta_+^* u (= (\Theta_+)^* u)$ and $\Theta^* u$ even for $u \in H^2(\mathfrak{F})$.

(observe that the Fourier representations $\Phi_+^{\mathfrak{F}}$ and $\Phi_+^{\mathfrak{F}_1}$ of the spaces $H^2(\mathfrak{F})$ and $H^2(\mathfrak{F}_1)$ are the identity transformations). Equations (4.3) and (4.4) imply for $v = \Theta_1 u$ $(u \in H^2(\mathfrak{E}))$:

$$\Theta u = \Theta_2 \Theta_1 u;$$

choosing $u(\lambda) \equiv h$ (constant function, $h \in \mathfrak{H}$), we obtain the relation (4.2).

Part (b): When equality holds in (4.1) a.e., then X is an isometry. By virtue of Lemma 3.2, $\Theta_2(\lambda)$ is then an inner function. If, moreover, $\Theta(\lambda)$ is an outer function, then

$$H^2(\mathfrak{F}) \supset \overline{\Theta_2 H^2(\mathfrak{F}_1)} \supset \overline{\Theta_2 \Theta_1 H^2(\mathfrak{E})} = \overline{\Theta H^2(\mathfrak{E})} = H^2(\mathfrak{F}),$$

so that $\Theta_2(\lambda)$ is also an outer function. By virtue of Proposition 2.3, $\Theta_2(\lambda)$ is then a unitary constant, and this concludes the proof.

2. Consider now a function $N(t)$ $(0 \le t \le 2\pi)$, whose values are self-adjoint operators on a (separable) Hilbert space $\mathfrak{E}$, and which is measurable (strongly or weakly, which amounts to the same thing because $\mathfrak{E}$ is separable); moreover, assume that

$$O \le N(t) \le I. \tag{4.5}$$

The formula $(Nv)(t) = N(t)v(t)$ defines a self-adjoint operator N on $L^2(\mathfrak{E})$, bounded by O and I.

Proposition 4.2. *There exists a contractive outer function* $\{\mathfrak{E}, \mathfrak{F}_1, \Theta_1(\lambda)\}$ *with the following properties:*

(i) $$N(t)^2 \ge \Theta_1(e^{it})^* \Theta_1(e^{it}) \quad a.e.$$

(ii) *For every other contractive analytic function* $\{\mathfrak{E}, \mathfrak{F}, \Theta(\lambda)\}$ *such that*

$$N(t)^2 \ge \Theta(e^{it})^* \Theta(e^{it}) \quad \text{a.e.} \tag{4.6}$$

we also have

$$\Theta_1(e^{it})^* \Theta_1(e^{it}) \ge \Theta(e^{it})^* \Theta(e^{it}) \quad \text{a.e.} \tag{4.7}$$

Moreover, these properties determine the outer function $\Theta_1(\lambda)$ *up to a constant unitary factor from the left. In order that equality hold in* (i) *a.e., it is necessary and sufficient that the condition*

$$\bigcap_{n \ge 0} e^{int} \overline{N H^2(\mathfrak{E})} = \{0\} \tag{4.8}$$

be satisfied.

Proof. Let $U^\times$ denote multiplication by e^{it} on $L^2(\mathfrak{E})$. As N commutes with $U^\times$, and as $H^2(\mathfrak{E})$ is invariant for $U^\times$, the subspace

$$\mathfrak{N} = \overline{N H^2(\mathfrak{E})} \tag{4.9}$$

of $L^2(\mathfrak{E})$ is also invariant for $U^\times$. Thus $U^\times$ induces an isometry on $\mathfrak{N}$. Let

$$\mathfrak{N} = M_+(\mathfrak{F}_1) \oplus \mathfrak{N}_0 \tag{4.10}$$

be the Wold decomposition of $\mathfrak{N}$ corresponding to this isometry:

$$\mathfrak{F}_1 = \mathfrak{N} \ominus U^\times \mathfrak{N}, \quad M_+(\mathfrak{F}_1) = \bigoplus_{n \geq 0} U^{\times n} \mathfrak{F}_1, \quad \mathfrak{N}_0 = \bigcap_{n \geq 0} U^{\times n} \mathfrak{N}. \tag{4.11}$$

Let P denote the orthogonal projection of $\mathfrak{N}$ onto $M_+(\mathfrak{F}_1)$ and N_+ denote the multiplication operator from $H^2(\mathfrak{E})$ into $L^2(\mathfrak{E})$; because $M_+(\mathfrak{F}_1)$ reduces $U^\times|\mathfrak{N}$ we have $U^\times P = P(U^\times|\mathfrak{N})$. Because N commutes with $U^\times$ we obtain: $U^\times P N_+ = P(U^\times|\mathfrak{N})N_+ = PU^\times N_+ = PN_+U_+^\times$. Hence

$$U_{1+}^\times X = XU_+^\times,$$

where

$$X = PN_+, \quad U_{1+}^\times = U^\times|M_+(\mathfrak{F}_1), \text{ and } U_+^\times = U^\times|H^2(\mathfrak{E}).$$

We can therefore apply Lemma 3.2 to the case

$$\mathfrak{R}_+ = H^2(\mathfrak{E}), \quad U_+ = U_+^\times, \quad \mathfrak{A} = \mathfrak{E};$$
$$\mathfrak{R}'_+ = M_+(\mathfrak{F}_1), \quad U'_+ = U_{1+}^\times, \quad \mathfrak{A}' = \mathfrak{F}_1, \text{ and } Q = X.$$

Because

$$\overline{Q\mathfrak{R}_+} = \overline{XH^2(\mathfrak{E})} = \overline{PNH^2(\mathfrak{E})} = \overline{P\mathfrak{N}} = M_+(\mathfrak{F}_1) = \mathfrak{R}'_+,$$

we obtain that there exists a contractive outer function $\{\mathfrak{E}, \mathfrak{F}_1, \Theta_1(\lambda)\}$ such that

$$\Phi_+^{\mathfrak{F}_1} Xu = \Theta_1 u \quad \text{for} \quad u \in H^2(\mathfrak{E}); \tag{4.12}$$

here we have also used the fact that $\Phi_+^{\mathfrak{E}} u = u$, because $\mathfrak{E}$ is the subspace of $H^2(\mathfrak{E})$ consisting of the constant functions.

The transformation $\Phi_+^{\mathfrak{F}_1}$ is unitary, thus (4.12) implies

$$\|Nu\| \geq \|PNu\| = \|Xu\| = \|\Theta_1 u\| \quad \text{for} \quad u \in H^2(\mathfrak{E}). \tag{4.13}$$

Now the elements of the form Nu ($u \in H^2(\mathfrak{E})$) are dense in $\mathfrak{N}$ (cf. (4.9)), so there is equality in (4.13) for all u if, and only if, $P = I_{\mathfrak{N}}$. This condition is equivalent to the condition $\mathfrak{N}_0 = \{0\}$, that is, to (4.8).

Let us set in (4.13) $u(\lambda) = p(\lambda)h$, where $h \in \mathfrak{E}$, and $p(\lambda)$ is a (scalar) polynomial of λ. We get:

$$\frac{1}{2\pi}\int_0^{2\pi} |p(e^{it})|^2 \|N(t)h\|^2\, dt = \|Nu\|^2$$
$$\geq \|\Theta_1 u\|^2 = \frac{1}{2\pi}\int_0^{2\pi} |p(e^{it})|^2 \|\Theta_1(e^{it})h\|^2\, dt.$$

Every trigonometric polynomial $\varphi(e^{it})$ is of the form $e^{-int}p(e^{it})$ for an (algebraic) polynomial $p(\lambda)$, therefore we have

$$\int_0^{2\pi} |\varphi(e^{it})|^2 \|N(t)h\|^2\,dt \geq \int_0^{2\pi} |\varphi(e^{it})|^2 \|\Theta_1(e^{it})h\|^2\,dt \tag{4.14}$$

for all trigonometric polynomials $\varphi(e^{it})$. Hence we deduce, just as in the proof of Lemma 3.1, that

$$\|N(t)h\|^2 \geq \|\Theta_1(e^{it})h\|^2 \quad \text{a.e.} \tag{4.15}$$

Because $\mathfrak{E}$ is separable, the exceptional set of zero measure can be chosen to be independent of h. Thus the inequality (i) holds a.e.

Let us observe, moreover, that equality in (4.13) for all $u \in H^2(\mathfrak{E})$ implies equality in (4.14) for all φ, and hence equality in (4.15) and in (i) a.e. By virtue of what we have already stated concerning equality in (4.13) it follows that the outer function $\Theta_1(\lambda)$ constructed above satisfies (i) with the equality sign a.e. if, and only if, condition (4.8) holds.

Now let $\{\mathfrak{E}, \mathfrak{F}, \Theta(\lambda)\}$ be an arbitrary contractive analytic function, satisfying (4.6). Then there exists a contraction Y of $\mathfrak{N}$ into $H^2(\mathfrak{F})$ such that

$$Y(Nu) = \Theta u \qquad (u \in H^2(\mathfrak{E})). \tag{4.16}$$

We obviously have

$$Y \cdot e^{it}Nu = Y \cdot Ne^{it}u = \Theta \cdot e^{it}u = e^{it} \cdot \Theta u = e^{it} \cdot YNu \quad (u \in H^2(\mathfrak{E}));$$

hence

$$Y \cdot e^{it}v = e^{it} \cdot Yv \quad (v \in \mathfrak{N}).$$

This implies

$$Y\mathfrak{N}_0 \subset \bigcap_{n\geq 0} Y \cdot e^{int}\mathfrak{N} = \bigcap_{n\geq 0} e^{int}Y\mathfrak{N} \subset \bigcap_{n\geq 0} e^{int}H^2(\mathfrak{F}) = \{0\},$$

and hence $Y\mathfrak{N}_0 = \{0\}$, $Y(I_{\mathfrak{N}} - P) = O$. It follows that

$$YNu = YPNu = YXu \qquad (u \in H^2(\mathfrak{E})); \tag{4.17}$$

using (4.12), (4.16), and the fact that $\Phi_+^{\mathfrak{F}_1}$ is unitary and Y is a contraction, we conclude:

$$\|\Theta_1 u\| = \|Xu\| \geq \|YXu\| = \|YNu\| = \|\Theta u\| \qquad (u \in H^2(\mathfrak{E})). \tag{4.18}$$

The method that has led us from (4.13) to (4.15) and to property (i) of $\Theta_1(\lambda)$, can be applied to (4.18) and leads to (4.7). This argument completes the proof of property (ii) for $\Theta_1(\lambda)$.

It remains to prove uniqueness. Let $\{\mathfrak{E},\mathfrak{F}_1',\Theta_1'(\lambda)\}$ be any contractive outer function with the properties (i) and (ii). Then we have

$$\Theta_1(e^{it})^*\Theta_1(e^{it}) = \Theta_1'(e^{it})^*\Theta_1'(e^{it}) \quad \text{a.e.}$$

By virtue of Proposition 4.1(b) this implies that $\Theta_1'(\lambda) = Z\cdot\Theta_1(\lambda)$, with a unitary operator Z from $\mathfrak{F}_1$ to $\mathfrak{F}_1'$.

This completes the proof of Proposition 4.2.

Remark. From Proposition 4.2 it follows at once that if the function $N(t)^2$ is factorable in the form $\Theta(e^{it})^*\Theta(e^{it})$ with a contractive analytic function $\Theta(\lambda)$, outer or not, then condition (4.8) is satisfied.

3. From Propositions 4.1 and 4.2 we can deduce that every contractive analytic function $\{\mathfrak{E},\mathfrak{E}_*,\Theta(\lambda)\}$ can be factored into the product of inner and outer factors. In fact, setting

$$N(t) = [\Theta(e^{it})^*\Theta(e^{it})]^{1/2},$$

we can apply Proposition 4.2, and obtain an outer function $\{\mathfrak{E},\mathfrak{F},\Theta_e(\lambda)\}$ with properties (i) and (ii). As $\Theta(e^{it})^*\Theta(e^{it}) = N(t)^2$, these properties imply that

$$N(t)^2 \geq \Theta_e(e^{it})^*\Theta_e(e^{it}) \geq \Theta(e^{it})^*\Theta(e^{it}) = N(t)^2 \quad \text{a.e.,}$$

and hence $\Theta_e(e^{it})^*\Theta_e(e^{it}) = \Theta(e^{it})^*\Theta(e^{it})$ a.e. Thus, by virtue of Proposition 4.1, there exists an inner function $\{\mathfrak{F},\mathfrak{E}_*,\Theta_i(\lambda)\}$ such that

$$\Theta(\lambda) = \Theta_i(\lambda)\Theta_e(\lambda) \qquad (\lambda \in D). \tag{4.19}$$

This is called the *canonical factorization* of $\Theta(\lambda)$ into the product of its outer factor $\Theta_e(\lambda)$ and inner factor $\Theta_i(\lambda)$. This factorization is unique in the sense that if $\Theta(\lambda) = \Theta_i'(\lambda)\Theta_e'(\lambda)$ is any factorization with some outer $\Theta_e'(\lambda)$ and inner $\Theta_i'(\lambda)$, and with some intermediary space $\mathfrak{F}'$, then there exists a unitary operator Z from $\mathfrak{F}$ to $\mathfrak{F}'$ such that

$$\Theta_e'(\lambda) = Z\cdot\Theta_e(\lambda) \quad \text{and} \quad \Theta_i'(\lambda) = \Theta_i(\lambda)\cdot Z^{-1} \quad (\lambda \in D).$$

This follows readily from Proposition 4.1(b). In particular, $\Theta_e'(\lambda)$ coincides with $\Theta_e(\lambda)$, and $\Theta_i'(\lambda)$ coincides with $\Theta_i(\lambda)$, in the sense of Sec. 2.4.

If we take the canonical factorization of the adjoint function $\Theta^\sim(\lambda)$ and then return to $\Theta(\lambda)$, we arrive at a factorization

$$\Theta(\lambda) = \Theta_{*e}(\lambda)\Theta_{*i}(\lambda) \quad (\lambda \in D) \tag{4.20}$$

with a $*$-outer factor $\Theta_{*e}(\lambda)$ and a $*$-inner factor $\Theta_{*i}(\lambda)$. This is called the $*$-*canonical factorization* of $\Theta(\lambda)$; uniqueness in the same sense follows as for the canonical factorization.

4. Let $u(\lambda)$ $(\lambda \in D)$ be a scalar-valued contractive analytic function, that is, $u \in H^\infty$, $|u(\lambda)| \leq 1$ on D. If $u(\lambda) \not\equiv 0$, then to every Borel subset α of the unit circle

C there corresponds a factorization

$$u(\lambda) = u_{2\alpha}(\lambda)u_{1\alpha}(\lambda)$$

into the product of two scalar-valued contractive analytic functions, such that $u_{1\alpha}(\lambda)$ is outer, and a.e.

$$|u_{1\alpha}(e^{it})| = \begin{cases} |u(e^{it})| & \text{for } e^{it} \in \alpha, \\ 1 & \text{for } e^{it} \in \alpha' = C \backslash \alpha, \end{cases} \tag{4.21}$$

and

$$u_{2\alpha}(e^{it}) = \begin{cases} 1 & \text{for } e^{it} \in \alpha, \\ |u(e^{it})| & \text{for } e^{it} \in \alpha'. \end{cases} \tag{4.22}$$

We can simply choose

$$u_{1\alpha}(\lambda) = \exp\left(\frac{1}{2\pi} \int_{(\alpha)} \frac{e^{it} + \lambda}{e^{it} - \lambda} \log|u(e^{it})| \, dt \right), \tag{4.23}$$

where $(\alpha) = \{t\colon\ 0 \leq t < 2\pi, e^{it} \in \alpha\}$. In fact, if $u_{1\alpha}(\lambda)$ is the function corresponding to α' in the same manner, then $u_{1\alpha'}(\lambda)u_{1\alpha}(\lambda)$ is equal to the outer factor of $u(\lambda)$, and hence

$$u_{2\alpha}(\lambda) = u(\lambda)/u_{1\alpha}(\lambda) \tag{4.24}$$

is equal to the inner factor of $u(\lambda)$ multiplied by $u_{1\alpha'}(\lambda)$; thus $u_{2\alpha}(\lambda)$ is a contractive analytic function. Equation (4.21) follows immediately from the definition (4.23), and (4.22) follows from (4.21) and (4.24), because $u(e^{it}) \neq 0$ a.e.

Applying our preceding results we can generalize these facts to operator-valued analytic functions as follows.

Proposition 4.3. *Let $\{\mathfrak{E}, \mathfrak{E}_*, \Theta(\lambda)\}$ be a contractive analytic function such that*

$$\Theta(e^{it})^{-1} \quad \textit{exists a.e. } \text{(not necessarily boundedly).} \tag{4.25}$$

(a) *To every Borel subset α of C there corresponds a factorization*

$$\Theta(\lambda) = \Theta_{2\alpha}(\lambda)\Theta_{1\alpha}(\lambda) \quad (\lambda \in D) \tag{4.26}$$

of $\Theta(\lambda)$ into the product of an outer function $\{\mathfrak{E}, \mathfrak{F}, \Theta_{1\alpha}(\lambda)\}$ and a contractive analytic function $\{\mathfrak{F}, \mathfrak{E}_*, \Theta_{2\alpha}(\lambda)\}$, such that

$$\Theta_{1\alpha}(e^{it}) \textit{ is unitary for almost every } t \in (\alpha'), \tag{4.27}$$

$$\Theta_{2\alpha}(e^{it}) \textit{ is isometric for almost every } t \in (\alpha). \tag{4.28}$$

(b) *The factorization* (4.26) *is unique in the sense that if* $\{\mathfrak{E}, \mathfrak{F}', \Theta'_{1\alpha}(\lambda)\}$ *and* $\{\mathfrak{F}', \mathfrak{E}_*, \Theta'_{2\alpha}(\lambda)\}$ *satisfy the same conditions, then we have*

$$\Theta'_{1\alpha}(\lambda) = Z \cdot \Theta_{1\alpha}(\lambda) \quad \textit{and} \quad \Theta'_{2\alpha}(\lambda) = \Theta_{2\alpha}(\lambda) \cdot Z^{-1}$$

with a constant unitary operator Z *from* $\mathfrak{F}$ *to* $\mathfrak{F}'$.

(c) *If* $\Theta(e^{it})$ *is not a.e. isometric on* (α) *as well as on* (α'), *then the factorization* (4.26) *is nontrivial (i.e., neither factor is a unitary constant).*

Remark. Equations (4.26) and (4.28) imply

$$\Theta_{1\alpha}(e^{it})^*\Theta_{1\alpha}(e^{it}) = \Theta(e^{it})^*\Theta(e^{it}) \quad \text{a.e. on } (\alpha). \tag{4.27'}$$

Proof. Part (a): Let us set

$$N_\alpha(t) = [\Theta(e^{it})^*\Theta(e^{it})]^{\frac{1}{2}} \text{ for } t \in (\alpha) \quad \text{and} \quad N_\alpha(t) = I_{\mathfrak{E}} \text{ for } t \in (\alpha'); \tag{4.29}$$

then we have

$$N_\alpha(t)^2 \geq \Theta(e^{it})^*\Theta(e^{it}) \quad \text{a.e.}, \tag{4.30}$$

and hence

$$\|N_\alpha u\|_{L^2(\mathfrak{E})} \geq \|\Theta u\|_{H^2(\mathfrak{E}_*)} \quad (u \in H^2(\mathfrak{E})).$$

Thus there exists a contraction Y of $\overline{N_\alpha H^2(\mathfrak{E})}$ into $H^2(\mathfrak{E}_*)$ such that

$$Y(N_\alpha u) = \Theta u \quad (u \in H^2(\mathfrak{E})).$$

As the operators N_α and Θ commute with multiplication by the function e^{it}, so does Y, and hence

$$Y \bigcap_{n \geq 0} e^{int}\overline{N_\alpha H^2(\mathfrak{E})} \subset \bigcap_{n \geq 0} e^{int}\overline{\Theta H^2(\mathfrak{E})} \subset \bigcap_{n \geq 0} e^{int} H^2(\mathfrak{E}_*) = \{0\}.$$

Thus, in order to prove the validity of condition (4.8) for the function $N_\alpha(t)$, it suffices to show that the only element w of $\overline{N_\alpha H^2(\mathfrak{E})}$ for which $Yw = 0$, is $w = 0$. In other words, we have to show that if a sequence of functions $v_n(t) \in H^2(\mathfrak{E})$ satisfies the conditions

$$N_\alpha v_n \to w, \quad \Theta v_n \to 0 \quad \text{(convergence in the mean)},$$

then $w = 0$. Replacing the sequence, if necessary, by a subsequence (so that $\sum_n \|N_\alpha v_n - w\|^2$ and $\sum_n \|\Theta v_n\|^2$ converge), we have

$$N_\alpha(t)v_n(t) \to w(t) \quad \text{and} \quad \Theta(e^{it})v_n(t) \to 0 \quad \text{a.e.}$$

in the norm of $\mathfrak{E}$ and $\mathfrak{E}_*$, respectively. Thus we have a.e. in (α')

$$\Theta(e^{it})w(t) = \Theta(e^{it})\lim_n N_\alpha(t)v_n(t) = \Theta(e^{it})\lim_n v_n(t) = \lim_n \Theta(e^{it})v_n(t) = 0$$

and a.e. in (α)

$$\|\Theta(e^{it})w(t)\| \leq \|w(t)\| = \lim_n \|N_\alpha(t)v_n(t)\| = \lim_n \|\Theta(e^{it})v_n(t)\| = 0,$$

and hence

$$\Theta(e^{it})w(t) = 0 \text{ a.e. in } (0,2\pi).$$

By virtue of the hypothesis (4.25) this implies $w(t) = 0$ a.e., that is, $w = 0$.

Thus $N_\alpha(t)$ satisfies condition (4.8). Hence, by virtue of Proposition 4.2, there exists a contractive outer function $\{\mathfrak{E}, \mathfrak{F}, \Theta_{1\alpha}(\lambda)\}$ such that

$$\Theta_{1\alpha}(e^{it})^* \Theta_{1\alpha}(e^{it}) = N_\alpha(t)^2 \quad \text{a.e.} \tag{4.31}$$

From (4.30) and (4.31) it follows by Proposition 4.1 that there exists a contractive analytic function $\{\mathfrak{F}, \mathfrak{E}_*, \Theta_{2\alpha}(\lambda)\}$ such that equation (4.26) holds. The factor $\Theta_{1\alpha}(\lambda)$ satisfies (4.27) by virtue of (4.29) and (4.31) and because, being outer, it verifies the relation

$$\overline{\Theta_{1\alpha}(e^{it})\mathfrak{E}} = \mathfrak{F} \quad \text{a.e. in } (0,2\pi); \tag{4.32}$$

see Proposition 2.4. In order to prove (4.28), we deduce from (4.26), (4.29), and (4.31) that for almost every $t \in (\alpha)$ and every $h \in \mathfrak{E}$,

$$\|\Theta_{2\alpha}(e^{it})\Theta_{1\alpha}(e^{it})h\| = \|\Theta(e^{it})h\| = \|N_\alpha(t)h\| = \|\Theta_{1\alpha}(e^{it})h\|.$$

Hence it follows by (4.32) that

$$\|\Theta_{2\alpha}(e^{it})h_*\| = \|h_*\|$$

for every $h_* \in \mathfrak{E}_*$, a.e. in (α).

Part (b): Recalling Remark (4.27′) we see that in the present situation

$$\Theta'_{1\alpha}(e^{it})^* \Theta'_{1\alpha}(e^{it}) = \Theta_{1\alpha}(e^{it})^* \Theta_{1\alpha}(e^{it}) \text{ a.e.;}$$

thus there exists by virtue of Proposition 4.1(b) a unitary operator Z from $\mathfrak{F}$ to $\mathfrak{F}'$ such that $\Theta'_{1\alpha}(\lambda) = Z \cdot \Theta_{1\alpha}(\lambda)$. So we have

$$\Theta_{2\alpha}(\lambda)\Theta_{1\alpha}(\lambda) = \Theta(\lambda) = \Theta'_{2\alpha}(\lambda)\Theta'_{1\alpha}(\lambda) = \Theta'_{2\alpha}(\lambda)Z \cdot \Theta_{1\alpha}(\lambda).$$

The space $\Theta_{1\alpha}(\lambda)\mathfrak{E}$ is dense in $\mathfrak{F}$ (cf. Proposition 2.4), therefore

$$\Theta_{2\alpha}(\lambda) = \Theta'_{2\alpha}(\lambda)Z, \quad \Theta'_{2\alpha}(\lambda) = \Theta_{2\alpha}(\lambda)Z^{-1}.$$

Part (c): This is obvious. In fact, if $\Theta_{1\alpha}(\lambda)$ is a unitary constant, (4.26) and (4.28) imply that $\Theta(e^{it})$ is isometric a.e. in (α). On the other hand, if $\Theta_{2\alpha}(\lambda)$ is a unitary constant, (4.26) and (4.27) imply that $\Theta(e^{it})$ is isometric a.e. in (α').

This concludes the proof of Proposition 4.3.

Remarks. (1) Condition (4.25) is equivalent to the condition

$$\overline{\Theta(e^{it})^*\mathfrak{E}_*} = \mathfrak{E} \quad \text{a.e.} \tag{4.33}$$

Now by virtue of Proposition 2.4, condition (4.33) holds for every $*$-*outer* function $\Theta(\lambda)$.

(2) If $\Theta(\lambda)$ is not an inner function, then $\Theta(e^{it})^*\Theta(e^{it})$ is different from $I_\mathfrak{E}$ on a set ρ of points e^{it}, of positive measure. Then we can choose α so that both $\alpha \cap \rho$ and $\alpha' \cap \rho$ are of positive measure.

By virtue of Proposition 4.3(c) we can state a corollary.

Corollary 4.4. *For every contractive analytic function, which is $*$-outer and not inner, there exist nontrivial factorizations of type* (4.26).

5. The following example is interesting for several reasons. Let A be a self-adjoint operator on the Hilbert space $\mathfrak{E}$, satisfying the inequalities $O \le A \le I$ and such that 0 and 1 are not eigenvalues of A. Define the function $\{\mathfrak{E}, \mathfrak{E}, \Theta(\lambda)\}$ by setting $\Theta(\lambda) \equiv A$. This function is obviously analytic and purely contractive. (Here we use the assumption that 1 is not an eigenvalue of A.) If u is an element of $H^2(\mathfrak{E})$ orthogonal to $AH^2(\mathfrak{E})$, then we have $Au(\lambda) = 0$ for $\lambda \in D$; because 0 is not an eigenvalue of A this implies that $u(\lambda) = 0$ (i.e., $u = 0$). Hence $AH^2(\mathfrak{E})$ is dense in $H^2(\mathfrak{E})$: the function $\Theta(\lambda) \equiv A$ is outer. Moreover, $\Theta^\sim(\lambda) \equiv A^* = A \equiv \Theta(\lambda)$, so that the function $\Theta(\lambda) \equiv A$ is *outer from both sides.*

We now determine the factorization (4.26) of this function for any Borel set α on C. To this end we introduce the function

$$\omega_\alpha(\lambda) = \frac{1}{2\pi} \int_{(\alpha)} \frac{e^{is} + \lambda}{e^{is} - \lambda}\, ds$$

which is holomorphic on D. For $\lambda = re^{it}$ $(0 \le r < 1)$ we can also write

$$\omega_\alpha(\lambda) = \frac{1}{2\pi} \int_{(\alpha)} P_r(t-s)\, ds + \frac{i}{2\pi} \int_{(\alpha)} Q_r(t-s)\, ds, \tag{4.34}$$

where P and Q denote the Poisson kernel and the conjugate Poisson kernel. By well-known properties of integrals with these kernels (cf. ZYGMUND [1] Secs. 3.4 and 7.1), we infer that the nontangential limit of $\omega_\alpha(\lambda)$ exists at a.e. $e^{it} \in C$ and equals $\chi_\alpha(t) + i\tilde{\chi}_\alpha(t)$, where $\chi_\alpha(t)$ denotes the characteristic function of the set (α), and $\tilde{\chi}_\alpha(t)$ the (trigonometric) conjugate function. From (4.34) it follows that Re $\omega_\alpha(\lambda) \ge 0$ (and even Re $\omega_\alpha(\lambda) > 0$ unless α is of zero measure). Hence $|a^{\omega_\alpha(\lambda)}| \le 1$ for $0 \le a \le 1$ and $\lambda \in D$.

Now we define

$$\Omega_\alpha(\lambda) = A^{\omega_\alpha(\lambda)} \quad (\lambda \in D), \tag{4.35}$$

the operator on the right hand side being understood as the integral of the function $a^{\omega_\alpha(\lambda)}$ on the interval $0 \le a \le 1$ with respect to the spectral measure of A. (Note that the points 0 and 1 have zero spectral measure.) We obtain readily that $\Omega_\alpha(\lambda)$ is a contractive analytic function. From the relations

$$\omega_\alpha(\lambda) + \omega_{\alpha'}(\lambda) = \frac{1}{2\pi}\int_0^{2\pi} \frac{e^{is}+\lambda}{e^{is}-\lambda}\, ds = 1 \quad (\lambda \in D)$$

and

$$\text{Re}\ \omega_\alpha(e^{it}) = \chi_\alpha(t) \quad \text{(a.e.)}$$

we infer by well-known properties of spectral integrals that

$$\Omega_\alpha(\lambda)\Omega_{\alpha'}(\lambda) = \Omega_{\alpha'}(\lambda)\Omega_\alpha(\lambda) = A \quad (\lambda \in D) \tag{4.36}$$

and, a.e.,

$$\Omega_\alpha(e^{it})^*\Omega_\alpha(e^{it}) = \Omega_\alpha(e^{it})\Omega_\alpha(e^{it})^* = A^{2\chi_\alpha(t)} = \begin{cases} A^2 & \text{if } t \in (\alpha), \\ I & \text{if } t \in (\alpha'). \end{cases} \tag{4.37}$$

From (4.36) we deduce that

$$AH^2(\mathfrak{E}) = \Omega_\alpha\Omega_{\alpha'}H^2(\mathfrak{E}) \subset \Omega_\alpha H^2(\mathfrak{E}),$$

and this shows that $\Omega_\alpha H^2(\mathfrak{E})$ is dense in $H^2(\mathfrak{E})$. Because the relation (4.36) implies $\Omega_{\tilde{\alpha}}(\lambda)\Omega_{\tilde{\alpha}'}(\lambda) = A$ we obtain similarly that $\Omega_{\tilde{\alpha}}H^2(\mathfrak{E})$ is also dense in $H^2(\mathfrak{E})$. Thus the function $\Omega_\alpha(\lambda)$ is *outer from both sides*. Relations (4.36) and (4.37) (when applied for α' as well as for α) prove that for $\Theta(\lambda) \equiv A$ *the factors in* (4.26) *are given by* $\Theta_{1\alpha}(\lambda) = \Omega_\alpha(\lambda)$ *and* $\Theta_{2\alpha}(\lambda) = \Omega_{\alpha'}(\lambda)$.

Some further remarks are of interest. From (4.37) we see that $\Omega_\alpha(e^{it})$ is *unitary* for almost every $t \in (\alpha')$. But if A is not boundedly invertible and α is not of zero measure then at every point $z = e^{it}$, where $\Omega_\alpha(z)$ is unitary, $\Omega_\alpha(\lambda)$ *does not tend to* $\Omega_\alpha(z)$ *in the operator norm* as $\lambda \to z$. Indeed, otherwise $\Omega_\alpha(\lambda)$ would be boundedly invertible for λ close enough to z, and this is impossible because $\Omega_\alpha(\lambda)^*\Omega_\alpha(\lambda) = A^{2\ \text{Re}\ \omega_\alpha(\lambda)}$ and Re $\omega_\alpha(\lambda) > 0$.

Hence it follows in particular that if A is not boundedly invertible and α is an arc of positive length of C, then $\Omega_\alpha(\lambda)$ has unitary (nontangential) limit almost everywhere on α', but cannot be extended analytically through α'. This remark is useful later in illuminating the role of Proposition 6.7, and we return to it in the Notes to Chap. VII.

5 Nontrivial factorizations

As we show in Chap. VII, the problem of finding the nontrivial invariant subspaces for operators on Hilbert space is equivalent to the problem of finding the factorizations of contractive analytic functions into the product of two functions of the same kind, neither of which is a unitary constant (i.e., *nontrivial* factorizations), and which, moreover, satisfy a certain regularity condition.

It is therefore natural to ask if every (nonconstant unitary) contractive analytic function can be factored nontrivially, without requiring first any additional property of this factorization. However, if $\{\mathfrak{E},\mathfrak{E}_*,\Theta(\lambda)\}$ is an arbitrary contractive analytic function that is not a constant unitary and $\{\mathfrak{E},\mathfrak{F},Z\}$ and $\{\mathfrak{E}_*,\mathfrak{F}_*,Z_*\}$ are constant nonunitary isometries then the factorizations $\Theta(\lambda)=(\Theta(\lambda)Z^*)Z=Z_*^*(Z_*\Theta(\lambda))$ are obviously nontrivial. Moreover when

$$\mathfrak{N}(\Theta)=\{e\in\mathfrak{E}:\ \Theta(\lambda)e=0 \text{ for } |\lambda|<1\}$$

is not $\{0\}$, then $\Theta(\lambda)$ has the obviously nontrivial factorization $\Theta(\lambda)=\Theta(\lambda)(I-P+\lambda P)$ where P denotes the orthogonal projection of $\mathfrak{E}$ onto $\mathfrak{N}(\Theta)$. Also if $\mathfrak{N}(\Theta^\sim)\neq\{0\}$ and P_* denotes the orthogonal projection of $\mathfrak{E}_*$ onto $\mathfrak{N}(\Theta^\sim)$, then $\Theta(\lambda)=(I-P_*+\lambda P_*)\Theta(\lambda)$ is again a nontrivial factorization. In view of these examples it is reasonable to call *strictly nontrivial* a factorization $\Theta(\lambda)=\Theta_2(\lambda)\Theta_1(\lambda)$, where $\{\mathfrak{E},\mathfrak{F},\Theta_1(\lambda)\}$ and $\{\mathfrak{F},\mathfrak{E}_*,\Theta(\lambda)\}$ are contractive analytic functions, such that $\mathfrak{N}(\Theta_1)=\mathfrak{N}(\Theta)$, $\mathfrak{N}(\Theta^\sim)=\mathfrak{N}(\Theta_2^\sim)$ and neither $\Theta_1(\lambda)$ nor $\Theta_2(\lambda)$ is constant as a function of λ. Note that if $\Theta(\lambda)=\Theta_2(\lambda)\Theta_1(\lambda)$ is strictly nontrivial then so is $\Theta^\sim(\lambda)=\Theta_2^\sim(\lambda)\Theta_1^\sim(\lambda)$.

If $\{\mathfrak{E},\mathfrak{E}_*,\Theta(\lambda)\}$ is a constant partial isometry, say W, then $\Theta(\lambda)$ has no strictly nontrivial factorization. Indeed if

$$W=\Theta_2(\lambda)\Theta_1(\lambda),$$

with $\{\mathfrak{E},\mathfrak{F},\Theta_1(\lambda)\}$ and $\{\mathfrak{F},\mathfrak{E}_*,\Theta_2(\lambda)\}$ contractive analytic functions, is a strictly nontrivial factorization then applying Proposition 2.1 to $\Theta_1(\lambda)$ and using the property that $\mathfrak{N}(\Theta_1)=\mathfrak{N}(\Theta)$ equals the kernel of W we obtain that $\Theta_1(\lambda)$ is a constant partial isometry W_1 with kernel equal to that of W. Applying the same argument to $\Theta^\sim(\lambda)$ we infer that $\Theta_2(\lambda)$ is a partial isometry W_2 such that the kernel of W_2^* coincides with that of W^*. Thus both $\Theta_1(\lambda)$ and $\Theta_2(\lambda)$ are constant (in λ) and this contradicts our initial assumption.

The aim of the present section is to prove that strictly nontrivial factorizations do exist for any contractive analytic function which is not a constant partial isometry.

1. We begin with some geometrical considerations.

Let us introduce the following definition. If V is a unilateral shift on the Hilbert space $\mathfrak{H}$, let us denote by $\pi(V)$ the *class* of those bounded operators Q on $\mathfrak{H}$, with which it is possible to associate a Hilbert space $\mathfrak{H}_Q$, a unilateral shift V_Q on $\mathfrak{H}_Q$, and a bounded operator A from $\mathfrak{H}_Q$ to $\mathfrak{H}$, in such a way that the following conditions hold,

$$VA=AV_Q, \tag{5.1}$$

$$AA^*=Q. \tag{5.2}$$

Proposition 5.1. *Let Q be a bounded self-adjoint operator on $\mathfrak{H}$, $Q\geq O$. In order that Q belong to $\pi(V)$, it is necessary and sufficient that*

$$Q-VQV^*\geq O. \tag{5.3}$$

Proof. If $Q \in \pi(V)$ then (5.1) and (5.2) imply

$$Q - VQV^* = AA^* - VAA^*V^* = AA^* - AV_QV_Q^*A^* = A(I - V_QV_Q^*)A^* \geq O,$$

because $I - V_QV_Q^* \geq O$.

Let us suppose, conversely, that Q satisfies (5.3). Let us denote by R the positive square root of the operator on the left hand side of (5.3). Then $Q = R^2 + VQV^*$; hence it follows by iteration

$$Q = R^2 + VR^2V^* + \cdots + V^nR^2V^{*n} + V^{n+1}QV^{*n+1} \qquad (n = 1, 2, \ldots).$$

Because V is a unilateral shift, V^{*n} converges strongly to O as $n \to \infty$. Thus we obtain

$$Q = \sum_{n=0}^{\infty} V^nR^2V^{*n}$$

and consequently

$$\|Q^{1/2}h\|^2 = \sum_{n=0}^{\infty} \|RV^{*n}h\|^2 \quad \text{for all} \quad h \in \mathfrak{H}. \tag{5.4}$$

Consider now the Hilbert space $\mathfrak{H}_Q$ of the sequences $\mathbf{x} = \{x_n\}_0^\infty$ such that $x_n \in \overline{R\mathfrak{H}}$ $(n = 0, 1, \ldots)$ and $\|\mathbf{x}\|^2 = \sum_0^\infty \|x_n\|^2 < \infty$. Let V_Q denote the unilateral shift on $\mathfrak{H}_Q$ defined by

$$V_Q\{x_0, x_1, \ldots\} = \{0, x_0, x_1, \ldots\}.$$

Let us set, for $h \in \mathfrak{H}$,

$$Bh = \{Rh, RV^*h, \ldots, RV^{*n}h, \ldots\}.$$

From (5.4) it follows that B is an operator from $\mathfrak{H}$ to $\mathfrak{H}_Q$ such that

$$\|Bh\|_{\mathfrak{H}_Q} = \|Q^{1/2}h\|_{\mathfrak{H}}. \tag{5.5}$$

Moreover,

$$V_Q^*\{x_0, x_1, \ldots\} = \{x_1, x_2, \ldots\},$$

implies

$$BV^*h = \{RV^*h, RV^{*2}h, \ldots\} = V_Q^*Bh,$$

and thus $BV^* = V_Q^*B$, $VB^* = B^*V_Q$. The operator $A = B^*$ therefore satisfies (5.1) and, on account of (5.5), it also satisfies (5.2).

Proposition 5.2. *Let Q be an operator on $\mathfrak{H}$, of class $\pi(V)$ and such that $O \leq Q \leq I$. The operator $Q_\alpha = \alpha Q + (1-\alpha)I$, where $0 < \alpha < 1$, is then also of class $\pi(V)$. If V is of infinite multiplicity, one can choose $\mathfrak{H}_Q = \mathfrak{H}$ and $V_{Q_\alpha} = V$, thus in this case there exists an operator A_α on $\mathfrak{H}$, commuting with V and such that $Q_\alpha = A_\alpha A_\alpha^*$.*

Proof. The first assertion follows immediately from Proposition 5.1. As to the second assertion, let us observe first that if $\mathfrak{A}$ is the generating subspace for V and if $P_{\mathfrak{A}}$

is the orthogonal projection onto $\mathfrak{A}$, then

$$R_\alpha^2 = Q_\alpha - VQ_\alpha V^* = \alpha(Q - VQV^*) + (1-\alpha)(I - VV^*) \geq (1-\alpha)P_{\mathfrak{A}}.$$

Hence it follows that $R_\alpha h = 0$ implies $P_{\mathfrak{A}} h = 0$ and consequently $\overline{R_\alpha \mathfrak{H}} \supset P_{\mathfrak{A}}\mathfrak{H} = \mathfrak{A}$; thus

$$\dim \mathfrak{H} \geq \dim \overline{R_\alpha \mathfrak{H}} \geq \dim \mathfrak{A}. \tag{5.6}$$

On the other hand, we have

$$\dim \mathfrak{H} = \aleph_0 \cdot \dim \mathfrak{A} = \dim \mathfrak{A}, \tag{5.7}$$

because $\dim \mathfrak{A}$ is infinite. Thus (5.6) and (5.7) imply that $\dim \overline{R_\alpha \mathfrak{H}} = \dim \mathfrak{A}$. Let φ be a unitary operator from $\overline{R_\alpha \mathfrak{H}}$ to $\mathfrak{A}$. This induces by

$$\Phi \mathbf{x} = \sum_0^\infty V^n(\varphi x_n)$$

a unitary operator Φ from the space $\mathfrak{H}_{Q_\alpha}$ of the sequences $\mathbf{x} = \{x_n\}_0^\infty$ $(x_n \in \overline{R_\alpha \mathfrak{H}})$ to the space $\mathfrak{H}$. We have

$$\Phi(V_{Q_\alpha}\mathbf{x}) = \sum_1^\infty V^n(\varphi x_{n-1}) = V \sum_1^\infty V^{n-1}(\varphi x_{n-1}) = V\Phi\mathbf{x}.$$

Thus, if A'_α is the operator from $\mathfrak{H}_{Q_\alpha}$ to $\mathfrak{H}$, associated with Q_α according to the proof of Proposition 5.1, then the operator $A_\alpha = A'_\alpha \Phi^*$ commutes with V and $A_\alpha A_\alpha^* = A'_\alpha \Phi^* \Phi A'^*_\alpha = A'_\alpha A'^*_\alpha = Q_\alpha$.

Proposition 5.3. *Let A and B be bounded operators on the space $\mathfrak{H}$, commuting with an isometry V on $\mathfrak{H}$. Let us suppose, moreover, that A is an isometry, B is a contraction, and*

$$BB^* \geq AA^*. \tag{5.8}$$

Then the operator $C = B^*A$ is an isometry on $\mathfrak{H}$, commuting with V and such that $A = BC$. Furthermore, if B has dense range and A^* does not commute with V then C^* does not commute with V.

Proof. Relation (5.8) and the fact that B is a contraction imply $I - AA^* \geq I - BB^* \geq O$; hence $(I - AA^*)h = 0$ (for an $h \in \mathfrak{H}$) implies $(I - BB^*)h = 0$. Because A is an isometry, we have $(I - AA^*)Ag = Ag - AA^*Ag = Ag - Ag = 0$ for every $g \in \mathfrak{H}$, and hence $(I - BB^*)Ag = 0$ or $Ag = BB^*Ag$. Thus, setting $C = B^*A$, we have $A = BC$. Because A is an isometry and B, C are contractions, C is necessarily an isometry, too.

As A and B commute with V, we have

$$V^*CV = V^*B^*AV = (BV)^*(AV) = (VB)^*(VA) = B^*V^*VA = B^*A = C,$$

and hence

$$(VV^*)CV = VC. \tag{5.9}$$

Now C and V being isometries, so are CV and VC too. On the other hand, VV^* is an orthogonal projection (namely to $V\mathfrak{H}$), thus we conclude from (5.9) that $CV = VC$. If C^* commutes with V, then $A^*VB = A^*BV = C^*V = VC^* = VA^*B$, whence $A^*V = VA^*$ follows under the assumption that B has dense range.

2. Let A be an isometry on $\mathfrak{H}$. Suppose that A commutes with a unilateral shift V on $\mathfrak{H}$, of *infinite* multiplicity, and suppose that A^* does not commute with V, in particular A is not unitary.

Set $Q = AA^*$ and $Q_\alpha = \alpha Q + (1-\alpha)I$ $(0 < \alpha < 1)$. By virtue of Proposition 5.2 there exists an operator B_α on $\mathfrak{H}$, commuting with V and such that $B_\alpha B_\alpha^* = Q_\alpha$; since $O \leq Q_\alpha \leq I$, B_α is a contraction on $\mathfrak{H}$. Because $Q_\alpha = \alpha Q + (1-\alpha)I \geq Q$, we have $B_\alpha B_\alpha^* \geq AA^*$. Then, by virtue of Proposition 5.3, the operator $C_\alpha = B_\alpha^* A$ is an isometry, commuting with V and such that $A = B_\alpha C_\alpha$.

We show that neither B_α nor C_α is a coisometry. In fact, equation $B_\alpha B_\alpha^* = I$ is impossible since it would imply $Q_\alpha = I$, and thus $Q = I$, $AA^* = I$, i.e., the isometry A would be unitary, which contradicts the hypothesis. Similarly, $C_\alpha C_\alpha^* = I$ would imply $B_\alpha B_\alpha^* = B_\alpha C_\alpha C_\alpha^* B^* = AA^*$, $Q_\alpha = Q$, and thus $Q = I$, which is impossible.

Let us show that $B_\alpha \mathfrak{H}$ is dense in $\mathfrak{H}$. In the contrary case there would exist an $h \neq 0$ such that $B_\alpha^* h = 0$, $Q_\alpha h = B_\alpha B_\alpha^* h = 0$. Since $Q_\alpha \geq (1-\alpha)I$, we should have $(1-\alpha)h = 0$, which implies $h = 0$, a contradiction.

Since A^* does not commute with V, we infer from Proposition 5.3 that neither does C_α^*. Assuming that B_α^* commutes with V, it follows that Q_α and so Q also commute with V. Then $AA^*V = QV = VQ = VAA^* = AVA^*$, and since A is an isometry, we obtain that $A^*V = VA^*$, which contradicts our assumption. Consequently, B_α^* does not commute with V.

So we have proved:

Proposition 5.4. *Every isometry A on $\mathfrak{H}$, commuting with a unilateral shift V on $\mathfrak{H}$ of infinite multiplicity, but not commuting with V^*, is the product $A = BC$ of two noncoisometric operators on $\mathfrak{H}$, both commuting with V, but neither commuting with V^*, C is an isometry and B is a contraction with dense range in $\mathfrak{H}$.*

It should be remarked that the Hilbert spaces considered in this paragraph are not necessarily separable.

3. Let us now consider a separable Hilbert space $\mathfrak{E}$ of infinite dimension, and a nonconstant inner function $\{\mathfrak{E}, \mathfrak{E}, \Theta(\lambda)\}$. The corresponding operator Θ_+ on $H^2(\mathfrak{E})$ is then isometric, Θ_+ commutes with the operator $U^\times$ of multiplication by λ in $H^2(\mathfrak{E})$, which is a unilateral shift of infinite multiplicity (because $\dim \mathfrak{E}$ is infinite). Moreover, by Lemma 3.2(f) Θ_+^* does not commute with $U^\times$.

Let us apply Proposition 5.4 to the case $A = \Theta_+$, $V = U^\times$. So we obtain that Θ_+ equals the product $B_2 B_1$ of two noncoisometric operators on $H^2(\mathfrak{E})$, commuting with $U^\times$, and such that B_1 is isometric and B_2 is a contraction with range dense in $H^2(\mathfrak{E})$, and neither of them constant. Applying then Lemma 3.2 we obtain that the operators B_k are generated by nonconstant contractive analytic functions $\{\mathfrak{E}, \mathfrak{E}, \Theta_k(\lambda)\}$ so that $B_k = \Theta_{k+}$ $(k = 1, 2)$, $\Theta_1(\lambda)$ being inner and $\Theta_2(\lambda)$ outer.

We have proved the following result.

Theorem 5.5. *If $\mathfrak{E}$ is a Hilbert space of dimension $\aleph_0$, then every nonconstant inner function $\{\mathfrak{E}, \mathfrak{E}, \Theta(\lambda)\}$, can be factored into the product*

$$\Theta(\lambda) = \Theta_2(\lambda)\Theta_1(\lambda) \quad (\lambda \in D) \tag{5.10}$$

of two nonconstant contractive analytic functions $\{\mathfrak{E}, \mathfrak{E}, \Theta_k(\lambda)\}$ $(k = 1,2)$, $\Theta_1(\lambda)$ being inner, $\Theta_2(\lambda)$ outer, and neither one is $*$-inner.

Remark 1. If $\mathfrak{E}$ is of finite dimension, such a factorization is impossible. In fact, because $\Theta(\lambda)$ and $\Theta_1(\lambda)$ are inner functions, $\Theta(e^{it})$ and $\Theta_1(e^{it})$ are isometries on $\mathfrak{E}$ a.e. (cf. Sec. 2.3); now every isometry on $\mathfrak{E}$ is unitary. It follows that $\Theta_2(e^{it}) = \Theta(e^{it}) \cdot \Theta_1(e^{it})^{-1}$ is also unitary on $\mathfrak{E}$, a.e., and hence the operator Θ_{2+} generated by the function $\Theta_2(\lambda)$ on $H^2(\mathfrak{E})$ is isometric. Thus $\Theta_2(\lambda)$ is an inner function, and it cannot be simultaneously a nonconstant unitary, outer function.

Thus Theorem 5.5 illustrates the essential difference one encounters in the study of factoring inner functions, when one passes from the case of a finite-dimensional space $\mathfrak{E}$ to the case of an infinite-dimensional one.

Remark 2. We note that Theorem 5.5 is valid also for any nonconstant inner function $\{\mathfrak{E}, \mathfrak{E}_*, \Theta(\lambda)\}$ such that both $\mathfrak{E}$ and $\mathfrak{E}_*$ are of dimension $\aleph_0$. In this case, the factorization (5.10) is strictly nontrivial if and only if $\mathfrak{N}(\Theta^\sim) = \{0\}$. If $\mathfrak{N}(\Theta^\sim) \neq \{0\}$, then $\mathfrak{E}'_* = \mathfrak{E}_* \ominus \mathfrak{N}(\Theta^\sim)$ is still of dimension $\aleph_0$ because $\Theta(e^{it})$ is isometric a.e. Moreover we can consider instead of $\{\mathfrak{E}, \mathfrak{E}_*, \Theta(\lambda)\}$ the function $\{\mathfrak{E}, \mathfrak{E}'_*, \Theta(\lambda)\}$. For this function, Theorem 5.5 provides the strictly nontrivial factorization $\Theta(\lambda) = \Theta_2(\lambda)\Theta_1(\lambda)$ where $\{\mathfrak{E}, \mathfrak{F}, \Theta_1(\lambda)\}$ is inner and $\{\mathfrak{F}, \mathfrak{E}'_*, \Theta_2(\lambda)\}$ is outer. By replacing this last function with $\{\mathfrak{F}, \mathfrak{E}_*, \Theta_2(\lambda)\}$ we obviously obtain a strictly nontrivial factorization of the initial function $\{\mathfrak{E}, \mathfrak{E}_*, \Theta(\lambda)\}$.

4. We are now able to prove the following result.

Theorem 5.6. *Every contractive analytic function $\{\mathfrak{E}, \mathfrak{E}_*, \Theta(\lambda)\}$ (with separable $\mathfrak{E}, \mathfrak{E}_*$), that is not a constant partial isometry, can be factored into the product of two contractive analytic functions, say $\{\mathfrak{E}, \mathfrak{F}, \Theta_1(\lambda)\}$ and $\{\mathfrak{F}, \mathfrak{E}_*, \Theta_2(\lambda)\}$ (with separable $\mathfrak{F}$), such that the factorization $\Theta(\lambda) = \Theta_2(\lambda)\Theta_1(\lambda)$ is strictly nontrivial.*

Proof. Let $\Theta(\lambda) = \Theta_i(\lambda)\Theta_e(\lambda) = \Theta_{*e}(\lambda)\Theta_{*i}(\lambda)$ be the canonical and the $*$-canonical factorizations of $\Theta(\lambda)$; see Sec. 4.3. There are two possibilities: (a) $\Theta(\lambda)$ has a nonconstant inner *or* $*$-inner factor, and (b) $\Theta(\lambda)$ is outer *and* $*$-outer. In case (b) it follows from Proposition 2.3 and from Corollary 4.4 that $\Theta(\lambda)$ has a strictly nontrivial factorization. In case (a) it suffices to get a factorization of the inner or of the $*$-inner factor, respectively. Strictly nontrivial factorizations $\Theta'(\lambda) = \Theta''(\lambda)\Theta'''(\lambda)$ give rise to strictly nontrivial factorizations $\Theta'^\sim(\lambda) = \Theta'''^\sim(\lambda)\Theta''^\sim(\lambda)$, thus our problem reduces to finding strictly nontrivial factorizations for a nonconstant inner function $\{\mathfrak{E}, \mathfrak{E}_*, \Theta(\lambda)\}$.

In this case $\dim \mathfrak{E} \leq \dim \mathfrak{E}_* \leq \aleph_0$ (cf. Sec. 2.3). Thus we can suppose that $\mathfrak{E}$ is a subspace of $\mathfrak{E}_*$, and we can embed $\mathfrak{E}_*$ in a space $\mathfrak{F}$ so that $\mathfrak{F} \ominus \mathfrak{E}_*$ is also of dimension $\aleph_0$. Then both $\mathfrak{F} \ominus \mathfrak{E}$ and $\mathfrak{F} \ominus \mathfrak{E}_*$ have dimension $\aleph_0$, so there exists a partially isometric operator Z on $\mathfrak{F}$ with initial domain $\mathfrak{F} \ominus \mathfrak{E}$ and range $\mathfrak{F} \ominus \mathfrak{E}_*$. Let us set

$$\widehat{\Theta}(\lambda) = \Theta(\lambda)P_{\mathfrak{E}} + Z, \tag{5.11}$$

where $P_{\mathfrak{E}} = I - Z^*Z$ denotes the orthogonal projection of $\mathfrak{F}$ onto $\mathfrak{E}$. Then we have for $f \in \mathfrak{F}$

$$\begin{aligned} \|\widehat{\Theta}(e^{it})f\|^2 &= \|\Theta(e^{it})P_{\mathfrak{E}}f + Zf\|^2 = \|\Theta(e^{it})P_{\mathfrak{E}}f\|^2 + \|Zf\|^2 \\ &= \|P_{\mathfrak{E}}f\|^2 + \|Zf\|^2 = ((I - Z^*Z)f, f) + (Z^*Zf, f) = \|f\|^2 \end{aligned}$$

at every point t where $\Theta(e^{it})$ is isometric, and hence almost everywhere. Thus $\{\mathfrak{F}, \mathfrak{F}, \widehat{\Theta}(\lambda)\}$ is a nonconstant inner function, so we can apply to it Remark 2 following Theorem 5.5 and obtain a strictly nontrivial factorization

$$\widehat{\Theta}(\lambda) = \widehat{\Theta}_2(\lambda) \cdot \widehat{\Theta}_1(\lambda) \quad (\lambda \in D) \tag{5.12}$$

with nonconstant contractive analytic factors $\{\mathfrak{F}, \mathfrak{F}, \widehat{\Theta}_k(\lambda)\}$ $(k = 1, 2)$.

In view of Proposition 2.1, the restriction $\widehat{\Theta}_1(\lambda)|\mathfrak{F} \ominus \mathfrak{E}$ is a constant isometry W_1 from $\mathfrak{F} \ominus \mathfrak{E}$ onto a subspace $\mathfrak{F}_1$ of $\mathfrak{F}$. It is clear that the restriction $\widehat{\Theta}_2(\lambda)|\mathfrak{F}_1$ is a constant isometry W_2 from $\mathfrak{F}_1$ onto the subspace $\mathfrak{F} \ominus \mathfrak{E}_*$, and $Z = W_2W_1$. Taking into account that $\widehat{\Theta}_1(\lambda)$ $(\lambda \in D)$ is a contraction, we obtain that $\widehat{\Theta}_1(\lambda)$ transforms $\mathfrak{E}$ into $\mathfrak{F}_0 = \mathfrak{F} \ominus \mathfrak{F}_1$, and in a similar way $\widehat{\Theta}_2(\lambda)$ transforms $\mathfrak{F}_0$ into $\mathfrak{E}_*$ Now it is easy to verify that the factorization

$$\Theta(\lambda) = \Theta_2(\lambda) \cdot \Theta_1(\lambda) \quad (\lambda \in D), \tag{5.13}$$

where the factors are defined by

$$\{\mathfrak{E}, \mathfrak{F}_0, \Theta_1(\lambda) = \widehat{\Theta}_1(\lambda)|\mathfrak{E}\} \quad \text{and} \quad \{\mathfrak{F}_0, \mathfrak{E}_*, \Theta_2(\lambda) = \widehat{\Theta}_2(\lambda)|\mathfrak{F}_0\},$$

is strictly nontrivial.

This concludes the proof of Theorem 5.6.

Remark. In a factorization of type (5.13) it is essential to allow intermediary spaces $\mathfrak{F}$ different from the spaces $\mathfrak{E}, \mathfrak{E}_*$. For example, the scalar function λ has no nontrivial scalar factorizations, but we have the factorization $\lambda = \Theta_2(\lambda)\Theta_1(\lambda)$ with

$$\Theta_2(\lambda) = \begin{bmatrix} \frac{1}{\sqrt{2}} & \frac{\lambda}{\sqrt{2}} \end{bmatrix} \quad \text{and} \quad \Theta_1(\lambda) = \begin{bmatrix} \lambda/\sqrt{2} \\ 1/\sqrt{2} \end{bmatrix}$$

(in this case $\mathfrak{E} = \mathfrak{E}_* = E^1$ and $\mathfrak{F} = E^2$).

6 Scalar multiples

1. An important case, when the study of the operator-valued function $\Theta(\lambda)$ can be reduced to the study of a scalar function, is indicated by the following

Definition. The contractive analytic function $\{\mathfrak{E}, \mathfrak{E}_*, \Theta(\lambda)\}$ is said to have the *scalar multiple* $\delta(\lambda)$, if $\delta(\lambda)$ is a scalar-valued analytic function, $\delta(\lambda) \not\equiv 0$, and there exists a contractive analytic function $\{\mathfrak{E}_*, \mathfrak{E}, \Omega(\lambda)\}$ such that

$$\Omega(\lambda)\Theta(\lambda) = \delta(\lambda)I_{\mathfrak{E}}, \quad \Theta(\lambda)\Omega(\lambda) = \delta(\lambda)I_{\mathfrak{E}_*} \quad (\lambda \in D). \tag{6.1}$$

A simple class of such functions $\Theta(\lambda)$ is exhibited by the following

Proposition 6.1. *If* $\dim \mathfrak{E} = \dim \mathfrak{E}_* = n < \infty$, *then every contractive analytic function* $\{\mathfrak{E}, \mathfrak{E}_*, \Theta(\lambda)\}$ *such that* $\Theta(\lambda)$ *is invertible for at least one* λ *in* D, *has a scalar multiple. In particular, the determinant* $d(\lambda)$ *of the matrix of* $\Theta(\lambda)$ *with respect to two orthonormal bases, in* $\mathfrak{E}$ *and in* $\mathfrak{E}_*$, *is a scalar multiple of* $\Theta(\lambda)$, *and the matrix of the corresponding function* $\Omega(\lambda)$ *is the algebraic adjoint of the matrix of* $\Theta(\lambda)$.

Proof. Let $\{e_i\}$ and $\{e_{i*}\}$ $(i = 1, \ldots, n)$ be orthonormal bases in $\mathfrak{E}$ and in $\mathfrak{E}_*$. The corresponding matrix $\vartheta(\lambda) = [\vartheta_{ij}(\lambda)]$ $(i, j = 1, \ldots, n)$ of $\Theta(\lambda)$ is defined by

$$\Theta(\lambda)e_j = \sum_{i=1}^{n} \vartheta_{ij}(\lambda)e_{*i} \qquad (j = 1, \ldots, n)$$

(i.e., $\vartheta_{ij}(\lambda) = (\Theta(\lambda)e_j, e_{*i})$.). Let $\omega(\lambda) = [\omega_{ij}(\lambda)]$ be the algebraic adjoint of $\vartheta(\lambda)$, that is, $\omega_{ij}(\lambda)$ is the determinant, multiplied by $(-1)^{i+j}$, of the matrix obtained from the matrix $\vartheta(\lambda)$ by deleting its ith column and its jth row. By their construction, the scalar-valued functions $\vartheta_{ij}(\lambda)$, $\omega_{ij}(\lambda)$, and $d(\lambda) = \det \vartheta(\lambda)$ are holomorphic and bounded on D; moreover, the matrix $\vartheta(\lambda)$ is regular for at least one value of λ in D, thus $d(\lambda) \not\equiv 0$. The operator-valued function $\{\mathfrak{E}_*, \mathfrak{E}, \Omega(\lambda)\}$ defined by

$$\Omega(\lambda)e_{*j} = \sum_{i=1}^{n} \omega_{ij}(\lambda)e_i \qquad (j = 1, \ldots, n)$$

is also holomorphic on D, and it follows from the matrix relations $\omega(\lambda)\vartheta(\lambda) = \vartheta(\lambda)\omega(\lambda) = d(\lambda)I_n$ (I_n is the identity matrix of order n) that the operator relations (6.1) are satisfied with $\delta = d$.

It remains to show that the function $\Omega(\lambda)$ is contractive. The self-adjoint operator $\Theta(\lambda)^*\Theta(\lambda)$ has an orthonormal set of eigenvectors $f_1(\lambda), \ldots, f_n(\lambda)$ corresponding to the eigenvalues $\rho_1(\lambda), \ldots, \rho_n(\lambda)$; as $\Theta(\lambda)$ and therefore $\Theta(\lambda)^*\Theta(\lambda)$ are contractive functions, we have $1 \geq \rho_i \geq 0$ for $i = 1, \ldots, n$. Let us observe next that

$$|d(\lambda)|^2 = \overline{d(\lambda)} \cdot d(\lambda) = \det \vartheta(\lambda)^* \cdot \det \vartheta(\lambda) = \det \vartheta(\lambda)^*\vartheta(\lambda).$$

Now $\vartheta(\lambda)^*\vartheta(\lambda)$ is the matrix of $\Theta(\lambda)^*\Theta(\lambda)$ with respect to the orthonormal basis $\{e_i\}$. Passing to the orthonormal basis $\{f_i(\lambda)\}$ by a unitary transformation $U(\lambda)$, the determinant of the matrix is not changed, and hence

$$|d(\lambda)|^2 = \rho_1(\lambda) \cdots \rho_n(\lambda) \leq \rho_j(\lambda) \qquad (j = 1, 2, \ldots, n).$$

Consequently, for every $e \in \mathfrak{E}$

$$|d(\lambda)|^2\|e\|^2 = |d(\lambda)|^2 \sum_{j=1}^{n} |(e,f_j(\lambda))|^2 \leq \sum_{j=1}^{n} \rho_j(\lambda)|(e,f_j(\lambda))|^2$$
$$= (\Theta(\lambda)^*\Theta(\lambda)e,e) = \|\Theta(\lambda)e\|^2.$$

On the other hand, (6.1) gives (for $\delta = d$) $\|\Omega(\lambda)\Theta(\lambda)e\| = \|d(\lambda)e\| = |d(\lambda)|\,\|e\|$, and thus we have

$$\|\Omega(\lambda)\Theta(\lambda)e\| \leq \|\Theta(\lambda)e\|. \tag{6.2}$$

For λ such that $d(\lambda) \neq 0$ the range of $\Theta(\lambda)$ equals $\mathfrak{E}_*$, and then (6.2) yields

$$\|\Omega(\lambda)e_*\| \leq \|e_*\| \quad \text{for every} \quad e_* \in \mathfrak{E}_*. \tag{6.3}$$

The zeros of $d(\lambda)$ in D form a discrete set, thus the validity of (6.3) extends by continuity to all λ in D. This concludes the proof of Proposition 6.1.

2. Now we consider a contractive analytic function $\{\mathfrak{E},\mathfrak{E}_*,\Theta(\lambda)\}$ with $\mathfrak{E}$ and $\mathfrak{E}_*$ not necessarily finite-dimensional.

Theorem 6.2. *If $\Theta(\lambda)$ has a scalar multiple $\delta(\lambda)$ and if*

$$\Theta(\lambda) = \Theta_i(\lambda)\Theta_e(\lambda), \quad \delta(\lambda) = \delta_i(\lambda)\delta_e(\lambda)$$

are the corresponding canonical factorizations, then $\delta_i(\lambda)$ is a scalar multiple of $\Theta_i(\lambda)$, and $\delta_e(\lambda)$ is a scalar multiple of $\Theta_e(\lambda)$. Particular consequences are:

(a) *If $\Theta(\lambda)$ is inner or outer, then $\delta_i(\lambda)$ or $\delta_e(\lambda)$ is also a scalar multiple of $\Theta(\lambda)$, respectively.*

(b) *If $\delta(\lambda)$ is inner or outer, then $\Theta(\lambda)$ is also inner or outer, respectively.*

(c) *If $\Theta(\lambda)$ is inner or outer, then it is so from both sides.*

Proof. We begin with the particular case (a), then we prove the general assertion, and finally we deduce the particular cases (b) and (c).

Case (a): (1) If $\Theta(\lambda)$ is inner, then (6.1) implies for almost every t

$$\|\Omega(e^{it})e_*\| = \|\Theta(e^{it})\Omega(e^{it})e_*\| = \|\delta(e^{it})e_*\| = |\delta_e(e^{it})|\,\|e_*\| \quad (e_* \in \mathfrak{E}_*). \tag{6.4}$$

Let $\{\mathfrak{E}_*,\mathfrak{E}_*,\Omega_1(\lambda)\}$ be the contractive analytic function defined by $\Omega_1(\lambda) = \delta_e(\lambda)I_{\mathfrak{E}_*}$. By virtue of (6.4) we have

$$\Omega(e^{it})^*\Omega(e^{it}) = \Omega_1(e^{it})^*\Omega_1(e^{it}) \quad \text{a.e.}$$

Because $\Omega_1(\lambda)$ is obviously outer, it follows from Proposition 4.1(b) that there exists an inner function $\{\mathfrak{E}_*,\mathfrak{E},\Omega_2(\lambda)\}$ such that

$$\Omega(\lambda) = \Omega_2(\lambda)\Omega_1(\lambda) = \delta_e(\lambda)\Omega_2(\lambda) \quad (\lambda \in D).$$

Introducing this expression of $\Omega(\lambda)$ into (6.1) we obtain

$$\delta_e(\lambda)[\Omega_2(\lambda)\Theta(\lambda)-\delta_i(\lambda)I_{\mathfrak{E}}]=O, \quad \delta_e(\lambda)[\Theta(\lambda)\Omega_2(\lambda)-\delta_i(\lambda)I_{\mathfrak{E}_*}]=O.$$

As $\delta_e(\lambda)\neq 0$ on D, this implies

$$\Omega_2(\lambda)\Theta(\lambda)=\delta_i(\lambda)I_{\mathfrak{E}}, \quad \Theta(\lambda)\Omega_2(\lambda)=\delta_i(\lambda)I_{\mathfrak{E}_*};$$

that is, $\delta_i(\lambda)$ is a scalar multiple of $\Theta(\lambda)$.

(2) If $\Theta(\lambda)$ is outer, then (6.1) implies

$$\overline{\Omega H^2(\mathfrak{E}_*)}=\overline{\Omega\overline{\Theta H^2(\mathfrak{E})}}=\overline{\Omega\Theta H^2(\mathfrak{E})}=\overline{\delta H^2(\mathfrak{E})}=\overline{\delta_i H^2(\mathfrak{E})}=\delta_i H^2(\mathfrak{E}). \tag{6.5}$$

Let $\Omega(\lambda)=\Omega_i(\lambda)\Omega_e(\lambda)$ be the canonical factorization of $\Omega(\lambda)$ with the inner factor $\{\mathfrak{G},\mathfrak{E},\Omega_i(\lambda)\}$ and the outer factor $\{\mathfrak{E}_*,\mathfrak{G},\Omega_e(\lambda)\}$. Then we have

$$\overline{\Omega H^2(\mathfrak{E}_*)}=\overline{\Omega_i\Omega_e H^2(\mathfrak{E}_*)}=\overline{\Omega_i\overline{\Omega_e H^2(\mathfrak{E}_*)}}=\overline{\Omega_i H^2(\mathfrak{G})}=\Omega_i H^2(\mathfrak{G}).$$

Compared with (6.5) this yields:

$$\Omega_i H^2(\mathfrak{G})=\delta_i H^2(\mathfrak{E}).$$

Thus to every $u\in H^2(\mathfrak{G})$ there corresponds a $v\in H^2(\mathfrak{E})$, and conversely, so that

$$\Omega_i u=\delta_i v,$$

and we have

$$\|u\|=\|\Omega_i u\|=\|\delta_i v\|=\|v\|.$$

Hence we see that $u\to v=Qu$ is a unitary transformation from $H^2(\mathfrak{G})$ to $H^2(\mathfrak{E})$, and

$$\Omega_i u=\delta_i Qu \quad \text{for} \quad u\in H^2(\mathfrak{G}). \tag{6.6}$$

It is obvious that Q commutes with multiplication by the variable λ: $Q\cdot\lambda u=\lambda\cdot Qu$ $(u\in H^2(\mathfrak{G}))$. We apply Lemma 3.2(d) to Q, and deduce the existence of a unitary transformation Z from $\mathfrak{G}$ to $\mathfrak{E}$ such that

$$(Qu)(\lambda)=Z\cdot u(\lambda) \qquad (\lambda\in D, u\in H^2(\mathfrak{G})).$$

We deduce, using (6.6), that $\Omega_i(\lambda)=\delta_i(\lambda)Z$ $(\lambda\in D)$. Substituting this value in (6.1) we obtain

$$\delta_i(\lambda)(Z\Omega_e(\lambda)\cdot\Theta(\lambda)-\delta_e(\lambda)I_{\mathfrak{E}})=O, \quad \delta_i(\lambda)(\Theta(\lambda)\cdot Z\Omega_e(\lambda)-\delta_e(\lambda)I_{\mathfrak{E}_*})=O.$$

The functions appearing in these relations are holomorphic on D, and $\delta_i(\lambda)$ has only a discrete set of zeros in D, therefore we conclude that

$$Z\Omega_e(\lambda)\cdot\Theta(\lambda)=\delta_e(\lambda)I_{\mathfrak{E}}, \quad \Theta(\lambda)\cdot Z\Omega_e(\lambda)=\delta_e(\lambda)I_{\mathfrak{E}_*} \quad (\lambda\in D).$$

This result shows that $\delta_e(\lambda)$ is a scalar multiple of $\Theta(\lambda)$.

The general case: Let $\mathfrak{F}$ be the intermediate space for the canonical factorization of $\Theta(\lambda)$ (cf. Sec. 4.3). By virtue of the characteristic property of the outer factors $\Theta_e(\lambda)$ and $\delta_e(\lambda)$ we have

$$\begin{aligned}\overline{\Theta_e\Omega\Theta_i H^2(\mathfrak{F})} &= \overline{\Theta_e\Omega\Theta_i\overline{\Theta_e H^2(\mathfrak{E})}} = \overline{\Theta_e\Omega\Theta H^2(\mathfrak{E})} = \overline{\Theta_e\delta H^2(\mathfrak{E})}\\ &= \overline{\delta\Theta_e H^2(\mathfrak{E})} = \overline{\delta\overline{\Theta_e H^2(\mathfrak{E})}} = \overline{\delta H^2(\mathfrak{F})} = \overline{\delta_i\overline{\delta_e H^2(\mathfrak{F})}}\\ &= \overline{\delta_i H^2(\mathfrak{F})} = \delta_i H^2(\mathfrak{F});\end{aligned}$$

the last equation follows from the fact that $\delta_i(\lambda)$ is an inner function. Hence for every $u \in H^2(\mathfrak{F})$ there exists a $v \in H^2(\mathfrak{F})$ such that

$$\Theta_e\Omega\Theta_i u = \delta_i v. \tag{6.7}$$

Applying Θ_i to both sides of (6.7) and taking account of (6.1) we obtain

$$\delta\Theta_i u = \Theta_i\delta_i v, \quad \Theta_i(\delta u - \delta_i v) = 0.$$

Because Θ_i is an isometry we deduce $\delta u - \delta_i v = 0$, and thus (6.7) implies

$$\Theta_e\Omega\Theta_i u = \delta u;$$

as u is arbitrary, it follows that

$$\Theta_e(\lambda)\Omega(\lambda)\Theta_i(\lambda) = \delta(\lambda)I_{\mathfrak{F}}. \tag{6.8}$$

On the other hand, relations (6.1) imply

$$\Theta_i(\lambda)\cdot\Theta_e(\lambda)\Omega(\lambda) = \Theta(\lambda)\Omega(\lambda) = \delta(\lambda)I_{\mathfrak{E}_*}$$

and

$$\Omega(\lambda)\Theta_i(\lambda)\cdot\Theta_e(\lambda) = \Omega(\lambda)\Theta(\lambda) = \delta(\lambda)I_{\mathfrak{E}}.$$

These relations, combined with (6.8), show that $\delta(\lambda)$ is a scalar multiple of both $\Theta_i(\lambda)$ and $\Theta_e(\lambda)$. The proof concludes by applying case (a) to $\Theta_i(\lambda)$ and to $\Theta_e(\lambda)$.

Case (b): This follows readily from the general case and from the obvious fact that a contractive analytic function cannot have the scalar multiple 1 unless it is a unitary constant.

Case (c): This is an immediate consequence of cases (a) and (b), and of the fact that if $\Theta(\lambda)$ has the scalar multiple $\delta(\lambda)$ then $\Theta^{\sim}(\lambda)$ has the scalar multiple $\delta^{\sim}(\lambda)$, which is inner or outer when $\delta(\lambda)$ is inner or outer, respectively.

This concludes the proof of the theorem.

Corollary 6.3. *Let $\{E^n, E^n, \vartheta(\lambda)\}$ be a matrix-valued contractive analytic function, such that $d(\lambda) = \det \vartheta(\lambda) \not\equiv 0$. In order that $\vartheta(\lambda)$ be inner or outer it is necessary and sufficient that $d(\lambda)$ be inner or outer, respectively.*

Proof. By virtue of Proposition 6.1, the sufficiency of the condition is contained in Theorem 6.2(b). Let us now suppose that $\vartheta(\lambda)$ is inner or outer. From Theorem 6.2(a) it follows that there exists a contractive analytic (matrix) function $\{E^n, E^n, \omega(\lambda)\}$ such that

$$\omega(\lambda)\vartheta(\lambda) = \vartheta(\lambda)\omega(\lambda) = \delta(\lambda)I_n \quad (\lambda \in D) \tag{6.9}$$

with a scalar function $\delta(\lambda)$, which is inner or outer, respectively. Denoting the determinant of $\omega(\lambda)$ by $d_*(\lambda)$ we obtain from (6.9), by taking determinants,

$$d_*(\lambda)d(\lambda) = \delta(\lambda)^n \quad (\lambda \in D).$$

Hence $d(\lambda)$ is a divisor (in H^∞) of $\delta(\lambda)^n$, and this is inner or outer when $\delta(\lambda)$ is inner or outer, respectively. Thus $d(\lambda)$ is of the same type (inner or outer) as $\delta(\lambda)$.

3. Suppose now that the contractive analytic function $\{\mathfrak{E}, \mathfrak{E}_*, \Theta(\lambda)\}$ is the product of the contractive analytic functions $\{\mathfrak{E}, \mathfrak{F}, \Theta_1(\lambda)\}$ and $\{\mathfrak{F}, \mathfrak{E}_*, \Theta_2(\lambda)\}$,

$$\Theta(\lambda) = \Theta_2(\lambda)\Theta_1(\lambda) \qquad (\lambda \in D). \tag{6.10}$$

It is obvious that *if $\Theta_1(\lambda)$ and $\Theta_2(\lambda)$ have the scalar multiples $\delta_1(\lambda)$ and $\delta_2(\lambda)$, then $\Theta(\lambda)$ has the scalar multiple $\delta(\lambda) = \delta_2(\lambda)\delta_1(\lambda)$.* The converse is in general not true: the existence of a scalar multiple of $\Theta(\lambda)$ does not imply the existence of scalar multiples of $\Theta_1(\lambda)$ and $\Theta_2(\lambda)$. For example, the function $\Theta(\lambda) \equiv \lambda$ has the trivial scalar multiple $\delta(\lambda) \equiv \lambda$, but the factors in its factorizations considered at the end of Sec. 5 have no scalar multiples at all. However, we can prove a partial converse.

Proposition 6.4. *If, in (6.10), $\Theta(\lambda)$ and one of the factors $\Theta_j(\lambda)$ $(j = 1,2)$ have scalar multiples then so has the other factor, and in this case every scalar multiple of $\Theta(\lambda)$ is also a scalar multiple of $\Theta_1(\lambda)$ and $\Theta_2(\lambda)$.*

Proof. Suppose $\Theta(\lambda)$ and $\Theta_1(\lambda)$ have the scalar multiple $\delta(\lambda)$ and $\delta_1(\lambda)$. Thus (6.1) holds, and

$$\Omega_1(\lambda)\Theta_1(\lambda) = \delta_1(\lambda)I_{\mathfrak{E}}, \quad \Theta_1(\lambda)\Omega_1(\lambda) = \delta_1(\lambda)I_{\mathfrak{F}} \tag{6.11}$$

hold. Multiplying (6.1) from the left by $\Theta_1(\lambda)$ and from the right by $\Omega_1(\lambda)$, and taking (6.11) into account, we obtain

$$\delta_1(\lambda)\Theta_1(\lambda)\Omega(\lambda)\Theta_2(\lambda) = \delta_1(\lambda)\delta(\lambda)I_{\mathfrak{F}} \quad (\lambda \in D).$$

Because $\delta_1(\lambda) \not\equiv 0$, the analyticity of the functions under consideration implies that

$$\Theta_1(\lambda)\Omega(\lambda)\Theta_2(\lambda) = \delta(\lambda)I_{\mathfrak{F}} \quad (\lambda \in D).$$

This relation, together with the second relation (6.1), shows that $\delta(\lambda)$ is a scalar multiple of $\Theta_2(\lambda)$ too. If it is the factor $\Theta_2(\lambda)$ which is supposed to have a scalar multiple $\delta_2(\lambda)$, we obtain by an analogous reasoning that $\Theta_1(\lambda)$ has the scalar multiple $\delta(\lambda)$. We conclude that both factors $\Theta_j(\lambda)$ $(j=1,2)$ have the scalar multiple $\delta(\lambda)$, as asserted.

4. For a matrix-valued contractive analytic function $\{E^n, E^n, \vartheta(\lambda)\}$, the function $d(\lambda) = \det \vartheta(\lambda)$ belongs to H^∞. Moreover, it follows from the relation

$$\overline{d(\lambda)}d(\lambda) = \det \vartheta(\lambda)^* \cdot \det \vartheta(\lambda) = \det[\vartheta(\lambda)^* \vartheta(\lambda)]$$

that $|d(e^{it})| = 1$ at any point e^{it} where $\vartheta(e^{it})$ is isometric (hence unitary, because the space is finite-dimensional). This fact extends to every contractive analytic function $\{\mathfrak{E}, \mathfrak{E}_*, \Theta(\lambda)\}$ with a scalar multiple $\delta(\lambda)$ in the following manner.

Proposition 6.5. *Let α be the set of points $\zeta = e^{it}$ at which $\Theta(\zeta)$ exists and is an isometry (from $\mathfrak{E}$ to $\mathfrak{E}_*$). Then*

(i) *$\Theta(\zeta)$ is even unitary a.e. in α.*
(ii) *$\Theta(\lambda)$ has a scalar multiple $\delta_{2\alpha}(\lambda)$ for which $|\delta_{2\alpha}(\zeta)| = 1$ a.e. in α; such is in particular the function $\delta_{2\alpha}(\lambda) = \delta(\lambda)/\delta_{1\alpha}(\lambda)$, where*

$$\delta_{1\alpha}(\lambda) = \exp\left[\frac{1}{2\pi}\int_{(\alpha)} \frac{e^{it}+\lambda}{e^{it}-\lambda} \log|\delta(e^{it})|\, dt\right] \quad (\lambda \in D).$$

(See the remark at the beginning of Sec. 4.4.)

Proof. Let $\{\mathfrak{E}, \mathfrak{F}, \Theta_e(\lambda)\}$ and $\{\mathfrak{F}, \mathfrak{E}_*, \Theta_i(\lambda)\}$ be the outer and inner factors of $\{\mathfrak{E}, \mathfrak{E}_*, \Theta(\lambda)\}$: $\Theta(\lambda) = \Theta_i(\lambda)\Theta_e(\lambda)$. By virtue of Theorem 6.2 these factors also have scalar multiples and $\Theta_i(\zeta)$ is unitary a.e. in C. Hence $\Theta_e(\zeta) = \Theta_i(\zeta)^{-1}\Theta(\zeta)$ is an isometry a.e. in α. Now as $\Theta_e(\lambda)$ is an outer function we have

$$\overline{\Theta_e(\zeta)\mathfrak{E}} = \mathfrak{F} \quad \text{a.e. in } C.$$

We conclude that $\Theta_e(\zeta)$ is unitary a.e. on α, and therefore so is the product $\Theta_i(\zeta)\Theta_e(\zeta) = \Theta(\zeta)$. This proves (i).

As to (ii), let us observe first that if $\Omega(\lambda)$ is the function associated with $\Theta(\lambda)$ and $\delta(\lambda)$ in the sense of (6.1), then for any fixed $e \in \mathfrak{E}$, $e_* \in \mathfrak{E}_*$ we have $|(\Omega(\zeta)e_*, e)| \leq \|e_*\|\|e\|$ a.e. in C, and

$$\begin{aligned} |(\Omega(\zeta)e_*, e)| &= |(\Theta(\zeta)\Omega(\zeta)e_*, \Theta(\zeta)e)| = |\delta(\zeta)||(e_*, \Theta(\zeta)e)| \\ &\leq |\delta(\zeta)| \cdot \|e_*\|\|e\| \quad \text{a.e. in } \alpha. \end{aligned}$$

Because $|\delta_{1\alpha}(\zeta)| = |\delta(\zeta)|$ a.e. in α and $|\delta_{1\alpha}(\zeta)| = 1$ a.e. in $\alpha' = C \backslash \alpha$, it follows that

$$|(\Omega(\zeta)e_*, e)| \leq |\delta_{1\alpha}(\zeta)| \cdot \|e_*\|\|e\| \quad \text{a.e. in } C. \tag{6.12}$$

The functions $(\Omega(\lambda)e_*, e)$ and $\delta_{1\alpha}(\lambda)$ are holomorphic on D and $\delta_{1\alpha}(\lambda)$ is an outer function, therefore (6.12) implies

$$|(\Omega(\lambda)e_*, e)| \le |\delta_{1\alpha}(\lambda)| \cdot \|e_*\| \|e\| \quad (\lambda \in D);$$

see HOFFMAN [1] p. 62. Hence

$$\Omega'(\lambda) = \delta_{1\alpha}(\lambda)^{-1} \cdot \Omega(\lambda)$$

is a contractive analytic function on D. From (6.1) it follows, dividing by $\delta_{1\alpha}(\lambda)$, that $\delta_{2\alpha}(\lambda) = \delta(\lambda)/\delta_{1\alpha}(\lambda)$ is a scalar multiple of $\Theta(\lambda)$. This concludes the proof.

5. We need the following lemma.

Lemma 6.6. *Let S be a domain in the plane of complex numbers, given by*

$$\{\lambda = \rho\, e^{it}: \ r_0 < \rho < R_0, t_1 < t < t_2\}$$

with $0 \le r_0 < 1 < R_0 \le \infty$, $0 < t_2 - t_1 \le 2\pi$. Let $\varphi_-(\lambda)$ and $\varphi_+(\lambda)$ be functions with values in a Banach space X, defined and holomorphic on S_- and S_+, respectively, where S_- denotes the part of S interior to the unit circle C, and S_+ the part exterior to C. Let us assume that the limits

$$\psi_-(t) = \lim_{r \to 1-0} \varphi_-(re^{it}), \quad \psi_+(t) = \lim_{R \to 1+0} \varphi_+(Re^{it}) \tag{6.13}$$

exist a.e. in the interval (t_1, t_2), and also in the mean L^1, that is,

$$\begin{aligned} &\int_{t_1}^{t_2} \|\varphi_-(re^{it}) - \psi_-(t)\|_X \, dt \to 0 \quad (r_0 < r \to 1-0), \\ &\int_{t_1}^{t_2} \|\varphi_+(Re^{it}) - \psi_+(t)\|_X \, dt \to 0 \quad (R_0 > R \to 1+0). \end{aligned} \tag{6.14}$$

Furthermore, let us suppose that

$$\psi_-(t) = \psi_+(t) \tag{6.15}$$

a.e. in the interval (t_1, t_2). Then the functions $\varphi_-(\lambda)$ and $\varphi_+(\lambda)$ are analytic continuations of each other through the arc $\alpha = \{e^{it}: \ t_1 < t < t_2\}$ of C.

Proof. Let $\beta = \{e^{it}: \ \tau_1 \le t \le \tau_2\}$ be a closed sub-arc of α such that the limits (6.13) exist for $t = \tau_1$ and for $t = \tau_2$, and are equal: on account of our hypotheses we can choose β as close to α as we wish. Choose r_1 and R_1 such that $r_0 < r_1 < 1 < R_1 < R_0$ and denote by Γ the contour $a_1 A_1 A_2 a_2$ indicated by the figure, where $a_k = r_1 e^{i\tau_k}$ and $A_k = R_1 e^{i\tau_k}$ $(k = 1, 2)$, and let Σ be the union of Γ and its interior. Let us denote by Σ_- and Σ_+ the parts of Σ situated in the interior and in the exterior of C, respectively. Let us define the function $F(\lambda)$ on $\Sigma \backslash \beta$ by setting

$$F(\lambda) = \begin{cases} \varphi_-(\lambda) \text{ for } \lambda \in \Sigma_-, \\ \varphi_+(\lambda) \text{ for } \lambda \in \Sigma_+; \end{cases}$$

due to the manner in which we have chosen β, the definition of $F(\lambda)$ extends to the points of intersection of Γ and C so that $F(\lambda)$ will be a continuous function on the whole contour Γ. Consequently, the integral

$$G(\lambda) = \frac{1}{2\pi i} \int_{\Gamma} \frac{F(\zeta)}{\zeta - \lambda} \, d\zeta \tag{6.16}$$

exists and is a holomorphic function of λ on the whole interior of Σ. We show that $G(\lambda) = \varphi_-(\lambda)$ in the interior of Σ_-, and $G(\lambda) = \varphi_+(\lambda)$ in the exterior of Σ_+; and this concludes the proof. To this end, let us fix a point λ_0 in the interior of Σ_-, and choose r and R such that $|\lambda_0| < r < 1 < R < R_1$. The arcs β_r and β_R, with radii r and R, cut Σ into three parts whose contours we denote by $\Gamma_1, \Gamma_2, \Gamma_3$; see the figure below. The integral (6.16) equals the sum of the integrals along these contours; the first of them is equal to $\varphi_-(\lambda_0)$, and the third equal to 0, because λ_0 is interior to Γ_1 and exterior to Γ_3 and because φ_- and φ_+ are holomorphic on S_- and on S_+, respectively. As regards the integral along Γ_2, it tends to 0 as $R - r \to 0$. This is obvious for the integrals along the two segments joining the extremities of β_r and β_R. On the other hand, it follows from (6.14) and (6.15) that if $r \to 1-0$ and $R \to 1+0$ then

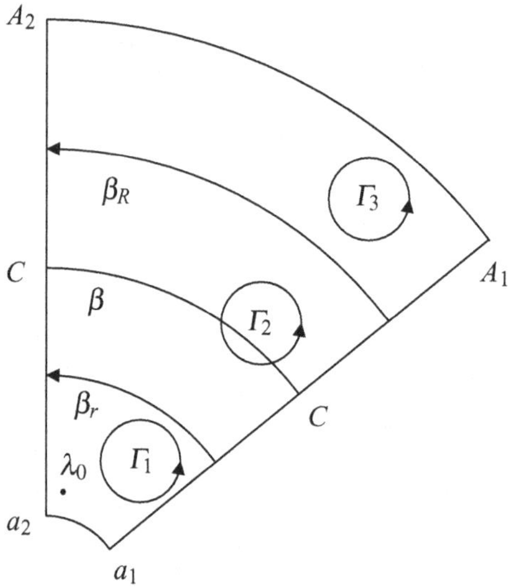

$$\frac{1}{2\pi}\left[\int_{\beta_R} - \int_{\beta_r}\right] \frac{F(\zeta)}{\zeta - \lambda_0} d\zeta = \frac{1}{2\pi} \int_{t_1}^{t_2} \left[\frac{Re^{it}}{Re^{it} - \lambda_0} \varphi_+(Re^{it}) - \frac{re^{it}}{re^{it} - \lambda_0} \varphi_-(re^{it})\right] dt$$

$$\to \frac{1}{2\pi} \int_{t_1}^{t_2} \frac{e^{it}}{e^{it} - \lambda_0} [\psi_+(t) - \psi_-(t)] \, dt = 0$$

(here we used the fact that $Re^{it}/(Re^{it}-\lambda_0)$ and $re^{it}/(re^{it}-\lambda_0)$ tend to $e^{it}/(e^{it}-\lambda_0)$ uniformly on (τ_1,τ_2)). We obtain that $G(\lambda_0)=\varphi_-(\lambda_0)$. Similar reasoning proves that for λ_0 in the interior of Σ_+ we have $G(\lambda_0)=\varphi_+(\lambda_0)$, thus concluding the proof.

We apply this lemma to prove the following result.

Proposition 6.7. *Let $\{\mathfrak{E},\mathfrak{E}_*,\Theta(\lambda)\}$ be a contractive outer function with a scalar multiple. Suppose $\Theta(\zeta)$ is an isometry (from $\mathfrak{E}$ into $\mathfrak{E}_*$) at almost every point $\zeta = e^{it}$ of an open arc α of the unit circle C. Then the function $\Theta(\lambda)$ has an analytic continuation through α to the exterior of C.*

Proof. By virtue of Theorem 6.2(a), $\Theta(\lambda)$ also has an *outer* scalar multiple $\delta(\lambda)$. On the other hand, by Proposition 6.5, $\Theta(\zeta)$ is unitary at almost every point $\zeta\in\alpha$, and $\Theta(\lambda)$ has as a scalar multiple the function $\delta_{2\alpha}(\lambda)=\delta(\lambda)/\delta_{1\alpha}(\lambda)$. Note that

$$\delta_{2\alpha}(\lambda)=\exp\left[\frac{1}{2\pi}\int_{(\alpha')}\frac{e^{it}+\lambda}{e^{it}-\lambda}\log|\delta(e^{it})|\,dt\right], \tag{6.17}$$

where $\alpha' = C\backslash\alpha$. If $\Omega(\lambda)$ is the function associated with $\Theta(\lambda)$ and $\delta_{2\alpha}(\lambda)$ in the sense of (6.1), we have

$$\Theta(\lambda)^{-1}=\frac{1}{\delta_{2\alpha}(\lambda)}\Omega(\lambda),\quad \|\Theta(\lambda)^{-1}\|\le\frac{1}{|\delta_{2\alpha}(\lambda)|}.$$

Using (6.17) we conclude that $\Theta(\lambda)^{-1}$ is defined and holomorphic on the interior D of C, and that it has bounded norm on every subset of D which is at a positive distance from α'. This implies that the function

$$\Phi_+(\lambda)=[\Theta(1/\bar{\lambda})^{-1}]^*$$

is defined and holomorphic on the exterior of C, and is bounded on every subset of this domain which is at positive distance from α'. Let α_1 be a closed arc in the interior of the arc α and let Δ be the domain limited by α_1 and by an arc β_1 having the same endpoints as α_1, but lying otherwise in the exterior of C. $\Phi_+(\lambda)$ is then holomorphic and bounded on Δ. Performing a conformal mapping of Δ onto D and applying Fatou's theorem in its generalized form (cf. Sec. 2.1) to the transform of the function $\Phi_+(\lambda)$, and then returning to the initial function, we obtain that $\Phi_+(\lambda)$ has a nontangential strong limit $\Phi_+(\zeta)$ at almost every point $\zeta\in\alpha_1$.

On the other hand we have as $r\to 1-0$,

$$I_{\mathfrak{E}}=\Theta(re^{it})^*\Theta(re^{it})^{*-1}=\Theta(re^{it})^*\Phi_+\left(\frac{1}{r}e^{it}\right)\to\Theta(e^{it})^*\Phi_+(e^{it})$$

strongly (cf. (2.10)) a.e. in α_1; thus

$$I_{\mathfrak{E}}=\Theta(\zeta)^*\Phi_+(\zeta)\quad\text{a.e. in }\alpha_1.$$

As $\Theta(\zeta)$ is unitary a.e. in α, we conclude that

$$\Phi_+(\zeta) = \Theta(\zeta) \quad \text{a.e. in } \alpha_1.$$

Let us set

$$\Phi_-(\lambda) = \Theta(\lambda) \quad \text{for} \quad |\lambda| < 1.$$

Then we can apply Lemma 6.6 to the functions $\Phi_-(\lambda)e$, $\Phi_+(\lambda)e$ and to the arc α_1 (for all $e \in \mathfrak{E}$). In fact, these functions being bounded on some domains adjacent to the arc α_1 (and lying in the interior and in the exterior of C, respectively), the pointwise (strong) convergences

$$\Phi_-(re^{it}) \to \Theta(e^{it}), \quad \Phi_+(Re^{it}) \to \Theta(e^{it}) \quad (r \to 1-0, R \to 1+0),$$

a.e. on α_1, imply the convergences in the mean of type (6.14).

Thus we obtain that, for $e \in \mathfrak{E}$, the functions $\Phi_-(\lambda)e = \Theta(\lambda)e$ ($|\lambda| < 1$) and $\Phi_+(\lambda)e$ ($|\lambda| > 1$) are analytic continuations of each other along α_1. This implies that the operator-valued function extends analytically across α_1 to $\Phi_+(\lambda)$. As α_1 is an arbitrary closed arc in the interior of α, this concludes the proof of Proposition 6.7.

Remark. As shown by the example of the functions $\Omega_\alpha(\lambda)$ considered in Sec. 4.5, Proposition 6.7 does not hold in general if we omit the assumption that the function admits a scalar multiple.

6. Here is a further useful remark on functions with scalar multiples.

Proposition 6.8. *If the contractive analytic function* $\{\mathfrak{E}, \mathfrak{E}_*, \Theta(\lambda)\}$ *has a scalar multiple* $\delta(\lambda)$*, then its purely contractive component* $\{\mathfrak{E}^0, \mathfrak{E}^0_*, \Theta^0(\lambda)\}$ *also has the scalar multiple* $\delta(\lambda)$*, and conversely.*

Proof. Let us set $\mathfrak{E}' = \mathfrak{E} \ominus \mathfrak{E}^0$, $\mathfrak{E}'_* = \mathfrak{E}_* \ominus \mathfrak{E}^0_*$; the function $\Theta(\lambda)|\mathfrak{E}'$ is then constant, its value being a unitary operator Z (from $\mathfrak{E}'$ to $\mathfrak{E}'_*$). Let us denote by P^0 the orthogonal projection of $\mathfrak{E}$ onto $\mathfrak{E}^0$, and by P^0_* the orthogonal projection of $\mathfrak{E}_*$ onto $\mathfrak{E}^0_*$.

Let us assume first that $\Theta(\lambda)$ has the scalar multiple $\delta(\lambda)$, thus there exists a contractive analytic function $\{\mathfrak{E}_*, \mathfrak{E}, \Omega(\lambda)\}$ satisfying

$$\Omega(\lambda)\Theta(\lambda) = \delta(\lambda)I_{\mathfrak{E}}, \quad \Theta(\lambda)\Omega(\lambda) = \delta(\lambda)I_{\mathfrak{E}_*}. \tag{6.18}$$

We have $\Theta(\lambda) = \Theta^0(\lambda)P^0 + Z(I_{\mathfrak{E}} - P^0)$, therefore the second relation (6.18) yields

$$\Theta^0(\lambda)P^0\Omega(\lambda)e^0_* - \delta(\lambda)e^0_* = -Z(I_{\mathfrak{E}} - P^0)\Omega(\lambda)e^0_* \quad \text{for} \quad e^0_* \in \mathfrak{E}^0_*.$$

For every fixed $\lambda \in D$, the left-hand side of this relation is an element of $\mathfrak{E}^0_*$, and the right-hand side is an element of $\mathfrak{E}'_*$. Therefore both equal 0, and hence we have

$\Theta^0(\lambda)P^0\Omega(\lambda)e^0_* = \delta(\lambda)e^0_*$ on D. On the other hand, the first relation (6.18) implies

$$P^0\Omega(\lambda)\Theta^0(\lambda)e^0 = P^0\Omega(\lambda)\Theta(\lambda)e^0 = P^0\delta(\lambda)e^0 = \delta(\lambda)e^0 \quad \text{for} \quad e^0 \in \mathfrak{E}^0.$$

Setting $\Omega_0(\lambda) = P^0\Omega(\lambda)|\mathfrak{E}^0_*$ we obtain therefore a contractive analytic function $\{\mathfrak{E}^0_*, \mathfrak{E}^0, \Omega_0(\lambda)\}$ satisfying the relations

$$\Omega_0(\lambda)\Theta^0(\lambda) = \delta(\lambda)I_{\mathfrak{E}^0}, \quad \Theta^0(\lambda)\Omega_0(\lambda) = \delta(\lambda)I_{\mathfrak{E}^0_*} \quad (\lambda \in D). \tag{6.19}$$

We have proved that every scalar multiple $\delta(\lambda)$ of $\Theta(\lambda)$ is also a scalar multiple of $\Theta^0(\lambda)$. Conversely, if $\delta(\lambda)$ is a scalar multiple of $\Theta^0(\lambda)$ (i.e., if there exists a contractive analytic function $\{\mathfrak{E}^0_*, \mathfrak{E}^0, \Omega_0(\lambda)\}$ satisfying relations (6.19)), then it is easy to verify that the function $\{\mathfrak{E}_*, \mathfrak{E}, \Omega(\lambda)\}$ defined by

$$\Omega(\lambda) = \Omega_0(\lambda)P^0_* + \delta(\lambda)Z^*(I_{\mathfrak{E}_*} - P^0_*)$$

is contractive analytic and satisfies (6.18). This completes the proof.

7 Factorization of functions with scalar multiple

1. For functions with a scalar multiple some of the results of Sec. 4 can be improved.

Let $N(t)$ be as in Sec. 4.2, that is, a strongly measurable function whose values are self-adjoint operators on a (separable) Hilbert space $\mathfrak{E}$, and such that $O \leq N(t) \leq I$ $(0 \leq t \leq 2\pi)$. Let $m(t)$ denote its lower bound function:

$$m(t) = \inf\{(N(t)e, e) \colon \ e \in \mathfrak{E}, \|e\| = 1\};$$

$m(t)$ is a scalar-valued, nonnegative, measurable function.

Proposition 7.1. (a) *If there exists a contractive analytic function $\{\mathfrak{E}, \mathfrak{F}, \Theta(\lambda)\}$ with a scalar multiple $\delta(\lambda)$, satisfying the inequality*

$$N(t)^2 \geq \Theta(e^{it})^*\Theta(e^{it}) \quad \text{a.e.,} \tag{7.1}$$

then we have

$$\log m(t) \in L^1(0, 2\pi). \tag{7.2}$$

(b) *Conversely, if* (7.2) *holds, then there exists a contractive analytic function, even an outer one $\{\mathfrak{E}, \mathfrak{F}_1, \Theta_1(\lambda)\}$, having a scalar multiple $\delta_1(\lambda)$ and satisfying the equality*

$$N(t)^2 = \Theta_1(e^{it})^*\Theta_1(e^{it}) \quad \text{a.e.;} \tag{7.3}$$

moreover, we can assume that

$$|\delta_1(e^{it})| = m(t) \quad a.e. \tag{7.4}$$

(c) *If* $\dim \mathfrak{E} < \infty$, *condition* (7.2) *is equivalent to the following one,*

$$\log \det[N(t)] \in L^1(0, 2\pi), \tag{7.5}$$

where $[N(t)]$ *denotes the matrix of the operator* $N(t)$ *with respect to any basis in* $\mathfrak{E}$.

Proof. Part (a): Suppose $\delta(\lambda)$ is a scalar multiple of $\Theta(\lambda)$; that is, $\delta \not\equiv 0$ and

$$\Omega(\lambda)\Theta(\lambda) = \delta(\lambda)I_{\mathfrak{E}}, \quad \Theta(\lambda)\Omega(\lambda) = \delta(\lambda)I_{\mathfrak{F}} \tag{7.6}$$

for some contractive analytic function $\{\mathfrak{F}, \mathfrak{E}, \Omega(\lambda)\}$. From (7.1) and (7.6) we deduce for every $h \in \mathfrak{E}$

$$\|h\| \geq \|N(t)h\| \geq \|\Theta(e^{it})h\| \geq \|\Omega(e^{it})\Theta(e^{it})h\| = |\delta(e^{it})|\|h\| \quad \text{a.e.};$$

hence

$$1 \geq m(t) \geq |\delta(e^{it})| \quad \text{a.e.} \tag{7.7}$$

Because $\log|\delta(e^{it})| \in L^1(0, 2\pi)$ (cf. Sec. III.1), we have (7.2).

Part (b): Condition (7.2) implies that there exists a scalar-valued outer function $\delta_1(\lambda)$ satisfying (7.4) (cf. Sec. III.1); because $m(t) \leq 1$, we have $|\delta_1(\lambda)| \leq 1$. Observe next that for $v \in L^2(\mathfrak{E})$,

$$\begin{aligned} \|Nv\|^2 &= \frac{1}{2\pi}\int_0^{2\pi} \|N(t)v(t)\|^2 \, dt \\ &\geq \frac{1}{2\pi}\int_0^{2\pi} m(t)^2 \|v(t)\|^2 \, dt = \frac{1}{2\pi}\int_0^{2\pi} |\delta_1(e^{it})|^2 \|v(t)\|^2 \, dt = \|\delta_1 v\|^2. \end{aligned}$$

Considering in particular the functions $u \in H^2(\mathfrak{E})$ we conclude that there exists a contraction X from $\mathfrak{N} = \overline{NH^2(\mathfrak{E})}$ into $H^2(\mathfrak{E})$ for which

$$X(Nu) = \delta_1 u \quad (u \in H^2(\mathfrak{E})).$$

Let us show that $Xw = 0$ (for a $w \in \mathfrak{N}$) implies $w = 0$. In fact, for such a w there exists a sequence $\{u_n\}$ in $H^2(\mathfrak{E})$ such that

$$Nu_n \to w \quad \text{and} \quad \delta_1 u_n = XNu_n \to Xw = 0$$

(convergence in $L^2(\mathfrak{E})$). Hence we obtain

$$\delta_1 N u_n \to \delta_1 w \quad \text{and} \quad N\delta_1 u_n \to N0 = 0;$$

from $\delta_1 N = N\delta_1$ we infer $\delta_1 w = 0$; that is,

$$\delta_1(\lambda)w(\lambda) = 0 \quad (\lambda \in D).$$

Because $\delta_1(\lambda) \not\equiv 0$, this implies $w(\lambda) \equiv 0$ $(\lambda \in D)$.

The operator X obviously commutes with $U^{\times}$ (multiplication by e^{it} in $L^2(\mathfrak{E})$), thus we have

$$X \bigcap_{n\geq 0} U^{\times n}\mathfrak{N} \subset \bigcap_{n\geq 0} U^{\times n} X\mathfrak{N} \subset \bigcap_{n\geq 0} U^{\times n} H^2(\mathfrak{E}) = \{0\};$$

as $Xw = 0$ implies $w = 0$ we obtain

$$\bigcap_{n\geq 0} U^{\times n}\mathfrak{N} = \{0\}.$$

By virtue of Proposition 4.2 there exists therefore a contractive outer function $\{\mathfrak{E}, \mathfrak{F}_1, \Theta(\lambda)\}$ satisfying (7.3). On the other hand, (7.4) implies $|\delta_1(e^{it})| I_{\mathfrak{E}} \leq N(t)$. Taking account of (7.3), we obtain

$$|\delta_1(e^{it})|^2 I_{\mathfrak{E}} \leq \Theta_1(e^{it})^* \Theta_1(e^{it}) \quad \text{a.e.}$$

By Proposition 4.1 (applied to $\Theta_1(\lambda)$ and to $\delta_1(\lambda) I_{\mathfrak{E}}$) there exists therefore a contractive analytic function $\{\mathfrak{F}_1, \mathfrak{E}, \Theta_1(\lambda)\}$ such that

$$\delta_1(\lambda) I_{\mathfrak{E}} = \Omega_1(\lambda)\Theta_1(\lambda) \quad (\lambda \in D).$$

Multiplying by $\Theta_1(\lambda)$ from the left yields

$$[\delta_1(\lambda) I_{\mathfrak{F}_1} - \Theta_1(\lambda)\Omega_1(\lambda)]\Theta_1(\lambda) = O \quad (\lambda \in D).$$

Because $\Theta_1(\lambda)$ is outer, we have $\overline{\Theta_1(\lambda)\mathfrak{E}} = \mathfrak{F}_1$ (cf. Proposition 2.4), so we conclude that

$$\delta_1(\lambda) I_{\mathfrak{F}_1} = \Theta_1(\lambda)\Omega_1(\lambda) \quad (\lambda \in D).$$

Thus $\delta_1(\lambda)$ is a scalar multiple of $\Theta_1(\lambda)$ and the proof of part (b) is complete.

Part (c): We simply observe that if $\rho_k(t)$ $(k = 1, \ldots, n; n = \dim \mathfrak{E})$ are the eigenvalues of $N(t)$ arranged in a nondecreasing order, we have $0 < \rho_1(t) \leq \rho_2(t) \leq \cdots \leq \rho_n(t) \leq 1$ a.e., and therefore

$$m(t) = \rho_1(t) \quad \text{and} \quad \det[N(t)] = \prod_{k=1}^{n} \rho_k(t) \begin{cases} \leq \rho_1(t), \\ \geq \rho_1(t)^n. \end{cases}$$

Thus

$$n \cdot \log m(t) \leq \log \det[N(t)] \leq \log m(t);$$

hence if one of the functions $\log m(t)$ and $\log \det[N(t)]$ belongs to $L^1(0, 2\pi)$ then so does the other. This completes the proof of Proposition 7.1.

2. If $\{\mathfrak{E}, \mathfrak{E}_*, \Theta(\lambda)\}$ is a contractive analytic function having a scalar multiple $\delta(\lambda)$, then (6.1) implies

$$\Theta(e^{it})^{-1} = \frac{1}{\delta(e^{it})}\Omega(e^{it}) \quad \text{a.e.}$$

Thus Proposition 4.3 applies to $\Theta(\lambda)$: for every Borel subset α of C there exists a factorization

$$\Theta(\lambda) = \Theta_{2\alpha}(\lambda)\Theta_{1\alpha}(\lambda) \quad (\lambda \in D)$$

with the properties (4.27) and (4.28); moreover $\Theta_{1\alpha}(\lambda)$ is outer. Set

$$N(t) = [\Theta(e^{it})^*\Theta(e^{it})]^{1/2}, \quad N_\alpha(t) = \begin{cases} N(t) & (t \in (\alpha)), \\ I_{\mathfrak{E}} & (t \in (\alpha')), \end{cases}$$

and let $m(t)$ and $m_\alpha(t)$ be the corresponding lower bounds; clearly

$$m_\alpha(t) = \begin{cases} m(t) & (t \in (\alpha)), \\ 1 & (t \in (\alpha')). \end{cases} \tag{7.8}$$

By Proposition 7.1(a) we have $\log m(t) \in L^1(0,2\pi)$; on account of (7.8) we have therefore $\log m_\alpha(t) \in L^1(0,2\pi)$ too. Applying Proposition 7.1(b) we obtain that there exists a contractive outer function $\Theta_\alpha(\lambda)$ satisfying the equation $N_\alpha(t)^2 = \Theta_\alpha(e^{it})^*\Theta_\alpha(e^{it})$ a.e., and admitting as a scalar multiple the outer function $\delta_\alpha(\lambda)$ determined by the relation $|\delta_\alpha(e^{it})| = m_\alpha(t)$ a.e., that is,

$$\delta_\alpha(\lambda) = \exp\left[\frac{1}{2\pi}\int_{(\alpha)} \frac{e^{it}+\lambda}{e^{it}-\lambda} \log m(t)\, dt\right]. \tag{7.9}$$

Let us remark that as $1 \geq m(t) \geq |\delta(e^{it})|$ a.e. (cf. (7.7)), the outer function

$$\delta_{1\alpha}(\lambda) = \exp\left[\frac{1}{2\pi}\int_{(\alpha)} \frac{e^{it}+\lambda}{e^{it}-\lambda} \log|\delta(e^{it})|\, dt\right] \tag{7.10}$$

is divisible (in H^∞) by $\delta_\alpha(\lambda)$ and we have $|\delta_{1\alpha}(\lambda)/\delta_\alpha(\lambda)| \leq 1$ on D; hence $\delta_{1\alpha}(\lambda)$ is also a scalar multiple of $\Theta_\alpha(\lambda)$.

Now from (4.27) and (4.27′) it follows that $\Theta_{1\alpha}(e^{it})^*\Theta_{1\alpha}(e^{it}) = N_\alpha(t)^2$ also holds a.e.; by Proposition 4.1(b), $\Theta_\alpha(\lambda)$ is therefore equal to $\Theta_{1\alpha}(\lambda)$ up to a constant unitary factor from the left. Consequently, $\delta_{1\alpha}(\lambda)$ is a scalar multiple of $\Theta_{1\alpha}(\lambda)$ also.

From Proposition 6.4 we get that $\Theta_{2\alpha}(\lambda)$ also has the scalar multiple $\delta(\lambda)$. Because $\Theta_{2\alpha}(e^{it})$ is isometric at almost every point e^{it} of α (cf. (4.28)), it follows from Proposition 6.5 that the function

$$\delta_{2\alpha}(\lambda) = \delta(\lambda)/\delta_{1\alpha}(\lambda) \tag{7.11}$$

is also a scalar multiple of $\Theta_{2\alpha}(\lambda)$.

Thus, Proposition 4.3 can be completed as follows.

Proposition 7.2. *Proposition* 4.3 *applies to every contractive analytic function* $\Theta(\lambda)$ *with a scalar multiple* $\delta(\lambda)$. *The factors* $\Theta_{1\alpha}(\lambda)$ *and* $\Theta_{2\alpha}(\lambda)$ *have the functions* $\delta_{1\alpha}(\lambda)$ *and* $\delta_{2\alpha}(\lambda)$ *defined by* (7.10) *and* (7.11) *as scalar multiples, respectively.*

8 Analytic kernels

Although the results of this section are not used in the sequel, we include them in this book because they are intimately connected with the subject of the preceding sections, in particular with Sec. 5.

Given a separable Hilbert space $\mathfrak{F}$ we call an *analytic kernel in* $\mathfrak{F}$, or in short a *kernel*, every function $\mathscr{K}(\mu,\lambda)$ defined on $D \times D$, whose values are bounded operators on $\mathfrak{F}$ and such that, for any fixed $\mu \in D$ and $f \in \mathfrak{F}$,

$$\mathscr{K}(\mu,\lambda)f \in H^2(\mathfrak{F}) \qquad \text{(as a function of } \lambda\text{)}. \tag{8.1}$$

We say that the kernel $\mathscr{K}(\mu,\lambda)$ is *positive definite*, and write $\mathscr{K}(\mu,\lambda) \gg \mathscr{O}$, if

$$\sum_{j=1}^{n} \sum_{k=1}^{n} (\mathscr{K}(\mu_j,\mu_k)f_j, f_k) \geq 0 \tag{8.2}$$

for every finite system of complex numbers $\mu_j \in D$ and vectors $f_j \in \mathfrak{F}$ $(j = 1,\ldots,n)$. For two kernels, the notation $\mathscr{K}_1 \gg \mathscr{K}_2$ or $\mathscr{K}_2 \ll \mathscr{K}_1$ indicates that $\mathscr{K}_1 - \mathscr{K}_2 \gg \mathscr{O}$. The particular kernel

$$\frac{1}{1-\bar{\mu}\lambda} I_{\mathfrak{F}}$$

is denoted by $\mathscr{J}(\mu,\lambda)$.

Proposition 8.1. (a) *For every contractive analytic function* $\{\mathfrak{E},\mathfrak{F},\Theta(\lambda)\}$ *(with separable* $\mathfrak{E},\mathfrak{F}$*)*

$$\mathscr{H}_\Theta(\mu,\lambda) = \Theta(\lambda)\Theta(\mu)^* \quad \textit{and} \quad \mathscr{K}_\Theta(\mu,\lambda) = (1-\bar{\mu}\lambda)^{-1}\Theta(\lambda)\Theta(\mu)^*$$

are analytic kernels in $\mathfrak{F}$, *and we have*

$$\mathscr{H}_\Theta \gg \mathscr{O}, \quad \mathscr{J} \gg \mathscr{K}_\Theta \gg \mathscr{O}.$$

(b) *In order that an analytic kernel* $\mathscr{K}(\mu,\lambda)$ *in* $\mathfrak{F}$ *admits a representation of the form* $\mathscr{K}_\Theta(\mu,\lambda)$ *with some contractive analytic function* $\Theta(\lambda)$, *it is sufficient* (and by virtue of (a) also necessary) *that it satisfies the conditions*

$$\mathscr{O} \ll \mathscr{K}(\mu,\lambda) \ll \mathscr{J}(\mu,\lambda) \quad \textit{and} \quad (1-\bar{\mu}\lambda)\mathscr{K}(\mu,\lambda) \gg \mathscr{O}. \tag{8.3}$$

Proof. Part (a): The assertion concerning $\mathscr{H}_\Theta(\mu,\lambda)$ is obvious. As to the others, let us observe first that the functions (of λ) of the form

$$f^\mu(\lambda) = (1-\bar{\mu}\lambda)^{-1}f \quad (\mu \in D; f \in \mathfrak{F}) \tag{8.4}$$

belong to $H^2(\mathfrak{F})$, and that

$$(u, f^\mu)_{H^2(\mathfrak{F})} = (u(\mu), f)_{\mathfrak{F}} \quad \text{for} \quad u \in H^2(\mathfrak{F}); \tag{8.5}$$

thus we have in particular

$$(g^\nu, f^\mu)_{H^2(\mathfrak{F})} = (1-\mu\bar{\nu})^{-1}(g,f)_{\mathfrak{F}}. \tag{8.6}$$

From (8.5) we also obtain that the elements of the form f^μ span $H^2(\mathfrak{F})$. Let us denote by L the set of the finite linear combinations

$$u(\lambda) = \sum_k f_k^{\mu_k}(\lambda) \quad \text{(with different } \mu_k\text{s);}$$

L is obviously a linear manifold. A function $u(\lambda) \in L$ extends to a meromorphic function in the whole complex plane and is determined by the $f_k^{\mu_k}$ (i.e., by its poles and residues). This implies that if R is a transformation in $H^2(\mathfrak{F})$ defined for the elements of the form f^μ, such that

$$R(cf)^\mu = c \cdot Rf^\mu \quad \text{and} \quad R(f_1+f_2)^\mu = Rf_1^\mu + Rf_2^\mu,$$

then R extends in a unique way to a linear transformation defined on the whole linear manifold L. We conclude in particular that the formula

$$\mathscr{K}(\mu,\lambda)f = (Rf^\mu)(\lambda) \quad (f \in \mathfrak{F}; \mu,\lambda \in D) \tag{8.7}$$

establishes a one-to-one and linear correspondence $\mathscr{K} \leftrightarrow R$ between the analytic kernels $\mathscr{K}(\mu,\lambda)$ in $\mathscr{F}$ and those linear transformations R in $H^2(\mathfrak{F})$ whose domain of definition is L. Relations (8.7) and (8.5) imply

$$(\mathscr{K}(\mu,\nu)f,g)_{\mathfrak{F}} = ((Rf^\mu)(\nu),g)_{\mathfrak{F}} = (Rf^\mu, g^\nu)_{H^2(\mathfrak{F})}, \tag{8.8}$$

and hence it follows readily that if $\mathscr{K} \leftrightarrow R$, then the conditions

$$\mathscr{K} \gg \mathscr{O} \quad \text{and} \quad R \geq O_L$$

are equivalent.[3]

If $R = I_L$, (8.7) gives that $\mathscr{K} = \mathscr{I}$; hence $\mathscr{I} \gg \mathscr{O}$.

Now let $\{\mathfrak{E},\mathfrak{F},\Theta(\lambda)\}$ be a contractive analytic function (with separable $\mathfrak{E}$ and $\mathfrak{F}$). For any $u \in H^2(\mathfrak{E})$ and $f \in \mathfrak{F}$ we have

$$\begin{aligned}(\Theta_+ u, f^\mu) &= \int_0^{2\pi} \frac{1}{1-\mu e^{-it}}(\Theta(e^{it})u(e^{it}),f)_{\mathfrak{F}}\frac{dt}{2\pi} = (\Theta(\mu)u(\mu),f)_{\mathfrak{F}} \\ &= (u(\mu),\Theta(\mu)^* f)_{\mathfrak{F}} = \int_0^{2\pi}\frac{1}{1-\mu e^{-it}}(u(e^{it}),\Theta(u)^* f)_{\mathfrak{F}}\frac{dt}{2\pi} \\ &= (u,(\Theta(\mu)^* f)^\mu),\end{aligned}$$

and hence

$$(\Theta_+^* f^\mu)(\lambda) = \frac{1}{1-\bar{\mu}\lambda}\Theta(\mu)^* f.$$

[3] The subscript indicates restriction to L.

From this we deduce

$$(\Theta_+\Theta_+^* f^\mu)(\lambda) = \frac{1}{1-\bar{\mu}\lambda}\Theta(\lambda)\Theta(\mu)^* f \quad (\mu,\lambda \in D),$$

thus the kernel corresponding to the operator $R = \Theta_+\Theta_+^*|L$ equals $\mathscr{K}_\Theta(\mu,\lambda)$. As $R \geq O_L$ and $I_L - R \geq O_L$, this kernel satisfies the conclusion in (a).

Part (b): Let us observe first that if V denotes multiplication by λ in $H^2(\mathfrak{F})$, then

$$(V^* f^\mu)(\lambda) = \frac{1}{\lambda}[f^\mu(\lambda) - f^\mu(0)] = \bar{\mu} f^\mu(\lambda)$$

and consequently

$$(VRV^* f^\mu)(\lambda) = \lambda\bar{\mu}(Rf^\mu)(\lambda)$$

for every R; hence if $\mathscr{K}(\mu,\lambda) \leftrightarrow R$ then

$$(1-\bar{\mu}\lambda)\mathscr{K}(\mu,\lambda) \leftrightarrow R - VRV^*.$$

Thus conditions (8.3) are equivalent to $O \leq R_\circ \leq I$ and $R_\circ - VR_\circ V^* \geq O$, where $R_\circ$ denotes the closure of R. By virtue of Proposition 5.1 these conditions imply that $R_\circ = AA^*$, where A is a bounded operator (indeed, a contraction) of a (separable) Hilbert space $\mathfrak{H}_R$ into $H^2(\mathfrak{F})$, such that $VA = AV_R$, and V_R is a unilateral shift on $\mathfrak{H}_R$. Taking the Fourier representation of $\mathfrak{H}_R$ with respect to V_R, A is transformed to the operator Θ_+ associated with a contractive analytic function $\{\mathfrak{E},\mathfrak{F},\Theta(\lambda)\}$ (with $\mathfrak{E} = \mathfrak{H}_R \ominus V_R\mathfrak{H}_R$); see Lemma 3.2. So we have $R_\circ = \Theta_+\Theta_+^*$. Now the kernel corresponding to $R = \Theta_+\Theta_+^*|L$ is $\mathscr{K}_\Theta(\mu,\lambda)$. We conclude that this kernel equals the given kernel $\mathscr{K}(\mu,\lambda)$. This completes the proof.

9 Notes

Operator-valued analytic functions, namely the resolvent of an operator, have long played a fundamental role in the study of operators, in particular concerning spectrum, functional calculus, spectral decompositions of Riesz–Dunford type, and so on. More recently, some other operator-valued analytic functions have gained importance in several domains of functional analysis. Let us mention the theory of characteristic functions inaugurated by LIVŠIC [1], [2], [3], the prediction theory for multivariate stochastic processes (cf. WIENER AND MASANI [1], [2]; MASANI [1], [2]; etc.), the description of the invariant subspaces of unilateral shifts of arbitrary multiplicity (cf. LAX [1] and HALMOS [2]), and finally the harmonic analysis of the unitary dilation of a contraction, which has led the authors of the present book to the functional models of contractions (cf. SZ.-N.–F. [2], [VIII]) and to various other results dealt with in the following chapters.

Proposition 2.1, concerning the decomposition of any contractive analytic function into its purely contractive part and its unitary constant part, was first proved by ŠTRAUS [1], [2]; SZ.-N.–F. [IX] found it independently and applied it to the study of the invariant subspaces of a contraction.

The notions of inner and outer functions can be extended in different ways to the operator-valued case (cf., e.g., HELSON AND LOWDENSLAGER [2]); the definitions in Sec. 2.2 were given by the authors in connection with the asymptotic behavior of the iterates of contractions (cf. SZ.-N.–F. [3], [VIII] and the following chapter).

Lemmas 3.1 and 3.2 on Fourier representations were formulated (as a single lemma) in SZ.-N.–F. [IX]; see also [X]. These lemmas determine in particular the general form of the contractions on $H^2(\mathfrak{E})$ that commute with multiplication by the variable λ; this form has been wellknown for some time.

Theorem 3.3 on the invariant subspaces of the unilateral shift is due to BEURLING [1] (case $\dim \mathfrak{E} = 1$), LAX [1] (case $\dim \mathfrak{E} < \infty$), and HALMOS [2] (general case).

The description of the hyperinvariant subspaces for a unilateral shift, stated in Sec. 3.4, appeared first in the original edition of this book (in French). A proof (simpler than the authors have had in mind, and concerning isometries of general type as well) was given later in DOUGLAS [5].

The problem of factorization of the type $N(t)^2 = \Theta(e^{it})^*\Theta(e^{it})$ for an operator-valued function $N(t) \geq O$ by means of an operator-valued analytic function $\Theta(\lambda)$, is a generalization, first considered by SZEGŐ [1], of the representation $f(t) = |q(e^{it})|^2$ of a scalar-valued function $f(t) \geq 0$ by means of a scalar-valued analytic function $q(\lambda)$. In its turn, Szegő's result generalizes a lemma of Fejér and Riesz, which states that if $f(t)$ is a trigonometric polynomial, $f(t) \geq 0$, then $q(\lambda)$ can be chosen as a polynomial of λ (cf. [*Func. Anal.*] Sec. 53). It was much later that ZASUHIN [1] and WIENER [1] observed the importance of this type of factorization of matrix-valued functions for the prediction theory of multivariate stochastic processes. Indeed, one of the problems of this theory can be formulated as follows. Let $\mathfrak{E}$ be a (separable) Hilbert space and let $N(t)$ be an operator-valued, measurable function on $(0, 2\pi)$, $O \leq N(t) \leq I$. Find for every $f \in \mathfrak{E}$ the distance

$$p(f) = \inf_u \left[\frac{1}{2\pi} \int_0^{2\pi} \|N(t)[e^{-it}f - u(e^{it})]\|_{\mathfrak{E}}^2 \, dt \right]^{1/2},$$

where u runs over $H^2(\mathfrak{E})$. If $N(t)^2$ admits of a factorization $\Theta(e^{it})^*\Theta(e^{it})$ by means of a contractive outer function $\{\mathfrak{E}, \mathfrak{E}_*, \Theta(\lambda)\}$, then, using the fact that $\overline{\Theta H^2(\mathfrak{E})} = H^2(\mathfrak{E}_*)$, we obtain

$$p(f) = \inf_v \left[\frac{1}{2\pi} \int_0^{2\pi} \|\Theta(e^{it})f - v(e^{it})\|_{\mathfrak{E}_*}^2 \, dt \right]^{1/2},$$

where v runs over the set of functions in $H^2(\mathfrak{E}_*)$ with $v(0) = 0$. So we have in this case

$$p(f) = \|\Theta(0)f\|.$$

If there is no such factorization for $N(t)$, then one can show that it is the outer function $\Theta_1(\lambda)$ appearing in Proposition 4.2 which furnishes the solution of the problem.

A general criterion for the existence of a factorization of the form $N(t)^2 = \Theta(e^{it})^*\Theta(e^{it})$ was obtained by LOWDENSLAGER [1] using the Wold decomposition. The method of this author, combined with Lemmas 3.1 and 3.2 on Fourier representation, leads us to our results in Sec. 4; see also SZ.-N.–F. [IX]. This combination of the two methods lends an evident unity to the reasoning in Sec. 4. The criterion of Lowdenslager (restricted to functions $N(t) \leq I$) is the one contained in Proposition 4.2, formula (4.8). (See also DOUGLAS [2].)

Proposition 4.1 was given in this form in SZ.-N.–F. [IX]. In the case that $\mathfrak{E}$ and $\mathfrak{E}_*$ are of finite dimension, Proposition 4.1(b) can also be found in HELSON AND LOWDENSLAGER [2] and ŠMUL'JAN [3].

The canonical factorization appeared in SZ.-N.–F. [3] as a consequence of the canonical triangulation of contractions (cf. Sec. II.4); the way it is obtained in Sec. 4.3 was indicated in SZ.-N.–F. [IX].

Proposition 4.3, due to the present authors, establishes the existence of a rich variety of nontrivial factorizations for any contractive analytic function, outer from both sides. It avoids the use of multiplicative integrals. For the use of these integrals see, for example, POTAPOV [1]; BRODSKIĬ AND LIVŠIC [1]; GINZBURG [1]; M.S. BORDSKIĬ [6]; V.M. BRODSKIĬ [1]; and V.M. BRODSKIĬ AND M.S. BRODSKIĬ [1].

Section 5 reproduces the paper SZ.-N.–F. [XII].

The notion of a scalar multiple, which generalizes to some extent that of the determinant, was introduced in SZ.-N.–F. [7]; the detailed exposition in Sec. 6 appeared for the first time in the French edition of this book. The results of this section are used in Chap. VIII. Corollary 6.3 coincides, for finite matrix-valued contractive outer functions, with a characterization given in HELSON AND LOWDENSLAGER [2] p. 204; see also HELSON [1] p. 125. Thus, a matrix-valued function which is outer in the sense of these authors, is outer from both sides in the sense adopted in the present book. Proposition 6.7 (on analytic continuation) was given (in the case of finite dimensional $\mathfrak{E}_*$) in SZ.-N.–F. [IX*].

Part (b) of Proposition 7.1, concerning the sufficiency of the condition

$$\log m(t) \in L^1(0, 2\pi)$$

in order that $N^2(t)$ admit a factorization $\Theta(e^{it})^*\Theta(e^{it})$, is due to DEVINATZ [1]. Part (a) shows that this condition is also necessary, if we also require that $\Theta(\lambda)$ have a scalar multiple. Part (c) was announced first in ZASUHIN [1], and proved independently by several authors, including WIENER AND MASANI [1]; WIENER AND AKUTOWICZ [1]; and HELSON AND LOWDENSLAGER [1].

Proposition 7.2 is new: it is used in Chap. VIII. It also allows us to compare the factorizations obtained in our Proposition 4.3 with those obtained by means of multiplicative integrals.

We have restricted our study throughout this chapter to functions $N(t)$, $\Theta(\lambda)$, and so on, which are *bounded* (indeed, contractive), the fundamental lemmas of Sec. 3 being established for *bounded* operators Q only. This restriction is justified by the nature of the problems to which we apply these results in the following chapters. However, the results obtained can be extended to some unbounded functions, namely to functions $N(t)$ such that $N(t)f \in L^2(\mathfrak{E})$ for the elements f of a dense linear manifold in $\mathfrak{E}$. We do not go into details of these generalizations here; for $\mathfrak{E} = E^n$ see HELSON [1].

Analytic kernels play an important role in the work announced by DE BRANGES and ROVNYAK [1]. The fact that

$$\frac{I - \Theta(\lambda)\Theta(\mu)^*}{1 - \bar{\mu}\lambda} \gg \mathcal{O}$$

(cf. Proposition 8.1(a)) was proved first by ROVNYAK [1]; another proof can be found in SZ.-NAGY [9]. Part (b) of Proposition 8.1 is contained (implicitly) in DE BRANGES AND ROVNYAK [1],[2].

In connection with this chapter see also DOUGLAS AND PEARCY [1]; DOUGLAS [1]; GINZBURG [4],[5]; and the factorization lemmas in SARASON [3] and SZ.-N–F. [11].

10 Further results

1. Consider a bounded domain Ω in the complex plane, with analytic boundary. For instance, the boundary of Ω could consist of a finite number of circles. As in the case of D, $H^\infty(\Omega)$ consists of the bounded analytic functions defined on Ω, and $H^p(\Omega)$ $(0 < p < \infty)$ consists of those analytic functions f in Ω such that $|f|^p$ has a harmonic majorant. We refer to FISHER [1] for the properties of these spaces. These definitions can be extended to spaces of functions with values in a Hilbert space. For applications to operator theory, it is very useful to consider functions f which at a point $\lambda \in \Omega$ take values in a Hilbert space $\mathfrak{E}_\lambda$ that depends analytically on λ. Formally, $\mathfrak{E} = (\mathfrak{E}_\lambda)_{\lambda \in \Omega}$ should be a Hermitian analytic vector bundle. When $p = 2$, one obtains a Hilbert space $H^2(\Omega, \mathfrak{E})$ on which one can define the *bundle shift* $S_{\mathfrak{E}}$ by setting

$$(S_{\mathfrak{E}} f)(\lambda) = \lambda f(\lambda), \quad f \in H^2(\Omega, \mathfrak{E}), \lambda \in \Omega.$$

Given two analytic vector bundles $\mathfrak{E}, \mathfrak{E}_*$, one can consider analytic operator-valued functions Θ such that $\Theta(\lambda) \in \mathscr{B}(\mathfrak{E}_\lambda, \mathfrak{E}_{*\lambda})$ for $\lambda \in \Omega$. In this context, one can define the notion of an inner function and prove a complete analogue of the BEURLING, LAX, AND HALMOS Theorem 3.3, with the operator U_+ replaced by $S_{\mathfrak{E}}$. We refer to ABRAHAMSE AND DOUGLAS [1] for these developments, and their applications to the theory of subnormal operators. This theory is instrumental in the study of the class C_0 associated with Ω; see ZUCCHI [1].

2. Given a Hilbert space $\mathfrak{H}$, we denote by $\mathfrak{H}^{\otimes n}$ the Hilbert space tensor product of n copies of $\mathfrak{H}$ $(n = 0, 1, \ldots)$. Thus, if $\{e_j\}_{j \in J}$ is an orthonormal basis for $\mathfrak{H}$, then the vectors $\{e_{j_1} \otimes e_{j_2} \otimes \cdots \otimes e_{j_n}\}_{j_1, j_2, \ldots, j_n \in J}$ form an orthonormal basis in $\mathfrak{H}^{\otimes n}$. The space $\mathfrak{H}^{\otimes 0}$ is a copy of the complex scalars. The *Fock space*

$$\mathscr{T}(\mathfrak{H}) = \bigoplus_{n=0}^{\infty} \mathfrak{H}^{\otimes n}$$

can be viewed as an analogue of the space H^2 corresponding to $k = \dim \mathfrak{H}$ noncommuting variables. In this context, the analogue of H^∞ is the weakly closed algebra $L^\infty_{\mathfrak{H}}$ generated by the left creation operators on $\mathscr{T}(\mathfrak{H})$. The left creation L_h operator associated with a vector $h \in \mathfrak{H}$ is defined by

$$L_h v = h \otimes v, \quad v \in \mathscr{T}(\mathfrak{H}).$$

This analogy can be carried surprisingly far. See, for instance, POPESCU [2] for a version of the BEURLING, LAX, AND HALMOS theorem. In its simplest form it states that the operators commuting with $L^\infty_{\mathfrak{H}}$ are precisely those in the algebra $R^\infty_{\mathfrak{H}}$ generated by the right creation operators. We refer to POPESCU [8] for further developments related to these ideas.

3. The existence of a bounded analytic left inverse for a bounded analytic function $\{\mathfrak{E}, \mathfrak{E}_*, \Theta(\lambda)\}$ is closely related to the classical Corona theorem. This appears in SZ.-N.–F. [27] in connection with similarity problems, and in TEODORESCU [1],[2],[4] in connection with the existence of invariant complements to an invariant subspace for a contraction. The Corona theorem in this context was studied by TREIL [1]–[4]. It is shown that, for spaces $\mathfrak{E}$ of infinite dimension, an extension of the Corona theorem holds only under some restrictions. See also TREIL AND WICK [1] for related results.

4. TAKAHASHI [1] gives necessary and sufficient conditions for a positive operator A to be of the form $A = B^*B$, where B is the operator of multiplication by a bounded analytic function.

Chapter VI

Functional Models

1 Characteristic functions

1. We recall the definition of the defect operators and defect spaces corresponding to a contraction T on the Hilbert space $\mathfrak{H}$:

$$D_T = (I - T^*T)^{1/2}, \quad D_{T^*} = (I - TT^*)^{1/2},$$
$$\mathfrak{D}_T = \overline{D_T\mathfrak{H}}, \qquad \mathfrak{D}_{T^*} = \overline{D_{T^*}\mathfrak{H}}.$$

Denote by Λ_T the set of complex numbers λ for which the operator $I - \lambda T^*$ is boundedly invertible.[1] For $\lambda \in \Lambda_T$ we define

$$\Theta_T(\lambda) = [-T + \lambda D_{T^*}(I - \lambda T^*)^{-1} D_T] | \mathfrak{D}_T. \tag{1.1}$$

From well-known properties of invertible operators (cf. [*Funct. Anal.*] Sec. 147) we infer that Λ_T is an open set containing the unit disc D and $\Theta_T(\lambda)$ is an analytic function on Λ_T; as a consequence of the relation $TD_T = D_{T^*}T$ (cf. (I.3.4)) the values of $\Theta_T(\lambda)$ are (bounded) operators from $\mathfrak{D}_T$ into $\mathfrak{D}_{T^*}$.

Moreover, by virtue of the same relation we obtain

$$\begin{aligned}\Theta_T(\lambda)D_T &= D_{T^*}[-T + \lambda(I - \lambda T^*)^{-1}(I - T^*T)] \\ &= D_{T^*}(I - \lambda T^*)^{-1}[-(I - \lambda T^*)T + \lambda(I - T^*T)],\end{aligned}$$

and hence

$$\Theta_T(\lambda)D_T = D_{T^*}(I - \lambda T^*)^{-1}(\lambda I - T) \qquad (\lambda \in \Lambda_T). \tag{1.2}$$

Replacing T by T^* we have analogously

$$\Theta_{T^*}(\mu)D_{T^*} = D_T(I - \mu T)^{-1}(\mu I - T^*) \qquad (\mu \in \Lambda_{T^*}). \tag{1.2*}$$

---

[1] For any operator T, the set Λ_T consists of the point $\lambda = 0$ and the symmetric image of $\rho(T)\setminus\{0\}$ with respect to the unit circle C, where $\rho(T)$ denotes the resolvent set for T. The set Λ_{T^*} is the symmetric image of Λ_T with respect to the real axis.

B.Sz.-Nagy et al., *Harmonic Analysis of Operators on Hilbert Space*, Universitext,
DOI 10.1007/978-1-4419-6094-8_6, © Springer Science+Business Media, LLC 2010

If $\lambda \in \Lambda_T$ and $\lambda^{-1} \in \Lambda_{T^*}$ (i.e., both λ and $\bar{\lambda}^{-1}$ belong to Λ_T), then equations (1.2) and (1.2)* imply

$$\Theta_T(\lambda)\Theta_{T^*}(\lambda^{-1})D_{T^*} = D_{T^*} \quad \text{and} \quad \Theta_{T^*}(\lambda^{-1})\Theta_T(\lambda)D_T = D_T;$$

that is, we have

$$\Theta_{T^*}(\lambda^{-1}) = \Theta_T(\lambda)^{-1} \quad \text{whenever} \quad \lambda, \bar{\lambda}^{-1} \in \Lambda_T. \tag{1.3}$$

This relation is used later. Now we set, for $\lambda, \mu \in \Lambda_T$,

$$A(\lambda,\mu) = I - T^*T - D_T\Theta_T(\mu)^*\Theta_T(\lambda)D_T$$

and using (1.2) we obtain

$$A(\lambda,\mu) = I - T^*T - (\bar{\mu}I - T^*)(I - \bar{\mu}T)^{-1}(I - TT^*)(I - \lambda T^*)^{-1}(\lambda I - T).$$

We observe that the following relations are valid for $\lambda \in \Lambda_T$:

$$(I - \lambda T^*)^{-1}(\lambda I - T) = -T + \lambda(I - \lambda T^*)^{-1}(I - T^*T),$$
$$(\lambda I - T)(I - \lambda T^*)^{-1} = -T + \lambda(I - TT^*)(I - \lambda T^*)^{-1};$$

to verify them one multiplies by $I - \lambda T^*$ from the left and from the right, respectively. From these relations we deduce, using the identity $(I - TT^*)T = T(I - T^*T)$, that

$$(I - TT^*)(I - \lambda T^*)^{-1}(\lambda I - T) = (\lambda I - T)(I - \lambda T^*)^{-1}(I - T^*T).$$

It follows that

$$\begin{aligned}
&A(\lambda,\mu)\\
&= [I - (\bar{\mu}I - T^*)(I - \bar{\mu}T)^{-1}(\lambda I - T)(I - \lambda T^*)^{-1}](I - T^*T)\\
&= [(I - \lambda T^*)(I - \bar{\mu}T) - (\bar{\mu}I - T^*)(\lambda I - T)](I - \bar{\mu}T)^{-1}(I - \lambda T^*)^{-1}(I - T^*T)\\
&= (1 - \lambda\bar{\mu})(I - T^*T)(I - \bar{\mu}T)^{-1}(I - \lambda T^*)^{-1}(I - T^*T).
\end{aligned}$$

Thus we have for any $h \in \mathfrak{H}$:

$$\begin{aligned}
&\|D_Th\|^2 - (\Theta_T(\lambda)D_Th, \Theta_T(\mu)D_Th) = (A(\lambda,\mu)h,h)\\
&= (1 - \lambda\bar{\mu})((I - \lambda T^*)^{-1}D_T^2h, (I - \mu T^*)^{-1}D_T^2h),
\end{aligned}$$

and consequently, for $f = D_Th$,

$$(f,f) - (\Theta_T(\lambda)f, \Theta_T(\mu)f) = (1 - \lambda\bar{\mu})((I - \lambda T^*)^{-1}D_Tf, (I - \mu T^*)^{-1}D_Tf);$$

thisrelation extends by continuity to every element f of $\mathfrak{D}_T = \overline{D_T\mathfrak{H}}$. When $\mu = \lambda$ we thus obtain

$$\|f\|^2 - \|\Theta_T(\lambda)f\|^2 = (1-|\lambda|^2)\|(I-\lambda T^*)^{-1}D_Tf\|^2 \qquad (f\in\mathfrak{D}_T, \lambda\in\Lambda_T). \quad (1.4)$$

As an immediate consequence of (1.4) we have

$$\|f\|^2 - \|\Theta_T(\lambda)f\|^2 \geq 0 \quad \text{for} \quad f\in\mathfrak{D}_T \quad \text{and} \quad \lambda\in D.$$

For $\lambda = 0$, (1.4) takes the form

$$\|f\|^2 - \|\Theta_T(0)f\|^2 = \|D_Tf\|^2 \qquad (f\in\mathfrak{D}_T).$$

Observe that $D_Tf = 0$ implies that f is orthogonal to every element of the form D_Th ($h\in\mathfrak{H}$) and hence orthogonal to $\mathfrak{D}_T$; if $f\in\mathfrak{D}_T$ this is impossible unless $f=0$. Thus we have

$$\|\Theta_T(0)f\| < \|f\| \quad \text{for every} \quad f\in\mathfrak{D}_T, \quad f\neq 0.$$

We have thereby proved that, when considered on the unit disc D, $\Theta_T(\lambda)$ *is a purely contractive analytic function* (*cf.* Sec. V.2).

Definition. The purely contractive analytic function $\{\mathfrak{D}_T, \mathfrak{D}_{T^*}, \Theta_T(\lambda)\}$ on D is called the *characteristic function* of the contraction T.

From (1.1) it follows that

$$\Theta_T(\lambda) = \Big[-T + \sum_{n=1}^{\infty}\lambda^n D_{T^*}T^{*n-1}D_T\Big]\Big|\mathfrak{D}_T \quad \text{for} \quad \lambda\in D, \qquad (1.1)'$$

the expansion being convergent in the norm.

Applying (1.1) to T^* as well as to T we obtain

$$\Theta_{T^*}(\lambda) = \Theta_T(\bar{\lambda})^* \quad \text{for} \quad \lambda\in\Lambda_{T^*}, \qquad (1.5)$$

and thus in particular

$$\Theta_{T^*}(\lambda) = \Theta_{\tilde{T}}(\lambda) \quad \text{on} \quad D. \qquad (1.6)'$$

If λ is a point on the unit circle C belonging to the resolvent set $\rho(T)$, then $\lambda = \bar{\lambda}^{-1}\in\Lambda_T$ so that we can apply (1.3). Using (1.5) we also obtain

$$\Theta_T(\lambda)^{-1} = \Theta_{T^*}(\lambda^{-1}) = \Theta_{T^*}(\bar{\lambda}) = \Theta_T(\lambda)^*;$$

that is, $\Theta_T(\lambda)$ is a unitary operator. Hence, *if α is an arc of C belonging to the resolvent set of T then the function $\Theta_T(\lambda)$ is analytic on α and its values on α are unitary operators* (from $\mathfrak{D}_T$ to $\mathfrak{D}_{T^*}$).

2. Let us now consider two contractions, T_1 on $\mathfrak{H}_1$ and T_2 on $\mathfrak{H}_2$, and suppose they are unitarily equivalent: $T_2 = \sigma T_1\sigma^{-1}$, where σ is a unitary operator from $\mathfrak{H}_1$

to $\mathfrak{H}_2$. From the definitions it follows readily that

$$\mathfrak{D}_{T_2} = \tau \mathfrak{D}_{T_1}, \quad \mathfrak{D}_{T_2^*} = \tau_* \mathfrak{D}_{T_1^*}, \quad \Theta_{T_2}(\lambda) = \tau_* \Theta_{T_1}(\lambda) \tau^{-1},$$

where τ and τ_* denote the restrictions of σ to $\mathfrak{D}_{T_1}$ and $\mathfrak{D}_{T_1^*}$, respectively.

Making use of the notion of coincidence introduced in Sec. V.2.4, we have therefore that *the characteristic functions of unitarily equivalent contractions coincide.*

The converse is not true, in this generality. Indeed, if $\mathfrak{H} = \mathfrak{H}_0 \oplus \mathfrak{H}_1$ is the decomposition of $\mathfrak{H}$ corresponding to the unitary part T_0 and the c.n.u part T_1 of T, then we have

$$D_T = O \oplus D_{T_1}, \quad D_{T^*} = O \oplus D_{T_1^*}, \quad \mathfrak{D}_T = \mathfrak{D}_{T_1}, \quad \mathfrak{D}_{T^*} = \mathfrak{D}_{T_1^*},$$

and hence $\Theta_T(\lambda) = \Theta_{T_1}(\lambda)$.

Therefore it suffices to consider c.n.u contractions. We show in Sec. 3 that for such contractions the above proposition has a complete converse.

3. If T is a contraction then so is

$$T_a = (T - aI)(I - \bar{a}T)^{-1}$$

for every complex a with $|a| < 1$; see Sec. I.4.4. For the characteristic functions of T and T_a we have that

$$\{\mathfrak{D}_{T_a}, \mathfrak{D}_{T_a^*}, \Theta_{T_a}(\lambda)\} \quad \textit{coincides with} \quad \left\{\mathfrak{D}_T, \mathfrak{D}_{T^*}, \Theta_T\left(\frac{\lambda + a}{1 + \bar{a}\lambda}\right)\right\}. \tag{1.6}$$

In fact, we obtain by elementary calculations that

$$I - T_a^* T_a = S^*(I - T^*T)S \quad \text{and} \quad I - T_a T_a^* = S(I - TT^*)S^*,$$

where

$$S = (1 - |a|^2)^{1/2}(I - \bar{a}T)^{-1},$$

and hence

$$\|D_{T_a}h\|^2 = \|D_T Sh\|^2, \quad \|D_{T_a^*}h\|^2 = \|D_{T^*}S^*h\|^2 \tag{1.7}$$

for $h \in \mathfrak{H}$. Because S and S^* map $\mathfrak{H}$ onto itself, relations (1.7) show that there exist a unitary operator Z from $\mathfrak{D}_{T_a}$ to $\mathfrak{D}_T$, and a unitary operator Z_* from $\mathfrak{D}_{T_a^*}$ to $\mathfrak{D}_{T^*}$, such that

$$ZD_{T_a} = D_T S \quad \text{and} \quad Z_* D_{T_a^*} = D_{T^*} S^*. \tag{1.8}$$

Using (1.2) and (1.8) we obtain

$$\begin{aligned}
Z_* \Theta_{T_a}(\lambda) Z^{-1} D_T &= Z_* \Theta_{T_a}(\lambda) D_{T_a} S^{-1} = Z_* D_{T_a^*}(I - \lambda T_a^*)^{-1}(\lambda I - T_a) S^{-1} \\
&= D_{T^*} S^* (I - \lambda T_a^*)^{-1}(\lambda I - T_a) S^{-1} \\
&= D_{T^*}(I - aT^*)^{-1}(I - \lambda T_a^*)^{-1}(\lambda I - T_a)(I - \bar{a}T) \\
&= D_{T^*}(I - \mu T^*)^{-1}(\mu I - T),
\end{aligned}$$

where

$$\mu = \frac{\lambda + a}{1 + \bar{a}\lambda}.$$

Thus

$$Z_* \Theta_{T_a}(\lambda) Z^{-1} D_T = \Theta_T(\mu) D_T;$$

the operators Z^{-1} and $\Theta_T(\mu)$ have the same domain $\mathfrak{D}_T$, hence

$$Z_* \Theta_{T_a}(\lambda) Z^{-1} = \Theta_T(\mu),$$

and (1.6) follows.

4. Let $\mathfrak{H} = \bigoplus_\alpha \mathfrak{H}_\alpha$ and, correspondingly, $T = \bigoplus_\alpha T_\alpha$, where T_α is a contraction on the space $\mathfrak{H}_\alpha$. Clearly,

$$D_T = \bigoplus_\alpha D_{T_\alpha}, \quad D_{T^*} = \bigoplus_\alpha D_{T_\alpha^*}, \quad \mathfrak{D}_T = \bigoplus_\alpha \mathfrak{D}_{T_\alpha}, \quad \mathfrak{D}_{T^*} = \bigoplus_\alpha \mathfrak{D}_{T_\alpha^*},$$

and hence

$$\Theta_T(\lambda) = \bigoplus_\alpha \Theta_{T_\alpha}(\lambda) \qquad (\lambda \in D).$$

As every contraction on a nonseparable Hilbert space is the orthogonal sum of contractions on separable Hilbert spaces, it suffices to restrict our further investigations on characteristic functions to the case of separable spaces. From now on we only consider separable Hilbert spaces. Then the defect spaces $\mathfrak{D}_T$ and $\mathfrak{D}_{T^*}$ are separable too, and as a consequence the results of Chap. V on contractive analytic functions can be applied to the characteristic functions. Therefore the strong operator limit

$$\Theta_T(e^{it}) = \lim \Theta_T(\lambda) \quad (\lambda \in D, \lambda \to e^{it} \text{ nontangentially to } C)$$

exists at almost every point e^{it} of the unit circle C; moreover, for every fixed $f \in \mathfrak{D}_T$ and for $r \to 1-0$ the function $\Theta_T(re^{it})f$ of t converges, as an element of $L^2(\mathfrak{D}_{T^*})$, to the function

$$\Theta_T(e^{it})f = -Tf + \sum_{n=1}^{\infty} e^{int} D_{T^*} T^{*n-1} D_T f; \tag{1.9}$$

the infinite sum is also to be understood in the sense of convergence in the mean, that is, in the space $L^2(\mathfrak{D}_{T^*})$ (cf. Secs. V.1 and 2).

2 Functional models for a given contraction

1. We have defined in Sec. 1 the characteristic function of a contraction T on the space $\mathfrak{H}$ without giving a motivation for this definition. We now show that this definition arises in a natural way in the context of our theory of unitary dilations.

As the space $\mathfrak{H}$ is separable, so is the space $\mathfrak{K}$ of the minimal unitary dilation U of T. Let $\mathfrak{K}_+$ be the space of the minimal isometric dilation U_+ of T: we always view $\mathfrak{K}_+$ as a subspace of $\mathfrak{K}$ and U_+ as the restriction of U to $\mathfrak{K}_+$.

By virtue of Theorem II.2.1, $\mathfrak{K}$ and $\mathfrak{K}_+$ admit the following decompositions

$$\mathfrak{K} = M(\mathfrak{L}_*) \oplus \mathfrak{R}, \quad \mathfrak{K}_+ = M_+(\mathfrak{L}_*) \oplus \mathfrak{R} = \mathfrak{H} \oplus M_+(\mathfrak{L}), \tag{2.1}$$

where

$$\mathfrak{L} = \overline{(U-T)\mathfrak{H}}, \quad \mathfrak{L}_* = \overline{(I-UT^*)\mathfrak{H}} \tag{2.2}$$

are subspaces of $\mathfrak{K}_+$ wandering for U_+ (and hence for U), and where $\mathfrak{R}$ reduces U and U_+ to their residual part R, which is unitary. The subspace $M(\mathfrak{L}_*)$ reduces U, and consequently $P^{\mathfrak{L}_*}$ (the orthogonal projection of $\mathfrak{K}$ onto $M(\mathfrak{L}_*)$) commutes with U and, invoking the same theorem once more,

$$P^{\mathfrak{L}_*} M_+(\mathfrak{L}) \subset M_+(\mathfrak{L}_*).$$

Thus we can apply Lemma V.3.1 to the bilateral shifts induced by U on $M(\mathfrak{L})$ and $M(\mathfrak{L}_*)$, and to the contraction

$$Q = P^{\mathfrak{L}_*} | M(\mathfrak{L})$$

of $M(\mathfrak{L})$ into $M(\mathfrak{L}_*)$. We obtain that there exists a contractive analytic function $\{\mathfrak{L}, \mathfrak{L}_*, \Theta_{\mathfrak{L}}(\lambda)\}$ such that

$$\Phi^{\mathfrak{L}_*} P^{\mathfrak{L}_*} f = \Theta_{\mathfrak{L}} \Phi^{\mathfrak{L}} f \quad \text{for} \quad f \in M(\mathfrak{L}), \tag{2.3}$$

where $\Phi^{\mathfrak{L}}$ and $\Phi^{\mathfrak{L}_*}$ denote the Fourier representations of $M(\mathfrak{L})$ and $M(\mathfrak{L}_*)$ on the functional spaces $L^2(\mathfrak{L})$ and $L^2(\mathfrak{L}_*)$; see Sec. V.3.1.

Because $\dim \mathfrak{L} = \mathfrak{d}_T$ and $\dim \mathfrak{L}_* = \mathfrak{d}_{T^*}$, $\mathfrak{L}$ and $\mathfrak{L}_*$ do not both equal $\{0\}$ unless T is unitary, a case which we exclude in the sequel. We also make the additional assumption that T is *completely nonunitary* on $\mathfrak{H}$ ($\neq \{0\}$). Then

$$M(\mathfrak{L}) \vee M(\mathfrak{L}_*) = \mathfrak{K}, \tag{2.4}$$

and consequently,

$$\overline{(I - P^{\mathfrak{L}_*}) M(\mathfrak{L})} = \mathfrak{R} \qquad \text{(cf. Theorem II.2.1).} \tag{2.5}$$

Set

$$\Delta_{\mathfrak{L}}(t) = [I_{\mathfrak{L}} - \Theta_{\mathfrak{L}}(e^{it})^* \Theta_{\mathfrak{L}}(e^{it})]^{1/2}$$

for those t at which $\Theta_{\mathfrak{L}}(e^{it})$ exists, thus a.e. For t fixed, $\Delta_{\mathfrak{L}}(t)$ is a self-adjoint operator on $\mathfrak{L}$, bounded by 0 and 1. As a function of t, $\Delta_{\mathfrak{L}}(t)$ is strongly measurable, and generates by

$$(\Delta_{\mathfrak{L}} v)(t) = \Delta_{\mathfrak{L}}(t) v(t) \qquad (v \in L^2(\mathfrak{L}))$$

a self-adjoint operator $\Delta_{\mathfrak{L}}$ on $L^2(\mathfrak{L})$, also bounded by 0 and 1.

For $f \in M(\mathfrak{L})$ we have

$$\begin{aligned}\|(I-P^{\mathfrak{L}_*})f\|^2 &= \|f\|^2 - \|P^{\mathfrak{L}_*}f\|^2 = \|\Phi^{\mathfrak{L}}f\|^2 - \|\Phi^{\mathfrak{L}_*}P^{\mathfrak{L}_*}f\|^2 \\ &= \|\Phi^{\mathfrak{L}}f\|^2 - \|\Theta_{\mathfrak{L}}\Phi^{\mathfrak{L}}f\|^2 \\ &= \frac{1}{2\pi}\int_0^{2\pi}[\|(\Phi^{\mathfrak{L}}f)(t)\|_{\mathfrak{L}}^2 - \|\Theta_{\mathfrak{L}}(e^{it})\cdot(\Phi^{\mathfrak{L}}f)(t)\|_{\mathfrak{L}_*}^2]\,dt \\ &= \frac{1}{2\pi}\int_0^{2\pi}\|\Delta_{\mathfrak{L}}(t)(\Phi^{\mathfrak{L}}f)(t)\|_{\mathfrak{L}}^2\,dt = \|\Delta_{\mathfrak{L}}\Phi^{\mathfrak{L}}f\|^2.\end{aligned}$$

Using (2.5) we deduce from this that there exists a unitary operator

$$\Phi_{\mathfrak{R}}:\ \mathfrak{R} \to \overline{\Delta_{\mathfrak{L}}L^2(\mathfrak{L})},$$

such that

$$\Phi_{\mathfrak{R}}(I-P^{\mathfrak{L}_*})f = \Delta_{\mathfrak{L}}\Phi^{\mathfrak{L}}f \quad \text{for} \quad f \in M(\mathfrak{L}). \tag{2.6}$$

Consequently,

$$\Phi = \Phi^{\mathfrak{L}_*} \oplus \Phi_{\mathfrak{R}} \tag{2.7}$$

is a unitary operator from the space

$$\mathfrak{K} = M(\mathfrak{L}_*) \oplus \mathfrak{R}$$

to the functional space

$$\mathbf{K} = L^2(\mathfrak{L}_*) \oplus \overline{\Delta_{\mathfrak{L}}L^2(\mathfrak{L})}. \tag{2.8}$$

Because U commutes with $P^{\mathfrak{L}_*}$ and because of the relation $\Phi^{\mathfrak{L}}U = e^{it}\Phi^{\mathfrak{L}}$ (cf. (V.3.2)), we obtain from (2.6):

$$\begin{aligned}\Phi_{\mathfrak{R}}U(I-P^{\mathfrak{L}_*})f &= \Phi_{\mathfrak{R}}(I-P^{\mathfrak{L}_*})Uf = \Delta_{\mathfrak{L}}\Phi^{\mathfrak{L}}Uf = \Delta_{\mathfrak{L}}\cdot e^{it}\Phi^{\mathfrak{L}}f \\ &= e^{it}\cdot\Delta_{\mathfrak{L}}\Phi^{\mathfrak{L}}f = e^{it}\cdot\Phi_{\mathfrak{R}}(I-P^{\mathfrak{L}_*})f\end{aligned}$$

for $f \in M(\mathfrak{L})$. All the operators occurring are continuous, and therefore we have $\Phi_{\mathfrak{R}}Ug = e^{it}\cdot\Phi_{\mathfrak{R}}g$ for every $g \in \mathfrak{R}$. On the other hand, $\Phi^{\mathfrak{L}_*}Uh = e^{it}\cdot\Phi^{\mathfrak{L}_*}h$ for $h \in M(\mathfrak{L}_*)$, thus

$$\Phi U = \mathbf{U}\Phi, \tag{2.9}$$

where $\mathbf{U}$ denotes the unitary operator on $\mathbf{K}$ defined by

$$\mathbf{U}(v_* \oplus v) = e^{it}v_*(t) \oplus e^{it}v(t) \quad (v_* \in L^2(\mathfrak{L}_*), v \in \overline{\Delta_{\mathfrak{L}}L^2(\mathfrak{L})}). \tag{2.10}$$

According to our convention of identifying the spaces of type $L^2_+(\mathfrak{A})$ with the corresponding spaces $H^2(\mathfrak{A})$ of analytic functions $u(\lambda)$ on D, it follows from the first of the decompositions (2.1) of $\mathfrak{K}_+$ that $\boldsymbol{\Phi}$ maps $\mathfrak{K}_+$ onto the space

$$\mathbf{K}_+ = H^2(\mathfrak{L}_*) \oplus \overline{\Delta_{\mathfrak{L}} L^2(\mathfrak{L})}. \tag{2.11}$$

The operator U_+ is thereby represented by the operator $\mathbf{U}_+$ on $\mathbf{K}_+$ defined by

$$\mathbf{U}_+(u_* \oplus v) = e^{it}u_*(e^{it}) \oplus e^{it}v(t) \quad (u_* \in H^2(\mathfrak{L}_*), \quad v \in \overline{\Delta_{\mathfrak{L}} L^2(\mathfrak{L})}). \tag{2.12}$$

The adjoint U_+^* will be represented by $\mathbf{U}_+^*$, which is given by the formula

$$\mathbf{U}_+^*(u_* \oplus v) = e^{-it}[u_*(e^{it}) - u_*(0)] \oplus e^{-it}v(t). \tag{2.13}$$

Let us find the image of the space $\mathfrak{H}$ by the representation $\boldsymbol{\Phi}$. Because $\mathfrak{H} = \mathfrak{K}_+ \ominus M_+(\mathfrak{L})$ (cf. (2.1)), we have $\boldsymbol{\Phi}\mathfrak{H} = \mathbf{K}_+ \ominus \boldsymbol{\Phi} M_+(\mathfrak{L})$. Now, for $g \in M_+(\mathfrak{L})$ we have

$$\boldsymbol{\Phi} g = \boldsymbol{\Phi}[P^{\mathfrak{L}_*}g + (I - P^{\mathfrak{L}_*})g] = \Phi^{\mathfrak{L}_*}P^{\mathfrak{L}_*}g \oplus \Phi_{\mathfrak{R}}(I - P^{\mathfrak{L}_*})g = \Theta_{\mathfrak{L}}\Phi^{\mathfrak{L}}g \oplus \Delta_{\mathfrak{L}}\Phi^{\mathfrak{L}}g,$$

and hence

$$\boldsymbol{\Phi} M_+(\mathfrak{L}) = \{\Theta_{\mathfrak{L}}u \oplus \Delta_{\mathfrak{L}}u \colon u \in H^2(\mathfrak{L})\}.$$

Thus $\boldsymbol{\Phi}\mathfrak{H} = \mathbf{H}$, where

$$\mathbf{H} = [H^2(\mathfrak{L}_*) \oplus \overline{\Delta L^2(\mathfrak{L})}] \ominus \{\Theta_{\mathfrak{L}}u \oplus \Delta_{\mathfrak{L}}u \colon u \in H^2(\mathfrak{L})\}. \tag{2.14}$$

Returning to the contraction T, recall that it is connected with its isometric dilation U_+ by the relation $T^* = U_+^*|\mathfrak{H}$; see (I.4.2). It follows that the transform of T by $\boldsymbol{\Phi}$, which we denote by $\mathbf{T}$, is connected with $\mathbf{U}_+$ by the relation

$$\mathbf{T}^* = \mathbf{U}_+^*|\mathbf{H}. \tag{2.15}$$

We study in more detail the contractive analytic function $\{\mathfrak{L}, \mathfrak{L}_*, \Theta_{\mathfrak{L}}(\lambda)\}$, using Lemma V.3.2 in the case

$$\mathfrak{R}_+ = M_+(\mathfrak{L}), \quad U_+ = U|\mathfrak{R}_+; \quad \mathfrak{R}'_+ = M_+(\mathfrak{L}_*), \quad U'_+ = U|\mathfrak{R}'_+; \quad Q = P^{\mathfrak{L}_*}|\mathfrak{R}_+.$$

We show first that $\Theta_{\mathfrak{L}}(\lambda)$ is *purely contractive*. Indeed, if $P_{\mathfrak{L}_*}$ denotes orthogonal projection onto $\mathfrak{L}_*$, we have $\|P_{\mathfrak{L}_*}P^{\mathfrak{L}_*}l\| < \|l\|$ for every $l \in \mathfrak{L}$, $l \neq 0$. Otherwise there would exist an $l \in \mathfrak{L}$, $l \neq 0$, such that $l = P_{\mathfrak{L}_*}P^{\mathfrak{L}_*}l$ (i.e. $l \in \mathfrak{L}_*$), and this contradicts the relation $\mathfrak{L} \cap \mathfrak{L}_* = \{0\}$ proved in Sec. II.2.

By virtue of point (b) of the same lemma, $\Theta_{\mathfrak{L}}(\lambda)$ is an inner function if and only if the operator $P^{\mathfrak{L}_*}|M_+(\mathfrak{L})$ is isometric. Because $P^{\mathfrak{L}_*}$ is the orthogonal projection onto $M(\mathfrak{L}_*)$, this condition means that $M_+(\mathfrak{L}) \subset M(\mathfrak{L}_*)$, or (which amounts to the same thing because $M(\mathfrak{L}_*)$ reduces U) that $M(\mathfrak{L}) \subset M(\mathfrak{L}_*)$. By (2.4) this is equivalent to the condition $M(\mathfrak{L}_*) = \mathfrak{K}$, and hence also to the condition $T^{*n} \to O$ $(n \to \infty)$; see Theorem II.1.2. We conclude that $\Theta_{\mathfrak{L}}(\lambda)$ is inner if and only if $T \in C_{\cdot 0}$.

We have proved the following result.

Proposition 2.1. *Let T be a completely nonunitary contraction on the space $\mathfrak{H}$. Let U be the minimal unitary dilation of T on the space $\mathfrak{K}$ and let $\mathfrak{L}$ and $\mathfrak{L}_*$ be the wandering subspaces for U defined by* (2.2). *Then there exists a purely contractive analytic function $\{\mathfrak{L}, \mathfrak{L}_*, \Theta_{\mathfrak{L}}(\lambda)\}$, satisfying condition* (2.3). *This function generates by* (2.7) *a unitary transformation Φ from $\mathfrak{K}$ to the functional space*

$$\mathbf{K} = L^2(\mathfrak{L}_*) \oplus \overline{\Delta_{\mathfrak{L}} L^2(\mathfrak{L})}, \quad \textit{where} \quad \Delta_{\mathfrak{L}}(t) = [I - \Theta_{\mathfrak{L}}(e^{it})^* \Theta_{\mathfrak{L}}(e^{it})]^{1/2}.$$

By means of this transformation, called the "Fourier representation" of $\mathfrak{K}$, U is represented by the operator $\mathbf{U}$ of multiplication by the function e^{it} in the space $\mathbf{K}$, and the subspace $\mathfrak{K}_+$ of $\mathfrak{K}$ is represented by the subspace

$$\mathbf{K}_+ = H^2(\mathfrak{L}_*) \oplus \overline{\Delta_{\mathfrak{L}} L^2(\mathfrak{L})}$$

of $\mathbf{K}$. Finally, the space $\mathfrak{H}$ and the contraction T are represented by the subspace $\mathbf{H}$ of $\mathbf{K}$ and the operator $\mathbf{T}$ on $\mathbf{H}$, defined by

$$\mathbf{H} = [H^2(\mathfrak{L}_*) \oplus \overline{\Delta_{\mathfrak{L}} L^2(\mathfrak{L})}] \ominus \{\Theta_{\mathfrak{L}} u \oplus \Delta_{\mathfrak{L}} u \colon u \in H^2(\mathfrak{L})\}$$

and

$$\mathbf{T}^*(u_* \oplus v) = e^{-it}[u_*(e^{it}) - u_*(0)] \oplus e^{-it} v(t) \quad (u_* \oplus v \in \mathbf{H}).$$

If the function $\Theta_{\mathfrak{L}}(\lambda)$ is inner (i.e. if $\Delta_{\mathfrak{L}}(t) = O$ a.e.), the above formulas for $\mathbf{K}$, $\mathbf{H}$, and $\mathbf{T}$ simplify to

$$\mathbf{K} = L^2(\mathfrak{L}_*), \quad \mathbf{H} = H^2(\mathfrak{L}_*) \ominus \Theta_{\mathfrak{L}} H^2(\mathfrak{L}),$$
$$(\mathbf{T}^* u_*)(\lambda) = \frac{1}{\lambda}[u_*(\lambda) - u_*(0)] \quad (u_* \in \mathbf{H}).$$

This is the case if and only if $T \in C_{\cdot 0}$.

2. Next we to establish a connection between the function $\Theta_{\mathfrak{L}}(\lambda)$ and the characteristic function of T.

Proposition 2.2. *The function $\Theta_{\mathfrak{L}}(\lambda)$ occurring in Proposition* 2.1 *coincides with the characteristic function $\Theta_T(\lambda)$ of T.*

Proof. Let us begin by observing that there is a unitary operator φ from $\mathfrak{L}$ to $\mathfrak{D}_T$ and a unitary operator φ_* from $\mathfrak{L}_*$ to $\mathfrak{D}_{T^*}$ such that

$$\varphi(U - T)h = D_T h \quad \text{and} \quad \varphi_*(I - UT^*)h = D_{T^*} h \quad (h \in \mathfrak{H}). \tag{2.16}$$

As to φ, this has been proved in Sec. II.1 (cf. (II.1.6)); for φ_* the proof is analogous. We shall prove that

$$\varphi_* \Theta_2(\lambda) \varphi^{-1} = \Theta_T(\lambda) \qquad (\lambda \in D), \tag{2.17}$$

so that $\Theta_{\mathfrak{L}}(\lambda)$ and $\Theta_T(\lambda)$ coincide (in the sense defined in Sec. V.2.4).

If $\Theta_{\mathfrak{L}}(\lambda) = \sum_0^\infty \lambda^n \Theta_n$ $(\lambda \in D)$ is the power series expansion of $\Theta_{\mathfrak{L}}(\lambda)$, with bounded operators Θ_n from $\mathfrak{L}$ into $\mathfrak{L}_*$ as coefficients, then we have

$$(\Theta_n l, l_*)_{\mathfrak{L}_*} = \frac{1}{2\pi}\int_0^{2\pi} e^{-int}(\Theta_{\mathfrak{L}}(e^{it})l, l_*)_{\mathfrak{L}_*}\, dt = (\Theta_{\mathfrak{L}}\Phi^{\mathfrak{L}} l, e^{int}\Phi^{\mathfrak{L}_*} l_*)_{L^2(\mathfrak{L}_*)}$$

for $l \in \mathfrak{L}$, $l_* \in \mathfrak{L}_*$. Now relation (2.3) defining $\Theta_{\mathfrak{L}}(\lambda)$ implies that $\Theta_{\mathfrak{L}}\Phi^{\mathfrak{L}} l = \Phi^{\mathfrak{L}_*} P^{\mathfrak{L}_*} l$; on the other hand we have $e^{int} \cdot \Phi^{\mathfrak{L}_*} l_* = \Phi^{\mathfrak{L}_*} U^n l_*$ from (V.3.2). As $\Phi^{\mathfrak{L}_*}$ is unitary, it follows that

$$(\Theta_n l, l_*)_{\mathfrak{L}_*} = (P^{\mathfrak{L}_*} l, U^n l_*)_{\mathfrak{K}},$$

and as

$$P^{\mathfrak{L}_*} U^n l_* = U^n P^{\mathfrak{L}_*} l_* = U^n l_*$$

we conclude that

$$(\Theta_n l, l_*)_{\mathfrak{L}_*} = (U^{*n} l, l_*)_{\mathfrak{K}} \qquad (l \in \mathfrak{L}; l_* \in \mathfrak{L}_*; n = 0, 1, \ldots).$$

Thus, denoting by $P_{\mathfrak{L}_*}$ the orthogonal projection of $\mathfrak{K}$ onto $\mathfrak{L}_*$, we have

$$\Theta_n = P_{\mathfrak{L}_*} U^{*n} | \mathfrak{L} \qquad (n = 0, 1, \ldots). \tag{2.18}$$

We now show by a straightforward calculation that for $l = (U - T)h$ $(h \in \mathfrak{H})$ we have $\Theta_n l = l_n$, with

$$l_0 = -(I - UT^*)Th \quad \text{and} \quad l_n = (I - UT^*)h_n,\ h_n = T^{*n-1}(I - T^*T)h \quad (n \geq 1).$$

These elements l_n $(n \geq 0)$ obviously belong to $\mathfrak{L}_*$. Therefore, by (2.18) we only have to show that

$$U^{*n} l - l_n \perp \mathfrak{L}_* \qquad (n \geq 0). \tag{2.19}$$

For $n = 0$ this is immediate. Indeed,

$$l - l_0 = U(I - T^*T)h \in U\mathfrak{H} \quad \text{and} \quad U\mathfrak{H} \perp U\mathfrak{L}^* = \mathfrak{L}_*.$$

For $n \geq 1$, (2.19) follows from the fact that, for every $h' \in \mathfrak{H}$,

$$\begin{aligned}
(U^{*n} l - l_n, (I - UT^*)h') &= ((U^{*n-1} - U^{*n}T)h - (I - UT^*)h_n, (I - UT^*)h') \\
&= ((U^{*n-1} - U^{*n}T)h - (I - UT^*)h_n, h') \\
&\quad - ((U^{*n} - U^{*n+1}T)h - (U^* - T^*)h_n, T^*h') \\
&= ((T^{*n-1} - T^{*n}T)h - (I - TT^*)h_n, h') \\
&\quad - ((T^{*n} - T^{*n+1}T)h - (T^* - T^*)h_n, T^*h') \\
&= ((I - TT^*)T^{*n-1}(I - T^*T)h, h') - ((I - TT^*)h_n, h') = 0.
\end{aligned}$$

Comparing these results with relations (2.16) we see that

$$\varphi_*\Theta_n\varphi^{-1}D_Th = \varphi_*\Theta_n l = \varphi_* l_n = \begin{cases} -D_{T^*}Th = -TD_Th & (n=0), \\ D_{T^*}h_n = D_{T^*}T^{*n-1}D_TD_Th & (n \geq 1). \end{cases}$$

The elements D_Th ($h \in \mathfrak{H}$) being dense in $\mathfrak{D}_T$, we conclude that

$$\varphi_*\Theta_n\varphi^{-1} = \begin{cases} -T|\mathfrak{D}_T & (n=0), \\ D_{T^*}T^{*n-1}D_T|\mathfrak{D}_T & (n \geq 1). \end{cases}$$

Comparing these formulas with (1.1)′ we obtain (2.17).

3. The above results allow us to construct functional models for c.n.u. contractions, in which the characteristic functions occur explicitly.

In fact, (2.17) implies $\Theta_T(\lambda)^*\Theta_T(\lambda) = \varphi\Theta_{\mathfrak{L}}(\lambda)^*\Theta_{\mathfrak{L}}(\lambda)\varphi^{-1}$, and thus, setting

$$\Delta_T(t) = [I - \Theta_T(e^{it})^*\Theta_T(e^{it})]^{1/2},$$

we have

$$\varphi\Delta_{\mathfrak{L}}(t)\varphi^{-1} = \Delta_T(t). \tag{2.20}$$

In this manner the unitary transformations

$$\varphi\colon \mathfrak{L} \to \mathfrak{D}_T, \quad \varphi_*\colon \mathfrak{L}_* \to \mathfrak{D}_{T^*}$$

generate, by

$$\widehat{\varphi}\colon u(e^{it}) \oplus v(t) \to \varphi_* u(e^{it}) \oplus \varphi v(t),$$

a unitary transformation

$$\widehat{\varphi}\colon H^2(\mathfrak{L}_*) \oplus \overline{\Delta_{\mathfrak{L}}L^2(\mathfrak{L})} \to H^2(\mathfrak{D}_{T^*}) \oplus \overline{\Delta_T L^2(\mathfrak{D}_T)},$$

commuting with multiplication by e^{it} and such that

$$\begin{aligned} \widehat{\varphi}(\Theta_{\mathfrak{L}}u \oplus \Delta_{\mathfrak{L}}u\colon u \in H^2(\mathfrak{L})\} &= \{\varphi_*\Theta_{\mathfrak{L}}u \oplus \varphi\Delta_{\mathfrak{L}}u\colon u \in H^2(\mathfrak{L})\} \\ &= \{\Theta_T\varphi u \oplus \Delta_T\varphi u\colon u \in H^2(\mathfrak{L})\} \\ &= \{\Theta_T w \oplus \Delta_T w\colon w \in H^2(\mathfrak{D}_T)\}. \end{aligned}$$

We proved the following result.

Theorem 2.3. *Every c.n.u. contraction T on the (separable) Hilbert space $\mathfrak{H}$ ($\neq \{0\}$) is unitarily equivalent to the operator $\mathbf{T}$ on the functional space*

$$\mathbf{H} = [H^2(\mathfrak{D}_{T^*}) \oplus \overline{\Delta_T L^2(\mathfrak{D}_T)}] \ominus \{\Theta_T u \oplus \Delta_T u\colon u \in H^2(\mathfrak{D}_T)\}$$

defined by

$$\mathbf{T}^*(u \oplus v) = e^{-it}[u(e^{it}) - u(0)] \oplus e^{-it}v(t) \quad (u \oplus v \in \mathbf{H}).$$

If $T \in C_{\cdot 0}$, and only in this case, $\Theta_T(\lambda)$ is inner, and then this model of T reduces to

$$\mathbf{H} = H^2(\mathfrak{D}_{T^*}) \ominus \Theta_T H^2(\mathfrak{D}_T), \quad \mathbf{T}^* u(\lambda) = \frac{1}{\lambda}[u(\lambda) - u(0)] \quad (u \in \mathbf{H}).$$

Interchanging the roles of T and T^* we obtain the following dual model.

Theorem 2.3*. *Under the conditions of Theorem 2.3, T is unitarily equivalent to the operator $\mathbf{T}'$ on the functional space*

$$\mathbf{H}' = [H^2(\mathfrak{D}_T) \oplus \overline{\Delta_{T^*} L^2(\mathfrak{D}_{T^*})}] \ominus \{\Theta_{T^*} u \oplus \Delta_{T^*} u \colon u \in H^2(\mathfrak{D}_{T^*})\}$$

defined by

$$\mathbf{T}'(u \oplus v) = e^{-it}[u(e^{it}) - u(0)] \oplus e^{-it} v(t) \quad (u \oplus v \in \mathbf{H}').$$

If $T \in C_{0 \cdot}$, and only in this case, $\Theta_T(\lambda)$ is $*$-inner, and then this model of T reduces to

$$\mathbf{H}' = H^2(\mathfrak{D}_T) \ominus \Theta_{T^*} H^2(\mathfrak{D}_{T^*}), \quad \mathbf{T}' u(\lambda) = \frac{1}{\lambda}[u(\lambda) - u(0)] \quad (u \in \mathbf{H}').$$

3 Functional models for analytic functions

1. The above theorems raise the problem of whether every contractive analytic function $\{\mathfrak{E}, \mathfrak{E}_*, \Theta(\lambda)\}$ generates, by analogous constructions, some c.n.u contractions $\mathbf{T}$ and $\mathbf{T}'$. As $\mathbf{T}$ and $\mathbf{T}'$ interchange roles if one passes to the function $\{\mathfrak{E}_*, \mathfrak{E}, \Theta^{\sim}(\lambda)\}$, it suffices to consider the case of $\mathbf{T}$.

Suppose we are given a contractive analytic function $\{\mathfrak{E}, \mathfrak{E}_*, \Theta(\lambda)\}$ with

$$\Theta(\lambda) = \sum_0^\infty \lambda^n \Theta_n,$$

and set

$$\mathbf{K} = L^2(\mathfrak{E}_*) \oplus \overline{\Delta L^2(\mathfrak{E})}, \quad \mathbf{K}_+ = H^2(\mathfrak{E}_*) \oplus \overline{\Delta L^2(\mathfrak{E})} \quad (\subset \mathbf{K}), \tag{3.1}$$

$$\mathbf{G} = \{\Theta w \oplus \Delta w \colon w \in H^2(\mathfrak{E})\} \quad (\subset \mathbf{K}_+), \tag{3.2}$$

where $\Delta(t) = [I_{\mathfrak{E}} - \Theta(e^{it})^* \Theta(e^{it})]^{1/2}$. By virtue of the relation

$$\begin{aligned}\|\Theta v\|^2_{L^2(\mathfrak{E}_*)} + \|\Delta v\|^2_{L^2(\mathfrak{E})} &= \frac{1}{2\pi}\int_0^{2\pi} ([\Theta(e^{it})^*\Theta(e^{it}) + \Delta(t)^2] v(t), v(t))_{\mathfrak{E}}\, dt \\ &= \frac{1}{2\pi}\int_0^{2\pi} (v(t), v(t))_{\mathfrak{E}}\, dt = \|v\|^2_{L^2(\mathfrak{E})} \quad (v \in L^2(\mathfrak{E})),\end{aligned}$$

$$\Omega \colon v \to \Theta v \oplus \Delta v \quad (v \in L^2(\mathfrak{E})) \tag{3.3}$$

is an isometry from $L^2(\mathfrak{E})$ into $\mathbf{K}$. As $\mathbf{G}$ is the image under Ω of the subspace $H^2(\mathfrak{E})$ of $L^2(\mathfrak{E})$, $\mathbf{G}$ is also a subspace of $\mathbf{K}_+$.

Finally, set

$$\mathbf{H} = \mathbf{K}_+ \ominus \mathbf{G}. \tag{3.4}$$

Denote by $\mathbf{U}$ the multiplication by e^{it} on $\mathbf{K}$ (i.e. simultaneous multiplication by e^{it} on the two component spaces). Clearly, $\mathbf{U}$ is a unitary operator on $\mathbf{K}$, and $\mathbf{K}_+$ is invariant for $\mathbf{U}$; let us set

$$\mathbf{U}_+ = \mathbf{U}|\mathbf{K}_+.$$

$\mathbf{G}$ is also invariant for $\mathbf{U}_+$, and consequently $\mathbf{H}$ is invariant for $\mathbf{U}_+^*$. One shows easily that

$$\mathbf{U}_+^*(u \oplus v) = e^{-it}[u(e^{it}) - u(0)] \oplus e^{-it}v(t) \qquad (u \oplus v \in \mathbf{K}_+).$$

The operator $\mathbf{U}_+$ is isometric, thus $\mathbf{U}_+^*$ is a contraction on $\mathbf{K}_+$, and so is the operator $\mathbf{T}$ on $\mathbf{H}$ defined by

$$\mathbf{T}^* = \mathbf{U}_+^*|\mathbf{H}. \tag{3.5}$$

Denote by $\mathbf{P}$ the orthogonal projection from $\mathbf{K}$ onto $\mathbf{H}$, and by $\mathbf{P}_+$ the orthogonal projection from $\mathbf{K}_+$ onto $\mathbf{H}$; then $\mathbf{P}_+ = \mathbf{P}|\mathbf{K}_+$. From (3.5) it follows that

$$\mathbf{T}^{*n} = \mathbf{U}_+^{*n}|\mathbf{H}, \tag{3.6}$$

and for $h, h' \in \mathbf{H}$,

$$(\mathbf{T}^n h, h')_{\mathbf{H}} = (h, \mathbf{T}^{*n}h')_{\mathbf{H}} = (h, \mathbf{U}_+^{*n}h')_{\mathbf{K}_+} = (\mathbf{U}_+^n h, h')_{\mathbf{K}_+} = (\mathbf{P}_+\mathbf{U}_+^n h, h')_{\mathbf{H}};$$

hence

$$\mathbf{T}^n = \mathbf{P}_+\mathbf{U}_+^n|\mathbf{H} = \mathbf{P}\mathbf{U}^n|\mathbf{H} \qquad (n \geq 0); \tag{3.7}$$

that is,$\mathbf{U}$ is *a unitary dilation of* $\mathbf{T}$.

2. Next we show that $\mathbf{T}$ is *completely nonunitary*. To this end we consider an element $u \oplus v \in \mathbf{H}$ such that for $n \geq 0$

$$\text{(a)} \quad \|\mathbf{T}^n(u \oplus v)\| = \|u \oplus v\|, \qquad \text{(b)} \quad \|\mathbf{T}^{*n}(u \oplus v)\| = \|u \oplus v\|.$$

As an element of $H^2(\mathfrak{E}_*)$, u has an expansion

$$u(\lambda) = \sum_0^\infty \lambda^k a_k \quad \left(a_k \in \mathfrak{E}_*, \sum_0^\infty \|a_k\|^2 < \infty\right).$$

By virtue of the relation

$$\begin{aligned} \|\mathbf{T}^{*n}(u \oplus v)\|^2 &= \left\| \sum_{k=n}^\infty e^{i(k-n)t} a_k \right\|^2_{L^2(\mathfrak{E}_*)} + \|v\|^2_{L^2(\mathfrak{E})} \\ &= \sum_n^\infty \|a_k\|^2_{\mathfrak{E}_*} + \|v\|^2_{L^2(\mathfrak{E})} \to \|v\|^2_{L^2(\mathfrak{E})} \quad (n \to \infty), \end{aligned} \tag{3.8}$$

assumptions (b) imply $u = 0$. On the other hand it follows from (3.7) that assumptions (a) mean that $\mathbf{U}_+^n(u \oplus v)$ is contained in $\mathbf{H}$ $(n \geq 0)$. Because $u = 0$, we have $\mathbf{U}_+^n(u \oplus v) = 0 \oplus e^{int}v$, and this has to be orthogonal to $\mathbf{G}$ for $n \geq 0$; that is,

$$0 = (0 \oplus e^{int}v, \Theta w \oplus \Delta w) = \frac{1}{2\pi}\int_0^{2\pi} e^{int}(v(t), \Delta(t)w(e^{it}))_{\mathfrak{E}}\, dt$$

for every $w \in H^2(\mathfrak{E})$, in particular for $w = e^{imt}f$ $(f \in \mathfrak{E}; m \geq 0)$; thus

$$\int_0^{2\pi} e^{i(n-m)t}(\Delta(t)v(t), f)_{\mathfrak{E}}\, dt = 0 \qquad (n, m \geq 0).$$

This implies $(\Delta(t)v(t), f)_{\mathfrak{E}} = 0$ a.e and, as $\mathfrak{E}$ is separable, $\Delta(t)v(t) = 0$ a.e.; thus $\Delta v = 0$, $v \perp \Delta L^2(\mathfrak{E})$. On the other hand, $v \in \overline{\Delta L^2(\mathfrak{E})}$, therefore $v = 0$.

This shows that (a) and (b) imply $u \oplus v = 0$; thus $\mathbf{T}$ is c.n.u.

3. We assume from now on that $\Theta(\lambda)$ is *purely* contractive, that is,

$$\|\Theta(0)f\| < \|f\| \quad \text{for} \quad f \in \mathfrak{E}, f \neq 0, \tag{3.9}$$

and we show that the characteristic function of $\mathbf{T}$ then coincides with $\Theta(\lambda)$. To this end we show first that $\mathbf{U}$ is the minimal unitary dilation of $\mathbf{T}$, that is,

$$\mathbf{K} = \bigvee_{-\infty}^{\infty} \mathbf{U}^n\mathbf{H}. \tag{3.10}$$

From the definition of $\mathbf{K}$ and $\mathbf{K}_+$ it follows immediately that $\mathbf{K} = \bigvee_{-\infty}^{0} \mathbf{U}^n\mathbf{K}_+$; hence to establish (3.10) it suffices to show that

$$\mathbf{K}_+ = \bigvee_0^{\infty} \mathbf{U}_+^n\mathbf{H} = \bigvee_0^{\infty} \mathbf{U}^n\mathbf{H}. \tag{3.11}$$

Suppose $u \oplus v$ is an element of $\mathbf{K}_+$ orthogonal to $\mathbf{U}_+^n\mathbf{H}$ $(n = 0, 1, \ldots)$, thus $\mathbf{U}_+^{*n}(u \oplus v)$ belongs to $\mathbf{G}$ for $n = 0, 1, \ldots$:

$$\mathbf{U}_+^{*n}(u \oplus v) = \Theta w^{(n)} \oplus \Delta w^{(n)} \qquad (w^{(n)} \in H^2(\mathfrak{E}); n = 0, 1, \ldots).$$

The recursive relation

$$\mathbf{U}_+^*(\Theta w^{(n)} \oplus \Delta w^{(n)}) = \Theta w^{(n+1)} \oplus \Delta w^{(n+1)} \qquad (n \geq 0)$$

gives

$$e^{-it}[\Theta w^{(n)} - (\Theta w^{(n)})(0)] \oplus e^{-it}\Delta w^{(n)} = \Theta w^{(n+1)} \oplus \Delta w^{(n+1)} \qquad (n \geq 0);$$

thus

$$\Theta\omega^{(n)} = \Theta(0)w^{(n)}(0), \quad \Delta\omega^{(n)} = 0 \quad (n \geq 0), \tag{3.12}$$

with

$$\omega^{(n)}(\lambda) = w^{(n)}(\lambda) - \lambda w^{(n+1)}(\lambda) \in H^2(\mathfrak{E}).$$

Now relations (3.12) imply

$$\omega^{(n)} = \Theta^* \Theta \omega^{(n)} = \Theta^* \Theta(0) w^{(n)}(0) = \sum_{k=0}^{\infty} e^{-ikt} \Theta_k^* \Theta_0 w^{(n)}(0).$$

Because $\omega^{(n)} \in H^2(\mathfrak{E})$, this is possible only if

$$\omega^{(n)}(\lambda) = \Theta_0^* \Theta_0 w^{(n)}(0) \tag{3.13}$$

for every λ, in particular for $\lambda = 0$. Because $\omega^{(n)}(0) = w^{(n)}(0)$ we obtain that

$$w^{(n)}(0) = \Theta_0^* \Theta_0 w^{(n)}(0), \quad \|w^{(n)}(0)\| = \|\Theta_0 w^{(n)}(0)\|.$$

As the function $\Theta(\lambda)$ is purely contractive, this implies $w^{(n)}(0) = 0$, and on account of (3.13) also $\omega^{(n)}(\lambda) \equiv 0$; hence $w^{(n)}(\lambda) = \lambda \cdot w^{(n+1)}(\lambda)$. This being true for $n \geq 0$, we obtain $w^{(0)}(\lambda) = \lambda^n \cdot w^{(n)}(\lambda)$ $(n \geq 0)$. Hence $w^{(0)}(\lambda)/\lambda^n$ belongs to $H^2(\mathfrak{E})$ for every $n \geq 0$, which is impossible unless $w^{(0)}(\lambda) \equiv 0$. So we have $u \oplus v = \Theta w^{(0)} \oplus \Delta w^{(0)} = 0$. This proves (3.11) and hence (3.10) also; that is the unitary dilation $\mathbf{U}$ of $\mathbf{T}$ is *minimal*. Using this fact we could continue our study of T by using Theorem II.2.1. However, we prefer to follow a more explicit analytic approach.

4. Our next step is to describe $\mathbf{L}_* = \overline{(\mathbf{I} - \mathbf{UT}^*)\mathbf{H}}$, where $\mathbf{I}$ denotes the identity operator on $\mathbf{K}$.

It is obvious that for $u \oplus v \in \mathbf{H}$ we have

$$(\mathbf{I} - \mathbf{UT}^*)(u \oplus v) = [u - (u - u(0))] \oplus [v - v] = u(0) \oplus 0, \tag{3.14}$$

where $u(0)$ is considered as a constant function in $L^2(\mathfrak{E}_*)$.

Let us choose in particular

$$\bar{u}(\lambda) = (I - \Theta(\lambda)\Theta_0^*)g, \quad \bar{v}(t) = -\Delta(t)\Theta_0^* g$$

with $g \in \mathfrak{E}_*$. It is obvious that $\bar{u} \oplus \bar{v} \in \mathbf{K}_+$; we show that actually $\bar{u} \oplus \bar{v} \in \mathbf{H}$. In fact, we have for $w \in H^2(\mathfrak{E})$,

$$(\bar{u} \oplus \bar{v}, \Theta w \oplus \Delta w) = \frac{1}{2\pi} \int_0^{2\pi} (\Theta(e^{it})^* \bar{u}(e^{it}) + \Delta(t)\bar{v}(t), w(e^{it}))_{\mathfrak{E}}\, dt = 0,$$

because

$$\Theta^* \bar{u} + \Delta \bar{v} = \Theta^* g - \Theta^* \Theta \Theta_0^* g - \Delta^2 \Theta_0^* g = (\Theta^* - \Theta_0^*)g = \sum_1^{\infty} e^{-int} \Theta_n^* g \perp H^2(\mathfrak{E}).$$

By virtue of (3.14) we have

$$(\mathbf{I} - \mathbf{UT}^*)(\bar{u} \oplus \bar{v}) = (I - \Theta_0 \Theta_0^*)g \oplus 0. \tag{3.15}$$

Now the elements of the form $(I-\Theta_0\Theta_0^*)g$ $(g\in\mathfrak{E}_*)$ are *dense* in $\mathfrak{E}_*$. Otherwise there would exist a $g'\in\mathfrak{E}_*$, $g'\neq 0$, such that

$$(I_{\mathfrak{E}_*}-\Theta_0\Theta_0^*)g'=0, \tag{3.16}$$

and hence $\|\Theta_0^*g'\|=\|g'\|=\|\Theta_0\Theta_0^*g'\|$. By virtue of (3.9) this implies $\Theta_0^*g'=0$, and by (3.16) $g'=0$. This contradiction proves our assertion.

This result, combined with (3.14) and (3.15), yields

$$\mathbf{L}_*=\mathfrak{E}_*\oplus\{0\} \tag{3.17}$$

(one identifies, as usual, constant functions in $L^2(\mathfrak{E}_*)$ with their values in $\mathfrak{E}_*$). We also have therefore

$$M(\mathbf{L}_*)=L^2(\mathfrak{E}_*)\oplus\{0\}. \tag{3.17$'$}$$

In other words, denoting by $P^{\mathbf{L}_*}$ the orthogonal projection from $\mathbf{K}$ onto $M(\mathbf{L}_*)$, we have for $u\oplus v\in\mathbf{K}$,

$$\begin{cases} P^{\mathbf{L}_*}(u\oplus v)=u\oplus 0,\\ \Phi^{\mathbf{L}_*}P^{\mathbf{L}_*}(u\oplus v)=\Phi^{\mathbf{L}_*}(u\oplus 0)=\Phi^{\mathfrak{E}_*}u\oplus 0=u\oplus 0.\end{cases} \tag{3.18}$$

5. The condition that the element $u\oplus v\in\mathbf{K}_+$ should belong to $\mathbf{H}$ can be expressed in a detailed form as

$$(u\oplus v,\Theta w\oplus\Delta w)_{\mathbf{K}_+}=0 \quad\text{or}\quad (\Theta^*u+\Delta v,w)_{L^2(\mathfrak{E})}=0,$$

for every $w\in H^2(\mathfrak{E})$. Thus our condition means that the function $\Theta^*u+\Delta v$ (which obviously belongs to $L^2(\mathfrak{E})$) should be orthogonal to $H^2(\mathfrak{E})$, so that their Fourier series is of the following form,

$$\Theta(e^{it})^*u(e^{it})+\Delta(t)v(t)=e^{-it}f_1+\cdots+e^{-int}f_n+\cdots, \tag{3.19}$$

where

$$f_n\in\mathfrak{E},\quad \sum_1^\infty\|f_n\|^2<\infty.$$

We calculate the explicit form of $\mathbf{T}$. By virtue of (3.7) we have

$$\mathbf{T}(u\oplus v)=\mathbf{PU}(u\oplus v)=\mathbf{P}(e^{it}u\oplus e^{it}v) \qquad (u\oplus v\in\mathbf{H});$$

that is,

$$\mathbf{T}(u\oplus v)=(e^{it}u\oplus e^{it}v)-(\Theta w\oplus\Delta w),$$

where the function $w\in H^2(\mathfrak{E})$ is defined by the condition

$$(e^{it}u\oplus e^{it}v)-(\Theta w\oplus\Delta w)\perp\Theta w'\oplus\Delta w' \quad\text{for all}\quad w'\in H^2(\mathfrak{E}),$$

or equivalently, by the condition

$$\Theta^*[e^{it}u - \Theta w] + \Delta[e^{it}v - \Delta w] = e^{it}(\Theta^* u + \Delta v) - w \perp H^2(\mathfrak{E}) \tag{3.20}$$

(in $L^2(\mathfrak{E})$). Using (3.19) we derive from (3.20) that w must equal f_1.

Thus *the explicit form of* $\mathbf{T}$ *is*

$$\mathbf{T}(u \oplus v) = (e^{it}u(e^{it}) - \Theta(e^{it})f_1) \oplus (e^{it}v(t) - \Delta(t)f_1) \qquad (u \oplus v \in \mathbf{H}), \tag{3.21}$$

where

$$f_1 = \frac{1}{2\pi}\int_0^{2\pi} e^{it}(\Theta(e^{it})^* u(e^{it}) + \Delta(t)v(t))\,dt. \tag{3.22}$$

Hence we obtain

$$(\mathbf{U} - \mathbf{T})(u \oplus v) = \Theta(e^{it})f_1 \oplus \Delta(t)f_1. \tag{3.23}$$

When $u \oplus v$ varies over $\mathbf{H}$ the corresponding elements f_1 vary over a set $\mathfrak{E}_1$ dense in $\mathfrak{E}$. To show this let us first observe that, on account of (3.9), the elements of the form

$$f = (I - \Theta_0^*\Theta_0)g \quad (g \in \mathfrak{E}, \Theta_0 = \Theta(0))$$

are dense in $\mathfrak{E}$, and setting

$$u \oplus v = e^{-it}[\Theta(e^{it}) - \Theta_0]g \oplus e^{-it}\Delta(t)g$$

we have

$$\begin{aligned}\Theta(e^{it})^* u(e^{it}) + \Delta(t)v(t) &= e^{-it}[\Theta(e^{it})^*\Theta(e^{it}) - \Theta(e^{it})^*\Theta_0 + \Delta(t)^2]g \\ &= e^{-it}[I - \Theta(e^{it})^*\Theta_0]g \\ &= e^{-it}(I - \Theta_0^*\Theta_0)g - e^{-2it}\Theta_1^*\Theta_0 g - \cdots;\end{aligned}$$

this shows that $u \oplus v$ belongs to $\mathbf{H}$ (cf. (3.19)), and the corresponding element f_1 is equal to the element f with which we started.

Let us recall definition (3.3) of the isometry Ω from $L^2(\mathfrak{E})$ into $\mathbf{K}$. Let ω be the restriction of Ω to the subspace $\mathfrak{E}$ of $L^2(\mathfrak{E})$ (formed by the constant functions); that is,

$$\omega : f \to \Theta f \oplus \Delta f \qquad (f \in \mathfrak{E}). \tag{3.24}$$

From (3.23), (3.24), and the fact that $\mathfrak{E}_1$ is dense in $\mathfrak{E}$, we deduce that

$$\mathbf{L} = \overline{(\mathbf{U} - \mathbf{T})\mathbf{H}} = \overline{\{\Theta f_1 \oplus \Delta f_1 : f_1 \in \mathfrak{E}_1\}} = \overline{\omega\mathfrak{E}_1} = \omega\overline{\mathfrak{E}_1} = \omega\mathfrak{E};$$

thus

$$\mathbf{L} = \{\Theta f \oplus \Delta f : f \in \mathfrak{E}\}. \tag{3.25}$$

Hence ω is a unitary transformation from $\mathfrak{E}$ to $\mathbf{L}$. On the other hand, we deduce from (3.17) that

$$\omega_*: f_* \to f_* \oplus 0_{\overline{\Delta L^2(\mathfrak{E})}} \qquad (f_* \in \mathfrak{E}_*) \tag{3.26}$$

is a unitary transformation from $\mathfrak{E}_*$ to $\mathbf{L}_*$. If we set

$$\boldsymbol{\Theta}(\lambda) = \omega_* \Theta(\lambda) \omega^{-1} \qquad (\lambda \in D), \tag{3.27}$$

we obtain a contractive analytic function $\{\mathbf{L}, \mathbf{L}_*, \boldsymbol{\Theta}(\lambda)\}$ coinciding with the function $\{\mathfrak{E}, \mathfrak{E}_*, \Theta(\lambda)\}$. We show that it satisfies the relation

$$\Phi^{\mathbf{L}_*} P^{\mathbf{L}_*} l = \boldsymbol{\Theta} \Phi^{\mathbf{L}} l \quad \text{for} \quad l \in M(\mathbf{L}). \tag{3.28}$$

We first consider the case when

$$l = \mathbf{U}^n l_n, \quad l_n = \omega f_n \quad (f_n \in \mathfrak{E}; n = 0, \pm 1, \ldots).$$

On account of (3.18) we have then

$$P^{\mathbf{L}_*} l = \mathbf{U}^n P^{\mathbf{L}_*} l_n = \mathbf{U}^n (\Theta f_n \oplus 0),$$

and hence

$$\begin{aligned}
[\Phi^{\mathbf{L}_*} P^{\mathbf{L}_*} l](\tau) &= (e^{in\tau} \Theta(e^{i\tau}) f_n \oplus 0) = \omega_* \Theta(e^{i\tau}) e^{in\tau} f_n \\
&= \omega_* \Theta(e^{i\tau}) \omega^{-1} e^{in\tau} \omega f_n = \boldsymbol{\Theta}(e^{i\tau}) [\Phi^{\mathbf{L}} \mathbf{U}^n \omega f_n](\tau) \\
&= \boldsymbol{\Theta}(e^{i\tau}) [\Phi^{\mathbf{L}} l](\tau) \quad \text{a.e.},
\end{aligned}$$

which proves (3.28) in the case $l \in \mathbf{U}^n \mathbf{L}$. The general case follows readily from this.

Comparing (3.28) with (2.3) we obtain, by virtue of Proposition 2.2, that the characteristic function of $\mathbf{T}$ coincides with $\boldsymbol{\Theta}(\lambda)$ and hence with the initial function $\Theta(\lambda)$ as well.

We have thus proved the following result.

Theorem 3.1. *Given an arbitrary contractive analytic function* $\{\mathfrak{E}, \mathfrak{E}_*, \Theta(\lambda)\}$ *and setting* $\Delta(t) = [I - \Theta(e^{it})^* \Theta(e^{it})]^{1/2}$, *the operator* $\mathbf{T}$ *defined on the Hilbert space*

$$\mathbf{H} = [H^2(\mathfrak{E}_*) \oplus \overline{\Delta L^2(\mathfrak{E})}] \ominus \{\Theta w \oplus \Delta w \colon w \in H^2(\mathfrak{E})\} \tag{3.29}$$

by

$$\mathbf{T}^*(u_* \oplus v) = e^{-it}[u_*(e^{it}) - u_*(0)] \oplus e^{-it} v(t) \qquad (u_* \oplus v \in \mathbf{H}), \tag{3.30}$$

will be a completely nonunitary contraction.

If the function $\{\mathfrak{E}, \mathfrak{E}_*, \Theta(\lambda)\}$ *is purely contractive, then it coincides with the characteristic function of* $\mathbf{T}$. *In this case, if we consider* $\mathbf{H}$ *in the natural way as a subspace of the spaces*

$$\mathbf{K} = L^2(\mathfrak{E}_*) \oplus \overline{\Delta L^2(\mathfrak{E})} \quad \textit{and} \quad \mathbf{K}_+ = H^2(\mathfrak{E}_*) \oplus \overline{\Delta L^2(\mathfrak{E})},$$

then the operators $\mathbf{U}$ *and* $\mathbf{U}_+$ *defined by*

$$\mathbf{U}(u_* \oplus v) = e^{it}u_* \oplus e^{it}v \quad (u_* \oplus v \in \mathbf{K}) \quad \textit{and} \quad \mathbf{U}_+ = \mathbf{U}|\mathbf{K}_+$$

are the minimal unitary and minimal isometric dilations of $\mathbf{T}$, *respectively.*

6. We complete this theorem in some respects.

Proposition 3.2. (a) *In order that the space* $\mathbf{H}$ *defined by* (3.29) *be* $\neq \{0\}$ *it is necessary and sufficient that the function* $\Theta(\lambda)$ *not be a unitary constant.*[2]

(b) *The characteristic function of the c.n.u contraction* $\mathbf{T}$ *on* $\mathbf{H}$, *defined by* (3.30), *coincides with the purely contractive part* $\{\mathfrak{E}^0, \mathfrak{E}^0_*, \Theta^0(\lambda)\}$ *of the function* $\{\mathfrak{E}, \mathfrak{E}_*, \Theta(\lambda)\}$ (cf Proposition V.2.1).

Proof. If $\Theta(\lambda) \equiv \Theta_0$, where Θ_0 is a unitary operator from $\mathfrak{E}$ to $\mathfrak{E}_*$, then $\Delta(t) \equiv 0$ and hence

$$\mathbf{H} = H^2(\mathfrak{E}_*) \ominus \Theta_0 H^2(\mathfrak{E}) = H^2(\mathfrak{E}_*) \ominus H^2(\Theta_0 \mathfrak{E}) = \{0\}.$$

If the function $\Theta(\lambda)$ is not a unitary constant, then assertion (b) implies that

$$\dim \mathfrak{D}_{\mathbf{T}} = \dim \mathfrak{E}^0 \quad \text{and} \quad \dim \mathfrak{D}_{\mathbf{T}^*} = \dim \mathfrak{E}^0_*,$$

where $\dim \mathfrak{E}^0$ and $\dim \mathfrak{E}^0_*$ are not both 0 (because $\Theta(\lambda)$ has a nontrivial purely contractive part). Because $\mathfrak{D}_{\mathbf{T}}$ and $\mathfrak{D}_{\mathbf{T}^*}$ are subspaces of $\mathbf{H}$ we conclude that $\dim \mathbf{H} > 0$.

Therefore it only remains to prove assertion (b). Let us denote by $\Delta^0(t)$ the function analogous to $\Delta(t)$ but formed from $\Theta^0(\lambda)$ instead of $\Theta(\lambda)$. The decomposition $\mathfrak{E} = \mathfrak{E}' \oplus \mathfrak{E}^0$ obviously implies the decompositions $L^2(\mathfrak{E}) = L^2(\mathfrak{E}') \oplus L^2(\mathfrak{E}^0)$ and $H^2(\mathfrak{E}) = H^2(\mathfrak{E}') \oplus H^2(\mathfrak{E}^0)$. Because $\Theta'(\lambda)$ is a unitary constant, we have $\Delta(t)v(t) = 0 \oplus \Delta^0(t)v^0(t)$ for any $v = v' \oplus v^0 \in L^2(\mathfrak{E})$. Consequently,

$$\{\Theta w \oplus \Delta w \colon w \in H^2(\mathfrak{E})\} = \{\Theta' w' \oplus \Theta^0 w^0 \oplus \Delta^0 w^0 \colon w' \in H^2(\mathfrak{E}'), w^0 \in H^2(\mathfrak{E}^0)\},$$

and hence

$$\{\Theta w \oplus \Delta w \colon w \in H^2(\mathfrak{E})\} = H^2(\mathfrak{E}'_*) \oplus \{\Theta^0 w^0 \oplus \Delta^0 w^0 \colon w^0 \in H^2(\mathfrak{E}^0)\},$$

because

$$\Theta' H^2(\mathfrak{E}') = H^2(\Theta' \mathfrak{E}') = H^2(\mathfrak{E}'_*).$$

It follows that for $u_* \oplus v \in \mathbf{H}$ we have $u_* \perp H^2(\mathfrak{E}'_*)$, and hence $u_* \in H^2(\mathfrak{E}^0_*)$. Consequently, the space $\mathbf{H}$ can be identified in a natural way with the space

$$\mathbf{H}^0 = [H^2(\mathfrak{E}^0_*) \oplus \overline{\Delta^0 L^2(\mathfrak{E}^0)}] \ominus \{\Theta^0 w^0 \oplus \Delta^0 w^0 \colon w^0 \in H^2(\mathfrak{E}^0)\},$$

and the operator $\mathbf{T}$ on $\mathbf{H}$ with the operator $\mathbf{T}^0$ on $\mathbf{H}^0$ defined by

$$\mathbf{T}^{0*}(u^0_* \oplus v^0) = e^{-it}[u^0_*(e^{it}) - u^0_*(0)] \oplus e^{-it}v^0(t) \qquad (u^0_* \oplus v^0 \in \mathbf{H}^0).$$

---

[2] We also admit the trivial transformation $0 \to 0$ from $\{0\}$ to $\{0\}$ as unitary.

Now, because the function $\{\mathfrak{E}^0, \mathfrak{E}^0_*, \Theta^0(\lambda)\}$ is *purely contractive*, we have by Theorem 3.1 that the characteristic function of $\mathbf{T}^0$—and hence the characteristic function of $\mathbf{T}$ as well—coincide with $\{\mathfrak{E}^0, \mathfrak{E}^0_*, \Theta^0(\lambda)\}$. The proof is complete.

Proposition 3.3. *If the contractive analytic functions*

$$\{\mathfrak{E}, \mathfrak{E}_*, \Theta(\lambda)\} \quad \textit{and} \quad \{\mathfrak{E}', \mathfrak{E}'_*, \Theta'(\lambda)\}$$

coincide, then the contractions $\mathbf{T}$ *and* $\mathbf{T}'$ *which they generate in the sense of Theorem* 3.1 *are unitarily equivalent.*

In fact, if

$$\tau\colon \mathfrak{E} \to \mathfrak{E}' \quad \text{and} \quad \tau_*\colon \mathfrak{E}_* \to \mathfrak{E}'_*$$

are unitary operators such that $\Theta'(\lambda) = \tau_* \Theta(\lambda)\tau^{-1}$ $(\lambda \in D)$, then

$$\hat{\tau}\colon u(e^{it}) \oplus v(t) \to \tau_* u(e^{it}) \oplus \tau v(t)$$

is a unitary operator from $\mathbf{H}$ to $\mathbf{H}'$ such that $\mathbf{T}' = \hat{\tau}\mathbf{T}\hat{\tau}^{-1}$. The proof is just the same as in the particular case considered in Sec. 2.3.

We apply this result to characteristic functions. In conjunction with Secs. 1.2 and 2.2, it yields the following theorem.

Theorem 3.4. *Two completely nonunitary contractions are unitarily equivalent if and only if their characteristic functions coincide.*

7. We have already seen (Theorem 2.3) that a c.n.u. contraction T is of class $C_{\cdot 0}$ if and only if its characteristic function is an inner function. Theorems 2.3 and 3.1 allow us to complete this result as follows.

Proposition 3.5. *For a c.n.u contraction T we have*

(i) $T \in C_{\cdot 0}$, (ii) $T \in C_{\cdot 1}$, (iii) $T \in C_{0 \cdot}$, (iv) $T \in C_{1 \cdot}$,

if and only if the characteristic function of T is

(i) *inner,* (ii) *outer,* (iii) $*$*-inner,* (iv) $*$*-outer, respectively.*

(For the definitions, cf. Secs. II.4 and V.2.3.)

Proof. The cases (iii) and (iv) reduce to the cases (i) and (ii) if one replaces T by T^* and recalls (1.5). Case (i) is contained in Theorem 2.3. It remains therefore to consider the case (ii). Because the classes $C_{\cdot 0}$, and so on, are obviously invariant under unitary equivalence, and because the property of a contractive analytic function of being outer (or inner) does not change if this function is replaced by a coinciding one, it suffices to prove our assertion for the functional model of T. Accordingly, let

$\mathbf{T}$ be the contraction defined by a purely contractive analytic function $\{\mathfrak{E}, \mathfrak{E}_*, \Theta(\lambda)\}$ in the sense of Theorem 3.1. By (3.8), we have

$$\lim_{n\to\infty} \|\mathbf{T}^{*n}(u \oplus v)\| = \|v\| \quad \text{for} \quad u \oplus v \in \mathbf{H}.$$

This shows that $\mathbf{T} \in C_{\cdot 1}$ if and only if $u \oplus 0 \in \mathbf{H}$ implies $u = 0$. Now $u \oplus 0 \in \mathbf{H}$ means that $u \perp \Theta H^2(\mathfrak{E})$, and the latter condition implies $u = 0$ if and only if $\Theta(\lambda)$ is an outer function.

8. In our functional models the minimal isometric dilations appear in an explicit form and this enables us to derive, by using Theorem II.2.3, explicit forms for the commutants also. For the sake of simplicity we only consider contractions of class $C_{\cdot 0}$, that is, we assume the functions $\Theta(\lambda)$ to be inner.

Thus suppose $\{\mathfrak{E}, \mathfrak{E}_*, \Theta(\lambda)\}$ is a purely contractive inner function, and let $\mathbf{T}$ be the contraction on the space

$$\mathbf{H} = H^2(\mathfrak{E}_*) \ominus \Theta H^2(\mathfrak{E}) \tag{3.31}$$

defined by

$$(\mathbf{T}^* u)(\lambda) = \frac{1}{\lambda}[u(\lambda) - u(0)] \qquad (u \in \mathbf{H}); \tag{3.32}$$

then, by Theorem 3.1, the operator $\mathbf{U}_+$ defined on $\mathbf{K}_+ = H^2(\mathfrak{E}_*)$ by

$$(\mathbf{U}_+ u)(\lambda) = \lambda \cdot u(\lambda)$$

is a minimal isometric dilation of $\mathbf{T}$. Let $\mathbf{H}'$, $\mathbf{T}'$, and so on, correspond similarly to a function $\{\mathfrak{E}', \mathfrak{E}'_*, \Theta'(\lambda)\}$ of the same kind. Because $\mathbf{U}_+$ and $\mathbf{U}'_+$ are unilateral shifts with $\mathfrak{E}_*$ and $\mathfrak{E}'_*$ as generating subspaces, Lemma V.3.2 implies that every bounded operator $\mathbf{Y}$ from $H^2(\mathfrak{E}'_*)$ to $H^2(\mathfrak{E}_*)$ such that $\mathbf{U}_+\mathbf{Y} = \mathbf{Y}\mathbf{U}'_+$, can be represented in the form

$$(\mathbf{Y}u)(\lambda) = Y(\lambda)u(\lambda) \qquad (u \in H^2(\mathfrak{E}'_*))$$

by means of some bounded analytic function $\{\mathfrak{E}'_*, \mathfrak{E}_*, Y(\lambda)\}$ such that

$$\|Y\|_\infty = \|\mathbf{Y}\|.$$[3]

Combining this fact with Theorem II.2.3 we obtain the following result.

Theorem 3.6. *Every bounded operator* $\mathbf{X}$ *from* $\mathbf{H}'$ *to* $\mathbf{H}$ *satisfying*

$$\mathbf{T}\mathbf{X} = \mathbf{X}\mathbf{T}' \tag{3.33}$$

can be represented in the form

$$\mathbf{X}u = \mathbf{P}_+(Yu) \qquad (u \in \mathbf{H}'), \tag{3.34}$$

[3] For a bounded analytic function $\{\mathfrak{A}, \mathfrak{A}_*, A(\lambda)\}$ we define $\|A\|_\infty$ as the supremum of the operator norm $\|A(\lambda)\|$ on D.

where $\mathbf{P}_+$ *denotes the orthogonal projection from* $H^2(\mathfrak{E}_*)$ *onto* $\mathbf{H}$, *and where* $\{\mathfrak{E}'_*, \mathfrak{E}_*, Y(\lambda)\}$ *is a bounded analytic function satisfying*

$$Y\Theta' H^2(\mathfrak{E}') \subset \Theta H^2(\mathfrak{E})^4 \tag{3.35}$$

and

$$\|Y\|_\infty = \|\mathbf{X}\|. \tag{3.36}$$

Conversely, every bounded analytic function $\{\mathfrak{E}'_*, \mathfrak{E}_*, Y(\lambda)\}$ *satisfying* (3.35) *yields by* (3.34) *a solution* $\mathbf{X}$ *of* (3.33) *with* $\|\mathbf{X}\| \leq \|Y\|_\infty$.

Observe that (3.35) is obviously satisfied if $\Theta(\lambda)$ and $\Theta'(\lambda)$ are scalar-valued and coincide; then $Y(\lambda)$ is also scalar-valued (i.e., we have $Y \in H^\infty$), and in this case the right-hand side of (3.34) is equal to $Y(\mathbf{T})u$; see Chapter III, in particular Theorem III.2.1(g). This result is obviously invariant under unitary equivalence so we get the following corollary.

Corollary 3.7. *Let* T *be a contraction of class* C_{00} *and with defect indices* $\mathfrak{d}_T = \mathfrak{d}_{T^*} = 1$. *Then every bounded operator* X *commuting with* T *is a function of* T, *say* $X = y(T)$, *where* $y \in H^\infty$. *Moreover, we can choose* y *such that*

$$\|y\|_\infty = \|X\|.$$

From the many possible applications of Theorem 3.6 we only mention one.

Proposition 3.8. *Let* $\mathbf{T}$ *be given by* (3.31) *and* (3.32) *and let* φ *be a function in* H^∞. *In order that the operator* $\varphi(\mathbf{T})$ *be boundedly invertible it is necessary and sufficient that there exist a bounded analytic function* $\{\mathfrak{E}_*, \mathfrak{E}_*, Y(\lambda)\}$ *such that for every* $u \in H^2(\mathfrak{E})$ *we have*

$$Y\Theta u \in \Theta H^2(\mathfrak{E}) \quad \textit{and} \quad u - \varphi Y u \in \Theta H^2(\mathfrak{E}).$$

The details of the proof may be supplied by the reader; we only remark that the inverse of $\varphi(\mathbf{T})$ and the function $Y(\lambda)$ are connected by the relation $\varphi(\mathbf{T})^{-1}u = \mathbf{P}_+(Yu)$, $u \in \mathbf{H}$.

4 The characteristic function and the spectrum

1. The following theorem establishes relations between the characteristic function of a c.n.u contraction T, and the spectrum $\sigma(T)$ or point spectrum $\sigma_p(T)$[5] of T. As before, C denotes the unit circle and D denotes the open unit disc.

---

[4] This condition is equivalent to the condition that there exists a bounded analytic function $\{\mathfrak{E}', \mathfrak{E}, Z(\lambda)\}$ such that

$$Y(\lambda)\Theta'(\lambda) = \Theta(\lambda)Z(\lambda). \tag{3.35$'$}$$

This follows easily by applying Lemma V.3.2.

[5] That is, the set of eigenvalues.

Theorem 4.1. *Let T be a c.n.u contraction on $\mathfrak{H}$. Denote by S_T the set of points $\mu \in D$ for which the operator $\Theta_T(\mu)$ is not boundedly invertible, together with those $\mu \in C$ not lying on any of the open arcs of C on which $\Theta_T(\lambda)$ is a unitary operator-valued analytic function of λ. Furthermore, denote by S_T^0 the set of points $\mu \in D$ for which $\Theta_T(\mu)$ is not invertible at all. Then*

$$\sigma(T) = S_T \text{ and } \sigma_p(T) = S_T^0.$$

Proof. Because T is a c.n.u contraction, it cannot have eigenvalues on C. Thus if $a \in \sigma_p(T)$ then $a \in D$; setting $T_a = (T - aI)(I - \bar{a}T)^{-1}$ we have therefore $0 \in \sigma_p(T_a)$. Consequently we have $T_a f = 0$ and hence $(I - T_a^* T_a)f = f$ for some nonzero f, which shows that $f \in \mathfrak{D}_{T_a}$ and, by virtue of (1.1),

$$\Theta_{T_a}(0)f = -T_a f = 0.$$

As $\Theta_{T_a}(0)$ differs from $\Theta_T(a)$ by unitary factors only (cf. Sec. 1.3) it follows that $\Theta_T(a)g = 0$ for some nonzero g, and hence $a \in S_T^0$. Thus we have $\sigma_p(T) \subset S_T^0$. The opposite inclusion can be proved by the same reasoning in reverse order. Thus $\sigma_p(T) = S_T^0$.

Now we consider the problem for $\sigma(T)$ and show first that

$$\sigma(T) \cap D = S_T \cap D; \tag{4.1}$$

that is, that a point $a \in D$ belongs either to both $\sigma(T)$ and S_T, or to neither of them. It suffices to show this for the point 0 because the case of a nonzero a can be reduced to this as in the above reasoning, replacing T by T_a.

For $a = 0$ we proceed as follows. First, we observe that $Z = T|(\mathfrak{H} \ominus \mathfrak{D}_T)$ is a unitary operator from $\mathfrak{H} \ominus \mathfrak{D}_T$ to $\mathfrak{H} \ominus \mathfrak{D}_{T^*}$ (see Sec. I.3.1). Next, from definition (1.1) of the characteristic function we infer that $T|\mathfrak{D}_T = -\Theta_T(0)$. Thus T is the orthogonal sum of $-\Theta_T(0)$ and the unitary operator Z. In order that T be boundedly invertible it is therefore necessary and sufficient that $\Theta_T(0)$ be boundedly invertible. This proves that 0 belongs to $\sigma(T)$ if and only if it belongs to S_T. This proves (4.1).

It remains to prove the equality

$$\sigma(T) \cap C = S_T \cap C. \tag{4.2}$$

The inclusion $\rho(T) \cap C \subset C \backslash S_T$ was proved at the end of Sec. 1.1. Taking complements with respect to C we obtain $\sigma(T) \cap C \supset S_T \cap C$. Thus to prove (4.2) it only remains to establish the reverse inclusion.

We have to prove, therefore, that if α is an arc of C on which the function $\Theta_T(\lambda)$ is analytic and unitary operator-valued, then $\alpha \subset \rho(T)$. Clearly it suffices to show this fact for our functional model for T. Thus let $\mathbf{T}$ be the c.n.u contraction generated in the sense of Theorem 3.1 by some purely contractive analytic function $\{\mathfrak{E}, \mathfrak{E}_*, \Theta(\lambda)\}$ such that $\Theta(\lambda)$ is analytic and unitary valued on some open arc α of C. We first prove the following result.

Lemma. *For every element* $u \oplus v$ *of*

$$\mathbf{H} = [H^2(\mathfrak{E}_*) \oplus \overline{\Delta L^2(\mathfrak{E})}] \ominus \{\Theta w \oplus \Delta w \colon w \in H^2(\mathfrak{E})\},$$

the function u $(\in H^2(\mathfrak{E}_*))$ *is analytic on* α.

Proof. By (3.19), the condition $u \oplus v \in \mathbf{H}$ is equivalent to the condition that the function

$$f(t) = \Theta(e^{it})^* u(e^{it}) + \Delta(t) v(t) \qquad (\in L^2(\mathfrak{E})) \tag{4.3}$$

be orthogonal to $H^2(\mathfrak{E})$, thus

$$f(t) = e^{-it} f_1 + e^{-2it} f_2 + \cdots \quad \text{with} \quad \sum_1^\infty \|f_k\|^2 < \infty. \tag{4.4}$$

Now the latter condition implies that the function

$$\varphi(\lambda) = \lambda f_1 + \lambda^2 f_2 + \cdots \qquad (\lambda \in D)$$

belongs to $H^2(\mathfrak{E})$. By virtue of Sec. V.1 we have

$$\varphi(re^{-it}) \to \varphi(e^{-it}) = f(t) \quad \text{a.e.} \quad (\text{convergence in } \mathfrak{E}) \tag{4.5}$$

and

$$\int_0^{2\pi} \|\varphi(re^{-it}) - f(t)\|^2_{\mathfrak{E}} \, dt \to 0, \tag{4.6}$$

as $r \to 1-0$. On the other hand, $u(\lambda) \in H^2(\mathfrak{E}_*)$, thus

$$u(re^{it}) \to u(e^{it}) \quad \text{a.e.} \quad (\text{convergence in } \mathfrak{E}_*) \tag{4.7}$$

and

$$\int_0^{2\pi} \|u(re^{it}) - u(e^{it})\|^2_{\mathfrak{E}_*} \, dt \to 0. \tag{4.8}$$

Observe that, because $\Theta(e^{it})$ is unitary for $e^{it} \in \alpha$, (4.3) implies

$$\Theta(e^{it}) f(t) = u(e^{it}) \text{ for almost every point } e^{it} \in \alpha. \tag{4.9}$$

Let G denote the domain where $\Theta(\lambda)$ is analytic; thus $G \supset D \cup \alpha$. Let G_+ be the part of G exterior to C, and let $G_- = D$. Set

$$\varphi_-(\lambda) = u(\lambda) \quad \text{for} \quad \lambda \in G_-$$

and

$$\varphi_+(\lambda) = \Theta(\lambda)\varphi(1/\lambda) \quad \text{for} \quad \lambda \in G_+.$$

We deduce from relations (4.5)–(4.9) that Lemma V.6.6 applies to these functions and to each arc α_1 with closure in α. It follows that $\varphi_-(\lambda)$ and $\varphi_+(\lambda)$ are analytic continuations of each other through the arc α_1, and as α_1 was arbitrary, through the whole arc α. This means in particular that $u(\lambda)$ is *analytic on* α, as asserted in the lemma.

Now we are able to complete the proof of Theorem 4.1. Let us observe that if $u(e^{it}) \oplus v(t) \in \mathbf{H}$ and if ν is any point of D then we also have $u_\nu \oplus v_\nu \in \mathbf{H}$, where

$$u_\nu(\lambda) = \frac{1}{\lambda - \nu}[\lambda u(\lambda) - \nu u(\nu)], \quad v_\nu(t) = \frac{1}{e^{it} - \nu} e^{it} v(t);$$

moreover,

$$(\mathbf{I} - \nu \mathbf{T}^*)(u_\nu \oplus v_\nu) = u \oplus v \qquad (\mathbf{I} = I_{\mathbf{H}}). \tag{4.10}$$

All this can be verified easily from the definition of $\mathbf{H}$ and $\mathbf{T}$, and from Theorem 3.1, also using the characterization (3.19) of the elements of $\mathbf{H}$. Now let ν tend to a point ε of the arc α. Because $\Theta(\lambda)$ is analytic and unitary-valued on α, it follows from our lemma that $u(\lambda)$ is analytic on α and hence in particular at $\lambda = \varepsilon$. On the other hand, we have $\Delta(t) = O$ for $t \in (\alpha)$, and consequently $v(t) = 0$ for $t \in (\alpha)$ (because $v \in \overline{\Delta L^2(\mathfrak{E})}$). Hence, by observing that $u_\nu(\lambda)$ is bounded when both ν and λ are in a small disc centered at ε, we easily conclude that $u_\nu \oplus v_\nu$ converges in $H^2(\mathfrak{E}_*) \oplus \overline{\Delta L^2(\mathfrak{E})}$ to a limit, which we denote by $u_\varepsilon \oplus v_\varepsilon$ and which also belongs to $\mathbf{H}$; relation (4.10) gives in the limit:

$$(\mathbf{I} - \varepsilon \mathbf{T}^*)(u_\varepsilon \oplus v_\varepsilon) = u \oplus v. \tag{4.10$'$}$$

Because $\mathbf{T}^*$ is c.n.u., it has no eigenvalue on C; consequently $\mathbf{I} - \varepsilon \mathbf{T}^*$ is invertible. By (4.10$'$) its inverse is defined everywhere on $\mathbf{H}$; thus $\mathbf{I} - \varepsilon \mathbf{T}^*$, and hence $\varepsilon \mathbf{I} - \mathbf{T}$ also, are boundedly invertible. This proves that every point of α belongs to the resolvent set $\rho(\mathbf{T})$. This ends the proof of Theorem 4.1.

2. We consider an example demonstrating the usefulness of our functional model in obtaining contractions with prescribed properties.

By Proposition II.3.5(iii) every contraction T of class C_{11} is quasi-similar to a unitary operator U, and by Proposition II.5.1 this quasi-similarity preserves in a certain sense the hyperinvariant subspaces. However, quasi-similarity does not generally preserve the spectrum. Indeed, we exhibit a contraction $T \in C_{11}$ whose spectrum consists of the closed disc $\overline{D}$. To this end recall the example treated in Sec V.4.5, namely the constant function $\Theta(\lambda) \equiv A$, where A is a self-adjoint operator satisfying the inequalities $O \le A \le I$ and such that 0 and 1 are not eigenvalues of A. This function is analytic, purely contractive, and outer from both sides, so that the c.n.u contraction $\mathbf{T}$ which it generates in the sense of Theorem 3.1 is of class C_{11}. Applying Theorem 4.1 we obtain that the spectrum of $\mathbf{T}$ coincides either with the unit circle C or with the closed unit disc $\overline{D}$, accordingly as the operator A is boundedly invertible or not. A complete description of the spectra of C_{11}-contractions is given in Chap. IX.

3. Theorem 4.1 suggests a further study of the relations between the resolvent $(\lambda I - T)^{-1}$ and the characteristic function of T when the spectrum of T does not cover the entire unit disc D.

As we already observed in 4.1, T is the orthogonal sum of $-\Theta_T(0)$ (from $\mathfrak{D}_T$ to $\mathfrak{D}_{T^*}$) and of a unitary operator Z (from $\mathfrak{H}\ominus\mathfrak{D}_T$ to $\mathfrak{H}\ominus\mathfrak{D}_{T^*}$). Hence, if T is boundedly invertible, then T^{-1} is the orthogonal sum of the operator $-\Theta_T(0)^{-1}$ and of the unitary operator Z^{-1}. As a consequence, $\|T^{-1}\|$ is then equal to the larger one of the values $\|\Theta_T(0)^{-1}\|$ and $\|Z^{-1}\| = 1$. Because $\Theta_T(0)$ is a contraction, its inverse has norm not less than 1. Therefore we have

$$\|T^{-1}\| = \|\Theta_T(0)^{-1}\|. \tag{4.11}$$

Consider now any point $a \in D\backslash\sigma(T)$ and the operator

$$T_a = (T - aI)(I - \bar{a}T)^{-1}; \tag{4.12}$$

T_a is a contraction and is boundedly invertible, with

$$T_a^{-1} = (I - \bar{a}T)(T - aI)^{-1}. \tag{4.13}$$

Applying (4.11) to T_a in place of T we obtain

$$\|T_a^{-1}\| = \|\Theta_{T_a}(0)^{-1}\|. \tag{4.14}$$

Now from 1.3 we know that $\Theta_{T_a}(0)$ and $\Theta_T(a)$ are equal up to unitary factors; hence we have

$$\|\Theta_{T_a}(0)^{-1}\| = \|\Theta_T(a)^{-1}\|. \tag{4.15}$$

From (4.13) we obtain

$$(T - aI)^{-1} = (I - \bar{a}T)^{-1}T_a^{-1},$$

and this implies that

$$\begin{aligned} \|(T - aI)^{-1}\| &\le \|(I - \bar{a}T)^{-1}\|\|T_a^{-1}\| \\ &\le (1 - |a|)^{-1}\|T_a^{-1}\|. \end{aligned} \tag{4.16}$$

Again from (4.13) we deduce

$$T_a^{-1} = -\bar{a}I + (1 - |a|^2)(T - aI)^{-1},$$

and this implies that

$$\begin{aligned} \|T_a^{-1}\| &\le |a| + (1 - |a|^2)\|(T - aI)^{-1}\| \\ &\le 1 + 2(1 - |a|)\|(T - aI)^{-1}\|. \end{aligned} \tag{4.17}$$

Relations (4.13)–(4.17) prove the following result.

Proposition 4.2. *If $\lambda \in D\backslash\sigma(T)$, then*

$$\|\Theta_T(\lambda)^{-1}\| = \|(\lambda I - T)^{-1}(I - \bar{\lambda}T)\|$$

and

$$(1-|\lambda|)\|(\lambda I-T)^{-1}\| \leq \|\Theta_T(\lambda)^{-1}\| \leq 1+2(1-|\lambda|)\|(\lambda I-T)^{-1}\|.$$

Let λ_0 be an isolated point of $\sigma(T)$ in D and let λ be variable in $D\backslash\sigma(T)$, tending to λ_0. From the above inequalities it follows that if there exists an integer $p \geq 1$ such that one of the values

$$\limsup_{\lambda\to\lambda_0} \|(\lambda-\lambda_0)^p(\lambda I-T)^{-1}\|, \quad \limsup_{\lambda\to\lambda_0} \|(\lambda-\lambda_0)^p\Theta_T(\lambda)^{-1}\|$$

is finite then so is the other. This remark proves that *if the point $\lambda_0 \in D$ is a pole of order p for one of the functions $(\lambda I-T)^{-1}$ and $\Theta_T(\lambda)^{-1}$ then it is a pole of order p for both.*

4. Consider now a c.n.u contraction T whose characteristic function $\Theta_T(\lambda)$ admits a scalar multiple $\delta(\lambda) \not\equiv 0$, that is, for which there exists a contractive analytic function $\{\mathfrak{D}_{T^*}, \mathfrak{D}_T, \Omega(\lambda)\}$ such that

$$\Omega(\lambda)\Theta_T(\lambda) = \delta(\lambda)I_{\mathfrak{D}_T}, \quad \Theta_T(\lambda)\Omega(\lambda) = \delta(\lambda)I_{\mathfrak{D}_{T^*}}.$$

Then we have

$$\Theta_T(\lambda)^{-1} = \frac{1}{\delta(\lambda)}\Omega(\lambda)$$

at every point $\lambda \in D$, where $\delta(\lambda) \neq 0$, and hence $\Theta_T(\lambda)^{-1}$ is a meromorphic function on *D*; the order of a pole *a* is at most equal to the multiplicity of *a* as a zero of δ.

By virtue of Theorem 4.1 and what we have just proved on the poles, we can state the following proposition.

Proposition 4.3. *If the characteristic function of a c.n.u contraction T admits a scalar multiple, then $(\lambda I-T)^{-1}$ is a meromorphic function on D.*

Let us make the additional assumption that $T \in C_{\cdot 1}$. As $\Theta_T(\lambda)$ is then an outer function its scalar multiple $\delta(\lambda)$ can be chosen to be outer also. In this case $\delta(\lambda) \neq 0$ on *D*, and thus $\Theta_T(\lambda)$ is boundedly invertible for every $\lambda \in D$. Consequently, $\sigma(T)$ is situated on the circle *C*. Suppose α is an arc of *C* such that $\Theta_T(e^{it})$ is isometric at almost every point $e^{it} \in \alpha$. By virtue of Proposition V.6.7, $\Theta_T(\lambda)$ then extends analytically through α to the whole exterior of *C* and, by virtue of Proposition V.6.5, $\Theta_T(e^{it})$ is, for $e^{it} \in \alpha$, even unitary. So it follows from Theorem 4.1 that α is contained in the resolvent set of T. Collecting these results we obtain the following corollary.

Proposition 4.4. *Let T be a c.n.u contraction of class $C_{\cdot 1}$ whose characteristic function admits a scalar multiple. The spectrum $\sigma(T)$ is then the complement on C of the union of the open arcs of C on which $\Theta_T(e^{it})$ is isometric a.e.*

We conclude this section by characterizing the similarity of a contraction T to a unitary operator in terms of the characteristic function of T.

Theorem 4.5. *In order that a contraction T on a (separable) Hilbert space $\mathfrak{H}$ be similar to a unitary operator, it is necessary and sufficient that $\Theta_T(\lambda)$ be boundedly invertible at every point $\lambda \in D$ and $\|\Theta_T(\lambda)^{-1}\|$ have a bound independent of λ. Equivalently, it is necessary and sufficient that $\Theta_T(\lambda)$ satisfy the conditions*

$$\|\Theta_T(\lambda)g\| \geq c\|g\| \qquad (g \in \mathfrak{D}_T, \lambda \in D) \quad \textit{and} \tag{4.18}$$

$$\Theta_T(\lambda)\mathfrak{D}_T = \mathfrak{D}_{T^*} \qquad (\lambda \in D), \tag{4.19}$$

the second at least at one (and then at every) *point of D. Under these conditions, T is similar in particular to the residual part R of its minimal unitary dilation. Moreover, the least upper bound of $\|\Theta_T(\lambda)^{-1}\|$ on D is equal to the minimum of $\|S\|\|S^{-1}\|$ for affinities S such that STS^{-1} is unitary.*

Proof. Taking into account the obvious fact that T is similar to a unitary operator if and only if its c.n.u part T_1 satisfies this property, and by recalling that $\Theta_{T_1}(\lambda) = \Theta_T(\lambda)$, we can assume that T is a c.n.u. contraction.

By virtue of Proposition II.3.6 similarity of T to a unitary operator is equivalent to the condition that the transformation $Q = P^{\mathfrak{L}_*}|M_+(\mathfrak{L})$ from $M_+(\mathfrak{L})$ to $M_+(\mathfrak{L}_*)$ be boundedly invertible, and then $\|Q^{-1}\|$ is the minimum of $\|S\|\|S^{-1}\|$ for affinities S such that STS^{-1} is unitary. Applying the Fourier representations $\Phi_+^{\mathfrak{L}}$ and $\Phi_+^{\mathfrak{L}_*}$, we obtain that the contractive analytic function $\{\mathfrak{L}, \mathfrak{L}_*, \Theta_{\mathfrak{L}}(\lambda)\}$ associated with T in the sense of Proposition 2.1 satisfies the intertwining relation $\Phi_+^{\mathfrak{L}_*}Q = (\Theta_{\mathfrak{L}})_+\Phi_+^{\mathfrak{L}}$. Recalling part (e) of Lemma V.3.2 we infer that Q is boundedly invertible if and only if $\Theta_{\mathfrak{L}}(\lambda)$ is boundedly invertible at every point $\lambda \in D$ and $\|\Theta_{\mathfrak{L}}(\lambda)^{-1}\|$ has a bound independent of λ; furthermore $\|Q^{-1}\|$ is the upper bound of $\|\Theta_{\mathfrak{L}}(\lambda)^{-1}\|$ on D. By virtue of Proposition 2.2 the function $\Theta_{\mathfrak{L}}(\lambda)$ coincides with the characteristic function $\Theta_T(\lambda)$, thus the previous conditions hold for $\Theta_{\mathfrak{L}}(\lambda)$ and $\Theta_T(\lambda)$ at the same time.

It remains to prove that if (4.18) is true for every $\lambda \in D$ and (4.19) is valid for at least one $\lambda_0 \in D$ then $\Theta_T(\lambda)$ is boundedly invertible on D. To show this, we begin by defining Λ as the set of points $\lambda \in D$ at which $\Theta_T(\lambda)$ is boundedly invertible. Because Λ is obviously an open set, and nonempty because $\lambda_0 \in \Lambda$, we have proved $\Lambda = D$ if we show that the assumptions

$$\lambda_n \in \Lambda, \quad \lambda_n \to \lambda \in D$$

imply $\lambda \in \Lambda$. Now, this can be seen from the relation

$$\Theta_T(\lambda) = \Theta_T(\lambda_n)[I + \Theta_T(\lambda_n)^{-1}[\Theta_T(\lambda) - \Theta_T(\lambda_n)]],$$

if we recall that $\|\Theta_T(\lambda_n)^{-1}\| \leq 1/c$ by (4.18), and that $\|\Theta_T(\lambda) - \Theta_T(\lambda_n)\| < c$ for λ_n sufficiently close to λ. Thus we have indeed $\Lambda = D$; that is, (4.19) holds for every $\lambda \in D$.

We recall (cf. (2.4) and the following discussion up to (2.10)) that if T is c.n.u., then R is unitarily equivalent to the operator of multiplication by e^{it} on the function

space
$$\overline{\Delta_T L^2(\mathfrak{D}_T)}, \quad \text{where} \quad \Delta_T(t) = [I - \Theta_T(e^{it})^* \Theta_T(e^{it})]^{1/2}. \tag{4.20}$$

We obtained the following corollary.

Corollary 4.6. *If the contraction T is similar to a unitary operator, then its c.n.u part is similar to multiplication by e^{it} on the space* (4.20).

It is possible to give Theorem 4.5 a form in which the characteristic function is replaced by the resolvent of T. One just has to recall the inequalities between $\|\Theta_T(\lambda)^{-1}\|$ and $(1-|\lambda|)\|(\lambda I - T)^{-1}\|$ established in Proposition 4.2. Thus Theorem 4.5 has the following consequence.

Corollary 4.7. *In order that the contraction T on $\mathfrak{H}$ be similar to a unitary operator, it is necessary and sufficient that the open unit disc D be contained in the resolvent set of T and that there exist a constant b such that*

$$\|(\lambda I - T)^{-1}\| \leq \frac{b}{1-|\lambda|} \text{ for } \lambda \in D. \tag{4.21}$$

Actually, it suffices to assume that at least one point of D belongs to the resolvent set and that we have

$$\|(\lambda I - T)h\| \geq \frac{1}{b}(1-|\lambda|)\|h\| \text{ for } \lambda \in D, h \in \mathfrak{H}. \tag{4.22}$$

The last statement can be proved in the same way as the analogous one for $\Theta_T(\lambda)$ in Theorem 4.5.[6]

5 The characteristic and the minimal functions

1. In Sec III.4 we have introduced the class C_0 of c.n.u. contractions T such that $\delta(T) = O$ for some function $\delta \in H^\infty$, $\delta(\lambda) \not\equiv 0$ (δ depending on T), and we have shown that $C_0 \subset C_{00}$; that is, $T \in C_0$ implies $T^n \to O$ and $T^{*n} \to O$ (strongly). Our aim in this section is to characterize the class C_0 by means of the characteristic function.

Theorem 5.1. *Let T be a contraction of class C_{00}. In order that T belong to the class C_0 it is necessary and sufficient that its characteristic function $\Theta_T(\lambda)$ admit a scalar multiple. To be more specific, a function $\delta \in H^\infty$ (with $\delta(\lambda) \not\equiv 0$ and $|\delta(\lambda)| \leq 1$) satisfies the equation $\delta(T) = O$ if and only if it is a scalar multiple of $\Theta_T(\lambda)$.*

Proof. We may consider instead of T its model $\mathbf{T}$ constructed with the aid of a contractive analytic function $\{\mathfrak{E}, \mathfrak{E}_*, \Theta(\lambda)\}$ coinciding with $\Theta_T(\lambda)$; see Theorem

[6] Using the last statement in Theorem 4.5 it is easy to extend Theorem 4.5 and its Corollary 4.7 to nonseparable spaces.

2.3. Because $T \in C_{00}$, the function $\Theta(\lambda)$ is inner (from both sides), and the model is of the simple form

$$\mathbf{H} = H^2(\mathfrak{E}_*) \ominus \Theta H^2(\mathfrak{E}), \quad \mathbf{T}^* u_*(\lambda) = \frac{1}{\lambda}[u_*(\lambda) - u_*(0)].$$

Let us assume that $\delta(\lambda)$ is a scalar multiple of $\Theta(\lambda)$, namely that there exists a contractive analytic function $\{\mathfrak{E}_*, \mathfrak{E}, \Omega(\lambda)\}$ for which $\Omega(\lambda)\Theta(\lambda) = \delta(\lambda) I_{\mathfrak{E}}$ and $\Theta(\lambda)\Omega(\lambda) = \delta(\lambda) I_{\mathfrak{E}_*}$. Then we have for $u_* \in \mathbf{H}$,

$$\delta(\mathbf{U})u_* = \delta u_* = \Theta\Omega u_* \in \Theta H^2(\mathfrak{E}) \perp \mathbf{H};$$

by virtue of the relation

$$\delta(\mathbf{T}) = P_{\mathbf{H}}\delta(\mathbf{U})|\mathbf{H} \quad \text{(cf. Theorem III.2.1(g))} \tag{5.1}$$

this yields $\delta(\mathbf{T}) = \mathbf{O}$.

Conversely, let us assume that δ is any function in H^∞ with $|\delta(\lambda)| \leq 1$, $\delta(\lambda) \not\equiv 0$, and $\delta(\mathbf{T}) = \mathbf{O}$. Let $\delta(\lambda) = \delta_e(\lambda)\delta_i(\lambda)$ be the canonical factorization of $\delta(\lambda)$, with inner factor $\delta_i(\lambda)$ and outer factor $\delta_e(\lambda)$; because $\delta_e(\mathbf{T})$ is invertible (cf. Sec. III.3) we also have $\delta_i(\mathbf{T}) = \mathbf{O}$. From relation (5.1) applied to δ_i we obtain $\delta_i(\mathbf{U})\mathbf{H} \perp \mathbf{H}$, and hence

$$\delta_i \mathbf{H} = \delta_i(\mathbf{U})\mathbf{H} \subset \mathbf{K}_+ \ominus \mathbf{H} = \Theta H^2(\mathfrak{E}).$$

On the other hand $\delta_i \Theta H^2(\mathfrak{E}) = \Theta\delta_i H^2(\mathfrak{E}) \subset \Theta H^2(\mathfrak{E})$, therefore we conclude:

$$\delta_i H^2(\mathfrak{E}_*) = \delta_i[\mathbf{H} \oplus \Theta H^2(\mathfrak{E})] \subset \Theta H^2(\mathfrak{E}).$$

From this inclusion we infer that to each $u_* \in H^2(\mathfrak{E}_*)$ there corresponds a unique $u \in H^2(\mathfrak{E})$ such that

$$\delta_i u_* = \Theta u. \tag{5.2}$$

Because $\delta_i(\lambda)$ and $\Theta(\lambda)$ are inner functions, (5.2) implies $\|u_*\| = \|u\|$; thus

$$u = Q u_*$$

defines an isometry Q from $H^2(\mathfrak{E}_*)$ into $H^2(\mathfrak{E})$. It is obvious that if $U_{*+}^\times$ and $U_+^\times$ denote the operators of multiplication by e^{it} on the spaces $H^2(\mathfrak{E}_*)$ and $H^2(\mathfrak{E})$, respectively, then $QU_{*+}^\times = U_+^\times Q$. Applying Lemma V.3.2 to the case

$$\mathfrak{R}_+ = H^2(\mathfrak{E}_*), \quad U_+ = U_{*+}^\times; \quad \mathfrak{R}'_+ = H^2(\mathfrak{E}), \quad U'_+ = U_+^\times; \quad Q,$$

and observing that in this case the Fourier transformations $\Phi^{\mathfrak{E}_*}$ and $\Phi^{\mathfrak{E}}$ are the identity transformations, we obtain that there exists an inner function $\{\mathfrak{E}_*, \mathfrak{E}, \Omega(\lambda)\}$ such that $Qu_* = \Omega u_*$. So we have

$$\delta_i u_* = \Theta\Omega u_* \quad (u_* \in H^2(\mathfrak{E}_*)). \tag{5.3}$$

Setting $u_* = \Theta u$ with $u \in H^2(\mathfrak{E})$ we derive $\Theta(\delta_i u - \Omega\Theta u) = 0$; the operator Θ is an isometry, therefore this relation implies

$$\delta_i u = \Omega\Theta u \qquad (u \in H^2(\mathfrak{E})). \tag{5.4}$$

Relations (5.3) and (5.4) say that $\Theta(\lambda)\Omega(\lambda) = \delta_i(\lambda)I_{\mathfrak{E}_*}$ and $\Omega(\lambda)\Theta(\lambda) = \delta_i(\lambda)I_{\mathfrak{E}}$; thus $\delta_i(\lambda)$ is a scalar multiple of $\Theta(\lambda)$. The same is then true for $\delta(\lambda) = \delta_e(\lambda)\delta_i(\lambda)$, because $\delta_e(\lambda)\Omega(\lambda)$ is also contractive (note that $\|\delta\|_\infty \le 1$ implies $\|\delta_e\|_\infty \le 1$).

This concludes the proof of Theorem 5.1.

An important particular case is as follows.

Theorem 5.2. *Let T be a contraction on a space $\mathfrak{H} \neq \{0\}$, of class C_{00}, and with finite defect indices $\mathfrak{d}_T$ and $\mathfrak{d}_{T^*}$. Then $T \in C_0$. Moreover, we have in this case $\mathfrak{d}_T = \mathfrak{d}_{T^*} = n \ge 1$ and*

$$d_T(T) = O,$$

where $d_T(\lambda)$ denotes the determinant of the matrix of $\Theta_T(\lambda)$ corresponding to some orthonormal bases in the defect spaces $\mathfrak{D}_T$ and $\mathfrak{D}_{T^*}$. The function $d_T(\lambda)$ is inner. If $n > 1$, the minimal function $m_T(\lambda)$ equals the quotient of $d_T(\lambda)$ by the greatest common inner divisor of the minors of order $n-1$ of the matrix of $\Theta_T(\lambda)$; if $n = 1$, then $m_T(\lambda) = d_T(\lambda)$. On the other hand, $d_T(\lambda)$ is always a divisor of $(m_T(\lambda))^n$.

Proof. According to Sec. 3.7, our assumption $T \in C_{00}$ implies that $\Theta_T(e^{it})$ is unitary a.e., and hence

$$\mathfrak{d}_T = \dim \mathfrak{D}_T = \dim \mathfrak{D}_{T^*} = \mathfrak{d}_{T^*}.$$

Because the defect indices were assumed finite, they must be equal to some integer $n \ge 1$ ($n = 0$ would mean that T is unitary, a case which is excluded by our assumptions). Let $\vartheta(\lambda)$ be the matrix of the operator $\Theta_T(\lambda)$ for some orthonormal bases in the defect spaces; the entries of $\vartheta(\lambda)$ are obviously functions of class H^∞, and so is

$$d_T(\lambda) = \det \vartheta(\lambda).$$

Under changes of the orthonormal bases, $\vartheta(\lambda)$ changes by constant unitary factors, and therefore its determinant changes by a constant numerical factor of modulus 1. Thus $d_T(\lambda)$ is determined by T up to a constant numerical factor of modulus 1. Moreover, because the operator $\Theta_T(e^{it})$ is unitary a.e., so is its matrix, and hence we have $|d_T(e^{it})| = 1$ a.e. Thus $d_T(\lambda)$ is an inner function. At every point $\lambda \in D$ such that $d_T(\lambda) \neq 0$, the operator $\Theta_T(\lambda)$ is (boundedly) invertible. By virtue of Proposition V.6.1, $d_T(\lambda)$ is a scalar multiple of $\Theta_T(\lambda)$, and the corresponding contractive analytic function $\Omega(\lambda)$ has the matrix $\vartheta^A(\lambda)$, namely the algebraic adjoint of the matrix $\vartheta(\lambda)$; if $n = 1$ then $\vartheta^A(\lambda)$ consists of the single entry 1. Obviously the entries of $\vartheta^A(\lambda)$ are also in H^∞.

Let us denote by $k(\lambda)$ the greatest common inner divisor (cf. Sec III.1) of the entries of $\vartheta^A(\lambda)$ (i.e., of the inner factors of the nonzero entries). Then we have $\Omega(\lambda) = k(\lambda)M(\lambda)$ and $d_T(\lambda) = k(\lambda)m(\lambda)$, where $m(\lambda)$ is a (scalar-valued) inner function and $M(\lambda)$ is an (operator-valued) *contractive*[7] *analytic* function the entries of whose matrix $\mu(\lambda)$ have no nonconstant common inner divisor (in H^∞). (For $n=1$ we have $k(\lambda)=1$, and hence $\Omega(\lambda)=M(\lambda)$, $d_T(\lambda)=m(\lambda)$.)

The relations $\Omega(\lambda)\Theta_T(\lambda)=d_T(\lambda)I_{\mathfrak{D}_T}$, $\Theta_T(\lambda)\Omega(\lambda)=d_T(\lambda)I_{\mathfrak{D}_{T^*}}$ imply

$$M(\lambda)\Theta_T(\lambda)=m(\lambda)I_{\mathfrak{D}_T},\quad \Theta_T(\lambda)M(\lambda)=m(\lambda)I_{\mathfrak{D}_{T^*}};\tag{5.5}$$

thus $m(\lambda)$ is a scalar multiple of $\Theta_T(\lambda)$, and consequently we have $m(T)=O$; cf. Theorem 5.1. Thus $T\in C_0$. By the same theorem, the minimal function $m_T(\lambda)$ is also a scalar multiple of T; thus there exists a contractive analytic function $\{\mathfrak{D}_{T^*},\mathfrak{D}_T,\Omega_T(\lambda)\}$ satisfying

$$\Omega_T(\lambda)\Theta_T(\lambda)=m_T(\lambda)I_{\mathfrak{D}_T},\quad \Theta_T(\lambda)\Omega_T(\lambda)=m_T(\lambda)I_{\mathfrak{D}_{T^*}}.\tag{5.6}$$

Because $m(\lambda)$ is inner and $m(T)=O$, we have $m(\lambda)=p(\lambda)m_T(\lambda)$ for some inner function $p(\lambda)$. From (5.5) and (5.6) we deduce

$$[p(\lambda)\Omega_T(\lambda)-M(\lambda)]\Theta_T(\lambda)=[p(\lambda)m_T(\lambda)-m(\lambda)]I_{\mathfrak{D}_T}=O\qquad(\lambda\in D).$$

Because $\Theta_T(\lambda)$ has a boundedly invertible value at every point where $d(\lambda)\neq 0$, and hence on a set dense in D, we infer that

$$p(\lambda)\Omega_T(\lambda)=M(\lambda)\qquad(\lambda\in D),\tag{5.7}$$

and hence, if we denote the matrix of $\Omega_T(\lambda)$ by $\omega(\lambda)$,

$$p(\lambda)\omega(\lambda)=\mu(\lambda).$$

This shows that the entries of the matrix $\mu(\lambda)$ have $p(\lambda)$ as a common inner divisor in H^∞, which is impossible unless $p(\lambda)$ is a constant (of modulus 1). Thus $m_T(\lambda)$ coincides with $m(\lambda)$.

Finally, it follows from (5.6) that

$$\omega(\lambda)\vartheta(\lambda)=m_T(\lambda)I_n=\vartheta(\lambda)\omega(\lambda);\tag{5.8}$$

taking determinants this implies $[\det\omega(\lambda)]\cdot d_T(\lambda)=[m_T(\lambda)]^n$. This shows that $d_T(\lambda)$ is a divisor of $[m_T(\lambda)]^n$, and hence concludes the proof of Theorem 5.2.

Remark. In the case $n=1$, $m_T(\lambda)$ coincides with $d_T(\lambda)$ and hence with $\Theta_T(\lambda)$. As the characteristic function of a c.n.u contraction T, considered up to coincidence, determines T up to unitary equivalence, it follows that *the contractions of class* C_{00} *with defect indices equal to* 1 *are determined by their minimal functions up to unitary equivalence.*

By Proposition III.4.6, two contractions of class C_0, one of which is a quasi-affine transform of the other, have the same minimal function. We obtain the following corollary.

[7] Consequence of relation (V.2.3).

Corollary 5.3. *Two contractions of class C_0, with defect indices equal to 1, are unitarily equivalent if one of them is a quasi-affine transform of the other.*

2. In order that a contraction belong to the class C_0 it is not necessary that the corresponding defect operators be of finite rank, or even that they be compact. In fact, we give an example for *a contraction T of class C_0, having the minimal function*

$$m_T(\lambda) = \exp\left(\frac{\lambda+1}{\lambda-1}\right), \tag{5.9}$$

and such that the defect operators D_T and D_{T^*} have no nonzero eigenvalue.

To this end, consider the purely contractive analytic function $\{\mathfrak{E}, \mathfrak{E}, \Theta(\lambda)\}$, where

$$\mathfrak{E} = L^2(0,1) \quad \text{and} \quad \Theta(\lambda)f(x) = \exp\left(x\frac{\lambda+1}{\lambda-1}\right)f(x) \qquad (f \in \mathfrak{E}).$$

Let $\mathbf{T}$ be the corresponding contraction in the sense of Theorem 3.1. Because

$$\left|\exp\left(x\frac{\lambda+1}{\lambda-1}\right)\right| = 1 \quad \text{for} \quad 0 \le x \le 1, \quad |\lambda| = 1, \quad \lambda \ne 1,$$

$\Theta(\lambda)$ is inner from both sides, and consequently $\mathbf{T} \in C_{00}$. Moreover, the function $\{\mathfrak{E}, \mathfrak{E}, \Omega(\lambda)\}$ defined by

$$\Omega(\lambda)f(x) = \exp\left[(1-x)\frac{\lambda+1}{\lambda-1}\right] \cdot f(x)$$

is also contractive and we have obviously

$$\Omega(\lambda)\Theta(\lambda) = \Theta(\lambda)\Omega(\lambda) = e_1(\lambda)I_{\mathfrak{E}},$$

where we again use the notation

$$e_s(\lambda) = \exp\left(s\frac{\lambda+1}{\lambda-1}\right) \qquad (s \ge 0).$$

Thus $e_1(\lambda)$ is a scalar multiple of $\Theta(\lambda)$. By Theorem 5.1 this implies $e_1(\mathbf{T}) = O$. Because $e_1(\lambda)$ admits no inner divisors other than $e_s(\lambda)$, $0 \le s \le 1$, the minimal function of $\mathbf{T}$ must be of the form $m_{\mathbf{T}}(\lambda) = e_s(\lambda)$ with some s such that $0 < s \le 1$. As $m_{\mathbf{T}}(\mathbf{T}) = O$, we see by invoking Theorem 5.1 again that $m_{\mathbf{T}}(\lambda)$ is a scalar multiple of $\Theta(\lambda)$, that is, there exists a contractive analytic function $\Omega'(\lambda)$ such that

$$\Omega'(\lambda)\Theta(\lambda) = \Theta(\lambda)\Omega'(\lambda) = m_{\mathbf{T}}(\lambda)I_{\mathfrak{E}}.$$

Hence

$$\|\Theta(0)f\| \ge \|\Omega'(0)\Theta(0)f\| = |m_{\mathbf{T}}(0)|\|f\|;$$

that is,

$$\left[\int_0^1 |e^{-x} f(x)|^2 \, dx\right]^{1/2} \geq e^{-s} \left[\int_0^1 |f(x)|^2 \, dx\right]^{1/2} \tag{5.10}$$

for every function $f \in \mathfrak{E} = L^2(0,1)$. Choosing $f(x)$ to be 0 except in a small neighborhood of $x = 1$, we conclude that s cannot be less than 1. Thus $m_T(\lambda) = e_1(\lambda)$, which proves (5.9).

It remains to prove that the defect operators D_T and D_{T^*} have no nonzero eigenvalues. As we have, up to coincidence,

$$\Theta_{\mathbf{T}^*}(\lambda) = \Theta_{\tilde{\mathbf{T}}}(\lambda) = \Theta^{\tilde{}}(\lambda) = \Theta(\lambda) = \Theta_{\mathbf{T}}(\lambda),$$

it follows from Proposition 3.3 that $\mathbf{T}$ and $\mathbf{T}^*$ are unitarily equivalent, and hence so are $D_{\mathbf{T}}$ and $D_{\mathbf{T}^*}$. It suffices therefore to show that $D_{\mathbf{T}^*}$ (or equivalently $D^2_{\mathbf{T}^*}$) has no nonzero eigenvalue.

For this purpose we use formula (3.14); it allows us to write

$$D^2_{\mathbf{T}^*} u = (\mathbf{I} - \mathbf{T}\mathbf{T}^*)u = \mathbf{P}(\mathbf{I} - \mathbf{U}\mathbf{T}^*)u = \mathbf{P}u(0), \qquad u \in \mathbf{H},$$

where

$$\mathbf{H} = H^2(\mathfrak{E}) \ominus \Theta H^2(\mathfrak{E}), \quad \mathfrak{E} = L^2(0,1),$$

and $\mathbf{P}$ denotes the orthogonal projection from $H^2(\mathfrak{E})$ onto $\mathbf{H}$. Let us make it explicit that we have here

$$u(\lambda) = u(\lambda; x) = \sum_0^\infty \lambda^n f_n(x) \quad (|\lambda| < 1; x \in (0,1); f_n \in L^2(0,1)).$$

The element $w = \mathbf{P}u(0)$ is determined by the conditions that $w \in \mathbf{H}$ and $u(0) - w \in \Theta H^2(\mathfrak{E})$; it follows that

$$w(\lambda) = [I_{\mathfrak{E}} - \Theta(\lambda)\Theta(0)^*]u(0) \qquad (u \in \mathbf{H}).$$

Suppose we have $(\mathbf{I} - \mathbf{T}\mathbf{T}^*)u = \rho u$ for some number $\rho \neq 0$ and some element $u \in \mathbf{H}$. It follows from the above that then

$$\rho u(\lambda) = [I_{\mathfrak{E}} - \Theta(\lambda)\Theta(0)^*]u(0), \tag{5.11}$$

and hence in particular

$$\rho u(0;x) = (1 - e^{-2x})u(0;x) \quad \text{a.e on} \quad 0 \leq x \leq 1;$$

therefore we have $u(0;x) = 0$ a.e., that is, $u(0) = 0$. Thus (5.11) implies $u(\lambda) = 0$ for every $\lambda \in D$, and hence $u = 0$ as an element of $\mathbf{H}$.

This proves that ρ is not an eigenvalue of $\mathbf{I} - \mathbf{T}\mathbf{T}^*$. All our assertions concerning the example have been established.

6 Spectral type of the minimal unitary dilation

1. Our functional model also allows us to solve completely the problem of finding the spectral type of the minimal unitary dilation U of an arbitrary c.n.u contraction T, namely determining the structure of U up to unitary equivalence.

Let us recall Theorem II.7.4, which says that if at least one of the defect indices of T is infinite, then U is a bilateral shift with multiplicity equal to $\mathfrak{d}_{\max} = \max\{\mathfrak{d}_T, \mathfrak{d}_{T^*}\}$; in other words, U is then unitarily equivalent to an orthogonal sum of $\mathfrak{d}_{\max}$ copies of the operator of multiplication by e^{it} on the space $L^2(0,2\pi)$ of scalar-valued functions $x(t)$. Note that this holds whenever $\|T\| < 1$ and the dimension of the Hilbert space is infinite.

We are now going to consider the remaining case when both defect indices are finite. We are concerned with spaces $L^2(S)$, where S is a measurable subset of $(0,2\pi)$, and scalar products and norms have to be taken with respect to the normalized Lebesgue measure $dt/2\pi$.

Theorem 6.1. *Let T be a c.n.u contraction with finite defect indices: $\mathfrak{d}_T = n$, $\mathfrak{d}_{T^*} = m$. The minimal unitary dilation U of T is then unitarily equivalent to the operator of multiplication by e^{it} on the space*

$$L^2(M_1) \oplus \cdots \oplus L^2(M_m) \oplus L^2(N_1) \oplus \cdots \oplus L^2(N_n), \tag{6.1}$$

where $M_1 = \cdots = M_m = (0,2\pi)$ and

$$N_k = \{t : t \in (0,2\pi), r(t) \geq k\} \qquad (k = 1,\ldots,n); \tag{6.2}$$

here $r(t)$ denotes the rank of the operator $\Delta_T(t) = [I - \Theta_T(e^{it})^*\Theta_T(e^{it})]^{1/2}$.

Proof. As the minimal unitary dilations of two unitarily equivalent contractions are obviously also unitarily equivalent, it suffices to consider our functional model $\mathbf{T}$ with the given defect indices n, m. Thus let $\{E^n, E^m, \Theta(\lambda)\}$ be a purely contractive analytic function and let $\mathbf{T}$ be the contraction it generates in the sense of Theorem 3.1. The characteristic function of $\mathbf{T}$ coincides with the given function $\Theta(\lambda)$, so $\Delta_T(t)$ is unitarily equivalent to $\Delta(t) = [I - \Theta(e^{it})^*\Theta(e^{it})]^{1/2}$ (by a constant unitary transformation). Hence

$$r(t) = \text{rank}\ \Delta_{\mathbf{T}}(t) = \text{rank}\ \Delta(t) \quad \text{for every } t.$$

By virtue of Theorem 3.1 the minimal unitary dilation $\mathbf{U}$ of $\mathbf{T}$ is equal to the multiplication by e^{it} on the space

$$\mathbf{K} = L^2(E^m) \oplus \overline{\Delta L^2(E^n)}.$$

Now it is obvious that $L^2(E^m)$ is the orthogonal sum of m copies of the space $L^2(0,2\pi)$ of scalar valued functions, with the multiplication by e^{it} on $L^2(E^m)$ corresponding to the multiplication by e^{it} on each of the component spaces $L^2(0,2\pi)$.

We study next the residual part of **U** on $\overline{\Delta L^2(E^n)}$. For each value of t at which it is defined, $\Delta(t)$ is a self-adjoint operator on E^n bounded by 0 and 1, and hence there exists an orthonormal base $\{\psi_k(t)\}_1^n$ of E^n composed of eigenvectors of $\Delta(t)$, that is,

$$\Delta(t)\psi_k(t) = \delta_k(t)\psi_k(t) \qquad (k = 1, \ldots n),$$

where the eigenvalues $\delta_k(t)$ are arranged in nonincreasing order:

$$1 \geq \delta_1(t) \geq \delta_2(t) \geq \cdots \geq \delta_n(t) \geq 0.$$

Because $\Delta(t)f$ is, for every $f \in E^n$, a measurable function of t, the eigenvalues $\delta_k(t)$ are also measurable functions of t; and because we obviously have

$$\{t\colon t \in (0, 2\pi), r(t) \geq k\} = \{t\colon t \in (0, 2\pi), \delta_k(t) > 0\},$$

the sets N_k defined in the theorem are also measurable. Moreover, the eigenvectors $\psi_k(t)$ can also be chosen so as to be measurable functions of t. The measurability of the greatest eigenvalue $\delta_1(t)$ is an easy consequence of the relation $\delta_1(t) = \sup(\Delta(t)f_k, f_k)$, where $\{f_k\}$ denotes a sequence dense on the unit sphere of E^n; for the other eigenvalues one can make use of the "minimax" theorem (cf. [*Func Anal.*] Sec. 95). The measurable choice of the eigenvectors is somewhat more involved; it can be established by manipulations with the minors of the matrix of $\Delta(t)$.

Setting $x_k(t) = (v(t), \psi_k(t))_{E^n}$ for $v \in L^2(E^n)$, we have

$$\Delta(t)v(t) = \Delta(t)\sum_1^n x_k(t)\psi_k(t) = \sum_1^n x_k(t)\delta_k(t)\psi_k(t),$$

and hence

$$\|\Delta(t)v(t)\|_{E^n}^2 = \sum_1^n |x_k(t)\delta_k(t)|^2$$

and

$$\|\Delta v\|_{L^2(E^n)}^2 = \frac{1}{2\pi}\sum_1^n \int_0^{2\pi} |x_k(t)\delta_k(t)|^2\, dt = \frac{1}{2\pi}\sum_1^n \int_{N_k} |x_k(t)\delta_k(t)|^2\, dt,$$

which shows that the transformation

$$\Delta v \to \{x_1(t)\delta_1(t), \ldots, x_n(t)\delta_n(t)\} \tag{6.3}$$

maps $\Delta L^2(E^n)$ isometrically into the space

$$L^2(N_1) \oplus \cdots \oplus L^2(N_n). \tag{6.4}$$

The obvious relation

$$(e^{it}v(t), \psi_k(t))_{E^n} = e^{it}x_k(t)$$

shows that multiplication by e^{it} on $\Delta L^2(E^n)$ is carried over by (6.3) into multiplication by e^{it} on each of the component spaces $L^2(N_k)$.

Let us choose, in particular, $v(t) = \varepsilon(t)\psi_k(t)$ for some fixed k $(1 \le k \le n)$ and some scalar-valued, bounded, measurable $\varepsilon(t)$; we have then $v \in L^2(E^n)$, and the vector corresponding to Δv by the transformation (6.3) will have its kth component equal to $\varepsilon(t)\delta_k(t)$ and its other components equal to 0. As the functions of type $\varepsilon(t)\delta_k(t)$ are dense in $L^2(N_k)$, we conclude that the transformation (6.3), when extended by continuity to the whole space $\overline{\Delta L^2(E^n)}$, maps this space isometrically onto the space (6.4). The assertion concerning multiplication by e^{it} will hold true, by continuity, after this extension also. Theorem 6.1 is proved.

2. It is obvious that $N_1 \supset N_2 \supset \cdots \supset N_n$. It is possible for the essential supremum $r_{\max}$ of the function $r(t)$ to be less than n; in this case the spaces $L^2(N_k)$ reduce, for $k > r_{\max}$, to the trivial space $\{0\}$, and hence can be omitted from (6.1).

There is an asymmetry in (6.1) with respect to the two defect indices, but this can easily be removed. Indeed, if we consider T^* instead of T we obtain that $U_T^* (= U_{T^*})$ is unitarily equivalent to multiplication by e^{it} on a space of type (6.1), but with the roles of m and n interchanged. The same is then true for U_T too (one has only to replace the sets occurring by their images under the transformation $t \to 2\pi - t$).[8] Thus Theorem 6.1 has the following consequence.

Corollary 6.2. *The minimal unitary dilation of a c.n.u contraction T with finite defect indices $d_T = n$ and $d_{T^*} = m$ is unitarily equivalent to multiplication by e^{it} on a space*

$$L^2(P_1) \oplus L^2(P_2) \oplus \cdots \oplus L^2(P_{n+m}),$$

where P_k $(k = 1, \ldots, n+m)$ are measurable subsets of $(0, 2\pi)$, with $P_k \supset P_{k+1}$ $(k = 1, \ldots, n+m-1)$ and with $P_k = (0, 2\pi)$ for $k = 1, \ldots, \max\{n, m\}$.

3. Now there arises the question if, conversely, multiplication by e^{it} on every such space is unitarily equivalent to the minimal unitary dilation of some c.n.u contraction. The answer is yes and we formulate it as follows.

Theorem 6.3. *Let n, m be nonnegative integers with $\nu = \max\{n, m\} \ge 1$. Let*

$$P_1 \supset P_2 \supset \cdots \supset P_{n+m}$$

[8] In the sum (6.1), the number of the terms different from $\{0\}$ is equal to $m + r_{\max}$. The roles of m and n are only seemingly asymmetric. This follows directly from the following: if Θ is a contraction from $\mathfrak{N}$ to $\mathfrak{M}$ (where $\mathfrak{N}$ and $\mathfrak{M}$ are Hilbert spaces of finite dimensions), then

$$\dim \mathfrak{M} + \dim \overline{(I - \Theta^*\Theta)^{1/2}\mathfrak{N}} = \dim \mathfrak{N} + \dim \overline{(I - \Theta\Theta^*)^{1/2}\mathfrak{M}}.$$

Indeed, the left-hand side is equal to $\dim \mathfrak{M} + \dim \mathfrak{N} - \dim \mathfrak{N}_0$, and the right-hand side to $\dim \mathfrak{N} + \dim \mathfrak{M} - \dim \mathfrak{M}_0$, where

$$\mathfrak{N}_0 = \{h \colon h \in \mathfrak{N}, (I - \Theta^*\Theta)h = 0\}, \quad \mathfrak{M}_0 = \{h \colon h \in \mathfrak{M}, (I - \Theta\Theta^*)h = 0\};$$

but $\dim \mathfrak{N}_0 = \dim \mathfrak{M}_0$ because Θ maps $\mathfrak{N}_0$ isometrically onto $\mathfrak{M}_0$.

be measurable subsets of $(0,2\pi)$ with $P_k = (0,2\pi)$ for $k = 1,\dots,\nu$. Then there exists a c.n.u contraction T with $\mathfrak{d}_T = n$ and $\mathfrak{d}_{T^*} = m$ such that the minimal unitary dilation of T is unitarily equivalent to multiplication by e^{it} on the space

$$L^2(P_1) \oplus L^2(P_2) \oplus \cdots \oplus L^2(P_{n+m}).$$

(We also allow the possibility that P_k equal $(0,2\pi)$ for some $k > \nu$, as well as the possibility that some P_k is of measure 0; in the latter case the corresponding spaces $L^2(P_k)$ reduce to $\{0\}$ and hence can be omitted.)

Proof. Let us assume first that $n \leq m$ so that $\nu = m$. The case $n = 0$ (i.e., $\nu = n+m$) is simple. Multiplication by e^{it} is then a bilateral shift with multiplicity ν, and hence it is the minimal unitary dilation of a unilateral shift with the same multiplicity.

In the case $1 \leq n \leq m$, let us consider the matrix-valued function

$$\Theta(\lambda) = [\vartheta_{jk}(\lambda)] \qquad (j = 1,\dots,m; k = 1,\dots,n; |\lambda| < 1)$$

with

$$\vartheta_{jk}(\lambda) \equiv 0 \text{ for } j \neq k,$$
$$\vartheta_{kk}(k) \equiv \lambda \cdot u_k(\lambda) \text{ for } k = 1,\dots,n,$$

where the function $u_k(\lambda)$ is determined in such a way that it belongs to H^∞ and satisfies

$$|u_k(e^{it})|^2 = 1 - \frac{1}{2}\chi_k(t) \quad \text{a.e.},$$

$\chi_k(t)$ denoting the characteristic function of the set P_{m+k}. The existence of such a function $u_k(\lambda)$ is guaranteed by the fact that $\log[1 - (1/2)\chi_k(t)] \in L^1(0,2\pi)$ (cf. (III.1.14–16)). As we have

$$|\vartheta_{kk}(\lambda)| \leq |u_k(\lambda)| \leq 1 \quad \text{and} \quad \vartheta_{kk}(0) = 0 \qquad (k = 1,\dots,n),$$

$\{E^n, E^m, \Theta(\lambda)\}$ will be a purely contractive analytic function.

The matrix $\Delta(t) = [I - \Theta(e^{it})^*\Theta(e^{it})]^{1/2}$ is of order n and of diagonal form; the entries on the diagonal are

$$\delta_{kk}(t) = \left[1 - \left(1 - \frac{1}{2}\chi_k(t)\right)\right]^{1/2} = \sqrt{\frac{1}{2}}\,\chi_k(t) \qquad (k = 1,\dots,n).$$

Thus we have

$$\operatorname{rank} \Delta(t) = \sum_{k=1}^{n} \chi_k(t).$$

The sets P_{m+k} $(k = 1,\dots,n)$ are arranged in nonincreasing order, thus the inequality $\operatorname{rank} \Delta(t) \geq k$ is satisfied (up to a set of 0 measure) exactly for the points of the set P_{m+k}.

Let $\mathbf{T}$ be the c.n.u contraction generated by the purely contractive analytic function $\{E^n, E^m, \Theta(\lambda)\}$ in the sense of Theorem 3.1. Applying Theorem 6.1 we see that the minimal unitary dilation of $\mathbf{T}$ is unitarily equivalent to multiplication by e^{it} on the space

$$\left[\bigoplus_1^m L^2(0,2\pi)\right] \oplus L^2(P_{m+1}) \oplus \cdots \oplus L^2(P_{m+n}).$$

This proves Theorem 6.3 in the case $n \leq m$.

In the case $n > m$ we first construct a c.n.u contraction S for which $\mathfrak{d}_S = m$ and $\mathfrak{d}_{S^*} = n$, and whose minimal unitary dilation is unitarily equivalent to multiplication by e^{it} on the space

$$L^2(P_1') \oplus L^2(P_2') \oplus \cdots \oplus L^2(P_{m+n}'),$$

where P_k' denotes the symmetric image of the set P_k with respect to the point $t = \pi$; such a contraction S exists by what we have just proved. Then $T = S^*$ satisfies the assertions of the theorem. Theorem 6.3 is proved.

Remark. Corollary 6.2 and Theorem 6.3 (for the case of finite defect indices) and Theorem II.7.4 (for the case of at least one infinite defect index) give a complete solution of the problem of determining the spectral type of the minimal unitary dilations of c.n.u contractions. Note that the contraction T constructed in the proof of Theorem 6.3 is, in general, reducible. It is natural to ask whether we can also find an irreducible T for this purpose. However, if for a contraction T we have $\max\{m,n\} > 1$ and $\min\{m,n\} = 0$, then, T or T^* is necessarily a shift of multiplicity > 1 and therefore always reducible. In the remaining case $\min\{m,n\} \geq 1$, an appropriate modification of the previous proof will yield an irreducible contraction T with the required properties. To present such a modification, first observe that, due to the last part of the proof of Theorem 6.3, we can assume $1 \leq n \leq m$. Also, observe that if $\{E^m, E^m, \Omega(\lambda)\}$ is an inner function and if $\widehat{\mathbf{T}}$ is the c.n.u contraction associated with the purely contractive analytic function $\{E^n, E^m, \hat{\Theta}(\lambda)\}$ with $\hat{\Theta}(\lambda) = \Omega(\lambda)\Theta(\lambda)$, where $\Theta(\lambda)$ and $\mathbf{T}$ are as in the proof above, then the minimal unitary dilation of $\widehat{\mathbf{T}}$ and $\mathbf{T}$ have the same spectral type. Thus it remains to construct $\Omega(\lambda)$ such that $\widehat{\mathbf{T}}$ be irreducible. To this end, choose an $m \times m$ unitary matrix $G = [g_{jk}]$ with nonzero constant entries (e.g $G = \exp(iA) = \sum_{l=0}^{\infty}(iA)^l/l!$ where all the entries of A are equal to $\pi\sqrt{2}$) and define $\Omega(\lambda) = [\omega_{jk}(\lambda)]$ by

$$\omega_{jk}(\lambda) = \lambda^{(j-1)m+(k-1)} g_{jk} \quad \text{for} \quad j,k = 1,2,\ldots,m.$$

With this choice, the entries $\hat{\vartheta}_{jk}(\lambda)$ $(j = 1,\ldots,m;\ k = 1,\ldots,n)$ of $\hat{\Theta}(\lambda)$ are linearly independent, a fact left to the reader to prove. Finally, if $\widehat{\mathbf{T}}$ were reducible then its characteristic function would be the orthogonal sum of the characteristic functions of the reducing components. Because it also coincides with $\hat{\Theta}(\lambda)$, there would exist nonzero vectors $c = (c_1,\ldots,c_n)$ in E^n and $d = (d_1,\ldots,d_m)$ in E^m such that

$$0 = (\hat{\Theta}(\lambda)c, d) = \sum_{j=1}^{m} \sum_{k=1}^{n} c_k d_j \hat{\vartheta}_{jk}(\lambda).$$

This contradicts the linear independence of the functions $\hat{\vartheta}_{jk}(\lambda)$. Thus the c.n.u contraction $\widehat{\mathbf{T}}$ is irreducible.

4. In Secs. II.1 and 7 we obtained several conditions upon a c.n.u contraction T which ensure that the minimal unitary dilation U of T be a bilateral shift. The above Theorem 6.3 shows that *there exist c.n.u contractions T for which U is not a bilateral shift*; indeed, we can choose some of the sets P_k such that both P_k and its complement $(0,2\pi)\backslash P_k$ have positive measure. However, it may be instructive to give here a concrete example based on Theorem 6.1.

Consider a numerical function $w(\lambda)$, holomorphic on D and such that $|w(\lambda)| \leq 1$ on D and $|w(0)| < 1$. Let $\mathbf{T}$ be the c.n.u contraction generated (in the sense of Theorem 3.1) by the purely contractive analytic function $\{E^1, E^1, w(\lambda)\}$. As the defect indices of $\mathbf{T}$ are equal to 1 it follows from Theorem 6.1 that the minimal unitary dilation of $\mathbf{T}$ is unitarily equivalent to multiplication by e^{it} on the space $L^2(0,2\pi) \oplus L^2(N)$, where $N = \{t \colon t \in (0,2\pi),\ 1 - |w(e^{it})|^2 > 0\}$. In particular, choose for $w(\lambda)$ the conformal map of D onto the half-disc $\{\lambda : |\lambda| < 1, \operatorname{Im}\lambda > 0\}$. Then we have $0 <$ meas $N < 2\pi$, and hence, obviously,

$$L^2(0,2\pi) \oplus \{0\} \subset L^2(0,2\pi) \oplus L^2(N) \subset L^2(0,2\pi) \oplus L^2(0,2\pi),$$

with proper inclusions. Now multiplication by e^{it} is a bilateral shift on the first as well as on the third space, with multiplicity 1 and 2, respectively. If it were a bilateral shift on the intermediate space $L^2(0,2\pi) \oplus L^2(N)$ also, then, by Proposition I.2.1, this space would coincide with one of the two other spaces, which is impossible because the measure of N differs from 0 and 2π.

7 Notes

1. The analysis of the resolvent $(T - \lambda I)^{-1}$ by methods of complex function theory has long since proved to be one of the principal means to study the structure of the operator T, at least if T is a normal operator or if the spectrum of T consists of several closed components. For other operators the analytical behavior of the resolvent yields but little information about the structure of T.

More recently, the school of M. G. Kreĭn in Odessa, inspired by Kreĭn's research on the extension of symmetric operators, began to associate with certain Hilbert space operators T a new type of operator-valued analytic function whose behavior reflects further details of the structure of T. The definition of these functions, called *characteristic functions*, developed gradually. First they appeared in LIVŠIC [1] (for operators such that $I - T^*T$ and $I - TT^*$ have rank 1) and [2] (for the case that these operators have a finite rank). The first general definition was given in ŠMUL′JAN [1]; in the case of a contraction T this definition coincides with our definition (1.1) or its equivalent form (1.2). The interest of this school shifted subsequently to bounded operators T such that $T^* - T$ is of finite rank or at least of finite trace. For these operators T a characteristic function was defined in an analogous way, and by means of it a far-reaching theory of these operators was achieved; see LIVŠIC [3]; M. S. BRODSKIĬ [1]–[3], and in particular BRODSKIĬ AND LIVŠIC [1], where one also finds further references. Interesting applications to some physical problems appeared recently in the book LIVŠIC [4]. One of the principal results is the construction of a concrete model for the operators considered, by means of Volterra-type integral operators and operators of multiplication by nondecreasing functions. Analogous results were obtained later for operators T such that $I-T^*T$ and $I-TT^*$ are of the same finite rank; see POLJACKIĬ [1],[2]. Later the interest of this school was directed mainly to the representation of the operators by integral operators that are no longer in immediate connection with the characteristic function; see, for example, SAHNOVIČ [2]–[4]; M. S. BRODSKIĬ [4],[7]; and GOHBERG AND KREĬN [1],[2],[6],[7].

The authors of this book came upon characteristic functions in 1962 in an entirely different way, namely as a result of the harmonic analysis of the minimal unitary dilation of c.n.u contractions T; moreover, they obtained simultaneously a functional model of T depending explicitly and exclusively on the characteristic function of T; see SZ.-N.–F. [2], [3], [VIII], and Secs. 2 and 3 of this chapter.

2. In the particular case of contractions $T \in C_{\cdot 0}$ and $T \in C_{0\cdot}$ the same functional model was also obtained by American mathematicians (cf. ROTA [1]; ROVNYAK [1]; and HELSON [1]), who arrived at it essentially in the following direct way.

Let T be a contraction on $\mathfrak{H}$ with $T^{*n} \to O$, and let $\mathbf{H}$ be the space of the sequences $\mathbf{h} = \{h_n\}_0^\infty$ with $h_n \in \mathfrak{D}_{T^*}$ and $\|\mathbf{h}\|^2 = \sum \|h_n\|^2 < \infty$. As we have $\|h\|^2 = \|D_{T^*}h\|^2 + \|T^*h\|^2$, and consequently

$$\|h\|^2 = \sum_0^\infty \|D_{T^*}T^{*j}h\|^2 \qquad (h \in \mathfrak{H}),$$

(cf. Sec. I.10), we can embed $\mathfrak{H}$ in $\mathbf{H}$ by identifying $h \in \mathfrak{H}$ with

$$\mathbf{h} = \{D_{T^*}h, D_{T^*}T^*h, D_{T^*}T^{*2}h, \ldots\} \in \mathbf{H}.$$

Denoting by V the unilateral shift $\{h_0, h_1, \ldots\} \to \{0, h_0, h_1, \ldots\}$ on $\mathbf{H}$, we have $T^* = V^*|\mathfrak{H}$. Take now the Fourier representation of $\mathbf{H}$ and V: identify $\mathbf{H}$ with $H^2(\mathfrak{D}_{T^*})$ and V with the operator $U^\times$ of multiplication by e^{it} on this space. Then the subspace $H^2(\mathfrak{D}_{T^*}) \ominus \mathfrak{H}$ is invariant for $U^\times$ and hence, by Theorem V.3.3, we have $H^2(\mathfrak{D}_{T^*}) \ominus \mathfrak{H} = \Theta H^2(\mathfrak{F})$ for some inner function $\{\mathfrak{F}, \mathfrak{D}_{T^*}, \Theta(\lambda)\}$. Thus we see that T is unitarily equivalent to the operator $\mathbf{T}$ defined on

$$H^2(\mathfrak{D}_{T^*}) \ominus \Theta H^2(\mathfrak{F}) \quad \text{by} \quad \mathbf{T}^* = U^{\times *}|[H^2(\mathfrak{D}_{T^*}) \ominus \Theta H^2(\mathfrak{F})]. \qquad (*)$$

This coincides with our functional model in the case $T \in C_{\cdot 0}$. However, in this nonelaborate form the model does not give any explicit information on $\Theta(\lambda)$ as a function of T, and in particular it does not state the relation between $\Theta(\lambda)$ and the characteristic function of T.

3. The functional model of Sec. 2 was originally given in SZ.-N.–F. [2], [VIII], and [IX] (the lemmas on Fourier representation were explicitly used in the third paper only). Section 3 reproduces part of the paper [VIII]. It should be noted that Theorem 3.4, which asserts that two c.n.u contractions are unitarily equivalent if and only if their characteristic functions coincide, is contained in a result of ŠTRAUS [3],[4]. However, our Theorem 3.4 is an immediate consequence of Theorem 3.1, that is, of the functional model for c.n.u contractions, for which no analogue is at present known for the more general operators considered by Štraus.

Our Proposition 3.5 was first stated in SZ.-N.–F [3] and proved in [VIII]; it establishes a correspondence between the nature (inner, outer, etc.) of the characteristic function and the class $C_{\alpha\beta}$ of the operator. This was made possible precisely by the manner in which we have extended the notions of inner and outer functions to operator-valued functions.

Theorem 3.6 was first proved in SZ.-N.–F [11] (for $\Theta(\lambda)$ and $\Theta'(\lambda)$ inner from both sides) using entirely different methods. The special case of scalar valued functions considered in Corollary 3.7 was proved, again by different methods, in SARASON [3]; the investigations on the dilation of commutants started with this paper. For the case of not necessarily inner $\Theta(\lambda)$ and $\Theta'(\lambda)$, see SZ.-N.–F. [16].

In connection with Proposition 3.8 see also FUHRMANN [1],[2], where the special case of finite matrix-valued functions $\Theta(\lambda)$ is considered.

Theorem 4.1 on the relations between the characteristic function and the spectrum of T was first proved in SZ.-N.–F. [VIII]. In the particular case of finite rectangular *matrix*-valued functions these relations were found earlier by some Soviet authors; see, for example LIVŠIC [1]–[3]; ŠMUL′JAN [1]; BRODSKIĬ AND LIVŠIC [1]; and POLJACKIĬ [2]. In the case of *operator*-valued inner functions related to an operator $T \in C_{\cdot 0}$ by the relations $(*)$ (and hence, by virtue of our Theorem 3.1, coinciding with the characteristic function of T), Theorem 4.1 was found independently and almost simultaneously by Srinivasan, Wang, and Helson (cf. HELSON [1] p. 74), and a little earlier for scalar valued inner functions by MOELLER [1]. Actually, the result of Moeller appears in the light of our Theorem 3.1 as a consequence of the corresponding result of LIVŠIC [1].

Proposition 4.2 is due to GOHBERG AND KREĬN [5]; the present proof is new.

The general criterion of Theorem 4.5 for a contraction to be similar to a unitary operator was originally published in SZ.-N.–F. [X]. This result illustrates the usefulness of both the characteristic functions and of the functional models related to them. Corollary 4.7 was found by GOHBERG–KREĬN as a consequence of Proposition 4.2.

The results in Sec. 4.4 are new, and so is Theorem 5.1 on the contractions of class C_0. Theorem 5.2, which states that every contraction T of class C_{00} with finite defect indices is also of class C_0, and shows how to calculate the minimal function $m_T(\lambda)$ from the characteristic function $\Theta_T(\lambda)$, was first given in SZ.-N.–F [VIII]. This theorem indicates a close analogy between the characteristic function $\Theta_T(\lambda)$ of such a contraction and the characteristic matrix $\mathscr{T}(\lambda)$ of a finite square matrix $\mathscr{T}$ defined as in linear algebra, namely by $\mathscr{T}(\lambda) = \lambda \mathscr{J} - \mathscr{T}$. Indeed, the minimal function of the contraction T and the minimal polynomial of the matrix $\mathscr{T}$ have to be calculated from $\Theta_T(\lambda)$ and from $\mathscr{T}(\lambda)$, respectively, in much the same way!

Section 6 reproduces SZ.-N.–F [VIII] with the only difference that in Subsec. 4 we give a simpler example for a c.n.u contraction whose minimal unitary dilation is not a bilateral shift. The first example of such a contraction was given in SZ.-N.–F

[V]; the construction there was based on a theorem of SAHNOVIČ [2] (cf. BRODSKIĬ AND LIVŠIC [1] p. 65). All the classes of c.n.u contractions which have been studied before yielded only examples where the minimal unitary dilation is a bilateral shift (cf. SCHREIBER [1]; DE BRUIJN [1]; and HALPERIN [3]).

4. By virtue of the relation $T(s) = e_s(T)$ between a continuous semigroup of contractions $\{T(s)\}_{s\geq 0}$ and its cogenerator T (cf. Sec. III.8), every model of T generates a model of $\{T(s)\}$. In particular, it follows from the results of Secs. 2 and 3 that *the completely nonunitary continuous semigroups admit the functional model* $\{\mathbf{H}, \mathbf{T}(s)\}$ *defined by*

$$\begin{cases} \mathbf{H} = [H^2(\mathfrak{E}_*) \oplus \overline{\Delta L^2(\mathfrak{E})}] \ominus \{\Theta w \oplus \Delta w\colon\, w \in H^2(\mathfrak{E})\}, \\ \mathbf{T}(s)(u_* \oplus v) = \mathbf{P}(e_s u_* \oplus e_s v) \quad (u_* \oplus v \in \mathbf{H}), \end{cases} \tag{a}$$

with $\mathbf{P}$ *denoting orthogonal projection onto* $\mathbf{H}$. Here $\{\mathfrak{E}, \mathfrak{E}_*, \Theta(\lambda)\}$ is a purely contractive, analytic, and otherwise arbitrary function on the unit disc; it coincides with the characteristic function of the cogenerator of the semigroup.

For semigroups of class $C_{\cdot 0}$, that is, for which $T(s)^* \to O$ as $s \to \infty$, or $T^{*n} \to O$ as $n \to \infty$ (the equivalence of these conditions having been established by Proposition III.9.1), the model simplifies to the following,

$$\begin{cases} \mathbf{H} = H^2(\mathfrak{E}_*) \ominus \Theta H^2(\mathfrak{E}), \\ \mathbf{T}(s)u_* = \mathbf{P}(e_s u_*) \quad (u_* \in \mathbf{H}); \end{cases} \tag{a$'$}$$

$\Theta(\lambda)$ is in this case an inner function.

We can give to this model an alternative form in which the roles of the unit disc D and its boundary C are played by the upper half-plane and the real axis. In fact, for an arbitrary (separable) Hilbert space $\mathfrak{A}$, the space $L^2(\mathfrak{A})$ of functions $u(e^{it})$ $(0 \leq t < 2\pi)$ is transformed unitarily onto the space $A(\mathfrak{A}) = L^2(-\infty, \infty; \mathfrak{A})$ of functions $f(x)$ $(-\infty < x < \infty)$[9] by the transformation $u \to f$, where

$$f(x) = \frac{1}{x+i} u\left(\frac{x-i}{x+i}\right).$$

The subspace $L^2_+(\mathfrak{A})$ of $L^2(\mathfrak{A})$, consisting of the limits on C of the functions analytic on D and of class $H^2(\mathfrak{A})$, is transformed thereby onto the subspace $A_+(\mathfrak{A})$ of $A(\mathfrak{A})$, consisting of the limits on the real axis of the functions $f(z)$ that are analytic on the upper half-plane and for which

$$\sup_{0<y<\infty} \int_{-\infty}^{\infty} \|f(x+iy)\|^2_{\mathfrak{A}}\, dx < \infty;$$

by the theorem of Paley and Wiener the functions in $A_+(\mathfrak{A})$ can also be characterized as the Fourier transforms of the functions of class $L^2(0, \infty; \mathfrak{A})$. see, for example, HOFFMAN [1] Chaps. 7 and 8.

[9] We take the measures $dt/d\pi$ and dx/π, respectively.

If we set

$$S(z) = \Theta\left(\frac{z-i}{z+i}\right) \quad \text{and} \quad D(x) = [I_{\mathfrak{E}} - S(x)^*S(x)]^{1/2}$$

the functional model of c.n.u continuous semigroups of contractions takes the following form

$$\begin{cases} \boldsymbol{H} = [A_+(\mathfrak{E}_*) \oplus \overline{DA(\mathfrak{E})}] \ominus \{Sf \oplus Df \colon \ f \in A_+(\mathfrak{E})\}, \\ T(s)(f_* \oplus g) = \boldsymbol{P}[e^{isx}f_*(x) \oplus e^{isx}g(x)] \quad (f_* \oplus g \in \boldsymbol{H}), \end{cases} \tag{b}$$

with $\boldsymbol{P}$ denoting orthogonal projection onto $\boldsymbol{H}$.

For semigroups of class $C_{\cdot 0}$ the values of the function $S(x)$ are a.e isometries on $\mathfrak{E}$, and the model (b) simplifies to

$$\begin{cases} \boldsymbol{H} = A_+(\mathfrak{E}_*) \ominus SA_+(\mathfrak{E}) \\ \boldsymbol{T}(s)f_* = \boldsymbol{P}[e^{isx}f_*(x)] \quad (f_* \in \boldsymbol{H}). \end{cases} \tag{b$'$}$$

The general model (b) was first formulated in FOIAŞ [7]. The particular case (b′) for the class $C_{\cdot 0}$ was obtained by LAX AND PHILLIPS [1]. These models can be used especially in "scattering theory": the function $S(x)$ appears there as the so-called "suboperator" of scattering. This was pointed out, in the case in which the particular model (b′) applies, in LAX AND PHILLIPS [1],[2], and in the general case in ADAMJAN AND AROV [1], [2].

In the sequel, we continue using functional models associated with the unit disk. Our results can be transferred to the setting of the upper half-plane.

In connection with this chapter see also ADAMJAN, AROV AND KREĬN [1]; DOUGLAS [3]; ŠVARCMAN [1]; and MUHLY [1].

8 Further results

1. In the first appearance of the characteristic function, LIVŠIC [1] studied a contraction operator with defect indices equal to 1 as a perturbation of a unitary operator. This also yields a model for the contraction on a function space which is useful in perturbation questions. This idea was developed independently in DE BRANGES [3]; note that the English translation of LIVŠIC [1] only appeared in 1960. The case of defect indices equal to 1 was considered again in CLARK [2], where an explicit unitary equivalence is constructed between the functional model described in this chapter, and the model arising from the spectral decomposition of a unitary rank one perturbation of a given contraction. This approach was extended to general functional models in BALL AND LUBIN [1].

2. Characteristic functions have been extended to commuting pairs of operators with finite rank imaginary parts. This extension and its function theoretical ramifications are discussed in detail in the monograph by LIVŠIC ET AL. [1].

3. In the context of noncommuting operators, characteristic functions were considered in FRAZHO [2]; see also POPESCU [2],[8] for the analysis of this situation, and for further extensions. See also BHATTACHARYYA, ESCHMEIER, AND SARKAR [1].

4. For other approaches to functional models and their extensions see BALL [1],[2]; BALL AND KRIETE [1]; BUNCE [1]; MAKAROV AND VASYUNIN [1]; KRIETE [2]; NIKOLSKIĬ AND HRUSŠČEV [1]; NIKOLSKIĬ AND TREIL [1]; and VASYUNIN [1].

5. The similarity of operators to unitary or selfadjoint ones can be studied in terms of characteristic functions or in terms of the resolvent of the operator. For a variety of results in this direction, see VAN CASTEREN [2]; NABOKO [1]; BENAMARA AND NIKOLSKIĬ [1]; NIKOLSKIĬ AND TREIL [1]; and KUPIN AND TREIL [1].

Chapter VII

Regular Factorizations and Invariant Subspaces

1 The fundamental theorem

1. We continue our study of the geometric structure of the space of the minimal unitary (or isometric) dilation of a contraction T, given in Secs. II.1 and II.2.1. We now consider decompositions of this space induced by invariant subspaces of T. Thus, let T be a contraction on $\mathfrak{H}$, U the minimal unitary dilation of T on $\mathfrak{K}$ ($\supset \mathfrak{H}$), and U_+ the minimal isometric dilation of T on $\mathfrak{K}_+$, where

$$\mathfrak{K}_+ = \bigvee_0^\infty U^n \mathfrak{H}$$

and $U_+ = U|\mathfrak{K}_+$. We recall that, according to (I.4.2),

$$T^* = U_+^*|\mathfrak{H}. \tag{1.1}$$

Let us suppose, furthermore, that $\mathfrak{H}_1$ is a subspace of $\mathfrak{H}$ invariant for T; $\mathfrak{H}_2 = \mathfrak{H} \ominus \mathfrak{H}_1$ is then invariant for T^*, and by (1.1) invariant for U_+^* also. This implies in turn that the orthogonal complement of $\mathfrak{H}_2$ in $\mathfrak{K}_+$,

$$\mathfrak{K}' = \mathfrak{K}_+ \ominus \mathfrak{H}_2, \tag{1.2}$$

is invariant for U_+. Let

$$\mathfrak{K}' = M_+(\mathfrak{F}) \oplus \mathfrak{R}_1 \tag{1.3}$$

be the Wold decomposition of $\mathfrak{K}'$ induced by the isometry $U_+|\mathfrak{K}'$; here we have

$$\mathfrak{F} = \mathfrak{K}' \ominus U_+\mathfrak{K}' \quad \text{and} \quad \mathfrak{R}_1 = \bigcap_0^\infty U_+^n \mathfrak{K}'. \tag{1.4}$$

B.Sz.-Nagy et al., *Harmonic Analysis of Operators on Hilbert Space*, Universitext,
DOI 10.1007/978-1-4419-6094-8_7, © Springer Science+Business Media, LLC 2010

As $\mathfrak{R}_1$ reduces U_+ to a unitary operator, $\mathfrak{R}_1$ is necessarily included in the space $\mathfrak{R}$ of the residual part of U_+; in fact it follows from the relation

$$\mathfrak{K}_+ = M_+(\mathfrak{L}_*) \oplus \mathfrak{R}, \quad \text{see (II.2.7)},$$

that $\mathfrak{R} = \bigcap_0^\infty U_+^n \mathfrak{K}_+$, which implies that $\mathfrak{R}$ is the maximal subspace of $\mathfrak{K}_+$ on which U_+ induces a unitary operator. Thus we have

$$\mathfrak{R} = \mathfrak{R}_1 \oplus \mathfrak{R}_2; \tag{1.5}$$

as $\mathfrak{R}$ and $\mathfrak{R}_1$ reduce U_+ to unitary operators, so does $\mathfrak{R}_2$.

From (1.2), (1.3), and (II.2.7) we deduce

$$\mathfrak{H}_2 = \mathfrak{K}_+ \ominus \mathfrak{K}' = [M_+(\mathfrak{L}_*) \oplus \mathfrak{R}] \ominus [M_+(\mathfrak{F}) \oplus \mathfrak{R}_1];$$

by (1.5) this becomes

$$\mathfrak{H}_2 = [M_+(\mathfrak{L}_*) \oplus \mathfrak{R}_2] \ominus M_+(\mathfrak{F}). \tag{1.6}$$

From (1.6) and from the analogous representation

$$\mathfrak{H} = \mathfrak{K}_+ \ominus M_+(\mathfrak{L}) = [M_+(\mathfrak{L}_*) \oplus \mathfrak{R}] \ominus M_+(\mathfrak{L}) \tag{1.7}$$

resulting from (II.2.5) and (II.2.7) we infer:

$$\mathfrak{H}_1 = \mathfrak{H} \ominus \mathfrak{H}_2 = \{[M_+(\mathfrak{L}_*) \oplus \mathfrak{R}] \ominus M_+(\mathfrak{L})\} \ominus \{[M_+(\mathfrak{L}_*) \oplus \mathfrak{R}_2] \ominus M_+(\mathfrak{F})\}.$$

Hence

$$\mathfrak{H}_1 = [M_+(\mathfrak{F}) \oplus \mathfrak{R}_1] \ominus M_+(\mathfrak{L}) = \mathfrak{K}' \ominus M_+(\mathfrak{L}). \tag{1.8}$$

These representations of $\mathfrak{H}_1$ and $\mathfrak{H}_2$ show that

$$M_+(\mathfrak{F}) \subset M_+(\mathfrak{L}_*) \oplus \mathfrak{R}_2 \quad \text{and} \quad M_+(\mathfrak{L}) \subset M_+(\mathfrak{F}) \oplus \mathfrak{R}_1. \tag{1.9}$$

For every wandering subspace $\mathfrak{A}$ for U we have

$$M(\mathfrak{A}) = \bigvee_{n \geq 0} U^{-n} M_+(\mathfrak{A})$$

and furthermore U^{-n} maps $\mathfrak{R}_1$ and $\mathfrak{R}_2$ onto themselves, thus (1.9) implies

$$M(\mathfrak{F}) \subset M(\mathfrak{L}_*) \oplus \mathfrak{R}_2 \quad \text{and} \quad M(\mathfrak{L}) \subset M(\mathfrak{F}) \oplus \mathfrak{R}_1. \tag{1.10}$$

Because $\mathfrak{R}_1 \subset \mathfrak{R} \perp M(\mathfrak{L}_*)$, it follows from the last relation that

$$[M(\mathfrak{L}) \vee M(\mathfrak{L}_*)] \subset [M(\mathfrak{F}) \vee M(\mathfrak{L}_*)] \oplus \mathfrak{R}_1. \tag{1.11}$$

If the contraction T is *completely nonunitary*, as we assume from now on, the left-hand side of (1.11) equals $\mathfrak{K}$; see (II.1.10). Thus (1.11) implies

$$\mathfrak{K} = [M(\mathfrak{F}) \vee M(\mathfrak{L}_*)] \oplus \mathfrak{R}_1. \tag{1.12}$$

On the other hand $\mathfrak{K} = M(\mathfrak{L}_*) \oplus \mathfrak{R}$ (cf. (II.2.1), thus (1.12) and (1.5) yield

$$\mathfrak{R}_2 = [M(\mathfrak{F}) \vee M(\mathfrak{L}_*)] \ominus M(\mathfrak{L}_*). \tag{1.13}$$

Let us return to the definition of $\mathfrak{F}$ by (1.4). Because $M_+(\mathfrak{L}) \subset \mathfrak{K}' \subset \mathfrak{K}_+$ (cf. (1.9)), we obtain, using (1.6) as well,

$$\mathfrak{F} = \mathfrak{K}' \ominus U_+\mathfrak{K}' \subset \mathfrak{K}_+ \ominus U_+M_+(\mathfrak{L}) = [\mathfrak{H} \oplus M_+(\mathfrak{L})] \ominus U_+M_+(\mathfrak{L}),$$

and thus

$$\mathfrak{F} \subset \mathfrak{H} \oplus \mathfrak{L}. \tag{1.14}$$

In analogy to the notations already adopted, let us denote by $P^{\mathfrak{L}}$, $P^{\mathfrak{L}_*}$, $P^{\mathfrak{F}}$, $P_{\mathfrak{R}}$, $P_{\mathfrak{R}_1}$, and $P_{\mathfrak{R}_2}$ the orthogonal projections of $\mathfrak{K}$ onto $M(\mathfrak{L})$, $M(\mathfrak{L}_*)$, $M(\mathfrak{F})$, $\mathfrak{R}$, $\mathfrak{R}_1$, and $\mathfrak{R}_2$, respectively.

Because of (1.14) we infer from the decomposition (II.1.4) of $\mathfrak{K}$ that $\mathfrak{F}$ is orthogonal to $U^\nu\mathfrak{L}$ and $U^{-\nu}\mathfrak{L}_*$ $(\nu \geq 1)$, consequently

$$U^n\mathfrak{L} \perp U^{-m}\mathfrak{F} \quad \text{and} \quad U^n\mathfrak{F} \perp U^{-m}\mathfrak{L}_* \quad \text{for} \quad n \geq 0, m \geq 1.$$

This implies

$$P^{\mathfrak{F}}M_+(\mathfrak{L}) \subset M_+(\mathfrak{F}) \quad \text{and} \quad P^{\mathfrak{L}_*}M_+(\mathfrak{F}) \subset M_+(\mathfrak{L}_*). \tag{1.15}$$

Observe now that (1.10) implies that for $f \in M(\mathfrak{F})$ and $l \in M(\mathfrak{L})$ we have

$$f = P^{\mathfrak{L}_*}f + P_{\mathfrak{R}_2}f, \quad l = P^{\mathfrak{F}}l + P_{\mathfrak{R}_1}l. \tag{1.16}$$

Let us choose in particular $f = P^{\mathfrak{F}}l$. The two relations (1.16) imply then

$$l = P^{\mathfrak{L}_*}P^{\mathfrak{F}}l + P_{\mathfrak{R}_1}l + P_{\mathfrak{R}_2}P^{\mathfrak{F}}l \quad (l \in M(\mathfrak{L})). \tag{1.17}$$

The first term of the right-hand side is an element of $M(\mathfrak{L}_*)$, and the sum of the two other terms is an element of $\mathfrak{R}_1 \oplus \mathfrak{R}_2 = \mathfrak{R}$. Because $M(\mathfrak{L}_*) \perp \mathfrak{R}$, we conclude for $l \in M(\mathfrak{L})$:

$$P^{\mathfrak{L}_*}l = P^{\mathfrak{L}_*}P^{\mathfrak{F}}l \tag{1.18}$$

$$P_{\mathfrak{R}}l = P_{\mathfrak{R}_1}l + P_{\mathfrak{R}_2}P^{\mathfrak{F}}l \tag{1.19}$$

On account of relation (II.2.1) defining $\mathfrak{R}$ and of relations (1.10) we have

$$P_{\mathfrak{R}}k = (I - P^{\mathfrak{L}_*})k \quad \text{for} \quad k \in \mathfrak{K}, \tag{1.20}$$

and

$$P_{\mathfrak{R}_1} l = (I - P^{\mathfrak{F}}) l \quad \text{for } l \in M(\mathfrak{L}), \qquad P_{\mathfrak{R}_2} f = (I - P^{\mathfrak{L}_*}) f \text{ for } f \in M(\mathfrak{F}). \tag{1.21}$$

Because

$$\overline{P_{\mathfrak{R}} M(\mathfrak{L})} = \overline{(I - P^{\mathfrak{L}_*}) M(\mathfrak{L})} = \mathfrak{R} \quad \text{(cf. (II.2.13))}, \tag{1.22}$$

relation (1.19) implies $\overline{P_{\mathfrak{R}_1} M(\mathfrak{L})} = \mathfrak{R}_1$ and $\overline{P_{\mathfrak{R}_2} P^{\mathfrak{F}} M(\mathfrak{L})} = \mathfrak{R}_2$, and hence *a fortiori*

$$\overline{P_{\mathfrak{R}_1} M(\mathfrak{L})} = \mathfrak{R}_1 \quad \text{and} \quad \overline{P_{\mathfrak{R}_2} M(\mathfrak{F})} = \mathfrak{R}_2. \tag{1.23}$$

2. If the space $\mathfrak{H}$ (and hence the space $\mathfrak{K}$ also) is separable, we can give the above relations a functional form, by using the Fourier representations $\Phi^{\mathfrak{L}_*}$, $\Phi^{\mathfrak{L}}$, and $\Phi^{\mathfrak{F}}$ of the spaces $M(\mathfrak{L}_*)$, $M(\mathfrak{L})$, and $M(\mathfrak{F})$ with respect to the bilateral shifts induced in these spaces by U. Namely, we choose the spaces and the contraction Q from one space into the other, in the three following ways (where, instead of the spaces themselves, we only indicate the generating spaces $\mathfrak{A}$ and $\mathfrak{A}'$).

(i) $\mathfrak{A} = \mathfrak{L}$, $\mathfrak{A}' = \mathfrak{L}_*$, $Q = P^{\mathfrak{L}_*} | M(\mathfrak{L})$ (case already considered in Sec. VI.2).
(ii) $\mathfrak{A} = \mathfrak{L}$, $\mathfrak{A}' = \mathfrak{F}$, $Q = P^{\mathfrak{F}} | M(\mathfrak{L})$.
(iii) $\mathfrak{A} = \mathfrak{F}$, $\mathfrak{A}' = \mathfrak{L}_*$, $Q = P^{\mathfrak{L}_*} | M(\mathfrak{F})$.

Q commutes with U in all three cases, because $M(\mathfrak{L}_*)$, $M(\mathfrak{L})$, and $M(\mathfrak{F})$ reduce U. Thus condition (V.3.5) is fulfilled. By virtue of (II.2.12) and of relations (1.15), condition (V.3.6) is also fulfilled.

Let

$$\{\mathfrak{L}, \mathfrak{L}_*, \Theta_{\mathfrak{L}}(\lambda)\}, \quad \{\mathfrak{L}, \mathfrak{F}, \Theta_1(\lambda)\}, \quad \{\mathfrak{F}, \mathfrak{L}_*, \Theta_2(\lambda)\} \tag{1.24}$$

be the corresponding contractive analytic functions, as in Lemma V.3.1. By virtue of Sec. VI.2 (Proposition VI.2.2), the first of these functions coincides with the characteristic function of T.

Relation (V.3.7) takes, respectively, the following forms:

$$\begin{cases} \Phi^{\mathfrak{L}_*} P^{\mathfrak{L}_*} l = \Theta_{\mathfrak{L}} \Phi^{\mathfrak{L}} l \text{ for } l \in M(\mathfrak{L}), \\ \Phi^{\mathfrak{F}} P^{\mathfrak{F}} l = \Theta_1 \Phi^{\mathfrak{L}} l \text{ for } l \in M(\mathfrak{L}), \\ \Phi^{\mathfrak{L}_*} P^{\mathfrak{L}_*} f = \Theta_2 \Phi^{\mathfrak{F}} f \text{ for } f \in M(\mathfrak{F}). \end{cases} \tag{1.25}$$

Let us apply $\Phi^{\mathfrak{L}_*}$ to both sides of (1.18). Because of (1.25) we obtain

$$\Theta_{\mathfrak{L}}(e^{it}) v(t) = \Theta_2(e^{it}) \Theta_1(e^{it}) v(t) \text{ (a.e.)} \quad \text{for} \quad v \in L^2(\mathfrak{L}),$$

in particular for every constant function $v(t) \equiv l$ ($l \in \mathfrak{L}$); as $\mathfrak{L}$ is separable this implies that $\Theta_{\mathfrak{L}}(e^{it}) = \Theta_2(e^{it}) \Theta_1(e^{it})$ (a.e.), and consequently

$$\Theta_{\mathfrak{L}}(\lambda) = \Theta_2(\lambda) \Theta_1(\lambda) \quad (\lambda \in D). \tag{1.26}$$

Using relations (1.20) and (1.22) we obtain, just as in Sec. VI.2.1, that there exists a unique unitary operator

$$\Phi_{\mathfrak{R}}\colon \mathfrak{R} \to \overline{\Delta_{\mathfrak{L}} L^2(\mathfrak{L})}, \tag{1.27}$$

such that

$$\Phi_{\mathfrak{R}} P_{\mathfrak{R}} l = \Delta_{\mathfrak{L}} \Phi^{\mathfrak{L}} l \quad (l \in M(\mathfrak{L})) \quad \text{(cf. (VI.2.6))}, \tag{1.28}$$

$\Delta_{\mathfrak{L}}$ denoting the operator generated on $L^2(\mathfrak{L})$ by the function

$$\Delta_{\mathfrak{L}}(t) = [I - \Theta_{\mathfrak{L}}(e^{it})^* \Theta_{\mathfrak{L}}(e^{it})]^{1/2}.$$

If we start with the relations (1.21) and (1.23) we obtain in the same way that there are two uniquely determined unitary operators,

$$\Phi_{\mathfrak{R}_1}\colon\ \mathfrak{R}_1 \to \overline{\Delta_1 L^2(\mathfrak{L})}, \quad \Phi_{\mathfrak{R}_2}\colon \mathfrak{R}_2 \to \overline{\Delta_2 L^2(\mathfrak{F})}, \tag{1.29}$$

such that

$$\Phi_{\mathfrak{R}_1} P_{\mathfrak{R}_1} l = \Delta_1 \Phi^{\mathfrak{L}} l \ (l \in M(\mathfrak{L})), \quad \Phi_{\mathfrak{R}_2} P_{\mathfrak{R}_2} f = \Delta_2 \Phi^{\mathfrak{F}} f \ (f \in M(\mathfrak{F})), \tag{1.30}$$

where the operators Δ_k are defined as multiplication by the functions

$$\Delta_k(t) = [I - \Theta_k(e^{it})^* \Theta_k(e^{it})]^{1/2} \quad (k = 1, 2).$$

Because $\mathfrak{R} = \mathfrak{R}_2 \oplus \mathfrak{R}_1$, the operator

$$Z = (\Phi_{\mathfrak{R}_2} \oplus \Phi_{\mathfrak{R}_1}) \Phi_{\mathfrak{R}}^{-1} \tag{1.31}$$

is also unitary; we have

$$Z\colon \overline{\Delta_{\mathfrak{L}} L^2(\mathfrak{L})} \to \overline{\Delta_2 L^2(\mathfrak{F})} \oplus \overline{\Delta_1 L^2(\mathfrak{L})}. \tag{1.32}$$

Combining relations (1.28), (1.19), (1.30), and (1.25) we obtain for $l \in M(\mathfrak{L})$:

$$\begin{aligned} Z \Delta_{\mathfrak{L}} \Phi^{\mathfrak{L}} l &= Z \Phi_{\mathfrak{R}} P_{\mathfrak{R}} l = (\Phi_{\mathfrak{R}_2} \oplus \Phi_{\mathfrak{R}_1}) P_{\mathfrak{R}} l = (\Phi_{\mathfrak{R}_2} \oplus \Phi_{\mathfrak{R}_1})(P_{\mathfrak{R}_2} P^{\mathfrak{F}} l \oplus P_{\mathfrak{R}_1} l) \\ &= \Phi_{\mathfrak{R}_2} P_{\mathfrak{R}_2} P^{\mathfrak{F}} l \oplus \Phi_{\mathfrak{R}_1} P_{\mathfrak{R}_1} l = \Delta_2 \Phi^{\mathfrak{F}} P^{\mathfrak{F}} l \oplus \Delta_1 \Phi^{\mathfrak{L}} l \\ &= \Delta_2 \Theta_1 \Phi^{\mathfrak{L}} l \oplus \Delta_1 \Phi^{\mathfrak{L}} l; \end{aligned}$$

as $\Phi^{\mathfrak{L}} l$ runs over $L^2(\mathfrak{L})$ it follows that

$$Z \Delta_{\mathfrak{L}} v = \Delta_2 \Theta_1 v \oplus \Delta_1 v \quad (v \in L^2(\mathfrak{L})). \tag{1.33}$$

Because Z is unitary, we conclude that

$$\overline{\{\Delta_2 \Theta_1 v \oplus \Delta_1 v\colon\ v \in L^2(\mathfrak{L})\}} = \overline{\Delta_2 L^2(\mathfrak{F})} \oplus \overline{\Delta_1 L^2(\mathfrak{L})}. \tag{1.34}$$

Observe also that Z commutes with multiplication by e^{it}. Indeed, it suffices to consider the elements $\Delta_{\mathfrak{L}} v$, which are dense in $\overline{\Delta_{\mathfrak{L}} L^2(\mathfrak{L})}$, and for these elements we have by (1.33)

$$Z(e^{it}\Delta_{\mathfrak{L}}v) = Z\Delta_{\mathfrak{L}}e^{it}v = \Delta_2\Theta_1 e^{it}v \oplus \Delta_1 e^{it}v = e^{it}\Delta_2\Theta_1 v \oplus e^{it}\Delta_1 v = e^{it}Z\Delta_{\mathfrak{L}}v.$$

Now consider the unitary operator

$$\Phi\colon \mathfrak{K} \to \mathbf{K} = L^2(\mathfrak{L}_*) \oplus \overline{\Delta_{\mathfrak{L}}L^2(\mathfrak{L})} \tag{1.35}$$

defined in Proposition VI.2.1 (i.e. the Fourier representation of $\mathfrak{K}$). By (VI.2.7) and (1.31) we have

$$\Phi = \Phi^{\mathfrak{L}_*} \oplus \Phi_{\mathfrak{R}} = \Phi^{\mathfrak{L}_*} \oplus Z^{-1}(\Phi_{\mathfrak{R}_2} \oplus \Phi_{\mathfrak{R}_1}). \tag{1.36}$$

By virtue of Proposition VI.2.1, Φ maps $\mathfrak{K}_+$ onto

$$\mathbf{K}_+ = H^2(\mathfrak{L}_*) \oplus \overline{\Delta_{\mathfrak{L}}L^2(\mathfrak{L})},$$

$M_+(\mathfrak{L})$ onto

$$\mathbf{G} = \{\Theta_{\mathfrak{L}}u \oplus \Delta_{\mathfrak{L}}u\colon u \in H^2(\mathfrak{L})\},$$

and $\mathfrak{H} = \mathfrak{K}_+ \ominus M_+(\mathfrak{L})$ onto

$$\mathbf{H} = \mathbf{K}_+ \ominus \mathbf{G} = [H^2(\mathfrak{L}_*) \oplus \overline{\Delta_{\mathfrak{L}}L^2(\mathfrak{L})}] \ominus \{\Theta_{\mathfrak{L}}u \oplus \Delta_{\mathfrak{L}}u\colon u \in H^2(\mathfrak{L})\}, \tag{1.37}$$

the contraction T being transformed into the contraction $\mathbf{T}$ defined by

$$\mathbf{T}^*(u_* \oplus v) = e^{-it}[u_*(e^{it}) - u_*(0)] \oplus e^{-it}v(t) \quad (u_* \oplus v \in \mathbf{H}).$$

Let us find the images $\mathbf{H}_1$ and $\mathbf{H}_2$ of the subspaces $\mathfrak{H}_1$ and $\mathfrak{H}_2$ by the Fourier representation $\Phi = \Phi^{\mathfrak{L}_*} \oplus \Phi_{\mathfrak{R}}$. We begin with $\mathfrak{H}_2$, for which we can use relation (1.6). Observe first that for $r_2 \in \mathfrak{R}_2$ we have

$$\Phi_{\mathfrak{R}}r_2 = Z^{-1}(\Phi_{\mathfrak{R}_2} \oplus \Phi_{\mathfrak{R}_1})(r_2 \oplus 0) = Z^{-1}(\Phi_{\mathfrak{R}_2}r_2 \oplus 0), \tag{1.38}$$

and hence

$$\Phi[M_+(\mathfrak{L}_*) \oplus \mathfrak{R}_2] = \Phi^{\mathfrak{L}_*}M_+(\mathfrak{L}_*) \oplus \Phi_{\mathfrak{R}}\mathfrak{R}_2 = H^2(\mathfrak{L}_*) \oplus Z^{-1}(\overline{\Delta_2 L^2(\mathfrak{F})} \oplus \{0\}).$$

On the other hand, the first relation (1.16) implies that

$$\Phi M_+(\mathfrak{F}) = \{\Phi^{\mathfrak{L}_*}P^{\mathfrak{L}_*}f \oplus \Phi_{\mathfrak{R}}P_{\mathfrak{R}_2}f\colon f \in M_+(\mathfrak{F})\},$$

whence, by virtue of (1.25), (1.30), and (1.38),

$$\Phi M_+(\mathfrak{F}) = \{\Theta_2 u \oplus Z^{-1}(\Delta_2 u \oplus 0)\colon u \in H^2(\mathfrak{F})\}.$$

Therefore, substituting in (1.6),

$$\begin{aligned}\mathbf{H}_2 =&[H^2(\mathfrak{L}_*)\oplus Z^{-1}(\overline{\Delta_2 L^2(\mathfrak{F})}\oplus\{0\})]\\ &\ominus\{\Theta_2 u\oplus Z^{-1}(\Delta_2 u\oplus 0)\colon u\in H^2(\mathfrak{F})\}.\end{aligned}\tag{1.39}$$

Finally, $\mathbf{H}_1=\mathbf{H}\ominus\mathbf{H}_2$ and $\overline{\Delta_{\mathfrak{L}}L^2(\mathfrak{L})}=Z^{-1}(\overline{\Delta_2 L^2(\mathfrak{F})}\oplus\overline{\Delta_1 L^2(\mathfrak{L})})$, thus (1.37) and (1.39) yield

$$\begin{aligned}\mathbf{H}_1 =&\{\Theta_2 u\oplus Z^{-1}(\Delta_2 u\oplus v)\colon\ u\in H^2(\mathfrak{F}), v\in\overline{\Delta_1 L^2(\mathfrak{L})}\}\\ &\ominus\{\Theta_{\mathfrak{L}}w\oplus\Delta_{\mathfrak{L}}w\colon\ w\in H^2(\mathfrak{L})\}.\end{aligned}\tag{1.40}$$

We note also that because the restriction of Φ, given by (1.35), to $\mathfrak{H}$ implements a unitary equivalence between the c.n.u. contraction T and the model operator $\mathbf{T}$ arising from the contractive analytic function $\{\mathfrak{L},\mathfrak{L}_*,\Theta_{\mathfrak{L}}(\lambda)\}$, it follows that every invariant subspace $\mathbf{H}_1$ of $\mathbf{T}$ is of the form (1.40) with an orthogonal complement $\mathbf{H}_2$ of the form (1.39), where $\Theta_{\mathfrak{L}}(\lambda)=\Theta_2(\lambda)\Theta_1(\lambda)$ is a regular factorization as defined below.

3. In order to give these results a compact form, let us make some remarks and formulate some definitions.

If $\Theta(\lambda)=\Theta_2(\lambda)\Theta_1(\lambda)$ is a factorization of a contractive analytic function $\{\mathfrak{E},\mathfrak{E}_*,\Theta(\lambda)\}$ into the product of contractive analytic functions $\{\mathfrak{E},\mathfrak{F},\Theta_1(\lambda)\}$ and $\{\mathfrak{F},\mathfrak{E}_*,\Theta_2(\lambda)\}$, then we obviously have

$$\begin{aligned}&I_{\mathfrak{E}}-\Theta(e^{it})^*\Theta(e^{it})\\ =&\Theta_1(e^{it})^*[I_{\mathfrak{F}}-\Theta_2(e^{it})^*\Theta_2(e^{it})]\Theta_1(e^{it})+[I_{\mathfrak{E}}-\Theta_1(e^{it})^*\Theta_1(e^{it})],\end{aligned}$$

and introducing the corresponding functions $\Delta(t)$, $\Delta_1(t)$, $\Delta_2(t)$ it follows that

$$Z(t)\colon\ \Delta(t)g\to\Delta_2(t)\Theta_1(e^{it})g\oplus\Delta_1(t)g\quad(g\in\mathfrak{E})\tag{1.41}$$

is an *isometry* of $\Delta(t)\mathfrak{E}$ into $\Delta_2(t)\mathfrak{F}\oplus\Delta_1(t)\mathfrak{E}$ a.e., indeed at every point such that the radial limits $\Theta(e^{it})$, $\Theta_1(e^{it})$, $\Theta_2(e^{it})$ exist. As a consequence,

$$Z\colon\ \Delta v\to\Delta_2\Theta_1 v\oplus\Delta_1 v\qquad(v\in L^2(\mathfrak{E}))\tag{1.42}$$

is an *isometry* of $\Delta L^2(\mathfrak{E})$ into $\Delta_2 L^2(\mathfrak{F})\oplus\Delta_1 L^2(\mathfrak{E})$. Completing by continuity we obtain *isometries* (denoted by the same letters):

$$Z(t)\colon\ \overline{\Delta(t)\mathfrak{E}}\to\overline{\Delta_2(t)\mathfrak{F}}\oplus\overline{\Delta_1(t)\mathfrak{E}}\quad\text{a.e., and}\tag{1.41'}$$

$$Z\colon\ \overline{\Delta L^2(\mathfrak{E})}\to\overline{\Delta_2 L^2(\mathfrak{F})}\oplus\overline{\Delta_1 L^2(\mathfrak{E})}.\tag{1.42'}$$

Definition. The factorization $\Theta(\lambda)=\Theta_2(\lambda)\Theta_1(\lambda)$ will be said to be *regular* if the operator Z that it generates (in the sense of (1.42) and (1.42′)) is unitary, that is, if

$$\overline{\{\Delta_2\Theta_1 u\oplus\Delta_1 u\colon u\in L^2(\mathfrak{E})\}}=\overline{\Delta_2 L^2(\mathfrak{F})}\oplus\overline{\Delta_1 L^2(\mathfrak{E})}.\tag{1.43}$$

By this definition and by (1.34), the factorization (1.26) is regular.

Now we formulate our results in terms of the functional model of c.n.u. contractions given by Theorem VI.3.1. Let **H** and **T** be the space and the c.n.u. contraction generated by the purely contractive analytic function $\{\mathfrak{E},\mathfrak{E}_*,\Theta(\lambda)\}$. The function $\{\mathfrak{L},\mathfrak{L}_*,\Theta_{\mathfrak{L}}(\lambda)\}$ corresponding to **T** coincides with $\{\mathfrak{E},\mathfrak{E}_*,\Theta(\lambda)\}$; in fact, each of them coincides with the characteristic function of **T**, (cf. Proposition VI.2.2 and Theorem VI.3.1). One concludes readily that part (a) of the following theorem holds.

Theorem 1.1. *Let $\{\mathfrak{E},\mathfrak{E}_*,\Theta(\lambda)\}$ be a purely contractive analytic function and let* **T** *be the contraction in the space*

$$\mathbf{H} = [H^2(\mathfrak{E}_*)\oplus\overline{\Delta L^2(\mathfrak{E})}]\ominus\{\Theta w\oplus\Delta w\colon w\in H^2(\mathfrak{E})\} \tag{1.44}$$

defined by

$$\mathbf{T}^*(u_*\oplus v) = e^{-it}[u_*(e^{it})-u_*(0)]\oplus e^{-it}v(t) \quad (u_*\oplus v\in\mathbf{H}). \tag{1.45}$$

(a) *To every subspace $\mathbf{H}_1$ of* **H**, *invariant for* **T**, *there corresponds a regular factorization*

$$\Theta(\lambda) = \Theta_2(\lambda)\Theta_1(\lambda) \tag{1.46}$$

of $\{\mathfrak{E},\mathfrak{E}_*,\Theta(\lambda)\}$ into the product of contractive analytic functions, $\{\mathfrak{E},\mathfrak{F},\Theta_1(\lambda)\}$ and $\{\mathfrak{F},\mathfrak{E}_*,\Theta_2(\lambda)\}$, such that if Z is the unitary operator $(1.42')$ corresponding to this factorization, then $\mathbf{H}_1$ and $\mathbf{H}_2=\mathbf{H}\ominus\mathbf{H}_1$ have the representations

$$\begin{aligned}\mathbf{H}_1 =&\{\Theta_2u\oplus Z^{-1}(\Delta_2u\oplus v)\colon u\in H^2(\mathfrak{F}), v\in\overline{\Delta_1L^2(\mathfrak{E})}\} &(1.47)\\ &\ominus\{\Theta w\oplus\Delta w\colon w\in H^2(\mathfrak{E})\},\\ \mathbf{H}_2 =&[H^2(\mathfrak{E}_*)\oplus Z^{-1}(\overline{\Delta_2L^2(\mathfrak{F})}\oplus\{0\})] &(1.48)\\ &\ominus\{\Theta_2u\oplus Z^{-1}(\Delta_2u\oplus 0)\colon u\in H^2(\mathfrak{F})\}.\end{aligned}$$

(b) *Every regular factorization of $\Theta(\lambda)$ generates in this way a subspace $\mathbf{H}_1$ of* **H**, *invariant for* **T**, *and its orthogonal complement $\mathbf{H}_2$.*

Proof. It only remains to prove (b). Observe that because Z is unitary, we have

$$\begin{aligned}\mathbf{G}_2 \equiv&\{\Theta_2u\oplus Z^{-1}(\Delta_2u\oplus v)\colon u\in H^2(\mathfrak{F}), v\in\overline{\Delta_1L^2(\mathfrak{E})}\}\\ &\supset\{\Theta_2\Theta_1w\oplus Z^{-1}(\Delta_2\Theta_1w\oplus\Delta_1w)\colon w\in H^2(\mathfrak{E})\}\\ =&\{\Theta w\oplus\Delta w\colon w\in H^2(\mathfrak{E})\}\equiv\mathbf{G}\end{aligned}$$

and

$$\begin{aligned}&[H^2(\mathfrak{E}_*)\oplus\overline{\Delta L^2(\mathfrak{E})}]\ominus\mathbf{G}_2 = [H^2(\mathfrak{E}_*)\oplus Z^{-1}(\overline{\Delta_2L^2(\mathfrak{F})}\oplus\overline{\Delta_1L^2(\mathfrak{E})})]\ominus\mathbf{G}_2\\ =&[H^2(\mathfrak{E}_*)\oplus Z^{-1}(\overline{\Delta_2L^2(\mathfrak{F})}\oplus\{0\})]\ominus\{\Theta_2u\oplus Z^{-1}(\Delta_2u\oplus 0)\colon\ u\in H^2(\mathfrak{F})\};\end{aligned}$$

the subspaces $\mathbf{H}_1, \mathbf{H}_2$ defined by (1.47) and (1.48) can therefore be expressed as

$$\mathbf{H}_1 = \mathbf{G}_2 \ominus \mathbf{G}, \quad \mathbf{H}_2 = [H^2(\mathfrak{E}_*) \oplus \overline{\Delta L^2(\mathfrak{E})}] \ominus \mathbf{G}_2.$$

On account of (1.44) it follows that $\mathbf{H} = \mathbf{H}_1 \oplus \mathbf{H}_2$.

From (1.48), the elements of $\mathbf{H}_2$ are the orthogonal sums $u_* \oplus Z^{-1}(v \oplus 0)$ with $u_* \in H^2(\mathfrak{E}_*)$ and $v \in \overline{\Delta_2 L^2(\mathfrak{F})}$ such that $\Theta_2^* u_* + \Delta_2 v \perp H^2(\mathfrak{F})$. The same conditions are then satisfied by $\bar{u}_* \oplus Z^{-1}(\bar{v} \oplus 0)$ with $\bar{u}_*(\lambda) = [u_*(\lambda) - u_*(0)]/\lambda$ and $\bar{v}(t) = e^{-it} v(t)$, indeed we have

$$\Theta_2^* \bar{u}_* + \Delta_2 \bar{v} = e^{-it}(\Theta_2^* u_* + \Delta_2 v) - e^{-it} \Theta_2(e^{it})^* u_*(0) \perp H^2(\mathfrak{F}).$$

Because Z commutes with multiplication by $e^{\pm it}$, it follows that $\mathbf{T}^* \mathbf{H}_2 \subset \mathbf{H}_2$. Hence $\mathbf{T} \mathbf{H}_1 \subset \mathbf{H}_1$. This concludes the proof of Theorem 1.1.

2 Some additional propositions

1. We begin with an addition to Theorem 1.1.

Proposition 2.1. *Under the conditions of Theorem* 1.1, *let*

$$\mathbf{T} = \begin{bmatrix} \mathbf{T}_1 & \mathbf{X} \\ \mathbf{O} & \mathbf{T}_2 \end{bmatrix}$$

be the triangulation of $\mathbf{T}$ *corresponding to the decomposition* $\mathbf{H} = \mathbf{H}_1 \oplus \mathbf{H}_2$ *induced by the regular factorization* $\Theta(\lambda) = \Theta_2(\lambda)\Theta_1(\lambda)$ *of* $\{\mathfrak{E}, \mathfrak{E}_*, \Theta(\lambda)\}$. *Then the characteristic functions of* $\mathbf{T}_1$ *and* $\mathbf{T}_2$ *coincide with the purely contractive parts of* $\Theta_1(\lambda)$ *and* $\Theta_2(\lambda)$, *respectively.*

Proof. Let us note first that

$$\mathbf{T}_2^* = \mathbf{T}^* | \mathbf{H}_2 \quad \text{and} \quad \mathbf{T}_1^* = \mathbf{P}_1 \mathbf{T}^* | \mathbf{H}_1,$$

where $\mathbf{P}_1$ denotes the orthogonal projection from $\mathbf{H}$ onto $\mathbf{H}_1$.

It is obvious that the operator Y defined by

$$Y\colon u_* \oplus Z^{-1}(v \oplus 0) \to u_* \oplus v \quad (u_* \in H^2(\mathfrak{E}_*), v \in \overline{\Delta_2 L^2(\mathfrak{F})}) \tag{2.1}$$

maps the space $H^2(\mathfrak{E}_*) \oplus Z^{-1}(\overline{\Delta_2 L^2(\mathfrak{F})} \oplus \{0\})$ unitarily onto $H^2(\mathfrak{E}_*) \oplus \overline{\Delta_2 L^2(\mathfrak{F})}$, and the subspace $\mathbf{H}_2$ of the first space onto the subspace

$$\mathscr{H}_2 = [H^2(\mathfrak{E}_*) \oplus \overline{\Delta_2 L^2(\mathfrak{F})}] \ominus \{\Theta_2 u \oplus \Delta_2 u \colon u \in H^2(\mathfrak{F})\} \tag{2.2}$$

of the second one. As Z commutes with multiplication by e^{it}, $\mathbf{T}^* | \mathbf{H}_2$ is transformed by Y to the operator $\mathscr{T}_2^*$ defined on $\mathscr{H}_2$ by

$$\mathscr{T}_2^*(u_* \oplus v_2) = e^{-it}[u_* - u_*(0)] \oplus e^{-it} v_2 \quad (u_* \oplus v_2 \in \mathscr{H}_2). \tag{2.3}$$

By Proposition VI.3.2(b) it follows from (2.2) and (2.3) that the characteristic function of $\mathscr{T}_2$ (hence that of $\mathbf{T}_2$ also) coincides with the purely contractive part of $\{\mathfrak{F}, \mathfrak{E}_*, \Theta_2(\lambda)\}$.

As regards $\mathbf{T}_1$, let us observe first that, because of the definition (1.47) of $\mathbf{H}_1$ and because $\Theta w \oplus \Delta w = \Theta_2\Theta_1 w \oplus Z^{-1}(\Delta_2\Theta_1 w \oplus \Delta_1 w)$, the elements of $\mathbf{H}$ belonging to $\mathbf{H}_1$ are those of the form $\Theta_2 u \oplus Z^{-1}(\Delta_2 u \oplus v)$ with $u \in H^2(\mathfrak{F})$, $v \in \overline{\Delta_1 L^2(\mathfrak{E})}$, and

$$\Theta_1^*\Theta_2^*\Theta_2 u + \Theta_1^*\Delta_2^2 u + \Delta_1 v \perp H^2(\mathfrak{E}); \tag{2.4}$$

the last condition obviously reduces to

$$\Theta_1^* u + \Delta_1 v \perp H^2(\mathfrak{E}). \tag{2.5}$$

For the elements of $\mathbf{H}_1$ we have

$$\begin{aligned}\mathbf{T}^*(\Theta_2 u \oplus Z^{-1}(\Delta_2 u \oplus v)) &= e^{-it}[\Theta_2 u - \Theta_2(0)u(0)] \oplus Z^{-1}(e^{-it}\Delta_2 u \oplus e^{-it} v)\\ &= [\Theta_2 u_1 \oplus Z^{-1}(\Delta_2 u_1 \oplus v_1)] + [u_2 \oplus Z^{-1}(v_2 \oplus 0)]\end{aligned}$$

with

$$\begin{aligned} u_1(\lambda) &= \frac{1}{\lambda}[u(\lambda) - u(0)], & u_2(\lambda) &= \frac{1}{\lambda}[\Theta_2(\lambda) - \Theta_2(0)]u(0),\\ v_1(t) &= e^{-it} v(t), & v_2(t) &= e^{-it}\Delta_2(t)u(0).\end{aligned}$$

On account of (2.5) we have

$$\Theta_1^* u_1 + \Delta_1 v_1 = e^{-it}(\Theta_1^* u + \Delta_1 v) - e^{-it}\Theta_1^* u(0) \perp H^2(\mathfrak{E}),$$

and hence $\Theta_2 u_1 \oplus Z^{-1}(\Delta_2 u_1 \oplus v_1) \in \mathbf{H}_1$. On the other hand, $u_2 \oplus Z^{-1}(v_2 \oplus 0) \in \mathbf{H}_2$ because, obviously,

$$\Theta_2^* u_2 + \Delta_2 v_2 = e^{-it} u(0) - e^{-it}\Theta_2^*\Theta_2(0)u(0) \perp H^2(\mathfrak{F}).$$

We conclude that

$$\mathbf{P}_1\mathbf{T}^*[\Theta_2 u \oplus Z^{-1}(\Delta_2 u \oplus v)] = \Theta_2 u_1 \oplus Z^{-1}(\Delta_2 u_1 \oplus v_1). \tag{2.6}$$

It is obvious that the operator W defined by

$$W\colon \Theta_2 u \oplus Z^{-1}(\Delta_2 u \oplus v) \to u \oplus v \quad (u \in H^2(\mathfrak{F}), v \in \overline{\Delta_1 L^2(\mathfrak{E})}) \tag{2.7}$$

maps the set of elements $\Theta_2 u \oplus Z^{-1}(\Delta_2 u \oplus v)$ *unitarily* onto $H^2(\mathfrak{F}) \oplus \overline{\Delta_1 L^2(\mathfrak{E})}$, and that for $w \in H^2(\mathfrak{E})$ we have

$$W(\Theta w \oplus \Delta w) = W(\Theta_2\Theta_1 w \oplus Z^{-1}(\Delta_2\Theta_1 w \oplus \Delta_1 w)) = \Theta_1 w \oplus \Delta_1 w.$$

It follows that the space $\mathbf{H}_1$ defined by (1.47) is mapped by W onto the space

$$\mathscr{H}_1 = [H^2(\mathfrak{F}) \oplus \overline{\Delta_1 L^2(\mathfrak{E})}] \ominus \{\Theta_1 w \oplus \Delta_1 w \colon w \in H^2(\mathfrak{E})\}. \tag{2.8}$$

By virtue of (2.6) the operator $\mathbf{T}_1^* (= \mathbf{P}_1 \mathbf{T}^* | \mathbf{H}_1)$ is transformed by W to the operator $\mathscr{T}_1^*$ defined on $\mathscr{H}_1$ by

$$\mathscr{T}_1^*(u \oplus v) = u_1 \oplus v_1 = e^{-it}[u - u(0)] \oplus e^{-it} v \quad (u \oplus v \in \mathscr{H}_1). \tag{2.9}$$

From (2.8) and (2.9) it follows by Proposition VI.3.2(b) that the characteristic function of $\mathscr{T}_1$ (hence that of $\mathbf{T}_1$ also) coincides with the purely contractive part of $\{\mathfrak{E}, \mathfrak{F}, \Theta_1(\lambda)\}$. This concludes the proof of Proposition 2.1.

We add that, as a consequence of formulas (2.8), (2.2), and of Proposition VI.3.2(a), $\mathscr{H}_1$ and $\mathscr{H}_2$ (and hence also $\mathbf{H}_1$ and $\mathbf{H}_2$) do not reduce to $\{0\}$ unless $\Theta_1(\lambda)$ or $\Theta_2(\lambda)$, respectively, is a unitary constant. Thus Theorem 1.1 can be completed as follows.

Theorem 2.2. *In order that the invariant subspace $\mathbf{H}_1$ generated by the regular factorization $\Theta(\lambda) = \Theta_2(\lambda)\Theta_1(\lambda)$ be nontrivial, it is necessary and sufficient that this factorization be nontrivial (i.e., neither factor be a unitary constant).*

Every c.n.u. contraction T is unitarily equivalent to the contraction $\mathbf{T}$ generated by the characteristic function of T, thus the preceding two results yield the following theorem.

Theorem 2.3. *In order that the completely nonunitary contraction T on a (separable) Hilbert space have a nontrivial invariant subspace, it is necessary and sufficient that the characteristic function of T admit a nontrivial regular factorization.*

2. As instructive examples, let us discuss the peculiar factorizations presented at the beginning of Sec. V.5. Recall that if $\{\mathfrak{E}, \mathfrak{E}_*, \Theta(\lambda)\}$ is a contractive analytic function and V is a nonunitary isometric operator from $\mathfrak{E}_*$ into a Hilbert space $\mathfrak{F}$, then the factorization $\Theta(\lambda) = \Theta_2(\lambda)\Theta_1(\lambda)$, where $\Theta_1(\lambda) = V\Theta(\lambda)$ and $\Theta_2(\lambda) = V^*$, is nontrivial. In this case, the operator-valued functions Δ_1 and Δ_2 defined in Sec. 1.3 are Δ and $I - VV^*$, respectively. It follows that

$$\begin{aligned}\overline{\{\Delta_2\Theta_1 u \oplus \Delta_1 u \colon u \in L^2(\mathfrak{E})\}} &= \overline{\{0 \oplus \Delta u \colon u \in L^2(\mathfrak{E})\}} = \{0\} \oplus \overline{\Delta L^2(\mathfrak{E})} \\ &\neq L^2((I - VV^*)\mathfrak{F}) \oplus \overline{\Delta L^2(\mathfrak{E})} = \\ &= \overline{\Delta_2 L^2(\mathfrak{F})} \oplus \overline{\Delta_1 L^2(\mathfrak{E})}\end{aligned}$$

and thus the factorization is not regular.

Recall also that if $\mathfrak{N}(\Theta^\sim) \neq \{0\}$ (see again Sec. V.5) then $\Theta(\lambda) = (I - P_* + \lambda P_*)\Theta(\lambda)$, where P_* denotes the orthogonal projection of $\mathfrak{E}_*$ onto $\mathfrak{N}(\Theta^\sim)$, is a nontrivial factorization. In this case $\Theta_2(\lambda) = I - P_* + \lambda P_*$, $\Theta_1(\lambda) = \Theta(\lambda)$, $\Delta_2 = 0$ and $\Delta_1 = \Delta$. Clearly, this factorization is regular. It is worth observing that in view of Proposition 2.1 the restriction $\mathbf{T}_1$ of $\mathbf{T}$ to the invariant subspace $\mathbf{H}_1$ corresponding to this factorization is unitarily equivalent to $\mathbf{T}$; moreover the compression $\mathbf{T}_2$ of $\mathbf{T}$ to the space $\mathbf{H}_2 = \mathbf{H} \ominus \mathbf{H}_1$ is the null operator. Note that the other two related examples in Sec. V.5 can be reduced to the ones already treated above by considering $\Theta^{\tilde{}}(\lambda)$ instead of $\Theta(\lambda)$ and by referring to the general Proposition 3.2 in the next section.

3. Another natural question related to Sec. V.5 is the relation between regular and strictly nontrivial factorizations. A partial answer is given by the following result.

Proposition 2.4. (a) *There exist nontrivial regular factorizations that are not strictly nontrivial.*

(b) *There exist strictly nontrivial factorizations that are not regular.*

Proof. Part (a): Take $\mathfrak{E}_* = \mathfrak{E} \oplus \mathfrak{E}$ and $\Theta(\lambda)e = e \oplus 0$, $e \in \mathfrak{E}$. Then

$$\Theta(\lambda) = \begin{bmatrix} I \\ 0 \end{bmatrix} = \begin{bmatrix} I & 0 \\ 0 & \lambda \end{bmatrix} \begin{bmatrix} I \\ 0 \end{bmatrix} = \Theta_2(\lambda)\Theta_1(\lambda) \ \text{ with } \Theta_1(\lambda) = \Theta(\lambda).$$

Note that $\Theta(\lambda)$ is a constant (in λ) nonunitary isometry. Therefore, as we pointed out in Sec. V.5, $\Theta(\lambda)$ has no strictly nontrivial factorization. Plainly, the above factorization is regular.

Part (b): Let $\Theta(\lambda) = \Theta_2(\lambda)\Theta_1(\lambda)$ be the strictly nontrivial factorization presented in Theorem V.5.5. Recall that Θ and Θ_1 are inner and that Θ_2 is outer. Therefore $\Delta = 0$, $\Delta_1 = 0$ and consequently (see Sec. 1.3)

$$\overline{\{\Delta_2\Theta_1 u \oplus \Delta_1 u \colon u \in L^2(\mathfrak{E})\}} = Z\{0\} = \{0\} \oplus \{0\},$$
$$\overline{\Delta_2 L^2(\mathfrak{E})} \oplus \overline{\Delta_1 L^2(\mathfrak{E})} = \overline{\Delta_2 L^2(\mathfrak{E})} \oplus \{0\}.$$

If this factorization is regular then we must have $\overline{\Delta_2 L^2(\mathfrak{E})} = \{0\}$, which is only possible if $\Delta_2(t) = 0$ a.e., that is, if Θ_2 is inner. Because Θ_2 is also outer it follows that Θ_2 is a constant unitary operator contradicting the fact that the factorization is strictly nontrivial.

4. Let us make a remark concerning the Beurling, Lax, and Halmos theorem (see Theorem V.3.3). Recall that this theorem states that if $U_+^\times$ is the unilateral shift on $H^2(\mathfrak{E})$ (i.e., $(U_+^\times h)(\lambda) = \lambda h(\lambda)$, $|\lambda| < 1$, $h \in H^2(\mathfrak{E})$) then all its invariant subspaces $\mathbf{H}_1$ are of the form $\Theta_2 H^2(\mathfrak{F})$ where $\{\mathfrak{F}, \mathfrak{E}, \Theta_2(\lambda)\}$ is some inner function. To derive this fact within the present approach observe first that the characteristic function of $U_+^\times$ is $\{\{0\}, \mathfrak{E}, \Theta(\lambda)\}$, where $\Theta(\lambda) \equiv 0$. The regular factorizations of this function are of the form

$$\Theta(\lambda) = 0 = \Theta_2(\lambda)\Theta_1(\lambda),$$

where $\{\mathfrak{F}, \mathfrak{E}, \Theta_2(\lambda)\}$ is inner and $\Theta_1(\lambda)$ is the zero operator from $\{0\}$ to $\mathfrak{F}$. It is easy to check that the formula (1.47) yields the invariant subspace $\mathbf{H}_1 = \Theta_2 H^2(\mathfrak{F})$.

5. Finally, we point out the existence of *strange factorizations*. A factorization $\Theta(\lambda) = \Theta_2(\lambda)\Theta_1(\lambda)$ is said to be *strange* if it is not regular, but there exists a *regular* factorization $\Theta(\lambda) = \Theta_2'(\lambda)\Theta_1'(\lambda)$ such that the pure part of Θ_j coincides with the pure part of Θ_j' for $j = 1,2$. Consider a separable, infinite-dimensional Hilbert space $\mathfrak{E}$, and two isometries U, V on $\mathfrak{E}$ with orthogonal ranges; thus $V^*U = O$. We have then $\Theta(\lambda) = \Theta_2(\lambda)\Theta_1(\lambda)$, where $\{\mathfrak{E}, \mathfrak{E}, \Theta(\lambda)\}$, $\{\mathfrak{E}, \mathfrak{E}, \Theta_1(\lambda)\}$, and

$\{\mathfrak{E}, \mathfrak{E}, \Theta_2(\lambda)\}$ are defined by $\Theta(\lambda) = O$, $\Theta_1(\lambda) = U$, and $\Theta_2(\lambda) = V^*$ for $\lambda \in D$. The pure part of Θ_1 is $\{\{0\}, \mathfrak{E} \ominus U\mathfrak{E}, O\}$, and the pure part of Θ_2 is $\{\mathfrak{E} \ominus V\mathfrak{E}, \{0\}, O\}$. Observe that

$$\overline{\Delta_2 L^2(\mathfrak{E})} \oplus \overline{\Delta_1 L^2(\mathfrak{E})} = L^2((I - VV^*)\mathfrak{E}) \oplus \{0\} = L^2((V\mathfrak{E})^{\perp}) \oplus \{0\}$$

and

$$\overline{\{\Delta_2 \Theta_1 u \oplus \Delta_1 u : u \in L^2(\mathfrak{E})\}} = L^2((I - VV^*)U\mathfrak{E}) \oplus \{0\} = L^2(U\mathfrak{E}) \oplus \{0\}.$$

We conclude that this factorization is regular if and only if $U\mathfrak{E} + V\mathfrak{E} = \mathfrak{E}$. Note that the pure part of Θ_1 coincides with $\{\{0\}, \mathfrak{E}, O\}$, and similarly the pure part of Θ_2 does not depend on V up to coincidence. Thus, these pure parts coincide with the pure parts of the factors of a regular factorization of Θ obtained when $U\mathfrak{E} + V\mathfrak{E} = \mathfrak{E}$. Therefore the factorization is strange if this equality does not hold. In either case we can construct, as in the proof of Theorem 1.1, spaces

$$\begin{aligned}
\mathbf{H} &= H^2(\mathfrak{E}) \oplus [L^2(\ker V^*) \ominus H^2(U\mathfrak{E})], \\
\mathbf{H}_1 &= H^2(\mathfrak{E}) \oplus [H^2(\ker V^*) \ominus H^2(U\mathfrak{E})], \\
\mathbf{H}_2 &= \{0\} \oplus [L^2(\ker V^*) \ominus H^2(\ker V^*)]
\end{aligned}$$

satisfying $\mathbf{H} = \mathbf{H}_1 \oplus \mathbf{H}_2$, and an operator $\mathbf{T}$ on $\mathbf{H}$ defined by

$$\mathbf{T}^*(f \oplus g)(e^{it}) = e^{-it}[f(e^{it}) - f(0)] \oplus e^{-it} g(e^{it}) \quad (e^{it} \in C, f \oplus g \in \mathbf{H}).$$

It is easy to see that, relative to the decomposition $\mathbf{H} = \mathbf{H}_1 \oplus \mathbf{H}_2$, we have the matrix representation

$$\mathbf{T} = \begin{bmatrix} \mathbf{T}_1 & \mathbf{X} \\ \mathbf{O} & \mathbf{T}_2 \end{bmatrix},$$

where $\mathbf{T}_1$ and $\mathbf{T}_2^*$ are unilateral shifts of infinite multiplicity. Observe that the unitary summand of $\mathbf{T}$ is multiplication by e^{it} on the space $\{0\} \oplus L^2(\ker V^* \ominus U\mathfrak{E})$. Hence $\mathbf{T}$ is completely nonunitary if and only if $U\mathfrak{E} + V\mathfrak{E} = \mathfrak{E}$, and in this case $\mathbf{T} = \mathbf{T}_1 \oplus \mathbf{T}_2$. Even when $U\mathfrak{E} + V\mathfrak{E} \neq \mathfrak{E}$, the completely nonunitary summand of $\mathbf{T}$ is unitarily equivalent to $\mathbf{T}_1 \oplus \mathbf{T}_2$, but different from it because $\mathbf{X} \neq \mathbf{O}$ in this case. (See also the notes concerning the general case of nonregular factorizations.)

3 Regular factorizations

1. Before continuing the investigation of the relations between invariant subspaces and regular factorizations, we need a closer look at the notion of regular factorization.

Beside the notion of regularity of a factorization of a *contractive analytic function*, defined in Sec. 1, we also introduce the notion of regularity of a factorization of an *individual contraction*.

Let $A = A_2A_1$ be an arbitrary factorization of a contraction A from a space $\mathfrak{A}$ into a space $\mathfrak{A}^*$, as a product of a contraction A_1 from $\mathfrak{A}$ into a space $\mathfrak{B}$ and of a contraction A_2 from $\mathfrak{B}$ into $\mathfrak{A}_*$. From the obvious relation

$$I_{\mathfrak{A}} - A^*A = A_1^*(I_{\mathfrak{B}} - A_2^*A_2)A_1 + (I_{\mathfrak{A}} - A_1^*A_1)$$

we infer that the transformation

$$Z\colon Da \to D_2A_1a \oplus D_1a \quad (a \in \mathfrak{A}) \tag{3.1}$$

is *isometric*; D, D_1, and D_2 denote here the defect operators:

$$D = D_A, \quad D_1 = D_{A_1}, \quad D_2 = D_{A_2}.$$

Completing by continuity we obtain an isometry (denoted by the same letter)

$$Z\colon \overline{D\mathfrak{A}} \to \overline{D_2\mathfrak{B}} \oplus \overline{D_1\mathfrak{A}}. \tag{3.2}$$

Definition. The factorization $A = A_2A_1$ is said to be *regular* if the corresponding operator Z is (not only isometric, but also) *unitary*, that is, if

$$\overline{\{D_2A_1a \oplus D_1a\colon a \in \mathfrak{A}\}} = \overline{D_2\mathfrak{B}} \oplus \overline{D_1\mathfrak{A}}. \tag{3.3}$$

Observe that there is an immediate relation between the two notions introduced above: *regularity of the factorization $\Theta(\lambda) = \Theta_2(\lambda)\Theta_1(\lambda)$ for contractive analytic functions means regularity of the factorization $\Theta = \Theta_2\Theta_1$ for the contractions on the respective L^2 spaces.*

A less immediate relation is established in the following proposition ("local characterization" of regular factorizations of functions).

Proposition 3.1. *In order that the (functional) factorization*

$$\Theta(\lambda) = \Theta_2(\lambda)\Theta_1(\lambda) \quad (|\lambda| < 1) \tag{3.4}$$

be regular it is necessary and sufficient that the (individual) factorization

$$\Theta(e^{it}) = \Theta_2(e^{it})\Theta_1(e^{it}) \quad (0 \le t \le 2\pi) \tag{3.5}$$

be regular a.e.

Proof. We set

$$\mathfrak{B} = \{\Delta_2\Theta_1 v \oplus \Delta_1 v\colon v \in L^2(\mathfrak{E})\}$$

and

$$\mathfrak{B}(t) = \{\Delta_2(t)\Theta_1(e^{it})e \oplus \Delta_1(t)e\colon e \in \mathfrak{E}\}.$$

Our proposition means then that the following two conditions are equivalent,

$$\overline{\mathfrak{B}} = \overline{\Delta_2 L^2(\mathfrak{F})} \oplus \overline{\Delta_1 L^2(\mathfrak{E})}, \tag{3.6}$$

$$\overline{\mathfrak{B}(t)} = \overline{\Delta_2(t)\mathfrak{F}} \oplus \overline{\Delta_1(t)\mathfrak{E}} \quad \text{a.e.}, \tag{3.7}$$

the closures being taken in $L^2(\mathfrak{F}) \oplus L^2(\mathfrak{E})$ and in $\mathfrak{F} \oplus \mathfrak{E}$, respectively.

Let us suppose first that (3.6) holds. Considering in particular two constant functions: $e(t) \equiv e$, $f(t) \equiv f$ $(e \in \mathfrak{E}, f \in \mathfrak{F})$, we deduce from (3.6) that there exists a sequence $\{v_n\}$ of elements of $L^2(\mathfrak{E})$ such that $\Delta_2\Theta_1 v_n \oplus \Delta_1 v_n$ tends to $\Delta_2 f \oplus \Delta_1 e$ in $L^2(\mathfrak{F}) \oplus L^2(\mathfrak{E})$; replacing this sequence if necessary by a suitable subsequence, we can also require that $\Delta_2(t)\Theta_1(e^{it})v_n(t) \oplus \Delta_1(t)v_n(t)$ tends to $\Delta_2(t)f \oplus \Delta_1(t)e$ a.e. (in the metric of $\mathfrak{F} \oplus \mathfrak{E}$). Thus $\Delta_2(t)f \oplus \Delta_1(t)e \in \overline{\mathfrak{B}(t)}$ a.e., the set of the exceptional points t depending on e and f. Let e and f run through two countable sets, say $\{e_n\}$ and $\{f_m\}$, dense in $\mathfrak{E}$ and in $\mathfrak{F}$, respectively. Taking the union of the corresponding exceptional sets we obtain that $\Delta_2(t)f_m \oplus \Delta_1(t)e_n \in \overline{\mathfrak{B}(t)}$ holds for every point t with the possible exception of a set of zero measure, independent of m and n. For the nonexceptional t we have thus

$$\overline{\Delta_2(t)\mathfrak{F}} \oplus \overline{\Delta_1(t)\mathfrak{E}} \subset \overline{\mathfrak{B}(t)},$$

and this implies (3.7).

Let us suppose now, conversely, that (3.7) holds a.e. Consider an element

$$v_2 \oplus v_1 \in \overline{\Delta_2 L^2(\mathfrak{F})} \oplus \overline{\Delta_1 L^2(\mathfrak{E})} \tag{3.8}$$

that is orthogonal to $\mathfrak{B}$, thus

$$(v_2, \Delta_2\Theta_1 v) + (v_1, \Delta_1 v) = 0 \quad \text{for each} \quad v \in L^2(\mathfrak{E}).$$

Then we have

$$\Theta_1(e^{it})^*\Delta_2(t)v_2(t) + \Delta_1(t)v_1(t) = 0 \quad \text{a.e.},$$

and consequently, for the nonexceptional t,

$$v_2(t) \oplus v_1(t) \perp \mathfrak{B}(t). \tag{3.9}$$

Now it is obvious that condition (3.8) implies

$$v_2(t) \oplus v_1(t) \in \overline{\Delta_2(t)\mathfrak{F}} \oplus \overline{\Delta_1(t)\mathfrak{E}} \quad \text{a.e.} \tag{3.10}$$

From (3.7), (3.9), and (3.10) we deduce that $v_2(t) \oplus v_1(t) = 0$ a.e. and thus $v_2 \oplus v_1 = 0$. This implies (3.6).

2. The following proposition lists some useful properties of regular factorizations of contractions.

Proposition 3.2. *Let $A = A_2A_1$ be a factorization of the contraction A (from $\mathfrak{A}$ into $\mathfrak{A}_*$) as a product of a contraction A_1 (from $\mathfrak{A}$ into $\mathfrak{B}$) and a contraction A_2 (from $\mathfrak{B}$ into $\mathfrak{A}_*$).*

(a) *If this factorization is regular then so is the factorization $A^* = A_1^*A_2^*$.*
(b) *The factorization $A = A_2A_1$ is regular whenever A_2 or A_1^* is isometric.*
(c) *If A is isometric (unitary) then the factorization $A = A_2A_1$ is regular if and only if A_1 and A_2 are isometric (unitary) also.*
(d) *We always have*

$$\dim \overline{D\mathfrak{A}} \leq \dim \overline{D_2\mathfrak{B}} + \dim \overline{D_1\mathfrak{A}}. \tag{3.11}$$

Equality holds for regular factorizations and, if $\dim \overline{D\mathfrak{A}} < \infty$, only for regular factorizations.

Proof. *Part* (a): The proof of this assertion is rather laborious. Let Z_* denote the (isometric) transformation generated by the factorization $A^* = A_1^*A_2^*$ (with the order of the component spaces reversed), that is,

$$Z_* : \overline{D_*\mathfrak{A}_*} \to \overline{D_{*2}\mathfrak{A}_*} \oplus \overline{D_{*1}\mathfrak{B}} \tag{3.12}$$

is obtained by completing the transformation

$$Z_* : D_*a_* \to D_{*2}a_* \oplus D_{*1}A_2^*a_* \quad (a_* \in \mathfrak{A}_*). \tag{3.13}$$

Here we have set $D_* = D_{A^*}$, $D_{*1} = D_{A_1^*}$, $D_{*2} = D^*_{A_2}$.

Let us set $S = Z_*AZ^*$. We have then for $a \in \mathfrak{A}$:

$$\begin{aligned} S(D_2A_1a \oplus D_1a) &= Z_*AZ^*ZDa = Z_*ADa = Z_*D_*Aa \\ &= D_{*2}Aa \oplus D_{*1}A_2^*Aa = D_{*2}A_2A_1a \oplus D_{*1}A_2^*A_2A_1a \\ &= A_2D_2A_1a \oplus [A_1D_1a - D_{*1}D_2(D_2A_1a)] \end{aligned}$$

and hence

$$S(u_2 \oplus u_1) = A_2u_2 \oplus (A_1u_1 - D_{*1}D_2u_2) \tag{3.14}$$

for $u_2 \oplus u_1 \in Z(D\mathfrak{A})$. Suppose that the factorization $A = A_2A_1$ is regular. The validity of (3.14) extends then by continuity to all elements $u_2 \oplus u_1$ of $\overline{D_2\mathfrak{B}} \oplus \overline{D_1\mathfrak{A}}$.

We recall that the relation

$$\mathfrak{D}_{T^*} = \overline{T\mathfrak{D}_T} \oplus \mathfrak{N}_{T^*} \quad \text{(cf. (I.3.7))} \tag{3.15}$$

holds for every contraction T from a space into itself or into some other space. This gives, for $T = A$,

$$\overline{D_*\mathfrak{A}_*} = \overline{AD\mathfrak{A}} \oplus \mathfrak{N}_{A^*};$$

from this we get, using (3.14),

$$
\begin{aligned}
Z_*\overline{D_*\mathfrak{A}_*} &= \overline{Z_*AD\mathfrak{A}} \oplus Z_*\mathfrak{N}_{A^*} = \overline{SZD\mathfrak{A}} \oplus Z_*\mathfrak{N}_{A^*} \\
&= \overline{S(\overline{D_2\mathfrak{B}} \oplus \overline{D_1\mathfrak{A}})} \oplus Z_*\mathfrak{N}_{A^*} \\
&= \overline{\{A_2u_2 \oplus (A_1u_1 - D_{*1}D_2u_2) \colon u_1 \in \overline{D_1\mathfrak{A}}, u_2 \in \overline{D_2\mathfrak{B}}\}} \oplus Z_*\mathfrak{N}_{A^*}.
\end{aligned}
\tag{3.16}
$$

Let $a'_* \oplus b'$ be an element of $\overline{D_{*2}\mathfrak{A}_*} \oplus \overline{D_{*1}\mathfrak{B}}$ orthogonal to $Z_*\overline{D_*\mathfrak{A}_*}$. As a consequence of (3.16), it is then orthogonal in particular (case $u_2 = 0$) to the elements of the form $0 \oplus A_1u_1$ with $u_1 \in \overline{D_1\mathfrak{A}}$. Applying (3.15) to $T = A_1^*$ we obtain $\overline{D_1\mathfrak{A}} = \overline{A_1^*D_{*1}\mathfrak{B}} \oplus \mathfrak{N}_{A_1}$ and hence $\overline{A_1D_1\mathfrak{A}} = \overline{A_1A_1^*D_{*1}\mathfrak{B}}$. We conclude that

$$
b' \perp A_1A_1^*D_{*1}\mathfrak{B} \quad \text{and hence} \quad A_1^*b' \perp A_1^*D_{*1}\mathfrak{B}.
$$

On the other hand $A_1^*b' \in \overline{A_1^*D_{*1}\mathfrak{B}}$, thus

$$
A_1^*b' = 0. \tag{3.17}
$$

This relation implies $D_{*1}^2b' = b' - A_1A_1^*b' = b'$ and consequently

$$
D_{*1}b' = b'. \tag{3.18}
$$

By (3.16), $a'_* \oplus b'$ is also orthogonal to $A_2u_2 \oplus (-D_{*1}D_2u_2)$ for every $u_2 \in \overline{D_2\mathfrak{B}}$ (case $u_1 = 0$). On account of (3.18) we deduce from this that

$$
A_2^*a'_* - D_2b' \perp \overline{D_2\mathfrak{B}}, \quad D_2(A_2^*a'_* - D_2b') \perp \mathfrak{B},
$$

and hence

$$
D_2A_2^*a'_* - D_2^2b' = 0. \tag{3.19}
$$

Finally, (3.16) shows that $a'_* \oplus b'$ is orthogonal to $Z_*\mathfrak{N}_{A^*}$, that is, to the elements $D_{*2}a_* \oplus D_{*1}A_2^*a_*$ with $a_* \in \mathfrak{N}_{A^*}$. From this and from (3.18),

$$
D_{*2}a'_* + A_2b' \perp \mathfrak{N}_{A^*}. \tag{3.20}
$$

But

$$
A^*(D_{*2}a'_* + A_2b') = A_1^*(A_2^*D_{*2}a'_* + A_2^*A_2b') = A_1^*(D_2A_2^*a'_* - D_2^2b' + b') = 0
$$

(cf. (3.17) and (3.19)); thus $D_{*2}a'_* + A_2b' \in \mathfrak{N}_{A^*}$. This result and (3.20) imply

$$
D_{*2}a'_* + A_2b' = 0, \tag{3.21}
$$

and consequently,

$$
A_2^*D_{*2}a'_* + A_2^*A_2b' = 0. \tag{3.22}
$$

Because $A_2^*D_{*2} = D_2A_2^*$, (3.19) and (3.22) give

$$b' = D_2^2 b' + A_2^* A_2 b' = 0.$$

So we obtain, from (3.21), $D_{*2}a'_* = 0$; as $a'_* \in \overline{D_{*2}\mathfrak{A}}$ we conclude that $a'_* = 0$.

Thus the only element $a'_* \oplus b'$ of $\overline{D_{*2}\mathfrak{A}_*} \oplus \overline{D_{*1}\mathfrak{B}}$ that is orthogonal to $\overline{Z_* D_* \mathfrak{A}_*}$, is zero. Hence Z_* maps $\overline{D_*\mathfrak{A}_*}$ onto $\overline{D_{*2}\mathfrak{A}_*} \oplus \overline{D_{*1}\mathfrak{B}}$, showing that the factorization $A^* = A_1^* A_2^*$ is regular.

Part (b): Due to the duality expressed by (a), it suffices to consider the case of an isometric A_2. Then $D_2 = O$, and hence

$$\begin{aligned}\overline{ZD\mathfrak{A}} = \overline{ZD\mathfrak{A}} &= \overline{\{D_2A_1a \oplus D_1a\colon a \in \mathfrak{A}\}} = \overline{\{0 \oplus D_1a\colon a \in \mathfrak{A}\}} \\ &= \{0\} \oplus \overline{D_1\mathfrak{A}} = \overline{D_2\mathfrak{B}} \oplus \overline{D_1\mathfrak{A}},\end{aligned}$$

and this proves our assertion.

Part (c): If A is isometric, then $D = O$. Now the transformation Z from $\overline{D\mathfrak{A}} = \{0\}$ into $\overline{D_2\mathfrak{B}} \oplus \overline{D_1\mathfrak{A}}$ is unitary if and only if $D_2 = O$ and $D_1 = O$, that is, if A_2 and A_1 are isometric. The assertion for unitary A follows because of the duality expressed by (a).

Part (d): Inequality (3.11) results immediately from the fact that the transformation Z generated by the factorization $A = A_2A_1$ is an isometry. If Z is even unitary (i.e., if the factorization is regular), then we have equality in (3.11). On the other hand, if we have

$$\dim \overline{D_2\mathfrak{B}} + \dim \overline{D_1\mathfrak{A}} = \dim \overline{D\mathfrak{A}} < \infty, \tag{3.23}$$

then Z is an isometric transformation from a finite dimensional space into a space of the same dimension, and hence it is necessarily unitary: the factorization is regular. This completes the proof of Proposition 3.2.

Here are some immediate consequences of Propositions 3.1 and 3.2.

Proposition 3.3. *Let*

$$\text{(F)} \quad \Theta(\lambda) = \Theta_2(\lambda)\Theta_1(\lambda)$$

be a factorization of the contractive analytic function $\{\mathfrak{E}, \mathfrak{E}_*, \Theta(\lambda)\}$ *into the product of the contractive analytic functions* $\{\mathfrak{E}, \mathfrak{F}, \Theta_1(\lambda)\}$ *and* $\{\mathfrak{F}, \mathfrak{E}_*, \Theta_2(\lambda)\}$, *and let*

$$\text{(F}^\sim\text{)} \quad \Theta^\sim(\lambda) = \Theta_1^\sim(\lambda)\Theta_2^\sim(\lambda)$$

be its dual.

(a) *If* (F) *is regular then so is* (F$^\sim$).
(b) (F) *is regular in each of the following cases.*

 (i) $\Theta_2(\lambda)$ *is an inner function.*
 (ii) $\Theta_1(\lambda)$ *is an* $*$*-inner function.*
 (iii) *At almost every point* t *where* $\Theta_2(e^{it})$ *is not isometric,* $\Theta_1(e^{it})^*$ *is isometric.*

(c) *When* $\Theta(\lambda)$ *is inner (inner from both sides),* (F) *is regular if and only if the factors are inner (inner from both sides) also.*

(d) *For* $\delta(t) = \dim \overline{\Delta(t)\mathfrak{E}}$, $\delta_1(t) = \dim \overline{\Delta_1(t)\mathfrak{E}}$, $\delta_2(t) = \dim \overline{\Delta_2(t)\mathfrak{F}}$ *we have*

$$\delta(t) \leq \delta_1(t) + \delta_2(t) \quad \text{a.e.} \tag{3.24}$$

Equality holds (a.e.) for regular (F) *and, if* $\delta(t) < \infty$ *a.e., only for regular* (F).

3. By virtue of part (b) of this proposition, the *canonical* and the $*$*-canonical factorizations* of $\Theta(\lambda)$ introduced in Sec. V.4.3, as well as the *factorizations considered in Proposition* V.4.3, *are regular factorizations*. In contrast, the factorization considered in Theorem V.5.5 is not regular; this follows from (c) and from Proposition V.2.3.

The regular factorizations considered in Proposition V.4.3 are used in Sec. 5.

Let us suppose now that $\Theta(\lambda)$ coincides with the characteristic function of the c.n.u. contraction T, and let $\Theta(\lambda) = \Theta_i(\lambda)\Theta_e(\lambda)$ be the canonical factorization of $\Theta(\lambda)$. This factorization is regular, therefore it induces, according to Theorem 1.1 and Proposition 2.1, a triangulation

$$\mathbf{T} = \begin{bmatrix} \mathbf{T}_1 & \mathbf{X} \\ \mathbf{O} & \mathbf{T}_2 \end{bmatrix}$$

of the functional model of T such that the characteristic functions of $\mathbf{T}_1$ and $\mathbf{T}_2$ coincide with the purely contractive parts of $\Theta_e(\lambda)$ and $\Theta_i(\lambda)$, respectively. Now it is obvious that the purely contractive part of an outer or inner function is also outer or inner, respectively. Thus by applying Proposition VI.3.5 we obtain that $\mathbf{T}_1 \in C_{\cdot 1}$ and $\mathbf{T}_2 \in C_{\cdot 0}$. Because T is unitarily equivalent to its functional model $\mathbf{T}$, this triangulation of $\mathbf{T}$ induces a triangulation of T of the same type.

Analogous reasoning applies to the $*$-canonical factorization of $\Theta(\lambda)$, which is also regular. We have proved the following proposition.

Proposition 3.4. *For a c.n.u. contraction* T *in the (separable) space* $\mathfrak{H}$, *the canonical and the* $*$*-canonical factorizations of the characteristic function of* T *induce the triangulations of type*

$$\begin{bmatrix} C_{\cdot 1} & * \\ O & C_{\cdot 0} \end{bmatrix} \quad \textit{and} \quad \begin{bmatrix} C_{0\cdot} & * \\ O & C_{1\cdot} \end{bmatrix},$$

respectively (cf. Sec. II.4).

4. Let us consider now, as an example, the case of a purely contractive analytic *scalar* function $\Theta(\lambda)$, namely the case $\mathfrak{E} = \mathfrak{E}_* = E^1$. Then $\delta(t) = 0$ or 1 according to $|\Theta(e^{it})| = 1$ or < 1. Let $\Theta(\lambda) = \Theta_2(\lambda)\Theta_1(\lambda)$ be a factorization of $\Theta(\lambda)$ as a product of the contractive analytic functions $\{E^1, \mathfrak{F}, \Theta_1(\lambda)\}$ and $\{\mathfrak{F}, E^1, \Theta_2(\lambda)\}$. By virtue of Proposition 3.3(d), this factorization is regular if and only if

$$\delta_1(t) + \delta_2(t) = \delta(t) \quad (\leq 1) \quad \text{a.e.} \tag{3.25}$$

Let us observe that, for $f \in \mathfrak{F}$ and for a.e. t,

$$\begin{aligned} f &= [I_{\mathfrak{F}} - \Theta_2(e^{it})^*\Theta_2(e^{it})]f + \Theta_2(e^{it})^*\Theta_2(e^{it})f \\ &\in [I_{\mathfrak{F}} - \Theta_2(e^{it})^*\Theta_2(e^{it})]\mathfrak{F} \vee \Theta_2(e^{it})^*E^1 \end{aligned} \tag{3.26}$$

and that

$$\dim \overline{[I_{\mathfrak{F}} - \Theta_2(e^{it})^*\Theta_2(e^{it})]\mathfrak{F}} = \delta_2(t), \quad \dim \overline{\Theta_2(e^{it})^*E^1} \leq 1. \tag{3.27}$$

From (3.25)–(3.27) we infer that $\dim \mathfrak{F} \leq 2$, and hence, up to an isomorphism, $\mathfrak{F} = E^0 (= \{0\})$, or $\mathfrak{F} = E^1$, or $\mathfrak{F} = E^2$.

The case $\mathfrak{F} = E^0$ can occur only if $\Theta(\lambda) \equiv O$, and then the factorization

$$O_{E^1 \to E^1} = O_{E^0 \to E^1} \cdot O_{E^1 \to E^0} \tag{3.28}$$

is obviously regular ($\delta(t) \equiv 1, \delta_1(t) \equiv 1, \delta_2(t) \equiv 0$). (We suggest to the reader as an interesting exercise to find the contraction T and the subspace invariant for T, corresponding to the characteristic function $\{E^1, E^1, O\}$ and to its regular factorization (3.28).)

When $\mathfrak{F} = E^1$ the functions $\Theta_1(\lambda)$ and $\Theta_2(\lambda)$ are also scalar. Because $\delta_k(t) = 0$ or 1 accordingly as $|\Theta_k(e^{it})| = 1$ or < 1 $(k = 1, 2)$, condition (3.25) means that, at almost every point t such that $|\Theta(e^{it})| = 1$, we have $|\Theta_k(e^{it})| = 1$ for $k = 1$ and $k = 2$, and at almost every point t such that $|\Theta(e^{it})| < 1$, we have $|\Theta_k(e^{it})| = 1$ for one of the indices $k = 1, 2$.

Finally, if $\mathfrak{F} = E^2$, the operator $\Theta_2(e^{it})$ (from E^2 into E^1) cannot be isometric; thus we have $\delta_2(t) > 0$. Hence in this case (3.25) means that $\delta_2(t) = 1$, $\delta_1(t) = 0$, $\delta(t) = 1$, a.e. Thus $\Theta_1(e^{it})$ is isometric a.e. If we represent $\Theta_2(\lambda)$ by its matrix $[\vartheta_{21}(\lambda)\ \vartheta_{22}(\lambda)]$, the operator $I_{\mathfrak{F}} - \Theta_2(e^{it})^*\Theta_2(e^{it})$ is represented by the matrix

$$\begin{bmatrix} 1 - |a|^2 & -\bar{a}b \\ -a\bar{b} & 1 - |b|^2 \end{bmatrix} \quad \text{with} \quad a = \vartheta_{21}(e^{it}), \quad b = \vartheta_{22}(e^{it}).$$

As

$$\dim[I_{\mathfrak{F}} - \Theta_2(e^{it})^*\Theta_2(e^{it})]\mathfrak{F} = \delta_2(t) = 1 < \dim \mathfrak{F},$$

this matrix has determinant 0 and hence $|a|^2 + |b|^2 = 1$; consequently $\Theta_2(e^{it})^*$ is isometric. Thus if $\mathfrak{F} = E^2$, condition (3.25) implies that $\Theta_1(\lambda)$ is inner and $\Theta_2(\lambda)$ is $*$-inner. Moreover, in this case $\delta(t) = 1$ and hence $|\Theta(e^{it})| < 1$ a.e. Conversely, the factorizations of this type (i.e., with $|\Theta(e^{it})| < 1$ a.e., $\Theta_1(\lambda)$ inner and $\Theta_2(\lambda)$ $*$-inner) are regular, because then $\delta(t) = 1$, $\delta_1(t) = 0$, and $\delta_2(t) = 1$ a.e.

So we have arrived at the results summarized below.

Proposition 3.5. *Let $\Theta(\lambda) = \Theta_2(\lambda)\Theta_1(\lambda)$ be a factorization of the scalar-valued contractive analytic function $\Theta(\lambda)$ as a product of two contractive analytic functions. This factorization is regular if and only if one of the following cases occurs.*

(i) *The trivial case* (3.28).

(ii) *The factors are scalar-valued and, a.e., at least one of the values* $|\Theta_k(e^{it})|$ $(k=1,2)$ *equals* 1.

(iii) $|\Theta(e^{it})| < 1$ *a.e., and the factors are of the form*

$$\Theta_1(\lambda) = \begin{bmatrix} \vartheta_{11}(\lambda) \\ \vartheta_{12}(\lambda) \end{bmatrix}, \quad \Theta_2(\lambda) = [\vartheta_{21}(\lambda)\ \vartheta_{22}(\lambda)]$$

with scalar-valued functions $\vartheta_{ik}(\lambda) \in H^\infty$ *such that*

$$|\vartheta_{11}(e^{it})|^2 + |\vartheta_{12}(e^{it})|^2 = 1 \quad \textit{and} \quad |\vartheta_{21}(e^{it})|^2 + |\vartheta_{22}(e^{it})|^2 = 1 \quad \textit{a.e.}$$

In the last case, the factorization is of the form

$$\Theta(\lambda) = \vartheta_{21}(\lambda)\vartheta_{11}(\lambda) + \vartheta_{22}(\lambda)\vartheta_{12}(\lambda).$$

An example for case (iii) is given by

$$\frac{\sqrt{3}}{2}\lambda = \begin{bmatrix} \frac{\sqrt{3}}{2} & \frac{\lambda}{2} \end{bmatrix} \cdot \begin{bmatrix} \frac{\lambda}{2} \\ \frac{\sqrt{3}}{2} \end{bmatrix}.$$

5. We conclude the section with a proposition, which does not deal with regular factorizations, but presents an interesting consequence of Propositions 2.1 and 3.3(d).

Proposition 3.6. *Let* T *be a c.n.u. contraction in* $\mathfrak{H}$, *with (finite or infinite) defect indices* $\mathfrak{d}_T$ *and* $\mathfrak{d}_{T^*}$. *Let* $\mathfrak{H}_1$ $(\neq \{0\})$ *be a subspace of* $\mathfrak{H}$, *invariant for* T. *Then the defect indices of the operator* $T_1 = T|\mathfrak{H}_1$ *satisfy the inequalities*

$$\mathfrak{d}_{T_1} \leq \mathfrak{d}_T \quad \textit{and} \quad \mathfrak{d}_{T_1^*} \leq \mathfrak{d}_T + \mathfrak{d}_{T^*}.$$

Proof. Let $\{\mathfrak{E}, \mathfrak{E}_*, \Theta(\lambda)\}$ be a contractive analytic function coinciding with the characteristic function of T. Let $\Theta(\lambda) = \Theta_2(\lambda)\Theta_1(\lambda)$ be the regular factorization of $\Theta(\lambda)$ into the product of the contractive analytic functions $\{\mathfrak{E}, \mathfrak{F}, \Theta_1(\lambda)\}$ and $\{\mathfrak{F}, \mathfrak{E}_*, \Theta_2(\lambda)\}$, corresponding to the invariant subspace $\mathfrak{H}_1$ in the sense of Theorem 1.1. According to Proposition 3.3(d) we have then $\dim \overline{\Delta(t)\mathfrak{E}} = \dim \overline{\Delta_2(t)\mathfrak{F}} + \dim \overline{\Delta_1(t)\mathfrak{E}}$ a.e., and consequently

$$\dim \overline{\Delta_2(t)\mathfrak{F}} \leq \dim \overline{\Delta(t)\mathfrak{E}} \leq \dim \mathfrak{E} = \mathfrak{d}_T \quad \text{a.e.} \tag{3.29}$$

On the other hand, the relation $I_{\mathfrak{F}} = \Delta_2(t)^2 + \Theta_2(e^{it})^*\Theta_2(e^{it})$ implies that, for almost every t, $\mathfrak{F}$ is spanned by $\Delta_2(t)\mathfrak{F}$ and $\Theta_2(e^{it})^*\Theta_2(e^{it})\mathfrak{F}$, and hence

$$\dim \mathfrak{F} \leq \dim \overline{\Delta_2(t)\mathfrak{F}} + \dim \overline{\Theta_2(e^{it})^*\Theta_2(e^{it})\mathfrak{F}}. \tag{3.30}$$

As $\Theta_2(e^{it})\mathfrak{F} \subset \mathfrak{E}_*$, we have $\dim \overline{\Theta_2(e^{it})\mathfrak{F}} \leq \dim \mathfrak{E}_* = \mathfrak{d}_{T^*}$; consequently,

$$\dim \overline{\Theta_2(e^{it})^* \Theta_2(e^{it})\mathfrak{F}} \leq \dim \overline{\Theta_2(e^{it})\mathfrak{F}} \leq \mathfrak{d}_{T^*}. \tag{3.31}$$

From (3.29)–(3.31) it follows that $\dim \mathfrak{F} \leq \mathfrak{d}_T + \mathfrak{d}_{T^*}$. Now we know (cf. Proposition 2.1) that the characteristic function of T_1 coincides with the purely contractive part $\{\mathfrak{E}^0, \mathfrak{F}^0, \Theta_1^0(\lambda)\}$ of $\{\mathfrak{E}, \mathfrak{F}, \Theta_1(\lambda)\}$. Because $\mathfrak{E}^0 \subset \mathfrak{E}$ and $\mathfrak{F}^0 \subset \mathfrak{F}$, we have

$$\mathfrak{d}_{T_1} = \dim \mathfrak{E}^0 \leq \dim \mathfrak{E} = \mathfrak{d}_T,$$
$$\mathfrak{d}_{T_1^*} = \dim \mathfrak{F}^0 \leq \dim \mathfrak{F} \leq \mathfrak{d}_T + \mathfrak{d}_{T^*},$$

and this completes the proof.

4 Arithmetic of regular divisors

1. We begin with a definition.

Definition. Let A and A_1 be contractions from the space $\mathfrak{A}$ into the spaces $\mathfrak{A}_*$ and $\mathfrak{B}$, respectively. We say that A_1 is a *divisor* of A if there exists a contraction A_2 from $\mathfrak{B}$ into $\mathfrak{A}_*$ such that

$$A = A_2 A_1.$$

If A_2 can be chosen so that this factorization is regular then we call A_1 a *regular divisor* of A.

This definition extends immediately to contractive analytic functions $\Theta(\lambda)$, $\Theta_1(\lambda), \Theta_2(\lambda)$ by considering the corresponding contractions $\Theta, \Theta_1, \Theta_2$, in the respective L^2 spaces.

One of the first problems in this connection is that of the transitivity of regular divisors. This requires finding regularity relations between several factorizations. The most important case is that of the factorizations

$$(\mathrm{F}_{21}) \quad A_{21} = A_2 A_1, \qquad (\mathrm{F}_{3(21)}) \quad A = A_3 A_{21},$$
$$(\mathrm{F}_{32}) \quad A_{32} = A_3 A_2, \qquad (\mathrm{F}_{(32)1}) \quad A = A_{32} A_1$$

induced by a factorization

$$(\mathrm{F}_{321}) \quad A = A_3 A_2 A_1$$

of a contraction $A\colon \mathfrak{A} \to \mathfrak{A}_*$ into the product of three contractions:

$$A_1\colon \mathfrak{A} \to \mathfrak{B}_1, \quad A_2\colon \mathfrak{B}_1 \to \mathfrak{B}_{21}, \quad A_3\colon \mathfrak{B}_{21} \to \mathfrak{A}_*.$$

We need a preliminary result.

Lemma 4.1. *The factorizations* $(\mathrm{F}_{21}), (\mathrm{F}_{3(21)}), (\mathrm{F}_{32}), (\mathrm{F}_{(32)1})$ *are all regular if one of the following conditions holds.*

(i) (F_{21}) *and* $(\mathrm{F}_{3(21)})$ *are regular.*

(ii) (F_{32}) *and* $(F_{(32)1})$ *are regular.*
(iii) $(F_{3(21)})$ *and* $(F_{(32)1})$ *are regular.*

Proof. For now let us denote by $D, D_1, \ldots, D_{32}$ the defect operators $D_A, D_{A_1}, \ldots, D_{A_{32}}$, and let us consider the isometric transformations $Z_{21}, Z_{3(21)}$, and so on, corresponding to the factorizations (F_{21}), $(F_{3(21)})$, and so on, in the sense of (3.1)–(3.2). Thus we have

$$\begin{aligned}
Z_{21}: & \quad D_{21}a \to D_2 A_1 a \oplus D_1 a & (a \in \mathfrak{A}), \\
Z_{3(21)}: & \quad Da \to D_3 A_{21} a \oplus D_{21} a & (a \in \mathfrak{A}), \\
Z_{32}: & \quad D_{32}b \to D_3 A_2 b \oplus D_2 b & (b \in \mathfrak{B}_1), \\
Z_{(32)1}: & \quad Da \to D_{32} A_1 a \oplus D_1 a & (a \in \mathfrak{A}).
\end{aligned}$$

Similarly, let us attach to the factorization (F_{321}) the transformation

$$Z_{321}: \quad Da \to D_3 A_{21} a \oplus D_2 A_1 a \oplus D_1 a \quad (a \in \mathfrak{A}),$$

which is also isometric. We use the same letters to denote the corresponding closed isometries. The relations between these isometric transformations are indicated by the following commutative diagram, the arrows denoting transformation from one space into the other.

$$(I_{3(21)} = I_{\overline{D_3 \mathfrak{B}_{21}}}, I_1 = I_{\overline{D_1 \mathfrak{A}}}).$$

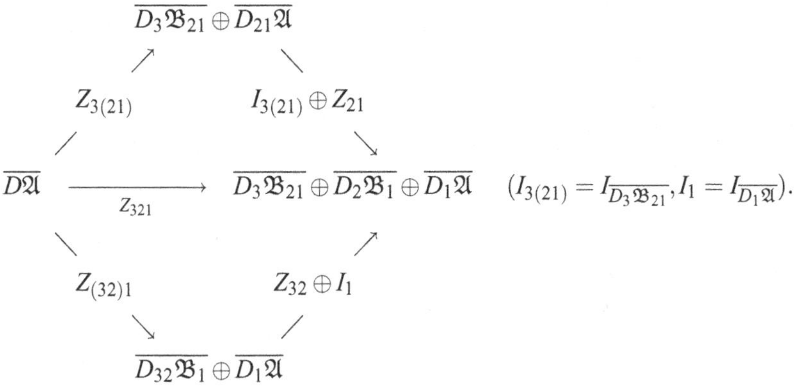

Commutativity of this diagram means

$$(I_{3(21)} \oplus Z_{21})Z_{3(21)} = Z_{321} = (Z_{32} \oplus I_1)Z_{(32)1};$$

these equations derive at once from the definitions.

Let us assume first that $(F_{3(21)})$ and (F_{21}) are regular; thus $Z_{3(21)}$ and Z_{21} (hence also $I_{3(21)} \oplus Z_{21}$) are unitary. By the commutativity of the diagram, Z_{321} is then also unitary and so are $Z_{(32)1}$ and $Z_{32} \oplus I_1$, and hence Z_{32} too. This proves that $(F_{(32)1})$ and (F_{32}) are regular. The case when $(F_{(32)1})$ and (F_{32}) are regular can be treated similarly.

Next we suppose that $(\mathrm{F}_{3(21)})$ and $(\mathrm{F}_{(32)1})$ are regular. These assumptions imply the relations

$$Z_{321}\overline{D\mathfrak{A}} = \begin{cases} (I_{3(21)} \oplus Z_{21})Z_{3(21)}\overline{D\mathfrak{A}} = (I_{3(21)} \oplus Z_{21})(\overline{D_3\mathfrak{B}_{21}} \oplus \overline{D_{21}\mathfrak{A}}) \\ \qquad = \overline{D_3\mathfrak{B}_{21}} \oplus Z_{21}\overline{D_{21}\mathfrak{A}}, \\ (Z_{32} \oplus I_1)Z_{(32)1}\overline{D\mathfrak{A}} = (Z_{32} \oplus I_1)(\overline{D_{32}\mathfrak{B}_1} \oplus \overline{D_1\mathfrak{A}}) \\ \qquad = Z_{32}\overline{D_{32}\mathfrak{B}_1} \oplus \overline{D_1\mathfrak{A}}, \end{cases} \tag{4.1}$$

where

$$Z_{21}\overline{D_{21}\mathfrak{A}} \subset \overline{D_2\mathfrak{B}_1} \oplus \overline{D_1\mathfrak{A}}, \quad Z_{32}\overline{D_{32}\mathfrak{B}_1} \subset \overline{D_3\mathfrak{B}_{21}} \oplus \overline{D_2\mathfrak{B}_1}. \tag{4.2}$$

From (4.1) and (4.2) we deduce that

$$Z_{321}\overline{D\mathfrak{A}} = \overline{D_3\mathfrak{B}_{21}} \oplus \mathfrak{B}_2 \oplus \overline{D_1\mathfrak{A}}, \tag{4.3}$$

with $\mathfrak{B}_2 \subset \overline{D_2\mathfrak{B}_1}$ and with

$$\overline{D_3\mathfrak{B}_{21}} \oplus \mathfrak{B}_2 = Z_{32}\overline{D_{32}\mathfrak{B}_1} = \overline{\{D_3A_2b \oplus D_2b \colon b \in \mathfrak{B}_1\}}.$$

Hence we obtain $\mathfrak{B}_2 = \overline{D_2\mathfrak{B}_1}$, and thus, from (4.3),

$$Z_{321}\overline{D\mathfrak{A}} = \overline{D_3\mathfrak{B}_{21}} \oplus \overline{D_2\mathfrak{B}_1} \oplus \overline{D_1\mathfrak{A}},$$

in other words, Z_{321} is unitary. This implies that all the isometric transformations occurring in our commutative diagram are unitary; hence all the factorizations are regular. This concludes the proof of Lemma 4.1.

Alternatively, we can formulate this lemma in the following manner.

Proposition 4.2. *Let A, A_{21}, and A_1 be contractions.*

(a) *If A_{21} is a regular divisor of A and if A_1 is a regular divisor of A_{21}, then A_1 is a regular divisor of A.*
(b) *If A_{21} and A_1 are regular divisors of A, if A_1 is a divisor of A_{21}, and if A_1 has dense range, then A_1 is a regular divisor of A_{21}.*

Proof. In fact, (a) corresponds to the case of Lemma 4.1 where the factorizations (F_{21}) and $(\mathrm{F}_{3(21)})$ are assumed to be regular.

Under the conditions of (b) there exist contractions A_2, A_3, and A' such that the factorizations $A = A_3A_{21}$, $A = A'A_1$ are regular, and $A_{21} = A_2A_1$. Because $A'A_1 = A = A_3A_{21} = A_3A_2A_1$ and A_1 has dense range, it follows that $A' = A_3A_2$. Thus the factorizations $(A_3A_2)A_1 = A = A_3(A_2A_1)$ of type $(\mathrm{F}_{(32)1})$ and $(\mathrm{F}_{3(21)})$, respectively, are regular. Hence Lemma 4.1 implies that the factorization $A_{21} = A_2A_1$ of type (F_{21}) is also regular; that is, A_1 is a regular divisor of A_{21}.

We note that the density of the range of A_1 cannot be omitted in (b). To construct an example we use the unilateral shift S on H^2 (i.e., the operator of multiplication by λ). Choose $A = S^3$, $A_1 = A_3 = S^2$, and $A_{21} = A_{23} = S$. The factorizations $A = A_3A_{21} = A_{23}A_1$ are obviously regular. Because $A_{21} = S^*A_1$, A_1 is a divisor of A_{21}.

We claim that no factorization $A_{21} = A'A_1$, with $\|A'\| \leq 1$, is regular. Assume to the contrary that such a factorization $S = A'S^2$ is regular. Proposition 3.2 implies that the factorization $S^* = S^{*2}A'^*$ is regular as well. The same proposition implies that

$$1 = \dim \overline{D_{S^*}H^2} = \dim \overline{D_{S^{*2}}H^2} + \dim \overline{D_{A^*}H^2} \geq 2,$$

a contradiction. Thus A_1 is not a regular divisor of A_{21}.

2. Let $\{\mathfrak{E}, \mathfrak{E}_*, \Theta(\lambda)\}$ be a purely contractive analytic function, and let **H** and **T** be the space and the contraction corresponding to $\Theta(\lambda)$ in the sense of (1.44) and (1.45).

Definition. Let $\Theta_1(\lambda)$ be a regular divisor of $\Theta(\lambda)$. A subspace $\mathbf{H}_1$ of $\mathbf{H}$ is called an *invariant subspace associated with* $\Theta_1(\lambda)$ if it derives by formula (1.47) from a regular factorization $\Theta_2(\lambda)\Theta_1(\lambda)$ of $\Theta(\lambda)$, with the given factor $\Theta_1(\lambda)$.

Remark. We show later that there can exist different invariant subspaces associated with the same regular divisor $\Theta_1(\lambda)$ of $\Theta(\lambda)$.

Here are the important relations between the regular divisors of $\Theta(\lambda)$ and the invariant subspaces associated with them.

Theorem 4.3. (a) *Let $\Theta_1(\lambda)$ and $\Theta_1'(\lambda)$ be two regular divisors of $\Theta(\lambda)$ such that we have $\mathbf{H}_1 \subset \mathbf{H}_1'$ for some invariant subspaces associated with $\Theta_1(\lambda)$ and with $\Theta_1'(\lambda)$, respectively. Then $\Theta_1(\lambda)$ is a regular divisor of $\Theta_1'(\lambda)$. If, moreover, $\mathbf{H}_1 = \mathbf{H}_1'$, then $\Theta_1(\lambda)$ and $\Theta_1'(\lambda)$ can only differ by a constant unitary left-factor.*

(b) *Let $\Theta_1'(\lambda)$ be a regular divisor of $\Theta(\lambda)$ and let $\Theta_1(\lambda)$ be a regular divisor of $\Theta_1'(\lambda)$* (hence of $\Theta(\lambda)$ as well, cf. Proposition 4.2(a)). *Every invariant subspace $\mathbf{H}_1'$ associated with $\Theta_1'(\lambda)$ includes then an invariant subspace $\mathbf{H}_1$ associated with $\Theta_1(\lambda)$*

Proof. Part (a): Let us assume that the invariant subspaces $\mathbf{H}_1$ and $\mathbf{H}_1'$ derive from the regular factorizations $\Theta(\lambda) = \Theta_2(\lambda)\Theta_1(\lambda)$ and $\Theta(\lambda) = \Theta_2'(\lambda)\Theta_1'(\lambda)$ with the intermediate spaces $\mathfrak{F}$ and $\mathfrak{F}'$, respectively, and that $\mathbf{H}_1 \subset \mathbf{H}_1'$. Denoting by Z and Z' the unitary operators generated by the above regular factorizations, we deduce from formula (1.47) that $\mathbf{H}_1 \subset \mathbf{H}_1'$ implies

$$\begin{aligned} \mathscr{K}_+ \equiv &\{\Theta_2 u \oplus Z^{-1}(\Delta_2 u \oplus v) \colon u \in H^2(\mathfrak{F}), v \in \overline{\Delta_1 L^2(\mathfrak{E})}\} \\ &\subset \{\Theta_2' u' \oplus Z'^{-1}(\Delta_2' u' \oplus v') \colon u' \in H^2(\mathfrak{F}'), v' \in \overline{\Delta_1' L^2(\mathfrak{E})}\} \equiv \mathscr{K}_+'. \end{aligned} \tag{4.4}$$

One of the consequences of the inclusion (4.4) is that to each $u \in H^2(\mathfrak{F})$ there corresponds a $u' \in H^2(\mathfrak{F}')$ and a $v' \in \overline{\Delta_1 L^2(\mathfrak{E})}$ such that

$$\Theta_2 u \oplus Z^{-1}(\Delta_2 u \oplus 0) = \Theta_2' u' \oplus Z'^{-1}(\Delta_2' u' \oplus v'). \tag{4.5}$$

Because

$$\|\Theta_2 u \oplus Z^{-1}(\Delta_2 u \oplus 0)\|^2 = \|\Theta_2 u\|^2 + \|\Delta_2 u\|^2 = \|u\|^2$$

and

$$\|\Theta_2'u' \oplus Z'^{-1}(\Delta_2'u' \oplus v')\|^2 = \|\Theta_2'u'\|^2 + \|\Delta_2'u'\|^2 + \|v'\|^2 = \|u'\|^2 + \|v'\|^2,$$

(4.5) implies that

$$\|u\|^2 = \|u'\|^2 + \|v'\|^2. \tag{4.6}$$

It follows that the operators Q and R defined by

$$Qu = u', \quad Ru = v'$$

are *contractions* from $H^2(\mathfrak{F})$ into $H^2(\mathfrak{F}')$ and $\overline{\Delta_1'L^2(\mathfrak{E})}$, respectively. All the operators figuring in formula (4.5) commute with multiplication by the function e^{it}, therefore so does Q. As multiplication by e^{it} is a unilateral shift on $H^2(\mathfrak{F})$ as well as on $H^2(\mathfrak{F}')$, with the respective generating subspaces consisting of the constant functions, it follows from Lemma V.3.2 that there exists a contractive analytic function $\{\mathfrak{F}, \mathfrak{F}', \Omega(\lambda)\}$ such that

$$u'(\lambda) = (Qu)(\lambda) = \Omega(\lambda)u(\lambda) \quad (u \in H^2(\mathfrak{F})). \tag{4.7}$$

Another consequence of (4.4) is the inclusion

$$\bigcap_{n\geq 0} e^{int}\mathscr{K}_+ \subset \bigcap_{n\geq 0} e^{int}\mathscr{K}_+'. \tag{4.8}$$

Let W be the unitary operator from $\mathscr{K}_+$ to $H^2(\mathfrak{F}) \oplus \overline{\Delta_1 L^2(\mathfrak{E})}$ defined by (2.7). It is obvious that W commutes with multiplication by e^{it}, therefore we have

$$\begin{aligned}\bigcap_{n\geq 0} e^{int}\mathscr{K}_+ &= W^{-1}\Big[\bigcap_{n\geq 0} e^{int}H^2(\mathfrak{F}) \oplus \bigcap_{n\geq 0} e^{int}\overline{\Delta_1 L^2(\mathfrak{E})}\Big] \\ &= W^{-1}[\{0\} \oplus \overline{\Delta_1 L^2(\mathfrak{E})}] = \{0 \oplus Z^{-1}(0 \oplus v) : v \in \overline{\Delta_1 L^2(\mathfrak{E})}\}\end{aligned}$$

and analogously for $\mathscr{K}_+'$. Thus (4.8) means that

$$\{0 \oplus Z^{-1}(0 \oplus v) : v \in \overline{\Delta_1 L^2(\mathfrak{E})}\} \subset \{0 \oplus Z'^{-1}(0 \oplus v') : v' \in \overline{\Delta_1' L^2(\mathfrak{E})}\}. \tag{4.9}$$

It follows that for every $v \in \overline{\Delta_1 L^2(\mathfrak{E})}$ there exists a $v' \in \overline{\Delta_1' L^2(\mathfrak{E})}$ such that

$$Z^{-1}(0 \oplus v) = Z'^{-1}(0 \oplus v'). \tag{4.10}$$

Because (4.10) implies $\|v\| = \|v'\|$, the operator V defined by

$$Vv = v'$$

is a well-defined *isometry* from $\overline{\Delta_1 L^2(\mathfrak{E})}$ into $\overline{\Delta_1' L^2(\mathfrak{E})}$.

Let $w \in H^2(\mathfrak{E})$. Then

$$\Theta w \oplus \Delta w = \Theta_2'\Theta_1' w \oplus Z'^{-1}(\Delta_2'\Theta_1' w \oplus \Delta_1' w), \tag{4.11}$$

and making use of the operators Q, R, V defined above, and of (4.5) and (4.7),

$$\begin{aligned}\Theta w \oplus \Delta w &= \Theta_2\Theta_1 w \oplus Z^{-1}(\Delta_2\Theta_1 w \oplus \Delta_1 w)\\ &= [\Theta_2\Theta_1 w \oplus Z^{-1}(\Delta_2\Theta_1 w \oplus 0)] + [0 \oplus Z^{-1}(0 \oplus \Delta_1 w)] \qquad (4.12)\\ &= [\Theta_2'\Omega\Theta_1 w \oplus Z'^{-1}(\Delta_2'\Omega\Theta_1 w \oplus R\Theta_1 w)] + [0 \oplus Z'^{-1}(0 \oplus V\Delta_1 w)]\\ &= \Theta_2'\Omega\Theta_1 w \oplus Z'^{-1}(\Delta_2'\Omega\Theta_1 w \oplus \bar{v}),\end{aligned}$$

where $\bar{v} = R\Theta_1 w + V\Delta_1 w \in \overline{\Delta_1' L^2(\mathfrak{E})}$. Comparing (4.11) with (4.12) we get

$$\Theta_2'\Theta_1' w = \Theta_2'\Omega\Theta_1 w, \quad \Delta_2'\Theta_1' w = \Delta_2'\Omega\Theta_1 w \quad (w \in H^2(\mathfrak{E})). \qquad (4.13)$$

The transformation $u' \to \Theta_2' u' \oplus \Delta_2' u'$ $(u' \in H^2(\mathfrak{F}'))$ being isometric, (4.13) implies $\Theta_1' w = \Omega\Theta_1 w$ and hence

$$\Theta_1'(\lambda) = \Omega(\lambda)\Theta_1(\lambda) \quad (\lambda \in D).$$

In view of the equations (4.5) and (4.7) we readily obtain that

$$\Theta_2 u = \Theta_2' u' = \Theta_2' Q u = \Theta_2'\Omega u \qquad (u \in H^2(\mathfrak{F})),$$

whence

$$\Theta_2(\lambda) = \Theta_2'(\lambda)\Omega(\lambda) \qquad (\lambda \in D)$$

follows. The factorizations

$$\Theta(\lambda) = (\Theta_2'(\lambda)\Omega(\lambda))\Theta_1(\lambda) = \Theta_2'(\lambda)(\Omega(\lambda)\Theta_1(\lambda))$$

of types $(\mathrm{F}_{(32)1})$ and $(\mathrm{F}_{3(21)})$, respectively, are regular because $\Theta_2'(\lambda)\Omega(\lambda) = \Theta_2(\lambda)$ and $\Omega(\lambda)\Theta_1(\lambda) = \Theta_1'(\lambda)$. Thus Lemma 4.1 implies that the factorization $\Theta_1'(\lambda) = \Omega(\lambda)\Theta_1(\lambda)$ of type (F_{21}) is also regular; that is, $\Theta_1(\lambda)$ is a regular divisor of $\Theta_1'(\lambda)$.

If $\mathbf{H}_1 = \mathbf{H}_1'$, we have equality in (4.4), (4.8), and (4.9) also. Now, equality in (4.4) implies (because of (4.9)) that if u runs over $H^2(\mathfrak{F})$, then $u' = Qu$ runs over $H^2(\mathfrak{F}')$, and thus $\overline{QH^2(\mathfrak{F})} = H^2(\mathfrak{F}')$. On the other hand, equality in (4.9) implies that if v runs over $\overline{\Delta_1 L^2(\mathfrak{E})}$, then $v' = Vv$ runs over $\overline{\Delta_1' L^2(\mathfrak{E})}$, and thus V is a unitary operator from $\overline{\Delta_1 L^2(\mathfrak{E})}$ onto $\overline{\Delta_1' L^2(\mathfrak{E})}$. On account of (4.10) the right-hand side of (4.5) is then equal to

$$\begin{aligned}&[\Theta_2' u' \oplus Z'^{-1}(\Delta_2' u' \oplus 0)] + [0 \oplus Z'^{-1}(0 \oplus v')]\\ =&[\Theta_2' u' \oplus Z'^{-1}(\Delta_2' u' \oplus 0)] + [0 \oplus Z^{-1}(0 \oplus V^{-1}v')];\end{aligned}$$

hence, subtracting $0 \oplus Z^{-1}(0 \oplus V^{-1}v')$ from both sides of (4.5), we obtain

$$\Theta_2 u \oplus Z^{-1}(\Delta_2 u \oplus (-V^{-1}v')) = \Theta_2' u' \oplus Z'^{-1}(\Delta_2' u' \oplus 0),$$

and taking norms,

$$\|u\|^2 + \|v'\|^2 = \|u'\|^2.$$

Together with (4.6) this gives $\|u\| = \|u'\|$; hence Q is isometric. Collecting the results obtained for Q, it follows that Q is unitary. By virtue of Lemma V.3.2(d) the function $\Omega(\lambda)$ is therefore a unitary constant.

Part (b): Let $\Theta_1'(\lambda)$ also be denoted by $\Theta_{21}(\lambda)$. Let $\mathbf{H}_{21}$ be an invariant subspace associated with the regular divisor $\Theta_{21}(\lambda)$ of $\Theta(\lambda)$: $\mathbf{H}_{21}$ corresponds to some regular factorization

$$(\mathrm{F}_{3(21)}) \quad \Theta(\lambda) = \Theta_3(\lambda)\Theta_{21}(\lambda),$$

where $\{\mathfrak{F}_1, \mathfrak{F}_{21}, \Theta_{21}(\lambda)\}$ is a contractive analytic function. By hypothesis, $\Theta_1(\lambda)$ is a regular divisor of $\Theta_{21}(\lambda)$, and thus there exists a regular factorization

$$(\mathrm{F}_{21}) \quad \Theta_{21}(\lambda) = \Theta_2(\lambda)\Theta_1(\lambda).$$

By virtue of Lemma 4.1, the factorizations

$$(\mathrm{F}_{32}) \quad \Theta_{32}(\lambda) = \Theta_3(\lambda)\Theta_2(\lambda), \qquad (\mathrm{F}_{(32)1}) \quad \Theta(\lambda) = \Theta_{32}(\lambda)\Theta_1(\lambda)$$

are then regular too. Consequently, the isometric transformations $Z_{3(21)}$, and so on, corresponding to these factorizations are all unitary; applying our commutative diagram to the present situation we obtain

$$Z_{(32)1}^{-1}(Z_{32}^{-1} \oplus I_1) = Z_{321}^{-1} = Z_{3(21)}^{-1}(I_{3(21)} \oplus Z_{21}^{-1}). \tag{4.14}$$

This is a unitary transformation

$$\overline{\Delta_3 L^2(\mathfrak{F}_{21})} \oplus \overline{\Delta_2 L^2(\mathfrak{F})} \oplus \overline{\Delta_1 L_2(\mathfrak{E})} \to \overline{\Delta L^2(\mathfrak{E})}.$$

Let $\mathbf{H}_1$ be the invariant subspace corresponding to the regular factorization $\Theta(\lambda) = \Theta_{32}(\lambda)\Theta_1(\lambda)$; the invariant subspace $\mathbf{H}_{21}$ corresponds to the regular factorization $\Theta(\lambda) = \Theta_3(\lambda)\Theta_{21}(\lambda)$, thus

$$\mathbf{H}_1 = \{\Theta_{32}u \oplus Z_{(32)1}^{-1}(\Delta_{32}u \oplus v) \colon u \in H^2(\mathfrak{F}_1),\ v \in \overline{\Delta_1 L^2(\mathfrak{E})}\} \ominus \mathbf{G}, \tag{4.15}$$
$$\mathbf{H}_{21} = \{\Theta_3 u' \oplus Z_{3(21)}^{-1}(\Delta_3 u' \oplus v') \colon u' \in H^2(\mathfrak{F}_{21}),\ v' \in \overline{\Delta_{21} L^2(\mathfrak{E})}\} \ominus \mathbf{G},$$

with $\mathbf{G} = \{\Theta w \oplus \Delta w \colon w \in H^2(\mathfrak{E})\}$; see (1.47).

With every pair of functions u, v occurring in the definition of $\mathbf{H}_1$ let us associate the functions

$$u' = \Theta_2 u, \quad v' = Z_{21}^{-1}(\Delta_2 u \oplus v);$$

these are among those occurring in the definition of $\mathbf{H}_{21}$. Because $\Theta_{32}(\lambda) = \Theta_3(\lambda)\Theta_2(\lambda)$, we have

$$\Theta_{32}u = \Theta_3 u'; \tag{4.16}$$

on the other hand (4.14) implies

$$\begin{aligned} Z_{(32)1}^{-1}(\Delta_{32}u \oplus v) &= Z_{(32)1}^{-1}(Z_{32}^{-1} \oplus I_1)((\Delta_3\Theta_2 u \oplus \Delta_2 u) \oplus v) \\ &= Z_{3(21)}^{-1}(I_{3(21)} \oplus Z_{21}^{-1})(\Delta_3\Theta_2 u \oplus (\Delta_2 u \oplus v)) \\ &= Z_{3(21)}^{-1}(\Delta_3 u' \oplus v'). \end{aligned} \tag{4.17}$$

From relations (4.15)–(4.17) it follows that $\mathbf{H}_1 \subset \mathbf{H}_{21}$. This concludes the proof of Theorem 4.3.

Notice that it turns out from the proof that the regular factorizations $\Theta(\lambda) = \Theta_2(\lambda)\Theta_1(\lambda) = \Theta_2'(\lambda)\Theta_1'(\lambda)$ provide the same invariant subspace if and only if $\Theta_1'(\lambda) = Z\Theta_1(\lambda)$ and $\Theta_2'(\lambda) = \Theta_2(\lambda)Z^*$ hold with a constant unitary operator Z from the intermediate space $\mathfrak{F}$ onto the intermediate space $\mathfrak{F}'$.

3. As already remarked, it is possible for more than one invariant subspace to be associated with the same regular divisor of $\Theta(\lambda)$. For an example let us consider the simplest of the factorizations discussed in Sec. 2.4. Namely let $\{\mathfrak{E}, \mathfrak{E}_*, \Theta(\lambda)\}$ be chosen with $\mathfrak{E} = \{0\}$, $\mathfrak{E}_* = E^1$, and $\Theta(\lambda) \equiv O$. Then the factorization

$$\Theta(\lambda) = \Theta_2(\lambda)\Theta(\lambda)$$

is regular for every inner function $\{E^1, E^1, \Theta_2(\lambda)\}$. Now if we choose in particular $\Theta_2(\lambda)$ equal to λ or λ^2, the invariant subspaces $\mathbf{H}_1$ corresponding to the two factorizations are λH^2 and $\lambda^2 H^2$, respectively, hence different, Moreover, the characteristic function of $\mathbf{T}_2 = \mathbf{P}_2\mathbf{T}|\mathbf{H}_2$ coincides with $\{E^1, E^1, \lambda\}$ in the first case and with $\{E^1, E^1, \lambda^2\}$ in the second case, and these functions clearly do not coincide. It is worthwhile to observe that in this example we have to do (up to unitary equivalence) with the unilateral shift S on H^2, and with the subspaces SH^2 and S^2H^2, respectively. This remark raises the problem of finding cases where uniqueness does hold. To this end we introduce a definition.

Definition. A contractive analytic function $\{\mathfrak{E}, \mathfrak{F}, \Theta_1(\lambda)\}$ is called a *strong regular divisor* of $\{\mathfrak{E}, \mathfrak{E}_*, \Theta(\lambda)\}$ if there exists a unique contractive analytic function $\{\mathfrak{F}, \mathfrak{E}_*, \Theta_2(\lambda)\}$ such that

$$\text{(F)} \quad \Theta(\lambda) = \Theta_2(\lambda)\Theta_1(\lambda)$$

is a regular factorization of $\Theta(\lambda)$.

Observe that if in a factorization (F), regular or not, the factor $\Theta_1(\lambda)$ is an outer function, then the factor $\Theta_2(\lambda)$ is uniquely determined by $\Theta(\lambda)$ and $\Theta_1(\lambda)$. This follows immediately from the fact that we have then $\overline{\Theta_1(\lambda)\mathfrak{E}} = \mathfrak{F}$ for every $\lambda \in D$; see Proposition V.2.4. Also, if the factor $\Theta_1(\lambda)$ in (F) is a $*$-inner function, that is, if $\Theta_1^{\sim}(e^{-it}) = \Theta_1(e^{it})^*$ is isometric a.e., then we have $\Theta_2(e^{it}) = \Theta_2(e^{it})\Theta_1(e^{it})\Theta_1(e^{it})^* = \Theta(e^{it})\Theta_1(e^{it})^*$ a.e.; thus $\Theta_2(e^{it})$ is determined by $\Theta(e^{it})$ and $\Theta_1(e^{it})$ a.e., and hence the function $\Theta_2(\lambda)$ is determined by $\Theta(\lambda)$ and $\Theta_1(\lambda)$. We have proved the following proposition.

Proposition 4.4. *Every outer or $*$-inner regular divisor of a contractive analytic function $\Theta(\lambda)$, is a strong regular divisor of $\Theta(\lambda)$.*

Thus, in particular, the outer factor in the canonical factorization of $\Theta(\lambda)$, and the $*$-inner factor in the $*$-canonical factorization of $\Theta(\lambda)$, are *strong* regular divisors of $\Theta(\lambda)$. Somewhat less immediate is the following result.

Proposition 4.5. *Let $\{\mathfrak{E},\mathfrak{E}_*,\Theta(\lambda)\}$ be a contractive analytic function for which the conditions*

$$u_* \in H^2(\mathfrak{E}_*), \quad v \in \overline{\Delta L^2(\mathfrak{E})}, \quad \Theta^* u_* + \Delta v = 0 \tag{4.18}$$

imply $u_* = 0, v = 0$. Then every regular divisor of $\Theta(\lambda)$ is strong.

Proof. We have to show that if $\Theta_2(\lambda)\Theta_1(\lambda)$ and $\Theta_2'(\lambda)\Theta_1(\lambda)$ are two regular factorizations of $\Theta(\lambda)$ with the same second factor $\{\mathfrak{E},\mathfrak{F},\Theta_1(\lambda)\}$ then $\Theta_2(\lambda) = \Theta_2'(\lambda)$.

Let $w \in H^2(\mathfrak{F})$. From the regularity property of the factorizations we deduce that there exist sequences $\{v_n\}$, $\{v_n'\}$ of elements of $L^2(\mathfrak{E})$ such that

$$\Delta_2\Theta_1 v_n \oplus \Delta_1 v_n \to \Delta_2 w \oplus 0, \quad \Delta_2'\Theta_1 v_n' \oplus \Delta_1 v_n' \to \Delta_2' w \oplus 0 \quad (n \to \infty).$$

As $\Delta_2\Theta_1 v_n \oplus \Delta_1 v_n = Z\Delta v_n$ and $\Delta_2'\Theta_1 v_n' \oplus \Delta_1 v_n' = Z'\Delta v_n'$, where Z and Z' are the isometries corresponding to the two factorizations, we conclude that the limits $l = \lim \Delta v_n$ and $l' = \lim \Delta v_n'$ also exist (in $L^2(\mathfrak{E})$). We have

$$\Delta^2 v_n = \Theta_1^*\Delta_2^2\Theta_1 v_n + \Delta_1^2 v_n \to \Theta_1^*\Delta_2^2 w = \Theta_1^* w - \Theta_1^*\Theta_2^*\Theta_2 w = \Theta_1^* w - \Theta^*\Theta_2 w$$

and similarly

$$\Delta^2 v_n' \to \Theta_1^* w - \Theta^*\Theta_2' w \quad (n \to \infty).$$

On the other hand $\Delta^2 v_n \to \Delta l$ and $\Delta^2 v_n' \to \Delta l'$, thus it follows that

$$\Theta^*(\Theta_2 w - \Theta_2' w) + \Delta(l - l') = 0,$$

which implies by our hypothesis that $\Theta_2 w - \Theta_2' w = 0$ and, because w was arbitrary, that $\Theta_2(\lambda) = \Theta_2'(\lambda)$.

Proposition 4.6. *The hypothesis in Proposition 4.5 is satisfied in particular if $\Theta(e^{it})$ is unitary at the points t of a set of positive measure.*

Proof. Condition (4.18) implies then that $u_*(e^{it}) = 0$ on this set of positive measure. As the numerical function $(u_*(\lambda), e_*)$ belongs to H^2 for every $e_* \in \mathfrak{E}_*$, we deduce that $(u_*(\lambda), e_*) \equiv 0$. It follows that $u_*(\lambda) \equiv 0$, and hence by (4.18) that $\Delta(t)v(t) = 0$ a.e. Because $v \in \overline{\Delta L^2(\mathfrak{E})}$, this implies $v(t) = 0$ a.e.

The cases considered in the last two propositions can be characterized in terms of contractions in the following manner:

Theorem 4.7. *Suppose $\Theta(\lambda)$ coincides with the characteristic function of a c.n.u. contraction T on the space $\mathfrak{H}$. In order that $\Theta(\lambda)$ satisfy the hypothesis of Proposition 4.5 it is necessary and sufficient that there exist no nonzero invariant subspace*

$\mathfrak{H}_1$ *for* T *such that* $T|\mathfrak{H}_1$ *is isometric. In order that* $\Theta(\lambda)$ *satisfy the hypothesis of Proposition* 4.6, *it suffices that the spectrum of* T *not contain the whole unit circle.*

Proof. The second assertion follows immediately from Theorem VI.4.1.

As to the first one, let us observe that a subspace $\mathfrak{H}_1$ is invariant for T and $T|\mathfrak{H}_1$ is isometric if and only if $\mathfrak{H}_1$ is invariant for any isometric dilation U_+ of T. Thus, in order that no such $\mathfrak{H}_1 \neq \{0\}$ exist, it is necessary and sufficient that there be no nonzero h in $\mathfrak{H}$ all of whose images under U_+^n $(n = 1,2,\ldots)$ belong to $\mathfrak{H}$. In terms of the functional model of T given by Theorem VI.3.1 this means that there exists no nonzero $u_* \oplus v \in H^2(\mathfrak{E}_*) \oplus \overline{\Delta L^2(\mathfrak{E})}$ such that

$$e^{int}u_* \oplus e^{int}v \in \mathbf{H} \qquad (n = 0,1,2,\ldots),$$

that is, such that

$$e^{int}(\Theta^* u_* + \Delta v) \perp H^2(\mathfrak{E}) \qquad (n = 0,1,2,\ldots); \tag{4.19}$$

but (4.19) is obviously equivalent to the condition $\Theta^* u_* + \Delta v = 0$.

4. Let $\{\mathfrak{E}, \mathfrak{E}_*, \Theta(\lambda)\}$ be a purely contractive analytic function, and let $\mathbf{H}$ and $\mathbf{T}$ be the space and the contraction generated by (1.44)–(1.45). With each strong regular divisor $\Theta_1(\lambda)$ of $\Theta(\lambda)$ there is associated a *unique* invariant subspace for $\mathbf{T}$; let us denote it by $\mathbf{H}(\Theta_1)$.

It is easy to show that if $\Theta_1(\lambda)$ is a strong regular divisor of $\Theta(\lambda)$, then so is every function $\Theta_1'(\lambda)$ which differs from $\Theta_1(\lambda)$ in a constant unitary left-factor only, and in this case $\mathbf{H}(\Theta_1') = \mathbf{H}(\Theta_1)$.

In particular, every constant unitary function $\{\mathfrak{E}, \mathfrak{F}, \Theta_0\}$ is a strong regular divisor of $\Theta(\lambda)$; for these functions (and only for these; cf. Theorem 4.3(a)) we have $\mathbf{H}(\Theta_0) = \{0\}$. On the other hand, assume the function $\Theta(\lambda)$ itself is a strong regular divisor of $\Theta(\lambda)$ (which is not always the case, as shown by the example given at the beginning of Subsec. 3); then we have $\mathbf{H}(\Theta_1) = \mathbf{H}$ exactly for those functions $\Theta_1(\lambda)$ that differ from $\Theta(\lambda)$ only by a constant unitary left-factor.

When *all* regular divisors of $\Theta(\lambda)$ are *strong* (e.g., in the cases considered in Propositions 4.5 and 4.6), *every set* $\{\Theta_\gamma(\lambda)\}$ *of regular divisors of* $\Theta_T(\lambda)$ *admits a greatest common regular divisor* $\Theta_\wedge(\lambda)$ *and a least common regular multiple* $\Theta_\vee(\lambda)$, *determined up to coincidence by the following conditions.*

(d_1) $\Theta_\wedge(\lambda)$ is a regular divisor of every $\Theta_\gamma(\lambda)$ (and consequently of $\Theta(\lambda)$).

(d_2) If a function $\Theta'(\lambda)$ is a regular divisor of every $\Theta_\gamma(\lambda)$, it is also a regular divisor of $\Theta_\wedge(\lambda)$.

(m_1) Every $\Theta_\gamma(\lambda)$ is a regular divisor of $\Theta_\vee(\lambda)$.

(m_2) If a regular divisor $\Theta''(\lambda)$ of $\Theta(\lambda)$ is such that every $\Theta_\gamma(\lambda)$ is a regular divisor of $\Theta''(\lambda)$, then $\Theta_\vee(\lambda)$ is a regular divisor of $\Theta''(\lambda)$ (in particular, $\Theta_\vee(\lambda)$ is a regular divisor of $\Theta(\lambda)$).

To prove this, let us consider the subspaces

$$\mathbf{H}_\wedge = \bigcap_\gamma \mathbf{H}(\Theta_\gamma) \quad \text{and} \quad \mathbf{H}_\vee = \bigvee_\gamma \mathbf{H}(\Theta_\gamma)$$

of $\mathbf{H}$. They are obviously invariant for $\mathbf{T}$ and hence associated with some regular divisors of $\Theta(\lambda)$, which we denote by $\Theta_\wedge(\lambda)$ and $\Theta_\vee(\lambda)$, respectively. These functions are determined up to constant unitary left-factors. Properties (d_1) and (m_1) follow readily from Theorem 4.3(a). If $\Theta'(\lambda)$ is of the type given in (d_2), $\Theta'(\lambda)$ is a regular divisor of $\Theta(\lambda)$ also, and by virtue of Theorem 4.3(b), $\mathbf{H}(\Theta') \subset \mathbf{H}(\Theta_\gamma)$ for every Θ_γ; hence $\mathbf{H}(\Theta') \subset \mathbf{H}_\wedge = \mathbf{H}(\Theta_\wedge)$. By virtue of Theorem 4.3(a) this implies that $\Theta'(\lambda)$ is a regular divisor of $\Theta_\wedge(\lambda)$, proving property (d_2). Property (m_2) of $\Theta_\vee(\lambda)$ can be established in an analogous manner.

5 Invariant subspaces for contractions of class C_{11}

1. By Proposition II.3.5 every contraction $T \in C_{11}$ is quasi-similar to a unitary operator, indeed to the residual part $R = R_T$ of the minimal unitary dilation $U = U_T$ of T. Using this fact, we were able to construct a large family of hyperinvariant subspaces for such a T. We now continue this study by applying the methods developed since Chapter V. Because for unitary operators we have spectral theory at our disposal, it does not essentially restrict generality if we only consider completely nonunitary contractions (of class C_{11}).

We begin by finding the form of R in the functional model. Thus let T be a c.n.u. contraction in a (separable) Hilbert space $\mathfrak{H}$ and let $\{\mathfrak{E}, \mathfrak{E}_*, \Theta(\lambda)\}$ coincide with the characteristic function of T. It follows from the results of Sec. VI.3 (cf. in particular (VI.3.1) and (VI.3.17′)) that R is unitarily equivalent to multiplication by e^{it} in $\overline{\Delta L^2(\mathfrak{E})}$. Consequently, if we denote by $E_R(\alpha)$ the spectral measure corresponding to the unitary operator R and to the Borel subset α of the unit circle C, then $E_R(\alpha)$ is unitarily equivalent to multiplication on the function space $\overline{\Delta L^2(\mathfrak{E})}$ by the function

$$\chi_\alpha(t) = \begin{cases} 1 \text{ for } t \in (\alpha), \\ 0 \text{ for } t \in (\alpha'), \end{cases}$$

where $\alpha' = C \backslash \alpha$ and, according to our convention, $(\beta) = \{t \in [0, 2\pi) : e^{it} \in \beta\}$.

So we obtain readily that $E_R(\alpha) = O$ if and only if $\Delta(t) = O$ a.e. on (α). The resolvent set of R on C is therefore composed of the open arcs on which the radial limit of $\Theta(\lambda)$ is isometric a.e.

Definition 1. For a c.n.u. contraction T, let $\varepsilon(T)$ denote the set of points $e^{it} \in C$ at which the radial limit of the characteristic function of T exists and is *not isometric*.

Definition 2. For any subset α of C, its *essential support*, denoted by

$$\text{ess supp } \alpha,$$

is defined as the complement, with respect to C, of the maximal open subset of C whose intersection with α is of zero Lebesgue measure.

We can now express the above result as follows.

Proposition 5.1. *For a c.n.u. contraction T on a (separable) space $\mathfrak{H}$, we have* $\sigma(R) = \text{ess supp } \varepsilon(T)$.

Let us recall in this connection Proposition II.6.2, which implies that for every contraction T,

$$\sigma(R) \subset \sigma(T) \cap C,$$

and Proposition VI.4.4, which states that

$$\sigma(T) = \sigma(T) \cap C = \text{ ess supp } \varepsilon(T)$$

at least for those c.n.u. contractions $T \in C_{\cdot 1}$ *whose characteristic function has a scalar multiple*. For these contractions we have thus

$$\sigma(R) = \sigma(T).$$

We add one more definition.

Definition 3. A Borel subset α of the unit circle C is said to be *residual* for the c.n.u. contraction T if $\Theta_T(e^{it})$ is isometric at almost every point of the complement α' of α.

2. After these preliminaries let us consider a c.n.u. contraction T of class $C_{1\cdot}$ on the (separable) space $\mathfrak{H}$, and let $\{\mathfrak{E}, \mathfrak{E}_*, \Theta(\lambda)\}$ be a function coinciding with the characteristic function of T. As this function is $*$-outer, we can apply to it Proposition V.4.3; see the Remark added to this proposition. Thus to every Borel subset α of C there corresponds a factorization

$$\Theta(\lambda) = \Theta_{2\alpha}(\lambda)\Theta_{1\alpha}(\lambda)$$

of $\{\mathfrak{E}, \mathfrak{E}_*, \Theta(\lambda)\}$ as a product of two contractive analytic functions, with some intermediate space $\mathfrak{F}_\alpha$, such that $\Theta_{1\alpha}(\lambda)$ is outer and we have

$$\Theta_{1\alpha}(e^{it})^*\Theta_{1\alpha}(e^{it}) = \begin{cases} \Theta(e^{it})^*\Theta(e^{it}) & \text{a.e. on } (\alpha), \\ I_{\mathfrak{E}} & \text{a.e. on } (\alpha'), \end{cases} \tag{5.1}$$

$$\Theta_{2\alpha}(e^{it})^*\Theta_{2\alpha}(e^{it}) = I_{\mathfrak{F}_\alpha} \quad \text{a.e. on } (\alpha). \tag{5.2}$$

We have already seen in Sec. 3.3 that this factorization is regular. Moreover, because $\Theta_{1\alpha}(\lambda)$ is outer, it is a strong regular divisor of $\Theta(\lambda)$. Thus if $\mathbf{T}$ is the functional model of T on the space $\mathbf{H}$ corresponding to $\Theta(\lambda)$ in the sense of Theorem VI.3.1, then there exists a unique invariant subspace $\mathbf{H}(\Theta_{1\alpha})$ associated with $\Theta_{1\alpha}(\lambda)$; for the sake of brevity we denote it by $\mathbf{H}_\alpha$, and set $\mathbf{T}_\alpha = \mathbf{T}|\mathbf{H}_\alpha$. Because $\Theta(\lambda)$ is $*$-outer, we have

$$H^2(\mathfrak{E}) = \overline{\Theta^{\tilde{}} H^2(\mathfrak{E}_*)} = \overline{\Theta_{1\alpha}^{\tilde{}}\Theta_{2\alpha}^{\tilde{}} H^2(\mathfrak{E}_*)} \subset \overline{\Theta_{1\alpha}^{\tilde{}} H^2(\mathfrak{F}_\alpha)} \subset H^2(\mathfrak{E}),$$

and hence we infer that $\Theta_{1\alpha}(\lambda)$ is $*$-outer too. Thus $\Theta_{1\alpha}(\lambda)$ is outer from both sides and so is its purely contractive part $\Theta_{1\alpha}^0(\lambda)$, hence also the characteristic function of $\mathbf{T}_\alpha$. This implies $\mathbf{T}_\alpha \in C_{11}$. Moreover, $\Theta_{1\alpha}(e^{it})$ is isometric (even unitary) a.e. in (α'), therefore $\Theta_{1\alpha}^0(e^{it})$ is also isometric a.e. in (α'); thus α is residual for $\mathbf{T}_\alpha$.

Now let $\mathbf{H}_1$ be any subspace of $\mathbf{H}$, invariant for $\mathbf{T}$ and such that $\mathbf{T}_1 = \mathbf{T}|\mathbf{H}_1 \in C_{11}$ and α is residual for $\mathbf{T}_1$. $\mathbf{H}_1$ is associated with a regular factor $\Theta_1(\lambda)$ of $\Theta(\lambda)$ (which is uniquely determined up to a constant unitary left-factor by Theorem 4.3(a)). Because $\mathbf{T}_1 \in C_{11}$, the purely contractive part $\Theta_1^0(\lambda)$ of $\Theta_1(\lambda)$ (coinciding with the characteristic function of $\mathbf{T}_1$) is outer from both sides. Consequently $\Theta_1(\lambda)$ is also outer, and hence a strong regular divisor of $\Theta(\lambda)$; thus $\mathbf{H}_1$ is the only invariant subspace associated with $\Theta_1(\lambda)$,

$$\mathbf{H}_1 = \mathbf{H}(\Theta_1).$$

On the other hand, the set α is residual for $\mathbf{T}_1$, thus $\Theta_1^0(e^{it})$ is isometric a.e. in (α'), and the same is then true for $\Theta_1(e^{it})$:

$$\Theta_1(e^{it})^*\Theta_1(e^{it}) = I_{\mathfrak{E}} \quad \text{a.e. in } (\alpha'). \tag{5.3}$$

As $\Theta_1(\lambda)$ is a divisor of $\Theta(\lambda)$, we have obviously

$$\Theta(e^{it})^*\Theta(e^{it}) \leq \Theta_1(e^{it})^*\Theta_1(e^{it}) \quad \text{a.e. on } (0,2\pi). \tag{5.4}$$

Comparing (5.3) and (5.4) with (5.1) we obtain

$$\Theta_{1\alpha}(e^{it})^*\Theta_{1\alpha}(e^{it}) \leq \Theta_1(e^{it})^*\Theta_1(e^{it}) \quad \text{a.e. on } (0,2\pi).$$

By Proposition V.4.1(a) it follows that $\Theta_1(\lambda)$ is a divisor of $\Theta_{1\alpha}(\lambda)$; and, taking into account that $\Theta_1(\lambda)$ is outer, by Proposition 4.2(b), $\Theta_1(\lambda)$ is even a regular divisor of $\Theta_{1\alpha}(\lambda)$. Applying Theorem 4.3(b) to the strong regular divisors $\Theta_{1\alpha}(\lambda)$ and $\Theta_1(\lambda)$ of $\Theta(\lambda)$ we obtain that $\mathbf{H}_1 \subset \mathbf{H}_\alpha$ (because these are the only invariant subspaces associated with $\Theta_1(\lambda)$ and $\Theta_{1\alpha}(\lambda)$, respectively).

This proves that $\mathbf{H}_\alpha$ is the maximal subspace of $\mathbf{H}$ satisfying the conditions

$$\text{(i) } \mathbf{T}\mathbf{H}_\alpha \subset \mathbf{H}_\alpha, \quad \text{(ii) } \mathbf{T}_\alpha = \mathbf{T}|\mathbf{H}_\alpha \in C_{11}, \quad \text{(iii) } \alpha \text{ is residual for } \mathbf{T}_\alpha.$$

As these properties are obviously invariant with respect to unitary equivalence, the existence of maximal subspaces $\mathfrak{H}_\alpha$ with these properties follows for every c.n.u. contraction $T \in C_{1\cdot}$ on a (separable) Hilbert space $\mathfrak{H}$. Maximality clearly implies uniqueness of $\mathfrak{H}_\alpha$. Let us show that $\mathfrak{H}_\alpha$ is even hyperinvariant for T, that is, invariant for every bounded operator X commuting with T. It suffices to consider such X that are boundedly invertible, because every other X could be replaced by $X - \mu I$ with

$|\mu| > \|X\|$. The space $\mathfrak{H}'_\alpha = X\mathfrak{H}_\alpha$ is then invariant for T, and $T'_\alpha = T|\mathfrak{H}'_\alpha$ is similar to $T_\alpha = T|\mathfrak{H}_\alpha$: indeed we have

$$T'_\alpha = ST_\alpha S^{-1} \quad \text{with} \quad S = X|\mathfrak{H}_\alpha,$$

S being an affinity from $\mathfrak{H}_\alpha$ onto $\mathfrak{H}'_\alpha$. Consequently, $T_\alpha \in C_{11}$ implies $T'_\alpha \in C_{11}$. Denoting by R_α and R'_α the residual parts of the minimal unitary dilations of T_α and T'_α, respectively, R_α is quasi-similar to T_α and R'_α is quasi-similar to T'_α; as T_α and T'_α are similar we conclude that R_α and R'_α are quasi-similar and hence unitarily equivalent; see Proposition II.3.4. This implies that the set α is residual for T'_α as well. By the maximality of $\mathfrak{H}_\alpha$ we have thus $\mathfrak{H}'_\alpha \subset \mathfrak{H}_\alpha, X\mathfrak{H}_\alpha \subset \mathfrak{H}_\alpha$. This proves that $\mathfrak{H}_\alpha$ is hyperinvariant for T.

Let us return to the model $\mathbf{T}$ of T on $\mathbf{H}$ and find conditions for $\mathbf{H}_{\alpha_1} \subset \mathbf{H}_{\alpha_2}$. A necessary and sufficient condition follows from Theorem 4.3, namely that $\Theta_{1\alpha_1}(\lambda)$ be a regular divisor of $\Theta_{1\alpha_2}(\lambda)$. Now if a contractive analytic $\Theta_1(\lambda)$ is a divisor of a contractive analytic $\Theta_2(\lambda)$, then we have

$$\Theta_2(e^{it})^*\Theta_2(e^{it}) \leq \Theta_1(e^{it})^*\Theta_1(e^{it}) \quad \text{a.e.} \tag{5.5}$$

If $\Theta_1(\lambda)$ is outer, this inequality implies, conversely, that $\Theta_1(\lambda)$ is a divisor of $\Theta_2(\lambda)$ (cf. Proposition V.4.1(a)), and if both of them are regular divisors of $\Theta(\lambda)$ then Proposition 4.2(b) yields that $\Theta_1(\lambda)$ is a regular divisor of $\Theta_2(\lambda)$. When applied to the case under consideration, these facts imply that for $\mathbf{H}_{\alpha_1} \subset \mathbf{H}_{\alpha_2}$ it is necessary and sufficient that

$$\Theta_{1\alpha_2}(e^{it})^*\Theta_{1\alpha_2}(e^{it}) \leq \Theta_{1\alpha_1}(e^{it})^*\Theta_{1\alpha_1}(e^{it}) \quad \text{a.e.}$$

By (5.1), this condition is equivalent to the condition that almost every point of $\alpha_1 \cap \alpha'_2$ belong to $\varepsilon(\mathbf{T})'$, that is, that the complementary set, $\alpha'_1 \cup \alpha_2$, be residual for $\mathbf{T}$. Thus for $\mathbf{H}_{\alpha_1} \subset \mathbf{H}_{\alpha_2}$ it is necessary and sufficient that $\alpha'_1 \cup \alpha_2$ be residual for $\mathbf{T}$. This condition is satisfied in particular if $\alpha_1 \subset \alpha_2$, because then $\alpha'_1 \cup \alpha_2 = C$.

We deduce from this result that $\mathbf{H}_{\alpha_1} = \mathbf{H}_{\alpha_2}$ if and only if both $\alpha'_1 \cup \alpha_2$ and $\alpha'_2 \cup \alpha_1$ are residual for $\mathbf{T}$. The intersection of these two sets being equal to the complement, in C, of the symmetric difference $\alpha_1 \triangle \alpha_2$, we conclude that $\mathbf{H}_{\alpha_1} = \mathbf{H}_{\alpha_2}$ if and only if the complement of $\alpha_1 \triangle \alpha_2$ in C is residual for $\mathbf{T}$ (i.e., if $\Theta_T(e^{it})$ is isometric a.e. in $\alpha_1 \triangle \alpha_2$).

Let us consider another problem. Let $\{\alpha_n\}$ be a (finite or infinite) sequence of Borel sets $\alpha_n \subset C$ and let $\alpha = \bigcup_n \alpha_n$. Because $\alpha_n \subset \alpha$, we have $\mathbf{H}_{\alpha_n} \subset \mathbf{H}_\alpha$ and hence

$$\mathbf{M} = \bigvee_n \mathbf{H}_{\alpha_n} \subset \mathbf{H}_\alpha. \tag{5.6}$$

The subspace $\mathbf{M}$, being invariant for $\mathbf{T}$, is associated with a regular divisor $\Omega(\lambda)$ of $\Theta(\lambda)$. Because $\mathbf{H}_{\alpha_n} \subset \mathbf{M}$, Theorem 4.3 a) implies that $\Theta_{1\alpha_n}(\lambda)$ is a divisor of $\Omega(\lambda)$. So we have

$$\Theta(e^{it})^*\Theta(e^{it}) \leq \Omega(e^{it})^*\Omega(e^{it}) \leq \Theta_{1\alpha_n}(e^{it})^*\Theta_{1\alpha_n}(e^{it}) \quad \text{a.e.;}$$

by virtue of (5.1) it follows that

$$\Omega(e^{it})^*\Omega(e^{it}) = \Theta(e^{it})^*\Theta(e^{it}) \tag{5.7}$$

a.e. in (α_n). This being true for $n \geq 1$, we conclude that (5.7) holds a.e. in (α). Using (5.1) again, we obtain

$$\Omega(e^{it})^*\Omega(e^{it}) \leq \Theta_{1\alpha}(e^{it})^*\Theta_{1\alpha}(e^{it}) \quad \text{a.e.}$$

This implies, by virtue of Proposition V.4.1(a), that $\Theta_{1\alpha}(\lambda)$ is a divisor of $\Omega(\lambda)$, and (because $\Theta_{1\alpha}(\lambda)$ is outer) by virtue of Proposition 4.2 b), even a regular divisor of $\Omega(\lambda)$. Applying Theorem 4.3 b) and recalling that $\mathbf{H}_\alpha$ is the only invariant subspace associated with $\Theta_{1\alpha}(\lambda)$, we obtain that $\mathbf{H}_\alpha \subset \mathbf{M}$. Together with (5.6), this gives $\mathbf{M} = \mathbf{H}_\alpha$.

Finally, consider the particular cases where α is either empty or the whole circle C. For α *empty*, relation (5.1) implies that $\Theta_{1\alpha}(\lambda)$ is an inner function. As it is also outer, it must be a unitary constant (cf. Proposition V.2.3). Direct application of formula (1.47) yields then $\mathbf{H}_\alpha = \{0\}$. For $\alpha = C$, (5.1) implies

$$\Theta_{1\alpha}(e^{it})^*\Theta_{1\alpha}(e^{it}) = \Theta(e^{it})^*\Theta(e^{it}) \quad \text{a.e.,}$$

and hence it follows that $\Theta_{1\alpha}(\lambda)$ is the outer factor in the canonical factorization of $\Theta(\lambda)$; see Proposition V.4.1(b). Consequently $\mathbf{H}_\alpha$ is equal to the space of the $C_{\cdot 1}$ part in the triangulation of $\mathbf{T}$ of type

$$\begin{bmatrix} C_{\cdot 1} & * \\ O & C_{\cdot 0} \end{bmatrix};$$

see Sec. 3.3. In particular, if $\mathbf{T} \in C_{11}$ we have necessarily $\mathbf{H}_\alpha = \mathbf{H}$ for $\alpha = C$.

We can summarize our results for $T \in C_{11}$ as follows.

Theorem 5.2. *Let T be a c.n.u. contraction of class C_{11} on a (separable) space $\mathfrak{H} \neq \{0\}$. Let α be a Borel subset of the unit circle C. Among the subspaces $\mathfrak{L}$ of $\mathfrak{H}$, invariant for T and such that*

$$T|\mathfrak{L} \in C_{11} \quad \textit{and} \quad \alpha \textit{ is residual for } T|\mathfrak{L},$$

there is a maximal one (i.e., containing all the others), which we denote by $\mathfrak{H}_\alpha$; $\mathfrak{H}_\alpha$ is even hyperinvariant for T. We have

(i) $\mathfrak{H}_\alpha = \{0\}$ *if α is empty;* $\mathfrak{H}_\alpha = \mathfrak{H}$ *if $\alpha = C$.*
(ii) $\bigvee_n \mathfrak{H}_{\alpha_n} = \mathfrak{H}_\alpha$ *for $\alpha = \bigcup_n \alpha_n$.*
(iii) $\mathfrak{H}_{\alpha_1} \subset \mathfrak{H}_{\alpha_2}$ *if and only if $\alpha_1' \cup \alpha_2$ is residual for T, thus in particular if $\alpha_1 \subset \alpha_2$.*

From these properties it follows, moreover, that

(a) $\mathfrak{H}_{\alpha_1} = \mathfrak{H}_{\alpha_2}$ *if and only if $C \backslash (\alpha_1 \triangle \alpha_2)$ is residual for T.*
(b) $\mathfrak{H}_\alpha = \{0\}$ *if and only if $C \backslash \alpha$ is residual for T.*
(c) $\mathfrak{H}_\alpha = \mathfrak{H}$ *if and only if α is residual for T.*

We recall that T, being of class C_{11}, is quasi-similar to R; hence we infer in particular that the space $\mathfrak{R}$ of R is different from $\{0\}$. On the other hand, because T is c.n.u., its minimal unitary dilation U, hence also R, has absolutely continuous spectra. It follows that the spectrum $\sigma(R)$ is a nonempty perfect subset of C. If α and β are any two disjoint open arcs of C both having nonempty intersection with the perfect set $\sigma(R)$, then (by virtue of Proposition 5.1) the corresponding subspaces $\mathfrak{H}_\alpha$ and $\mathfrak{H}_\beta$ are *nontrivial* (by (b) and (c)), *different* (by (a)), and *neither of them includes the other* (by (iii)). Thus, Theorem 5.2 implies in particular the following statement.

For every c.n.u. $T \in C_{11}$ there is an infinity of hyperinvariant subspaces, and T is not unicellular (cf. Sec. III.7.2).

6 Spectral decomposition of contractions of class C_{11} whose characteristic function admits a scalar multiple

1. For a c.n.u. contraction $T \in C_{11}$, whose characteristic function admits a scalar multiple, Theorem 5.2 can be given a more complete form. We begin with a lemma.

Lemma 6.1. *Let $\{\mathfrak{E}, \mathfrak{E}_*, \Theta(\lambda)\}$ be a contractive analytic function admitting a scalar multiple $\delta(\lambda)$, and let $\Theta(\lambda) = \Theta_2(\lambda)\Theta_1(\lambda)$ be a factorization of $\Theta(\lambda)$ as the product of contractive analytic functions $\{\mathfrak{E}, \mathfrak{F}, \Theta_1(\lambda)\}$ and $\{\mathfrak{F}, \mathfrak{E}_*, \Theta_2(\lambda)\}$ such that $\Theta_2(e^{it})$ is isometric at the points t of a set of positive measure. Then $\Theta_1(\lambda)$ admits the scalar multiple $\delta(\lambda)$ also.*

Proof. By hypothesis, there exists a contractive analytic function $\{\mathfrak{E}_*, \mathfrak{E}, \Omega(\lambda)\}$ such that

$$\Omega(\lambda)\Theta_2(\lambda)\Theta_1(\lambda) = \delta(\lambda)I_{\mathfrak{E}}, \quad \Theta_2(\lambda)\Theta_1(\lambda)\Omega(\lambda) = \delta(\lambda)I_{\mathfrak{E}_*}. \tag{6.1}$$

The second equation yields:

$$\Theta_2(\lambda)\Phi(\lambda) \equiv O \quad \text{with} \quad \Phi(\lambda) = \Theta_1(\lambda)\Omega(\lambda)\Theta_2(\lambda) - \delta(\lambda)I_{\mathfrak{F}}.$$

Hence $\Theta_2(e^{it})\Phi(e^{it}) = O$ a.e., and consequently $\Phi(e^{it}) = O$ at almost every point t where $\Theta_2(e^{it})$ is isometric. Thus $\Phi(e^{it}) = O$ on a set of positive measure. Because $\Phi(\lambda)$ is a bounded analytic function, we have then necessarily $\Phi(\lambda) \equiv O$, and thus

$$\Theta_1(\lambda)\Omega(\lambda)\Theta_2(\lambda) = \delta(\lambda)I_{\mathfrak{F}}. \tag{6.2}$$

Equation (6.2) and the first equation (6.1) prove that $\delta(\lambda)$ is a (scalar) multiple of $\Theta_1(\lambda)$.

Now we are able to prove the following theorem which together with Theorem 5.2 establishes a *spectral decomposition* of these operators.

Theorem 6.2. *For a c.n.u. contraction T on $\mathfrak{H}$ of class C_{11} and such that the characteristic function of T admits a scalar multiple, the invariant subspaces $\mathfrak{H}_\alpha$ considered in Theorem 5.2 have the following additional properties,*

(iv) $\sigma(T_\alpha) \subset \overline{\alpha}$ (the closure of α).
(v) $\mathfrak{H}_\alpha = \bigcap_n \mathfrak{H}_{\alpha_n}$ *for* $\alpha = \bigcap \alpha_n$.

Proof. Part (iv): By virtue of (5.2), $\Theta_{2\alpha}(e^{it})$ is isometric a.e. in (α). If α has positive measure, it follows from Lemma 6.1 that $\Theta_{1\alpha}(\lambda)$ also admits a scalar multiple. The same is true if α is of zero measure. Indeed, in this case (5.1) implies $\Theta_{1\alpha}(e^{it})^*\Theta_{1\alpha}(e^{it}) = I_{\mathfrak{E}}$ a.e., so the outer function $\Theta_{1\alpha}(\lambda)$ is also inner, and hence a unitary constant (cf. Proposition V.2.3), consequently it admits the scalar multiple $\delta(\lambda) \equiv 1$. Thus $\Theta_{1\alpha}(\lambda)$ admits in every case a scalar multiple, and the same is then true for the purely contractive part of $\Theta_{1\alpha}(\lambda)$ (cf. Proposition V.6.8). The latter function coincides with the characteristic function of T_α (observe that T_α is c.n.u. and of class C_{11}); thus we have, by virtue of Proposition VI.4.4,

$$\sigma(T_\alpha) = \text{ess supp}\, \varepsilon(T_\alpha) \quad \text{(cf. Sec. 5.1).}$$

On the other hand, because α is residual for T_α (cf. Theorem 5.2), $\varepsilon(T_\alpha) \cap \alpha'$ is of zero measure and *a fortiori* so is $\varepsilon(T_\alpha) \cap \overline{\alpha}'$. Because $\overline{\alpha}$ is an open set, this implies

$$\overline{\alpha}' = [\text{ess supp}\, \varepsilon(T_\alpha)]', \quad \overline{\alpha} \supset \text{ess supp}\, \varepsilon(T_\alpha),$$

and therefore $\sigma(T_\alpha) \subset \overline{\alpha}$.

Part (v): In this proof we apply the functional model $\mathbf{T}$ of T on $\mathbf{H}$. Because $\alpha = \bigcap_n \alpha_n$ implies $\alpha \subset \alpha_n$ $(n = 1, 2, \ldots)$, it follows from Theorem 5.2 that $\mathbf{H}_\alpha \subset \mathbf{H}_{\alpha_n}$ $(n = 1, 2, \ldots)$, and hence

$$\mathbf{H}_{\alpha_m} \supset \mathbf{M} = \bigcap_n \mathbf{H}_{\alpha_n} \supset \mathbf{H}_\alpha. \tag{6.3}$$

As $\mathbf{M}$ is invariant for $\mathbf{T}$, it is associated with a regular divisor $\Omega(\lambda)$ of $\Theta(\lambda)$. By virtue of Theorem 4.3(a), it follows from (6.3) that $\Theta_{1\alpha}(\lambda)$ is a regular divisor of $\Omega(\lambda)$ and $\Omega(\lambda)$ is a regular divisor of $\Theta_{1\alpha_m}(\lambda)$ $(m = 1, 2, \ldots)$. This implies, a.e.,

$$\Theta_{1\alpha}(e^{it})^*\Theta_{1\alpha}(e^{it}) \geq \Omega(e^{it})^*\Omega(e^{it}) \geq \Theta_{1\alpha_m}(e^{it})^*\Theta_{1\alpha_m}(e^{it}) \quad (m = 1, 2, \ldots);$$

taking (5.1) into account we deduce that

$$\Omega(e^{it})^*\Omega(e^{it}) = \Theta_{1\alpha}(e^{it})^*\Theta_{1\alpha}(e^{it}) \quad \text{a.e.} \tag{6.4}$$

On the other hand, $\Omega(\lambda)$ being a *regular divisor of* $\Theta_{1\alpha_m}(\lambda)$, the factors in the regular factorization $\Theta_{1\alpha_m}(\lambda) = \Omega_m(\lambda)\Omega(\lambda)$ must be such that $\Omega_m(e^{it})$ is isometric a.e. where $\Theta_{1\alpha_m}(e^{it})$ is isometric (cf. Propositions 3.1 and 3.2(c)), i.e. on (α'_m). Suppose α'_m is of positive measure for at least one m. We have observed while proving assertion (iv) that the functions $\Theta_{1\alpha}(\lambda)$ (for arbitrary α) admit scalar multiples. Lemma 6.1 then tells us that $\Omega(\lambda)$ also admits a scalar multiple. As $\Omega(\lambda)$ is $*$-outer (because $\Omega^{\tilde{}}(\lambda)\Omega_{\tilde{m}}(\lambda) = \Theta_{1\tilde{\alpha}_m}(\lambda)$ and $\Theta_{1\tilde{\alpha}_m}(\lambda)$ is outer), it follows that $\Omega(\lambda)$ is outer from both sides (cf. Theorem V.6.2(c)). Now the relation (6.4) between the outer functions $\Omega(\lambda)$ and $\Theta_{1\alpha}(\lambda)$ implies that they are equal up to constant unitary left-factors (cf. Proposition V.4.1); hence $\mathbf{M} = \mathbf{H}_\alpha$. This was proved under the assumption that at least one of the sets α'_m has positive measure. In the remaining case all the sets $\alpha_1, \alpha_2, \ldots$ as well as their intersection α are obviously residual for $\mathbf{T}$; by Theorem 5.2(c) we have therefore $\mathbf{H}_{\alpha_1} = \mathbf{H}_{\alpha_2} = \cdots = \mathbf{H}$, $\mathbf{H}_\alpha = \mathbf{H}$, and thus $\mathbf{M} = \mathbf{H}_\alpha$ holds in this case too. This completes the proof.

2. Let us consider now another problem, closely related to the previous one. This is to study c.n.u. contractions T of class $C_{\cdot 1}$, whose defect indices are not both infinite. The characteristic function $\Theta_T(\lambda)$ is then an outer function, thus

$$\mathfrak{d}_{T^*} = \dim \mathfrak{D}_{T^*} \leq \dim \mathfrak{D}_T = \mathfrak{d}_T$$

(cf. the remark at the end of Sec. V.2.3). Therefore our hypothesis means that $\mathfrak{d}_{T^*}$ is finite.

Theorem 6.3. *Let T be a c.n.u. contraction of class $C_{\cdot 1}$ such that $\mathfrak{d}_{T^*} < \infty$. Then either every point of the open unit disc D is an eigenvalue of T, or no point of D belongs to the spectrum of T. In order that the second case occur it is necessary and sufficient that one of the following (equivalent) conditions be satisfied.*

(i) $T \in C_{11}$.
(ii) $\mathfrak{d}_T = \mathfrak{d}_{T^*}$.
(iii) $\Theta_T(\lambda)$ *admits a scalar multiple.*

Remark. In Sec. VI.4.2 we constructed an example of a $T \in C_{11}$ for which $\sigma(T) = \overline{D}$, but then we had $\mathfrak{d}_T = \mathfrak{d}_{T^*}$ $(= \dim \mathfrak{E}) = \infty$.

Proof. No higher-dimensional space can be applied linearly, continuously, and one-to-one into a finite-dimensional space, so in the case $\mathfrak{d}_T > \mathfrak{d}_{T^*}$ there is no $\lambda \in D$ for which the operator $\Theta_T(\lambda)$ (from $\mathfrak{D}_T$ into $\mathfrak{D}_{T^*}$) is invertible. The characteristic function $\Theta_T(\lambda)$ is outer because T is of class $C_{\cdot 1}$. Hence, if $\mathfrak{d}_T = \mathfrak{d}_{T^*}$ then $\Theta_T(\lambda)$ is invertible for every $\lambda \in D$, and so $\Theta_T(\lambda)$ admits a scalar multiple by Proposition V.6.1; that is, (ii) implies (iii). On the other hand, if (iii) holds then the outer function $\Theta_T(\lambda)$ is also $*$-outer by Theorem V.6.2(c), and hence $T \in C_{11}$. Finally, (i) implies that $\Theta_T(\lambda)$ is outer from both sides, whence $\mathfrak{d}_T = \mathfrak{d}_{T^*}$ immediately follows.

Examples. Both cases indicated in the theorem actually occur, as shown by the following examples.

(1) The function $\{E^1, E^1, \frac{1}{2}(\lambda - 1)\}$ is purely contractive analytic and, by virtue of Proposition III.1.3, outer (clearly from both sides). The c.n.u. contraction T generated by this function is therefore of class C_{11}, it has the defect indices $\mathfrak{d}_T = \mathfrak{d}_{T^*} = 1$, and $\sigma(T)$ is contained in the unit circle (in fact, $\sigma(T) = C$).

(2) The function $\{E^2, E^1, \Theta(\lambda)\}$ defined by

$$\Theta(\lambda) \begin{bmatrix} x_1 \\ x_2 \end{bmatrix} = \frac{1}{\sqrt{2}} \frac{1}{2}(\lambda - 1)x_1 + \frac{1}{\sqrt{2}} \lambda x_2$$

is evidently purely contractive and analytic. Moreover, it is outer, because

$$\Theta(\lambda) \begin{bmatrix} -2\sqrt{2}\, u(\lambda) \\ \sqrt{2}\, u(\lambda) \end{bmatrix} = u(\lambda) \quad (u \in H^2),$$

and hence

$$\Theta H^2(E^2) = H^2 = H^2(E^1).$$

The contraction T generated by this function is thus of class $C_{\cdot 1}$ and has $\mathfrak{d}_T = 2$, $\mathfrak{d}_{T^*} = 1$. So every point of D is an eigenvalue of T.

7 Notes

1. The fundamental Theorem 1.1, stating the relation between the invariant subspaces for a c.n.u. contraction T on the (separable) space $\mathfrak{H}$ and the regular factorizations of the characteristic function $\Theta_T(\lambda)$, was first proved in Sz.-N.–F. [4] and [IX]. However, there is a difference in the presentation that needs some explanation.

Here, as well as in the papers referred to, we make use of the functional model $\{\mathbf{H},\mathbf{T}\}$ of $\{\mathfrak{H},T\}$, generated by a contractive analytic function $\{\mathfrak{E},\mathfrak{E}_*,\Theta(\lambda)\}$ coinciding with the characteristic function of T. Let $\Theta(\lambda) = \Theta_2(\lambda)\Theta_1(\lambda)$ be a factorization of $\Theta(\lambda)$ as a product of two contractive analytic functions, $\{\mathfrak{E},\mathfrak{F},\Theta_1(\lambda)\}$ and $\{\mathfrak{F},\mathfrak{E}_*,\Theta_2(\lambda)\}$, and let Z be the corresponding isometry

$$Z\colon \overline{\Delta L^2(\mathfrak{E})} \to \overline{\Delta_2 L^2(\mathfrak{F})} \oplus \overline{\Delta_1 L^2(\mathfrak{E})};$$

see (1.42). If Z is unitary (i.e., if the factorization is regular), then we can identify the elements of the two spaces corresponding to Z and so we can pass from the model $\{\mathbf{H},\mathbf{T}\}$ to the (unitarily equivalent) model $\{\mathscr{H},\mathscr{T}\}$ defined by the formulas:

$$\begin{aligned}\mathscr{H} =& [H^2(\mathfrak{E}_*) \oplus \overline{\Delta_2 L^2(\mathfrak{F})} \oplus \overline{\Delta_1 L^2(\mathfrak{E})}] \\ & \ominus \{\Theta_2\Theta_1 u \oplus \Delta_2\Theta_1 u \oplus \Delta_1 u\colon u \in H^2(\mathfrak{E})\},\end{aligned} \tag{1}$$

$$\begin{aligned}\{\mathscr{T}^*(u_* \oplus v_2 \oplus v_1) =& e^{-it}[u_*(e^{it}) - u_*(0)] \oplus e^{-it}v_2(t) \oplus e^{-it}v_1(t) \\ & (u_* \oplus v_2 \oplus v_1 \in \mathscr{H});\end{aligned} \tag{2}$$

this was done in the papers mentioned. To the decomposition $\mathbf{H} = \mathbf{H}_1 \oplus \mathbf{H}_2$ (cf. (1.47) and (1.48)) corresponds the decomposition $\mathscr{H} = \mathscr{H}_1 \oplus \mathscr{H}_2$, where

$$\begin{aligned}\mathscr{H}_1 =& \{\Theta_2 u \oplus \Delta_2 u \oplus v_1\colon u \in H^2(\mathfrak{F}), v_1 \in \overline{\Delta_1 L^2(\mathfrak{E})}\} \\ & \ominus \{\Theta_2\Theta_1 w \oplus \Delta_2\Theta_1 w \oplus \Delta_1 w\colon w \in H^2(\mathfrak{E})\},\end{aligned} \tag{3}$$

$$\mathscr{H}_2 = [H^2(\mathfrak{E}_*) \oplus \overline{\Delta_2 L^2(\mathfrak{F})} \oplus \{0\}] \ominus \{\Theta_2 u \oplus \Delta_2 u \oplus 0\colon u \in H^2(\mathfrak{F})\}. \tag{4}$$

When we have to do with a fixed regular factorization, the two manners of representation are equally useful. However, if we wish to study the relations between the invariant subspaces corresponding to different regular factorizations, then the model $\{\mathscr{H},\mathscr{T}\}$ cannot be used because it depends on the factorization chosen. This is why we have preferred in this book to construct the invariant subspaces in the model $\{\mathbf{H},\mathbf{T}\}$. This enabled us to obtain the results stated in Theorem 4.3, and in particular those in Secs. 5 and 6. Nevertheless, the space $\mathscr{H}$ and its operator $\mathscr{T}$

have their own advantage: they can be constructed by the formulas (1) and (2) even if the factorization under consideration is not regular. $\mathscr{T}$ will always be a contraction on $\mathscr{H}$, and the space $\mathscr{H}_2$ defined by (4) will be equal to $\mathscr{H} \ominus \mathscr{H}_1$. Moreover one can show in analogy to Proposition 2.1 that if

$$\mathscr{T} = \begin{bmatrix} \mathscr{T}_1 & \mathscr{X} \\ O & \mathscr{T}_2 \end{bmatrix}$$

is the triangulation of $\mathscr{T}$ corresponding to the decomposition $\mathscr{H} = \mathscr{H}_1 \oplus \mathscr{H}_2$, then the characteristic functions of $\mathscr{T}_1$ and $\mathscr{T}_2$ coincide with the purely contractive parts of $\Theta_1(\lambda)$ and $\Theta_2(\lambda)$, respectively. But *if the factorization is not regular then $\mathscr{T}$ will not be unitarily equivalent to* $\mathbf{T}$ (or to T, which amounts to the same thing). In fact, since Z is then nonunitary, the space

$$\mathbf{Z} = [\overline{\Delta_2 L^2(\mathfrak{F})} \oplus \overline{\Delta_1 L^2(\mathfrak{E})}] \ominus Z \cdot \overline{\Delta L^2(\mathfrak{E})}$$

is nonzero. Thus it follows that $\mathscr{T}$ is the orthogonal sum of a contraction unitarily equivalent to $\mathbf{T}$ (and to T), and of a unitary operator: the multiplication by the function e^{it} on the space $\mathbf{Z}$. We can formulate the following result.

Theorem. *To every factorization $\Theta_T(\lambda) = \Theta_2(\lambda)\Theta_1(\lambda)$ of the characteristic function of a c.n.u. contraction T as a product of two contractive analytic functions, there corresponds an invariant subspace (if not for T, at least) for $T' = T \oplus U$ (where U is some unitary operator with absolutely continuous spectrum), such that if*

$$T' = \begin{bmatrix} T_1' & X' \\ O & T_2' \end{bmatrix}$$

is the triangulation of T' induced by this invariant subspace and its orthogonal complement, then the characteristic functions of T_1' and T_2' coincide with the purely contractive parts of the factors $\Theta_1(\lambda)$ and $\Theta_2(\lambda)$, respectively.

See SZ.-N.–F. [IX] p. 300. A different proof, entirely independent of dilation theory and using only the definition of the characteristic function, was given in SZ.-N.–F. [3] and [8].

2. The fact that to every nontrivial invariant subspace for an operator there corresponds a nontrivial factorization of the characteristic function of the operator, was discovered by Soviet authors including LIVŠIC AND POTAPOV [1], M. S. BRODSKIĬ [1], and BRODSKIĬ AND LIVŠIC [1], among others. The operators T considered by them were those that differ from a unitary or a self-adjoint one by an additive term of finite rank or at least of finite trace. These authors also noticed that, conversely, every factorization of the characteristic function generates an invariant subspace, if not of T at least of some operator $T' = T \oplus S$, where S is a unitary or self-adjoint operator, respectively. They did not find criteria for the factorization to generate an invariant subspace of the given T itself, except in some particular cases; see BRODSKIĬ AND LIVŠIC [1] Theorem 16, and ŠMUL'JAN [2] Theorem 2.7; and even in these particular cases their criteria are more involved than our general criterion of regularity of the factorization. Let us mention that ŠMUL′JAN [2] (§4) obtained a result equivalent to case (ii) of our Proposition 3.5. (For the further development of this research see BRODSKIĬ AND ŠMUL′JAN [1] and the monograph of M. S. BRODSKIĬ [9].)

The research presented in this chapter, although influenced in part by the problems treated by the authors cited above, follows an entirely different path, which is related in a natural manner to the study of the structure of the unitary dilation and of the corresponding functional model. Sections 1 and 2 reproduce essentially parts of the papers SZ.-N.–F. [4] and [IX]. SZ.-NAGY [14] asked whether strange factorizations exist. The first examples of such factorizations were constructed in FOIAŞ [10], where it was shown that Θ_T has no strange factorizations if the defect indices of T are finite. The entire Sec. 4 constitutes an elaborated and precise form of some results sketched in SZ.-N.–F. [7]. Proposition 3.2 is new. The spectral decomposition of the operators of class C_{11} has been indicated in SZ.-N.–F. [IX] and [IX*], but its study in Secs. 5 and 6 is much more complete.

The reader may ask whether Theorem 6.2, which we have established for contractions of class C_{11} whose characteristic function admits a scalar multiple, does hold for every c.n.u. contraction of class C_{11}. As to property (iv) in this theorem, this is certainly not valid in the general case. Indeed, choose for T the contraction on $\mathfrak{H}$ so that its characteristic function coincides with the constant function $\Theta(\lambda) \equiv A$, where A is a self-adjoint operator for which (a) $O \leq A \leq I$, (b) 0 and 1 are not eigenvalues of A, and (c) A^{-1} is not bounded. Let α be a proper subarc of C, of positive length. We have seen in Sec. V.4.5 that in the corresponding factorization $\Theta(\lambda) = \Theta_{2\alpha}(\lambda)\Theta_{1\alpha}(\lambda)$ the function $\Theta_{1\alpha}(\lambda)$ has a unitary limit $\Theta_{1\alpha}(e^{it})$ a.e. in α', and $\Theta_{1\alpha}(\lambda)$ cannot be extended analytically through α'. Thus by Theorem VI.4.1 we have $\sigma(T_\alpha) \supset \alpha'$. (Actually $\sigma(T_\alpha) = \overline{D}$.) This shows that in this case property (iv) does not hold.

As to property (v) in the same theorem, the problem of its general validity is related to the following question. Does every contractive analytic function $\Theta(\lambda)$, outer from both sides and such that $\Theta(e^{it})$ is isometric on a set of positive measure, have the property that every regular divisor of $\Theta(\lambda)$ is also outer from both sides? The negative answer to both these question is given in Chap. IX (see Sec. 4).

Theorem 6.3 was stated in SZ.-N.–F. [IX] (Theorem 2) in a more general form, but the proof there was incomplete and it is still unknown whether in that general form the theorem holds. The restricted form given in the text was indicated in SZ.-N.–F. [IX*]. The fundamental problem, left open, is *to elucidate the structure of the contractive analytic functions that are outer from one side and inner from the other side.*

Chapter VIII

Weak Contractions

1 Scalar multiples

1. According to the usual definition, a self-adjoint operator A on a Hilbert space $\mathfrak{H}$, $A \geq O$, is said to be of *finite trace* if A is compact (i.e., completely continuous) and the sum of its eigenvalues $\neq 0$ (each counted with the respective multiplicity) is finite. This sum is the *trace* of A and is denoted by $\operatorname{tr}(A)$.

If A is of finite trace and if X is a bounded operator on $\mathfrak{H}$, then $A' = X^*AX$ is also of finite trace. In fact, A' is compact, $A' \geq O$, and if

$$Ah = \sum_n \mu_n (h, \varphi_n)\varphi_n, \quad A'h = \sum_m \mu'_m (h, \varphi'_m)\varphi'_m \quad (h \in \mathfrak{H})$$

are the spectral decompositions of A and A' with respect to the orthonormal systems $\{\varphi_n\}$, $\{\varphi'_m\}$ of eigenvectors (with the respective eigenvalues $\mu_n > 0$, $\mu'_m > 0$), then we have

$$\begin{aligned}\sum_m \mu'_m &= \sum_m (A'\varphi'_m, \varphi'_m) = \sum_m (AX\varphi'_m, X\varphi'_m) = \sum_m \sum_n \mu_n |(X\varphi'_m, \varphi_n)|^2 \\ &= \sum_n \mu_n \sum_m |(\varphi'_m, X^*\varphi_n)|^2 \leq \sum_n \mu_n \|X^*\varphi_n\|^2 \leq \|X^*\|^2 \sum_n \mu_n,\end{aligned}$$

so that

$$\operatorname{tr}(X^*AX) \leq \|X\|^2 \cdot \operatorname{tr}(A).$$

It follows in particular that if $P_{\mathfrak{L}}$ is the orthogonal projection from $\mathfrak{H}$ onto a subspace $\mathfrak{L}$, and if A is of finite trace, then $P_{\mathfrak{L}}AP_{\mathfrak{L}}$ and therefore $A_{\mathfrak{L}} = P_{\mathfrak{L}}A|\mathfrak{L}$ are also of finite trace.

Definition. A contraction T on Hilbert space $\mathfrak{H}$ is called a *weak contraction* if (i) its spectrum $\sigma(T)$ does not fill the unit disc D, and (ii) $I - T^*T$ is of finite trace.

Thus in particular all contractions T with finite defect index $\mathfrak{d}_T$ and with $\sigma(T) \neq \overline{D}$ (among them all unitary operators) are weak contractions.

B.Sz.-Nagy et al., *Harmonic Analysis of Operators on Hilbert Space*, Universitext,
DOI 10.1007/978-1-4419-6094-8_8, © Springer Science+Business Media, LLC 2010

Let us recall that if T is a contraction, then so is

$$T_a = (T - aI)(I - \bar{a}T)^{-1} \quad (|a| < 1),$$

and we have

$$I - T_a^* T_a = S^*(I - T^*T)S \quad \text{with} \quad S = (1 - |a|^2)^{1/2}(I - \bar{a}T)^{-1};$$

see Sec. VI.1.3. On the other hand, $\sigma(T_a)$ is the image of $\sigma(T)$ by the homography $\lambda \to \lambda_a = (\lambda - a)/(1 - \bar{a}\lambda)$, we conclude that if T is a weak contraction, then so is T_a for every $a \in D$.

Let T be a weak contraction and let us fix a point $a \in D$ not belonging to $\sigma(T)$, that is, such that T_a is boundedly invertible. Let

$$(I - T_a^* T_a)h = \sum \mu_n (h, \varphi_n) \varphi_n \tag{1.1}$$

be the spectral representation of $I - T_a^* T_a$ by means of an orthonormal sequence of its eigenvectors φ_n corresponding to the respective eigenvalues $\mu_n \neq 0$ (if there is no nonzero eigenvalue, the right-hand side of (1.1) should be taken to be 0). We have

$$\begin{aligned}(T_a \varphi_m, T_a \varphi_n) &= (\varphi_m, \varphi_n) - ((I - T_a^* T_a)\varphi_m, \varphi_n) \\ &= (1 - \mu_m)(\varphi_m, \varphi_n) = (1 - \mu_m)\delta_{mn}.\end{aligned}$$

As $\varphi_n \neq 0$ implies $T_a \varphi_n \neq 0$, we have $1 - \mu_n > 0$, and it follows that the vectors

$$\psi_n = (1 - \mu_n)^{-1/2} T_a \varphi_n \tag{1.2}$$

also form an orthonormal sequence. Moreover, we have

$$T_a^* \psi_n = (1 - \mu_n)^{-1/2} T_a^* T_a \varphi_n = (1 - \mu_n)^{-1/2} [\varphi_n - (I - T_a^* T_a)\varphi_n] = (1 - \mu_n)^{1/2} \varphi_n,$$

and hence

$$\varphi_n = (1 - \mu_n)^{-1/2} T_a^* \psi_n. \tag{1.3}$$

By (1.1)–(1.3) we have, for $g \in \mathfrak{H}$,

$$\begin{aligned}(I - T_a T_a^*) T_a g = T_a (I - T_a^* T_a) g &= \sum \mu_n (g, \varphi_n) T_a \varphi_n \\ &= \sum \mu_n (g, T_a^* \psi_n) \psi_n = \sum \mu_n (T_a g, \psi_n) \psi_n.\end{aligned}$$

As $T_a \mathfrak{H} = \mathfrak{H}$, this gives

$$(I - T_a T_a^*)h = \sum \mu_n (h, \psi_n) \psi_n \tag{1.4}$$

for every $h \in \mathfrak{H}$. This shows that $I - T_a T_a^*$ has the same nonzero eigenvalues as $I - T_a^* T_a$; hence $\operatorname{tr}(I - T_a^* T_a) = \operatorname{tr}(I - T_a T_a^*)$. Moreover, $\sigma(T_a^*)$ does not contain the point 0, thus T_a^* is a weak contraction. Finally, the obvious relation

$$T^* = (T_a^*)_{-\bar{a}}$$

implies that T^* is also a weak contraction. Thus, *if T is a weak contraction, then so is T^*.*

An important property of weak contractions is established in the following theorem.

Theorem 1.1. *The characteristic function of every weak contraction T has a scalar multiple.*

Proof. Let us choose a point a as above and consider the characteristic function of T_a. The defect spaces of T_a are of the same dimension, because the sequences $\{\varphi_n\}$ and $\{\psi_n\}$ of eigenvectors of $I - T_a^* T_a$ and $I - T_a T_a^*$, connected by (1.2) and (1.3), form orthonormal bases for these defect spaces.

If this dimension is a finite number N, then the characteristic function $\Theta_{T_a}(\lambda)$ has with respect to these bases the matrix

$$[(\Theta_{T_a}(\lambda)\varphi_i, \psi_j)] \qquad (j,i = 1,\ldots,N). \tag{1.5}$$

As $\Theta_{T_a}(0) = -T_a|\mathfrak{D}_T$ is invertible, we can apply Proposition V.6.1, and obtain that the determinant $d(\lambda)$ of the matrix (1.5) is a scalar multiple of $\Theta_T(\lambda)$. Moreover, we have

$$(\Theta_{T_a}(0)\varphi_i, \psi_j) = -(T_a\varphi_i, \psi_j) = -(1-\mu_i)^{1/2}(\psi_i, \psi_j) = -(1-\mu_i)^{1/2}\delta_{ij},$$

and hence

$$|d(0)|^2 = \prod_1^N (1-\mu_i) > 0. \tag{1.6}$$

If the defect spaces of T_a are infinite-dimensional, denote by $P^{(n)}$ and $P_*^{(n)}$ the orthogonal projections of these spaces onto the subspaces spanned by the vectors $\varphi_1,\ldots,\varphi_n$ and $\psi_1,\ldots,\psi_n$, respectively. Define functions

$$\{\mathfrak{D}_{T_a}, \mathfrak{D}_{T_a^*}, \Theta_n(\lambda)\} \quad (n = 1,2,\ldots)$$

by

$$\Theta_n(\lambda)f = P_*^{(n)}\Theta_{T_a}(\lambda)P^{(n)}f + \sum_{n+1}^{\infty}(f,\varphi_k)\psi_k \quad (f \in \mathfrak{D}_{T_a}).$$

These functions are contractive analytic and such that

$$\Theta_n(\lambda)^*g = P^{(n)}\Theta_{T_a}(\lambda)^*P_*^{(n)}g + \sum_{n+1}^{\infty}(g,\psi_k)\varphi_k \quad (g \in \mathfrak{D}_{T_a^*}).$$

As $n \to \infty$, we have thus

$$\Theta_n(\lambda) \to \Theta_{T_a}(\lambda), \quad \Theta_n(\lambda)^* \to \Theta_{T_a}(\lambda)^* \quad \text{(strongly)}. \tag{1.7}$$

Let $d^{(n)}(\lambda)$ be the determinant of the matrix

$$[(\Theta_{T_a}(\lambda)\varphi_i, \psi_j)] \qquad (i,j = 1,\ldots,n).$$

We have, for the same reason as (1.6),

$$|d^{(n)}(0)|^2 = \prod_1^n (1-\mu_i), \tag{1.8}$$

and $d^{(n)}(\lambda)$ is a scalar multiple of the contractive function $\{\mathfrak{D}^{(n)}, \mathfrak{D}_*^{(n)}, \Theta^{(n)}(\lambda)\}$, where

$$\mathfrak{D}^{(n)} = P^{(n)}\mathfrak{D}_{T_a}, \quad \mathfrak{D}_*^{(n)} = P_*^{(n)}\mathfrak{D}_{T_a^*}, \quad \Theta^{(n)}(\lambda) = P_*^{(n)}\Theta_{T_a}(\lambda)|\mathfrak{D}^{(n)}.$$

Thus there exist contractive analytic functions $\{\mathfrak{D}_*^{(n)}, \mathfrak{D}^{(n)}, \Omega^{(n)}(\lambda)\}$ $(n = 1, \ldots)$ such that

$$\Omega^{(n)}(\lambda)\Theta^{(n)}(\lambda) = d^{(n)}(\lambda)I_{\mathfrak{D}^{(n)}}, \quad \Theta^{(n)}(\lambda)\Omega^{(n)}(\lambda) = d^{(n)}(\lambda)I_{\mathfrak{D}_*^{(n)}}.$$

Setting

$$\Omega_n(\lambda)g = \Omega^{(n)}(\lambda)P_*^{(n)}g + d^{(n)}(\lambda)\sum_{n+1}^{\infty}(g, \psi_k)\varphi_k \quad (g \in \mathfrak{D}_{T_a^*})$$

we obtain contractive analytic functions $\{\mathfrak{D}_{T_a^*}, \mathfrak{D}_{T_a}, \Omega_n(\lambda)\}$ $(n = 1, \ldots)$ such that

$$\Omega_n(\lambda)\Theta_n(\lambda) = d^{(n)}(\lambda)I_{\mathfrak{D}_{T_a}}, \quad \Theta_n(\lambda)\Omega_n(\lambda) = d^{(n)}(\lambda)I_{\mathfrak{D}_{T_a^*}}. \tag{1.9}$$

As the defect spaces of T_a are separable and the functions $d^{(n)}(\lambda)$ and $\Omega^{(n)}(\lambda)$ are analytic on D with

$$|d^{(n)}(\lambda)| \leq 1, \quad \|\Omega_n(\lambda)\| \leq 1 \quad (\lambda \in D),$$

we know from the Vitali–Montel theorem that there exists a sequence of integers $n_q \to \infty$ for which $d^{(n_q)}(\lambda)$ converges on D to an analytic function $d(\lambda)$, $|d(\lambda)| \leq 1$, and $\Omega_{n_q}(\lambda)$ converges on D (weakly) to a contractive analytic function $\Omega(\lambda)$. Using also of (1.7) we deduce from (1.9) that

$$\Omega(\lambda)\Theta_{T_a}(\lambda) = d(\lambda)I_{\mathfrak{D}_{T_a}}, \quad \Theta_{T_a}(\lambda)\Omega(\lambda) = d(\lambda)I_{\mathfrak{D}_{T_a^*}}.$$

Moreover, (1.8) implies

$$|d(0)|^2 = \prod_1^{\infty}(1-\mu_i);$$

as $1-\mu_i > 0$ $(i = 1, \ldots)$ and $\sum \mu_i = \operatorname{tr}(I - T_a^* T_a) < \infty$, we have thus $d(0) \neq 0$. Hence $d(\lambda) \not\equiv 0$, and consequently $d(\lambda)$ is a scalar multiple of the characteristic function of T_a.

The relation between the characteristic functions of T and T_a (cf. Sec. VI.1.3) implies that T also has equal defect indices and that $\delta(\lambda) = d((\lambda - a)(1 - \bar{a}\lambda))$ is a scalar multiple of $\Theta_T(\lambda)$.

2. Let the contraction T on $\mathfrak{H}$ be such that $I - T^*T$ is of finite trace, and let $\mathfrak{L}$ be an invariant subspace for T. For $T_{\mathfrak{L}} = T|\mathfrak{L}$ we have then

$$I_{\mathfrak{L}} - T_{\mathfrak{L}}^* T_{\mathfrak{L}} = P_{\mathfrak{L}}(I - T^*T)|\mathfrak{L},$$

$P_{\mathfrak{L}}$ denoting the orthogonal projection onto $\mathfrak{L}$. Thus $I_{\mathfrak{L}} - T_{\mathfrak{L}}^* T_{\mathfrak{L}}$ is also of finite trace. If $\sigma(T_{\mathfrak{L}}) \neq \overline{D}$, then $T_{\mathfrak{L}}$ is also a weak contraction. But, in general, $T_{\mathfrak{L}}$ may not be a weak contraction if T is a weak contraction. An example is furnished by the regular factorization

$$\frac{\sqrt{3}}{2}\lambda = \begin{bmatrix} \frac{\sqrt{3}}{2} & \frac{1}{2}\lambda \end{bmatrix} \cdot \begin{bmatrix} \frac{1}{2}\lambda \\ \frac{\sqrt{3}}{2} \end{bmatrix} \qquad (\lambda \in D) \tag{1.10}$$

of the scalar-valued function $\Theta(\lambda) \equiv \sqrt{3}\lambda/2$; see Sec. VII.3.4. $\Theta(\lambda)$ is the characteristic function of a c.n.u. contraction T with defect indices equal to 1, and with $\sigma(T) \cap D = \{0\}$. Thus T is a weak contraction. The regular factorization (1.10) induces, in the sense of Theorem VII.1.1, an invariant subspace $\mathfrak{L}$ for T. As the second factor in (1.10) is a *purely* contractive analytic function $\{E^1, E^2, \Theta_1(\lambda)\}$, it coincides with the characteristic function of $T_{\mathfrak{L}} = T|\mathfrak{L}$ (cf. Proposition VII.2.1). $\Theta_1(\lambda)$ maps, for every fixed λ, the space E^1 into the space E^2, so it cannot have an inverse defined on the whole of E^2, and hence (by virtue of Theorem VI.4.1) we have $\sigma(T_{\mathfrak{L}}) = \overline{D}$; thus $T_{\mathfrak{L}}$ is not a weak contraction.

2 Decomposition C_0–C_{11}

1. Let T be any contraction on the space $\mathfrak{H}$, and let $T = T^{(0)} \oplus T^{(u)}$ be its canonical decomposition into its unitary part $T^{(u)}$ and c.n.u. part $T^{(0)}$. Then we have

$$I - T^*T = [I^{(0)} - T^{(0)*}T^{(0)}] \oplus O$$

and $\sigma(T) \cap D = \sigma(T^{(0)}) \cap D$. Hence we infer that *if T is a weak contraction then so is $T^{(0)}$, and conversely*. Therefore it does not restrict generality if we suppose in the sequel that T is c.n.u.

Thus let T be a c.n.u. weak contraction on $\mathfrak{H}$. Let

$$T = \begin{bmatrix} T_0 & X \\ O & T_1' \end{bmatrix}, \qquad \mathfrak{H} = \mathfrak{H}_0 \oplus \mathfrak{H}_1' \tag{2.1}$$

be the triangulation of T of type

$$\begin{bmatrix} C_{0\cdot} & * \\ O & C_{1\cdot} \end{bmatrix}. \tag{2.2}$$

This corresponds to the $*$-canonical factorization of $\Theta_T(\lambda)$ (cf. Sec. VII.3.3); thus the characteristic functions of T_0 and T_1' coincide with the purely contractive parts of the $*$-inner and $*$-outer factors of $\Theta_T(\lambda)$, respectively. We have just proved that

$\Theta_T(\lambda)$ admits a scalar multiple (Theorem 1.1); this implies, by virtue of Theorem V.6.2 and Proposition V.6.8, that the factors in the $*$-canonical factorization of $\Theta_T(\lambda)$, as well as their purely contractive parts, also admit scalar multiples. Hence it follows that these purely contractive parts are, respectively, inner and outer from both sides (cf. Theorem V.6.2). Consequently, T_0 belongs to C_{00} and hence also to C_0, whereas T_1' belongs to C_{11} (cf. Proposition VI.3.5 and Theorem VI.5.1).

Let $m_{T_0}(\lambda)$ be the minimal function of T_0 ($m_{T_0}(\lambda) \equiv 1$ if and only if $\mathfrak{H}_0 = \{0\}$). By virtue of Theorem III.5.1, the spectrum of T_0 in D consists of the zeros of $m_{T_0}(\lambda)$ and hence is a discrete set in D. Hence $T_0 = T|\mathfrak{H}_0$ is also a weak contraction.

Moreover, if $\mathfrak{L}$ is an arbitrary subspace of $\mathfrak{H}$, invariant for T and such that $T_{\mathfrak{L}} = T|\mathfrak{L} \in C_0$, then we have

$$T^n l = T_{\mathfrak{L}}^n l \to 0 \quad \text{for} \quad l \in \mathfrak{L}, n \to \infty$$

(because $T_{\mathfrak{L}} \in C_0 \subset C_{00}$), and this implies $\mathfrak{L} \subset \mathfrak{H}_0$ by virtue of the relation

$$\mathfrak{H}_0 = \{h\colon h \in \mathfrak{H}, T^n h \to 0\};$$

see (II.4.3). Hence $\mathfrak{H}_0$ is the maximal invariant subspace for T on which T induces a contraction of class C_0.

As to T_1', because it belongs to C_{11} and its characteristic function has a scalar multiple, its spectrum must lie entirely on the circle C (cf. Proposition VI.4.4). Moreover, $T_1'^* = T^*|\mathfrak{H}_1'$ and T^* is a weak contraction, thus we conclude that $T_1'^*$, and hence T_1' as well, are weak contractions.

We now show that $\mathfrak{H}_1'$ is maximal among those subspaces $\mathfrak{L}_*$ invariant for T^* for which $T^*|\mathfrak{L}_* \in C_{11}$. In fact, if $\mathfrak{L}_*$ is such a subspace, set

$$T_* = (T^*|\mathfrak{L}_*)^* \quad \text{and} \quad \mathfrak{L}_*' = \{l\colon l \in \mathfrak{L}_*, m_{T_0}(T_*)l = 0\}.$$

$\mathfrak{L}_*'$ is invariant for T_* and we have $m_{T_0}(T_*') = O$ for $T_*' = T_*|\mathfrak{L}_*'$. It follows that

$$T_*^n|\mathfrak{L}_*' = T_*'^n \to O \quad \text{as} \quad n \to \infty.$$

Because $T_* \in C_{11}$, this implies $\mathfrak{L}_*' = \{0\}$. Hence

$$l \in \mathfrak{L}_* \quad \text{and} \quad m_{T_0}(T_*)l = 0 \quad \text{imply} \quad l = 0,$$

and this statement is obviously equivalent to the following,

$$\overline{m_{T_0}(T_*)^*\mathfrak{L}_*} = \mathfrak{L}_*. \tag{2.3}$$

Now we have for $u \in H^\infty$

$$u(T_*)^* = u\tilde{\ }(T_*^*) = u\tilde{\ }(T^*|\mathfrak{L}_*) = u\tilde{\ }(T^*)|\mathfrak{L}_* = u(T)^*|\mathfrak{L}_*,$$

and hence for $l \in \mathfrak{L}_*$ and $h \in \mathfrak{H}_0$

$$(u(T_*)^* l, h) = (u(T)^* l, h) = (l, u(T)h) = (l, u(T_0)h).$$

In particular, taking $u = m_{T_0}$ this gives

$$(m_{T_0}(T_*)^* l, h) = 0 \quad (l \in \mathfrak{L}_*, h \in \mathfrak{H}_0);$$

thus $m_{T_0}(T_*)^* \mathfrak{L}_* \perp \mathfrak{H}_0$. By virtue of (2.3) this implies $\mathfrak{L}_* \perp \mathfrak{H}_0$ and hence $\mathfrak{L}_* \subset \mathfrak{H}'_1$.

In addition to the triangulation (2.1) of type (2.2) let us also consider the triangulation

$$T = \begin{bmatrix} T_1 & Y \\ O & T'_0 \end{bmatrix}, \quad \mathfrak{H} = \mathfrak{H}_1 \oplus \mathfrak{H}'_0, \tag{2.4}$$

of type

$$\begin{bmatrix} C_{\cdot 1} & * \\ O & C_{\cdot 0} \end{bmatrix}. \tag{2.5}$$

By arguments similar to the above, or simply observing that (2.4) is equivalent to the triangulation of T^* of type (2.2):

$$T^* = \begin{bmatrix} T_0'^* & Y^* \\ O & T_1^* \end{bmatrix}, \quad \mathfrak{H} = \mathfrak{H}'_0 \oplus \mathfrak{H}_1,$$

one sees that T_1 and T'_0 are weak contractions of classes C_{11} and C_0, respectively, that $\mathfrak{H}_1$ includes all the subspaces $\mathfrak{M}$ invariant for T such that $T|\mathfrak{M} \in C_{11}$, and $\mathfrak{H}'_0$ includes all the subspaces $\mathfrak{M}_*$ invariant for T^* such that $T^*|\mathfrak{M}_* \in C_0$.

We call T_0 *the* C_0 *part* and T_1 *the* C_{11} *part of* T. Obviously, the C_0 and C_{11} parts of T^* are then $T_0'^*$ and $T_1'^*$, respectively.

The subspaces $\mathfrak{H}_0$ and $\mathfrak{H}_1$ are invariant for T, therefore so is $\mathfrak{H}_0 \cap \mathfrak{H}_1$. The restriction of T to this intersection (being at the same time a restriction of $T_0 \in C_0$ and a restriction of $T_1 \in C_{11}$) belongs to $C_{0\cdot}$ and to $C_{1\cdot}$, which is impossible unless $\mathfrak{H}_0 \cap \mathfrak{H}_1 = \{0\}$. Hence $\mathfrak{H}_0 \cap \mathfrak{H}_1 = \{0\}$. This result, applied to T^* instead of T, yields $\mathfrak{H}'_0 \cap \mathfrak{H}'_1 = \{0\}$; by virtue of the obvious relation

$$\mathfrak{H} \ominus (\mathfrak{H}_0 \vee \mathfrak{H}_1) = \mathfrak{H}'_0 \cap \mathfrak{H}'_1,$$

this implies $\mathfrak{H}_0 \vee \mathfrak{H}_1 = \mathfrak{H}$.

We turn now to the study of the relations between the spectra. We observe first that an operator matrix

$$M = \begin{bmatrix} X_1 & X \\ O & X_2 \end{bmatrix},$$

whose entries are bounded operators, is boundedly invertible with the inverse

$$M^{-1} = \begin{bmatrix} X_1^{-1} & -X_1^{-1} X X_2^{-1} \\ O & X_2^{-1} \end{bmatrix} \tag{2.6}$$

provided X_1 and X_2 are boundedly invertible. From (2.4) follows the relation

$$T-\lambda I=\begin{bmatrix} T_1-\lambda I_1 & Y \\ O & T_0'-\lambda I_0' \end{bmatrix},$$

which together with (2.6) gives

$$\sigma(T)\subset\sigma(T_1)\cup\sigma(T_0'). \tag{2.7}$$

We show next that the contractions T_0 and T_0', of class C_0, have the same minimal function, that is,

$$m_{T_0}(\lambda)=m_{T_0'}(\lambda). \tag{2.8}$$

This implies, by virtue of Theorem III.5.1, that $\sigma(T_0)=\sigma(T_0')$; thus (2.7) takes the form

$$\sigma(T)\subset\sigma(T_1)\cup\sigma(T_0). \tag{2.9}$$

To this end, let us consider the canonical and $*$-canonical factorizations of $\Theta_T(\lambda)$,

$$\Theta_T(\lambda)=\Theta_0'(\lambda)\Theta_1(\lambda),\quad \Theta_T(\lambda)=\Theta_1'(\lambda)\Theta_0(\lambda); \tag{2.10}$$

they correspond to the triangulations (2.4) and (2.1) of T, respectively. $m_{T_0}(\lambda)$ is a scalar multiple of $\Theta_{T_0}(\lambda)$ (cf. Theorem VI.5.1) and hence of $\Theta_0(\lambda)$ too (because the purely contractive part of $\Theta_0(\lambda)$ coincides with $\Theta_{T_0}(\lambda)$). On the other hand, the purely contractive part of $\Theta_1'(\lambda)$ coincides with $\Theta_{T_1'}(\lambda)$, which is outer from both sides (because $T_1'\in C_{11}$). $\Theta_{T_1'}(\lambda)$ admits a scalar multiple (because T_1' is a weak contraction), and by virtue of Theorem V.6.2 it admits even an outer scalar multiple $\delta_e(\lambda)$. Thus $\Theta_1'(\lambda)$ also admits $\delta_e(\lambda)$ as a scalar multiple. It follows from these results that the function $\Theta_T(\lambda)$ $(=\Theta_1'(\lambda)\Theta_0(\lambda))$ has the scalar multiple $\delta(\lambda)=\delta_e(\lambda)m_{T_0}(\lambda)$. Therefore the inner factor $\Theta_0'(\lambda)$ in the canonical factorization $\Theta_T(\lambda)=\Theta_0'(\lambda)\Theta_1(\lambda)$ will have as a scalar multiple the inner factor of $\delta(\lambda)$, that is, $m_{T_0}(\lambda)$ (cf. Proposition V.6.2). The purely contractive part of $\Theta_0'(\lambda)$ coincides with $\Theta_{T_0'}(\lambda)$, therefore $\Theta_{T_0'}(\lambda)$ will also have the scalar multiple $m_{T_0}(\lambda)$; as $T_0'\in C_0$ this implies $m_{T_0}(T_0')=O$ (cf. Theorem VI.5.1). Therefore $m_{T_0'}$ is an inner divisor of m_{T_0}. If we repeat these arguments, interchanging the roles of the two factorizations (2.10), we arrive at the result that m_{T_0} is an inner divisor of $m_{T_0'}$. The two results together imply (2.8).

The inclusion (2.9) being thereby established, we now prove its opposite, namely that

$$\sigma(T_0)\subset\sigma(T)\quad\text{and}\quad\sigma(T_1)\subset\sigma(T). \tag{2.11}$$

The points of $\sigma(T_0)$ in the interior of the unit circle C are eigenvalues of T_0 (because $T_0\in C_0$; see Proposition III.7.1) and hence also of T. On the other hand, $\sigma(T_1)$ is confined to C: this follows from Proposition VI.4.4 because $T_1\in C_{11}$ and $\Theta_{T_1}(\lambda)$ admits a scalar multiple. Thus, in order to obtain (2.11), it remains only to consider the parts of the spectra on C.

Now the inclusion

$$\sigma(T|\mathfrak{L}) \cap C \subset \sigma(T) \tag{2.12}$$

holds for every contraction T on $\mathfrak{H}$ and for every invariant subspace $\mathfrak{L}$ for T.

In fact, we infer from the expansion

$$(\mu I - T)^{-1} = \sum_{0}^{\infty} \mu^{-n-1} T^n,$$

valid for $|\mu| > 1$, that $\mathfrak{L}$ is invariant for $(\mu I - T)^{-1}$ too and that

$$(\mu I - T)^{-1}|\mathfrak{L} = (\mu I_{\mathfrak{L}} - T|\mathfrak{L})^{-1} \qquad (|\mu| > 1). \tag{2.13}$$

If ε is a point of C belonging to the resolvent set of T, then $(\mu I - T)^{-1}$ converges in norm to $(\varepsilon I - T)^{-1}$ as $\mu \to \varepsilon$, and thus the right-hand side of (2.13) also converges in norm to a limit, this being necessarily equal to $(\varepsilon I - T|\mathfrak{L})^{-1}$. Thus every point $\varepsilon \in C$ that belongs to the resolvent set of T, also belongs to the resolvent set of $T|\mathfrak{L}$. This implies (2.12).

We conclude that the relations (2.11) are valid. Together with (2.9), they imply the equality $\sigma(T) = \sigma(T_1) \cup \sigma(T_0)$.

Summarizing, we have the following result.

Theorem 2.1. (Decomposition $C_0 - C_{11}$.) *Let T be a completely nonunitary weak contraction on $\mathfrak{H}$. Among the subspaces $\mathfrak{L}$, invariant for T and such that $T|\mathfrak{L} \in C_0$, there exists a maximal one, denoted by $\mathfrak{H}_0$. Also, among the subspaces $\mathfrak{M}$, invariant for T and such that $T|\mathfrak{M} \in C_{11}$, there exists a maximal one, denoted by $\mathfrak{H}_1$. The contractions*

$$T_0 = T|\mathfrak{H}_0 \quad \text{and} \quad T_1 = T|\mathfrak{H}_1,$$

called the C_0 part and the C_{11} part of T, are equal to those appearing in the triangulations

$$T = \begin{bmatrix} T_0 & X \\ O & T_1' \end{bmatrix} \quad \text{and} \quad T = \begin{bmatrix} T_1 & Y \\ O & T_0' \end{bmatrix}$$

of type

$$\begin{bmatrix} C_{0\cdot} & * \\ O & C_{1\cdot} \end{bmatrix} \quad \text{and} \quad \begin{bmatrix} C_{\cdot 1} & * \\ O & C_{\cdot 0} \end{bmatrix},$$

respectively. $T_0'^*$ and $T_1'^*$ are then the C_0 and C_{11} parts of T^*. All these contractions are weak ones; moreover, T_0 and T_0' have the same minimal function. We have

$$\mathfrak{H}_0 \vee \mathfrak{H}_1 = \mathfrak{H}, \quad \mathfrak{H}_0 \cap \mathfrak{H}_1 = \{0\}, \tag{2.14}$$

$$\sigma(T) = \sigma(T_0) \cup \sigma(T_1), \tag{2.15}$$

$\sigma(T_1)$ lying on the unit circle C.

Remark. Together with Theorem III.5.1, this implies that *for every weak contraction* (c.n.u. or not) *the part of the spectrum interior to C is discrete.*

2. We have shown by an example in Sec. 1 that the restriction of a weak contraction to an invariant subspace need not be a weak contraction. But if it is, some important relations hold.

Proposition 2.2. *Under the conditions of Theorem* 2.1, *let* $\mathfrak{L}$ *be an invariant subspace for* T *such that* $T_{\mathfrak{L}} = T|\mathfrak{L}$ *is also a weak contraction. For the spaces* $\mathfrak{L}_0$ *and* $\mathfrak{L}_1$ *of the* C_0 *and* C_{11} *parts of* $T_{\mathfrak{L}}$ *we have then*

$$\mathfrak{L}_0 = \mathfrak{L} \cap \mathfrak{H}_0, \quad \mathfrak{L}_1 = \mathfrak{L} \cap \mathfrak{H}_1. \tag{2.16}$$

Proof. As $\mathfrak{L}_0$ is invariant for T, with $T|\mathfrak{L}_0 = T_{\mathfrak{L}}|\mathfrak{L}_0 \in C_0$, it follows from the maximality of $\mathfrak{H}_0$ that $\mathfrak{L}_0 \subset \mathfrak{H}_0$. Similarly, we have $\mathfrak{L}_1 \subset \mathfrak{H}_1$. Thus

$$\mathfrak{L}_0 \subset \mathfrak{L} \cap \mathfrak{H}_0, \quad \mathfrak{L}_1 \subset \mathfrak{L} \cap \mathfrak{H}_1. \tag{2.17}$$

On the other hand, as $\mathfrak{L}_0$ is the space of the $C_{0\cdot}$ part in the triangulation of $T_{\mathfrak{L}}$ of type (2.2), we have

$$\mathfrak{L}_0 = \{l\colon l \in \mathfrak{L}, T_{\mathfrak{L}}^n l \to 0\ (n \to \infty)\}. \tag{2.18}$$

As $T_{\mathfrak{L}}^n l = T^n l$ for $l \in \mathfrak{L}$, it follows hence that $\mathfrak{L}_0 \supset \mathfrak{L} \cap \mathfrak{H}_0$. The first of the relations (2.16) is thereby proved.

As regards the second one, we know (2.17) and it remains to prove that $\mathfrak{L}_1 \supset \mathfrak{L} \cap \mathfrak{H}_1$. Recalling the maximality property of $\mathfrak{L}_1$ it suffices to prove to this effect that

$$T_{\wedge} = T|\mathfrak{H}_{\wedge} \in C_{11}, \quad \text{where} \quad \mathfrak{H}_{\wedge} = \mathfrak{L} \cap \mathfrak{H}_1. \tag{2.19}$$

Now because $T_{\wedge} = (T|\mathfrak{H}_1)|\mathfrak{H}_{\wedge} = T_1|\mathfrak{H}_{\wedge}$ and $T_1 \in C_{11}$, we have $T_{\wedge} \in C_{1\cdot}$. In order to prove (2.19) it suffices therefore to prove that $T_{\wedge}$ is a weak contraction (indeed, its characteristic function then admits a scalar multiple, thus this characteristic function is not only $*$-outer, but outer from both sides, and hence $T_{\wedge}$ is not only of class $C_{1\cdot}$, but also of class C_{11}). Because $T_{\wedge}$ is the restriction of the weak contraction T to the invariant subspace $\mathfrak{H}_{\wedge}$, we have proved that $T_{\wedge}$ is also a weak contraction if we show that $\sigma(T_{\wedge})$ does not include the whole disc D.

To show this let us consider the triangulation of $T_{\wedge}$ with respect to the invariant subspace $\mathfrak{L}_1$,

$$T_{\wedge} = \begin{bmatrix} T' & X \\ O & T'' \end{bmatrix}, \quad \mathfrak{H}_{\wedge} = \mathfrak{L}_1 \oplus \mathfrak{L}''. \tag{2.20}$$

As $T' = T_{\wedge}|\mathfrak{L}_1 = T_{\mathfrak{L}}|\mathfrak{L}_1$, T' is the C_{11} part of $T_{\mathfrak{L}}$; hence $\sigma(T') \subset C$. As to T'', we observe that $\mathfrak{L}''$ is contained in $\mathfrak{L} \ominus \mathfrak{L}_1$ (i.e., in the space of the C_0 part of $T_{\mathfrak{L}}^*$); let us denote this part of $T_{\mathfrak{L}}^*$ simply by S. Let $u(\lambda)$ be any function in H^∞. From the facts that (i) $\mathfrak{L}''$ is invariant for $T_{\wedge}^*$, (ii) $\mathfrak{H}_{\wedge}$ is invariant for $T_{\mathfrak{L}}$, (iii) $\mathfrak{L} \ominus \mathfrak{L}_1$ is invariant for $T_{\mathfrak{L}}^*$, (iv) $\mathfrak{H}_{\wedge} \supset \mathfrak{L}''$, and (v) $\mathfrak{L} \ominus \mathfrak{L}_1 \supset \mathfrak{L}''$, we deduce:

$$\begin{aligned} u(T''^*) &= u(T_{\wedge}^*)|\mathfrak{L}'' = [P_{\wedge} u(T_{\mathfrak{L}}^*)|\mathfrak{H}_{\wedge}]|\mathfrak{L}'' = P_{\wedge} u(T_{\mathfrak{L}}^*)|\mathfrak{L}'' \\ &= [P_{\wedge} u(T_{\mathfrak{L}}^*)|\mathfrak{L} \ominus \mathfrak{L}_1]|\mathfrak{L}'' = P_{\wedge} u(S)|\mathfrak{L}'', \end{aligned}$$

where $P_\wedge$ denotes the orthogonal projection of $\mathfrak{H}$ onto $\mathfrak{H}_\wedge$. Choosing for u in particular the minimal function of S, we obtain

$$m_S(T''^*) = O, \quad m_{\tilde{S}}(T'') = O.$$

Hence $T'' \in C_0$, and so the part of $\sigma(T'')$ in D is a discrete set. Recalling (2.7), we see that

$$\sigma(T_\wedge) \cap D \subset (\sigma(T') \cup \sigma(T'')) \cap D = \sigma(T'') \cap D \subset \sigma(T'');$$

hence $\sigma(T_\wedge)$ does not cover D.

The proof is complete.

3. The following two propositions give useful complements to the preceding results.

Proposition 2.3. *If the spectrum of the weak contraction T does not include the whole unit circle C, then the restriction of T to any invariant subspace $\mathfrak{L}$ is a weak contraction.*

Proof. We only have to show that the spectrum of $T_\mathfrak{L} = T|\mathfrak{L}$ does not include the whole disc D. To this end, we start with the relations

$$[(\mu I - T)^{-1}|\mathfrak{L}] \cdot [\mu I_\mathfrak{L} - T_\mathfrak{L}] = [\mu I_\mathfrak{L} - T_\mathfrak{L}] \cdot [(\mu I - T)^{-1}|\mathfrak{L}] = I_\mathfrak{L}, \tag{2.21}$$

valid for $|\mu| > 1$ by (2.13). Next we observe that our assumptions on T imply that the resolvent set of T is a domain G containing all the exterior of C, one or more arcs of C, and all the interior of C except a discrete set of points. Relations (2.21), valid on the exterior of C, extend by analyticity to the whole domain G (indeed, $\mu I_\mathfrak{L} - T_\mathfrak{L}$ is an analytic function of μ on the whole plane, whereas $(\mu I - T)^{-1}$ is analytic on G). Hence the resolvent set of $T_\mathfrak{L}$ includes G; consequently $\sigma(T_\mathfrak{L}) \cap D$ is a discrete set.

Proposition 2.4. *Under the assumptions of Theorem* 2.1 *we have*

$$\mathfrak{H}_0 = \{h : h \in \mathfrak{H}, m_{T_0}(T)h = 0\} \tag{2.22}$$

and

$$\mathfrak{H}_1 = \overline{m_{T_0}(T)\mathfrak{H}}. \tag{2.23}$$

As a consequence $\mathfrak{H}_0$ and $\mathfrak{H}_1$ are even hyperinvariant for T.

Proof. Denoting the subspace defined by the right-hand side of (2.22) by $\mathfrak{L}_0$, it is obvious that $\mathfrak{H}_0 \subset \mathfrak{L}_0$; indeed we have $m_{T_0}(T)h = m_{T_0}(T_0)h = 0$ for $h \in \mathfrak{H}_0$. On the other hand, $\mathfrak{L}_0$ is invariant for T and we have $m_{T_0}(T|\mathfrak{L}_0) = O$. Thus $T|\mathfrak{L}_0 \in C_0$, which implies by the maximality property of $\mathfrak{H}_0$ that $\mathfrak{L}_0 \subset \mathfrak{H}_0$. This proves the equation (2.22).

As regards (2.23), let us observe first that

$$\mathfrak{H} \ominus \overline{m_{T_0}(T)\mathfrak{H}} = \{h\colon h \in \mathfrak{H}, m_{T_0}(T)^* h = 0\}, \tag{2.24}$$

and that, because T_0 and T_0' have the same minimal function,

$$m_{T_0}(T)^* = m_{T_0'}(T)^* = m_{\tilde{T_0'}}(T^*) = m_{T_0'^*}(T^*).$$

Now, because $T_0'^*$ is the C_0 part of T^*, relation (2.22) applied to T^* instead of T shows that the right-hand side of (2.24) is equal to $\mathfrak{H}_0'$, that is, to $\mathfrak{H} \ominus \mathfrak{H}_1$. This implies the validity of (2.23).

3 Spectral decomposition of weak contractions

1. Let T be a c.n.u. weak contraction on $\mathfrak{H}$, and let T_0 and T_1 be the C_0 part and the C_{11} part of T $(T_0 = T|\mathfrak{H}_0, T_1 = T|\mathfrak{H}_1)$. Let

$$m_{T_0}(\lambda) = B(\lambda) \cdot \exp\left[-\int_0^{2\pi} \frac{e^{it} + \lambda}{e^{it} - \lambda} d\mu_t\right] \tag{3.1}$$

be the parametric representation of the minimal function as an inner function: $B(\lambda)$ is a Blaschke product and μ is a nonnegative, bounded, singular measure defined for the Borel subsets of $(0, 2\pi)$.

With every Borel subset ω of the plane of complex numbers we associate the inner function

$$m_\omega(\lambda) = B_\omega(\lambda) \cdot \exp\left[-\int_{(C\cap\omega)} \frac{e^{it} + \lambda}{e^{it} - \lambda} d\mu_t\right], \tag{3.2}$$

where $B_\omega(\lambda)$ means the product of those factors of $B(\lambda)$ whose zeros lie in ω.

We define two new subspaces:

$$\mathfrak{H}_0(\omega) = \{h\colon h \in \mathfrak{H}_0, m_\omega(T_0)h = 0\}, \tag{3.3}$$

and $\mathfrak{H}_1(\omega)$ is the subspace of $\mathfrak{H}_1$ associated with the contraction T_1 (the C_{11} part of T) and the Borel set

$$\alpha = C \cap \omega$$

in the sense of Theorem VII.5.2. The space $\mathfrak{H}_0(\omega)$ is hyperinvariant for T_0 (evident from (3.3)) and $\mathfrak{H}_1(\omega)$ is hyperinvariant for T_1 (by Theorem VII.5.2). In particular, $\mathfrak{H}_0(\omega)$ is invariant for T_0, $\mathfrak{H}_1(\omega)$ is invariant for T_1, and hence both are invariant for T. The subspace

$$\mathfrak{H}(\omega) = \mathfrak{H}_0(\omega) \vee \mathfrak{H}_1(\omega) \tag{3.4}$$

is then also invariant for T.

We show that $\mathfrak{H}(\omega)$ is even hyperinvariant for T. To this effect let us consider an arbitrary bounded operator X commuting with T. As $\mathfrak{H}_0$ and $\mathfrak{H}_1$ are hyperinvariant for T (cf. Proposition 2.4), we have $X\mathfrak{H}_j \subset \mathfrak{H}_j$, moreover $X_j = X|\mathfrak{H}_j$ commutes with T_j $(j = 0,1)$. Because $\mathfrak{H}_j(\omega)$ is hyperinvariant for T_j, we have $X\mathfrak{H}_j(\omega) = X_j\mathfrak{H}_j(\omega) \subset \mathfrak{H}_j(\omega)$ $(j = 0,1)$, and hence $X\mathfrak{H}(\omega) \subset \mathfrak{H}(\omega)$, which proves the hyperinvariance of $\mathfrak{H}(\omega)$ for T.

Next we show that $T(\omega) = T|\mathfrak{H}(\omega)$ is a weak contraction. To this effect let us also consider the operators $T_0(\omega) = T_0|\mathfrak{H}_0(\omega)$ $(= T|\mathfrak{H}_0(\omega))$ and $T_1(\omega) = T_1|\mathfrak{H}_1(\omega)$ $(= T|\mathfrak{H}_1(\omega))$. From the definition (3.3) of $\mathfrak{H}_0(\omega)$ and from Theorem III.6.3 it follows that $T_0(\omega)$ has its minimal function equal to m_ω. Because the zeros of m_ω in D are a subset of the zeros of m_{T_0}, it follows from Theorem III.5.1 and from (2.15) that

$$\sigma(T_0(\omega)) \cap D \subset \sigma(T_0) \cap D = \sigma(T) \cap D. \tag{3.5}$$

As to $T_1(\omega)$, its spectrum lies entirely on the unit circle C (cf. Theorem VII.6.2, applied to $T_1 \in C_{11}$ and to $\alpha = C \cap \omega$). Let a be a point of D not belonging to $\sigma(T)$. Then it belongs neither to $\sigma(T_0(\omega))$ nor to $\sigma(T_1(\omega))$, so we have $(T_j(\omega) - aI_j)\mathfrak{H}_j(\omega) = \mathfrak{H}_j(\omega)$ $(j = 0,1$; I_j denoting the identity operator on $\mathfrak{H}_j)$. This implies

$$\begin{aligned}\overline{(T-aI)\mathfrak{H}(\omega)} &= \overline{(T-aI)[\mathfrak{H}_0(\omega) \vee \mathfrak{H}_1(\omega)]}\\ &= (T_0(\omega) - aI_0)\mathfrak{H}_0(\omega) \vee (T_1(\omega) - aI_1)\mathfrak{H}_1(\omega)\\ &= \mathfrak{H}_0(\omega) \vee \mathfrak{H}_1(\omega) = \mathfrak{H}(\omega).\end{aligned}$$

As $T - aI$ is an affinity on $\mathfrak{H}$, we infer that $T(\omega) - aI_{\mathfrak{H}(\omega)}$ is an affinity on $\mathfrak{H}(\omega)$: the point a does not belong to the spectrum of $T(\omega)$. Therefore $T(\omega)$ is a weak contraction.

Let $\mathfrak{H}(\omega)_0$ and $\mathfrak{H}(\omega)_1$, respectively, denote the subspaces of $\mathfrak{H}(\omega)$ on which the C_0 and C_{11} parts of $T(\omega)$ act. We are going to prove that

$$\mathfrak{H}(\omega)_j = \mathfrak{H}_j(\omega) \qquad (j = 0,1). \tag{3.6}$$

To being with, we observe that, by virtue of Proposition 2.2,

$$\mathfrak{H}(\omega)_j = \mathfrak{H}(\omega) \cap \mathfrak{H}_j \qquad (j = 0,1). \tag{3.7}$$

We have $\mathfrak{H}_j(\omega) \subset \mathfrak{H}_j$ by the definition of $\mathfrak{H}_j(\omega)$, and $\mathfrak{H}_j(\omega) \subset \mathfrak{H}(\omega)$ by virtue of the definition (3.4) of $\mathfrak{H}(\omega)$, thus (3.7) implies

$$\mathfrak{H}(\omega)_j \supset \mathfrak{H}_j(\omega) \qquad (j = 0,1). \tag{3.8}$$

To establish (3.6) it remains to prove the opposite inclusion.

Now, every $h \in \mathfrak{H}(\omega)$ is the limit of a sequence $\{h_{0n} + h_{1n}\}$ with $h_{0n} \in \mathfrak{H}_0(\omega)$ and $h_{1n} \in \mathfrak{H}_1(\omega)$, thus

$$m_\omega(T)h = \lim_n m_\omega(T)(h_{0n} + h_{1n}) = \lim_n m_\omega(T)h_{1n} \in \mathfrak{H}_1(\omega) \subset \mathfrak{H}_1.$$

If $h \in \mathfrak{H}_0$ we have, on the other hand, $m_\omega(T)h = m_\omega(T_0)h \in \mathfrak{H}_0$. As $\mathfrak{H}_0 \cap \mathfrak{H}_1 = \{0\}$, it follows that $m_\omega(T)h = 0$ for $h \in \mathfrak{H}(\omega) \cap \mathfrak{H}_0$, that is, $h \in \mathfrak{H}_0(\omega)$. Hence we have $\mathfrak{H}(\omega) \cap \mathfrak{H}_0 \subset \mathfrak{H}_0(\omega)$. Recalling (3.7) (case $j=0$) we obtain $\mathfrak{H}(\omega)_0 \subset \mathfrak{H}_0(\omega)$. Together with (3.8) (case $j=0$) this yields (3.6) for $j=0$.

As regards the case $j=1$, let us first recall that the minimal function of $T_0(\omega)$ equals m_ω. Applying Proposition 2.4 to $T(\omega)$ we obtain:

$$\begin{aligned}\mathfrak{H}(\omega)_1 &= \overline{m_\omega(T(\omega))\mathfrak{H}(\omega)} = \overline{m_\omega(T)\mathfrak{H}(\omega)} = \overline{m_\omega(T)(\mathfrak{H}_0(\omega) \vee \mathfrak{H}_1(\omega))} \\ &= m_\omega(T_0)\mathfrak{H}_0(\omega) \vee m_\omega(T_1)\mathfrak{H}_1(\omega) = \overline{m_\omega(T_1)\mathfrak{H}_1(\omega)} \subset \mathfrak{H}_1(\omega);\end{aligned}$$

together with (3.8) (case $j=1$) this yields (3.6) for $j=1$.

We now show that the spaces $\mathfrak{H}(\omega)$ enjoy some maximality properties. Let $\mathfrak{L} \subset \mathfrak{H}$ be a hyperinvariant subspace for T and let $T_\mathfrak{L} = T|\mathfrak{L}$. Then $\sigma(T_\mathfrak{L}) \subset \sigma(T)$ and hence $T_\mathfrak{L}$ is a weak contraction (see Sect. 1.2). Let $\mathfrak{L}_0$ and $\mathfrak{L}_1$ denote the spaces of the C_0 part $T_{\mathfrak{L}_0}$ of $T_\mathfrak{L}$ and the C_{11} part $T_{\mathfrak{L}_1}$ of $T_\mathfrak{L}$, respectively. By Proposition 2.2 we know that $\mathfrak{L}_0 \subset \mathfrak{H}_0$ and $\mathfrak{L}_1 \subset \mathfrak{H}_1$. Assume that $m_{T_{\mathfrak{L}_0}}(\lambda)$ divides $m_\omega(\lambda)$ and that the set $\omega \cap C$ is residual for $T_{\mathfrak{L}_1}$ (see Definition 3 in Sec. VII.5.1). Then due to Theorems III.6.3(ii) and VII.5.2, we have $\mathfrak{L}_0 \subset \mathfrak{H}_0(\omega)$ and $\mathfrak{L}_1 \subset \mathfrak{H}_1(\omega)$. By virtue of (2.14) (applied to $T_\mathfrak{L}$ and $\mathfrak{L}$) and of Definition (3.4) we have

$$\mathfrak{L} = \mathfrak{L}_0 \vee \mathfrak{L}_1 \subset \mathfrak{H}_0(\omega) \vee \mathfrak{H}_1(\omega) = \mathfrak{H}(\omega).$$

From the relations (3.6) established above we obtain, applying (2.15) to $T(\omega)$ instead of T, that

$$\sigma(T(\omega)) = \sigma(T_0(\omega)) \cup \sigma(T_1(\omega)).$$

Now from Theorem III.5.1 it follows that $\sigma(T_0(\omega)) \subset \overline{\omega}$, and from Theorem VII.6.2 it follows that $\sigma(T_1(\omega)) \subset \overline{\alpha} \subset \overline{\omega}$ (where $\alpha = C \cap \omega$). So we obtain

$$\sigma(T(\omega)) \subset \overline{\omega} \tag{3.9}$$

for every Borel subset ω of the set of complex numbers.

If moreover ω is a *closed* set, we show that $\mathfrak{H}(\omega)$ contains all the subspaces $\mathfrak{L}$ of $\mathfrak{H}$ which are invariant for T and such that (a) $T_\mathfrak{L} = T|\mathfrak{L}$ is a weak contraction, and (b) $\sigma(T_\mathfrak{L}) \subset \omega$.

If $\mathfrak{L}_0$ and $\mathfrak{L}_1$ are the spaces of the C_0 and C_{11} parts of $T_\mathfrak{L}$, we derive from (2.15) that

$$\sigma(T_{\mathfrak{L}_0}) \cup \sigma(T_{\mathfrak{L}_1}) = \sigma(T_\mathfrak{L}) \subset \omega, \ \text{ where } T_{\mathfrak{L}_j} = T_\mathfrak{L}|\mathfrak{L}_j = T|\mathfrak{L}_j \ (j=0,1). \tag{3.10}$$

As $T_{\mathfrak{L}_0} \in C_0$, we have $\mathfrak{L}_0 \subset \mathfrak{H}_0$ by the maximality property of $\mathfrak{H}_0$. This implies $m_{T_0}(T_{\mathfrak{L}_0}) = m_{T_0}(T_0)|\mathfrak{L}_0 = O$; thus, denoting the minimal function of $T_{\mathfrak{L}_0}$ by l_0, l_0 is a divisor of m_{T_0}. On the other hand, (3.10) implies $\sigma(T_{\mathfrak{L}_0}) \subset \omega$; thus, by Theorem III.5.1, l_0 has the following properties. (i) Its zeros in D belong to the set ω, and (ii) it is analytic on every open arc of C, disjoint from the closed set $\alpha = C \cap \omega$. Now we deduce from (3.1) and (3.2) that the inner divisors of m_{T_0} with these properties are also divisors of m_ω. Because $\mathfrak{L}_0 \subset \mathfrak{H}_0$, we have for $h \in \mathfrak{L}_0$

$$l_0(T_0)h = l_0(T_{\mathfrak{L}_0})h = 0;$$

as l_0 is a divisor of m_ω this implies $m_\omega(T_0)h = 0$; as a consequence of (3.3) we have therefore $h \in \mathfrak{H}_0(\omega)$. Thus

$$\mathfrak{L}_0 \subset \mathfrak{H}_0(\omega). \tag{3.11}$$

On the other hand, because (3.10) also implies $\sigma(T_{\mathfrak{L}_1}) \subset \omega$, the Borel set $\alpha = C \cap \omega$ is residual for the operator $T_{\mathfrak{L}_1} \in C_{11}$ (cf. Theorem VI.4.1); by virtue of the maximality property of the subspaces of type $\mathfrak{H}_\alpha$ (established in Theorem VII.5.2) we conclude that

$$\mathfrak{L}_1 \subset \mathfrak{H}_1(\omega). \tag{3.12}$$

Applying (2.14) to $\mathfrak{L}$ (instead of $\mathfrak{H}$) and using (3.11) and (3.12), we obtain

$$\mathfrak{L} = \mathfrak{L}_0 \vee \mathfrak{L}_1 \subset \mathfrak{H}_0(\omega) \vee \mathfrak{H}_1(\omega) = \mathfrak{H}(\omega),$$

thus establishing the maximality property of $\mathfrak{H}(\omega)$.

The above arguments prove Parts (i) and (ii) of the following theorem.

Theorem 3.1. (Spectral decomposition) *Let T be a c.n.u. weak contraction on $\mathfrak{H}$. Then to every Borel subset ω of the plane of complex numbers there corresponds a unique subspace $\mathfrak{H}(\omega)$ of $\mathfrak{H}$ with the following properties.*

(i) *$\mathfrak{H}(\omega)$ is the maximal hyperinvariant subspace $\mathfrak{L}$ for T satisfying the following conditions. The minimal function of $T_{\mathfrak{L}_0}$ divides m_ω and $\omega \cap C$ is residual for $T_{\mathfrak{L}_1}$ (see the discussion above).*

(ii) *$T(\omega) = T|\mathfrak{H}(\omega)$ is a weak contraction and $\sigma(T(\omega)) \subset \overline{\omega}$; moreover, if ω is closed, then $\mathfrak{H}(\omega)$ is maximal under these conditions, that is, $\mathfrak{H}(\omega)$ includes all the subspaces $\mathfrak{L}$, invariant for T and such that $T_{\mathfrak{L}} = T|\mathfrak{L}$ is a weak contraction with $\sigma(T_{\mathfrak{L}}) \subset \omega$.*

(iii) *$\mathfrak{H}(\bigcup_n \omega_n) = \bigvee_n \mathfrak{H}(\omega_n), \mathfrak{H}(\bigcap_n \omega_n) = \bigcap_n \mathfrak{H}(\omega_n)$ for any sequence $\{\omega_n\}$.*

(iv) $\mathfrak{H}(\omega) = \begin{cases} \{0\} & \textit{if } \omega \textit{ is empty,} \\ \mathfrak{H} & \textit{if } \omega \supset \sigma(T). \end{cases}$

(v) *$\mathfrak{H}(\omega) \neq \{0\}$ if ω is open and $\omega \cap \sigma(T)$ is not empty.*

Proof. It remains to prove Properties (iii), (iv) and (v). The first property (iii), respectively, property (iv) are simple, respectively, trivial consequences of the corresponding properties of the spaces $\mathfrak{H}_0(\omega)$ and $\mathfrak{H}_1(\omega)$ and the definition (3.4) of $\mathfrak{H}(\omega)$. For the first one we notice that $m_{\cup \omega_n}$ is the smallest common inner multiple of the functions m_{ω_n} and then apply Theorem III.6.3(iii), as for the spaces $\mathfrak{H}_1(\omega_n)$ we apply Theorem VII.5.2. Concerning the second property (iii), we observe that $m_{\cap \omega_n}$ is the largest common inner divisor of the functions m_{ω_n} and thus by virtue of Theorem III.6.3(iii) we have

$$\mathfrak{H}_0\left(\bigcap_n \omega_n\right) = \bigcap_n \mathfrak{H}_0(\omega_n).$$

On the other hand by applying Theorem VII.6.2 (v) we also have

$$\mathfrak{H}_1\left(\bigcap_n \omega_n\right) = \bigcap_n \mathfrak{H}_1(\omega_n).$$

It follows that

$$\begin{aligned}\mathfrak{L} &= \bigcap_n \mathfrak{H}(\omega_n) = \bigcap_n (\mathfrak{H}_0(\omega_n) \vee \mathfrak{H}_1(\omega_n)) \\ &\supset \left(\bigcap_n \mathfrak{H}_0(\omega_n)\right) \vee \left(\bigcap_n \mathfrak{H}_1(\omega_n)\right) = \mathfrak{H}\left(\bigcap_n \omega_n\right).\end{aligned}$$

But the space $\mathfrak{L}$ satisfies the conditions in Property (i) with respect to $\omega = \bigcap_n \omega_n$ hence $\mathfrak{L} \subset \mathfrak{H}(\bigcap_n \omega_n)$. This completes the proof of Property (iii).

Let us now consider an open set ω. If $\mathfrak{H}_0(\omega) = \{0\}$, then $m_{T_0} = m_{T_0}/m_\omega$ by Theorem III.6.3. Hence $m_\omega = 1$ and therefore no zero of the Blaschke product B lies in $\omega \cap D$, and $\mu(\beta) = 0$ for the open arcs β of C of which $\alpha = C \cap \omega$ is composed. But $\mu(\beta) = 0$ implies that m_{T_0} is analytic on β. It follows from Theorem III.5.1 that ω is contained in the resolvent set of T_0. On the other hand, if $\mathfrak{H}_1(\omega) = \mathfrak{H}_\alpha = \{0\}$, it follows from Theorem VII.5.2 that $C \backslash \alpha$ is residual for T_1 and hence $\Theta_{T_1}(e^{it})$ is isometric a.e. in (α); by virtue of Proposition VI.4.4 the set α (composed of open arcs) is contained in the resolvent set of T_1.

From these results it follows that if $\mathfrak{H}(\omega) = \{0\}$, then ω is contained in the resolvent sets of T_0 and T_1, so its intersection with $\sigma(T) = \sigma(T_0) \cup \sigma(T_1)$ is void. This implies the validity of (v) and thus concludes the proof of Theorem 3.1.

2. By virtue of this theorem the subspaces $\mathfrak{H}(\omega)$ have properties analogous to those of the spectral subspaces for a normal operator.

In particular, if $\sigma(T)$ has more than one point, this theorem establishes a nontrivial spectral decomposition of the space $\mathfrak{H}$, and, consequently, *the existence of nontrivial hyperinvariant subspaces for* T. Consider indeed two disjoint open sets, say ω_1 and ω_2, both having nonempty intersections with $\sigma(T)$. The corresponding spaces $\mathfrak{H}(\omega_1)$ and $\mathfrak{H}(\omega_2)$ will then be different from $\{0\}$, and because $\mathfrak{H}(\omega_1) \cap \mathfrak{H}(\omega_2) = \{0\}$, neither of them will equal $\mathfrak{H}$; thus $\mathfrak{H}(\omega_1)$ and $\mathfrak{H}(\omega_2)$ will be nontrivial disjoint subspaces of $\mathfrak{H}$, both hyperinvariant for T.

Let us consider now a weak contraction T on $\mathfrak{H}$, whose spectrum consists of a single point τ. We distinguish two cases accordingly as $|\tau| < 1$ or $|\tau| = 1$.

(1) If $|\tau| < 1$, then we have, by the spectral radius theorem, $\|T^n\|^{1/n} \to |\tau| < 1$ (cf. [*Func. Anal.*] Sec. 149), hence $\|T^n\| \to 0$. Consequently, T has no unitary part and even no C_{11} part; thus $T \in C_0$. From Theorem III.5.1 it follows that the minimal function is of the form

$$m_T(\lambda) = \left(\frac{\lambda - \tau}{1 - \bar{\tau}\lambda}\right)^n \quad (n \text{ a positive integer}). \tag{3.13}$$

Now *for a weak contraction* T, (3.13) *implies* $\dim \mathfrak{H} < \infty$. In fact, denoting by $\mathfrak{L}_\tau$ the subspace formed by the solutions h of the equation $Th = \tau h$, $\mathfrak{L}_\tau$ is invariant for T and $T_\tau = T|\mathfrak{L}_\tau$ satisfies the equation $I_\tau - T_\tau^* T_\tau = (1 - |\tau|^2) I_\tau$ (I_τ: the identity operator on $\mathfrak{L}_\tau$). Since $(1 - |\tau|^2) I_\tau$ has to be of finite trace, we have $\dim \mathfrak{L}_\tau = d_\tau < \infty$. Now let $\mathfrak{L}$ be an arbitrary finite dimensional subspace of $\mathfrak{H}$, say of dimension d.

Because (3.13) implies

$$(T-\tau I)^n = O, \text{ and thus } T^n = c_0 I + c_1 T + \cdots + c_{n-1} T^{n-1}, \tag{3.14}$$

the subspace $\mathfrak{M}$ spanned by $\mathfrak{L}$, $T\mathfrak{L}, \ldots, T^{n-1}\mathfrak{L}$ is invariant for T. As $\mathfrak{M}$ has finite dimension ($\leq nd$), we can choose in $\mathfrak{M}$ a basis with respect to which the matrix of $T|\mathfrak{M}$ has Jordan normal form. From (3.14) it follows that the *orders* of the Jordan cells in this matrix do not exceed n; hence the *number* v of these cells is $\geq \dim \mathfrak{M}/n$. To every cell corresponds an eigenvector of T, and the eigenvectors so obtained are linearly independent, thus we must have $v \leq d_\tau$. Consequently, $d \leq \dim \mathfrak{M} \leq nv \leq nd_\tau$. This implies obviously that $\mathfrak{H}$ itself has dimension $\leq nd_\tau$.

(2) For *any* contraction T such that $\sigma(T) = \{\tau\}$, $|\tau| = 1$, a necessary and sufficient condition in order that T be c.n.u., is that τ not be an eigenvalue of T. This follows readily from the fact, implied by the spectral theorem, that every isolated point of the spectrum of a unitary operator is an eigenvalue.

Let us assume therefore that T is a contraction on $\mathfrak{H}$ with $\sigma(T) = \{\tau\}$, $|\tau| = 1$, and that τ is not an eigenvalue of T. As T is then c.n.u., we deduce from Proposition II.6.7 that $T \in C_{00}$. If, moreover, the characteristic function of T has a scalar multiple—in particular, if T is a weak contraction on a separable space $\mathfrak{H}$—then $T \in C_{00}$ implies $T \in C_0$. From Theorem III.5.1 we deduce that the minimal function of T is of the form

$$m_T(\lambda) = e_a(\bar{\tau}\lambda) \quad (a > 0), \tag{3.15}$$

where $e_a(\lambda) = \exp(a(\lambda+1)/(\lambda-1))$; see the proof of Proposition III.7.3. It is clear that this case can only occur if $\mathfrak{H}$ *is of infinite dimension*. We come back to the study of such contractions in Chapter X.

4 Dissipative operators. Class (Ω_0^+)

1. By virtue of the relations between a one-parameter continuous semigroup of contractions and its cogenerator, and between a maximal accretive or dissipative operator and its Cayley transform, our results obtained from Chap. VI on, concerning functional models, invariant subspaces, spectral decompositions, and so on, carry over in a more or less immediate manner from individual contractions to one-parameter continuous semigroups of contractions, or to accretive and dissipative operators (as regards the models of semigroups see the Notes to Chap. VI).

Let us consider, for example, operators on the space $\mathfrak{H}$ of the form

$$A = R + iQ, \tag{4.1}$$

where R and Q are self-adjoint operators, with

$$O \leq Q \leq 2q \cdot I. \tag{4.2}$$

It is obvious that A is dissipative. The resolvent set of A contains in particular the points $z = x + iy$ with $y < 0$ and $y > 2q$. In fact, we have

$$A - zI = M + N, \quad \text{where} \quad M = (R - xI) + i(q - y)I, \quad N = i(Q - qI);$$

condition (4.2) implies that $\|N\| \le q$, and if $y \ne q$ then M is boundedly invertible, with $\|M^{-1}\| \le |q - y|^{-1}$. As a consequence we have

$$A - zI = M(I + M^{-1}N), \quad \text{where} \quad \|M^{-1}N\| \le q \cdot |q - y|^{-1}.$$

The right-hand side of the last inequality is < 1 if $y < 0$ or $y > 2q$; hence, in these cases, $A - zI$ also is boundedly invertible. It follows in particular that the Cayley transform of A, that is, the operator

$$T = (A - iI)(A + iI)^{-1} = I - 2i(A + iI)^{-1} \quad \text{(cf. (IV.4.12))} \tag{4.3}$$

is defined everywhere on $\mathfrak{H}$, which implies that A is *maximal* dissipative.

Moreover, because the homography

$$z \to \lambda = \frac{z - i}{z + i} \tag{4.4}$$

transforms the points $z(\ne -i)$ of the resolvent set of A to points of the resolvent set of T, it follows that the spectrum of T does not cover the unit disc $|\lambda| < 1$. Moreover, if A is bounded, the spectrum of T does not even cover the circle $|\lambda| = 1$.

We obtain by a simple calculation:

$$I - T^*T = J^*QJ, \quad I - TT^* = JQJ^*, \tag{4.5}$$

where

$$J = 2i(A + iI)^{-1} = I - T;$$

thus it follows that the transformations

$$\tau\colon D_T h \to Q^{1/2}Jh, \quad \tau_*\colon D_{T^*}h \to Q^{1/2}J^*h \quad (h \in \mathfrak{H}) \tag{4.6}$$

are isometric. On account of the relation

$$\overline{Q^{1/2}J\mathfrak{H}} = \overline{Q^{1/2}\overline{J\mathfrak{H}}} = \overline{Q^{1/2}\mathfrak{H}} = \overline{Q\mathfrak{H}},$$

and the analogous one with J^* in place of J, the transformations (4.6) extend by continuity to *unitary* ones:

$$\tau\colon \mathfrak{D}_T \to \mathfrak{Q}, \quad \tau_*\colon \mathfrak{D}_{T^*} \to \mathfrak{Q}, \text{ where } \mathfrak{Q} = \overline{Q\mathfrak{H}}. \tag{4.7}$$

It follows that the characteristic function $\{\mathfrak{D}_T, \mathfrak{D}_{T^*}, \Theta_T(\lambda)\}$ of T coincides with $\{\mathfrak{Q}, \mathfrak{Q}, \tau_*\Theta_T(\lambda)\tau^{-1}\}$.

Let us set

$$\Theta_A(z) = \tau_* \Theta_T(\lambda) \tau^{-1} \quad \left(z = i\frac{1+\lambda}{1-\lambda}, \quad |\lambda| < 1 \right) \tag{4.8}$$

and calculate $\Theta_A(z)$ explicitly. On account of the relation

$$\Theta_T(\lambda) D_T = D_{T^*}(I - \lambda T^*)^{-1}(\lambda I - T), \text{ see (VI.1.2),}$$

and the definition (4.6) of τ and τ_*, we have for $h \in \mathfrak{H}$:

$$\begin{aligned}
\Theta_A(z) Q^{1/2} J h &= \tau_* \Theta_T(\lambda) D_T h = Q^{1/2} J^* (I - \lambda T^*)^{-1} (\lambda I - T) h = \\
&= Q^{1/2} J^* \left[I - \frac{z-i}{z+i} (A^* + iI)(A^* - iI)^{-1} \right]^{-1} \\
&\quad \cdot \left[\frac{z-i}{z+i} I - (A - iI)(A + iI)^{-1} \right] h \\
&= -2iQ^{1/2} [(z+i)(A^* - iI) - (z-i)(A^* + iI)]^{-1} \\
&\quad \cdot [(z-i)(A+iI) - (z+i)(A-iI)](A+iI)^{-1} h \\
&= Q^{1/2} (A^* - zI)^{-1} (A - zI) J h = Q^{1/2} [I + 2i(A^* - zI)^{-1} Q] J h \\
&= [I + 2iQ^{1/2}(A^* - zI)^{-1} Q^{1/2}] Q^{1/2} J h;
\end{aligned}$$

hence we deduce that

$$\Theta_A(z) = [I + 2iQ^{1/2}(A^* - zI)^{-1} Q^{1/2}] | \mathfrak{Q}. \tag{4.9}$$

When Q is compact, with spectral decomposition

$$Qh = \sum \omega_k (h, \varphi_k) \varphi_k, \tag{4.10}$$

where $\{\varphi_k\}$ is an orthonormal system of eigenvectors corresponding to the eigenvalues $\omega_k > 0$, this system is a basis of the subspace $\mathfrak{Q}$ of $\mathfrak{H}$, and the operator $\Theta_A(z)$ has, with respect to this basis, the (finite or infinite) matrix with the entries

$$(\Theta_A(z)\varphi_j, \varphi_k) = \delta_{jk} + 2i(\omega_j \omega_k)^{1/2} ((A^* - zI)^{-1} \varphi_j, \varphi_k). \tag{4.11}$$

2. We now show that if A is bounded, then *A and T have the same invariant subspaces $\mathfrak{L}$, and $T_\mathfrak{L} = T|\mathfrak{L}$ is the Cayley transform of the dissipative operator $A_\mathfrak{L} = A|\mathfrak{L}$.* To this end, first consider a subspace $\mathfrak{L}$ invariant for A. Choose a circle having $\sigma(A)$ in its interior and the point $-i$ in its exterior. If z_0 is the center and r the radius of this circle, we obtain from the spectral radius theorem that

$$r \geq \lim_{n \to \infty} \|(A - z_0 I)^n\|^{1/n}.$$

As a consequence we have for $|z-z_0|>r$ the expansion, convergent in norm,

$$(A-zI)^{-1}=[(A-z_0I)-(z-z_0)I]^{-1}=-\sum_0^\infty (z-z_0)^{-n-1}(A-z_0I)^n, \tag{4.12}$$

which implies that $\mathfrak{L}$ is invariant for $(A-zI)^{-1}$ also. As we can choose in particular $z=-i$, we obtain that $\mathfrak{L}$ is invariant for $(A+iI)^{-1}$ and, as a consequence of (4.3), of T also. Moreover, relation (4.12) implies for $z=-i$

$$(A+iI)^{-1}|\mathfrak{L}=(A_{\mathfrak{L}}+iI_{\mathfrak{L}})^{-1}$$

and hence

$$T_{\mathfrak{L}}=(A-iI)(A+iI)^{-1}|\mathfrak{L}=(A_{\mathfrak{L}}-iI_{\mathfrak{L}})(A_{\mathfrak{L}}+iI_{\mathfrak{L}})^{-1};$$

this proves that $T_{\mathfrak{L}}$ is the Cayley transform of $A_{\mathfrak{L}}$.

If we suppose, conversely, that $\mathfrak{L}$ is invariant for T, then the expansion

$$(T-\mu I)^{-1}=-\sum_0^\infty \mu^{-n-1}T^n \quad (|\mu|>1)$$

implies that $\mathfrak{L}$ is invariant for $(T-\mu I)^{-1}$ also. Because $T-I$ is boundedly invertible, with $(T-I)^{-1}=(1/2i)(A+iI)$ (cf. (4.3)), $(T-I)^{-1}$ is the limit (in norm) of $(T-\mu I)^{-1}$ as $\mu\to 1$; thus we conclude that $\mathfrak{L}$ is invariant for $(T-I)^{-1}$ and hence for $A=i(I+T)(I-T)^{-1}$ as well.

One of the consequences of the statement just proved is that if A is bounded, *A is unicellular if and only if T is unicellular.*

3. Suppose A is bounded, $\sigma(A)$ consists of the single point 0, and 0 is not an eigenvalue of A. Then the spectrum of the contraction $T'=-T$ consists of the single point 1, which is not an eigenvalue of T'. Moreover, T' is completely nonunitary (cf. the remark made at the end of Sec. 3.2) and hence A is completely nonselfadjoint. T' is the cogenerator of a continuous one-parameter semigroup of contractions $\{T'(s)\}_{s\geq 0}$,

$$T'(s)=e_s(T') \quad (s\geq 0); \tag{4.13}$$

see Theorem III.8.1. From (4.3) we deduce that the generator A' of this semigroup is given by $A'=(iA)^{-1}$ (A^{-1} exists and has dense domain, because -1 is not an eigenvalue of T, and hence not of T^* either). The functional calculus introduced in Sec. IV.4.4 justifies the notation

$$T'(s)=\exp(sA'). \tag{4.14}$$

Under the additional hypothesis that Q is of finite trace, it follows from (4.5) that T, and hence T' also, are weak contractions. By virtue of Theorem 1.1 the characteristic function of T' admits a scalar multiple. Because $\sigma(T')=\{1\}$, we have $T'\in C_{00}$ (cf. Proposition II.6.7). It follows that $T'\in C_0$ (cf. Theorem VI.5.1) and

$m_{T'}(\lambda) = e_a(\lambda)$, with $a = a_{T'} > 0$. This means that $e_a(T') = O$, whereas $e_b(T') \neq O$ for $0 \leq b < a$; using (4.13) and (4.14) we can express this in the alternative form:

$$\exp(aA') = O \quad \text{and} \quad \exp(bA') \neq O \quad \text{for} \quad 0 \leq b < a.$$

Let us introduce the notation (Ω_0^+) for the class of bounded dissipative operators $A = R + iQ$ for which Q is of finite trace, $\sigma(A) = \{0\}$, and 0 is not an eigenvalue of A. The above results prove the following proposition (except the final statement (4.15)).

Proposition 4.1. *For $A \in (\Omega_0^+)$ the operator $A' = (iA)^{-1}$ is the generator of a continuous one-parameter semigroup of contractions $\{T'(s)\}_{s\geq 0}$, such that $T'(s) = O$ for s large enough. The smallest of these values, denoted by s_A, is equal to the value $a = a_{T'}$ occurring in the minimal function $e_a(\lambda)$ of $T' = (iI - A)(iI + A)^{-1}$. The following relation also holds.*

$$s_A = \limsup_{z \to 0} [|z| \cdot \log \|(A + zI)^{-1}\|]. \tag{4.15}$$

Proof. In order to prove (4.15) let us observe first that, setting $z' = (iz)^{-1}$ $(z \neq 0)$, we have $A' + z'I = (zI + A)(izA)^{-1}$, and hence

$$(A' + z'I)^{-1} = izA(A + zI)^{-1}. \tag{4.16}$$

We also have $A'^{-1} = iA$, thus every (finite) complex number belongs to the resolvent set of A'. By virtue of relation (IV.4.17) we have

$$(-A' - z'I)^{-1} = \int_0^{s_A} e^{tz'} T'(t)\, dt \tag{4.17}$$

if $\operatorname{Re} z' < 0$; as both sides of (4.17) are entire functions of z', they must coincide for every complex z'. An immediate consequence is the inequality

$$\|(A' + z'I)^{-1}\| \leq \int_0^{s_A} e^{t|z'|}\, dt < \exp(s_A|z'|) \quad \text{for} \quad |z'| \geq 1;$$

hence

$$\limsup_{z' \to \infty} \left[\frac{1}{|z'|} \cdot \log \|(A' + z'I)^{-1}\| \right] \leq s_A. \tag{4.18}$$

On the other hand, it follows from the well-known theorem of PALEY AND WIENER [1] that for every scalar function $\varphi(t) \in L^2(-\sigma, \sigma)$ we have

$$\limsup_{z' \to \infty} \left[\frac{1}{|z'|} \cdot \log |\widehat{\varphi}(z')| \right] = \alpha_\varphi, \quad \text{where} \quad \widehat{\varphi}(z') = \int_{-\sigma}^{\sigma} e^{tz'} \varphi(t)\, dt$$

and α_φ denotes the smallest of the numbers $\alpha \geq 0$ such that $\varphi(t)$ vanishes a.e. on the set $(-\sigma, \sigma) \backslash (-\alpha, \alpha)$. By the definition of s_A we have $T'(s) \neq O$ for $0 \leq s < s_A$, thus there exist for every $\varepsilon > 0$ elements $h, g \in \mathfrak{H}$ such that the function $\varphi(t)$ defined by $(T'(t)h,g)$ for $0 \leq t < s_A$ and by 0 for $-\infty < t < 0$, satisfies $\alpha_\varphi > s_A - \varepsilon$. In this case we have, by virtue of (4.17),

$$\widehat{\varphi}(z') = ((-A' - z'I)^{-1}h, g),$$

we obtain that

$$\limsup_{z'\to\infty} \left[\frac{1}{|z'|} \cdot \log|((A' + z'I)^{-1}h, g)|\right] > s_A - \varepsilon,$$

and this implies that in (4.18) the equality sign holds. Using (4.16) we can express this equation also in the form

$$s_A = \limsup_{z\to 0}[|z| \cdot \log\|zA(A+zI)^{-1}\|]. \tag{4.19}$$

THe operator A is bounded, thus (4.19) implies

$$s_A \leq \limsup_{z\to 0}[|z| \cdot \log\|z(A+zI)^{-1}\|]. \tag{4.20}$$

Let $\{z_n\}$ be a sequence ($z_n \to 0$) for which this upper limit is approached. As $s_A > 0$, we have necessarily $\|z_n(A + z_nI)^{-1}\| \to \infty$, and hence, on account of the relation $z(A+zI)^{-1} = I - A(A+zI)^{-1}$, we obtain

$$\begin{aligned}\lim_n[|z_n| \cdot \log\|z_n(A+z_nI)^{-1}\|] &\leq \limsup_n[|z_n| \cdot \log(1 + \|A(A+z_nI)^{-1}\|)] \\ &= \limsup_n[|z_n| \cdot \log\|A(A+z_nI)^{-1}\|] \leq s_A\end{aligned}$$

(cf. (4.19)). Therefore in (4.20) and hence in (4.15), the equality sign holds.

4. Let $A = R + iQ \in (\Omega_0^+)$ be such that Q is of finite rank N. It follows from (4.5) that the defect indices of T are equal to the rank of Q, that is, to N. The determinant $d_T(\lambda)$ is an inner function, divisible by $m_T(\lambda)$, and a divisor of $m_T(\lambda)^N$ (cf. Theorem VI.5.2). Because $m_T(\lambda) = m_{T'}(-\lambda) = e_a(-\lambda) = \exp(a\zeta)$, where

$$\zeta = \frac{-\lambda+1}{-\lambda-1} = -\frac{i}{z},$$

it follows that

$$d_T(\lambda) = \varkappa \cdot \exp(c\zeta), \quad \text{where} \quad |\varkappa| = 1, a \leq c \leq Na. \tag{4.21}$$

(Observe that $|\lambda| < 1$ implies Re $\zeta < 0$, and conversely.) The value of c can be calculated using (4.8) and (4.11). By virtue of these formulas, the matrix of $\Theta_T(\lambda)$ corresponding to the orthonormal basis $\{\tau^{-1}\varphi_k\}$ of $\mathfrak{D}_T$ and the orthonormal basis $\{\tau_*^{-1}\varphi_j\}$ of $\mathfrak{D}_{T^*}$ has the entries

$$\vartheta_{kj}(\lambda) = (\Theta_A(-i/\zeta)\varphi_j, \varphi_k) = \delta_{jk} - 2(\omega_j\omega_k)^{1/2}\zeta((I + i\zeta A^*)^{-1}\varphi_j, \varphi_k). \tag{4.22}$$

As $d_T(\lambda)$ is the determinant of this matrix, (4.21) and (4.22) imply

$$\frac{d}{d\zeta}\det[\vartheta_{kj}(\lambda)]|_{\zeta=0} = \varkappa c. \tag{4.23}$$

Now from (4.22) we get

$$\vartheta_{kj}(\lambda)|_{\zeta=0} = \delta_{kj} \quad \text{and} \quad \frac{d}{d\zeta}\vartheta_{kj}(\lambda)|_{\zeta=0} = -2(\omega_j\omega_k)^{1/2}\delta_{jk},$$

which shows that the left-hand side of (4.23) equals $-2\sum\omega_j$. As $\omega_j > 0$, we conclude that

$$c = 2\sum\omega_j. \tag{4.24}$$

So we have proved the following theorem.

Theorem 4.2. *For $A \in (\Omega_0^+)$, with $Q = \operatorname{Im} A$ of finite rank, the value s_A defined in the preceding proposition satisfies the inequality*

$$s_A \le 2 \operatorname{tr} Q. \tag{4.25}$$

In Chap. X we show that equality holds in (4.25) if and only if A is unicellular.

5. If the defect indices of T are equal to 1, then T is determined up to unitary equivalence by its minimal function $e_a(-\lambda)$, where $a = a_T = s_A$ (cf. the remark following Theorem VI.5.2), and hence by s_A. On account of the obvious relation $s_{rA} = r \cdot s_A$ $(r > 0)$ we conclude that *if $A_0 = R_0 + iQ_0 \in (\Omega_0^+)$ with* rank $Q_0 = 1$ *and $s_{A_0} = 1$, then every operator $A = R + iQ \in (\Omega_0^+)$ with* rank $Q = 1$ *is unitarily equivalent to $s_A \cdot A_0$.* Consider, for example, the operator A_0 defined on $L^2(0,1)$ by

$$A_0 f(x) = i\int_0^x f(t)\,dt. \tag{4.26}$$

As A_0 is a Volterra-type integral operator, we have $\sigma(A_0) = \{0\}$, and it is obvious that 0 is not an eigenvalue of A_0. For $Q = \operatorname{Im} A_0$ we have

$$Qf(x) = \frac{1}{2}\int_0^1 f(t)\,dt;$$

hence $Q \ge O$ and rank $Q = 1$. Thus $A_0 \in (\Omega_0^+)$. It is easy to show that $A_0' = (iA_0)^{-1}$ is the generator of the continuous one-parameter semigroup of contractions defined by

$$(\exp sA_0')f(x) = \begin{cases} f(x-s) & \text{for} \quad x \in [s,\infty] \cap [0,1] \\ 0 & \text{otherwise,} \end{cases}$$

and hence $s_{A_0} = 1$. Thus the following statement is proved.

Proposition 4.3. *The operator A_0 on $L^2(0,1)$, defined by* (4.26)*, is unicellular. Every operator $A \in (\Omega_0^+)$ for which* $\operatorname{Im} A$ *is of rank* 1*, is unitarily equivalent to a positive multiple of A_0, namely to $s_A \cdot A_0$.*

5 Dissipative operators similar to self-adjoint ones

1. We now consider operators $A = R + iQ$ on the (separable) space $\mathfrak{H}$, where R and Q are self-adjoint, and Q is bounded and positive; such an operator A is maximal dissipative; see Sec. 4.1.

It is obvious that A is similar to a self-adjoint operator[1] if its Cayley transform T is similar (by the same affinity S) to a unitary operator. As we know that $\sigma(T)$ does not cover the unit disc D, we infer from Theorem VI.4.1 that there exist points $\lambda \in D$ at which $\Theta_T(\lambda)$ is boundedly invertible. From Theorem VI.4.5 we deduce therefore that T is similar to a unitary operator if and only if

$$\|\Theta_T(\lambda)g\| \geq c\|g\| \qquad (g \in \mathfrak{D}_T, \lambda \in D) \tag{5.1}$$

for some constant $c > 0$. By virtue of relations (4.8) and (4.9), and because τ and τ_* are unitary, we obtain, setting

$$q = \tau g \text{ and } z = i\frac{1+\lambda}{1-\lambda} \quad (g \in \mathfrak{D}_T, |\lambda| < 1, \operatorname{Im} z > 0),$$

that

$$\|\Theta_T(\lambda)g\| = \|\Theta_A(z)q\| = \|q + 2iQ^{1/2}(A^* - zI)^{-1}Q^{1/2}q\| \quad \text{and} \tag{5.2}$$

$$\|g\| = \|q\|. \tag{5.3}$$

If, in particular, rank $Q = 1$, that is,

$$Qh = \omega(h, q_0)q_0 \qquad (\omega > 0, \|q_0\| = 1), \tag{5.4}$$

then $\mathfrak{Q} = \overline{Q\mathfrak{H}}$ consists of the numerical multiples of q_0, thus $q = \omega^{-1/2}Q^{1/2}q$ for $q \in \mathfrak{Q}$. As, moreover,

$$\|Q^{1/2}h\| = (Qh, h)^{1/2} = [\omega(h, q_0)(q_0, h)]^{1/2} = |(q_0, \omega^{1/2}h)|,$$

the right-hand side of (5.2) can also be written in the form

$$|(q_0, q + 2i\omega(A^* - zI)^{-1}q)|,$$

and hence we have

$$\|\Theta_A(z)q\| = |\vartheta(z)| \cdot \|q\| \quad (q \in \mathfrak{Q}), \tag{5.5}$$

with

$$\vartheta(z) = 1 - 2i\omega((A - \bar{z}I)^{-1}q_0, q_0) \quad (\operatorname{Im} z > 0). \tag{5.6}$$

By virtue of (5.2)–(5.5), condition (5.1) takes the form

$$|\vartheta(z)| \geq c(>0) \quad (\operatorname{Im} z > 0). \tag{5.7}$$

[1] For unbounded operators A_1, A_2 similarity is defined as for bounded ones: existence of an affinity S such that $A_1 S = S A_2$.

2. As an example consider the operator A on $\mathfrak{H} = L^2(0,1)$ defined by

$$Ah(x) = a(x)h(x) + i\int_0^x h(t)\,dt, \tag{5.8}$$

where $a(x)$ is a real-valued, a.e. finite, measurable function. We have then $A = R + iQ$ with

$$Rh(x) = a(x)h(x) + \frac{i}{2}\left[\int_0^x - \int_x^1\right] h(t)\,dt, \quad Qh(x) = \frac{1}{2}\int_0^1 h(t)\,dt = \frac{1}{2}(h,e_0)e_0$$

and $e_0(x) \equiv 1$; thus Q satisfies (5.4) with $q_0 = e_0$ and $\omega = \frac{1}{2}$.

To determine $\vartheta(z)$ let us set

$$u_\zeta = (A + \zeta I)^{-1}e_0 \quad (\zeta = -\bar{z}).$$

Then

$$(a(x)+\zeta)u_\zeta(x) + i\int_0^x u_\zeta(t)\,dt = 1 \qquad (0 \le x \le 1),$$

or, setting $(a(x)+\zeta)u_\zeta(x) = v_\zeta(x)$,

$$v_\zeta(x) + i\int_0^x \frac{1}{a(t)+\zeta} v_\zeta(t)\,dt = 1. \tag{5.9}$$

This equation has the unique solution

$$v_\zeta(x) = \exp\left(-i\int_0^x \frac{1}{a(t)+\zeta}\,dt\right). \tag{5.10}$$

By virtue of (5.6), (5.9), and (5.10) we have therefore

$$\vartheta(z) = 1 - i(u_\zeta, e_0) = v_\zeta(1) = \exp\left(-i\int_0^1 \frac{dt}{a(t)+\zeta}\right),$$

or, if we introduce the distribution function of $a(x)$, that is, the function $\sigma(a) = \text{meas}\{x\colon 0 \le x \le 1, a(x) \le a\}$ $(-\infty < a < \infty)$, then

$$\vartheta(z) = \exp\left(-i\int_{-\infty}^{\infty} \frac{d\sigma(a)}{a+\zeta}\right) \quad (\zeta = -\bar{z}), \tag{5.11}$$

and hence

$$|\vartheta(z)| = \exp\left(-\int_{-\infty}^{\infty} \frac{\beta}{(a-\alpha)^2+\beta^2} d\sigma(a)\right) \quad (z = \alpha + i\beta, \beta > 0). \tag{5.12}$$

Thus, condition (5.7) takes the form

$$F(\alpha,\beta) \equiv \int_{-\infty}^{\infty} \frac{\beta}{(a-\alpha)^2+\beta^2} d\sigma(a) \le M \quad (-\infty < \alpha < \infty, \beta > 0), \tag{5.13}$$

with $M = \log(1/c) < \infty$.

If (5.13) is satisfied, we have for every finite interval (α_1, α_2) and for every $\eta > 0$:

$$M(\alpha_2 - \alpha_1) \geq \int_{\alpha_1}^{\alpha_2} F(\alpha, \beta)\, d\alpha = \int_{-\infty}^{\infty} \left[\int_{\alpha_1}^{\alpha_2} \frac{\beta\, d\alpha}{(a-\alpha)^2 + \beta^2} \right] d\sigma(a)$$
$$= \int_{-\infty}^{\infty} \left[\arctan \frac{\alpha_2 - a}{\beta} - \arctan \frac{\alpha_1 - a}{\beta} \right] d\sigma(a).$$

If α_1 and α_2 are points of continuity of $\sigma(a)$, then we can let $\beta \to 0$ and obtain, using Fatou's lemma,

$$M(\alpha_2 - \alpha_1) \geq \pi \int_{\alpha_1}^{\alpha_2} d\sigma(a) = \pi[\sigma(\alpha_2) - \sigma(\alpha_1)]; \tag{5.14}$$

thus $\sigma(a)$ satisfies the Lipschitz condition with constant M/π.

Conversely, for any $\sigma(a)$ satisfying (5.14) we have

$$F(\alpha, \beta) \leq \frac{M}{\pi} \int_{-\infty}^{\infty} \frac{\beta\, da}{(a-\alpha)^2 + \beta^2} = \frac{M}{\pi} \left[\arctan \frac{a-\alpha}{\beta} \right]_{a=-\infty}^{a=+\infty} = M,$$

and thus (5.13) holds. We have proved part (a) of the following proposition.

Proposition 5.1. *Let A be the operator defined by* (5.8).

(a) *A is similar to a self-adjoint operator if and only if $\sigma(a)$, the distribution function of $a(x)$, satisfies a Lipschitz condition.*

(b) *If $\sigma(a)$ satisfies a Lipschitz condition, then the completely nonself-adjoint part of A is similar to the self-adjoint operator A_0 defined on $L^2(\Omega)$ by $A_0 f(\xi) = \xi \cdot f(\xi)$, where $\Omega = \{a\colon \sigma'(a) > 0\}$.*

Proof. It remains to prove part (b). From (5.2), (5.3), and (5.5) we obtain

$$\|\Theta_T(e^{it})g\| = \lim_{z \to \xi} |\vartheta(z)| \cdot \|g\| \quad (g \in \mathfrak{D}_T), \tag{5.15}$$

where ξ denotes that point of the real axis which is the image of $\lambda = e^{it}$ under the homography $\lambda \to z = i(1+\lambda)/(1-\lambda)$, that is, $\xi = -\cot(t/2)$, and where z tends to ξ from the upper half-plane, nontangentially to the real axis. Now if $\sigma(a)$ satisfies a Lipschitz condition, we have for almost every $\xi \in (-\infty, \infty)$,

$$\lim_{z \to \xi} \int_{-\infty}^{\infty} \frac{\beta\, d\sigma(a)}{(a-\alpha)^2 + \beta^2} = \pi \sigma'(\xi) \tag{5.16}$$

as $z = \alpha + i\beta$ tends to ξ nontangentially to the real axis (cf., e.g., HOFFMAN [11] p. 123). It follows from (5.12), (5.15), and (5.16) that

$$\|\Theta_T(e^{it})g\| = \exp(-\pi\sigma'(\xi)) \cdot \|g\| \quad (g \in \mathfrak{D}_T)$$

for almost every point e^{it} of the unit circle, and hence

$$(\Delta_T(t)^2 g, g) = [1 - \exp(-2\pi\sigma'(\xi))] \cdot (g, g) \quad (g \in \mathfrak{D}_T)$$

for almost every point t of $(0, 2\pi)$. This implies, for these t,

$$\Delta_T(t) g = [1 - \exp(-2\pi\sigma'(\xi))]^{1/2} g \quad (g \in \mathfrak{D}_T).$$

Because $\mathfrak{D}_T = \tau^{-1}\mathfrak{Q}$ has dimension 1, the space $\overline{\Delta_T L_0^2(\mathfrak{D}_T)}$ can be identified with the space $\overline{\delta L_0^2(0, 2\pi)}$,[2] where

$$\delta(t) = \eta(-\cot(t/2)), \quad \eta(\xi) = [1 - \exp(-2\pi\sigma'(\xi))]^{1/2}.$$

In this way we see that $T^{(1)}$, the c.n.u. part of T, is similar to multiplication by the function e^{it} on the space $\overline{\delta L_0^2(0, 2\pi)}$. As a consequence, the completely nonself-adjoint part of A (i.e. $A^{(1)} = i(I + T^{(1)})(I - T^{(1)})^{-1}$), is similar to multiplication by $i(1 + e^{it})/(1 - e^{it}) = -\cot(t/2)$ in the same space.

Consider the transformation

$$f(t) \equiv F(e^{it}) \to \frac{1}{[\pi(1 + \xi^2)]^{1/2}} F\left(\frac{\xi - i}{\xi + i}\right) \quad (0 \le t \le 2\pi; -\infty < \xi < \infty).$$

It maps $L_0^2(0, 2\pi)$ unitarily onto $L^2(-\infty, \infty)$ in such a way that the space $\overline{\delta L_0^2(0, 2\pi)}$ is mapped onto the space $\overline{\eta L^2(-\infty, \infty)}$. Moreover, to multiplication by $-\cot(t/2)$ in the first subspace corresponds multiplication by ξ in the second subspace. Now the space $\overline{\eta L^2(-\infty, \infty)}$ can be identified in a natural way with the space $L^2(\Omega)$, where

$$\Omega = \{\xi : \eta(\xi) \neq 0\} = \{\xi : \sigma'(\xi) \neq 0\},$$

and this concludes the proof of Proposition 5.1.

3. When the operator A is completely non-self-adjoint, it is itself similar to A_0 (if $\sigma(a)$ is Lipschitzian). This is the case if, for example,

$$a(x) \equiv x, \quad \text{and hence} \quad \sigma(a) = \begin{cases} 0 \text{ for } a \le 0, \\ a \text{ for } 0 \le a \le 1 \\ 1 \text{ for } 1 \le a < \infty; \end{cases} \tag{5.17}$$

then we have $\Omega = (0, 1)$. In fact, suppose there exists a subspace $\mathfrak{H}'$ of $L^2(0, 1)$, which reduces A to a self-adjoint operator A'. Because A is bounded, we have for $h \in \mathfrak{H}'$,

$$0 = (A' - A'^*)h = (A - A^*)h = 2iQh = i(h, e_0)e_0$$

[2] We use in this section the subscript 0 in L_0^2 to indicate that integration has been taken with respect to the measure $dt/(2\pi)$.

and hence $(h,e_0)=0$. As a consequence, we also have

$$(h,A^n e_0)=(A^{*n}h,e_0)=0 \quad \text{for} \quad h\in\mathfrak{H}' \quad \text{and} \quad n=1,2,\dots .$$

Now,

$$Ae_0 = x+ix=(1+i)x,$$
$$A^2e_0=(1+i)\left(1+\frac{i}{2}\right)x^2,\dots,A^ne_0=(1+i)\left(1+\frac{i}{2}\right)\dots\left(1+\frac{i}{n}\right)x^n,\dots$$

so that $h\perp x^n$ $(n=0,1,\dots)$, which implies $h=0$. Hence $\mathfrak{H}'=\{0\}$, which proves that A is completely non-self-adjoint. We obtain the following corollary.

Corollary 5.2. *The operator A on $L^2(0,1)$ defined by*

$$Ah(x)=x\cdot h(x)+i\int_0^x h(t)\,dt$$

is completely nonself-adjoint, but similar to the self-adjoint operator A_0 on $L^2(0,1)$ defined by

$$A_0 g(x)=x\cdot g(x).$$

6 Notes

The results of Secs. 1–3 of this chapter were announced in SZ.-N.–F. [7], where the contractions in question were called "almost unitary." The term "weak contraction" was proposed by M.G. KREĬN [1]. (Thus we avoid such paradoxical expressions as "completely nonunitary, almost unitary contraction.") The operators T for which $I-T^*T$ and $I-TT^*$ are of finite rank have been called by M.S. Livšic and "quasi-unitary" by others, see POLJACKIĬ [1]–[3]. In their general context, our theorems seem to be new even in the case of finite defect indices.

The first systematic study of operators in the class (Ω_0^+) was undertaken by M.S. Brodskiĭ and M.S. Livšic (cf. BRODSKIĬ [1]) as well as by their collaborators.

Recall that for an operator A in the class (Ω_0^+), the Cayley transform T of A is a weak contraction (cf. Sec. 4). Thus the spectral decomposition of T generates an analogous spectral decomposition of A. In the particular case that A is bounded and ω is either the intersection of $\sigma(A)$ with the real line, or the part of $\sigma(A)$ in the interior of the upper half-plane, the corresponding subspace $\mathfrak{H}(\omega)$ was constructed earlier by M.S. BRODSKIĬ [8], using another method. GINZBURG [3] arrived at results similar to our Theorems 2.1 and 3.1, making use of certain parts of the paper SZ.-N.–F. [IX], but applying the method of multiplicative integrals.

Proposition 5.1 appeared first in SZ.-N.–F. [X]. Particular cases were considered earlier in BRODSKIĬ AND LIVŠIC [1] (e.g., Corollary 5.2, which is due to SAHNOVIČ). GOHBERG AND KREĬN [5] generalized Proposition 5.1 to integral operators of a more general type.

Consider two completely nonunitary contractions T_1, T_2 with defect indices equal to one, and let Θ_1, Θ_2 be the corresponding characteristic functions, which are simply scalar contractive analytic functions. KRIETE [3] shows that T_1 and T_2 are similar if and only if $\Theta_1/\Theta_2, \Theta_2/\Theta_1$ are bounded and $\{z \in C : |\Theta_1(z)| = 1\} = \{z \in C : |\Theta_2(z)| = 1\}$ up to sets of measure zero. For an extension of this result to contractions with finite defect indices see SZ.-N.–F. [22].

In connection with this chapter see also M.S. BRODSKIĬ [5]; BRODSKIĬ ET AL. [1]; FRIEDRICHS [1]; KALISCH [1]; KISILEVS′KIĬ [1]; SAHNOVIČ [1]–[10]; SARASON [2]; and ŠMUL′JAN [2].

Chapter IX

The Structure of $C_{1\cdot}$-Contractions

1 Unitary and isometric asymptotes

1. We systematically exploit the operators intertwining a given contraction with an isometry or unitary operator. Given operators T on $\mathfrak{H}$ and T' on $\mathfrak{H}'$, we denote by $\mathscr{I}(T,T')$ the set of all *intertwining operators*; these are the bounded linear transformations $X\colon \mathfrak{H} \to \mathfrak{H}'$ such that $XT = T'X$. We also use the notation $\{T\}' = \mathscr{I}(T,T)$ for the *commutant* of T. Fix a contraction T on $\mathfrak{H}$, an isometry (resp., unitary operator) V on $\mathfrak{K}$, and $X \in \mathscr{I}(T,V)$ such that $\|X\| \leq 1$. The pair (X,V) is called an *isometric* (resp., *unitary*) *asymptote* of T if for every isometry (resp., unitary operator) V', and every $X' \in \mathscr{I}(T,V')$ with $\|X'\| \leq 1$, there exists a unique $Y \in \mathscr{I}(V,V')$ such that $X' = YX$ and $\|Y\| \leq 1$.

Assume that (X,V) is an isometric or unitary asymptote of T. If the operator X is zero then we also have $\mathfrak{K} = \{0\}$. Indeed, we have $I_{\mathfrak{K}}X = OX$, and hence $I_{\mathfrak{K}} = O$ by uniqueness. If $X \neq O$, we must have $\|X\| = 1$. Indeed, setting $X' = X/\|X\|$, we deduce the existence of $Y \in \{V\}'$ such that $YX = X'$ and $\|Y\| \leq 1$. The desired conclusion follows because clearly $\|Y\| \geq 1/\|X\|$.

The following result demonstrates a uniqueness property of asymptotes.

Lemma 1.1. *For any two isometric (resp., unitary) asymptotes $(X,V),(X',V')$ of a contraction T, there exists a unique unitary transformation $Y \in \mathscr{I}(V,V')$ such that $YX = X'$.*

Proof. The existence of a contractive $Y \in \mathscr{I}(V,V')$ such that $YX = X'$ follows because (X,V) is an asymptote. Similarly, because (X',V') is an asymptote, there is a contractive $Y' \in \mathscr{I}(V',V)$ such that $Y'X' = X$. The relation

$$(Y'Y)X = Y'(YX) = Y'X' = X$$

and the fact that $Y'Y$ is a contraction imply that $Y'Y = I$ by the uniqueness clause in the definition of asymptotes. Similarly, $YY' = I$ so that Y is the desired unitary transformation.

Corollary 1.2. *Let T be a contraction on $\mathfrak{H}$, and let (X,V) be a unitary asymptote of T, with V acting on $\mathfrak{K}$. The smallest reducing subspace for V containing $X\mathfrak{H}$ is $\mathfrak{K}$.*

B.Sz.-Nagy et al., *Harmonic Analysis of Operators on Hilbert Space*, Universitext,
DOI 10.1007/978-1-4419-6094-8_9, © Springer Science+Business Media, LLC 2010

Proof. It suffices to consider the case $X \neq O$. Let $\mathfrak{K}'$ be the smallest reducing subspace for V containing $X\mathfrak{H}$, denote by Y_1 the orthogonal projection of $\mathfrak{K}$ onto $\mathfrak{K}'$, and by Y_2 the identity operator on $\mathfrak{K}$. We have $Y_1X = Y_2X = X$ and $\|Y_1\| = \|Y_2\| \leq 1$. The definition of unitary asymptotes implies $Y_1 = Y_2$, and therefore $\mathfrak{K}' = \mathfrak{K}$.

The following result allows us to deal with intertwining operators without worrying about their norms.

Lemma 1.3. *Let T be a contraction on $\mathfrak{H}$, and let (X,V) be a unitary asymptote of T, with V acting on $\mathfrak{K}$.*

(1) *The pair $(X,V|\mathfrak{K}_+)$ is an isometric asymptote for T, where $\mathfrak{K}_+ = \overline{X\mathfrak{H}}$.*
(2) *Assume that V' is an isometry and $X' \in \mathscr{I}(T,V')$. There exists a unique $Y \in \mathscr{I}(V|\mathfrak{K}_+,V')$ such that $X' = YX$. This operator satisfies $\|Y\| = \|X'\|$.*
(3) *Assume that V' is a unitary operator and $X' \in \mathscr{I}(T,V')$. There exists a unique $Y \in \mathscr{I}(V,V')$ such that $X' = YX$. This operator satisfies*

$$\|Y\| = \|Y|\mathfrak{K}_+\| = \|X'\|.$$

Proof. It suffices to consider the case $X \neq O$. To prove (1), assume that V' is an isometry on $\mathfrak{K}'$, and W is the minimal unitary extension of V'. Given $X' \in \mathscr{I}(T,V')$ with $\|X'\| \leq 1$, we also have $X' \in \mathscr{I}(T,W)$, and therefore there exists a unique $Y \in \mathscr{I}(V,W)$ such that $YX = X'$ and $\|Y\| \leq 1$. Clearly $Y\mathfrak{K}_+ \subset \overline{YX\mathfrak{H}} = \overline{X'\mathfrak{H}} \subset \mathfrak{K}'$, so that the operator $Z = Y|\mathfrak{K}_+$ belongs to $\mathscr{I}(V|\mathfrak{K}_+,V')$, $ZX = X'$, and $\|Z\| \leq 1$. In fact, Z is unique with these properties. Indeed, assume that $Z_1 \in \mathscr{I}(V|\mathfrak{K}_+,V')$, $Z_1X = X'$, and $\|Z_1\| \leq 1$. The range of X is dense in $\mathfrak{K}$, therefore the equation $Z_1X = X' = ZX$ yields $Z_1 = Z$, and this concludes the proof of (1).

Let now V' and X' be as in (2), and assume $Y_1, Y_2 \in \mathscr{I}(V|\mathfrak{K}_+,V')$ satisfy $Y_1X = Y_2X = X'$. Then $Y_1Xh = Y_2Xh = X'h$, $h \in \mathfrak{H}$, and hence $Y_1 = Y_2$ because $X\mathfrak{H}$ is dense in $\mathfrak{K}_+$. To prove the existence of Y, it suffices to consider the case $X' \neq O$. There exists then $Y_0 \in \mathscr{I}(V|\mathfrak{K}_+,V')$ such that $Y_0X = X'/\|X'\|$ and $\|Y_0\| \leq 1$; in fact $\|Y_0\| = 1$ because $X'/\|X'\|$ has norm one. It suffices to take $Y = \|X'\|Y_0$.

Finally, let V' and X' be as in (3), and assume $Y_1, Y_2 \in \mathscr{I}(V,V')$ satisfy $Y_1X = Y_2X = X'$. Then

$$Y_1V^{-n}Xh = V'^{-n}Y_1Xh = V'^{-n}X'h = V'^{-n}Y_2Xh = Y_2V^{-n}Xh$$

for all $h \in \mathfrak{H}$ and $n \geq 0$. The equality $Y_1 = Y_2$ follows then from Corollary 1.2. The existence of Y is proved as in case (2). The equality $\|Y\| = \|Y|\mathfrak{K}_+\|$ follows because

$$\|YV^{-n}Xh\| = \|YXh\| \quad (x \in \mathfrak{H}),$$

and the vectors $\{V^{-n}Xh : h \in \mathfrak{H}, n \geq 0\}$ are dense in $\mathfrak{K}$.

Isometric and unitary asymptotes also have a useful commutant extension property.

Lemma 1.4. *Given isometric (resp., unitary) asymptotes (X,V) and (X',V') of T and T', respectively, and $A \in \mathscr{I}(T,T')$, there exists a unique transformation $B =$*

$\gamma(A) \in \mathscr{I}(V,V')$ such that $BX = X'A$. This operator satisfies $\|B\| \le \|A\|$. When $T = T', X = X'$, and $V = V'$, the map γ is a unital algebra homomorphism. It follows in particular that $\sigma(\gamma(A)) \subset \sigma(A)$ for every $A \in \{T\}'$.

Proof. The existence of B follows immediately from an application of the definition of asymptotes to the operator $X'A \in \mathscr{I}(T,V')$. The uniqueness of B, and the fact that γ is a contractive homomorphism follow easily from the uniqueness properties in Lemma 1.3. For instance, given $A, A' \in \{T\}'$, the relation

$$XAA' = \gamma(A)XA' = \gamma(A)\gamma(A')X$$

implies $\gamma(AA') = \gamma(A)\gamma(A')$. The spectral inclusion follows from the fact that γ is a unital homomorphism. Indeed, if $A \in \{T\}'$ is invertible then $A^{-1} \in \{T\}'$ and $\gamma(A^{-1})\gamma(A) = \gamma(A)\gamma(A^{-1}) = \gamma(I) = I$.

We now provide a direct construction proving the existence of isometric and unitary asymptotes. Later we identify them with parts of the unitary dilation of a given operator. Fix a contraction T on $\mathfrak{H}$, and note that for every $x \in \mathfrak{H}$ the sequence $\{\|T^n x\|\}_{n=1}^{\infty}$ is decreasing, hence convergent. The polar identity

$$(Tx, Ty) = \frac{1}{4}\sum_{k=1}^{4} i^k \|T(x + i^k y)\|^2$$

shows that the sequence $\{(T^n x, T^n y)\}_{n=1}^{\infty}$ also converges for every $x, y \in \mathfrak{H}$. The form defined by

$$w_T(x,y) = \lim_{n\to\infty} (T^n x, T^n y) \quad (x, y \in \mathfrak{H})$$

is linear in x, conjugate linear in y, and $0 \le w_T(x,x) \le \|x\|^2$ for $x \in \mathfrak{H}$. Therefore there exists a unique operator A_T on $\mathfrak{H}$ such that

$$(A_T x, y) = w_T(x,y) \quad (x, y \in \mathfrak{H})$$

and $O \le A_T \le I$. The obvious identity $w_T(Tx,Ty) = w_T(x,y)$ implies $T^*A_T T = A_T$, and in particular $\|A_T^{1/2}Tx\| = \|A_T^{1/2}x\|$ for $x \in \mathfrak{H}$. Consider now the space $\mathfrak{K}_T^+ = \overline{A_T^{1/2}\mathfrak{H}}$. The preceding identity implies the existence of an isometry V_T on $\mathfrak{K}_T^+$ such that $V_T A_T^{1/2} x = A_T^{1/2} Tx$ for every $x \in \mathfrak{H}$. We also define an operator $X_T^+ : \mathfrak{H} \to \mathfrak{K}_T^+$ by setting $X_T^+ x = A_T^{1/2} x$ for $x \in \mathfrak{H}$. The operator V_T has a minimal unitary extension W_T acting on a space $\mathfrak{K}_T \supset \mathfrak{K}_T^+$. Define $X_T \in \mathscr{I}(T, W_T)$ by setting $X_T x = X_T^+ x$ for $x \in \mathfrak{H}$.

Proposition 1.5. *The pair (X_T^+, V_T) (resp., (X_T, W_T)) is an isometric (resp., unitary) asymptote of T.*

Proof. Clearly $\|X_T^+\| \le 1$ and $X_T^+ \in \mathscr{I}(T, V_T)$. Let V be an arbitrary isometry, and $X \in \mathscr{I}(T,V)$. The inequality

$$\|Xx\| = \|V^n Xx\| = \|XT^n x\| \le \|X\| \|T^n x\|$$

implies that

$$\|Xx\| \leq \|X\| \lim_{n\to\infty} \|T^n x\| = \|X\|\|X_T^+ x\| \quad (x \in \mathfrak{H}).$$

It follows that the map $Y_0 : X_T^+ x \mapsto Xx$ is well defined, and it extends uniquely to a linear transformation Y satisfying $\|Y\| \leq \|X\|$ and $YX_T^+ = X$. The equations

$$VYX_T^+ = VX = XT = YX_T^+T = YV_TX_T^+,$$

and the fact that X_T^+ has dense range, imply $Y \in \mathscr{I}(V_T, V)$. The uniqueness of such an operator Y is obvious. This proves the statement concerning (X_T^+, V_T).

Assume now that U is an arbitrary unitary operator, and $X \in \mathscr{I}(T,U)$. It follows from the first part of the proof that there exists a unique $Y \in \mathscr{I}(V_T, U)$ such that $YX_T^+ = X$. To conclude the proof it suffices to show that Y has a unique extension $Y_1 \in \mathscr{I}(W_T, U)$ with the same norm. Note that Y_1 must satisfy

$$Y_1(W_T^{-n}x) = U^{-n}(Yx) \quad (x \in \mathfrak{H}, n = 1,2,\ldots).$$

This formula defines Y_1 as a continuous linear transformation on the linear manifold $\bigcup_{n\geq 1} W_T^{-n}\mathfrak{K}_T^+$, and the desired conclusion follows from the density of this set in $\mathfrak{K}_T$.

Assume now that the operator T on $\mathfrak{H}$ has an invariant subspace $\mathfrak{H}'$, and consider the triangulation

$$T = \begin{bmatrix} T' & * \\ O & T'' \end{bmatrix}$$

associated with the orthogonal decomposition $\mathfrak{H} = \mathfrak{H}' \oplus \mathfrak{H}''$. The subspace $\mathfrak{K}' = \bigvee_{n=1}^{\infty} W_T^{-n} X_T \mathfrak{H}'$ is reducing for W_T, say $W_T = W' \oplus W''$ relative to the decomposition $\mathfrak{K}_T = \mathfrak{K}' \oplus \mathfrak{K}''$. Because $X_T\mathfrak{H}' \subset \mathfrak{K}'$, we can define operators $X' \in \mathscr{I}(T', W')$ and $X'' \in \mathscr{I}(T'', W'')$ by setting

$$X'x = X_T x \quad (x \in \mathfrak{H}')$$

and

$$X''x = P_{\mathfrak{K}''} X_T x \quad (x \in \mathfrak{H}'').$$

Theorem 1.6. *With the above notation, we have*

(1) *The pair (X', W') is a unitary asymptote of T'.*
(2) *The pair (X'', W'') is a unitary asymptote of T''.*
(3) *W_T is unitarily equivalent to $W_{T'} \oplus W_{T''}$.*

Proof. The fact that $(X_{T'}, W_{T'})$ is a unitary asymptote for T' implies the existence of a contraction $Y \in \mathscr{I}(W_{T'}, W')$ such that $YX_{T'} = X'$. Note that

$$\|X_{T'}x\| = \lim_{n\to\infty} \|T^n x\| = \|X_T x\| = \|X'x\| \quad (x \in \mathfrak{H}')$$

and therefore Y is isometric on the range of $X_{T'}$, and its range contains $X'\mathfrak{H}'$. Moreover,

$$YW_{T'}^{-n}X_{T'} = W'^{-n}YX_{T'} = W'^{-n}X' \quad (n = 1,2,\ldots),$$

and this implies that Y is an isometry and its range is dense, and hence Y is unitary. This proves (1). To prove (2), consider an arbitrary unitary operator U, and $X \in \mathscr{I}(T'',U)$. Observe that $XP_{\mathfrak{H}''} \in \mathscr{I}(T,U)$ and therefore we can find a unique $Y_0 \in \mathscr{I}(W_T,U)$ such that $\|Y_0\| \leq \|XP_{\mathfrak{H}''}\| = \|X\|$ and $Y_0X_T = XP_{\mathfrak{H}''}$. The operator Y_0 is zero on $X_T\mathfrak{H}'$, and the identity

$$Y_0W_T^{-n} = U^{-n}Y_0 \quad (n = 1,2,\dots)$$

implies that Y_0 is zero on $\mathfrak{K}'$. Therefore we can write $Y_0 = YP_{\mathfrak{K}''}$, with $Y \in \mathscr{I}(W'',U)$ and $\|Y\| = \|Y_0\| \leq \|X\|$. Moreover, we clearly have $YX'' = X$. The uniqueness of Y follows from the uniqueness of Y_0. The last assertion follows from the fact that $W_T = W' \oplus W''$, because W' and W'' are unitarily equivalent to $W_{T'}$ and $W_{T''}$, respectively.

Note that V_T need not be unitarily equivalent to $V_{T'} \oplus V_{T''}$. Indeed, take T to be a bilateral shift of multiplicity one, and $\mathfrak{H}'$ a nonreducing invariant subspace.

2. We now relate the unitary asymptote to the minimal unitary dilation of an operator. Let T be a contraction on $\mathfrak{H}$, and let U on $\mathfrak{K} \supset \mathfrak{H}$ be the minimal unitary dilation of T. Consider the $*$-residual part R_* of U on the reducing space $\mathfrak{R}_* \subset \mathfrak{K}$, and the operator $X\colon \mathfrak{H} \to \mathfrak{R}_*$ defined by

$$Xx = P_{\mathfrak{R}_*}x \quad (x \in \mathfrak{H}).$$

By Proposition II.3.1 and (II.3.6), we have $X \in \mathscr{I}(T,R_*)$ and

$$\|Xx\| = \lim_{n\to\infty} \|T^nx\| = \|X_T^+x\| \quad (x \in \mathfrak{H}).$$

It follows from the construction of V_T that the pair $(X, R_*|\overline{X\mathfrak{H}})$ is an isometric asymptote of T.

Proposition 1.7. *The pair (X,R_*) is a unitary asymptote of T.*

Proof. We only have to verify that R_* is the minimal unitary extension of $R_*|\overline{X\mathfrak{H}}$. The space $\mathfrak{R}_*^0 = \bigvee_{n=1}^{\infty} R_*^{-n}X\mathfrak{H}$ reduces R_*, and thus $\mathfrak{R}_*^0 \oplus M(\mathfrak{L})$ reduces U. This reducing space for U contains $\mathfrak{H}$, and therefore it must equal $\mathfrak{K}$ by the minimality of U. We deduce that $\mathfrak{R}_*^0 = \mathfrak{R}_*$, as desired.

Proposition 1.8. *Let (X,W) be a unitary asymptote of the contraction T. The transformation X is one-to-one if and only if T is of class $C_1.$. If T is of class C_{11} then X is a quasi-affinity, and T is quasi-similar to W.*

Proof. The first assertion is obvious. The second one follows from Proposition II.3.5.

2 The spectra of $C_1.$-contractions

1. Let T be a contraction on the Hilbert space $\mathfrak{H}$, and let (X_T,W_T) be its unitary asymptote. It follows from Lemma 1.4 that $\sigma(W_T) \subset \sigma(T)$. We show that, even if T is of class C_{11} or C_{10}, there are very few restrictions on the set $\sigma(T)$ beside this inclusion. We restrict ourselves to the case in which T is c.n.u. In this case, the spectral measure of the minimal unitary dilation U of T is absolutely continuous relative to normalized Lebesgue measure m on the unit circle C (cf. Theorem II.6.4), and therefore so is the spectral measure of the $*$-residual part R_*, and that of W_T which is unitarily equivalent to R_*. In other words, there exists a Borel subset $\omega_T \subset C$ such that the spectral measure E of W_T is mutually absolutely continuous with the scalar measure $\chi_{\omega_T}\,dm$; here we use χ_ω to denote the characteristic function of the set ω. We emphasize that ω_T is only determined up to sets of measure zero. Theorem VI.2.3 and Proposition 1.7 imply that

$$\omega_T = \{\zeta \in C : \Delta_*(\zeta) \neq O\},$$

where $\Delta_*(\zeta) = (I - \Theta_T(\zeta)\Theta_T(\zeta)^*)^{1/2}$. This follows from the representation of T^* as a functional model associated with its characteristic function $\Theta_{T^*}(\lambda)$, and from the fact that $\Theta_{T^*}(\lambda) = \Theta_{\tilde{T}}(\lambda)$. Thus, R_* is unitarily equivalent to multiplication by ζ on $\overline{\Delta_* L^2(\mathfrak{D}_{T^*})}$. If T is of class C_{11} then, for a.e. $\zeta \in C$, $\Theta_T(\zeta)$ is isometric if and only if it is unitary (cf. Propositions VI.3.5 and V.2.4), and therefore ω_T is equal to the smallest Borel set that is residual for T in the sense of Definition VII.5.3.

We recall from Definition VII.5.2 that the *essential support* of a Borel set $\alpha \subset C$ is the complement of the largest open set $\omega \subset C$ such that $m(\omega \cap \alpha) = 0$. We use the short notation α^- for the essential support of α, which should not be confused with the closure $\overline{\alpha}$. Observe that the equality $\alpha_1^- = \alpha_2^-$ can occur even when the sets α_1 and α_2 differ by a set of positive measure. We say that α is *essentially closed* if $\alpha = \alpha^-$. Clearly the set

$$\sigma(W_T) = \omega_T^-$$

is essentially closed.

We are ready to describe a further condition the sets $\sigma(T)$ and $\sigma(W_T)$ must satisfy if $T \in C_{1\cdot}$. We say that an essentially closed set $\alpha \subset C$ is *neatly contained* in a compact set $\sigma \subset \overline{D}$ if $\alpha \subset \sigma$ and each nonempty closed subset $\sigma' \subset \sigma$, such that $\sigma \setminus \sigma'$ is also closed, satisfies $m(\sigma' \cap \alpha) > 0$.

Remark. It is useful to note that, given a Borel set $\alpha \subset C$ of positive measure and an open arc $\beta \subset C$, then $m(\alpha \cap \beta) = 0$ if and only if $m(\alpha^- \cap \beta) = 0$.

Proposition 2.1. *For every c.n.u. contraction $T \in C_{1\cdot}$, the spectrum $\sigma(W_T)$ is neatly contained in $\sigma(T)$.*

Proof. Let $\sigma' \subset \sigma(T)$ be a nonempty closed set such that $\sigma(T) \setminus \sigma'$ is closed. The Riesz–Dunford functional calculus provides an invariant subspace $\mathfrak{H}'$ for T such that $\sigma(T|\mathfrak{H}') = \sigma'$ (cf. Sec. 148 in [*Func. Anal.*]). Because $T' \in C_{1\cdot}$, the operator $W_{T|\mathfrak{H}'}$ acts on a nonzero space, and hence the essentially closed set $\sigma(W_{T|\mathfrak{H}'})$ has positive measure. By Theorem 1.6.(3), $\sigma(W_{T|\mathfrak{H}'}) \subset \sigma(W_T)$, and therefore $m(\sigma' \cap \sigma(W_T)) \geq m(\sigma(W_{T|\mathfrak{H}'})) > 0$, as claimed.

Remark. Every contraction T can be written as $T = T_0 \oplus T_1$, with T_0 c.n.u. and T_1 unitary. We say that T is an *absolutely continuous contraction* if the spectral measure of T_1 is absolutely continuous relative to m. If T is absolutely continuous, the unitary asymptote $W_T \simeq W_{T_0} \oplus T_1$ is absolutely continuous, and ω_T can be defined as in the c.n.u. case. Proposition 2.1 easily extends to absolutely continuous contractions.

2. We now show that Proposition 2.1 provides the only constraint on the spectrum of an operator T of class C_{11} and that of its unitary asymptote, even if T is assumed cyclic (i.e., $\bigvee_{n=0}^{\infty} T^n h = \mathfrak{H}$ for some $h \in \mathfrak{H}$). The examples we construct must have infinite defect indices. Indeed, Theorem VII.6.3 indicates that $\sigma(T) \subset C$ if $T \in C_{11}$ and T has finite defect indices. An example of $T \in C_{11}$ with $\sigma(T) = \overline{D}$ was presented in VI.4.2.

Theorem 2.2. *Assume that $\omega_0 \subset C$ is a Borel set of positive measure, $\omega = \omega_0^-$, and ω is neatly contained in a compact set $\sigma \subset \overline{D}$. Then there exists a cyclic c.n.u. contraction $T \in C_{11}$ such that $\sigma(T) = \sigma$ and $\omega_T = \omega_0$. In particular, $\omega = \sigma(W_T)$.*

We need a few preliminaries. If $\beta \subset C$ is a Borel set, recall that $L^2(\beta)$ can be identified with $\chi_\beta L^2$, where χ_β denotes the characteristic function of β. We denote by M_β the unitary operator of multiplication by ζ on $L^2(\beta)$. It is well known that M_β has a cyclic vector. For instance, the function $u(e^{it}) = \chi_\beta(e^{it})e^{-1/t}$ $(0 < t \leq 2\pi)$ is cyclic for M_β.

Lemma 2.3. *For every Borel set $\alpha \subset C$ with positive measure, and for every $c > 0$, there exists a cyclic c.n.u. contraction $T \in C_{11}$ such that $\sigma(T) = \alpha^-$, $\|T^{-1}\| > c$, and W_T is unitarily equivalent to M_α.*

Proof. Fix $\varepsilon \in (0,1)$, and an outer function $\vartheta \in H^\infty$ such that $|\vartheta(\zeta)| = \chi_{C\setminus\alpha}(\zeta) + \varepsilon\chi_\alpha(\zeta)$ for a.e. $\zeta \in C$. Let T be a c.n.u. contraction whose characteristic function coincides with $\{E^1, E^1, \vartheta(\lambda)\}$. Because ϑ is outer, we deduce that $T \in C_{11}$ and $\|T^{-1}\| = |\vartheta(0)|^{-1} = \varepsilon^{-m(\alpha)} > c$ if ε is sufficiently small (cf. (VI.4.11) and (III.1.14)). The operator W_T is unitarily equivalent to R_*, and this operator is unitarily equivalent to M_α because, up to sets of measure zero, $\alpha = \{\zeta \in C : |\vartheta(\zeta)| < 1\}$. Finally, because $T \in C_{11}$, it is quasi-similar to M_α, and hence it has a cyclic vector.

Next is a spectral mapping theorem.

Proposition 2.4. *Let T be a contraction on $\mathfrak{H}$, and assume that the function $u \in H^\infty$ can be extended continuously to $\overline{D}$. Then*

$$\sigma(u(T)) = u(\sigma(T)).$$

Proof. Let A, B be two commuting operators on $\mathfrak{H}$. We claim that for every $\lambda \in \sigma(A)$ (resp., $\mu \in \sigma(B)$) there exists $\mu \in \sigma(B)$ (resp., $\lambda \in \sigma(A)$) such that $|\lambda - \mu| \leq \|A - B\|$. Indeed, assume that a complex number λ is at a distance greater than $\|A - B\|$ from every $\mu \in \sigma(B)$. Then the operator $(B - \lambda I)^{-1}$ has a spectral radius less than

$1/\|A-B\|$, and by commutativity it follows that $(B-\lambda I)^{-1}(A-B)$ has a spectral radius less than one. Indeed,

$$\begin{aligned}\|[(B-\lambda I)^{-1}(A-B)]^n\|^{1/n} &= \|(B-\lambda I)^{-n}(A-B)^n\|^{1/n}\\ &\le \|(B-\lambda I)^{-n}\|^{1/n}\|A-B\|\end{aligned}$$

for every n, and the desired conclusion follows by letting $n\to\infty$. Therefore the operator

$$A-\lambda I=(B-\lambda I)(I+(B-\lambda I)^{-1}(A-B))$$

is invertible.

Observe that the functions $u_r(\lambda)=u(r\lambda)$ converge uniformly to u, and therefore $\|u_r(T)-u(T)\|\to 0$ as $r\to 1-0$. Because the operators $u_r(T)$ commute with $u(T)$, the preceding observation implies that the compact sets $\sigma(u_r(T))$ converge to $\sigma(u(T))$ in the Hausdorff metric. Also, the sets $u_r(\sigma(T))=u(r\sigma(T))$ converge to $u(\sigma(T))$. Now, the operators $u_r(T)$ can also be calculated by the Riesz–Dunford functional calculus because u_r is analytic in a neighborhood of $\sigma(T)$. Therefore the desired conclusion follows from the spectral mapping theorem for the Riesz–Dunford functional calculus: $\sigma(u_r(T))=u_r(\sigma(T))$.

The construction of C_{11}-contractions with complicated spectra relies essentially on the following lemma.

Lemma 2.5. *Let $\Omega\subset D$ be a simply connected open set bounded by a rectifiable Jordan curve Γ. Assume that $\Gamma\cap C$ contains a nontrivial arc J. Let $\alpha\subset J$ be a Borel set of positive measure, $\mu_0\in\Omega$, and $c>0$. There exists a c.n.u. contraction $T\in C_{11}$ with the following properties.*

(1) $\sigma(T)=\alpha^-$.
(2) W_T *is unitarily equivalent to* M_α.
(3) $\|(T-\mu_0 I)^{-1}\|\ge c$.
(4) $\|(T-\mu I)^{-1}\|\le 1/\mathrm{dist}(\mu,\overline{\Omega})$ *for every* μ *in the complement of* $\overline{\Omega}$.

Proof. Fix a homeomorphism $u\colon\overline{D}\to\overline{\Omega}$ such that $u|D$ is holomorphic and $u(0)=\mu_0$. The existence of the conformal map $u|D$ follows from the Riemann mapping theorem. The fact that u extends to $\overline{D}$ is due to Carathéodory (cf. Theorem II.4 in GOLUZIN [1]) because Γ is a Jordan curve. By results of F. and M. Riesz, because Γ is rectifiable, a set $\omega\subset C$ has Lebesgue measure zero if and only if $u(\omega)$ has arclength zero (cf. Sec. X.1, Theorem 2 in GOLUZIN [1]). Therefore the set $\beta=u^{-1}(\alpha)\subset C$ has positive Lebesgue measure, and

$$\beta^-=u^{-1}(\alpha^-). \tag{2.1}$$

We can factor the difference

$$u(\lambda)-\mu_0=u(\lambda)-u(0)=\lambda v(\lambda)\quad(\lambda\in D), \tag{2.2}$$

with $v \in H^\infty$. Apply now Lemma 2.3 (with β and $c\|v\|_\infty$ in place of α and c) to produce a cyclic c.n.u. contraction $T_1 \in C_{11}$ such that

$$\sigma(T_1) = \beta^-, \quad \|T_1^{-1}\| \geq c\|v\|_\infty, \tag{2.3}$$

and W_{T_1} is unitarily equivalent to M_β. The required operator T is defined as $T = u(T_1)$. Theorem III.2.1(e) implies that T is a c.n.u. contraction, and the equality

$$\sigma(T) = u(\sigma(T_1)) = u(\beta^-) = \alpha^-$$

follows from Proposition 2.4 and (2.1). Because T_1 is quasi-similar to M_β, the operator $T = u(T_1)$ is quasi-similar to $u(M_\beta)$. We claim that $u(M_\beta)$ is unitarily equivalent to M_α. Indeed, an explicit unitary equivalence $Z\colon L^2(\alpha) \to L^2(\beta)$ is provided by the formula

$$(Zg)(\zeta) = \begin{cases} |u'(\zeta)|^{1/2} g(u(\zeta)) & \text{if } \zeta \in \beta \\ 0 & \text{if } \zeta \in C \setminus \beta, \end{cases}$$

and this is well defined for $g \in L^2(\alpha)$ due to the F. and M. Riesz theorem referred to above. We conclude that T is quasi-similar to M_α, so that T is cyclic, of class C_{11}, and $\omega_T = \alpha$. To verify the other properties of T, we apply (2.2) to deduce that $T - \mu_0 I = u(T_1) - u(0)I = T_1 v(T_1)$ so that

$$c\|v\|_\infty \leq \|T_1^{-1}\| \leq \|v(T_1)\|\|(T - \mu_0 I)^{-1}\| \leq \|v\|_\infty \|(T - \mu_0 I)^{-1}\|,$$

where we use (2.3) and Theorem III.2.1(b) to estimate the norm of $v(T_1)$. The inequality (3) follows. For (4), note that the function $v_\mu(\lambda) = 1/(u(\lambda) - \mu)$ belongs to H^∞ if $\mu \notin \overline{\Omega}$, and $\|v_\mu\|_\infty = 1/\mathrm{dist}(\mu, \overline{\Omega})$. Moreover, for such μ we have $(T - \mu I)^{-1} = v_\mu(T_1)$, and (4) follows from another application of Theorem III.2.1(b). The lemma is proved.

We are now ready to prove Theorem 2.2.

Proof. Let $\{\mu_n\}_{n=1}^\infty$ be a dense sequence in σ, such that each of its terms appears infinitely many times. We construct a sequence $\{\alpha_n\}_{n=1}^\infty$ of pairwise disjoint subsets of ω_0 with $m(\alpha_n) > 0$, and a sequence $\{T_n\}_{n=1}^\infty$ of cyclic c.n.u. C_{11}-contractions with the following properties.

(1) $\sigma(T_n) = \alpha_n^-$.
(2) W_{T_n} is unitarily equivalent to M_{α_n}.
(3) If $(T_n - \mu_n I)^{-1}$ exists then $\|(T_n - \mu_n I)^{-1}\| > n$.
(4) $\|(T_n - \mu I)^{-1}\| \leq 1/[\mathrm{dist}(\mu, \sigma) - 1/n]$ if $\mathrm{dist}(\mu, \sigma) > 1/n$.

Once these operators are constructed, we set $\alpha_0 = \omega_0 \setminus (\bigcup_{n=1}^\infty \alpha_n)$, and construct by Lemma 2.3 a c.n.u. contraction $T_0 \in C_{11}$ such that $\sigma(T_0) = \alpha_0^-$ and W_{T_0} is unitarily equivalent to M_{α_0}; if α_0 has measure zero, we can take T_0 to act on the trivial space $\{0\}$. The desired operator is the c.n.u. contraction defined as $T = \bigoplus_{n=0}^\infty T_n$. Each T_n is quasi-similar to W_{T_n}, and therefore T is quasi-similar to M_{ω_0}. In particular, T is of class C_{11}, it is cyclic, and $\omega_T = \omega_0$. The spectrum of T is determined by observing that, given a complex number λ, the operator $T-\lambda I$ is invertible if and only if $T_n-\lambda I$ is invertible for all n, and the sequence $\{\|(T_n-\lambda I)^{-1}\|\}_{n=0}^{\infty}$ is bounded. Condition (3) then indicates that μ_i belongs to $\sigma(T)$ for all i, and thus $\sigma\subset\sigma(T)$ by the density of $\{\mu_n\}_{n=1}^{\infty}$. On the other hand, condition (4) shows that $\sigma(T)\subset\sigma$.

To conclude the proof we need to construct sets α_n and operators T_n satisfying the above conditions. Let us set

$$G_n=\{\lambda:\operatorname{dist}(\lambda,\sigma)<1/n\}\quad(n=1,2,\dots),$$

and denote by G'_n the connected component of G_n which contains μ_n. The set $G'_n\cap\sigma$ thus contains μ_n, it is closed, and $\sigma\setminus(G'_n\cap\sigma)$ is closed as well. Because ω_0 is neatly contained in σ, we must have $m(G'_n\cap\omega_0)>0$ (cf. the remark preceding Proposition 2.1). We can find inductively Borel subsets $\beta_n\subset G'_n\cap\omega_0$ such that

$$0<m(\beta_n)\le\frac{1}{3}m(\beta_{n-1})\quad(n\ge 2).$$

The Borel set $\alpha_n=\beta_n\setminus(\bigcup_{k=1}^{\infty}\beta_{n+k})$ has positive measure because

$$m(\alpha_n)\ge m(\beta_n)-\sum_{k=1}^{\infty}m(\beta_{n+k})\ge\left(1-\sum_{k=1}^{\infty}3^{-k}\right)m(\beta_n)=\frac{1}{2}m(\beta_n).$$

There is a closed arc $J_n\subset G'_n\cap C$ such that $m(\alpha_n\cap J_n)>0$. Replacing each α_n by $\alpha_n\cap J_n$ we can also assume that $\alpha_n\subset J_n$. The sets $\{\alpha_n\}_{n=1}^{\infty}$ are pairwise disjoint.

Next we construct the operators T_n. Fix $n\ge1$, and choose a point $\mu'_n\in G'_n\cap D$ such that

$$|\mu'_n-\mu_n|<\frac{1}{2n};$$

we can take $\mu'_n=\mu_n$ if $\mu_n\in D$. Assume that ζ_1 and ζ_2 are the two endpoints of the arc J_n. The set $G'_n\cap D$ is connected, therefore we can find a simple rectifiable curve $\Gamma_n\subset(G'_n\cap D)\cup\{\zeta_1,\zeta_2\}$, with endpoints ζ_1 and ζ_2, such that the simply connected region Ω_n bounded by $J_n\cup\Gamma_n$ is entirely contained in $G'_n\cap D$, and $\mu'_n\in\Omega_n$. We now apply Lemma 2.5 with $\Omega_n,\alpha_n,\mu'_n,3n$ in place of Ω,α,μ_0,c, respectively. We obtain a cyclic c.n.u. C_{11}-contraction T_n satisfying conditions (1) and (2) above, such that

$$\|(T_n-\mu'_nI)^{-1}\|\ge 3n$$

and

$$\|(T_n-\mu I)^{-1}\|\le 1/\operatorname{dist}(\mu,\Omega_n)\quad(\mu\notin\overline{\Omega_n}).$$

The last condition implies (4) because $\Omega_n\subset G_n$, and therefore

$$\operatorname{dist}(\mu,\sigma)\le\operatorname{dist}(\mu,G_n)+\frac{1}{n}\le\operatorname{dist}(\mu,\Omega_n)+\frac{1}{n}$$

for all scalars μ. Finally choose a unit vector x such that $\|(T_n - \mu_n' I)x\| \le 1/2n$. We have

$$\|(T_n - \mu_n I)x\| \le \|(T_n - \mu_n' I)x\| + |\mu_n' - \mu_n| < \frac{1}{n},$$

and this in turn implies condition (3). The proof is complete.

3. We consider now the case of a contraction $T \in C_{10}$, which is necessarily c.n.u. If T has a finite defect index, then the two defect indices must be different; indeed, Θ_T is inner, but not $*$-inner. In this case it follows that $\sigma(T) = \overline{D}$. Surprisingly, however, the general form of the spectrum for the class C_{10} is the same as for the class C_{11}.

Theorem 2.6. *Assume that $\omega_0 \subset C$ is a Borel set of positive measure, $\omega = \omega_0^-$, and ω is neatly contained in a compact set $\sigma \subset \overline{D}$. Then there exists a cyclic contraction $T \in C_{10}$ such that $\sigma(T) = \sigma$ and $\omega_T = \omega_0$. In particular, $\omega = \sigma(W_T)$.*

The proof of this result depends on an analogue of Lemma 2.3 for the class C_{10}. This is considerably more difficult because the contractions involved must have infinite defect indices. Our construction depends on identifying appropriate invariant subspaces of bilateral weighted shifts.

Consider a sequence $\beta = \{\beta(n)\}_{n=-\infty}^{\infty}$ of positive numbers such that $\beta(n) \ge \beta(n+1) \ge 1$ for all n. We denote by L^2_β the Hilbert space consisting of those functions $f \in L^2$ whose Fourier series $\sum_{n=-\infty}^{\infty} u_n \zeta^n$ is such that

$$\sum_{n=-\infty}^{\infty} \beta(n)^2 |u_n|^2 < \infty,$$

where the norm of f is given by

$$\|f\|_\beta^2 = \sum_{n=-\infty}^{\infty} \beta(n)^2 |u_n|^2.$$

The functions $e_n(\zeta) = \zeta^n$ form an orthonormal basis in L^2, and $\beta(n)^{-1} e_n$ form an orthonormal basis in L^2_β. Denote by U the bilateral shift on L^2, and observe that $UL^2_\beta \subset L^2_\beta$. Moreover, the restriction U_β of U to L^2_β is a contraction satisfying

$$U_\beta(\beta(n)^{-1} e_n) = \beta(n)^{-1} e_{n+1} = \frac{\beta(n+1)}{\beta(n)} (\beta(n+1)^{-1} e_{n+1})$$

for all integers n. Thus U_β is a weighted bilateral shift. The inclusion operator $X_\beta : L^2_\beta \to L^2$ is obviously a contraction, and $X_\beta \in \mathscr{I}(U_\beta, U)$.

Lemma 2.7. *Assume that* $\lim_{n\to+\infty} \beta(n) = 1$. *Then the pair (X_β, U) is an isometric asymptote for U_β.*

Proof. It suffices to observe that the range of X_β is dense, U is unitary, and $\|X_\beta f\| = \lim_{n\to\infty} \|U_\beta^n f\|_\beta$ whenever f is a finite linear combination of the vectors $\{e_k\}_{k=-\infty}^{\infty}$.

The following results show that restrictions of U_β provide operators T such that $W_T = W_T^+$ is unitarily equivalent to M_α for suitable choices of the sequence β.

Lemma 2.8. *Let $\beta = \{\beta(n)\}_{n=-\infty}^{\infty}$ be a sequence of positive numbers such that $\beta(n) \geq \beta(n+1)$ and $\lim_{n\to\infty}\beta(n) = 1$. Assume that $\alpha \subset C$ is a Borel set with the property that the functions $\zeta^n\chi_\alpha(\zeta)$ belong to L^2_β for all integers n. Denote by $\mathfrak{H}$ the closed linear subspace of L^2_β generated by $\{\zeta^n\chi_\alpha(\zeta)\}_{n=-\infty}^{\infty}$, set $T = U_\beta|\mathfrak{H}$, and define $X \in \mathscr{I}(T, M_\alpha)$ by $X = X_\beta|\mathfrak{H}$. Then the pair (X, M_α) is an isometric asymptote for T.*

Proof. It is clear that $\mathfrak{H}$ is invariant for U_β and $X_\beta\mathfrak{H}$ is dense in $L^2(\alpha)$, so that the result follows from Theorem 1.6.

The operator X_β is one-to-one, therefore the operator T constructed in the preceding lemma is of class $C_{1\cdot}$, and we have $\alpha^- = \sigma(M_\alpha) \subset \sigma(T)$ by Lemma 1.4. We show that sequences β can be found so that the hypothesis of Lemma 2.8 is satisfied, and in addition $T \in C_{10}$ and $\sigma(T) = \alpha^-$. The sequences we need consist entirely of powers of 2. More precisely,

$$\beta(n) = \begin{cases} 1 & \text{for } n \geq 0 \\ 2^p & \text{for } r_p \leq -n < r_{p+1}, \end{cases}$$

where $1 = r_1 < r_2 < \cdots$ are integers such that the sequence $\{r_{p+1} - r_p\}_{p=1}^{\infty}$ is increasing and unbounded. Such a sequence β is called a *simple weight sequence.*

Lemma 2.9. *If β is a simple weight sequence, then $U_\beta \in C_{10}$ and $\sigma(U_\beta) = C$. Moreover, $\|U_\beta^{-n}\| = \beta(-n)$ for $n \geq 0$.*

Proof. Note that $U_\beta^*(\beta(k)^{-1}e_k) = (\beta(k)/\beta(k-1))(\beta(k-1)^{-1}e_{k-1})$, so that

$$U_\beta^{*n}(\beta(k)^{-1}e_k) = \frac{\beta(k)}{\beta(k-n)}(\beta(k-n)^{-1}e_{k-n})$$

tends to zero because $\beta(k-n) \to \infty$ as $n \to \infty$, thus $U_\beta \in C_{10}$. We already know that $C = \sigma(U) \subset \sigma(U_\beta)$, and it is clear that U_β is invertible with $\|U_\beta^{-1}\| = 2$. More generally,

$$\|U_\beta^{-n}\| = \sup_k \frac{\beta(k)}{\beta(k+n)} = \beta(-n) = 2^p$$

for $r_p \leq n < r_{p+1}$. Therefore

$$\|U_\beta^{-n}\|^{1/n} = 2^{p/n} \leq 2^{p/r_p},$$

and it follows that the spectral radius of U^{-1} is equal to 1 because

$$\lim_{p\to\infty} \frac{p}{r_p} = 0.$$

It follows that $\sigma(U_\beta) \subset C$, and the lemma is proved.

Lemma 2.10. *For any sequence $\{f_k\}_{k=1}^\infty \subset L^2$ there exists a simple weight sequence β such that $\{f_k\}_{k=1}^\infty \subset L^2_\beta$.*

Proof. There is no loss of generality in assuming that $\sum_{k=1}^\infty \|f_k\| < \infty$. Consider the fourier series $\sum_{n=-\infty}^\infty u_n^{(k)}\zeta^n$ of f_k, and set $u_n = \left(\sum_{k=1}^\infty |u_n^{(k)}|^2\right)^{1/2}$. It suffices to find β such that the function $f \in L^2$ with Fourier series $\sum_{n=-\infty}^\infty u_n\zeta^n$ belongs to L^2_β. Choose integers $1 = r_1 < r_2 < \cdots$ such that $\sum_{n=r_p}^\infty |u_{-n}|^2 \le 5^{-p}$ for $p \ge 2$. Increasing r_p if necessary, we may assume that the sequence $r_{p+1} - r_p$ tends increasingly to infinity. We have then

$$\begin{aligned}\|f\|_\beta^2 &= \sum_{p=2}^\infty 4^p \sum_{n=r_p}^{r_{p+1}-1} |u_{-n}|^2 + 4\sum_{n=r_1}^{r_2-1} |u_{-n}|^2 + \sum_{n=0}^\infty |u_n|^2 \\ &\le \sum_{p=1}^\infty \left(\frac{4}{5}\right)^p + 4\|f\|^2 < \infty,\end{aligned}$$

as desired.

In order to control $\sigma(T)$ in Lemma 2.8 we need one more result.

Lemma 2.11. *If T is an invertible C_1.-contraction such that*

$$\sum_{n=1}^\infty n^{-p}\|T^{-n}\| < \infty \tag{2.4}$$

for some integer p, then $\sigma(T) = \sigma(W_T)$.

Proof. Lemma 1.4 implies the inclusion $\sigma(W_T) \subset \sigma(T)$. Because the sequence $\{n^{-p}\|T^{-n}\|\}_{n=1}^\infty$ is bounded, the spectral radius of T^{-1} is at most 1; thus $\sigma(T) \subset C$. To conclude the proof, we must show that $T - \zeta_0 I$ is invertible for each $\zeta_0 \in C \setminus \sigma(W_T)$. Fix such a scalar ζ_0, and construct an infinitely differentiable function g on C such that $g(\zeta) = -1$ in a neighborhood of ζ_0, and $g(\zeta) = 0$ in an open set containing $\sigma(W_T)$. If we factor

$$g(\zeta) - g(\zeta_0) = (\zeta - \zeta_0)h(\zeta) \quad (\zeta \in C),$$

the function h is also infinitely differentiable. Let $\sum_{n=-\infty}^\infty u_n\zeta^n$ be the Fourier series of h, so that $\sum_{n=-\infty}^\infty (in)^k u_n\zeta^n$ is the Fourier series of the kth derivative of h. The sequence $\{n^k u_n\}_n$ must thus be bounded for every $k \ge 1$, and (2.4) implies that $\sum_{n=-\infty}^\infty |u_n|\|T^n\| < \infty$. We can then define the operator $Y \in \{T\}'$ by setting

$$Y = \sum_{n=-\infty}^\infty u_n T^n.$$

Note that $I = g(W_T) - g(\zeta_0)I = (W_T - \zeta_0 I)h(W_T)$, and $X_T \in \mathscr{I}(Y, h(W_T))$. We deduce that

$$X_T(T - \zeta_0 I)Y = (W_T - \zeta_0 I)h(W_T)X_T = X_T.$$

Finally, because $T \in C_{1\cdot}$, the operator X_T is one-to-one, and the last identity implies that Y is the inverse of $T - \zeta_0 I$.

We are now ready for an analogue of Lemma 2.3, but without cyclicity.

Proposition 2.12. *For every Borel set $\alpha \subset C$ with positive measure, and for every $c > 0$, there exists a C_{10}-contraction A such that $\sigma(A) = \alpha^-$, $\omega_A = \alpha$, and $\|A^{-1}\| \geq c$.*

Proof. Let β be a simple weight sequence provided by Lemma 2.10 applied to the functions $f_n(\zeta) = \zeta^n \chi_\alpha(\zeta)$ $(n = 0, \pm 1, \pm 2, \dots)$. Lemma 2.8 produces an operator $T \in C_{1\cdot}$ such that W_T is unitarily equivalent to M_α. The operator $U_\beta \in C_{10}$ is invertible and T has dense range. We conclude that $T \in C_{10}$ is invertible as well, and $\|T^{-n}\| \leq \|U_\beta^{-n}\|$ for $n \geq 1$. Therefore the equality $\sigma(T) = \alpha^-$ follows from Lemma 2.11 if $\sum_{n=1}^\infty n^{-3}\|U_\beta^{-n}\| < \infty$. This condition is achieved if $\|U_\beta^{-n}\| = \beta(-n) \leq 2n$ for $n \geq 1$, and for this it suffices to take $r_p \geq 2^{p-1}$ in the definition of β. (Note that enlarging the numbers r_p also enlarges the space L^2_β.) Observe, however, that $\|T^{-1}\| \leq 2$. In order to construct the required operator, consider the characteristic function $\{\mathfrak{D}_T, \mathfrak{D}_{T^*}, \Theta_T(\lambda)\}$. The properties of T imply that this function is inner and $*$-outer, $\Theta_T(\lambda)$ is invertible for every $\lambda \in D$, the set $\{\zeta \in C : \Theta_T(\zeta) \text{ is not unitary}\}$ coincides with α up to sets of measure zero, and for every $\zeta \in C \setminus \alpha^-$ the function Θ_T extends analytically to a neighborhood of ζ (cf. Theorem VI.4.1). Fix an arbitrary unit vector $x_0 \in \mathfrak{D}_T$, and choose a unitary transformation $Z\colon \mathfrak{D}_{T^*} \to \mathfrak{D}_T$ such that $Z\Theta_T(0)x_0 = \|\Theta_T(0)x_0\|x_0$. Such a unitary transformation exists because the spectral conditions on T imply that $\mathfrak{d}_T = \mathfrak{d}_{T^*} = \infty$. The function $\{\mathfrak{D}_T, \mathfrak{D}_T, Z\Theta_T(\lambda)\}$ coincides with Θ_T. For each natural number n, denote by A_n a c.n.u. contraction whose characteristic function coincides with $\{\mathfrak{D}_T, \mathfrak{D}_T, (Z\Theta_T(\lambda))^n\}$. The function $(Z\Theta_T(\lambda))^n$ is also inner, $*$-outer, invertible for $\lambda \in D$, it extends analytically at points in $C \setminus \alpha^-$, and $\alpha = \{\zeta \in C : (Z\Theta_T(\zeta))^n \text{ is not unitary}\}$. It follows that A_n is of class C_{10}, $\omega_{A_n} = \alpha$, and $\sigma(A_n) = \alpha^-$. Now,

$$\|A_n^{-1}\| = \|(Z\Theta_T(0))^{-n}\| \geq \|(Z\Theta_T(0))^{-n}x_0\| = \|\Theta_T(0)x_0\|^{-n} \to \infty$$

as $n \to \infty$ (cf. formula (VI.4.11)). Therefore the conclusion of the proposition is satisfied by A_n for sufficiently large n.

The following result shows the application of functional calculus (as in the proof of Lemma 2.5) produces operators of class C_{10}. Then we show how to produce a cyclic operator.

Lemma 2.13. *Let T be a C_{10}-contraction such that $\sigma(T)$ does not contain D, and let $u \in H^\infty$ be a nonconstant function which extends continuously to $\overline{D}$ such that $\|u\|_\infty = 1$ and $|u| = 1$ on ω_T. Then the contraction $u(T)$ also belongs to C_{10} and $\omega_{u(T)} = u(\omega_T)$.*

Proof. First note that $X_T u(T) = u(W_T)X_T$, and $u(W_T)$ is unitary. Because X_T is one-to-one, we deduce that $u(T) \in C_{1\cdot}$. Setting $\mu = u(0) \in D$, we know that $u(T) \in C_{\cdot 0}$ if and only if the operator

$$A = (u(T) - \mu)(I - \bar{\mu}u(T))^{-1}$$

belongs to $C_{\cdot 0}$ (cf. Theorem III.2.1, Sec. VI.1.3, and Proposition VI.3.5). Now, there exists a function $v \in H^\infty$ such that

$$\frac{u(\lambda) - \mu}{1 - \bar{\mu}u(\lambda)} = \lambda v(\lambda) \quad (\lambda \in D).$$

Considering the values of v on C we see that $\|v\|_\infty \leq 1$. We conclude that for every $x \in \mathfrak{H}$ we have

$$\|A^{*n}x\| = \|v(T)^{*n}T^{*n}x\| \leq \|T^{*n}x\| \to 0$$

as $n \to \infty$, and it follows that $A \in C_{\cdot 0}$. Thus $u(T) \in C_{10}$, as claimed. To verify the last assertion, observe that the adjoint of W_T^+ is a quasi-affine transform of T^*. The requirement on $\sigma(T)$, and the Wold decomposition, imply that $W_T^+ = W_T$, and therefore X_T is a quasi-affinity. Because $X_T \in \mathscr{I}(u(T), u(W_T))$ and $u(W_T)$ is unitary, there exists $Y \in \mathscr{I}(W_{u(T)}, u(W_T))$ such that $X_T = YX_{u(T)}$. The range of Y is necessarily dense and therefore $u(W_T)$ is unitarily equivalent to a direct summand of $W_{u(T)}$ (cf. the proof of Proposition II.3.4). An application of Proposition 2.4 shows that $X_{u(T)}$ is also a quasi-affinity. By Lemma 1.4, there exists a contraction $\widetilde{T} \in \{W_{u(T)}\}'$ satisfying $X_{u(T)}T = \widetilde{T}X_{u(T)}$. The equalities

$$W_{u(T)}X_{u(T)} = X_{u(T)}u(T) = u(\widetilde{T})X_{u(T)}$$

yield then $u(\widetilde{T}) = W_{u(T)}$. Considering the canonical decomposition of $\widetilde{T}$, we infer by Theorem III.2.1 that $\widetilde{T}$ is in fact a unitary operator. Thus there exists an operator $Z \in \mathscr{I}(W_T, \widetilde{T})$ such that $X_{u(T)} = ZX_T$. Because $Z \in \mathscr{I}(u(W_T), u(\widetilde{T}))$ has dense range, it follows that $W_{u(T)}$ is unitarily equivalent to a direct summand of $u(W_T)$. We conclude that $W_{u(T)}$ and $u(W_T)$ are unitarily equivalent, and therefore $\omega_{u(T)} = u(\omega_T)$ (Cf. KADISON AND SINGER [1] and the proof of Lemma 2.5).

To see that the construction in the proof of Theorem 2.6 produces a cyclic operator, we need some auxiliary results. The first one is a general observation about direct sums. In the proof we use a well known theorem of Runge. In our context it simply says that for every proper compact subset $A \subset C$, every continuous function f on C, and every $\varepsilon > 0$, there exists a polynomial p such that $|f - p| < \varepsilon$ in a neighborhood of A.

Lemma 2.14. *Let $\{T_n\}_{n=1}^\infty$ be a sequence of cyclic contractions with pairwise disjoint spectra contained in the unit circle C. Then the direct sum $T = \bigoplus_{n=1}^\infty T_n$ is cyclic as well.*

Proof. Assume that T_n acts on $\mathfrak{H}_n$. For $n \leq N$, we denote by $Q_{N,n}$ the orthogonal projection of $\bigoplus_{k=1}^N \mathfrak{H}_k$ onto $\mathfrak{H}_n$. The spectra are assumed disjoint, therefore Runge's theorem mentioned above, and the Riesz–Dunford functional calculus imply the existence of a polynomial $p_{N,n}$ satisfying

$$\left\| p_{N,n}\left(\bigoplus_{k=1}^{N} T_k\right) - Q_{N,n} \right\| < \frac{1}{N}.$$

Choose positive numbers δ_N such that

$$N \sup\{\|p_{N',n}(T)\| : 1 \leq n \leq N' \leq N\} < \delta_N^{-1} \quad (N = 1, 2, \ldots).$$

Choose for each n, a cyclic vector $e_n \in \mathfrak{H}_n$ for T_n such that $\|e_n\| = \delta_n$. We show that the vector $e = \bigoplus_{n=1}^{\infty} e_n$ is cyclic for T. For $n \leq N$ we have

$$\begin{aligned} \|p_{N,n}(T)e - e_n\| &\leq \left\| p_{N,n}\left(\bigoplus_{k=1}^{N} T_k\right)\left(\bigoplus_{k=1}^{N} e_k\right) - Q_{N,n}\left(\bigoplus_{k=1}^{N} e_k\right) \right\| \\ &\quad + \left\| p_{N,n}\left(\bigoplus_{k=N+1}^{\infty} T_k\right)\left(\bigoplus_{k=N+1}^{\infty} e_k\right) \right\| \\ &\leq \frac{\|e\|}{N} + \|p_{N,n}(T)\| \left(\sum_{k=N+1}^{\infty} \delta_k^2\right)^{1/2} \\ &\leq \frac{\|e\|}{N} + \left(\sum_{k=N+1}^{\infty} \frac{1}{k^2}\right)^{1/2} < \frac{\|e\|}{N} + \frac{1}{\sqrt{N}}. \end{aligned}$$

Letting $N \to \infty$, we see that e_n belongs to the cyclic space for T generated by e, and therefore e is a cyclic vector for T.

For the following lemma, denote by $\omega_{T,x}$ the set $\omega_{T|\mathfrak{H}_x}$, where

$$\mathfrak{H}_x = \bigvee_{n=0}^{\infty} T^n x$$

is the cyclic space for T generated by x. Recall that the sets $\omega_{T,x}$ and ω_T are only determined up to sets of measure zero.

Lemma 2.15. *Let T be an a. c. contraction on the separable Hilbert space $\mathfrak{H}$. The set $\{x \in \mathfrak{H} : \omega_{T,x} = \omega_T\}$ is a dense G_δ in $\mathfrak{H}$.*

Proof. The case in which $m(\omega_T) = 0$ is trivial, so we assume that $m(\omega_T) > 0$. Let (X, V) be a unitary asymptote of T. We can take $V = \bigoplus_{k=1}^{\infty} M_{\omega_k}$ with $C \supset \omega_1 \supset \omega_2 \supset \cdots$, and $Xx = \bigoplus_{k=1}^{\infty} X_k x$, with $X_k \in \mathscr{I}(T, M_{\omega_k})$. The equality $\omega_{T,x} = \omega_T$ holds if the set $\{\zeta \in \omega_1 : (X_1 x)(\zeta) = 0\}$ has measure zero (cf. Theorem 1.6).

The complement of the set in the statement is

$$\bigcup_{n=1}^{\infty} \{x \in \mathfrak{H} : m(\omega_T \setminus \omega_{T,x}) \geq 1/n\}.$$

Thus it suffices to prove that the set

$$S_\delta = \{x \in \mathfrak{H} : m(\omega_T \setminus \omega_{T,x}) \geq \delta\}$$

is closed and nowhere dense for $\delta > 0$. Let $\{x_n\}_{n=1}^\infty \subset S_\delta$ be a sequence with limit x, and set $\omega_n = \omega_T \setminus \omega_{T,x_n}$. Passing to a subsequence, we may assume that the sequence $\{\chi_{\omega_n}\}_{n=1}^\infty$ converges weakly in $L^2(\omega_T)$ to a function f. Note that $0 \leq f \leq \chi_{\omega_T}$ and $\int f\,dm = \lim_{n\to\infty} m(\omega_n) \geq \delta$, so that $m(\{\zeta : f(\zeta) > 0\}) \geq \delta$. We have

$$\int |(X_k x)(\zeta)| f(\zeta)\,dm(\zeta) = \lim_{n\to\infty} \int |(X_k x_n)(\zeta)| \chi_{\omega_n}(\zeta)\,dm(\zeta) = 0$$

for all k, so that

$$m(\{\zeta \in \omega_1 : (X_k x)(\zeta) = 0, k \geq 1\}) \geq m(\{\zeta : f(\zeta) > 0\}) \geq \delta,$$

and hence $x \in S_\delta$. This proves that S_δ is closed.

To show that S_δ has empty interior it suffices to produce a single vector x such that $\omega_{T,x} = \omega_T$. Indeed, if $x \in \mathfrak{H}$ is such a vector, and $y \in \mathfrak{H}$ is an arbitrary vector, the set $E_\varepsilon = \{\zeta \in \omega_1 : (X_1 y)(\zeta) + \varepsilon (X_1 x)(\zeta) = 0\}$ has positive measure for at most countably many values of the scalar ε. It follows that $\omega_{T,y+\varepsilon x} = \omega_T$ for values of ε arbitrarily close to zero, in particular $y + \varepsilon x \notin S_\delta$. To prove the existence of such a vector x, choose a total sequence $\{x_n\}_{n=1}^\infty \subset \mathfrak{H}$ such that $\|x_n\| < 2^{-n}$, and set $f_n = X_1 x_n$. The inequalities $\|f_n\|_1 \leq \|f_n\|_2 \leq \|x_n\| < 2^{-n}$ imply that $\sum_{n=1}^\infty |f_n(\zeta)| < \infty$ for a.e. $\zeta \in \omega_1$. Setting $\alpha_n = \{\zeta \in \omega_1 : f_n(\zeta) \neq 0\}$, we have $\bigcup_{n=1}^\infty \alpha_n = \omega_1$ a.e. We can construct inductively nonzero scalars $c_n \in D$ such that, for every $n \geq 1$,

$$\sum_{k=1}^n c_k f_k(\zeta) \neq 0 \quad \text{for a.e. } \zeta \in \bigcup_{k=1}^n \alpha_k,$$

and the set

$$\alpha_n' = \left\{\zeta \in \bigcup_{k=1}^{n-1} \alpha_k : |c_n f_n(\zeta)| \geq 3^{-n} \left|\sum_{k=1}^{n-1} c_k f_k(\zeta)\right|\right\}$$

satisfies $m(\alpha_n') < 2^{-n}$ (here $\alpha_1' = \varnothing$). Indeed, set $c_1 = 1$, and assume that $n > 1$ and c_j has been defined for $j < n$. Choosing $\delta_n \in (0,1)$ sufficiently small, we have $m(\alpha_n') < 2^{-n}$ if $|c_n| = \delta_n$. Choose next a complex number λ_n with absolute value 1 such that

$$\sum_{k=1}^{n-1} c_k f_k(\zeta) + \lambda_n \delta_n f_n(\zeta) \neq 0 \quad \text{for a.e. } \zeta \in \bigcup_{k=1}^n \alpha_k,$$

and set $c_n = \lambda_n \delta_n$. Because $\sum_{n=2}^\infty m(\alpha_n') < \infty$, we conclude that a.e. $\zeta \in \omega_1$ is only contained in finitely many of the sets α_n'. It follows then that the function $\sum_{n=1}^\infty c_n f_n$ is different from zero a.e. on ω_1, and hence the vector $x = \sum_{n=1}^\infty c_n x_n$ satisfies the equality $\omega_{T,x} = \omega_1 = \omega_T$.

We have now the necessary ingredients to prove an analogue of Lemma 2.5 for the class C_{10}.

Lemma 2.16. *Let $\Omega \subset D$ be a simply connected open set bounded by a rectifiable Jordan curve Γ. Assume that that the set $\Gamma \cap C$ contains a nontrivial arc J with $0 < m(J) < 1$. Let $\alpha \subset J$ be a Borel set of positive measure, $\mu_0 \in \Omega$, and $c > 0$. There exists a cyclic contraction $T \in C_{10}$ with the following properties.*

(1) $\sigma(T) = \alpha^-$.
(2) *W_T is unitarily equivalent to M_α.*
(3) $\|(T - \mu_0 I)^{-1}\| \geq c$.
(4) $\|(T - \mu I)^{-1}\| \leq 1/\text{dist}(\mu, \overline{\Omega})$ *for every μ in the complement of $\overline{\Omega}$.*

Proof. Fix a homeomorphism u, a set β, and a factorization

$$u(\lambda) - \mu_0 = \lambda v(\lambda) \quad (\lambda \in D)$$

as in the proof of Lemma 2.5. Proposition 2.12 provides an operator $T_1 \in C_{10}$ such that

$$\sigma(T_1) = \beta^-, \quad \|T_1^{-1}\| > c\|v\|_\infty,$$

and $\omega_{T_1} = \beta$. (Actually, W_{T_1} is unitarily equivalent to the orthogonal sum of some copies of M_β by the proof of Proposition 2.12.) Proposition 2.4 and Lemma 2.13 now show that the operator $T_2 = u(T_1)$ is of class C_{10},

$$\sigma(T_2) = u(\sigma(T_1)) = u(\beta^-) = \alpha^-, \quad \omega_{T_2} = \alpha,$$

and

$$\|(T_2 - \mu_0 I)^{-1}\| > c.$$

The operator T_2 also satisfies condition (4) by virtue of Theorem III.2.1(b). The required operator T is obtained as the restriction $T = T_2|\mathfrak{M}$, where $\mathfrak{M}$ is the cyclic space generated by a vector x such that $\omega_{T_2,x} = \omega_{T_2} = \alpha$. Such vectors are dense, therefore T satisfies (3) for an appropriate choice of x, provided $T - \mu_0 I$ is invertible. Property (2) is verified by virtue of the equality $\omega_{T,x} = \omega_T = \alpha$. Observe now that the operators $(T_2 - \lambda I)^{-1}$ $(\lambda \notin \sigma(T_2))$ can be approximated in norm by polynomials in T_2; this follows from Runge's theorem via the Riesz–Dunford functional calculus. It follows that $\mathfrak{M}$ is invariant for $(T_2 - \lambda I)^{-1}$ $(\lambda \notin \sigma(T_2))$, and this immediately implies that $\sigma(T) \subset \sigma(T_2) = \alpha^-$ and condition (4) is satisfied. Finally, condition (1) is also satisfied because $\alpha^- = \sigma(W_T) \subset \sigma(T)$.

The following result is a slight variation of Lemma 2.16.

Lemma 2.17. *Let $\beta \subset C$, $\varnothing \neq \beta \neq C$, be an essentially closed set, and fix $\varepsilon \in (0,1)$. There exists a cyclic C_{10}-contraction T with the following properties.*

(1) $\sigma(T) = \beta$.
(2) *W_T is unitarily equivalent to M_β.*
(3) $\|(T - \mu I)^{-1}\| \leq 1/(\text{dist}(\mu, \beta) - \varepsilon)$ *whenever* $\text{dist}(\mu, \beta) > \varepsilon$.

Proof. There exist a finite number of pairwise disjoint open arcs

$$J_1, J_2, \ldots, J_n \subset C$$

such that

$$\beta \subset \bigcup_{k=1}^{n} J_k \subset \{\mu : \text{dist}(\mu, \beta) < \varepsilon/2\}.$$

The sets

$$\Omega_k = \{r\zeta : \zeta \in J_k, r \in (1 - (\varepsilon/2), 1)\}$$

are contained in $\{\mu : \text{dist}(\mu, \beta) < \varepsilon\}$. Lemma 2.16 provides then cyclic C_{10}-contractions T_k such that $\sigma(T_k) = \beta \cap J_k$, W_{T_k} is unitarily equivalent to $M_{\beta \cap J_k}$, and $\|(T_k - \mu I)^{-1}\| \leq 1/\text{dist}(\mu, \Omega_k)$ for $\mu \notin \overline{\Omega_k}$. The operator $T = \bigoplus_{k=1}^{n} T_k$ satisfies all the requirements of the lemma. Its cyclicity is guaranteed by Lemma 2.14.

We can now prove Theorem 2.6.

Proof. Let $\{\mu_n\}_{n=1}^{\infty}$ be a dense sequence in σ, such that each of its terms appears infinitely many times. Using Lemma 2.16 in place of Lemma 2.5, and applying the regularity of Lebesgue measure, we construct a sequence $\{\alpha_n\}_{n=1}^{\infty}$ of pairwise disjoint essentially closed subsets of ω_0 with $m(\alpha_n) > 0$, and a sequence $\{T'_n\}_{n=1}^{\infty}$ of cyclic C_{10}-contractions with the following properties.

(1) $\sigma(T'_n) = \alpha_n^- = \alpha_n$.
(2) $W_{T'_n}$ is unitarily equivalent to M_{α_n}.
(3) if $(T'_n - \mu_n I)^{-1}$ exists then $\|(T'_n - \mu_n I)^{-1}\| > n$.
(4) $\|(T'_n - \mu I)^{-1}\| \leq 1/[\text{dist}(\mu, \sigma) - 1/n]$ if $\text{dist}(\mu, \sigma) > 1/n$.

We can then form the C_{10}-contraction $T' = \bigoplus_{n=1}^{\infty} T'_n$ that satisfies $\omega_{T'} = \bigcup_{n=1}^{\infty} \alpha_n$ and $\sigma(T') = \sigma$.

The set $\omega_1 = \omega_0 \setminus \omega_{T'}$ may have positive measure. If that is the case, regularity of Lebesgue measure implies the existence of pairwise disjoint essentially closed sets $\beta_n \subset \omega_1$ such that $\bigcup_{n=1}^{\infty} \beta_n = \omega_1$ (up to sets of measure zero). By Lemma 2.17, there exist cyclic C_{10}-contractions T''_n such that:

(a) $\sigma(T''_n) = \beta_n$.
(b) $W_{T''_n}$ is unitarily equivalent to M_{β_n}.
(c) $\|(T''_n - \mu I)^{-1}\| \leq 1/[\text{dist}(\mu, \beta_n) - 1/n]$ if $\text{dist}(\mu, \beta_n) > 1/n$.

Condition (c) implies, of course,

(c′) $\|(T''_n - \mu I)^{-1}\| \leq 1/[\text{dist}(\mu, \sigma) - 1/n]$ if $\text{dist}(\mu, \sigma) > 1/n$.

These conditions imply that the C_{10}-contraction

$$T'' = \bigoplus_{n=1}^{\infty} T''_n$$

satisfies $\sigma(T'') = \omega_1^-$ and $\omega_{T''} = \omega_1$. Moreover, note that the family $\{\alpha_n, \beta_n\}_{n=1}^{\infty}$ consists of pairwise disjoint closed sets. Therefore Lemma 2.14 implies that $T = T' \oplus T''$ is cyclic. Thus T satisfies all the requirements of Theorem 2.6.

3 Intertwining with unilateral shifts

1. In the preceding sections we studied intertwinings in $\mathscr{I}(T,W)$, where W is an isometric or unitary operator. In this section we focus on $\mathscr{I}(W,T)$, where T is an absolutely continuous contraction, and W is isometric. The results are particularly useful for contractions of class $C_1.$, but they can be applied whenever T has a nontrivial $*$-residual part, which is why we chose to prove them in greater generality. Let $W = W_0 \oplus W_1$ on $\mathfrak{K}_0 \oplus \mathfrak{K}_1$ be the Wold decomposition of W, with W_1 unitary and W_0 a unilateral shift. If $T \in C_{.0}$ and $X \in \mathscr{I}(W,T)$, then obviously $X|\mathfrak{K}_1 = 0$. Therefore we restrict ourselves to the case of a unilateral shift W. We start by explaining in rough outline the idea of this section. Fix a contraction T on a Hilbert space $\mathfrak{H}$, and let U be its minimal unitary dilation on $\mathfrak{K} \supset \mathfrak{H}$. If the $*$-residual space $\mathfrak{R}_*$ has nonzero intersection with $\mathfrak{H}$, then $\mathfrak{H} \cap \mathfrak{R}_*$ is invariant for T. Indeed, for $x \in \mathfrak{H} \cap \mathfrak{R}_*$,

$$\|P_{\mathfrak{R}_*}(Tx)\| = \|R_* P_{\mathfrak{R}_*} x\| = \|R_* x\| = \|x\| \geq \|Tx\|,$$

and therefore $P_{\mathfrak{R}_*}(Tx) = Tx$. If, in addition, T is c.n.u., then $T|(\mathfrak{H} \cap \mathfrak{R}_*)$ is a unilateral shift. We show that the space $\mathfrak{R}_*$ always contains vectors which are arbitrarily close to $\mathfrak{H}$ in the L^∞ norm, and this provides restrictions of T similar to unilateral shifts when $\omega_T = C$. The meaning of L^∞ approximation is explicated by Lemma 3.1. We focus first on the c.n.u. case, so let us assume that $\{\mathfrak{E}, \mathfrak{E}_*, \Theta(\lambda)\}$ is a purely contractive analytic function, and that $\mathbf{T}$ is the model operator associated with this function (as in Sec. VI.3) acting on $\mathbf{H}$, whereas $\mathbf{U}$ and $\mathbf{U}_+$ are the operators of multiplication by ζ on the spaces

$$\mathbf{K} = L^2(\mathfrak{E}_*) \oplus \overline{\Delta L^2(\mathfrak{E})}, \quad \mathbf{K}_+ = H^2(\mathfrak{E}_*) \oplus \overline{\Delta L^2(\mathfrak{E})},$$

respectively, where $\Delta(\zeta) = (I - \Theta(\zeta)^*\Theta(\zeta))^{1/2}$ for a.e. $\zeta \in C$. The spaces $\mathfrak{E}$ and $\mathfrak{E}_*$ are assumed to be separable. The crucial approximation result is the following purely function theoretical result.

Lemma 3.1. *Fix an essentially bounded function $u \in L^2(\mathfrak{E}_*)$ and a positive number ε. There exist $u' \in H^2(\mathfrak{E}_*)$ and an inner function $\varphi \in H^\infty$ such that $\|u'(\zeta) - \varphi(\zeta)u(\zeta)\| < \varepsilon$ for a.e. $\zeta \in C$.*

Proof. We may assume that $\varepsilon < 1$ and $\|u(\zeta)\| \leq 1$ a.e. There exists a measurable partition $C = \bigcup_{n=1}^\infty \sigma_n$, and there are vectors $x_n \in \mathfrak{E}_*$ such that $\|x_n\| \leq 1$,

$$\|u(\zeta) - x_n\| < \frac{\varepsilon}{2} \quad (\text{a.e. } \zeta \in \sigma_n, n \geq 1).$$

This is easily seen by constructing a Borel partition of the unit ball of $\mathfrak{E}_*$ into sets of diameter less than $\varepsilon/2$, and defining σ_n to be preimages under u of elements of this partition. Construct outer functions $\psi_n \in H^\infty$ such that

$$|\psi_n| = \chi_{\sigma_n} + \frac{\varepsilon}{10^n} \chi_{C \setminus \sigma_n} \quad (\text{a.e. on } C).$$

The function $\psi = \sum_{n=1}^{\infty} \psi_n$ belongs to H^∞, and

$$||\psi(\zeta)| - 1| \le \sum_{n=1}^{\infty} \frac{\varepsilon}{10^n} = \frac{\varepsilon}{9} \quad (\text{a.e. } \zeta \in C).$$

Factor now $\psi = g\varphi$, with g outer and φ inner. Because

$$||g(\zeta)| - 1| = ||\psi(\zeta)| - 1| \le \frac{\varepsilon}{9},$$

the function g is invertible in H^∞, and $\|g^{-1}\|_\infty < 2$. Finally, we define

$$u' = \sum_{n=1}^{\infty} g^{-1} \psi_n x_n.$$

The series above converges a.e. and in $L^2(\mathfrak{E}_*)$, and therefore its sum belongs to $H^2(\mathfrak{E}_*)$. For almost every $\zeta \in \sigma_n$ we have

$$\begin{aligned} \|u'(\zeta) - \varphi(\zeta)u(\zeta)\| &< \frac{\varepsilon}{2} + \|u'(\zeta) - \varphi(\zeta)x_n\| \\ &\le \frac{\varepsilon}{2} + \|g^{-1}\|_\infty \|g(\zeta)u'(\zeta) - g(\zeta)\varphi(\zeta)x_n\| \\ &\le \frac{\varepsilon}{2} + 2|\psi_n(\zeta) - \psi(\zeta)| + 2\sum_{k\neq n} \|\psi_k(\zeta)x_k\| \\ &\le \frac{\varepsilon}{2} + 4\sum_{k=1}^{\infty} \frac{\varepsilon}{10^k} < \varepsilon. \end{aligned}$$

Thus the functions u' and φ verify the conclusion of the lemma.

In this functional representation we have

$$\mathbf{R}_* = \mathbf{K} \ominus M(\mathbf{L}) = \mathbf{K} \ominus \{\Theta u \oplus \Delta u : u \in L^2(\mathfrak{E})\},$$

and the projection onto $M(\mathfrak{L})$ is the multiplication operator by the operator-valued function

$$\begin{bmatrix} \Theta \\ \Delta \end{bmatrix} [\Theta^*, \Delta] = \begin{bmatrix} \Theta\Theta^* & \Theta\Delta \\ \Delta\Theta^* & \Delta^2 \end{bmatrix}.$$

It follows that $P_{\mathbf{R}_*}$ is the operator of multiplication by the projection-valued function

$$P(\zeta) = \begin{bmatrix} I - \Theta(\zeta)\Theta(\zeta)^* & -\Theta(\zeta)\Delta(\zeta) \\ -\Delta(\zeta)\Theta(\zeta)^* & \Theta(\zeta)^*\Theta(\zeta) \end{bmatrix},$$

where $P(\zeta)$ must be viewed as an operator on the space $\mathfrak{E}_* \oplus \overline{\Delta(\zeta)\mathfrak{E}}$.

The operator R_* is determined, up to unitary equivalence, by the measurable spectral multiplicity function

$$\mu(\zeta) = \text{rank}(P(\zeta)) \quad (\zeta \in C),$$

whose possible values are nonnegative integers or $\aleph_0$. More precisely, if we set

$$\omega_n = \{\zeta : \mu(\zeta) \geq n\} \quad (n = 1, 2, \ldots),$$

then R_* is unitarily equivalent to $\bigoplus_{n=1}^{\infty} M_{\omega_n}$. Let $X \in \mathscr{I}(\bigoplus_{n=1}^{\infty} M_{\omega_n}, R_*)$ be a unitary operator, and consider the sequence $\{\chi_{\omega_n}\}_{n=1}^{\infty}$, where χ_{ω_n} is viewed as an element in the nth component of $\bigoplus_{n=1}^{\infty} L^2(\omega_n)$. The sequence $\{w_n\}_{n=1}^{\infty} \subset \mathbf{R}_*$ defined by $w_n = X\chi_{\omega_n}$ has the property that $w_n(\zeta) = 0$ for $n > \mu(\zeta)$ and $\{w_n(\zeta)\}_{0 \leq n-1 < \mu(\zeta)}$ is an orthonormal basis in the range of $P(\zeta)$ for a.e. $\zeta \in C$. Such a sequence $\{w_n\}_{n=1}^{\infty}$ is called a *basic sequence* for R_*. Conversely, given a basic sequence $\{w_n\}_{n=1}^{\infty}$ for R_*, there exists a unitary operator $X \in \mathscr{I}(\bigoplus_{n=1}^{\infty} M_{\omega_n}, R_*)$ such that $X\chi_{\omega_n} = w_n$. This discussion, including the concept of a basic sequence, can be applied to any a.c. unitary operator on a separable Hilbert space. We refer to DUNFORD AND SCHWARTZ [2], where the more general case of normal operators is considered.

Fix a separable, infinite-dimensional Hilbert space $\mathfrak{F}$ with orthonormal basis $\{e_n\}_{n=1}^{\infty}$, and a basic sequence $\{w_n\}_{n=1}^{\infty}$ for an a.c. unitary operator W. Denote by $S_{\mathfrak{F}}$ the unilateral shift on $H^2(\mathfrak{F})$. We can then construct an operator $J \in \mathscr{I}(S_{\mathfrak{F}}, W)$ satisfying $Je_n = w_n$; here e_n is regarded as a constant function in $H^2(\mathfrak{F})$. Such an operator is called a *basic operator*. The existence of J is seen most easily when $W = \bigoplus_{n=1}^{\infty} M_{\omega_n}$ and $w_n = \chi_{\omega_n}$. If $f = \sum_{n=1}^{\infty} f_n e_n$ is an arbitrary element of $H^2(\mathfrak{F})$, with $f_n \in H^2$, the corresponding basic operator is defined by

$$Jf = \bigoplus_{n=1}^{\infty} \chi_{\omega_n} f_n.$$

The following theorem is the main result in this section.

Theorem 3.2. *Let T be an a.c. contraction on a separable Hilbert space $\mathfrak{H}$, let U be its minimal unitary dilation on $\mathfrak{K} \supset \mathfrak{H}$, and let $\varepsilon \in (0,1)$. Denote by R_* the $*$-residual part of U. There exist operators $Y, Y' \in \mathscr{I}(S_{\mathfrak{F}}, T)$ and $Z, Z' \in \mathscr{I}(T, R_*)$ such that*

(i) *The norms $\|Y\|, \|Y'\|, \|Z\|, \|Z'\|$ are less than $1 + \varepsilon$.*
(ii) *ZY and $Z'Y'$ are basic operators.*
(iii) *$YH^2(\mathfrak{F}) \vee Y'H^2(\mathfrak{F}) = \mathfrak{H}$.*

Proof. Let $T = T_0 \oplus T_1$ be the decomposition of T such that T_0 is c.n.u. and T_1 is an absolutely continuous unitary operator. We may assume that T_0 is the model operator $\mathbf{T}$ associated with the purely contractive analytic function $\{\mathfrak{E}, \mathfrak{E}_*, \Theta(\lambda)\}$, and $T_1 = \bigoplus_{k=1}^{\infty} M_{\alpha_k}$, where $\alpha_1 \supset \alpha_2 \supset \cdots$ are Borel subsets of C. Setting $\mathfrak{H}_1 = \bigoplus_{n=1}^{\infty} L^2(\alpha_k)$, the minimal unitary dilation of T acts on $\mathfrak{K} = \mathbf{K} \oplus \mathfrak{H}_1$, and $\mathfrak{R}_* = \mathbf{R}_* \oplus \mathfrak{H}_1$. Consider a basic sequence $\{w_n\}_{n=1}^{\infty}$ for R_*. Writing $w_n = u_n \oplus v_n \oplus g_n$ with $u_n \in L^2(\mathfrak{E}_*), v_n \in \overline{\Delta L^2(\mathfrak{E})}$, and $g_n \in \bigoplus_{k=1}^{\infty} L^2(\alpha_k)$, we deduce from Lemma 3.1 that there exist inner functions $\varphi_n \in H^\infty$, and functions $u'_n \in H^2(\mathfrak{E}_*)$, such that

$$\|u'_n(\zeta) - \varphi_n(\zeta) u_n(\zeta)\| < \frac{\varepsilon}{10^n} \quad (\text{a.e. } \zeta \in C, n \geq 1).$$

The functions $w_n' = u_n' \oplus \varphi_n v_n \oplus \varphi_n g_n$ belong then to $\mathfrak{K}_+$, and

$$\|w_n'(\zeta) - \varphi_n(\zeta) w_n(\zeta)\| < \frac{\varepsilon}{10^n} \quad (\text{a.e. } \zeta \in C, n \geq 1). \tag{3.1}$$

Replacing $\{w_n\}_{n=1}^\infty$ by the basic sequence $\{\varphi_n w_n\}_{n=1}^\infty$ we can assume that $\varphi_n = 1$ for every n. Let us also set

$$w_n'' = P_{\mathfrak{R}_*} w_n' \quad (n \geq 1).$$

As in the c.n.u. case, $P_{\mathfrak{R}_*}$ is a multiplication operator by a projection valued function $Q(\zeta)$, and therefore by (3.1) and by the assumption that $\varphi_n = 1$, we also have

$$\|w_n''(\zeta) - w_n(\zeta)\| < \frac{\varepsilon}{10^n} \quad (\text{a.e. } \zeta \in C, n \geq 1).$$

The construction of the desired operators proceeds as follows:

(a) Show that there is an operator $A \in \{R_*\}'$ such that $\|A\| < 1 + \varepsilon$ and $Aw_n'' = w_n$.
(b) Define $Z = AP_{\mathfrak{R}_*}|\mathfrak{H}$.
(c) Show that there is an operator $X \in \mathscr{I}(S_{\mathfrak{F}}, U_+)$ such that $\|X\| < 1 + \varepsilon$ and $Xe_n = w_n'$ for all n.
(d) Define $Y = P_{\mathfrak{H}} X$.
(e) Verify that $ZYe_n = w_n$, and therefore ZY is basic.

We start with the construction of A. The vectors $\{w_n(\zeta)\}_{0 \leq n-1 < \operatorname{rank}(Q(\zeta))}$ form an orthonormal basis of the range of $Q(\zeta)$. There exists an operator $B(\zeta)$ on this space such that $B(\zeta) w_n(\zeta) = w_n''(\zeta)$ for $n - 1 < \operatorname{rank}(Q(\zeta))$, and

$$\|B(\zeta) - I\| \leq \|B(\zeta) - I\|_2 \leq \left(\sum_{n=1}^\infty \frac{\varepsilon^2}{100^n} \right)^{1/2} < \frac{\varepsilon}{9}.$$

Here we use $\|D\|_2$ to denote the Hilbert–Schmidt norm: $\|D\|_2^2 = \sum_n \|Dw_n(\zeta)\|^2$ for an operator on the range of $Q(\zeta)$. The function $B(\zeta)$ is strongly measurable, and therefore the operator B of pointwise multiplication by $B(\zeta)$ belongs to $\{R_*\}'$ and $\|B - I\| < \varepsilon/9$. It follows that B is invertible, and the operator $A = B^{-1}$ satisfies condition (a). Analogously, for a.e. $\zeta \in C$ there are operators $X(\zeta)$ defined on $\mathfrak{F}$ such that $X(\zeta)e_n = w_n'(\zeta)$ and $\|X(\zeta)\| < 1 + \varepsilon$. Indeed, $X(\zeta)$ differs from the partially isometric operator $X_0(\zeta) : e_n \mapsto w_n(\zeta)$ by a Hilbert–Schmidt operator with norm less than ε. The operator X is defined as pointwise multiplication by the function $X(\zeta)$.

To verify (e), note that $\mathfrak{K}_+ \ominus \mathfrak{H} \subset M(\mathfrak{L}) = \mathfrak{R}_*^\perp$, thus $P_{\mathfrak{R}_*} P_{\mathfrak{H}} | \mathfrak{K}_+ = P_{\mathfrak{R}_*} | \mathfrak{K}_+$. Therefore

$$ZYe_n = ZP_{\mathfrak{H}} w_n' = AP_{\mathfrak{R}_*} P_{\mathfrak{H}} w_n' = AP_{\mathfrak{R}_*} w_n' = Aw_n'' = w_n$$

for $n \geq 1$. Furthermore, because $P_{\mathfrak{H}} | \mathfrak{K}_+ \in \mathscr{I}(U_+, T)$ and $P_{\mathfrak{R}_*} | \mathfrak{H} \in \mathscr{I}(T, R_*)$, the operators Z and Y defined in (b) and (d) satisfy the required commutation and norm requirements.

To prove (iii), it suffices to construct a second operator X' such that

$$XH^2(\mathfrak{F}) \vee X'H^2(\mathfrak{F}) = \mathfrak{K}_+.$$

Fix then a total sequence $\{g_n\}_{n=1}^{\infty}$ in $\mathfrak{K}_+$ consisting of essentially bounded functions. Replace the vector w'_n in the above construction by $w'_n + \delta_n g_n$, where $\delta_n \neq 0$ is chosen so that the condition

$$\|w_n(\zeta) - (w'_n(\zeta) + \delta_n g_n(\zeta))\| < \frac{\varepsilon}{9^n}$$

is satisfied a.e. The above construction of X shows that there exists a unique operator $X' \in \mathscr{I}(S_{\mathfrak{F}}, U_+)$ satisfying $X'e_n = w'_n + \delta_n g_n$ $(n \geq 1)$ and $\|X'\| < 1 + \varepsilon$ Then $\delta_n g_n = X'e_n - Xe_n$, so that the ranges of X and X' span the whole space $\mathfrak{K}_+$. The theorem is proved.

2. Fix an a.c. contraction T on $\mathfrak{H}$, and let U on $\mathfrak{K} \supset \mathfrak{H}$ be the minimal unitary dilation of T. As noted before, there exist Borel sets $C \supset \omega_1 \supset \omega_2 \supset \cdots$ such that R_* is unitarily equivalent to $\bigoplus_{n=1}^{\infty} M_{\omega_n}$. The operator R_* is unitarily equivalent to W_T, thus it follows that $\omega_1 = \omega_T$. We also use the notation $\omega_{\aleph_0} = \bigcap_{n=1}^{\infty} \omega_n$. Given a Borel set $\alpha \subset C$, and a cardinal number $n \leq \aleph_0$, we denote by $M_\alpha^{(n)}$ the direct sum of n copies of M_α. Fix also a separable Hilbert space $\mathfrak{F}$ with orthonormal basis $\{e_k\}_{k=1}^{\infty}$, and denote by $\mathfrak{F}_n$ the closed linear space generated by $\{e_k : k-1 < n\}$; note that $\mathfrak{F}_{\aleph_0} = \mathfrak{F}$.

Proposition 3.3. *With the notation above, let $\alpha \subset C$ be a Borel set. The following are equivalent for every n:*

(1) $m(\alpha \setminus \omega_n) = 0$.
(2) *There exist operators $Y \in \mathscr{I}(S_{\mathfrak{F}_n}, T)$ and $Z \in \mathscr{I}(T, M_\alpha^{(n)})$ such that ZY is a basic operator.*

If these conditions are satisfied and $\varepsilon > 0$, the operators Y and Z can be chosen such that $\|Y\| < 1 + \varepsilon$ and $\|Z\| < 1 + \varepsilon$.

Proof. Theorem 3.2 provides operators $Y_0 \in \mathscr{I}(S_{\mathfrak{F}}, T)$ and $Z_0 \in \mathscr{I}(T, R_*)$ such that $\|Y_0\| < 1 + \varepsilon, \|Z_0\| < 1 + \varepsilon$, and $Z_0 Y_0$ is a basic operator. Denote by $\{w_k = Z_0 Y_0 e_k\}_{k=1}^{\infty}$ the corresponding basic sequence for R_*. Assume first that (1) is true. The reducing space $\mathfrak{N}$ for R_* generated by the functions $\{\chi_\alpha w_k : k-1 < n\}$ is such that $R_*|\mathfrak{N}$ is unitarily equivalent to $M_\alpha^{(n)}$. The required operators Y, Z are obtained simply as $Y = Y_0|H^2(\mathfrak{F}_n)$ and $Z = AP_{\mathfrak{N}} Z_0$, where $A \in \mathscr{I}(R_*|\mathfrak{N}, M_\alpha^{(n)})$ is unitary. The operators Y, Z thus constructed also satisfy $\|Y\|, \|Z\| < 1 + \varepsilon$.

Conversely, assume that (2) holds, and observe that the smallest reducing subspace for $M_\alpha^{(n)}$ which contains the range of Z is the entire space. This follows from the fact that $Z\mathfrak{H}$ contains the range of the basic operator ZY. Because $(P_{\mathfrak{R}_*}|\mathfrak{H}, R_*)$ is a unitary asymptote of T, there exists $A \in \mathscr{I}(R_*, M_\alpha^{(n)})$ such that $Z = AP_{\mathfrak{R}_*}|\mathfrak{H}$. We also have $A \in \mathscr{I}(R_*^{-1}, M_\alpha^{(n)-1})$, and this implies that $\overline{A\mathfrak{R}_*}$ is a reducing subspace for $M_\alpha^{(n)}$, and therefore A has dense range because $\overline{A\mathfrak{R}_*} \supset Z\mathfrak{H}$. Equivalently,

$A^* \in \mathcal{I}(M_\alpha^{(n)}, R_*)$ is one-to-one, and we see as above that the closure of the range of A^* is reducing R_*. The restriction of R_* to this reducing subspace is unitarily equivalent to $M_\alpha^{(n)}$. Spectral multiplicity decreases when passing to a reducing subspace, thus (1) follows.

When $n = \aleph_0$, the argument of Theorem 3.2 shows that the preceding construction actually produce two operators Y, Y' such that

$$YH^2(\mathfrak{F}_{\aleph_0}) \vee Y'H^2(\mathfrak{F}_{\aleph_0}) = \mathfrak{H}$$

where, of course, $\mathfrak{F}_{\aleph_0} = \mathfrak{F}$. For finite n we have the following result.

Proposition 3.4. *With the notation of the preceding proposition, assume that $m(\alpha \setminus \omega_n) = 0$. Then, for every $\varepsilon > 0$, there exist sequences $\{Y_k\}_{k=1}^\infty \subset \mathcal{I}(S_{\mathfrak{F}_n}, T)$ and $\{Z_k\}_{k=1}^\infty \subset \mathcal{I}(T, M_\alpha^{(n)})$ such that*

(1) *$Z_k Y_k$ is a basic operator for $k \geq 1$.*
(2) *$\|Y_k\|, \|Z_k\| < 1 + \varepsilon$ for $k \geq 1$.*
(3) *$\bigvee_{k=1}^\infty Y_k H^2(\mathfrak{F}_n) = \mathfrak{H}$.*

Proof. We adapt the last part of the argument of Theorem 3.2. Using the notation in that proof, we construct operators $X_k \in \mathcal{I}(S_{\mathfrak{F}_n}, U_+)$ such that $X_k e_j = w_j'$ for $j = 2, 3, \ldots, n$, and $X_k e_1 = w_1' + \delta_k g_k$, where $\delta_k > 0$ is sufficiently small. It is clear that the ranges of the resulting operators $X, X_1, X_2, \ldots$ generate a dense subspace of $\mathfrak{K}_+$. The proposition follows easily from this observation.

3. The preceding results are most useful when the set α can be chosen to be equal to C. In this case, basic operators are isometric. Thus, if Y, Z satisfy the conclusion of Proposition 3.3, then

$$(1+\varepsilon)^{-1}\|h\| \leq \|Yh\| \leq (1+\varepsilon)\|h\| \quad (h \in H^2(\mathfrak{F}_n)). \tag{3.2}$$

In particular, $YH^2(\mathfrak{F}_n)$ is closed and invariant for T, and $T|(YH^2(\mathfrak{F}_n))$ is similar to $S_{\mathfrak{F}_n}$. We have proved one of the main results of this chapter.

Theorem 3.5. *Assume that T is an a.c. contraction and $\omega_n = C$ for some n, $1 \leq n \leq \aleph_0$. Then there exists an invariant subspace $\mathfrak{M}$ for T such that $T|\mathfrak{M}$ is similar to $S_{\mathfrak{F}_n}$.*

This result can be strengthened by using Proposition 3.4. For this purpose, we denote by $\mathrm{Lat}(T, n, \varepsilon)$ the collection of those invariant subspaces $\mathfrak{M}$ for T for which there is an invertible operator $Y \in \mathcal{I}(S_{\mathfrak{F}_n}, T|\mathfrak{M})$ satisfying (3.2). Proposition 3.4 yields then the following result. The sets ω_n in the statement are such that R_* or W_T is unitarily equivalent to $\bigoplus_{n=1}^\infty M_{\omega_n}$.

Theorem 3.6. *Let T be an a.c. contraction on a separable Hilbert space $\mathfrak{H}$, and let n be a cardinal number such that $\omega_n = C$. For every $\varepsilon > 0$, there exists a sequence $\{\mathfrak{M}_k\}_{k=1}^\infty \subset \mathrm{Lat}(T, n, \varepsilon)$ such that $\bigvee_{k=1}^\infty \mathfrak{M}_k = \mathfrak{H}$. If $n = \aleph_0$, there exist subspaces $\mathfrak{M}_1, \mathfrak{M}_2 \in \mathrm{Lat}(T, n, \varepsilon)$ such that $\mathfrak{M}_1 \vee \mathfrak{M}_2 = \mathfrak{H}$. Conversely, if $\mathrm{Lat}(T, n, \varepsilon) \neq \varnothing$, it follows that $\omega_n = C$.*

Proof. We only need to verify the last statement. Assume therefore that $\mathfrak{M}$ is an invariant subspace for T, and $X \in \mathscr{I}(T|\mathfrak{M}, S_{\mathfrak{F}_n})$ is an invertible operator. Because $W_{T|\mathfrak{M}}$ is a direct summand of W_T, it suffices to prove the result for $T|\mathfrak{M}$. Assume therefore that $\mathfrak{M} = \mathfrak{H}$. Inasmuch as $H^2(\mathfrak{F}_n) \subset L^2(\mathfrak{F}_n)$, X can also be viewed as an operator in $\mathscr{I}(T,W)$, where W denotes multiplication by ζ in $L^2(\mathfrak{F}_n)$. Therefore there exists $A \in \mathscr{I}(W_T, W)$ such that $AX_T = X$. The range of A contains $H^2(\mathfrak{F}_n)$, and therefore it must be dense in $L^2(\mathfrak{F}_n)$. The conclusion $m(C \setminus \omega_n) = 0$ is reached as in the proof of Proposition 3.3.

When the set $\omega_T \neq C$, we can still obtain a result about invariant subspaces for powers of T. We use the notation $T \prec S$ to indicate that T is a quasi-affine transform of S.

Theorem 3.7. *Let $T \in C_1.$ be an a.c. contraction on a separable Hilbert space $\mathfrak{H}$, and let $n > 2\pi/m(\omega_T)$ be an integer. Then T^n has no cyclic vectors.*

Proof. Let (X, V_T) be the isometric asymptote of T. The operator X is a quasi-affinity because $T \in C_1.$, so that $T \prec V_T$. Without loss of generality, we may assume that T has a cyclic vector. In this case V_T must have a cyclic vector so that either V_T is a unilateral shift of multiplicity one, or V_T is unitarily equivalent to M_{ω_T}. Because $T^n \prec V_T^n$, it suffices to show that V_T^n has no cyclic vectors for $n > 2\pi/m(\omega_T)$. If V_T is a shift of multiplicity one then $\omega_T = C$, and V_T^n does not have cyclic vectors because it is a shift of multiplicity $n \geq 2$. Assume then that V_T is unitarily equivalent to M_{ω_T} and $n > 2\pi/m(\omega_T)$. The operator $M_{\omega_T}^n$ is unitarily equivalent to $\bigoplus_{k=1}^n M_{\alpha_k}$, where

$$\alpha_k = \left\{ e^{int} : e^{it} \in \omega_T, t \in \left[\frac{2(k-1)\pi}{n}, \frac{2k\pi}{n}\right) \right\} \quad (k = 1, 2, \ldots, n).$$

Because $\sum_{k=1}^n m(\alpha_k) = nm(\omega_T) > 2\pi$, these sets cannot be disjoint. Thus $M_{\omega_T}^n$ has spectral multiplicity greater than one, and hence it has no cyclic vectors.

4. We conclude this section with a reflexivity result. Given an operator T on a Hilbert space $\mathfrak{H}$, we denote by Lat(T) the collection of all invariant subspaces of T. Given a collection $\mathscr{L}$ of closed subspaces of $\mathfrak{H}$, we denote by Alg$(\mathscr{L})$ the algebra consisting of all operators T such that $T\mathfrak{M} \subset \mathfrak{M}$ for every $\mathfrak{M} \in \mathscr{L}$. It is clear that the algebra AlgLat(T)=Alg(Lat(T)) contains T and is closed in the weak operator topology. The operator T is said to be *reflexive* if AlgLat(T) is the smallest algebra containing T and I which is closed in the weak operator topology. If T is an a.c. contraction and $u \in H^\infty$, then $u(T)$ is a strong limit of polynomials in T. Therefore the following result implies the reflexivity of T provided that $\omega_T = C$.

Theorem 3.8. *Let T be an a.c. contraction on a separable space, and assume that $\omega_T = C$. Then* AlgLat$(T) = \{u(T) : u \in H^\infty\}$.

Proof. We prove the result first when T is the unilateral shift S on H^2. The functions $k_\mu \in H^2$ defined for $|\mu| < 1$ by

$$k_\mu(\lambda) = (1 - \bar{\mu}\lambda)^{-1} = \sum_{k=0}^{\infty} \bar{\mu}^k \lambda^k \quad (\lambda \in D)$$

satisfy $S^*k_\mu = \overline{\mu}k_\mu$ and $(f,k_\mu) = f(\mu)$ for $f \in H^2$. In particular, the set $\{k_\mu : \mu \in D\}$ is total in H^2. Consider now an operator $A \in \text{AlgLat}(S)$, and note that $A^* \in \text{AlgLat}(S^*)$. In particular, $A^*k_\mu = \alpha_\mu k_\mu$ for some scalar α_μ. We have then $A^*S^*k_\mu = S^*A^*k_\mu$ and, because the vectors k_μ form a total set, A commutes with S. An application of Lemma V.3.2 (with $U_+ = U'_+ = S$) implies that $A = u(S)$ for some $u \in H^\infty$. Thus the theorem is true for $T = S$, and it therefore be true for any a.c. contraction that is similar to S.

Consider now the general case, and let $A \in \text{AlgLat}(T)$. Theorem 3.6 implies the existence of operators $\{Y_n\}_{n=1}^\infty \subset \mathscr{I}(S,T)$ such that

$$\frac{1}{2}\|h\| \le \|Y_n h\| \le 2\|h\| \quad (n \ge 1, h \in H^2),$$

and $\bigvee_{n=1}^\infty Y_n H^2 = \mathfrak{H}$. The operator $T|(Y_n H^2)$ is then similar to S, and

$$A|(Y_n H^2) \in \text{AlgLat}(T|(Y_n H^2)).$$

The first part of the proof implies the existence of functions $u_n \in H^\infty$ such that $Ax = u_n(T)x$ for $n \ge 1$ and $x \in Y_n H^2$. Replacing A by $A - u_1(T) \in \text{AlgLat}(T)$, we may assume that $u_1 = 0$. It suffices to prove that $u_n = 0$ for $n \ge 2$. Assume to the contrary that u_n is not zero for some $n \ge 2$. In this case the operator $A|(Y_n H^2) = u_n(T)|(Y_n H^2)$ is similar to $u_n(S)$, and is therefore one-to-one; hence $(Y_n H^2) \cap (Y_1 H^2) = \{0\}$. The operator $Y = Y_1 + (1/8)Y_n \in \mathscr{I}(S,T)$ satisfies the inequalities

$$\frac{1}{4}\|h\| \le \|Yh\| \le \frac{9}{4}\|h\| \quad (h \in H^2),$$

and therefore $T|(YH^2)$ is similar to S. There is thus $u \in H^\infty$ such that $Ax = u(T)x$ for $x \in YH^2$. Note then that for $f \in H^2 \setminus \{0\}$

$$u(T)Y_1 f + \frac{1}{8}u(T)Y_n f = u(T)Yf = AYf = AY_1 f + \frac{1}{8}AY_n f = \frac{1}{8}u_n(T)Y_n f,$$

so $Y_1(uf) = u(T)Y_1 f = (1/8)(u_n(T) - u(T))Y_n f \in (Y_n H^2) \cap (Y_1 H^2) = \{0\}$. We deduce that $u = 0$, and hence $u_n(T)Y_n = O$ as well. This implies the desired conclusion that $u_n = 0$.

4 Hyperinvariant subspaces of C_{11}-contractions

1. In this section we use the unitary asymptotes of T and T^* to carry out a thorough investigation of the lattice of hyperinvariant subspaces of a contraction $T \in C_{11}$. A different approach was used in Sec. VII.5.

Given a unitary asymptote (X,W) for T^*, the pair (W^*,X^*) is called a *unitary $*$-asymptote* for T. We record for further use the properties of a $*$-asymptote; these follow directly from the definition of unitary asymptotes, and from their concrete construction.

Lemma 4.1. *Let T on $\mathfrak{H}$ be a contraction, and let (V,Y) be a unitary $*$-asymptote for T.*

1. *For every unitary operator U, and every $X \in \mathscr{I}(U,T)$, there exists a unique $Z \in \mathscr{I}(U,V)$ such that $X = YZ$. Moreover, we have $\|Z\| = \|X\|$.*
2. *There is a contractive homomorphism $\gamma_*\colon \{T\}' \to \{V\}'$ such that $AY = Y\gamma_*(A)$ for $A \in \{T\}'$.*
3. *We have $\|Y^*x\| = \lim_{n\to\infty} \|T^{*n}x\|$ for $x \in \mathfrak{H}$.*
4. *If $T \in C_{11}$ then T is quasi-similar to V.*

Assume now that $T \in C_{1\cdot}$ on $\mathfrak{H}$. As seen in Sec. II.4, there exists a largest invariant subspace $\mathfrak{H}_1 \in \operatorname{Lat}(T)$ such that the restriction $T|\mathfrak{H}_1$ is in C_{11} (cf. Proposition II.4.2). We have

$$\mathfrak{H} \ominus \mathfrak{H}_1 = \{x \in \mathfrak{H} : \lim_{n\to\infty} \|T^{*n}x\| = 0\} = \ker Y^*,$$

where (V,Y) is a unitary $*$-asymptote for T. Therefore $\mathfrak{H}_1$ is precisely the closure of the range of Y. More generally, assume that $\mathfrak{M} \in \operatorname{Lat}(T)$. Then the restriction $T|\mathfrak{M}$ is also in $C_{1\cdot}$, and therefore there exists a largest space $\mathfrak{M}_1 \in \operatorname{Lat}(T)$ such that $\mathfrak{M}_1 \subset \mathfrak{M}$ and $T|\mathfrak{M}_1$ is in C_{11}. We use the notation $\mathfrak{M}_1 = \Psi_{11}(\mathfrak{M})$. Thus Ψ_{11} is a map from $\operatorname{Lat}(T)$ to the collection $\operatorname{Lat}_1(T)$ of those invariant subspaces $\mathfrak{N}$ such that $T|\mathfrak{N} \in C_{11}$. When $T \in C_{11}$, the elements of $\operatorname{Lat}_1(T)$ are also called *quasi-reducing subspaces* for T. The quasi-reducing subspaces of a unitary operator are obviously the same as the reducing subspaces, but this is generally not true for general C_{11}-contractions.

It is easy to check that $\operatorname{Lat}_1(T)$ is a lattice. More precisely, if $\mathfrak{M}_1, \mathfrak{M}_2 \in \operatorname{Lat}_1(T)$, then $\mathfrak{M}_1 \vee \mathfrak{M}_2 \in \operatorname{Lat}_1(T)$ is the least upper bound of the two spaces, and $\Psi_{11}(\mathfrak{M}_1 \cap \mathfrak{M}_2)$ is their greatest lower bound. The examples at the end of this section show that $\Psi_{11}(\mathfrak{M}_1 \cap \mathfrak{M}_2)$ may be different from $\mathfrak{M}_1 \cap \mathfrak{M}_2$.

We use the more general notation $\operatorname{Lat}\mathscr{S}$ for the collection of all closed subspaces of $\mathfrak{H}$ that are invariant for every element A in a family $\mathscr{S}$ of operators on $\mathfrak{H}$. Thus, $\operatorname{Lat}\{T\}'$ represents the collection of hyperinvariant subspaces of T. Denote by $\{T\}'' = \bigcap_{A\in\{T\}'} \{A\}'$ the *double commutant* of T.

Proposition 4.2. *Let T on $\mathfrak{H}$ be a C_{11}-contraction.*

1. *For every $\mathfrak{M} \in \operatorname{Lat}_1(T)$ there exists $A \in \{T\}'$ such that $\mathfrak{M} = \overline{A\mathfrak{H}} = \overline{A\mathfrak{M}}$.*
2. $\operatorname{Lat}_1(T) \subset \operatorname{Lat}\{T\}''$.

Proof. Let (X,W) be a unitary asymptote of T, and let (V,Y) be a unitary $*$-asymptote of $T|\mathfrak{M}$. Note that the range of Y is dense in $\mathfrak{M}$ and $W|\overline{X\mathfrak{M}}$ is unitary because $T|\mathfrak{M} \in C_{11}$. We know that $T|\mathfrak{M}$ is quasi-similar to its unitary asymptote and $*$-asymptote, and the unitary asymptote of $T|\mathfrak{M}$ is unitarily equivalent to the direct summand $W|\overline{X\mathfrak{M}}$ of W. It follows that there exists a map $B \in \mathscr{I}(W,V)$ such that $BX|\mathfrak{M}$ has dense range. The desired operator is then defined as $A = YBX$. Part (2) obviously follows from (1).

We later show that the opposite inclusion to (2) is not true in general.

Let us set

$$\operatorname{Lat}_1\{T\}' = \operatorname{Lat}_1(T) \cap \operatorname{Lat}\{T\}'.$$

Proposition 4.3. *Let $T \in C_{11}$ be a cyclic operator. Then $\{T\}' = \{T\}''$ and hence* $\mathrm{Lat}_1(T) \subset \mathrm{Lat}\{T\}'$.

Proof. Let (X,W) be a unitary asymptote of T, and let γ: $\{T\}' \to \{W\}'$ be the homomorphism satisfying $\gamma(A)X = XA$ for $A \in \{T\}'$. Because W is quasi-similar to T, it is also cyclic, and therefore $\{W\}'$ is commutative (cf. Theorems IX.3.4 and IX.6.6 in CONWAY [1]). The fact that X is one-to-one implies then that γ is one-to-one, and therefore $\{T\}'$ is commutative. Thus $\{T\}' = \{T\}''$, and the final inclusion follows from the preceding corollary.

Quasi-reducing hyperinvariant subspaces have a basic maximality property. We write $T \sim S$ to indicate that T and S are quasi-similar.

Proposition 4.4. *Let $T \in C_{11}$, $\mathfrak{M} \in \mathrm{Lat}_1\{T\}'$, and $\mathfrak{N} \in \mathrm{Lat}_1(T)$. If $T|\mathfrak{M} \sim T|\mathfrak{N}$ then $\mathfrak{M} \supset \mathfrak{N}$. If in addition $\mathfrak{N} \in \mathrm{Lat}_1\{T\}'$, we must have $\mathfrak{M} = \mathfrak{N}$.*

Proof. It is enough to prove the first assertion. Fix an operator $A \in \{T\}'$ such that $\mathfrak{M} = \overline{A\mathfrak{M}} = \overline{A\mathfrak{H}}$, and a quasi-affinity $Q \in \mathscr{I}(T|\mathfrak{M}, T|\mathfrak{N})$. The operator QA commutes with T and therefore it leaves $\mathfrak{M}$ invariant. Therefore

$$\mathfrak{N} = \overline{Q\mathfrak{M}} = \overline{QA\mathfrak{M}} \subset \mathfrak{M},$$

as claimed.

2. We now relate the set $\mathrm{Lat}_1\{T\}'$ to the corresponding set for the unitary asymptote of T. First, a preliminary result.

Lemma 4.5. *Let T and T' be two contractions such that $T \prec T'$. If $T \in C_{\cdot 1}$ then $T' \in C_{\cdot 1}$ as well. If T and T' are in C_{11} then $T \sim T'$.*

Proof. Let $A \in \mathscr{I}(T,T')$ be a quasi-affinity. We have $A^*T'^{*n}x = T^{*n}A^*x$ for $n \geq 1$. If $\lim_{n\to\infty} \|T'^{*n}x\| = 0$ we deduce that $\lim_{n\to\infty} T^{*n}A^*x = 0$. If $T \in C_{\cdot 1}$ this is possible only for $x = 0$, and therefore $T' \in C_{\cdot 1}$. Finally, if both T and T' are of class C_{11}, they are quasi-similar to their unitary asymptotes $W_T, W_{T'}$, and therefore $W_T \prec W_{T'}$. It follows that W_T and $W_{T'}$ are unitarily equivalent, and the desired conclusion $T \sim T'$ follows.

Theorem 4.6. *Let T be a contraction of class C_{11}, let (X,W) be a unitary asymptote of T, and let (W,Y) be a unitary $*$-asymptote of T. There exists a bijection φ: $\mathrm{Lat}_1\{T\}' \to \mathrm{Lat}\{W\}'$ such that $T|\mathfrak{M} \sim W|\varphi(\mathfrak{M})$ for every $\mathfrak{M} \in \mathrm{Lat}_1\{T\}'$. We have $\varphi(\mathfrak{M}) = \overline{X\mathfrak{M}}$ for $\mathfrak{M} \in \mathrm{Lat}_1\{T\}'$ and $\varphi^{-1}(\mathfrak{N}) = \overline{Y\mathfrak{N}}$ for $\mathfrak{N} \in \mathrm{Lat}\{W\}'$.*

Proof. Assume that T acts on $\mathfrak{H}$ and W acts on $\mathfrak{H}'$. We know from Proposition 4.4 applied to W (resp., T) that for every $\mathfrak{M} \in \mathrm{Lat}_1\{T\}'$ (resp., $\mathfrak{N} \in \mathrm{Lat}\{W\}'$) there exists at most one subspace $\mathfrak{N} \in \mathrm{Lat}\{W\}'$ (resp., $\mathfrak{M} \in \mathrm{Lat}_1\{T\}'$) such that $T|\mathfrak{M} \sim$

$W|\mathfrak{N}$. It suffices therefore to show that such spaces do exist. We start with $\mathfrak{M} \in \mathrm{Lat}_1\{T\}'$, and construct the space $\mathfrak{N} \in \mathrm{Lat}\{W\}'$ by setting

$$\mathfrak{N} = \bigvee\{AX\mathfrak{M} : A \in \{W\}'\}.$$

Note that $YAX \in \{T\}'$ for every $A \in \{W\}'$, and because $\mathfrak{M}$ is hyperinvariant we have

$$XY\mathfrak{N} = \bigvee\{XYAX\mathfrak{M} : A \in \{W\}'\} \subset \overline{X\mathfrak{M}}.$$

On the other hand, $XY \in \{W\}', P_{\mathfrak{N}} \in \{W\}''$, and XY has dense range. Thus

$$\overline{XY\mathfrak{N}} = \overline{XYP_{\mathfrak{N}}\mathfrak{H}'} = \overline{P_{\mathfrak{N}}XY\mathfrak{H}'} = P_{\mathfrak{N}}\mathfrak{H}' = \mathfrak{N},$$

and the last two relations imply $\overline{X\mathfrak{M}} = \mathfrak{N}$. Note that

$$X|\mathfrak{M} \in \mathscr{I}(T|\mathfrak{M}, W|\overline{X\mathfrak{M}}),$$

hence $T|\mathfrak{M} \prec W|\overline{X\mathfrak{M}}$. Because $W|\overline{X\mathfrak{M}}$ is unitary, the preceding lemma shows that $T|\mathfrak{M} \sim W|\overline{X\mathfrak{M}}$.

Finally, fix a space $\mathfrak{N} \in \mathrm{Lat}\{W\}'$, and set $\mathfrak{M} = \overline{Y\mathfrak{N}}$. As in the preceding argument, we have $W|\mathfrak{N} \prec T|\mathfrak{M}$, and $W|\mathfrak{N}$ is unitary. The preceding lemma implies then that $T|\mathfrak{M} \in C_{\cdot 1}$, and therefore $T|\mathfrak{M} \in C_{11}$. The same lemma implies now that $W|\mathfrak{N} \sim T|\mathfrak{M}$. To conclude the proof, it suffices to show that $\mathfrak{M}$ is hyperinvariant. Consider then an arbitrary operator $A \in \{T\}'$, and denote by $B \in \{W\}'$ the unique operator satisfying $AY = YB$ provided by Lemma 4.1(2). We have

$$A\mathfrak{M} \subset \overline{AY\mathfrak{N}} = \overline{YB\mathfrak{N}} \subset \overline{Y\mathfrak{N}} = \mathfrak{M},$$

where we used the fact that $\mathfrak{N}$ is hyperinvariant. The theorem is proved.

The preceding result shows, in particular, that $\mathrm{Lat}_1\{T\}'$ is a lattice, isomorphic to $\mathrm{Lat}\{W\}'$. Indeed, the map φ and its inverse preserve inclusions.

As noted already, quasi-similar C_{11}-operators have unitarily equivalent unitary asymptotes. The preceding theorem implies they also have isomorphic lattices of quasi-reducing hyperinvariant subspaces. The following result relates this isomorphism to the construction in the proof of Proposition II.5.1.

Recall that $\Psi_{11}(\mathfrak{M})$ denotes the C_{11} part of an invariant subspace $\mathfrak{M}$ for a $C_{1\cdot}$-operator.

Corollary 4.7. *Let $T_1 \sim T_2$ be two C_{11}-contractions, and fix quasi-affinities $X \in \mathscr{I}(T_1, T_2), Y \in \mathscr{I}(T_2, T_1)$. There exists a unique bijection $\varphi : \mathrm{Lat}_1\{T_1\}' \to \mathrm{Lat}_1\{T_2\}'$ such that $T_1|\mathfrak{M}_1 \sim T_2|\varphi(\mathfrak{M}_2)$ for every $\mathfrak{M}_1 \in \mathrm{Lat}_1\{T_1\}'$, namely*

$$\varphi(\mathfrak{M}_1) = \bigvee\{AX\mathfrak{M}_1 : A \in \{T_2\}'\} = \Psi_{11}(Y^{-1}\mathfrak{M}_1) \quad (\mathfrak{M}_1 \in \mathrm{Lat}_1\{T_1\}').$$

Its inverse is given by

$$\varphi^{-1}(\mathfrak{M}_2) = \bigvee\{BY\mathfrak{M}_2 : B \in \{T_1\}'\} = \Psi_{11}(X^{-1}\mathfrak{M}_2) \quad (\mathfrak{M}_2 \in \mathrm{Lat}_1\{T_2\}').$$

Proof. The existence and uniqueness of φ follow from the preceding theorem. To establish the formula for φ, fix $\mathfrak{M}_1 \in \mathrm{Lat}_1\{T_1\}'$, and set $\mathfrak{M}_2 = \varphi(\mathfrak{M}_1)$. Note first that $T_2|\mathfrak{M}_2 \prec T_1|\overline{Y\mathfrak{M}_2}$. An application of Lemma 4.5 shows that $T_1|\overline{Y\mathfrak{M}_2}$ is of class C_{11}, and

$$T_1|\overline{Y\mathfrak{M}_2} \sim T_2|\mathfrak{M}_2 \sim T_1|\mathfrak{M}_1.$$

Proposition 4.4 shows then that $Y\mathfrak{M}_2 \subset \mathfrak{M}_1$, and therefore $\mathfrak{M}_2 \subset Y^{-1}\mathfrak{M}_1$. Set $\mathfrak{N} = \Psi_{11}(Y^{-1}\mathfrak{M}_1)$. Because $\mathfrak{N}$ is the largest subspace of $Y^{-1}\mathfrak{M}_1$ where the restriction of T_2 is C_{11}, we have

$$\mathfrak{M}_2 \subset \mathfrak{N}.$$

Obviously, $Y\mathfrak{N} \subset \mathfrak{M}_1$. By symmetry, we also have $X\mathfrak{M}_1 \subset \mathfrak{M}_2$, and therefore $\mathfrak{N}' = \overline{XY\mathfrak{N}}$ is contained in $\mathfrak{M}_2$. Because XY is a quasi-affinity, Lemma 4.6 yields $T_2|\mathfrak{N}' \sim T_2|\mathfrak{N}$, and thus there exists a quasi-affinity $Z \in \mathscr{I}(T_2|\mathfrak{N}', T_2|\mathfrak{N})$. By Proposition 4.2, there exists $B \in \{T_2\}'$ such that $\mathfrak{N}' = \overline{B\mathfrak{N}'} = \overline{B\mathfrak{H}_2}$. Because $ZB \in \{T_2\}'$ and $\mathfrak{M}_2$ is hyperinvariant for T_2, we conclude that

$$\mathfrak{N} = \overline{ZB\mathfrak{N}'} \subset \overline{ZB\mathfrak{M}_2} \subset \mathfrak{M}_2.$$

The opposite inclusion was verified earlier.

Finally, the space $q(\mathfrak{M}_1) = \bigvee\{AX\mathfrak{M}_1 : A \in \{T_2\}'\}$ is clearly in $\mathrm{Lat}\{T_2\}'$, and

$$X\mathfrak{M}_1 \subset \mathfrak{M}_2 \in \mathrm{Lat}\{T_2\}'$$

implies that $q(\mathfrak{M}) \subset \mathfrak{M}_2$. Because

$$T_2|\overline{X\mathfrak{M}_1} \sim T_1|\mathfrak{M}_1 \sim T_2|\mathfrak{M}_2,$$

there exists a quasi-affinity $Q \in \mathscr{I}(T_2|\overline{X\mathfrak{M}_1}, T_2|\mathfrak{M}_2)$. Proposition 4.2 yields an operator $D \in \{T_2\}'$ such that $\overline{X\mathfrak{M}_1} = \overline{DX\mathfrak{M}_1} = \overline{D\mathfrak{H}_2}$. It follows that $QD \in \{T_2\}'$, and

$$\mathfrak{M}_2 = \overline{QDX\mathfrak{M}_1} \subset \overline{QDq(\mathfrak{M}_1)} \subset q(\mathfrak{M}_1).$$

The formula for φ^{-1} is obtained by interchanging the roles of T_1 and T_2. The corollary is proved.

If W is a unitary operator on a separable space, the projections in $\{W\}''$ are of the form $\chi_\alpha(W)$, and therefore the spaces in $\mathrm{Lat}\{W\}'$ are precisely the ranges of these operators (cf. Sec. IX.8 in CONWAY [1]). Assume that W is absolutely continuous, and is therefore unitarily equivalent to an operator of the form $\bigoplus_{n=1}^{\infty} M_{\omega_n}$ for some Borel sets $C \supset \omega_1 \supset \omega_2 \supset \cdots$. Then $\chi_{\alpha_1}(W) = \chi_{\alpha_2}(W)$ if and only if $\alpha_1 \cap \omega_1 = \alpha_2 \cap \omega_1$. If T is a C_{11}-contraction quasi-similar to W, and $\alpha \subset C$ is a Borel set, we denote by $\mathfrak{H}_{T,\alpha}$ the space that corresponds to the range of $\chi_\alpha(W)$ under the isomorphism between $\mathrm{Lat}\{W\}'$ and $\mathrm{Lat}_1\{T\}'$. These are easily seen to be precisely the spaces considered in Theorem VII.5.2.

We can give a new description of the spaces $\mathfrak{H}_{T,\alpha}$ when $T = \mathbf{T}$ is a model operator. This description does not use regular factorizations, and depends explicitly on the characteristic function and on the set α. Thus, assume that $\mathfrak{E}$ and $\mathfrak{E}_*$

are separable Hilbert spaces, $\{\mathfrak{E},\mathfrak{E}_*,\Theta(\lambda)\}$ is a purely contractive analytic function, and the spaces $\mathbf{K},\mathbf{K}_+,\mathbf{H}$ are constructed as usual. Let us also set $\Delta_*(\zeta)=(I-\Theta(\zeta)\Theta(\zeta)^*)^{1/2}$ for a.e. $\zeta\in C$.

Corollary 4.8. *Assume that* $\mathbf{T}$ *is of class* C_{11} *and* $\alpha\subset C$ *is a Borel set. We have then*

$$\mathbf{H}_{\mathbf{T},\alpha}=\Psi_{11}(\{u\oplus v\in\mathbf{H}:-\Delta_*(\zeta)u(\zeta)+\Theta(\zeta)v(\zeta)=0 \text{ for a.e. } \zeta\in C\setminus\alpha\}).$$

Proof. The pair $(P_{\mathfrak{R}_*}|\mathbf{H},R_*)$ is a unitary asymptote for $\mathbf{T}$, and the preceding corollary implies that $\mathbf{H}_{\mathbf{T},\alpha}=\Psi_{11}((P_{\mathfrak{R}_*}|\mathbf{H})^{-1}(\chi_\alpha\mathfrak{R}_*))$. Now, $\chi_\alpha\mathfrak{R}_*=\{w\in\mathfrak{R}_*:w(\zeta)=0$ for a.e. $\zeta\in C\setminus\alpha\}$, and $P_{\mathfrak{R}_*}$ is the operator of pointwise multiplication by the projection-valued function

$$P(\zeta)=\begin{bmatrix} I-\Theta(\zeta)\Theta(\zeta)^* & -\Theta(\zeta)\Delta(\zeta)\\ -\Delta(\zeta)\Theta(\zeta)^* & \Theta(\zeta)^*\Theta(\zeta)\end{bmatrix}.$$

The corollary follows because $P(\zeta)$ can also be written as

$$P(\zeta)=\begin{bmatrix}\Delta_*(\zeta)^2 & -\Delta_*(\zeta)\Theta(\zeta)\\ -\Theta(\zeta)^*\Delta_*(\zeta) & \Theta(\zeta)^*\Theta(\zeta)\end{bmatrix}=\begin{bmatrix}-\Delta_*(\zeta)\\ \Theta(\zeta)^*\end{bmatrix}[-\Delta_*(\zeta),\Theta(\zeta)],$$

and

$$\begin{bmatrix}-\Delta_*(\zeta)\\ \Theta(\zeta)^*\end{bmatrix}$$

is an isometry for a.e. $\zeta\in C$.

3. We conclude with a few observations about the lattice structure of $\operatorname{Lat}(T)$ and $\operatorname{Lat}_1(T)$. For an arbitrary set $\mathscr{S}$ of operators on $\mathfrak{H}$, any family $\{\mathfrak{M}_j\}_{j\in J}$ of subspaces in $\operatorname{Lat}\mathscr{S}$ has a least upper bound and a greatest lower bound, namely $\bigvee_{j\in J}\mathfrak{M}_j$ and $\bigcap_{j\in J}\mathfrak{M}_j$. If $T\in C_{11}$ and $\{\mathfrak{M}_j\}_{j\in J}\subset\operatorname{Lat}_1(T)$, the family $\{\mathfrak{M}_j\}_{j\in J}$ still has a least upper bound and a greatest lower bound in $\operatorname{Lat}_1(T)$, namely the spaces $\mathfrak{M}=\bigvee_{j\in J}\mathfrak{M}_j$ and $\Psi_{11}(\bigcap_{j\in J}\mathfrak{M}_j)$, respectively. We show that the latter space can in fact be different from $\bigcap_{j\in J}\mathfrak{M}_j$.

Example 1. There exist a cyclic absolutely continuous contraction $T\in C_{11}$, and two subspaces $\mathfrak{M}_1,\mathfrak{M}_2\in\operatorname{Lat}_1\{T\}'$ such that $\mathfrak{M}=\mathfrak{M}_1\cap\mathfrak{M}_2\notin\operatorname{Lat}_1(T)$. In particular, $\mathfrak{M}\in\operatorname{Lat}\{T\}'\setminus\operatorname{Lat}_1(T)$.

In order to facilitate the construction, observe that an operator of the form

$$\begin{bmatrix}AD_T\\ T\end{bmatrix}$$

is always a contraction if T and A are contractions, where A could even act between different spaces. Similarly, $[T,D_{T^*}A]$ is a contraction if T and A are contractions.

We start with c.n.u. contractions $T_j\in C_{11}$ on $\mathfrak{H}_j$ $(j=0,1,2)$ that are not boundedly invertible; the existence of such operators was proved in Sec. VI.4.2. Fix vectors $x_j \in \mathfrak{H}_j$ with $\|x_j\| = 1/2$ ($j = 0,1,2$), and denote by E^1 the one-dimensional space of complex scalars. Define an operator T on $\mathfrak{H} = \mathfrak{H}_0 \oplus E^1 \oplus \mathfrak{H}_1 \oplus \mathfrak{H}_2$ by setting

$$T = \begin{bmatrix} T_0 & X_0 & O & O \\ O & O & Y_1 & Y_2 \\ O & O & T_1 & O \\ O & O & O & T_2 \end{bmatrix},$$

where $X_0 \colon E^1 \to \mathfrak{H}_0$ is defined by $X_0\lambda = \lambda D_{T_0^*}x_0$, $\lambda \in E^1$, and $Y_j \colon \mathfrak{H}_j \to E^1$ is defined by $Y_j v = (D_{T_j}v, x_j)$ for $v \in \mathfrak{H}_j$ and $j = 1,2$. In order to see that T is a contraction, note that T maps $\mathfrak{H}_0 \oplus E^1 \oplus \{0\} \oplus \{0\}$ and $\{0\} \oplus \{0\} \oplus \mathfrak{H}_1 \oplus \mathfrak{H}_2$ into orthogonal subspaces. Thus it is enough to verify contractivity separately on these two spaces. The restrictions of T to these two spaces are contractions of the form just discussed above. For instance,

$$\begin{bmatrix} Y_1 & Y_2 \\ T_1 & O \\ O & T_2 \end{bmatrix} = \begin{bmatrix} AD_{T_1 \oplus T_2} \\ T_1 \oplus T_2 \end{bmatrix},$$

where $A \colon \mathfrak{H}_1 \oplus \mathfrak{H}_2 \to E^1$ is a contraction defined by $A(h_1 \oplus h_2) = (h_1, x_1) + (h_2, x_2)$.

Obviously the only vectors $x \in \mathfrak{H}$ satisfying $\lim_{n\to\infty} \|T^n x\| = 0$ belong to $\mathfrak{H}_0 \oplus E^1 \oplus \{0\} \oplus \{0\}$, and these vectors x are in the kernel of T. Choosing x_0 such that $D_{T_0^*}x_0$ does not belong to $T_0\mathfrak{H}_0$ will then ensure that $T \in C_{1\cdot}$. Analogously, if $D_{T_j}x_j$ is not in $T_j^*\mathfrak{H}_j$ for $j = 1,2$, then that $T \in C_{\cdot 1}$. Such choices are in fact possible. For instance, the operator T_0 is not onto, and therefore we can choose $x_0 = D_{T_0^*}y_0$ for some $y_0 \in \mathfrak{H}_0 \setminus T_0\mathfrak{H}_0$. Therefore $D_{T_0^*}x_0 = y_0 - T_0T_0^*y_0 \notin T_0\mathfrak{H}_0$. Making analogous choices for x_1 and x_2, we have constructed an operator of class C_{11}.

Next, repeated application of Theorem 1.6 shows that W_T is unitarily equivalent to $W_{T_0} \oplus W_{T_1} \oplus W_{T_2}$. By Theorem 2.2 we can choose W_{T_j} to be cyclic unitary operators with disjoint spectra. With this choice, W_T is also a cyclic a.c. operator, and therefore so is $T \sim W_T$.

Observe now that the spaces $\mathfrak{M}_1 = \mathfrak{H}_0 \oplus E^1 \oplus \mathfrak{H}_1 \oplus \{0\}$ and $\mathfrak{M}_2 = \mathfrak{H}_0 \oplus E^1 \oplus \{0\} \oplus \mathfrak{H}_2$ are invariant for T, and the above arguments show that $T|\mathfrak{M}_j \in C_{11}$ for $j = 1,2$. Proposition 4.4 shows that in fact $\mathfrak{M}_j \in \mathrm{Lat}_1\{T\}'$ for $j = 1,2$. Finally, $\mathfrak{M} = \mathfrak{M}_1 \cap \mathfrak{M}_2 = \mathfrak{H}_0 \oplus E^1 \oplus \{0\} \oplus \{0\}$, and obviously $(T|\mathfrak{M})^*$ has nonzero kernel. Thus $\mathfrak{M} \notin \mathrm{Lat}_1(T)$.

The interested reader be able to verify that T is in fact c.n.u. A more elaborate construction yields the following example.

Example 2. There exist a cyclic a.c. contraction $T \in C_{11}$, and a sequence $\mathfrak{M}_1 \supset \mathfrak{M}_2 \supset \cdots$ in $\mathrm{Lat}_1\{T\}'$ such that $\bigcap_{n=1}^{\infty} \mathfrak{M}_n \notin \mathrm{Lat}_1(T)$.

We only sketch the construction briefly. Start with a sequence of c.n.u. C_{11}-contractions $T_0, T_1, \ldots$ that are not boundedly invertible, and construct

$$T = \begin{bmatrix} T_0 & X_0 & O & O & O & \cdots \\ O & O & Y_1 & Y_2 & Y_3 & \cdots \\ O & O & T_1 & O & O & \cdots \\ O & O & O & T_2 & O & \cdots \\ O & O & O & O & T_3 & \cdots \\ \vdots & \vdots & \vdots & \vdots & \vdots & \ddots \end{bmatrix}$$

on $\mathfrak{H}_0 \oplus E^1 \oplus \mathfrak{H}_1 \oplus \mathfrak{H}_2 \oplus \cdots$, where the operators X_0 and Y_j are defined as in the preceding example. With proper choices, this is a cyclic c.n.u. C_{11}-contraction. We then set $\mathfrak{M}_n = \mathfrak{H}_0 \oplus E^1 \oplus \bigoplus_{j=1}^{\infty} \mathfrak{K}_j$, with $\mathfrak{K}_j = \mathfrak{H}_j$ for $j > n$ and $\mathfrak{K}_j = \{0\}$ for $j \leq n$. The intersection of these spaces is $\mathfrak{H}_0 \oplus E^1 \oplus \{0\} \oplus \{0\} \oplus \cdots$, and it does not belong to $\mathrm{Lat}_1(T)$ as seen in the preceding example.

Remark. The situation discussed in Example 1 shows that part (v) of Theorem VII.6.2 is not necessarily true if the characteristic function is not assumed to have a scalar multiple. Example 2 shows that, in the absence of scalar multiples, Theorem VII.6.2(v) may fail even if the sequence $\{\alpha_n\}$ is decreasing. In terms of characteristic functions, the subspace $\mathfrak{M}$ in Example 1 shows that there exists a regular factorization $\Theta(\lambda) = \Theta_2(\lambda)\Theta_1(\lambda)$ of a function $\Theta(\lambda)$, outer from both sides, such that $\Theta_1(\lambda)$ is not outer, although $\Theta(\zeta)$ is isometric on a set of positive measure (cf. Theorem VII.1.1 and Propositions VII.2.1 and VI.3.5).

Remark. It is interesting to note that for a subspace $\mathfrak{M} \in \mathrm{Lat}\{T\}'$, the C_{11} part $\Psi_{11}(\mathfrak{M})$ is also hyperinvariant. Consider indeed an operator $X \in \{T\}'$, and consider the matrices

$$T|\mathfrak{M} = \begin{bmatrix} T_{11} & T_{12} \\ O & T_{22} \end{bmatrix}, \quad X|\mathfrak{M} = \begin{bmatrix} X_{11} & X_{12} \\ X_{21} & X_{22} \end{bmatrix}$$

relative to the decomposition $\mathfrak{M} = \Psi_{11}(\mathfrak{M}) \oplus (\mathfrak{M} \ominus \Psi_{11}(\mathfrak{M}))$. These operators commute, thus $X_{21} \in \mathscr{I}(T_{11}, T_{22})$. Because $T_{11} \in C_{\cdot 1}$ and $T_{22} \in C_{\cdot 0}$, it follows immediately that $X_{21} = O$. Thus $\Psi_{11}(\mathfrak{M})$ is invariant for an arbitrary $X \in \{T\}'$. Consider now a family $\{\mathfrak{M}_j\}_{j \in J} \subset \mathrm{Lat}_1\{T\}'$. It follows that the greatest lower bound $\Psi_{11}(\bigcap_{j \in J} \mathfrak{M}_j)$ of this family in $\mathrm{Lat}_1(T)$ actually belongs to $\mathrm{Lat}_1\{T\}'$. On the other hand, $\mathrm{Lat}_1\{T\}'$ is obviously closed under the usual operation of taking the closed linear span of a family of subspaces.

5 Notes

As seen in Sec. II.5, power-bounded operators of class C_{11} are quasi-similar to unitary operators. This idea was developed in KÉRCHY [8], where isometric and unitary asymptotes were constructed for arbitrary power-bounded operators. Section 1 has been organized so as to suggest this more general development. Theorem 1.6 is proved in KÉRCHY [8] in this general setting. In the context of contractions, part

(3) of this theorem also appears in BERCOVICI AND KÉRCHY [1] where the tool is regular factorization. The identification of the unitary asymptote with the $*$-residual part of the minimal unitary dilation, and the explicit identification of the corresponding intertwiners is from KÉRCHY [5].

Isometric and unitary asymptotes were introduced for some operators that are not even power-bounded. For instance, if $\{\|T^n\|\}_{n=1}^{\infty}$ is a regular sequence (or more generally, for certain semigroups with regular norm behavior) this was done in KÉRCHY [10],[12] and in KÉRCHY AND LÉKA [1]. The regularity property was characterized in KÉRCHY [11] and KÉRCHY AND MÜLLER [1]. A study of $C_{1\cdot}$-contractions based on the unitary asymptote can also be found in BEAUZAMY [2, Chapter XII].

The first example of a C_{11}-contraction such that $\overline{D} \neq \sigma(T) \not\subset C$ is given in ECKSTEIN [1]. An example of a cyclic C_{11}-contraction with $\sigma(T) = \overline{D}$ appears in BERCOVICI AND KÉRCHY [1]; a noncyclic example is in VI.4.2. Theorem 2.2 is from BERCOVICI AND KÉRCHY [2]. Proposition 2.4 is a special form of a result of FOIAŞ AND MLAK [1]. GILFEATHER [1] presented a weighted bilateral shift $T \in C_{10}$ such that $\sigma(T) = C$. BEAUZAMY [1] constructed a contraction $T \in C_{10}$ whose spectrum contains a nontrivial arc of C disjoint from $\sigma(W_T)$. The complete description of the spectra of C_{10}-contractions in Theorem 2.6 is from BERCOVICI AND KÉRCHY [3], which extends results in KÉRCHY [4]. Lemma 2.11 also follows from the fact that T is a generalized scalar operator; see COLOJOARĂ AND FOIAŞ [1, Theorem 5.1.4]. A spectral mapping theorem generalizing the relation $\omega_{u(T)} = u(\omega_T)$ (cf. Lemma 2.13) can be found in KÉRCHY [15]. The proof of Lemma 2.14 is inspired by NIKOLSKIĬ [1].

The material of Sec. 3 is from KÉRCHY [9],[16]. However, the approximation Lemma 3.1 is based on ideas from KÉRCHY [6],[7], and Theorem 3.7 is due to SZ.-N AND F. [31].

The concept of reflexivity was introduced by SARASON [4] who proved that normal operators and unilateral shifts are reflexive. An overview of this area is in HADWIN [1]. WU [3],[5] proved that C_{11}-contractions with finite defect indices are reflexive. Theorem 3.8 was proved in TAKAHASHI [4]. The case when $T \in C_{11}$ was done earlier by KÉRCHY [6]. The reflexivity of such operators also follows from BROWN AND CHEVREAU [1], where it was shown that an a.c. contraction T is reflexive if $\|u(T)\| = \|u\|_\infty$ for every $u \in H^\infty$.

The isomorphism of $\mathrm{Lat}_1\{T\}'$ to $\mathrm{Lat}\{W_T\}'$ was proved in KÉRCHY [8] for power-bounded operators; this is an extension of Theorem 4.6. Theorem 4.6 can also be derived, at least in the c.n.u. case, from Theorem VII.5.2 and a result of TEODORESCU [3] on regular factorizations.

Examples 1, 2 in Sec. 4, and the subsequent remarks are from KÉRCHY [1]. The examples also use some ideas from BERCOVICI AND KÉRCHY [1].

The isomorphisms of various invariant subspace lattices, and their implementation, are studied in KÉRCHY [3]. In particular, Corollary 4.8 is from that paper. Proposition 4.3, along with a more detailed discussion of cyclic C_{11}-contractions, is in KÉRCHY [2]. The classification of lattices of invariant subspaces of isometries is discussed in CONWAY AND GILLESPIE [1] and KÉRCHY [14]. The existence of hyperinvariant subspaces for $C_{1\cdot}$-contractions T was discussed in BEAUZAMY

[2, Theorem XII.8.1]. It is shown there that an invertible $T \in C_1.$ has a nontrivial hyperinvariant subspace if it is not a scalar multiple of the identity, and $\sum_{n=1}^{\infty} n^{-2} \log \|T^{-n}x\| < \infty$ for some $x \neq 0$. This result was extended by KÉRCHY [13] to $C_1.$-operators with regular norm sequences, and it was shown that there is in fact an infinite family of completely disjoint hyperinvariant subspaces.

An interesting connection between unitary asymptotes and the existence of disjoint invariant subspaces is discussed in TAKAHASHI [6].

TAKAHASHI [2] shows that a contraction T of class $C_1.$, whose defect operators are Hilbert–Schmidt, is completely injection-similar to an isometry. The relation of injection-similarity was introduced in SZ.-N.–F. [24].

Chapter X

The Structure of Operators of Class C_0

1 Local maximal functions and maximal vectors

1. Let T be a c.n.u. contraction on the Hilbert space $\mathfrak{H}$, and $h \in \mathfrak{H}$. Denote by $\mathfrak{M}_h$ the cyclic space for T generated by h. Observe that for a function $u \in H^\infty$, we have $u(T)h = 0$ if and only if $u(T|\mathfrak{M}_h) = O$.

Definition. The operator T is said to be *locally of class* C_0 if, for every $h \in \mathfrak{H}$, there exists a function $u \in H^\infty$ (depending, generally, on h) such that $u(T)h = 0$. The minimal function of $T|\mathfrak{M}_h$ is denoted m_h if T is locally of class C_0.

The purpose of this section is to prove that operators T which are locally of class C_0 are actually of class C_0; that is, the function u in the definition above can be chosen independently of h (cf. Sec. III.4). The proof follows from the existence of maximal vectors, defined below.

Definition. Assume that T is locally of class C_0, and $h \in \mathfrak{H}$. The vector h is said to be T-*maximal* (or simply *maximal* when no confusion may arise) if m_g divides m_h for every $g \in \mathfrak{H}$.

Observe that, provided that T has a maximal vector h, then T is of class C_0, and $m_T = m_h$.

2. For the purposes of this chapter, we need to extend some of the concepts in Sec. III.1 as follows. Consider functions $\varphi, \psi \in H^2$, not both identically zero. We denote by $\varphi \wedge \psi$ the *greatest common inner divisor* of the functions φ and ψ. More generally, $\bigwedge_i \varphi_i$ denotes the greatest common inner divisor of a family $\{\varphi_i\}$ of functions, not all identically zero. Analogously, $\varphi \vee \psi$ stands for the *least common inner multiple* of φ and ψ, with a corresponding notation for inner multiples of families of functions. In this chapter, an equality $u = v$ between two inner functions is understood to hold only up to a constant factor of absolute value one.

Proposition 1.1. *Let $\{\varphi_i : i \in I\}$ be a family of nonconstant inner divisors of the inner function $\varphi \in H^\infty$. If $\varphi_i \wedge \varphi_j = 1$ for $i \neq j$, then the set I is at most countable.*

B.Sz.-Nagy et al., *Harmonic Analysis of Operators on Hilbert Space*, Universitext, DOI 10.1007/978-1-4419-6094-8_10, © Springer Science+Business Media, LLC 2010

Proof. Choose $\lambda \in D$ such that $\varphi(\lambda) \neq 0$. If $i_1, i_2, \ldots, i_n \in I$ are distinct, then

$$\varphi_{i_1}\varphi_{i_2}\cdots\varphi_{i_n} = \varphi_{i_1} \vee \varphi_{i_2} \vee \cdots \vee \varphi_{i_n}$$

divides φ and, in particular,

$$\sum_{k=1}^{n} -\log|\varphi_{i_k}(\lambda)| \leq -\log|\varphi(\lambda)|.$$

We conclude that $\sum_{i\in I} -\log|\varphi_i(\lambda)| < \infty$, and hence the set $I_1 = \{i \in I : |\varphi_i(\lambda)| \neq 0\}$ is at most countable. By assumption, $I = I_1$ and the proposition follows.

3. Assume that T is locally of class C_0, and $\mathfrak{K} \subset \mathfrak{H}$ is a subspace of dimension two with basis $\{h_1, h_2\}$. Let us set

$$m_{\mathfrak{K}} = m_{h_1} \vee m_{h_2},$$

and note that $m_{\mathfrak{K}}$ does not depend on the particular basis. Indeed, $m_{\mathfrak{K}}$ is the greatest common inner divisor of all functions $u \in H^\infty$ satisfying $u(T)\mathfrak{K} = \{0\}$.

Lemma 1.2. *Let T be locally of class C_0, and let $\mathfrak{K} \subset \mathfrak{H}$ be a subspace of dimension two. Then the set*

$$\{h \in \mathfrak{K} : m_h \neq m_{\mathfrak{K}}\}$$

is the union of an at most countable family of subspaces of dimension one.

Proof. Denote by A the set in the statement, and observe that $0 \in A$, and $m_{\lambda h} = m_h$ whenever λ is a nonzero scalar. We conclude that A is the union of a family of subspaces of dimension one:

$$A = \bigcup_{i \in I} E^1 h_i,$$

where E^1 denotes, as usual, the complex numbers, and h_i, h_j are linearly independent for $i \neq j$. Define $\varphi_i = m_{\mathfrak{K}}/m_{h_i}$, and note that φ_i is not constant because $h_i \in A$ for $i \in I$. If $i \neq j$, the vectors h_i and h_j form a basis of $\mathfrak{K}$, and therefore

$$\varphi_i \wedge \varphi_j = m_{\mathfrak{K}}/(m_{h_i} \vee m_{h_j}) = m_{\mathfrak{K}}/m_{\mathfrak{K}} = 1.$$

The lemma follows now immediately from Proposition 1.1.

Lemma 1.3. *Let T be locally of class C_0. For each $\lambda_0 \in D$ and every $\alpha > 0$, the set*

$$\sigma = \{h \in \mathfrak{H} : |m_h(\lambda_0)| \geq \alpha\}$$

is closed in $\mathfrak{H}$.

Proof. Let $\{h_n\} \subset \sigma$ be a sequence converging to h. An application of the Vitali and Montel theorem allows us to assume, upon dropping to a subsequence, that the sequence $\{m_{h_n}\}$ converges uniformly on the compact subsets of D to a function

$u \in H^\infty$. We certainly have $|u(\lambda)| \leq 1$ for $\lambda \in D$, and $|u(\lambda_0)| \geq \alpha$. By Theorem III.2.1, $\{m_{h_n}(T)\}$ converges weakly to $u(T)$, and therefore for $k \in \mathfrak{H}$ we have

$$\begin{aligned}|(u(T)h,k)| &\leq |((u(T) - m_{h_n}(T))h,k)| + |(m_{h_n}(T)(h-h_n),k)| \\ &\leq |((u(T) - m_{h_n}(T))h,k)| + \|h-h_n\|\|k\| \to 0,\end{aligned}$$

as $n \to \infty$; here we made use of the relation $m_{h_n}(T)h_n = 0$. Because k is arbitrary, we conclude that $u(T)h = 0$, and therefore $m_h|u$. We can thus write $u = m_h\varphi$ with $\varphi \in H^\infty$, and $|\varphi(e^{it})| = |u(e^{it})|$ a.e. It follows that $|\varphi(\lambda_0)| \leq 1$, and

$$\alpha \leq |u(\lambda_0)| \leq |m_h(\lambda_0)|$$

so that $h \in \sigma$, as desired.

The next result follows from an application of the Baire category theorem.

Lemma 1.4. *Assume that T is locally of class C_0. The set*

$$\{k \in \mathfrak{H} : |m_k(\lambda_0)| = \inf_{h \in \mathfrak{H}} |m_h(\lambda_0)|\}$$

is a dense G_δ set in $\mathfrak{H}$ for each $\lambda_0 \in D$.

Proof. Fix $\lambda_0 \in D$, and set

$$\alpha = \inf_{h \in \mathfrak{H}} |m_h(\lambda_0)|.$$

The complement of the set in the statement can be written as $\bigcup_{j=1}^\infty \sigma_j$, where

$$\sigma_j = \left\{h \in \mathfrak{H} : |m_h(\lambda_0)| \geq \alpha + \frac{1}{j}\right\}.$$

The preceding lemma implies that each σ_j is a closed set, and to finish the proof it suffices to show that each σ_j has empty interior. Suppose to the contrary that σ_j contains the open ball $B = \{h : \|h - h_0\| < \varepsilon\}$. Because $\sigma_j \neq \mathfrak{H}$, we can consider a linear space $\mathfrak{K}$ generated by h_0 and some vector $k \notin \sigma_j$, $k \neq 0$. Lemma 1.2 implies the existence of $f \in \mathfrak{K} \cap B$ such that $m_f = m_{\mathfrak{K}}$; in particular $m_k|m_f$, from which we infer $|m_f(\lambda_0)| \leq |m_k(\lambda_0)| < \alpha + (1/j)$ because $k \notin \sigma_j$. On the other hand, $f \in B \subset \sigma_j$, a contradiction. The lemma follows.

Theorem 1.5. *Assume that T is locally of class C_0 on $\mathfrak{H}$. Then there exist T-maximal vectors, and the set of T-maximal vectors is a dense G_δ in $\mathfrak{H}$. In particular, T is of class C_0 and $m_T = m_h$ for each T-maximal vector h.*

Proof. The intersection of countably many G_δ sets is still a dense G_δ, and therefore the set

$$M = \{h \in \mathfrak{H} : |m_h(\lambda_n)| = \inf_{k \in \mathfrak{H}} |m_k(\lambda_n)|, n \geq 0\}$$

is a dense G_δ for any choice of sequence $\{\lambda_n\} \subset D$. Choose this sequence to be dense in D. If $h \in M$ and $k \in \mathfrak{H}$, we have $|m_h(\lambda_n)| \leq |m_k(\lambda_n)|$ for all n, and by continuity

$$|m_h(\lambda)| \le |m_k(\lambda)|, \quad \lambda \in D.$$

We conclude that $m_k|m_h$, and thus every element of M is T-maximal. The other assertions of the theorem are now obvious.

4. We need the following variation of Theorem 1.5 on the existence of maximal vectors.

Theorem 1.6. *Let T be an operator of class C_0 on $\mathfrak{H}$, $\mathfrak{B}$ a Banach space, and $X\colon \mathfrak{B} \to \mathfrak{H}$ a bounded linear operator. If*

$$\mathfrak{H} = \bigvee_{n \ge 0} T^n X \mathfrak{B},$$

then the set

$$\{k \in \mathfrak{B} : m_{Xk} = m_T\}$$

is a dense G_δ in $\mathfrak{B}$.

Proof. The proof closely imitates that of Theorem 1.5. We provide the relevant details. Fix $\lambda_0 \in D$, and set

$$\alpha = \inf_{k \in \mathfrak{B}} |m_{Xk}(\lambda_0)|.$$

The sets

$$\sigma_j = \left\{k \in \mathfrak{B} : |m_{Xk}(\lambda_0)| \ge \alpha + \frac{1}{j}\right\} = X^{-1}\left\{h \in \mathfrak{H} : |m_h(\lambda_0)| \ge \alpha + \frac{1}{j}\right\}$$

are closed by Lemma 1.3. We then proceed as in the proof of Lemma 1.4 to show that each σ_j has empty interior. It follows that the set $\{k \in \mathfrak{B} : |m_{Xk}(\lambda_0)| = \alpha\}$ is a dense G_δ in $\mathfrak{B}$. Then the argument of Theorem 1.5 shows that the set

$$M = \{k \in \mathfrak{B} : |m_{Xk}(\lambda)| = \inf_{h \in \mathfrak{B}} |m_{Xh}(\lambda)|, \lambda \in D\}$$

is a dense G_δ. For $k \in M$ it follows that m_{Xk} is a multiple of m_{Xh} for all $h \in \mathfrak{B}$, and hence $m_{Xk}(T)(X\mathfrak{B}) = \{0\}$. This last relation implies

$$m_{Xk}(T)\left(\bigvee_{n \ge 0} T^n X \mathfrak{B}\right) = \{0\},$$

and hence $m_{Xk}(T) = O$. Therefore, for such k we have $m_{Xk} = m_T$. The theorem follows.

2 Jordan blocks

1. As usual, we denote by S the unilateral shift of multiplicity one acting on H^2.

Definition. For each inner function $\varphi \in H^\infty$, the *Jordan block* $S(\varphi)$ is the operator defined on $\mathfrak{H}(\varphi) = H^2 \ominus \varphi H^2$ by $S(\varphi) = P_{\mathfrak{H}(\varphi)} S|\mathfrak{H}(\varphi)$ or, equivalently, $S(\varphi)^* = S^*|\mathfrak{H}(\varphi)$.

We have already seen in Proposition III.4.4 that the Jordan block $S(\varphi)$ is an operator of class C_0 with minimal function φ. These operators can be viewed as the basic building blocks of arbitrary operators of class C_0, and it is worthwhile to study their properties in more detail.

Lemma 2.1. *If φ is a nonconstant inner function, then S is the minimal isometric dilation of $S(\varphi)$.*

Proof. This follows immediately from Theorem VI.3.1. Indeed, if φ is not constant, then $\{E^1, E^1, \varphi(\lambda)\}$ is a purely contractive analytic function.

Corollary 2.2. *We have $\mathfrak{d}_{S(\varphi)} = \mathfrak{d}_{S(\varphi)^*} = 1$ for every nonconstant inner function φ.*

Proof. By Theorem VI.3.1, $\{E^1, E^1, \varphi\}$ coincides with the characteristic function of $S(\varphi)$. The corollary follows at once.

Proposition 2.3. *Let T be a contraction of class $C_{\cdot 0}$ on $\mathfrak{H}$ such that $\mathfrak{d}_{T^*} = 1$. Then one of the following mutually exlusive possibilities holds.*

(1) *T is unitarily equivalent to S.*
(2) *T is unitarily equivalent to $S(\varphi)$ for some nonconstant inner function φ.*

Proof. The minimal isometric dilation of T is a unilateral shift of multiplicity one; in other words, it is unitarily equivalent to S. Thus we may assume that $\mathfrak{H} \subset H^2$ is invariant for S^*, and $T^* = S^*|\mathfrak{H}$. The proposition clearly follows from the classification of invariant subspaces of S. Indeed, either $\mathfrak{H} = H^2$, or $H^2 \ominus \mathfrak{H} = \varphi H^2$ for some inner function φ. In this last case, $\mathfrak{H} = \mathfrak{H}(\varphi)$ and $T = S(\varphi)$. The function φ cannot be constant because $\dim(\mathfrak{H}) \geq \dim(\mathfrak{D}_{T^*}) = 1$.

Recall that the adjoint of a function φ is defined by $\varphi^\sim(\lambda) = \overline{\varphi(\bar{\lambda})}$, $\lambda \in D$.

Corollary 2.4. *Let φ be an inner function in H^∞. The adjoint $S(\varphi)^*$ is unitarily equivalent to $S(\varphi^\sim)$.*

Proof. As noted above, $\{E^1, E^1, \varphi(\lambda)\}$ coincides with the characteristic function of $S(\varphi)$. Therefore the characteristic function of $S(\varphi)^*$ coincides with $\{E^1, E^1, \varphi^\sim(\lambda)\}$, and this yields the desired unitary equivalence by virtue of Proposition VI.1.1.

2. We study next the invariant subspaces and maximal vectors of $S(\varphi)$. It is convenient to denote by $\operatorname{ran} X$ the range of an operator X.

Proposition 2.5. *Let φ be a nonconstant inner function.*

(1) *For every $h \in \mathfrak{H}(\varphi)$ we have $m_h = \varphi/(h \wedge \varphi)$.*

(2) *Every invariant subspace of $S(\varphi)$ has the form $\psi H^2 \ominus \varphi H^2$ for some inner divisor ψ of φ. We have*

$$\psi H^2 \ominus \varphi H^2 = \ker(\varphi/\psi)(S(\varphi)) = \operatorname{ran}\psi(S(\varphi))$$

for each inner divisor ψ of φ.

(3) *If $\mathfrak{M} = \psi H^2 \ominus \varphi H^2$ is invariant for $S(\varphi)$ then $S(\varphi)|\mathfrak{M}$ is unitarily equivalent to $S(\varphi/\psi)$, and the compression of $S(\varphi)$ to $\mathfrak{H}(\varphi) \ominus \mathfrak{M} = \mathfrak{H}(\psi)$ is precisely $S(\psi)$.*

(4) *A vector $h \in \mathfrak{H}(\varphi)$ is cyclic for $S(\varphi)$ if and only if $\varphi \wedge h = 1$, that is, if and only if h is maximal. The set of cyclic vectors for $S(\varphi)$ is a dense G_δ in $\mathfrak{H}(\varphi)$.*

Proof. (1) Set $u = m_h$ and $v = \varphi/(h \wedge \varphi)$. We have

$$v(S(\varphi))h = P_{\mathfrak{H}(\varphi)}v(S)h = P_{\mathfrak{H}(\varphi)}(vh) = P_{\mathfrak{H}(\varphi)}\left[\varphi\frac{h}{h \wedge \varphi}\right] = 0,$$

and consequently $u|v$. Conversely, we know that $u(S(\varphi))h = 0$, so that $uh = \varphi g$ for some $g \in H^2$. Because u divides φ, it follows that $h = (\varphi/u)g$, and hence $(\varphi/u)|h$. Obviously $(\varphi/u)|\varphi$, thus $(\varphi/u)|(h \wedge \varphi)$ or, equivalently, $v|u$. We deduce that $v = u$, as desired.

(2) The description of the invariant subspaces of $S(\varphi)$ is part b) of Proposition III.4.3. Let ψ be an inner divisor of φ. We have $(\varphi/\psi)(S(\varphi))h = 0$ if and only if $m_h|(\varphi/\psi)$ or, equivalently by (1), $\psi|h$. Because

$$\{h \in \mathfrak{H}(\varphi) : \psi|h\} = \psi H^2 \ominus \varphi H^2,$$

we proved that $\ker(\varphi/\psi)(S(\varphi)) = \psi H^2 \ominus \varphi H^2$. For the second equality we note that

$$\begin{aligned}\psi(S(\varphi))\mathfrak{H}(\varphi) &= P_{\mathfrak{H}(\varphi)}\psi(S)\mathfrak{H}(\varphi)\\ &= P_{\mathfrak{H}(\varphi)}\psi(S)H^2\\ &= P_{\mathfrak{H}(\varphi)}\psi H^2 = \psi H^2 \ominus \varphi H^2.\end{aligned}$$

Part (3) follows from Theorem VI.1.1.

(4) If h is cyclic, we must have $m_h = m_{S(\varphi)}$, so that $h \wedge \varphi = 1$ by (1). Conversely, if $h \wedge \varphi = 1$, (2) shows that h does not belong to any proper invariant subspace of $S(\varphi)$, and hence h is a cyclic vector. The last statement follows from Theorem 1.5. The proposition is proved.

Corollary 2.6. *Every invariant subspace of $S(\varphi)$ is hyperinvariant.*

Proof. This follows from the equality

$$\psi H^2 \ominus \varphi H^2 = \operatorname{ran}(\psi(S(\varphi)))$$

if ψ is an inner divisor of φ.

Corollary 2.7. *Assume that $\varphi, u \in H^\infty$, and φ is inner. Then*

$$\ker u(S(\varphi)) = (\varphi/(u\wedge\varphi))H^2 \ominus \varphi H^2, (\operatorname{ran} u(S(\varphi)))^- = (u\wedge\varphi)H^2 \ominus \varphi H^2,$$

$S(\varphi)|\ker u(S(\varphi))$ is unitarily equivalent to $S(u\wedge\varphi)$, and $S(\varphi)|(\operatorname{ran} u(S(\varphi)))^-$ is unitarily equivalent to $S(\varphi/(u\wedge\varphi))$.

Proof. Observe that $u(S(\varphi))h = 0$ if and only if $m_h|u$. Because m_h always divides φ, we see that $m_h|u$ if and only if $m_h|(u\wedge\varphi)$. In other words, we have $\ker u(S(\varphi)) = \ker(u\wedge\varphi)(S(\varphi))$. Now Proposition 2.5(2) proves the first equality in the statement.

Analogously,

$$\begin{aligned}\mathfrak{H}(\varphi) \ominus (\operatorname{ran} u(S(\varphi)))^- &= \ker u^\sim(S(\varphi)^*)\\ &= \ker(u\wedge\varphi)^\sim(S(\varphi)^*)\\ &= \mathfrak{H}(\varphi) \ominus (\operatorname{ran}(u\wedge\varphi)(S(\varphi))),\end{aligned}$$

so that

$$(\operatorname{ran} u(S(\varphi)))^- = \operatorname{ran}(u\wedge\varphi)(S(\varphi)) = (u\wedge\varphi)H^2 \ominus \varphi H^2.$$

The last two assertions of the corollary follow from Part (3) of Proposition 2.5.

Corollary 2.8. *The set of cyclic vectors for $S(\varphi)$ is a dense G_δ in $\mathfrak{H}(\varphi)$.*

Proof. Proposition 2.5 implies that $h \in \mathfrak{H}(\varphi)$ is cyclic if and only if $m_h = \varphi$. The corollary follows now from Theorem 1.5.

3. Quite interestingly, Theorem 1.6 has the following consequence of intrinsic interest for the arithmetic of Hardy spaces. This is used in Sec. 6. We denote by ℓ^1 the Banach space of absolutely summable sequences of complex scalars.

Theorem 2.9. *Let $\{f_j : j \geq 0\}$ be a bounded sequence of functions in H^2,and let φ be an inner function. The set of those sequences $\alpha = \{\alpha_j\}$ in ℓ^1 satisfying the relation*

$$\left(\sum_{j=0}^{\infty} \alpha_j f_j\right) \wedge \varphi = \left(\bigwedge_{j=0}^{\infty} f_j\right) \wedge \varphi$$

is a dense G_δ in ℓ^1.

Proof. We may assume without loss of generality that

$$\left(\bigwedge_{j=0}^{\infty} f_j\right) \wedge \varphi = 1.$$

Indeed, we can replace f_j by f_j/ψ and φ by φ/ψ, with $\psi = \left(\bigwedge_{j=0}^{\infty} f_j\right) \wedge \varphi$. Under this additional assumption, the invariant subspace for $S(\varphi)$ generated by the vectors $\{P_{\mathfrak{H}(\varphi)} f_j : j \geq 0\}$ is $\mathfrak{H}(\varphi)$. Indeed, if this invariant subspace is $\psi H^2 \ominus \varphi H^2$, it follows that $\psi | \left(\bigwedge_{j=0}^{\infty} f_j\right) \wedge \varphi$, and hence $\psi = 1$. We can then apply Theorem 1.6 to the space $\mathfrak{B} = \ell^1$ and the linear operator $X\colon \mathfrak{B} \to \mathfrak{H}(\varphi)$ defined by

$$X\alpha = P_{\mathfrak{H}(\varphi)}\left(\sum_{j=0}^{\infty} \alpha_j f_j\right) \quad (\alpha = \{\alpha_j\} \in \ell^1).$$

We deduce that the set of those sequences $\alpha \in \ell^1$ for which $m_{X\alpha} = \varphi$ is a dense G_δ in ℓ^1. Finally, the condition $m_{X\alpha} = \varphi$ is equivalent to $(X\alpha) \wedge \varphi = 1$, and this is in turn equivalent to $\left(\sum_{j=0}^{\infty} \alpha_j f_j\right) \wedge \varphi = 1$.

4. We conclude this section with a few facts about operators that intertwine Jordan blocks.

Theorem 2.10. *Let φ be an inner function. For every operator $X \in \{S(\varphi)\}'$ there exists a function $u \in H^\infty$ such that $X = u(S(\varphi))$ and $\|u\| = \|X\|$.*

Proof. We may assume that φ is not constant, in which case S is the minimal isometric dilation of $S(\varphi)$. Given $X \in \{S(\varphi)\}'$, Theorem II.2.3 implies the existence of an operator $Y \in \{S\}'$ such that $\|Y\| = \|X\|$ and $X = P_{\mathfrak{H}(\varphi)}Y|\mathfrak{H}(\varphi)$. Apply now Lemma V.3.2 to deduce that $Y = u(S)$ for some $u \in H^\infty$. This function satisfies the conclusion of the theorem.

A more general form of Theorem 2.10 is useful in applications.

Theorem 2.11. *Let φ, φ' be inner functions, and let $X\colon \mathfrak{H}(\varphi) \to \mathfrak{H}(\varphi')$ satisfy the intertwining relation $XS(\varphi) = S(\varphi')X$. There exists a function $u \in H^\infty$ such that $\varphi' | u\varphi$, $\|u\| = \|X\|$, and*

$$X = P_{\mathfrak{H}(\varphi')}u(S)|\mathfrak{H}(\varphi).$$

Conversely, if $u \in H^\infty$ is such that $\varphi' | u\varphi$, then the above formula defines an operator such that $XS(\varphi) = S(\varphi')X$, and $X = O$ if and only if $\varphi' | u$.

Proof. As in the preceding argument, we may assume that φ and φ' are not constant, and then the commutant lifting theorem yields $Y \in \{S\}'$ such that $Y(\varphi H^2) \subset \varphi' H^2$ and $X = P_{\mathfrak{H}(\varphi')}Y|\mathfrak{H}(\varphi)$. If we write $Y = u(S)$, we see that the above inclusion is equivalent to $\varphi' | u\varphi$. The remaining assertions are easily verified.

3 Quasi-affine transforms and multiplicity

1. Let T be an operator on the complex Hilbert space $\mathfrak{H}$.

Definition. The *cyclic multiplicity μ_T of T* is the smallest cardinality of a subset $M \subset \mathfrak{H}$ with the property that the set $\{T^n h : h \in M, n \geq 0\}$ generates $\mathfrak{H}$. The operator T is said to be *multiplicity-free* if $\mu_T = 1$.

Note that T is multiplicity-free if and only if it has a cyclic vector.

Proposition 3.1. *Let V be a unilateral shift with wandering space $\mathfrak{F}$. We have $\mu_V =$* $\dim(\mathfrak{F})$.

Proof. If M is an orthonormal basis in $\mathfrak{F}$, then $\bigvee_{n=0}^{\infty} V^n M = \mathfrak{H}$, and therefore $\mu_V \leq \dim(\mathfrak{F})$. To prove the opposite inequality, let M be an arbitrary set such that $\mu_V = \operatorname{card}(M)$ and $\bigvee_{n=0}^{\infty} V^n M = \mathfrak{H}$. Then

$$\mathfrak{F} = \mathfrak{H} \ominus V\mathfrak{H} = \left(\bigvee_{n=0}^{\infty} V^n M\right) \ominus \left(\bigvee_{n=1}^{\infty} V^n M\right),$$

and it follows that $\mathfrak{F}$ is spanned as a closed space by the set $P_{\mathfrak{F}}M$. Consequently, $\dim(\mathfrak{F}) \leq \operatorname{card}(P_{\mathfrak{F}}M) \leq \operatorname{card}(M) = \mu_V$.

Lemma 3.2. *Let T and T' act on $\mathfrak{H}$ and $\mathfrak{H}'$, respectively, and let $X\colon \mathfrak{H}' \to \mathfrak{H}$ be a bounded linear transformation such that $XT' = TX$. If X has dense range then $\mu_T \leq \mu_{T'}$.*

Proof. Choose $M' \subset \mathfrak{H}'$ with $\operatorname{card}(M') = \mu_{T'}$ and $\bigvee_{n=0}^{\infty} T'^n M' = \mathfrak{H}'$. If X has dense range then

$$\bigvee_{n=0}^{\infty} T^n (XM') = \bigvee_{n=0}^{\infty} XT'^n M' = (X\mathfrak{H}')^- = \mathfrak{H},$$

and therefore

$$\mu_T \leq \operatorname{card}(XM') \leq \operatorname{card}(M') = \mu_{T'}.$$

Corollary 3.3. *If T is a contraction of class $C_{\cdot 0}$ then $\mu_T \leq \mathfrak{d}_{T^*}$.*

Proof. Let T act on $\mathfrak{H}$, and let U_+ on $\mathfrak{K}_+$ be the minimal isometric dilation of T. By Proposition II.3.1, we have $\mathfrak{R} = \{0\}$, and hence U_+ is a unilateral shift of multiplicity $\mathfrak{d}_{T^*}$. Because $P_{\mathfrak{H}}\mathfrak{K}_+ = \mathfrak{H}$ and $TP_{\mathfrak{H}} = P_{\mathfrak{H}}U_+$, we deduce

$$\mu_T \leq \mu_{U_+} = \mathfrak{d}_{T^*}$$

from the preceding results.

2. Contractions of class $C_{\cdot 0}$ with small defect $\mathfrak{d}_{T^*}$, particularly with $\mathfrak{d}_{T^*} = 1$, are relatively easy to understand (see Proposition 2.3). It is natural to reduce problems related to a contraction of class $C_{\cdot 0}$ to operators T with a small defect index $\mathfrak{d}_{T^*}$. This is achieved in the following two results.

Lemma 3.4. *Let T be an operator of class $C_{\cdot 0}$ on $\mathfrak{H}$. There exist a unilateral shift U on $\mathfrak{H}_1$, and a bounded linear transformation $X\colon \mathfrak{H}_1 \to \mathfrak{H}$ such that X has dense range, $XU = TX$, and $\mu_U = \mu_T$.*

Proof. Let U_+ on $\mathfrak{K}_+$ be the minimal isometric dilation of T; as noted above, U_+ is a unilateral shift. Fix a set $M \subset \mathfrak{H}$ with $\operatorname{card}(M) = \mu_T$ such that $\bigvee_{n=0}^{\infty} T^n M = \mathfrak{H}$. We define the space $\mathfrak{H}_1$ and the operators U, X as follows:

$$\mathfrak{H}_1 = \bigvee_{n=0}^{\infty} U_+^n M, \quad U = U_+|\mathfrak{H}_1, \quad X = P_{\mathfrak{H}}|\mathfrak{H}_1.$$

The relation $TX = XU$ follows because $TP_{\mathfrak{H}} = P_{\mathfrak{H}}U_+$. Next we see that

$$(X\mathfrak{H}_1)^- = \bigvee_{n=0}^{\infty} XU_+^n M = \bigvee_{n=0}^{\infty} T^n XM = \bigvee_{n=0}^{\infty} T^n M = \mathfrak{H},$$

and therefore X has dense range. Finally, U is a unilateral shift,

$$\mu_U \leq \operatorname{card}(M) = \mu_T,$$

and the opposite inequality $\mu_U \geq \mu_T$ follows from Lemma 3.2.

Recall that $T_1 \prec T_2$ indicates that T_1 is a quasi-affine transform of T_2, and $T_1 \sim T_2$ indicates that the T_1 and T_2 are quasi-similar.

Theorem 3.5. *For every contraction T of class $C_{\cdot 0}$ there exists a contraction T' of class $C_{\cdot 0}$ such that $T' \prec T$, and*

$$\mu_{T'} = \mathfrak{d}_{T'^*} = \mu_T.$$

Proof. Let $\mathfrak{H}, \mathfrak{H}_1, U$, and X be as in the preceding lemma, and set

$$\mathfrak{H}' = \mathfrak{H}_1 \ominus \ker X, \quad Y = X|\mathfrak{H}', \quad T' = P_{\mathfrak{H}'}U|\mathfrak{H}'.$$

Because $TX = XU$, $\ker X$ is an invariant subspace for U, and therefore T' is of class $C_{\cdot 0}$ as $T'^* = U^*|\mathfrak{H}'$. Clearly Y is one-to-one, and $Y\mathfrak{H}' = X\mathfrak{H}_1$, so that Y has dense range and is therefore a quasi-affinity. For every vector $x' \in \mathfrak{H}'$ we have

$$\begin{aligned} TYx' &= TXx' = XUx' \\ &= X(Ux' - P_{\ker X}Ux') \\ &= XP_{\mathfrak{H}'}Ux' = XT'x' = YT'x'. \end{aligned}$$

Thus $TY = YT'$, and this proves the relation $T' \prec T$. The inequalities

$$\mu_T \leq \mu_{T'} \leq \mathfrak{d}_{T'^*}$$

are obvious from Lemma 3.2 and its corollary. Finally, the wandering space $\mathfrak{F}$ of U has dimension μ_T by Lemma 3.4, and

$$\begin{aligned} I_{\mathfrak{H}'} - T'T'^* &= I_{\mathfrak{H}'} - T'U^*|\mathfrak{H}' \\ &= P_{\mathfrak{H}'}(I - UU^*)|\mathfrak{H}' \\ &= P_{\mathfrak{H}'}P_{\mathfrak{F}}|\mathfrak{H}'. \end{aligned}$$

We conclude that

$$\mathfrak{d}_{T'^*} = \operatorname{rank}(I - T'T'^*) \leq \operatorname{rank}(P_{\mathfrak{F}}) = \dim(\mathfrak{F}) = \mu_T,$$

and this completes the proof of the theorem.

4 Multiplicity-free operators and splitting

1. The adjoint of a multiplicity-free operator is not generally multiplicity-free; for example, the adjoint of a unilateral shift of countably infinite multiplicity has a cyclic vector. An easy way to see this is to show that $U^* \prec (S^*)^{(\aleph_0)}$, where U is the bilateral shift on L^2 and S is the unilateral shift on H^2. For this purpose, consider a partition $C = \bigcup_{n=1}^{\infty} \alpha_n$ into Borel sets of positive measure, and define $X\colon H^{2(\aleph_0)} \to L^2$ by $X(\{f_n\}_{n=1}^{\infty}) = \sum_{n=1}^{\infty} f_n \chi_{\alpha_n}$ for $\{f_n\}_{n=1}^{\infty} \in H^{2(\aleph_0)}$. Clearly X is a quasi-affinity in $\mathscr{I}(S^{(\aleph_0)}, U)$ so that $S^{(\aleph_0)} \prec U$, or equivalently $U^* \prec (S^*)^{(\aleph_0)}$, as claimed. Thus $\mu_{(S^*)^{(\aleph_0)}} \leq \mu_{U^*} = 1$ by Lemma 3.2 and the remark preceding Lemma IX.2.3.

We show that for operators T of class C_0 we have in fact $\mu_T = 1$ if and only if $\mu_{T^*} = 1$. First we prove an auxiliary result.

Proposition 4.1. *Let T be an operator of class C_0. If T is multiplicity-free, then $S(m_T) \prec T$. If T^* is multiplicity-free, then $T \prec S(m_T)$.*

Proof. Assume first that $\mu_T = 1$. It follows from Theorem 3.5 that there exists an operator T' of class $C_{\cdot 0}$ such that $T' \prec T$ and $\mathfrak{d}_{T'^*} = \mu_T = 1$. The operator T' is of class C_0 by Proposition III.4.6, and therefore it cannot be unitarily equivalent to S. Then Proposition 2.3 shows that T' is unitarily equivalent to $S(\varphi)$ for some inner function $\varphi \in H^\infty$. Thus we have $S(\varphi) \prec T$, and because $\varphi = m_{S(\varphi)} = m_T$, we conclude that $S(m_T) \prec T$, as desired.

If $\mu_{T^*} = 1$, the preceding proof shows that

$$S(m_{\tilde{T}}) = S(m_{T^*}) \prec T^*,$$

and hence $T \prec S(m_{\tilde{T}})^*$. The proposition follows thus from Corollary 2.4.

Theorem 4.2. *Let T be an operator of class C_0. The following conditions are equivalent.*

(1) *T is multiplicity-free.*
(2) *T^* is multiplicity-free.*
(3) *T is quasi-similar to $S(m_T)$.*

Proof. It suffices to prove that (2) implies (1). Indeed, it follows then by symmetry that (1) implies (2). Furthermore, if (1) and (2) are satisfied, then $T \sim S(m_T)$ by the preceding proposition. Conversely, if $S(m_T) \prec T$, then $\mu_T \leq \mu_{S(m_T)} = 1$, and (1) follows.

Assume therefore that T acts on $\mathfrak{H}$ and T^* is multiplicity-free. By Proposition 4.1 we can choose a quasi-affinity X such that $XT = S(m_T)X$. Theorem 1.5 allows us to choose a T-maximal vector $h \in \mathfrak{H}$. Denote by $\mathfrak{K}$ the cyclic space $\bigvee\{T^n h : n \geq 0\}$ generated by h. Thus we have $T\mathfrak{K} \subset \mathfrak{K}$ and $m_{T|\mathfrak{K}} = m_T$. The operator $T|\mathfrak{K}$ is multiplicity-free, therefore a second application of Proposition 4.1 yields an injective operator $Y\colon \mathfrak{H}(m_T) \to \mathfrak{H}$ such that $Y\mathfrak{H}(m_T)$ is dense in $\mathfrak{K}$ and $YS(m_T) = TY$. We have then

$$XYS(m_T) = XTY = S(m_T)XY,$$

so that $XY \in \{S(m_T)\}'$ and, of course, XY is injective. By Theorem 2.10, we have $XY = u(S(m_T))$ for some $u \in H^\infty$, and $u \wedge m_T = 1$ because

$$\ker u(S(m_T)) = \{0\}$$

(cf. Corollary 2.7). Observe now that

$$\begin{aligned} X(YX - u(T)) &= XYX - Xu(T) \\ &= XYX - u(S(m_T))X \\ &= (XY - u(S(m_T)))X = O, \end{aligned}$$

and hence $YX = u(T)$ because X is injective. Now, the relation $u \wedge m_T = 1$ implies via Proposition III.4.7 that $u(T)$ is a quasi-affinity. In particular,

$$\mathfrak{H} = (u(T)\mathfrak{H})^- \subset (Y\mathfrak{H}(m_T))^- \subset \mathfrak{K},$$

so that $\mathfrak{K} = \mathfrak{H}$ and h is a cyclic vector for T. The theorem is proved.

The preceding argument also yields the following result.

Corollary 4.3. *Let T be a multiplicity-free operator of class C_0 acting on $\mathfrak{H}$. A vector $h \in \mathfrak{H}$ is cyclic for T if and only if it is T-maximal. The set of cyclic vectors for T is a dense G_δ in $\mathfrak{H}$.*

Corollary 4.4. *Every restriction of a multiplicity-free operator of class C_0 to an invariant subspace is multiplicity-free.*

Proof. Let T be a multiplicity-free operator of class C_0, and let $\mathfrak{K}$ be an invariant subspace for T. If h is cyclic for T^* then $P_{\mathfrak{K}}h$ is cyclic for $(T|\mathfrak{K})^*$. Thus $(T|\mathfrak{K})^*$ is multiplicity-free, and therefore so is $T|\mathfrak{K}$.

2. Some of the results concerning operators intertwining Jordan blocks can be transferred to general multiplicity-free operators of class C_0.

Proposition 4.5. *Let T and T' be two multiplicity-free operators of class C_0, and let A satisfy the equation $AT = T'A$. If $m_T = m_{T'}$ then A is one-to-one if and only if it has dense range.*

Proof. Set $\varphi = m_T = m_{T'}$, so that T and T' are quasi-similar to $S(\varphi)$ by Theorem 4.2. Choose quasi-affinities X, Y satisfying $XS(\varphi) = T'X$ and $YT = TS(\varphi)$. The product XAY is easily seen to commute with $S(\varphi)$, and hence $XAY = u(S(\varphi))$ for some $u \in H^\infty$ by Theorem 2.10. If A is either one-to-one or has dense range, then XAY has the same property, and hence $u \wedge \varphi = 1$ in either case (cf. Corollary 2.7). Next we note that

$$\begin{aligned} X(AYX - u(T')) &= XAYX - Xu(T) \\ &= XAYX - u(S(\varphi))X \\ &= (XAY - u(S(\varphi)))X = O, \end{aligned}$$

and hence $u(T') = AYX$ because X is one-to-one. If $u \wedge \varphi = 1$, it follows that $u(T')$ is a quasi-affinity, and consequently A has dense range. Analogously, one can show that $YXA = u(T)$, and hence A is one-to-one if $u \wedge \varphi = 1$. The proposition follows easily from these observations.

Proposition 4.6. *Let T be a multiplicity-free operator of class C_0. There exists a function $v \in K_T^\infty$ such that every operator $A \in \{T\}'$ can be written as $A = (u/v)(T)$ for some $u \in H^\infty$.*

Proof. Let φ,X, and Y be as in the proof of the preceding proposition, with $T' = T$. If $A = I$, that proof implies the existence of $v \in H^\infty$ such that $v \wedge \varphi = 1$ and $YX = v(T)$; note that $v \in K_T^\infty$ by Proposition III.4.7. Now, if A is arbitrary in $\{T\}'$, we deduce the existence of $u \in H^\infty$ such that $YXA = u(T)$, so that $v(T)A = u(T)$. This means precisely that $A = (u/v)(T)$, as desired.

3. We prove next a result about invariant subspaces that justifies in particular the terminology "multiplicity-free".

Theorem 4.7. *Let T be an operator of class C_0. The following assertions are equivalent.*

(1) *T is multiplicity-free.*
(2) *For every inner divisor φ of m_T, there exists a unique invariant subspace $\mathfrak{K}$ for T satisfying the relation $m_{T|\mathfrak{K}} = \varphi$.*
(3) *If $\mathfrak{K}$ and $\mathfrak{K}'$ are invariant for T, and $T|\mathfrak{K} \prec T|\mathfrak{K}'$, then $\mathfrak{K} = \mathfrak{K}'$.*
(4) *There are no proper invariant subspaces $\mathfrak{K}$ for T such that $m_{T|\mathfrak{K}} = m_T$.*

Moreover, if T is multiplicity-free, the unique invariant subspace considered in (2) *is given by*

$$\mathfrak{K} = \ker\varphi(T) = [\operatorname{ran}(m_T/\varphi)(T)]^-.$$

Proof. Assume that T acts on $\mathfrak{H}$, it is multiplicity-free, $\mathfrak{K}$ is invariant for T, and $\varphi = m_{T|\mathfrak{K}}$. The operators $T' = T|\mathfrak{K}$ and $T'' = T|\ker\varphi(T)$ are multiplicity-free by Corollary 4.4, and they satisfy the relation $JT' = T''J$, where $J\colon \mathfrak{K} \to \ker\varphi(T)$ is the inclusion operator. Because T' and T'' both have minimal function φ, Proposition 4.5 implies that J must have dense range, and therefore $\mathfrak{K} = J\ker\varphi(T) = \ker\varphi(T)$. Thus (1) implies (2). It is obvious that (2) implies (4). Assume next that (4) holds, and h is a T-maximal vector. If we define $\mathfrak{K} = \bigvee\{T^n h : n \geq 0\}$, we have $m_{T|\mathfrak{K}} = m_T$, and hence $\mathfrak{K} = \mathfrak{H}$ by (4). Thus h is a cyclic vector, and we conclude that (4) implies (1). It remains to show that (3) is equivalent to the other three conditions. The fact that (2) implies (3) is obvious because $T|\mathfrak{K} \prec T|\mathfrak{K}'$ implies, in particular, the equality $m_{T|\mathfrak{K}} = m_{T|\mathfrak{K}'}$. Conversely, assume that (3) holds and h, h' are T-maximal vectors. Denote by $\mathfrak{K}, \mathfrak{K}'$ the cyclic spaces generated by h, h', respectively, and note that $T|\mathfrak{K} \sim T|\mathfrak{K}'$ by Theorem 4.2 and the transitivity of of quasi-similarity. In particular $T|\mathfrak{K} \prec T|\mathfrak{K}'$, so that $\mathfrak{K} = \mathfrak{K}'$ and therefore $h' \in \mathfrak{K}$. We conclude that $\mathfrak{K}$ contains all maximal vectors, so that $\mathfrak{K} = \mathfrak{H}$ because the set of maximal vectors is dense. Thus (3) implies (1).

The last assertion of the theorem follows because both $T|[\operatorname{ran}(m_T/\varphi)(T)]^-$ and $T|\ker\varphi(T)$ have minimal function φ.

The last assertion of the preceding theorem yields the following result.

Corollary 4.8. *Every invariant subspace of a multiplicity-free operator of class C_0 is hyperinvariant.*

We can now complete the characterization of unicellular operators of class C_0.

Corollary 4.9. *A contraction T of class C_0 is unicellular if and only if it is multiplicity-free and its spectrum consists of a single point.*

Proof. Assume first that T is unicellular. We already know from the corollary to Proposition III.7.3 that $\sigma(T)$ is a singleton. If T is not cyclic, Theorem 4.7.(4) implies the existence of a maximal vector h such that $\mathfrak{M}_h$ is not the whole space. Then Theorem 1.5 implies the existence of a maximal vector $k \notin \mathfrak{M}_h$. Neither of the spaces $\mathfrak{M}_h$ and $\mathfrak{M}_k$ is contained in the other, contradicting unicellularity. Thus T must be multiplicity-free. Conversely, assume that T is multiplicity-free, and $\sigma(T)$ is a singleton. In this case we have $T \sim S(m_T)$, and Theorem 4.7.(2) shows that T is unicellular if and only if the divisors of m_T are totally ordered by divisibility. Thus m_T is either a Blaschke product with a single zero, or a singular inner function determined by a measure supported by a single point. By Theorem III.5.1, this happens precisely when $\sigma(T)$ is a singleton. The corollary follows.

4. We now show how multiplicity-free operators can be used in the study of operators with larger multiplicity.

Theorem 4.10. *Let T be an operator of class C_0 on $\mathfrak{H}$, $h \in \mathfrak{H}$ a T-maximal vector, and $\mathfrak{K} = \bigvee\{T^n h : n \geq 0\}$. There exists an invariant subspace $\mathfrak{M}$ for T such that $\mathfrak{K} \vee \mathfrak{M} = \mathfrak{H}$ and $\mathfrak{K} \cap \mathfrak{M} = \{0\}$.*

Proof. The operator $T_1 = T|\mathfrak{K}$ is multiplicity-free, and by Theorem 4.2 there exists a vector $k \in \mathfrak{K}$ cyclic for T_1^*. We now set

$$\mathfrak{K}' = \bigvee_{n=0}^{\infty} T^{*n}k, \quad \mathfrak{M} = \mathfrak{H} \ominus \mathfrak{K}',$$

and define T_2 on $\mathfrak{K}'$ by $T_2^* = T^*|\mathfrak{K}'$. Because $\mathfrak{K}'$ is invariant for T^*, we have $P_{\mathfrak{K}'}T = T_2 P_{\mathfrak{K}'}$, and therefore the operator $X\colon \mathfrak{K} \to \mathfrak{K}'$ defined by $X = P_{\mathfrak{K}'}|\mathfrak{K}$ satisfies the intertwining relation $XT_1 = T_2 X$. Observe that $\mathfrak{K} \cap \mathfrak{M} = \ker X$, and

$$\mathfrak{H} \ominus (\mathfrak{K} \vee \mathfrak{M}) = (\mathfrak{H} \ominus \mathfrak{K}) \cap \mathfrak{K}' = \mathfrak{K}' \cap \ker P_{\mathfrak{K}} = \ker X^*.$$

To conclude the proof, it suffices to show that X is a quasi-affinity. To do this we first note that

$$(\operatorname{ran} X^*)^- = \bigvee_{n=0}^{\infty} X^* T_2^{*n} k = \bigvee_{n=0}^{\infty} T_1^{*n} X^* k = \bigvee_{n=0}^{\infty} T_1^{*n} k = \mathfrak{K},$$

and thus X^* has dense range. If $\varphi = m_{T_2}$, we have

$$(\varphi(T_1))^* X^* = X^* (\varphi(T_2))^* = O,$$

so that $(\varphi(T_1))^*$ vanishes on a dense set. We conclude that $m_T = m_{T_1}$ must divide φ, so that in fact $\varphi = m_T$. The fact that X is a quasi-affinity follows now from Proposition 4.5 because X^* has dense range, and the operators T_1, T_2 are multiplicity-free and have the same minimal function.

Corollary 4.11. *Let T be an operator of class C_0. There exist operators T' and T'' of class C_0 such that*

$$S(m_T) \oplus T' \prec T \prec S(m_T) \oplus T''.$$

Proof. Let $\mathfrak{K}$ and $\mathfrak{M}$ be as in Theorem 4.10. Then we have

$$(T|\mathfrak{K}) \oplus (T|\mathfrak{M}) \prec T$$

with the intertwining quasi-affinity $X\colon \mathfrak{K} \oplus \mathfrak{M} \to \mathfrak{H}$ defined by $X(u \oplus v) = u + v$. Thus

$$S(m_T) \oplus (T|\mathfrak{M}) \prec T$$

because $S(m_T) \sim T|\mathfrak{K}$. It follows that $T' = T|\mathfrak{M}$ satisfies the required relation. The existence of T'' is deduced similarly replacing T by T^*.

Theorem 4.10 can be used to prove a converse to Proposition 4.6, hence yet another characterization of multiplicity-free operators.

Theorem 4.12. *The following assertions are equivalent for an operator of class C_0.*

(1) *T is multiplicity-free.*
(2) *$\{T\}'$ is commutative.*
(3) *$\{T\}'$ consists of the bounded operators of the form $f(T)$ with $f \in N_T$.*

Proof. We already know from Proposition 4.6 that (1) implies (3), and (3) trivially implies (2). It remains to show that the commutant of T is not commutative if $\mu_T \geq 2$. Let $\mathfrak{K}, \mathfrak{M}$, and $\mathfrak{H}$ be as in Theorem 4.10; if $\mu_T \geq 2$ we must have $\mathfrak{K} \neq \mathfrak{H}$, and hence $\mathfrak{M} \neq \{0\}$. Define now $\mathfrak{K}'$, T_1 and T_2 by

$$\mathfrak{K}' = \mathfrak{H} \ominus \mathfrak{M}, \quad T_1 = T|\mathfrak{K}, \quad T_2^* = T^*|\mathfrak{K}'.$$

The operator $X = P_{\mathfrak{K}'}|\mathfrak{K}$ is a quasi-affinity, and $XT_1 = T_2X$. Both T_1 and T_2 are multiplicity-free, thus T_1 and T_2 are quasi-similar; indeed, both are quasi-similar to $S(m_T)$. Let Y be a quasi-affinity satisfying $YT_2 = T_1Y$, and define the operator $A \in \{T\}'$ by $A = YP_{\mathfrak{K}'}$. We clearly have

$$\ker A = \ker P_{\mathfrak{K}'} = \mathfrak{M},$$

and $(A\mathfrak{H})^- = \mathfrak{K}$. Assume that we can find a nonzero operator $Z\colon \mathfrak{K}' \to \mathfrak{M}$ such that $ZT_2 = (T|\mathfrak{M})Z$. Then the operator $B \in \{T\}'$ defined by $B = ZP_{\mathfrak{K}'}$ is such that $AB = O$ and

$$(BA\mathfrak{H})^- = (ZP_{\mathfrak{K}'}Y\mathfrak{K}')^- = (ZP_{\mathfrak{K}'}\mathfrak{K})^- = (Z\mathfrak{K}')^- \neq \{0\},$$

so that A and B do not commute. Thus, to conclude the proof, it suffices to produce such an operator Z. Because $\mathfrak{M} \neq \{0\}$, $T|\mathfrak{M}$ has a nonzero cyclic subspace $\mathfrak{M}_1$, and it would suffice to find a nonzero operator $Z\colon \mathfrak{K}' \to \mathfrak{M}_1$ such that $ZT_2 = (T|\mathfrak{M})Z$. Set now $\varphi = m_T = m_{T_2}$, and $\varphi' = m_{T|\mathfrak{M}_1}$, so that $T|\mathfrak{M}_1 \sim S(\varphi')$ and $T_2 \sim S(\varphi)$. The operator $R = P_{\mathfrak{H}(\varphi')}|\mathfrak{H}(\varphi)$ is not zero, and $RS(\varphi) = S(\varphi')R$. The desired operator Z can now be constructed by composing R with the appropriate quasi-affinities.

5 Jordan models

1. We have seen in the preceding section that multiplicity-free operators of class C_0 are uniquely determined, up to quasi-similarity, by their minimal functions. We have thus a complete classification of multiplicity-free operators of class C_0 up to quasi-similarity. In this section we extend this classification to all operators of class C_0 acting on a separable space.

Definition. Let $\Phi = \{\varphi_j : j \geq 0\} \subset H^\infty$ be a sequence of inner functions such that $\varphi_{j+1}|\varphi_j$ for all $j \geq 0$. The operator

$$S(\Phi) = \bigoplus_{j=0}^{\infty} S(\varphi_j)$$

is called a *Jordan operator*.

Note that some of the functions φ_j in the above definition may be constant. If this happens, the Jordan operator $S(\Phi)$ is unitarily equivalent to $\bigoplus_{j=0}^{k-1} S(\varphi_j)$, where k is the first integer such that φ_k is a constant function. Clearly, a Jordan operator is of class C_0, and $m_{S(\Phi)} = \varphi_0$.

Proposition 5.1. *For every operator T of class C_0 acting on a separable Hilbert space, there exists a Jordan operator $S(\Phi)$ such that $S(\Phi) \prec T$.*

Proof. Assume that T acts on the separable space $\mathfrak{H}$, choose a dense sequence $\{h_n : n \geq 0\}$ in $\mathfrak{H}$, and let $\{k_n : n \geq 0\}$ be a sequence in which each h_i is repeated infinitely many times. We inductively construct vectors $f_0, f_1, f_2, \dots$ in $\mathfrak{H}$, and invariant subspaces $\mathfrak{M}_{-1}, \mathfrak{M}_0, \mathfrak{M}_1, \dots$ for T with the following properties:

(1) $\mathfrak{M}_{-1} = \mathfrak{H}$;
(2) $f_j \in \mathfrak{M}_{j-1}, m_{f_j} = m_{T|\mathfrak{M}_{j-1}}$;
(3) $\mathfrak{K}_j \vee \mathfrak{M}_j = \mathfrak{M}_{j-1}, \mathfrak{K}_j \cap \mathfrak{M}_j = \{0\}$, where $\mathfrak{K}_j = \bigvee\{T^n f_j : j \geq 0\}$;
(4) $\|k_j - P_{\mathfrak{K}_0 \vee \mathfrak{K}_1 \vee \cdots \vee \mathfrak{K}_j} k_j\| \leq 2^{-j}$

for $j = 0, 1, 2 \dots$. Assume, indeed that f_j and $\mathfrak{M}_j$ have already been defined for $j < n$, and let us construct f_n and $\mathfrak{M}_n$. (Note that if $n = 0$, only $\mathfrak{M}_{-1}$ needs to be constructed, and there is no f_{-1}.) A repeated application of (3) yields

$$\begin{aligned} \mathfrak{H} &= \mathfrak{M}_{-1} = \mathfrak{K}_0 \vee \mathfrak{M}_0 = \mathfrak{K}_0 \vee \mathfrak{K}_1 \vee \mathfrak{M}_1 = \cdots \\ &= \mathfrak{K}_0 \vee \mathfrak{K}_1 \vee \cdots \vee \mathfrak{K}_{n-1} \vee \mathfrak{M}_{n-1}, \end{aligned}$$

so that we can find vectors $u_n \in \mathfrak{K}_0 \vee \mathfrak{K}_1 \vee \cdots \vee \mathfrak{K}_{n-1}$ and $v_n \in \mathfrak{M}_{n-1}$ such that

$$\|k_n - u_n - v_n\| \leq 2^{-n-1}.$$

By Theorem 1.5, we can then find a vector $f_n \in \mathfrak{M}_{n-1}$ such that $m_{f_n} = m_{T|\mathfrak{M}_{n-1}}$ (in other words, f_n is a $T|\mathfrak{M}_{n-1}$-maximal vector), and

$$\|v_n - f_n\| \leq 2^{-n-1}.$$

An application of Theorem 4.10 to the operator $T|\mathfrak{M}_{n-1}$ proves the existence of an invariant subspace $\mathfrak{M}_n$ satisfying (3) for $j = n$. It remains to verify (4) for $j = n$, and this follows because

$$\begin{aligned}\|k_n - P_{\mathfrak{K}_0\vee\mathfrak{K}_1\vee\cdots\vee\mathfrak{K}_n}k_n\| &\le \|k_n - u_n - f_n\| \\ &\le \|k_n - u_n - v_n\| + \|v_n - f_n\| \le 2^{-n}.\end{aligned}$$

Thus the existence of the vectors f_j and of the spaces $\mathfrak{M}_j$ is established by induction. A useful consequence of (4) is that

$$\mathfrak{H} = \bigvee_{j=0}^{\infty} \mathfrak{K}_j.$$

Indeed, this follows from the equality

$$\lim_{n\to\infty} \operatorname{dist}\left(k_n, \bigvee_{j=0}^{\infty} \mathfrak{K}_j\right) = 0$$

and the fact that each h_i is repeated infinitely many times among the k_n, so that $h_i \in \bigvee_{j=0}^{\infty} \mathfrak{K}_j$ for all i.

We define now $\Phi = \{\varphi_j : j \ge 0\}$ by setting $\varphi_j = m_{f_j}$. Relation (4), and the fact that $\mathfrak{M}_{j+1} \subset \mathfrak{M}_j$, easily imply that $\varphi_{j+1}|\varphi_j$ for all j, and hence $S(\Phi)$ is a Jordan operator. We now prove that $S(\Phi) \prec T$. The operator $T|\mathfrak{K}_j$ is multiplicity-free with minimal function φ_j. Therefore Proposition 4.1 implies the existence of a quasi-affinity X_j such that $XS(\varphi_j) = (T|\mathfrak{K}_j)X_j$. We can then define an operator X satisfying $XS(\Phi) = TX$ by the formula

$$X\left(\bigoplus_{j=1}^{\infty} g_j\right) = \sum_{j=0}^{\infty} \frac{2^{-j}}{\|X_j\|} X_j g_j \quad \text{for } \bigoplus_{j=0}^{\infty} g_j \in \bigoplus_{j=0}^{\infty} \mathfrak{H}(\varphi_j).$$

The reader will verify without difficulty that X is bounded. The range of X_j is dense in $\mathfrak{K}_j$, and the spaces $\mathfrak{K}_j$ span $\mathfrak{H}$, thus X has dense range. To prove that X is one-to-one, suppose that

$$g = \bigoplus_{j=0}^{\infty} g_j \in \ker X,\ g \ne 0,$$

and n is the first integer such that $g_n \ne 0$. By the definition of X, we have

$$\frac{X_n}{\|X_n\|} g_n = -\sum_{j=1}^{\infty} \frac{2^{-j}}{\|X_{n+j}\|} X_{n+j} g_{n+j}.$$

Thus $X_n g_n$, a nonzero element of $\mathfrak{K}_n$, belongs to $\bigvee_{j=1}^{\infty} \mathfrak{K}_{n+j} \subset \mathfrak{M}_n$. By (3), we must have $X_n g_n = 0$, and this contradiction implies that X is one-to-one. We thus determined a quasi-affinity X such that $TX = XS(\Phi)$, and this concludes the proof.

Corollary 5.2. *For every operator of class C_0 acting on a separable Hilbert space, there exists a Jordan operator $S(\Phi')$ such that $T \prec S(\Phi')$.*

Proof. Proposition 5.1, applied to T^*, shows the existence of a Jordan operator $S(\Psi)$, $\Psi = \{\psi_j : j \geq 0\}$, such that $S(\Psi) \prec T^*$. We have then $T \prec S(\Psi)^*$, and $S(\Psi)^*$ is unitarily equivalent to the Jordan operator $S(\Phi')$, where $\Phi' = \{\tilde{\psi_j} : j \geq 0\}$.

2. In order to complete the classification theorem, we prove that the operators $S(\Phi)$ and $S(\Phi')$ constructed above are necessarily identical. If T is an operator acting on $\mathfrak{H}$, and n is a natural number, we denote by $T^{(n)}$ the orthogonal sum of n copies of T acting on the orthogonal sum $\mathfrak{H}^{(n)}$ of n copies of $\mathfrak{H}$.

Lemma 5.3. *Let n and k be natural numbers, and φ a nonconstant inner function. If there exists an injective operator $X\colon \mathfrak{H}(\varphi)^{(k)} \to \mathfrak{H}(\varphi)^{(n)}$ such that $XS(\varphi)^{(k)} = S(\varphi)^{(n)}X$, then $k \leq n$.*

Proof. The operator X is represented by a matrix $[X_{ij}]_{1\leq i\leq n, 1\leq j\leq k}$ in the sense that

$$X\left(\bigoplus_{j=1}^{k} h_j\right) = \bigoplus_{i=1}^{n}\left(\sum_{j=1}^{k} X_{ij}h_j\right) \quad \text{for } \bigoplus_{j=1}^{k} h_j \in \mathfrak{H}(\varphi)^{(k)}.$$

The condition $XS(\varphi)^{(k)} = S(\varphi)^{(n)}X$ implies that the operators X_{ij} commute with $S(\varphi)$. By Theorem 2.10, we have

$$X_{ij}h = P_{\mathfrak{H}(\varphi)}(a_{ij}h), \quad h \in \mathfrak{H}(\varphi),$$

where $a_{ij} \in H^\infty$ for all i, j. Now, the operator X is one-to-one and, in particular, it is not zero. Therefore φ cannot divide all the functions a_{ij}. There exists then a minor of maximal rank of the matrix $[a_{ij}]_{i,j}$ that is not divisible by φ, and there is no loss of generality in assuming that this minor is $|a_{ij}|_{1\leq i,j\leq r}$, with $r \leq \min\{k, n\}$. Assuming now that $k > n$, consider the determinant

$$\det\begin{bmatrix} a_{11} & a_{12} & \cdots & a_{1r} & a_{1,r+1} \\ a_{21} & a_{22} & \cdots & a_{2r} & a_{2,r+1} \\ \vdots & \vdots & \ddots & \vdots & \vdots \\ a_{r1} & a_{r2} & \cdots & a_{rr} & a_{r,r+1} \\ x_1 & x_2 & \cdots & x_r & x_{r+1} \end{bmatrix} = \sum_{j=1}^{r+1} x_j u_j.$$

The sum $\sum_{j=1}^{r+1} a_{ij}u_j$ is zero if $1 \leq i \leq r$, and it equals a minor of order $r+1$ if $i > r$; therefore all of these sums are divisible by φ. We deduce that the vector $h = \bigoplus_{j=1}^{k} h_j \in \mathfrak{H}(\varphi)^{(k)}$ defined by

$$h_j = \begin{cases} P_{\mathfrak{H}(\varphi)}u_j & \text{for } 1 \leq j \leq r+1 \\ 0 & \text{for } j > r+1, \end{cases}$$

satisfies the relation $Xh = 0$. The injectivity of X implies that $h = 0$. In particular, $P_{\mathfrak{H}(\varphi)}u_{r+1} = 0$, or $u_{r+1} \in \varphi H^2$. However, the function $u_{r+1} = \det[a_{ij}]_{1\leq i,j\leq r}$ was chosen not to be divisible by φ. This contradiction shows that necessarily $k \leq n$, thus concluding the proof.

Corollary 5.4. *Let n and k be natural numbers, and φ a nonconstant inner function. If there exists an operator $X\colon \mathfrak{H}(\varphi)^{(k)} \to \mathfrak{H}(\varphi)^{(n)}$ with dense range such that $XS(\varphi)^{(k)} = S(\varphi)^{(n)}X$, then $k \geq n$.*

Proof. The operator X^* is one-to-one, and $X^*S(\varphi)^{*(n)} = S(\varphi)^{*(k)}X^*$. Because $S(\varphi)^*$ is unitarily equivalent to $S(\varphi^{\sim})$, the corollary follows immediately from Lemma 5.3.

We recall that, given an integer $N \geq 1$, an operator T of class C_0 belongs to $C_0(N)$ if $\mathfrak{d}_{T^*} = N$.

Lemma 5.5. *Let T be an operator of class $C_0(N)$ on $\mathfrak{H}$ with minimal function φ. There exists a surjective operator $X\colon \mathfrak{H}(\varphi)^{(N)} \to \mathfrak{H}$ such that $XS(\varphi)^{(N)} = TX$.*

Proof. The minimal isometric dilation of T is a unilateral shift of multiplicity N. We may assume without loss of generality that $\mathfrak{H} \subset (H^2)^{(N)}$ and $T^* = S^{*(N)}|\mathfrak{H}$. We have

$$O = \varphi^{\sim}(T^*) = \varphi^{\sim}(S^{*(N)})|\mathfrak{H},$$

so that

$$\mathfrak{H} \subset \ker \varphi^{\sim}(S^{*(N)}) = \mathfrak{H}(\varphi)^{(N)}.$$

We simply define then $X = P_{\mathfrak{H}}|\mathfrak{H}(\varphi)^{(N)}$.

Proposition 5.6. *Let $\varphi_0, \varphi_1, \ldots, \varphi_{n-1}$ be inner functions with a nonconstant common inner divisor φ. The cyclic multiplicity of the operator $T = \bigoplus_{j=0}^{n-1} S(\varphi_j)$ equals n.*

Proof. Each $S(\varphi_j)$ has a cyclic vector, and hence the cyclic multiplicity of their direct sum is at most n. On the other hand, by Proposition 2.5(3) we have

$$P_{\mathfrak{H}(\varphi)^{(n)}}T = S(\varphi)^{(n)}P_{\mathfrak{H}(\varphi)^{(n)}},$$

and this implies that the multiplicity of $\bigoplus_{j=0}^{n-1} S(\varphi_j)$ is at least equal to the multiplicity of $S(\varphi)^{(n)}$. Thus, it suffices to prove that this last operator has multiplicity $\geq n$. Set $N = \mu_{S(\varphi)^{(n)}}$, and use Theorem 3.5 to find an operator T of class $C_0(N)$ such that $T \prec S(\varphi)^{(n)}$. Fix a quasi-affinity Y satisfying $YT = S(\varphi)^{(n)}Y$. Next observe that $m_T = \varphi$, and Lemma 5.5 provides a surjective operator X such that $XS(\varphi)^{(N)} = TX$. We now have

$$(YX)S(\varphi)^{(N)} = S(\varphi)^{(n)}(YX),$$

and YX has dense range. The inequality $N \geq n$ follows from Corollary 5.4.

We are now ready for the classification theorem.

Theorem 5.7. *Let T be an operator of class C_0 acting on a separable Hilbert space. There exists a Jordan operator $S(\Phi)$ such that $T \sim S(\Phi)$. Moreover, $S(\Phi)$ is uniquely determined by either $S(\Phi) \prec T$ or $T \prec S(\Phi)$.*

The operator $S(\Phi)$ is called the *Jordan model* of T.

Proof. By Proposition 5.1 and Corollary 5.2, there exist Jordan operators $S(\Phi)$ and $S(\Phi')$ such that $S(\Phi) \prec T \prec S(\Phi')$; in particular, $S(\Phi) \prec S(\Phi')$. It suffices then to prove that the relation $S(\Phi) \prec S(\Phi')$ between two Jordan operators implies $S(\Phi) = S(\Phi')$. Assume therefore that $S(\Phi')X = XS(\Phi)$ for some quasi-affinity X. If u is an arbitrary function in H^∞, then clearly

$$[X\operatorname{ran} u(S(\Phi))]^- = [\operatorname{ran} u(S(\Phi'))X]^- = [\operatorname{ran} u(S(\Phi'))]^-,$$

so that

$$X|[\operatorname{ran} u(S(\Phi))]^- : [\operatorname{ran} u(S(\Phi))]^- \to [\operatorname{ran} u(S(\Phi'))]^-$$

is a quasi-affinity intertwining the restrictions of the two Jordan operators to these invariant subspaces. By Corollary 2.7, the operators

$$S(\Phi)|[\operatorname{ran} u(S(\Phi))]^- \text{ and } S(\Phi')|[\operatorname{ran} u(S(\Phi'))]^-$$

are unitarily equivalent to

$$A = \bigoplus_{j=0}^{\infty} S\left(\frac{\varphi_j}{u \wedge \varphi_j}\right) \text{ and } A' = \bigoplus_{j=0}^{\infty} S\left(\frac{\varphi'_j}{u \wedge \varphi'_j}\right),$$

respectively. We conclude that $A \prec A'$, and hence $\mu_{A'} \le \mu_A$. When $u = \varphi_n$, we have

$$A = \bigoplus_{j=0}^{n-1} S(\varphi_j/\varphi_n),$$

and therefore $\mu_{A'} \le n$. Proposition 5.6 implies in particular that the nth summand in A' must be trivial. We deduce that $\varphi'_n = \varphi'_n \wedge \varphi_n$ and hence $\varphi'_n|\varphi_n$ for all n. To conclude the proof, it suffices to show that φ_n also divides φ'_n. But we have $S(\Phi')^* \prec S(\Phi)^*$, and $S(\Phi')^*, S(\Phi)^*$ are unitarily equivalent to the Jordan operators $\bigoplus_{j=0}^{\infty} S(\varphi'^{\sim}_j), \bigoplus_{j=0}^{\infty} S(\varphi^{\sim}_j)$. By the first part of the argument we deduce that $\varphi^{\sim}_n|\varphi'^{\sim}_n$ for all n, and this is equivalent to $\varphi_n|\varphi'_n$.

6 The quasi-equivalence of matrices over H^∞

1. It is well known that the classical theorem of Jordan, concerning the classification of linear transformations on a finite-dimensional space, can be obtained as a consequence of a diagonalization theorem for polynomial matrices. One may ask whether the classification theorem for operators of class C_0 can be proved in a similar fashion. We show that this is indeed the case for operators of class C_0 with finite defect indices, and this follows from a diagonalization theorem for matrices over H^∞.

Let $\mathfrak{F}$ be a separable Hilbert space, and $\{\mathfrak{F}, \mathfrak{F}, \Theta(\lambda)\}$ a bounded analytic function. Let $\{e_j : 0 \le j < \dim(\mathfrak{F})\}$ be an orthonormal basis in $\mathfrak{F}$. With respect to this basis, $\Theta(\lambda)$ is represented by a matrix $[\theta_{ij}(\lambda)]_{0\le i,j<\dim(\mathfrak{F})}$, with $\theta_{ij} \in H^\infty$. A minor of order k, $1 \le k < \dim(\mathfrak{F})$, of Θ is a determinant of the form

$$\det \begin{bmatrix} \theta_{i_1 j_1} & \theta_{i_1 j_2} & \cdots & \theta_{i_1 j_k} \\ \theta_{i_2 j_1} & \theta_{i_2 j_2} & \cdots & \theta_{i_2 j_k} \\ \vdots & \vdots & \ddots & \vdots \\ \theta_{i_k j_1} & \theta_{i_k j_2} & \cdots & \theta_{i_k j_k} \end{bmatrix},$$

where $0 \le i_1 < i_2 < \cdots < i_k < \dim(\mathfrak{F})$ and $0 \le j_1 < j_2 < \cdots < j_k < \dim(\mathfrak{F})$.

Definition. Assume that not all minors of order k of Θ are identically zero. then $\mathscr{D}_k(\Theta)$ is defined as the greatest common inner divisor of all minors of order k of Θ. If all minors of order k are equal to zero, we set $\mathscr{D}_k(\Theta) = 0$.

Observe that each minor of order $k+1$ (with $k+1 \le \dim(\mathfrak{F})$) is a linear combination, with coefficients in H^∞, of minors of order k. Thus, if $\mathscr{D}_{k+1}(\Theta) \ne 0$, then $\mathscr{D}_k(\Theta) \ne 0$, and $\mathscr{D}_k(\Theta)$ divides $\mathscr{D}_{k+1}(\Theta)$.

Definition. The *invariant factors* of Θ are defined as follows:

$$\mathscr{E}_k(\Theta) = \begin{cases} \mathscr{D}_1(\Theta) & \text{if } k = 1, \\ \mathscr{D}_k(\Theta)/\mathscr{D}_{k-1}(\Theta) & \text{if } 2 \le k \le \dim(\mathfrak{F}) \text{ and } \mathscr{D}_k(\Theta) \ne 0, \\ 0 & \text{if } 2 \le k \le \dim(\mathfrak{F}) \text{ and } \mathscr{D}_k(\Theta) = 0. \end{cases}$$

Next, we introduce the equivalence relation between matrices over H^∞ which allows us to prove a diagonalization theorem. Let us recall that a bounded analytic function $\{\mathfrak{F}, \mathfrak{F}, \Phi(\lambda)\}$ has a scalar multiple $\varphi \in H^\infty$ if there exists a bounded analytic function $\{\mathfrak{F}, \mathfrak{F}, \Phi'(\lambda)\}$ satisfying the relations

$$\Phi'(\lambda)\Phi(\lambda) = \Phi(\lambda)\Phi'(\lambda) = \varphi(\lambda) I_{\mathfrak{F}} \quad (\lambda \in D).$$

Definition. Let $\{\mathfrak{F}, \mathfrak{F}, \Theta_1(\lambda)\}$ and $\{\mathfrak{F}, \mathfrak{F}, \Theta_2(\lambda)\}$ be bounded analytic functions, and $\omega \in H^\infty$ an inner function. Then Θ_1 and Θ_2 are said to be *ω-equivalent* if there exist bounded analytic functions $\{\mathfrak{F}, \mathfrak{F}, \Phi(\lambda)\}$, $\{\mathfrak{F}, \mathfrak{F}, \Psi(\lambda)\}$ with scalar multiples φ, ψ, respectively, such that

$$\varphi \wedge \omega = \psi \wedge \omega = 1,$$

and

$$\Phi(\lambda)\Theta_1(\lambda) = \Theta_2(\lambda)\Psi(\lambda) \quad (\lambda \in D).$$

The functions $\{\mathfrak{F}, \mathfrak{F}, \Theta_1(\lambda)\}$ and $\{\mathfrak{F}, \mathfrak{F}, \Theta_2(\lambda)\}$ are said to be *quasi-equivalent* if they are ω-equivalent for every inner function $\omega \in H^\infty$.

The fact that ω-equivalence (and hence quasi-equivalence) is reflexive and transitive is obvious. Symmetry is proved as follows. Let $\Phi, \Psi, \varphi, \psi$ be as in the above definition, and let Φ', Ψ' satisfy

$$\Phi\Phi' = \Phi'\Phi = \varphi I, \quad \Psi\Psi' = \Psi'\Psi = \psi I.$$

The functions $\{\mathfrak{F},\mathfrak{F},\varphi(\lambda)\Psi'(\lambda)\}$ and $\{\mathfrak{F},\mathfrak{F},\psi(\lambda)\Phi'(\lambda)\}$ both have the scalar multiple $\varphi\psi$ relatively prime to ω, and

$$(\psi\Phi')\Theta_2 = \Phi'\Theta_2\psi = \Phi'\Theta_2\Psi\Psi' = \Phi'\Phi\Theta_1\Psi' = \Theta_1(\varphi\Psi').$$

This shows that $\{\mathfrak{F},\mathfrak{F},\Theta_2(\lambda)\}$ is ω-equivalent to $\{\mathfrak{F},\mathfrak{F},\Theta_1(\lambda)\}$. Thus both ω-equivalence and quasi-equivalence are indeed equivalence relations.

Lemma 6.1. *Let the functions $\Theta_1,\Theta_2,\Phi,\Phi'\Psi,\Psi',\varphi$, and ψ satisfy the relations*

$$\Phi\Theta_1 = \Theta_2\Psi, \quad \Phi'\Phi = \Phi\Phi' = \varphi I, \text{ and } \Psi'\Psi = \Psi\Psi' = \psi I.$$

For every integer k, $1 \le k \le \dim(\mathfrak{F})$, we have

$$\mathscr{D}_k(\Theta_1)|\psi_0^k\mathscr{D}_k(\Theta_2), \quad \mathscr{D}_k(\Theta_2)|\varphi_0^k\mathscr{D}_k(\Theta_1),$$

where φ_0,ψ_0 denote the inner factors of φ,ψ, respectively.

Proof. Observe that

$$\varphi\Theta_1 = \Phi'\Phi\Theta_1 = \Phi'\Theta_2\Psi,$$

and clearly $\mathscr{D}_k(\varphi\Theta_1) = \varphi_0^k\mathscr{D}_k(\Theta_1)$. If we show that

$$\mathscr{D}_k(\Theta_2)|\mathscr{D}_k(\Phi'\Theta_2\Psi), \tag{6.1}$$

then we obtain the relation $\mathscr{D}_k(\Theta_2)|\varphi_0^k\mathscr{D}_k(\Theta_1)$. Now, to prove (6.1) it suffices to prove the general fact that $\mathscr{D}_k(A)|\mathscr{D}_k(AB)$ whenever $\{\mathfrak{F},\mathfrak{F},A(\lambda)\}$ and $\{\mathfrak{F},\mathfrak{F},\mathfrak{B}(\lambda)\}$ are bounded analytic functions. When $\dim(\mathfrak{F})$ is finite, this divisibility is obvious. Indeed, each minor of order k of AB is a finite sum of terms, each term being the product of a minor of order k of A with a minor of order k of B. If $\mathfrak{F}$ is infinite-dimensional, let us denote by P_n the orthogonal projection of $\mathfrak{F}$ onto the space generated by $\{e_1,e_2,\dots,e_n\}$. Then clearly

$$\mathscr{D}_k(A)|\mathscr{D}_k(P_nAP_n)|\mathscr{D}_k(P_nAP_nBP_n) \quad (n \ge 1),$$

by the finite-dimensional case. Now, each minor of AB is the pointwise limit of the corresponding minors in $P_nAP_nBP_n$ as $n \to \infty$. It follows easily that $\mathscr{D}_k(A)$ divides each minor of order k of AB, and therefore $\mathscr{D}_k(A)|\mathscr{D}_k(AB)$, as desired. The relation $\mathscr{D}_k(\Theta_1)|\psi_0^k\mathscr{D}_k(\Theta_2)$ is proved in an analogous manner.

Corollary 6.2. *If Θ_1 and Θ_2 are quasi-equivalent, then $\mathscr{D}_k(\Theta_1) = \mathscr{D}_k(\Theta_2)$ for every integer k, $1 \le k \le \dim(\mathfrak{F})$.*

Proof. Choose $\omega = \mathscr{D}_k(\Theta_1)\mathscr{D}_k(\Theta_2)$ and apply Lemma 6.1 to the functions provided by ω-equivalence.

2. The following result shows the relationship between quasi-equivalence and quasi-similarity. We are using the fact that $\mathfrak{D}_T = \mathfrak{D}_{T^*}$ if T is of class C_{00}, and thus Θ_T coincides with a function of the form $\{\mathfrak{F},\mathfrak{F},\Theta(\lambda)\}$.

Proposition 6.3. *Let T_1 and T_2 be operators of class C_0, and assume that the characteristic function of T_i coincides with $\{\mathfrak{F}, \mathfrak{F}, \Theta_i(\lambda)\}$ for $i = 1,2$. If Θ_1 and Θ_2 are quasi-equivalent then T_1 and T_2 are quasi-similar.*

Proof. We assume without loss of generality that

$$T_i = U_+^* | (H^2(\mathfrak{F}) \ominus \Theta_i H^2(\mathfrak{F})) \quad (i = 1,2),$$

where U_+ denotes the unilateral shift on $H^2(\mathfrak{F})$. Let θ_i denote the minimal function of T_i, and set $\omega = \theta_1 \theta_2$. By hypothesis, we can find bounded operator-valued analytic functions Φ, Φ', Ψ, Ψ', and scalar functions $\varphi, \psi \in H^\infty$, such that

$$\Phi\Phi' = \Phi'\Phi = \varphi I, \quad \Psi\Psi' = \Psi'\Psi = \psi I, \quad \Phi\Theta_1 = \Theta_2\Psi,$$

and $\varphi \wedge \omega = \psi \wedge \omega = 1$. We can then define a bounded linear transformation

$$X \colon H^2(\mathfrak{F}) \ominus \Theta_1 H^2(\mathfrak{F}) \to H^2(\mathfrak{F}) \ominus \Theta_2 H^2(\mathfrak{F})$$

by

$$Xu = P_2 \Phi u \quad \text{for } u \in H^2(\mathfrak{F}) \ominus \Theta_1 H^2(\mathfrak{F}),$$

where P_2 denotes the orthogonal projection onto $H^2(\mathfrak{F}) \ominus \Theta_2 H^2(\mathfrak{F})$. Because

$$\Phi\Theta_1 H^2(\mathfrak{F}) = \Theta_2 \Psi H^2(\mathfrak{F}) \subset \Theta_2 H^2(\mathfrak{F}),$$

we see that X satisfies the relation $XT_1 = T_2 X$. In an analogous manner, the operator

$$Y \colon H^2(\mathfrak{F}) \ominus \Theta_2 H^2(\mathfrak{F}) \to H^2(\mathfrak{F}) \ominus \Theta_1 H^2(\mathfrak{F})$$

defined by

$$Yv = P_1(\psi \Phi' v), \quad v \in H^2(\mathfrak{F}) \ominus \Theta_2 H^2(\mathfrak{F}),$$

satisfies $YT_2 = T_1 Y$ because

$$\begin{aligned} \psi\Phi'\Theta_2 H^2(\mathfrak{F}) &= \Phi'\Theta_2 \psi H^2(\mathfrak{F}) \\ &= \Phi'\Theta_2 \Psi\Psi' H^2(\mathfrak{F}) \\ &= \Phi'\Phi\Theta_1 \Psi' H^2(\mathfrak{F}) \\ &= \varphi\Theta_1 \Psi' H^2(\mathfrak{F}) \\ &\subset \Theta_1 H^2(\mathfrak{F}). \end{aligned}$$

Moreover, for $v \in H^2(\mathfrak{F}) \ominus \Theta_2 H^2(\mathfrak{F})$ and $u \in H^2(\mathfrak{F}) \ominus \Theta_1 H^2(\mathfrak{F})$ we have

$$XYv = P_2(\Phi\psi\Phi' v) = P_2(\varphi\psi v) = (\varphi\psi)(T_2)v$$

and

$$YXu = P_1(\psi\Phi'\Phi u) = P_1(\varphi\psi u) = (\varphi\psi)(T_1)u.$$

Thus $XY = (\varphi\psi)(T_2)$ and $YX = (\varphi\psi)(T_1)$, and both of these operators are quasi-affinities because $(\varphi\psi) \wedge \theta_1 = (\varphi\psi) \wedge \theta_2 = 1$. We conclude that X and Y are also quasi-affinities, thus establishing the quasi-similarity of T_1 and T_2.

3. Proposition 6.3 allows us to calculate the Jordan model of a given C_0-contraction directly from its characteristic function. This is achieved by proving a diagonalization theorem relative to quasi-equivalence. For this purpose, we first introduce some notation.

Given an integer k, $1 \leq k < \dim(\mathfrak{F})$, we denote by $\mathfrak{F}_k$ the subspace of $\mathfrak{F}$ generated by $\{e_j : j \geq k\}$. Thus $\dim(\mathfrak{F} \ominus \mathfrak{F}_k) = k$. If $\theta_1, \theta_2, \dots, \theta_k$ are functions in H^∞, and $\{\mathfrak{F}_k, \mathfrak{F}_k, \Theta_0(\lambda)\}$ is a bounded analytic function, then we can define a bounded analytic function $\{\mathfrak{F}, \mathfrak{F}, \Theta(\lambda)\}$ such that

$$\Theta(\lambda)e_j = \begin{cases} \theta_{j+1}(\lambda)e_j, & 0 \leq j \leq k-1, \\ \Theta_0(\lambda)e_j, & j \geq k. \end{cases}$$

We use the notation $\Theta = \mathrm{Diag}(\theta_1, \theta_2, \dots, \theta_k, \Theta_0)$ for this function.

The key step in the diagonalization process is as follows.

Lemma 6.4. *Let $\{\mathfrak{F}, \mathfrak{F}, \Theta(\lambda)\}$ be a bounded analytic function, and let $\omega \in H^\infty$ be an inner function. There exists a bounded analytic function $\{\mathfrak{F}_1, \mathfrak{F}_1, \Theta_1(\lambda)\}$ such that Θ is ω-equivalent to* $\mathrm{Diag}(\mathscr{D}_1(\Theta), \Theta_1)$, *and $\mathscr{D}_1(\Theta) | \mathscr{D}_1(\Theta_1)$. We have $\Theta_1 = O$ if and only if $\mathscr{D}_2(\Theta) = 0$.*

Proof. The lemma is trivial when $\Theta = O$, so we assume that $\Theta \neq O$. It is also clear that, provided that ω' is an inner multiple of ω, ω'-equivalence implies ω-equivalence. Therefore there is no loss of generality in assuming that $\mathscr{D}_1(\Theta)^2 | \omega$; simply replace ω by $\mathscr{D}_1(\Theta)^2\omega$. Let $[\theta_{ij}]_{1 \leq i,j < \dim(\mathfrak{F})}$ be the matrix of Θ. An application of Theorem 2.9 provides a sequence $\{\alpha_j : 0 \leq j < \dim(\mathfrak{F})\}$ of scalars such that

$$\alpha_0 \neq 0, \qquad \sum_{0 \leq j < \dim(\mathfrak{F})} |\alpha_j| < \infty, \tag{6.2}$$

and

$$\left(\sum_{0 \leq j < \dim(\mathfrak{F})} \alpha_j \theta_{ij} \right) \wedge \omega = \left(\bigwedge_{0 \leq j < \dim(\mathfrak{F})} \theta_{ij} \right) \wedge \omega \tag{6.3}$$

for $0 \leq i < \dim(\mathfrak{F})$. In order to realize simultaneously the conditions in (6.3), we must use the fact that a countable intersection of dense G_δ sets is also dense (Baire's theorem). If we set

$$\theta_i = \sum_{0 \leq j < \dim(\mathfrak{F})} \alpha_j \theta_{ij},$$

we have

$$\begin{aligned}\left(\bigwedge_{0\le i<\dim(\mathfrak{F})}\theta_i\right)\wedge\omega &= \bigwedge_{0\le i<\dim(\mathfrak{F})}(\theta_i\wedge\omega)\\ &= \bigwedge_{0\le i<\dim(\mathfrak{F})}\left[\left(\bigwedge_{0\le j<\dim(\mathfrak{F})}\theta_{ij}\right)\wedge\omega\right]\\ &= \left(\bigwedge_{0\le i,j<\dim(\mathfrak{F})}\theta_{ij}\right)\wedge\omega\\ &= \mathscr{D}_1(\Theta)\wedge\omega=\mathscr{D}_1(\Theta).\end{aligned}$$

because of the assumption that $\mathscr{D}_1(\Theta)^2|\omega$. One further application of Theorem 2.9 provides a sequence $\{\beta_j : 0\le j<\dim(\mathfrak{F})\}$ such that

$$\beta_0\neq 0,\qquad \sum_{0\le i<\dim(\mathfrak{F})}|\beta_i|<\infty, \tag{6.4}$$

and

$$\left(\sum_{0\le i<\dim(\mathfrak{F})}\beta_i\theta_i\right)\wedge\omega=\mathscr{D}_1(\theta). \tag{6.5}$$

We now construct boundedly invertible operators A,B on $\mathfrak{F}$ with matrices given by

$$a_{ij}=\begin{cases}\alpha_j & \text{if } i=0\\ \delta_{ij} & \text{if } i\ge 1\end{cases}\qquad b_{ij}=\begin{cases}\beta_i & \text{if } j=0\\ \delta_{ij} & \text{if } j\ge 1\end{cases}\qquad (0\le i,j<\dim(\mathfrak{F})),$$

where, as usual, $\delta_{ii}=1$ and $\delta_{ij}=0$ for $i\neq j$. Thus the function Θ is trivially quasi-equivalent to $\Theta'=B\Theta A=[\theta'_{ij}]$, and we have

$$\theta'_{00}=\sum_{0\le i<\dim(\mathfrak{F})}\beta_i\theta_i,$$

so that $\theta'_{00}\wedge\omega=\mathscr{D}_1(\Theta)=\mathscr{D}_1(\Theta')$ by (6.5). We have then

$$(\theta'_{00}/\mathscr{D}_1(\Theta))\wedge(\omega/\mathscr{D}_1(\Theta))=1, \tag{6.6}$$

and because $\mathscr{D}_1(\Theta)|(\omega/\mathscr{D}_1(\Theta))$, we also have

$$(\theta'_{00}/\mathscr{D}_1(\Theta))\wedge\mathscr{D}_1(\Theta)=1. \tag{6.7}$$

Relations (6.6) and (6.7) imply that $(\theta'_{00}/\mathscr{D}_1(\Theta))\wedge\omega=1$. We can therefore write

$$\theta'_{00}=\varphi\mathscr{D}_1(\Theta),\quad \varphi\wedge\omega=1.$$

We perform next an ω-equivalence to remove the factor φ from θ'_{00}. Let us set $\Theta'' = [\theta''_{ij}]_{0\le i,j<\dim(\mathfrak{F})}$, where

$$\theta''_{ij} = \begin{cases} \mathscr{D}_1(\Theta) & \text{if } i=j=0, \\ \theta'_{ij} & \text{if } ij=0 \text{ but } i\neq j, \\ \varphi\theta'_{ij} & \text{if } ij\neq 0. \end{cases}$$

We also set $\Phi = \mathrm{Diag}(1,\varphi I_{\mathfrak{F}_1})$ and $\Psi = \mathrm{Diag}(\varphi, I_{\mathfrak{F}_1})$. Both Φ and Ψ have the scalar multiple φ, and clearly $\Phi\Theta' = \Theta''\Psi$, so that Θ and Θ'' are ω-equivalent. Note that $\mathscr{D}_1(\Theta)$ continues to divide all the entries of Θ''. The final step of the proof eliminates the entries θ''_{i0} and θ''_{0i} for $i\neq 0$. We define two boundedly invertible matrices $\Phi_1 = [\varphi_{ij}]_{0\le i,j<\dim(\mathfrak{F})}$ and $\Psi_1 = [\psi_{ij}]_{0\le i,j<\dim(\mathfrak{F})}$ as follows.

$$\varphi_{ij} = \begin{cases} -\theta''_{ij}/\mathscr{D}_1(\Theta) & \text{if } j=0 \text{ and } i\ge 1, \\ \delta_{ij} & \text{if } j>0 \text{ or } i=j=0, \end{cases}$$

and

$$\psi_{ij} = \begin{cases} -\theta''_{ij}/\mathscr{D}_1(\Theta) & \text{if } i=0 \text{ and } j\ge 1, \\ \delta_{ij} & \text{if } i>0 \text{ or } i=j=0. \end{cases}$$

The boundedness of Ψ is verified as follows. If $f=\sum_{0\le j<\dim(\mathfrak{F})} f_j e_j$ is an element of $\mathfrak{F}$, then

$$\begin{aligned}\left|\sum_{0\le j<\dim(\mathfrak{F})} \psi_{0j}(\zeta)f_j\right| &= \left|\sum_{0\le j<\dim(\mathfrak{F})} \theta''_{0j}(\zeta)f_j\right| \\ &\le \|\Theta''(\zeta)f\| \le \|\Theta''(\zeta)\|\|f\|\end{aligned}$$

for almost every $\zeta\in C$, and hence

$$\|\Psi(\zeta)\| \le 1+\operatorname{ess\,sup}\{\|\Theta''(\zeta)\| : \zeta\in C\} \quad (\text{a.e. } \zeta\in C).$$

The boundedness of Φ is proved analogously, using its transpose matrix. It is also clear that Φ and Ψ are invertible; for instance, $\Phi^{-1} = 2I-\Phi$. Therefore Θ'' is quasi-equivalent to $\Theta''' = \Phi\Theta''\Psi$ so that

$$\mathscr{D}_1(\Theta''') = \mathscr{D}_1(\Theta'') = \mathscr{D}_1(\Theta).$$

Clearly Θ''' has the form $\mathrm{Diag}(\mathscr{D}_1(\Theta),\Theta_1)$, and $\mathscr{D}_1(\Theta)$ divides all the entries of Θ_1. The last assertion of the lemma follows because $\mathscr{D}_2(\Theta)=0$ if and only if $\mathscr{D}_2(\Theta''')=0$, and $\mathscr{D}_2(\Theta''') = \mathscr{D}_1(\Theta)\mathscr{D}_1(\Theta_1)$.

Theorem 6.5. *Assume that $\{\mathfrak{F},\mathfrak{F},\Theta(\lambda)\}$ is a bounded analytic function, and $\omega\in H^\infty$ is inner.*

(1) *We have $\mathscr{E}_j(\Theta)|\mathscr{E}_{j+1}(\Theta)$ for $0\le j<\dim(\mathfrak{F})$.*

(2) *If k is an integersuch that $1 \le k \le \dim(\mathfrak{F})$, there exists a bounded analytic function $\{\mathfrak{F}_k, \mathfrak{F}_k, \Theta_k(\lambda)\}$ such that $\mathscr{E}_k(\Theta)|\mathscr{D}_1(\Theta_k)$, and Θ is ω-equivalent to* $\mathrm{Diag}(\mathscr{E}_1(\Theta), \mathscr{E}_2(\Theta), \ldots, \mathscr{E}_k(\Theta), \Theta_k)$. *In order that $\Theta_k = O$ it is necessary and sufficient that $\mathscr{D}_{k+1}(\Theta) = 0$.*

Proof. We prove (2) first. As in the proof of Lemma 6.4, there is no loss of generality in assuming that $\mathscr{D}_j(\Theta)|\omega$ whenever $j \le k$ and $\mathscr{D}_j(\Theta) \ne 0$. We set $\Theta_0 = \Theta$. An inductive application of Lemma 6.4 shows the existence of bounded analytic functions $\{\mathfrak{F}_j, \mathfrak{F}_j, \Theta_j(\lambda)\}$, $i \le j \le k$, with the following properties: Θ_j is ω-equivalent to $\mathrm{Diag}(\mathscr{D}_1(\Theta_j), \Theta_{j+1})$ and $\mathscr{D}_1(\Theta_j)|\mathscr{D}_1(\Theta_{j+1})$ for $0 \le j < k$. Define functions $\delta_j \in H^\infty$ by $\delta_j = \mathscr{D}_1(\Theta_{j-1})$ for $1 \le j \le k$. Then Θ is ω-equivalent to $\Theta' = \mathrm{Diag}(\delta_1, \delta_2, \ldots, \delta_k, \Theta_k)$, we have $\delta_j|\delta_{j+1}$ for $1 \le j < k$, and $\delta_k|\mathscr{D}_1(\Theta_k)$. We now want to relate the functions δ_j to $\mathscr{D}_j(\Theta)$ and $\mathscr{E}_j(\Theta)$. To do this, we note that by Lemma 6.1 there exist inner functions $\varphi_0, \psi_0 \in H^\infty$ satisfying the relations

$$\varphi_0 \wedge \omega = \psi_0 \wedge \omega = 1, \tag{6.8}$$

and

$$\mathscr{D}_j(\Theta)|\psi_0^j \mathscr{D}_j(\Theta'), \quad \mathscr{D}_j(\Theta')|\varphi_0^j \mathscr{D}_j(\Theta) \tag{6.9}$$

for $1 \le j \le \dim(\mathfrak{F})$. Now, it is clear that $\mathscr{D}_j(\Theta') = \delta_1 \delta_2 \cdots \delta_j$ for $1 \le j \le k$, and thus (6.9) can be rewritten as

$$\mathscr{D}_j(\Theta)|\psi_0^j \delta_1 \delta_2 \cdots \delta_j, \quad \delta_1 \delta_2 \cdots \delta_j|\varphi_0^j \mathscr{D}_j(\Theta). \tag{6.10}$$

These relations show, in particular, that the first j for which $\delta_j = 0$ coincides with the first j for which $\mathscr{D}_j(\Theta) = 0$. Assume that j is such that these two functions are not zero. Because $\mathscr{D}_j(\Theta)|\omega$, (6.8) implies

$$\mathscr{D}_j(\Theta) \wedge \psi_0^j = \mathscr{D}_j(\Theta) \wedge \varphi_0^j = 1.$$

An application of the second relation in (6.10) (with $j-1$ in place of j) shows that $\mathscr{D}_j(\Theta)|\varphi_0^{j-1} \mathscr{D}_{j-1}(\Theta)\delta_j$. Hence $\mathscr{D}_j(\Theta)|\mathscr{D}_{j-1}(\Theta)$ or, equivalently, $\mathscr{E}_j(\Theta)|\delta_j$ so that we can write

$$\delta_j = \mathscr{E}_j(\Theta)\eta_j \quad (\eta_j \in H^\infty, j = 1, 2, \ldots, k). \tag{6.11}$$

Observe that

$$\begin{aligned}\mathscr{D}_k(\Theta) &= (\mathscr{E}_1(\Theta)\eta_1)(\mathscr{E}_2(\Theta)\eta_2) \cdots (\mathscr{E}_k(\Theta)\eta_k) \\ &= \delta_1 \delta_2 \cdots \delta_k|\varphi_0^k \mathscr{D}_k(\Theta),\end{aligned}$$

so that $\eta_1 \eta_2 \cdots \eta_k|\varphi_0^k$, and therefore $\eta_j \wedge \omega = 1$ for all j. Relation (6.11) can be extended to all $j \le k$ by setting $\eta_j = 1$ if $\mathscr{D}_j(\Theta) = \delta_j = 0$.

We apply one further ω-equivalence to eliminate the factors η_j. Define $\Psi = \mathrm{Diag}(\eta_1, \eta_2, \ldots, \eta_k, I_{\mathfrak{F}_k})$, and note that

$$\Theta' = \mathrm{Diag}(\mathscr{E}_1(\Theta), \mathscr{E}_2(\Theta), \ldots, \mathscr{E}_k(\Theta), \Theta_k)\Psi.$$

Moreover, Ψ has the scalar multiple $\eta_1\eta_2\cdots\eta_k$ which is relatively prime with ω. Thus the ω-equivalence in (2) is proved. The relation $\mathscr{E}_k(\Theta)|\mathscr{D}_1(\Theta_k)$ follows from (6.11) and $\delta_k|\mathscr{D}_1(\Theta_k)$. Finally, $\mathscr{D}_{k+1}(\mathrm{Diag}(\mathscr{E}_1(\Theta),\mathscr{E}_2(\Theta),\ldots,\mathscr{E}_k(\Theta),\Theta_k))=0$ if and only if $\mathscr{D}_{k+1}(\Theta)=0$, and this happens if and only if $\Theta_k=O$.

In order to prove (1) for some integer $j<\dim(\mathfrak{F})$, choose an integer k such that $j<k\leq\dim(\mathfrak{F})$, and perform the above construction. In (6.11) we have $\mathscr{E}_j(\Theta)\wedge\eta_{j+1}=1$ because $\mathscr{E}_j(\Theta)|\omega$ if $\mathscr{E}_j(\Theta)\neq 0$. Thus the relations $\mathscr{E}_j(\Theta)|\delta_j|\delta_{j+1}$ and $\delta_{j+1}=\mathscr{E}_{j+1}(\Theta)\eta_{j+1}$ imply that $\mathscr{E}_j(\Theta)|\mathscr{E}_{j+1}(\Theta)$, as desired.

The preceding theorem yields actual quasi-equivalence if $\mathscr{D}_k(\Theta)=0$, or if $\mathscr{F}$ is finite-dimensional and $k=\dim(\mathfrak{F})$. In particular, we obtain the following result.

Theorem 6.6. *Assume that T is an operator of class $C_0(N)$, and its characteristic function coincides with $\{E^N,E^N,\Theta(\lambda)\}$. Then*

$$T\sim\bigoplus_{j=1}^{N}S(\mathscr{E}_j(\Theta)).$$

Proof. This follows immediately from Theorem 6.5 and Proposition 6.3.

We note that, under the conditions of the preceding theorem, the Jordan model of T is $\bigoplus_{j=0}^{\infty}S(\varphi_j)$, where $\varphi_j=\mathscr{E}_{n-j-1}(\Theta)$ for $j<N$, and $\theta_j=1$ for $j\geq N$. Theorem 6.6 actually yields a new proof of the existence of Jordan models for the class $C_0(N)$. This method for calculating the Jordan model generally fails if T has infinite defect numbers. Consider for instance a Jordan operator $S(\Phi)$, where the sequence $\Phi=\{\varphi_j: j\geq 0\}$ satisfies $\bigwedge_j\varphi_j=1$. The characteristic function of $S(\Phi)$ coincides with $\Theta=\mathrm{diag}(\varphi_j)_{j=0}^{\infty}$, and it is easy to show that $\mathscr{E}_j(\Theta)=\mathscr{D}_j(\Theta)=1$ for all j.

7 Scalar multiples and Jordan models

1. The natural context for studying determinants and minors is that of exterior powers, which we discuss now in the case of complex Hilbert spaces. Given Hilbert spaces $\mathfrak{F}_1$ and $\mathfrak{F}_2$, we denote by $\mathfrak{F}_1\otimes\mathfrak{F}_2$ their Hilbert space tensor product. This is a Hilbert space generated by tensors $x_1\otimes x_2$ with $x_1\in\mathfrak{F}_1, x_2\in\mathfrak{F}_2$ whose scalar product satisfies

$$(x_1\otimes x_2,y_1\otimes y_2)=(x_1,y_1)(x_2,y_2),\quad x_1,y_1\in\mathfrak{F}_1,\quad x_2,y_2\in\mathfrak{F}_2.$$

This tensor product can be realized concretely as the Hilbert space of Hilbert and Schmidt operators from the dual $\mathfrak{F}_1^*$ of $\mathfrak{F}_1$ to $\mathfrak{F}_2$; $\mathfrak{F}_1^*$ is simply the Hilbert space consisting of all linear functionals $\varphi\colon\mathfrak{F}_1\to E^1$. The tensor $x_1\otimes x_2$ corresponds then to the rank-one operator

$$\varphi\to\varphi(x_1)x_2\quad(\varphi\in\mathfrak{F}_1^*).$$

Given bounded linear operators T_j on $\mathfrak{F}_j$, $j = 1, 2$, there exists a unique bounded linear operator $T_1 \otimes T_2$ on $\mathfrak{F}_1 \otimes \mathfrak{F}_2$ such that

$$(T_1 \otimes T_2)(x_1 \otimes x_2) = (T_1 x_1) \otimes (T_2 x_2) \quad (x_1 \in \mathfrak{F}_1, x_2 \in \mathfrak{F}_2).$$

The norm of this operator is $\|T_1 \otimes T_2\| = \|T_1\| \cdot \|T_2\|$.

Tensor products can be iterated to yield a space $\mathfrak{F}_1 \otimes \mathfrak{F}_2 \otimes \cdots \otimes \mathfrak{F}_n$ associated with a finite number of Hilbert spaces $\mathfrak{F}_j$ $(j = 1, 2, \ldots, n)$. In particular, for any integer $n \geq 1$ we can form the tensor power

$$\mathfrak{F}^{\otimes n} = \underbrace{\mathfrak{F} \otimes \mathfrak{F} \otimes \cdots \otimes \mathfrak{F}}_{n \text{ times}}$$

of a Hilbert space $\mathfrak{F}$, and the tensor power

$$T^{\otimes n} = \underbrace{T \otimes T \otimes \cdots \otimes T}_{n \text{ times}}$$

of an operator acting on $\mathfrak{F}$. The map $T \to T^{\otimes n}$ is multiplicative, contractive, $(T^*)^{\otimes n} = (T^{\otimes n})^*$, and $T^{\otimes n}$ is isometric if T is isometric. Moreover, sequential strong convergence $T_k \to T$ implies $T_k^{\otimes n} \to T^{\otimes n}$.

We are interested in a subspace $\mathfrak{F}^{\wedge n} \subset \mathfrak{F}^{\otimes n}$, called the nth exterior power of $\mathfrak{F}$. This space is the range of the orthogonal projection P_n on $\mathfrak{F}^{\otimes n}$ which satisfies

$$P_n(x_1 \otimes x_2 \otimes \cdots \otimes x_n) = \frac{1}{n!} \sum_{\sigma \in \mathfrak{S}_n} \varepsilon(\sigma) x_{\sigma(1)} \otimes x_{\sigma(2)} \otimes \cdots \otimes x_{\sigma(n)},$$

where $\mathfrak{S}_n$ denotes the group of permutations of $\{1, 2, \ldots, n\}$, and $\varepsilon(\sigma)$ is the sign of the permutation σ, that is, $\varepsilon(\sigma) = 1$ if σ is even and $\varepsilon(\sigma) = -1$ if σ is odd. It is obvious that P_n commutes with every tensor power $T^{\otimes n}$, and therefore $\mathfrak{F}^{\wedge n}$ is a reducing space for such operators. We write $T^{\wedge n} = T^{\otimes n} | \mathfrak{F}^{\wedge n}$. The map $T \to T^{\wedge n}$ is multiplicative.

We use the notation

$$x_1 \wedge x_2 \wedge \cdots \wedge x_n = \sqrt{n!} P_n(x_1 \otimes x_2 \otimes \cdots \otimes x_n).$$

For the remainder of this chapter, the symbol $\wedge$ is only used in relation to exterior powers, and no longer designates greatest common inner divisors, even when applied to functions. Observe that

$$x_{\sigma(1)} \wedge x_{\sigma(2)} \wedge \cdots \wedge x_{\sigma(n)} = \varepsilon(\sigma) x_1 \wedge x_2 \wedge \cdots \wedge x_n \quad (\sigma \in \mathfrak{S}_n),$$

and

$$(x_1 \wedge x_2 \wedge \cdots \wedge x_n, y_1 \wedge y_2 \wedge \cdots \wedge y_n) = \det[(x_i, y_j)]_{1 \leq i,j \leq n}.$$

If $\mathfrak{F}$ is separable and $\{e_i : 0 \leq i < \dim(\mathfrak{F})\}$ is an orthonormal basis for $\mathfrak{F}$, the system

$$\{e_{i_1} \wedge e_{i_2} \wedge \cdots \wedge e_{i_n} : 0 \leq i_1 < i_2 < \cdots < i_n < \dim(\mathfrak{F})\}$$

is an orthonormal basis for $\mathfrak{F}^{\wedge n}$. In particular, $\mathfrak{F}^{\wedge n} = \{0\}$ if $\dim(\mathfrak{F}) < n$. There is a continuous bilinear map

$$\wedge\colon \mathfrak{F}^{\wedge n} \times \mathfrak{F}^{\wedge m} \to \mathfrak{F}^{\wedge n+m}$$

such that

$$(x_1 \wedge x_2 \wedge \cdots \wedge x_n) \wedge (x_{n+1} \wedge x_{n+2} \wedge \cdots \wedge x_{n+m}) = x_1 \wedge x_2 \wedge \cdots \wedge x_{n+m} \quad (x_j \in \mathfrak{F}).$$

Clearly we have

$$(T^{\wedge n} u) \wedge (T^{\wedge m} v) = T^{\wedge n+m}(u \wedge v) \quad (u \in \mathfrak{F}^{\wedge n}, v \in \mathfrak{F}^{\wedge m}),$$

for every linear operator T on $\mathfrak{F}$.

Given a vector $e \in \mathfrak{F}$, the operator

$$u \to e \wedge u \quad (u \in \mathfrak{F}^{\wedge n}),$$

is a multiple of a partial isometry from $\mathfrak{F}^{\wedge n}$ to $\mathfrak{F}^{\wedge n+1}$. Its adjoint is denoted

$$v \to e^* \vee v, \quad v \in \mathfrak{F}^{\wedge n+1}.$$

This is sometimes called a *contraction operation*, and

$$e^* \vee (f_1 \wedge f_2 \wedge \cdots \wedge f_{n+1}) = \sum_{j=1}^{n+1} (-1)^{j-1} (f_j, e) f_1 \wedge \cdots \wedge \widehat{f_j} \wedge \cdots \wedge f_{n+1}$$

for $f_1, \ldots, f_{n+1} \in \mathfrak{F}$, where the symbol $\widehat{\ }$ indicates that the corresponding factor is omitted. This formula is easily verified using an orthonormal basis (in $\mathfrak{F}$) including the vector e. If T is a bounded operator on $\mathfrak{F}$, we have

$$\begin{aligned} & e^* \vee T^{\wedge n+1}(f_1 \wedge f_2 \wedge \cdots \wedge f_{n+1}) \\ &= T^{\wedge n} \sum_{j=1}^{n+1} (-1)^{j-1} (T f_j, e) f_1 \wedge \cdots \wedge \widehat{f_j} \wedge \cdots \wedge f_{n+1}. \end{aligned} \tag{7.1}$$

2. The continuity of the multilinear operations described above implies that, given an inner function $\{\mathfrak{F}, \mathfrak{F}, \Theta(\lambda)\}$ in D, then $\{\mathfrak{F}^{\wedge n}, \mathfrak{F}^{\wedge n}, \Theta(\lambda)^{\wedge n}\}$ is also an inner function. Analogously, if $f_j\colon D \to \mathfrak{F}$, $j = 1, 2, \ldots, n$, are analytic functions, then the function $\lambda \to f_1(\lambda) \wedge f_2(\lambda) \wedge \cdots \wedge f_n(\lambda)$ is analytic as well, and it is denoted $f_1 \wedge f_2 \wedge \cdots \wedge f_n$. Such products do not necessarily belong to $H^2(\mathfrak{F}^{\wedge n})$ if $f_j \in H^2(\mathfrak{F})$. Let us denote by $H^\infty(\mathfrak{F})$ the collection of all bounded functions in $H^2(\mathfrak{F})$. The following lemma is easily verified.

Lemma 7.1. *Given functions $f_2, f_3, \ldots, f_n \in H^\infty(\mathfrak{F})$, the map*

$$f \to f \wedge f_2 \wedge \cdots \wedge f_n$$

is a bounded linear operator from $H^2(\mathfrak{F})$ to $H^2(\mathfrak{F}^{\wedge n})$.

3. Now let $\mathfrak{F}$ be a separable Hilbert space, and let $\{\mathfrak{F},\mathfrak{F},\Theta(\lambda)\}$ be an inner function. We denote by $\mathbf{T}_\Theta$ the functional model associated with this function. Thus $\mathbf{T}_\Theta$ is a c.n.u. contraction whose characteristic function coincides with Θ. According to Theorem VI.5.1, $\mathbf{T}_\Theta$ is of class C_0 if and only if Θ has a scalar multiple, that is,

$$\Theta(\lambda)\Omega(\lambda) = \Omega(\lambda)\Theta(\lambda) = u(\lambda)I_{\mathfrak{F}} \quad (\lambda \in D),$$

where $\{\mathfrak{F},\mathfrak{F},\Omega(\lambda)\}$ is another inner function, and $u \in H^\infty$ is inner as well.

Lemma 7.2. *If $u \in H^\infty$ is a scalar multiple of $\{\mathfrak{F},\mathfrak{F},\Theta(\lambda)\}$, and $n \le \dim(\mathfrak{F})$ is a positive integer, then u^n is a scalar multiple of $\{\mathfrak{F}^{\wedge n},\mathfrak{F}^{\wedge n},\Theta(\lambda)^{\wedge n}\}$.*

Proof. If Ω satisfies the relation above, we have

$$\Theta(\lambda)^{\wedge n}\Omega(\lambda)^{\wedge n} = \Omega(\lambda)^{\wedge n}\Theta(\lambda)^{\wedge n} = u(\lambda)^n I_{\mathfrak{F}^{\wedge n}} \quad (\lambda \in D).$$

The lemma follows.

The least inner scalar multiple of Θ is precisely the minimal function of $\mathbf{T}_\Theta$ (cf. Theorem VI.5.1). We denote by $d_{n,\Theta} \in H^\infty$ the least inner scalar multiple of $\Theta^{\wedge n}$. Note that $\Theta^{\wedge n}$ acts on the space $\{0\}$ if $n > \dim(\mathfrak{F})$, so we do not use the notation $d_{n,\Theta}$ for such values of n.

Proposition 7.3. *Assume that the inner function $\{\mathfrak{F},\mathfrak{F},\Theta(\lambda)\}$ has a scalar multiple. Then $d_{n,\Theta}$ divides $d_{n+1,\Theta}$ for all $n < \dim(\mathfrak{F})$.*

Proof. The function $u = d_{n+1,\Theta}$ is a scalar multiple of $\Theta^{\wedge n+1}$. Equivalently, $uf \in \Theta^{\wedge n+1}H^2(\mathfrak{F}^{\wedge n+1})$ for every $f \in H^2(\mathfrak{F}^{\wedge n+1})$. We need to prove that this requirement is also satisfied with n in place of $n+1$. In fact, because $\Theta^{\wedge n}H^2(\mathfrak{F}^{\wedge n})$ is invariant for the unilateral shift on $H^2(\mathfrak{F}^{\wedge n})$, it suffices to show that

$$ue_1 \wedge e_2 \wedge \cdots \wedge e_n \in \Theta^{\wedge n}H^2(\mathfrak{F}^{\wedge n})$$

for any orthonormal system $\{e_j : 1 \le j \le n\}$ in $\mathfrak{F}$. Consider such an orthonormal system, and choose a unit vector $e_0 \in \mathfrak{F}$ orthogonal to e_j for $1 \le j \le n$. The hypothesis implies the existence of a function $f \in H^2(\mathfrak{F}^{\wedge n+1})$ such that

$$u(\lambda)e_0 \wedge e_1 \wedge \cdots \wedge e_n = \Theta(\lambda)^{\wedge n+1} f(\lambda) \quad (\lambda \in D).$$

We have then

$$\begin{aligned} u(\lambda)e_1 \wedge \cdots \wedge e_n &= e_0^* \vee (u(\lambda)e_0 \wedge e_1 \wedge \cdots \wedge e_n) \\ &= e_0^* \vee \Theta(\lambda)^{\wedge n+1} f(\lambda), \end{aligned}$$

and (7.1) shows that this function is of the form $\Theta(\lambda)^{\wedge n}g(\lambda)$ for some analytic function g. The function g must in fact be bounded because $\Theta^{\wedge n}$ is inner, and therefore $g \in H^2(\mathfrak{F}^{\wedge n})$, as desired.

We now show that all the functions $d_{n,\Theta}$ can be calculated in terms of the Jordan model of $\mathbf{T}_\Theta$.

Theorem 7.4. *Let* $\{\mathfrak{F},\mathfrak{F},\Theta(\lambda)\}$ *and* $\{\mathfrak{F}',\mathfrak{F}',\Theta'(\lambda)\}$ *be inner functions, both having scalar multiples. If* $\mathbf{T}_\Theta \prec \mathbf{T}_{\Theta'}$ *then* $d_{n,\Theta'}$ *divides* $d_{n,\Theta}$ *provided that* $n \le \dim(\mathfrak{F})$ *and* $n \le \dim(\mathfrak{F}')$.

Proof. Let $X\colon \mathbf{H} \to \mathbf{H}'$ be a quasi-affinity such that $X\mathbf{T}_\Theta = \mathbf{T}_{\Theta'}X$; here

$$\mathbf{H} = H^2(\mathfrak{F}) \ominus \Theta' H^2(\mathfrak{F}), \quad \mathbf{H}' = H^2(\mathfrak{F}') \ominus \Theta H^2(\mathfrak{F}').$$

By Theorem VI.3.6, there exists a bounded analytic function $\{\mathfrak{F},\mathfrak{F}',\Psi(\lambda)\}$ such that

$$\Psi\Theta H^2(\mathfrak{F}) \subset \Theta' H^2(\mathfrak{F}')$$

and

$$Xf = P_{\mathbf{H}'}\Psi f \quad (f \in \mathfrak{H}(\Theta)).$$

The first requirement on Ψ implies the existence of a bounded analytic function $\{\mathfrak{F},\mathfrak{F}',\Psi'(\lambda)\}$ such that

$$\Psi\Theta = \Theta'\Psi'.$$

The range of X is equal to $(\Psi H^2(\mathfrak{F}) + \Theta' H^2(\mathfrak{F}')) \ominus \Theta' H^2(\mathfrak{F}')$, and therefore the space $\Psi H^2(\mathfrak{F}) + \Theta' H^2(\mathfrak{F}')$ is dense in $H^2(\mathfrak{F}')$. Fix now

$$n \le \min\{\dim(\mathfrak{F}), \dim(\mathfrak{F}')\},$$

and set $u = d_{n,\Theta}$, so that $uH^2(\mathfrak{F}^{\wedge n}) \subset \Theta^{\wedge n} H^2(\mathfrak{F}^{\wedge n})$. It suffices to show that

$$ux_1 \wedge x_2 \wedge \cdots \wedge x_n \in \Theta'^{\wedge n} H^2(\mathfrak{F}'^{\wedge n})$$

for any choice of vectors $x_1, x_2, \ldots, x_n \in \mathfrak{F}'$. For each $j = 1,2,\ldots,n$, there exist functions $f_{jk} \in H^2(\mathfrak{F})$ and $g_{jk} \in H^2(\mathfrak{F}')$ such that

$$\lim_{k\to\infty} (\Psi f_{jk} + \Theta' g_{jk}) = x_j \quad (j = 1,2,\ldots,n),$$

in the H^2 norm. Moreover, the functions f_{jk} and g_{jk} can be assumed to be bounded. It follows that for each k, the function

$$h_{k_1k_2\ldots k_n} = (\Psi f_{1k_1} + \Theta' g_{1k_1}) \wedge (\Psi f_{2k_2} + \Theta' g_{2k_2}) \wedge \cdots \wedge (\Psi f_{nk_n} + \Theta' g_{nk_n})$$

belongs to $H^2(\mathfrak{F}'^{\wedge n})$. Moreover, Lemma 7.1 allow us to let the indices k_j tend to infinity one at a time, allowing us to conclude that the vector $x_1 \wedge x_2 \wedge \cdots \wedge x_n$ belongs to the norm closure of the set $S = \{h_{k_1k_2\ldots k_n} : k_1, k_2, \ldots, k_n = 1,2,\ldots\}$. Thus it suffices to show that $uh \in \Theta'^{\wedge n} H^2(\mathfrak{F}'^{\wedge n})$ when $h \in S$. Any such function h is a finite sum of functions of the form

$$F_0 = \Theta' g_1 \wedge \Theta' g_2 \wedge \cdots \wedge \Theta' g_n, \quad F_n = \Psi f_1 \wedge \Psi f_2 \wedge \cdots \wedge \Psi f_n$$

or

$$F_j = \Psi f_1 \wedge \Psi f_2 \wedge \cdots \wedge \Psi f_j \wedge \Theta' g_1 \wedge \Theta' g_2 \wedge \cdots \wedge \Theta' g_{n-j},$$

where $1 \le j \le n-1$, $f_1, f_2, \ldots, f_n \in H^\infty(\mathfrak{F})$, and $g_1, g_2, \ldots, g_n \in H^\infty(\mathfrak{F}')$. It therefore suffices to show that $uF \in \Theta'^{\wedge n} H^2(\mathfrak{F}'^{\wedge n})$ for such functions F. When $j = 0$ there is nothing to prove. Assume therefore that $j \ge 1$. By Proposition 7.3, the function $d_{j,\Theta}$ divides u, and therefore

$$uf_1 \wedge f_2 \wedge \cdots \wedge f_j = \Theta^{\wedge j} f$$

for some $f \in H^\infty(\mathfrak{F}^{\wedge j})$. For $j < n$, we conclude that

$$\begin{aligned} uF_j &= \Psi^{\wedge j}(uf_1 \wedge f_2 \wedge \cdots \wedge f_j) \wedge \Theta'^{\wedge n-j}(g_1 \wedge g_2 \wedge \cdots \wedge g_{n-j}) \\ &= \Psi^{\wedge j} \Theta^{\wedge j} f \wedge \Theta'^{\wedge n-j}(g_1 \wedge g_2 \wedge \cdots \wedge g_{n-j}) \\ &= \Theta'^{\wedge j} \Psi'^{\wedge j} f \wedge \Theta'^{\wedge n-j}(g_1 \wedge g_2 \wedge \cdots \wedge g_{n-j}) \\ &= \Theta'^{\wedge n}(\Psi'^{\wedge j} f \wedge (g_1 \wedge g_2 \wedge \cdots \wedge g_{n-j})). \end{aligned}$$

For $j = n$ the same equation holds, except that there are no factors of the form g_i. The theorem is proved.

We can now calculate the Jordan model of $\mathbf{T}_\Theta$ in terms of the functions $d_{n,\Theta}$.

Theorem 7.5. *Let $\mathfrak{F}$ be a separable, infinite-dimensional Hilbert space, and let $\{\mathfrak{F}, \mathfrak{F}, \Theta(\lambda)\}$ be an inner function such that the operator $T = \mathbf{T}_\Theta$ is of class C_0. Define inner functions $\varphi_0 = d_{1,\Theta}$ and $\varphi_j = d_{j+1,\Theta}/d_{j,\Theta}$ for $j \ge 1$.*

(1) *We have $\varphi_{j+1} | \varphi_j$ for all j.*
(2) *The Jordan model of T is $\bigoplus_{j=0}^\infty S(\varphi_j)$.*

Proof. Let $\bigoplus_{j=0}^\infty S(\psi_j)$ be the Jordan model of T, and denote by Θ' the inner function $\operatorname{diag}(\psi_j)_{j=0}^\infty$. The function $\Theta'^{\wedge n}$ is diagonal as well, with diagonal entries $\{\psi_{j_1}\psi_{j_2}\cdots\psi_{j_n} : 0 \le j_1 < j_2 < \cdots < j_n\}$. Clearly, $d_{n,\Theta'}$ is the least common multiple of these entries; that is, $d_{n,\Theta'} = \psi_0\psi_1\cdots\psi_{n-1}$. Theorem 7.4 implies the equality $d_{n,\Theta} = d_{n,\Theta'}$ for all $n \ge 1$. The theorem follows immediately from this observation.

Corollary 7.6. *Let T be a contraction of class C_0 on a Hilbert space, and let*

$$T = \begin{bmatrix} T' & * \\ O & T'' \end{bmatrix}$$

be the triangulation associated with an invariant subspace of T. Denote by $S(\Phi)$, $S(\Phi')$, and $S(\Phi'')$ the Jordan models of T, T', and T'', respectively, where $\Phi = \{\varphi_j : j \ge 0\}$, $\Phi' = \{\varphi'_j : j \ge 0\}$, and $\Phi'' = \{\varphi''_j : j \ge 0\}$. Then we have

$$\varphi_0\varphi_1\cdots\varphi_{n-1} | \varphi'_0\varphi'_1\cdots\varphi'_{n-1} \cdot \varphi''_0\varphi''_1\cdots\varphi''_{n-1}$$

for all $n \ge 1$.

Proof. According to Theorem VII.1.1, there exist inner functions $\{\mathfrak{F}, \mathfrak{F}, \Theta(\lambda)\}$, $\{\mathfrak{F}, \mathfrak{F}, \Theta'(\lambda)\}$, and $\{\mathfrak{F}, \mathfrak{F}, \Theta''(\lambda)\}$ whose pure parts coincide with the characteristic functions of T, T', and T'', respectively, and such that $\Theta(\lambda) = \Theta''(\lambda)\Theta'(\lambda)$. By the preceding theorem, the corollary is equivalent to the statement that $d_{n,\Theta} | d_{n,\Theta'} d_{n,\Theta''}$.

Thus we must prove that $d_{n,\Theta'}d_{n,\Theta''}$ is a scalar multiple of $\Theta^{\wedge n}$. Choose inner functions Ω', Ω'' such that $\Theta'^{\wedge n}\Omega' = \Omega'\Theta'^{\wedge n} = d_{n,\Theta'}I_{\mathfrak{F}^{\wedge n}}$ and $\Theta''^{\wedge n}\Omega'' = \Omega''\Theta''^{\wedge n} = d_{n,\Theta''}I_{\mathfrak{F}^{\wedge n}}$, and set $\Omega = \Omega'\Omega''$. It is easy to see that $\Theta^{\wedge n}\Omega = d_{n,\Theta'}d_{n,\Theta''}I_{\mathfrak{F}^{\wedge n}}$, as desired.

8 Weak contractions of class C_0

1. We have seen in Proposition V.6.1 how algebraic adjoints can be used to show that $\det(\Theta(\lambda))$ is a scalar multiple of the function $\{\mathfrak{F}, \mathfrak{F}, \Theta(\lambda)\}$ when $\dim(\mathfrak{F}) < \infty$. There is an extension of the notion of algebraic adjoint that allows us to show that $\det(\Theta(\lambda))$ is a scalar multiple of $\Theta^{\wedge n}$ as well.

Proposition 8.1. *Assume that $\mathfrak{F}$ is a Hilbert space of finite dimension N. For $1 \leq n < N$, there exists a continuous map $A \to A^{\mathrm{Ad}n}$ which associates with each operator A on $\mathfrak{F}$ an operator $A^{\mathrm{Ad}n}$ on $\mathfrak{F}^{\wedge n}$ such that:*

(1) $A^{\mathrm{Ad}n}A^{\wedge n} = A^{\wedge n}A^{\mathrm{Ad}n} = \det(A)I_{\mathfrak{F}^{\wedge n}}$.
(2) $(AB)^{\mathrm{Ad}n} = B^{\mathrm{Ad}n}A^{\mathrm{Ad}n}$.
(3) $\|A^{\mathrm{Ad}n}\| = \|A^{\wedge N-n}\|$.

Proof. Fix n, and an orthonormal basis $\{e_j : 1 \leq j \leq N\}$ in $\mathfrak{F}$. The bilinear form

$$B(h,g) = (h \wedge g, e_1 \wedge e_2 \wedge \cdots \wedge e_N) \quad (h \in \mathfrak{F}^{\wedge n}, g \in \mathfrak{F}^{\wedge N-n}),$$

is nondegenerate; that is, the equality $B(h,g) = 0$ for every g (resp., every h) implies that $h = 0$ (resp., $g = 0$). In fact $B(e_{\sigma(1)} \wedge \cdots \wedge e_{\sigma(n)}, e_{\sigma(n+1)} \wedge \cdots \wedge e_{\sigma(N)}) = \varepsilon(\sigma)$ if $\sigma \in \mathfrak{S}_N$, and this implies that there is an isometric transformation C from $\mathfrak{F}^{\wedge N-n}$ onto the dual space $(\mathfrak{F}^{\wedge n})^*$ such that

$$(Cg)(h) = B(h,g) \quad (h \in \mathfrak{F}^{\wedge n}, g \in \mathfrak{F}^{\wedge N-n}).$$

Given an operator A on $\mathfrak{F}$, we can then define the operator X on $(\mathfrak{F}^{\wedge n})^*$ by $X = CA^{\wedge N-n}C^{-1}$. The operator $A^{\mathrm{Ad}n}$ is then defined so that its dual (in the Banach space sense) is equal to X. In other words,

$$B(A^{\mathrm{Ad}n}h, g) = B(h, A^{\wedge N-n}g) \quad (h \in \mathfrak{F}^{\wedge n}, g \in \mathfrak{F}^{\wedge N-n}).$$

Clearly, the map $A \to A^{\mathrm{Ad}n}$ is continuous, and equality (3) follows immediately. Equality (2) is due to the fact that taking duals reverses the order of the factors. To prove (1) note that for every $h \in \mathfrak{F}^{\wedge n}, g \in \mathfrak{F}^{\wedge N-n}$ we have

$$\begin{aligned} B(A^{\mathrm{Ad}n}A^{\wedge n}h, g) &= B(A^{\wedge n}h, A^{\wedge N-n}g) \\ &= (A^{\wedge N}(h \wedge g), e_1 \wedge e_2 \wedge \cdots \wedge e_N) \\ &= \det(A)B(h,g), \end{aligned}$$

where we used the fact that $A^{\wedge N} = \det(A)I_{\mathfrak{F}^{\wedge N}}$. Because B is a nondegenerate form, we obtain $A^{\mathrm{Ad}n}A^{\wedge n} = \det(A)I_{\mathfrak{F}^{\wedge n}}$. Equality (1) follows immediately for invertible A, and it extends to arbitrary A by continuity.

Observe that the construction given above depends on the choice of basis, but (1) shows that the result is independent of this choice. It is implicit in the proof above that $A^{\mathrm{Ad}n}$ is unitarily equivalent to $A^{\wedge N-n}$, so its matrix entries of $A^{\mathrm{Ad}n}$ are minors of order $N-n$ of the matrix A.

Corollary 8.2. *Let A be a linear operator on the Hilbert space $\mathfrak{F}$ of dimension N, and let $\{e_1, e_2, \dots, e_n\}$ be an orthonormal system in $\mathfrak{F}$. If P denotes the orthogonal projection onto the space generated by $\{e_1, e_2, \dots, e_n\}$, we have*

$$(A^{\mathrm{Ad}n}(e_1 \wedge e_2 \wedge \cdots \wedge e_n), e_1 \wedge e_2 \wedge \cdots \wedge e_n) = \det(P + (I-P)A(I-P)).$$

Proof. Complete the given vectors to an orthonormal basis $\{e_j : 1 \le j \le N\}$ for $\mathfrak{F}$, and set $f = e_1 \wedge e_2 \wedge \cdots \wedge e_n$, $g = e_{n+1} \wedge \cdots \wedge e_N$. Using the notation in the preceding proof, we have

$$\begin{aligned} (A^{\mathrm{Ad}n}f, f) &= ((A^{\mathrm{Ad}n}f) \wedge g, f \wedge g) \\ &= B(A^{\mathrm{Ad}n}f, g) \\ &= B(f, A^{\wedge N-n}g) \\ &= (f \wedge A^{\wedge N-n}g, f \wedge g). \end{aligned}$$

Because $(P + A(I-P))e_j = e_j$ for $j = 1, 2, \dots, n$ and $(P + A(I-P))e_j = Ae_j$ for $j = n+1, n+2, \dots, N$, we have

$$(P + A(I-P))^{\wedge N}(f \wedge g) = f \wedge A^{\wedge N-n}g,$$

and thus

$$\begin{aligned} (A^{\mathrm{Ad}n}f, f) &= ((P + A(I-P))^{\wedge N}(f \wedge g), f \wedge g) \\ &= \det(P + A(I-P)). \end{aligned}$$

To conclude the proof, observe that

$$\det(P + A(I-P)) = \det(P + (I-P)A(I-P))$$

because $P + A(I-P)$ has a block upper triangular form.

Remark. The above formula can be extended as follows. Let $\{e_1, e_2, \dots, e_n\}$ and $\{f_1, f_2, \dots, f_n\}$ be orthonormal systems in $\mathfrak{F}$, with $1 \le n < \dim(\mathfrak{F})$. Choose a unitary operator U on $\mathfrak{F}$ such that $\det(U) = 1$ and $Uf_j = e_j$ for $j = 1, 2, \dots, n$. Denote by P the orthogonal projection onto the space generated by $\{e_1, e_2, \dots, e_n\}$, and set $e = e_1 \wedge e_2 \wedge \cdots \wedge e_n$, $f = f_1 \wedge f_2 \wedge \cdots \wedge f_n$. We have $U^{\wedge n}f = e$, so that $U^{\mathrm{Ad}n}e = f$, and thus

$$(A^{\mathrm{Ad}n}f, e) = ((UA)^{\mathrm{Ad}n}e, e) = \det(P + (I-P)UA(I-P)). \tag{8.1}$$

2. In order to extend algebraic adjoints to infinite dimensions, we need a few additional facts about the *trace class* whose definition we now recall. An operator X on a Hilbert space $\mathfrak{F}$ is of trace class if the positive operator $|X| = (X^*X)^{1/2}$ has finite trace as defined in Section VIII.1. In other words, $|X|$ is compact with summable eigenvalues. We use the notation $\mathfrak{S}_1(\mathfrak{F})$ for the collection of trace class operators, and we set

$$\|X\|_1 = \operatorname{tr}(|X|) \quad (X \in \mathfrak{S}_1(\mathfrak{F})).$$

Then $\mathfrak{S}_1(\mathfrak{F})$ is a Banach space with this norm, and the operators with finite rank are dense in $\mathfrak{S}_1(\mathfrak{F})$. With each $X \in \mathfrak{S}_1(\mathfrak{F})$ one can associate a complex number $\det(I+X)$ that behaves like the usual determinant. More precisely,

(a) $\det(AB) = \det(A)\det(B)$ if $A, B \in I + \mathfrak{S}_1(\mathfrak{F})$.
(b) A is invertible if and only if $\det(A) \neq 0$.
(c) $|\det(A)| = 1$ if A is unitary.
(d) $|\det(A)| \leq 1$ if $\|A\| \leq 1$.
(e) If $A(\lambda)$ is analytic (resp., continuous), then $\det(A(\lambda))$ is analytic (resp., continuous).
(f) If $\mathfrak{F}$ has an orthonormal basis $\{e_j : j = 1, 2, \ldots\}$, then

$$\det(A) = \lim_{n\to\infty} \det[(Ae_i, e_j)]_{i,j=1}^n.$$

(g) If $B \in \mathfrak{S}_1(\mathfrak{F})$ then $e^B \in I + \mathfrak{S}_1(\mathfrak{F})$ and

$$\det(e^B) = e^{\operatorname{tr} B}.$$

For these facts the reader can consult GOHBERG AND KREĬN [4]. The following result is essentially proved in Sec. VIII.4 for the finite-dimensional case.

Corollary 8.3. *Let $A(\lambda) \in I + \mathfrak{S}_1(\mathfrak{F})$ be an analytic function in a neighborhood of 0 such that $A(0) = I$. Then*

$$\left.\frac{d}{d\lambda}\det(A(\lambda))\right|_{\lambda=0} = \operatorname{tr} A'(0).$$

Proof. Property (g) above shows that $\det(A(\lambda)) = \exp(\operatorname{tr}\log A(\lambda))$ for λ close to zero. Here $\log A(\lambda)$ is calculated using the Riesz and Dunford calculus for the principal branch of the logarithm. The lemma follows immediately from this observation via the chain rule for differentiation.

Lemma 8.4. *Assume that* $\dim(\mathfrak{F}) < \infty$*, and let A be a linear operator on $\mathfrak{F}$. Then we have*

$$\|A^{\mathrm{Ad}\,n}\| \leq \exp((1 + \|A - I\|_1)^2 - 1).$$

for $1 \leq n < \dim(\mathfrak{F})$.

Proof. Assume first that $A \geq O$, and $\lambda_1 \geq \lambda_2 \geq \cdots \geq \lambda_N \geq 0$ are its eigenvalues, repeated according to their multiplicities. Then $A^{\mathrm{Ad}n}$ is positive, and its eigenvalues are $\{\lambda_{i_1}\lambda_{i_2}\cdots\lambda_{i_{N-n}} : 1 \leq i_1 < i_2 < \cdots < i_{N-n} \leq N\}$. Therefore

$$\begin{aligned}\|A^{\mathrm{Ad}n}\| &= \lambda_1\lambda_2\cdots\lambda_{N-n}\\ &\leq (1+|\lambda_1-1|)(1+|\lambda_2-1|)\cdots(1+|\lambda_N-1|)\\ &\leq \exp(|\lambda_1-1|)\exp(|\lambda_2-1|)\cdots\exp(|\lambda_N-1|)\\ &= \exp(\mathrm{tr}(|A-I|)) = \exp(\|A-I\|_1).\end{aligned}$$

In the general case we use the polar decomposition $A = U|A|$:

$$\|A^{\mathrm{Ad}n}\| = \|U^{\mathrm{Ad}n}|A|^{\mathrm{Ad}n}\| \leq \||A|^{\mathrm{Ad}n}\| \leq \exp(\||A|-I\|_1),$$

and the easy estimate

$$\begin{aligned}\||A|-I\|_1 &\leq \|A^*A-I\|_1 \leq \|A^*(A-I)\|_1 + \|A-I\|_1\\ &\leq (\|A\|+1)\|A-I\|_1\\ &\leq (\|A-I\|_1+2)\|A-I\|_1\\ &= (1+\|A-I\|_1)^2-1.\end{aligned}$$

The lemma follows.

Theorem 8.5. *Assume that $\mathfrak{F}$ is a separable, infinite dimensional Hilbert space, and $n \geq 1$. For every operator $A \in I+\mathfrak{S}_1(\mathfrak{F})$ there exists an operator $A^{\mathrm{Ad}n}$ on $\mathfrak{F}^{\wedge n}$ with the following properties.*

(1) $A^{\mathrm{Ad}n}A^{\wedge n} = A^{\wedge n}A^{\mathrm{Ad}n} = \det(A)I_{\mathfrak{F}^{\wedge n}}$.
(2) $(AB)^{\mathrm{Ad}n} = B^{\mathrm{Ad}n}A^{\mathrm{Ad}n}$.
(3) *If P denotes the orthogonal projection onto the space generated by an orthonormal system $\{e_1,e_2,\dots,e_n\}$, we have*

$$(A^{\mathrm{Ad}n}(e_1\wedge e_2\wedge\cdots\wedge e_n), e_1\wedge e_2\wedge\cdots\wedge e_n) = \det(P+(I-P)A(I-P)).$$

(4) $\|A^{\mathrm{Ad}n}\| \leq \exp((1+\|A-I\|_1)^2-1)$.
(5) $\|A^{\mathrm{Ad}n}\| \leq 1$ *if* $\|A\| \leq 1$.
(6) *If $A(\lambda)$ is an analytic function, then $A(\lambda)^{\mathrm{Ad}n}$ is also analytic.*

Proof. Fix n, choose an orthonormal basis $\{f_j : j \geq 1\}$ for $\mathfrak{F}$, and denote by P_k the orthogonal projection onto the space $\mathfrak{F}_k$ generated by $\{f_j : 1 \leq j \leq k\}$. Given $A \in I+\mathfrak{S}_1(\mathfrak{F})$, set $A_k = P_kAP_k$, and define an operator X_k on $\mathfrak{F}^{\wedge n}$ by setting

$$X_k|\mathfrak{F}_k^{\wedge n} = (A_k|\mathfrak{F}_k)^{\mathrm{Ad}n} \quad \text{and} \quad X_k|(\mathfrak{F}_k^{\wedge n})^{\perp} = O.$$

Because $\|A_k - I_{\mathfrak{F}_k}\|_1 \leq \|A-I\|_1$, Lemma 8.4 implies that the operators X_k are uniformly bounded. We show that they converge weakly. Given integers $1 \leq i_1 < i_2 < \cdots < i_n$ and $1 \leq j_1 < j_2 < \cdots < j_n$, there exists a unitary operator U on $\mathfrak{F}$ such that

$Uf_k = f_k$ for $k > \max\{i_n, j_n\}$ and $Uf_{j_\ell} = f_{i_\ell}$ for $\ell = 1,2,\dots,n$. We deduce from (8.1) that

$$(X_k(f_{j_1} \wedge \cdots \wedge f_{j_n}), f_{i_1} \wedge \cdots \wedge f_{i_n}) = \det(P + (I-P)UA_k(I-P))$$

for $k > \max\{i_n, j_n\}$, where P denotes the orthogonal projection onto the space generated by $\{f_{i_\ell} : 1 \le \ell \le n\}$. Now, $\lim_{k\to\infty} \|A - A_k\|_1 = 0$, and property (e) of the determinants implies that the determinants in the right-hand side of the above equation tend to $\det(P + (I-P)UA(I-P))$ as $k \to \infty$.

We denote by $A^{\mathrm{Ad}\,n}$ the weak limit of the operators X_k. Properties (4) and (5) are trivially verified, and property (6) follows from Dunford's theorem because the above argument, along with property (e) of determinants, shows that the matrix entries of $A(\lambda)^{\mathrm{Ad}\,n}$ are analytic functions. To verify (1), observe that

$$X_k A_k^{\mathrm{Ad}\,n} h = A_k^{\mathrm{Ad}\,n} X_k h = \det(A_k + I - P_k)h \quad (h \in \mathfrak{F}_k^{\wedge n}),$$

and let $k \to \infty$.

Let $\{A_m\}_{m=1}^\infty \subset I + \mathfrak{S}(\mathfrak{F})$ be a sequence such that $\lim_{m\to\infty} \|A - A_m\|_1 = 0$. The above considerations show that $A_m \to A$ in the weak operator topology.

Property (2) follows from (1) if A and B are invertible. For the general case, choose invertible operators $A_m, B_m \in I + \mathfrak{S}(\mathfrak{F})$ such that $\lim_{m\to\infty} \|A - A_m\|_1 = \lim_{m\to\infty} \|B - B_m\|_1 = 0$. The equality $(A_m B_\ell)^{\mathrm{Ad}\,n} = B_\ell^{\mathrm{Ad}\,n} A_m^{\mathrm{Ad}\,n}$ yields the desired result as $m \to \infty$ and $\ell \to \infty$.

Finally, observe that the operators $A^{\mathrm{Ad}\,n}$ do not depend on the basis $\{f_j\}_{j=1}^\infty$. This follows from (1) when A is invertible, and from approximation by invertible operators in general. Note also that (3) has already been verified if $e_j = f_j$ for $1 \le j \le n$. The general case follows if we construct $A^{\mathrm{Ad}\,n}$ using a basis for $\mathfrak{F}$ that contains the elements $\{e_j\}_{j=1}^n$.

For applications to weak contractions, the following result is very useful.

Theorem 8.6. *Let $\{\mathfrak{F}, \mathfrak{F}, \Theta(\lambda)\}$ be an inner function such that $I - \Theta(\lambda) \in \mathfrak{S}_1(\mathfrak{F})$ for all $\lambda \in D$. If $\det(\Theta(\lambda))$ is not identically zero, then it is an inner scalar multiple of $\Theta^{\wedge n}$ for all integers n, $1 \le n \le \dim \mathfrak{F}$.*

Proof. Assume that the determinant is not identically zero. The relation

$$\Theta(\lambda)^{\wedge n} \Theta(\lambda)^{\mathrm{Ad}\,n} = \Theta(\lambda)^{\mathrm{Ad}\,n} \Theta(\lambda)^{\wedge n} = \det(\Theta(\lambda)) I_{\mathfrak{F}^{\wedge n}}$$

shows that $\det(\Theta(\lambda))$ is a scalar multiple of $\Theta(\lambda)^{\wedge n}$. Write $\det(\Theta(\lambda)) = d_i d_o$, with d_i inner and d_o outer. Then d_i is also a scalar multiple of $\Theta(\lambda)^{\wedge n}$. Thus $\Theta(\lambda)^{\wedge n} \Omega_n(\lambda) = d_i I_{\mathfrak{F}^{\wedge n}}$ for some contractive analytic function $\Omega_n(\lambda)$. The relation

$$\Theta(\lambda)^{\wedge n} (\Theta(\lambda)^{\mathrm{Ad}\,n} - d_o(\lambda)\Omega_n(\lambda)) = O \quad (\lambda \in D),$$

and the fact that $\Theta(\lambda)^{\wedge n}$ is inner, imply the equality $\Theta(\lambda)^{\mathrm{Ad}n} = d_o(\lambda)\Omega_n(\lambda)$. In particular, $\|\Theta(0)^{\mathrm{Ad}n}\| \leq |d_o(0)|$, and therefore

$$|(\Theta(0)^{\mathrm{Ad}n} e_1 \wedge e_2 \wedge \cdots \wedge e_n, e_1 \wedge e_2 \wedge \cdots \wedge e_n)| \leq |d_o(0)|,$$

where $\{e_j : j = 1, 2, \ldots\}$ is any orthonormal basis in $\mathfrak{F}$. Denoting by P_n the orthogonal projection onto the space generated by $\{e_j : 1 \leq j \leq n\}$, Theorem 8.5(3) implies that

$$|\det(P_n + (I - P_n)\Theta(0)(I - P_n)| \leq |d_o(0)| \quad (n \geq 1).$$

The operators $P_n + (I - P_n)\Theta(0)(I - P_n)$ tend to I in the trace norm, and therefore their determinants tend to 1. We deduce that $|d_o(0)| \geq 1$, and the maximum modulus principle implies that d_o is constant. Thus $\det(\Theta(\lambda)) = d_o(0)d_i(\lambda)$ is inner, as claimed.

We need one more result about the trace class.

Proposition 8.7. *Let A be an invertible operator on $\mathfrak{F}$ such that $O \leq A \leq I$. If the sequence $\{\|(A^{\wedge n})^{-1}\| : n \geq 1\}$ is bounded, then $I - A \in \mathfrak{S}_1(\mathfrak{F})$.*

Proof. Write the spectral integral $I - A = \int_0^1 t\, dE(t)$. We show first that $I - A$ is compact. Indeed, in the contrary case there exists $\varepsilon > 0$ such that the space $E((\varepsilon, 1])\mathfrak{F}$ is infinite dimensional. Choose an orthonormal system $\{f_j : j \geq 1\}$ in this space, and note that $\|A^{\wedge n}(f_1 \wedge f_2 \wedge \cdots \wedge f_n)\| \leq (1 - \varepsilon)^n$. Thus $\|(A^{\wedge n})^{-1}\| \geq (1 - \varepsilon)^{-n}$, contrary to the hypothesis. Hence $I - A$ is compact. If $I - A$ has finite rank, we are done. If not, let $\lambda_1 \geq \lambda_2 \geq \cdots$ be the eigenvalues of $I - A$, repeated according to multiplicity, and let $\{e_j : j \geq 1\}$ be an orthonormal system such that $Ae_j = (1 - \lambda_j)e_j$ for $j \geq 1$. We have then

$$\|A^{\wedge n}(e_1 \wedge e_2 \wedge \cdots \wedge e_n)\| = (1 - \lambda_1)(1 - \lambda_2) \cdots (1 - \lambda_n) \leq \exp\left(-\sum_{j=1}^{n} \lambda_j\right).$$

The hypothesis implies that $\sum_{j=1}^{\infty} \lambda_j < \infty$, that is, $I - A$ is of trace class.

3. We can now characterize weak contractions of class C_0 in terms of their Jordan models.

Theorem 8.8. *Let T be a contraction of class C_0 on a separable Hilbert space, and let $S(\Phi)$, $\Phi = \{\varphi_j : j \geq 0\}$, be its Jordan model. The following conditions are equivalent.*

(1) *T is a weak contraction.*
(2) *$S(\Phi)$ is a weak contraction.*
(3) *The functions $\{\varphi_0\varphi_1 \cdots \varphi_n : n \geq 0\}$ have a common inner multiple.*

Proof. Without loss of generality, we can assume that T is invertible because otherwise we replace T by $T_a = (I - \bar{a}T)^{-1}(T - aI)$. Then $\Theta_T(0)$ is invertible and hence, in the polar decomposition $\Theta_T(0) = U|\Theta_T(0)|$, U is a unitary operator. The function $\Theta_T(\lambda)$ coincides then with $\Theta(\lambda) = U^*\Theta_T(\lambda)$.

Assume now that T is a weak contraction, and let $\mu_1 \geq \mu_2 \geq \cdots$ be the nonzero eigenvalues of $I - T^*T$, repeated according to multiplicity. By assumption, we have $\sum_{j=1}^{\infty} \mu_j < \infty$. The eigenvalues of the operator $\Theta(0) = |\Theta_T(0)| = |T||\mathfrak{D}_T$ are then $\{(1-\mu_j)^{1/2} : j \geq 1\}$. Because

$$1-(1-\mu_j)^{1/2} = \frac{\mu_j}{1+(1-\mu_j)^{1/2}} \leq \mu_j,$$

the operator $I - \Theta(0)$ is of trace class. Now, the operators D_{T^*} and D_T are Hilbert and Schmidt, and therefore the product $D_{T^*}(I - \lambda T^*)^{-1} D_T$ is of trace class. We conclude that $\Theta(\lambda) \in I + \mathfrak{S}_1(\mathfrak{D}_T)$ for every $\lambda \in D$, and therefore $\det(\Theta(\lambda))$ is an inner function by Theorem 8.6. Theorems 8.5 and 8.6 imply that the function $d_{n,\Theta}$ (introduced before Proposition 7.3) divides $\det(\Theta(\lambda))$ for all integers n, $1 \leq n \leq \dim(\mathfrak{F})$. We conclude that (3) is true because $d_{n,\Theta} = \varphi_0\varphi_1 \cdots \varphi_{n-1}$ by Theorem 7.5.

Assume next that (3) is true, and let $u \in H^\infty$ be an inner common multiple of $\{\varphi_0\varphi_1 \cdots \varphi_n : n \geq 0\}$. Thus $u = u_n\varphi_0\varphi_1 \cdots \varphi_n$ for some inner function u_n, so that

$$\sum_{j=0}^{n} -\log|\varphi_j(0)| \leq -\log|u(0)| \quad (n \geq 0).$$

Thus $\sum_{j=0}^{\infty} -\log|\varphi_j(0)|$ converges, or equivalently $\sum_{j=0}^{\infty}(1 - |\varphi_j(0)|) < \infty$. We can now calculate

$$\operatorname{tr}(I - S(\Phi)^*S(\Phi)) = \sum_{j=0}^{\infty}(1 - |\varphi_j(0)|^2) \leq 2\sum_{j=0}^{\infty}(1 - |\varphi_j(0)|) < \infty$$

to conclude that (2) is true.

Assume now that (2) is true, so that

$$\sum_{j=0}^{\infty}(1 - |\varphi_j(0)|) \leq \sum_{j=0}^{\infty}(1 - |\varphi_j(0)|^2) = \operatorname{tr}(I - S(\Phi)^*S(\Phi)) < \infty.$$

In this case the numbers

$$|\varphi_0(0)\varphi_1(0)\cdots\varphi_n(0)| = \exp\left(\sum_{j=0}^{n} \log|\varphi_j(0)|\right)$$

are bounded away from zero. An argument similar to that in the proof of Lemma 1.3 shows that some subsequence of $\{\varphi_0\varphi_1 \cdots \varphi_n : n \geq 0\}$ converges to a nonzero function $v \in H^\infty$ such that

$$|v(\lambda)| \leq |\varphi_0(\lambda)\varphi_1(\lambda)\cdots\varphi_n(\lambda)| \quad (\lambda \in D, n \geq 0).$$

It follows that v, and hence its inner factor, is a multiple of $\varphi_0\varphi_1 \cdots \varphi_n$ for all n, and therefore (3) is true.

It remains to prove that (3) implies (1). Assume then that (3) is true, and let u be the least common inner multiple of $\{\varphi_0\varphi_1 \cdots \varphi_n : n \geq 0\}$; note that $u(0) \neq 0$. Then

u is a multiple of $d_{n,\Theta}$; that is,

$$\Theta(\lambda)^{\wedge n}\Omega_n(\lambda) = u(\lambda)I_{\mathfrak{D}_T^{\wedge n}} \quad (\lambda \in D),$$

for some contractive analytic function $\Omega_n(\lambda)$. In particular,

$$\|[\Theta(0)^{\wedge n}]^{-1}\| = |u(0)|^{-1}\|\Omega_n(0)\| \leq |u(0)|^{-1}.$$

Proposition 8.7 implies that $I - \Theta(0)$ is of trace class, and therefore $I - T^*T = (I + |T|)^{-1}(I - |T|)$ is of trace class as well. The theorem is proved.

Corollary 8.9. *Every C_0-contraction with finite multiplicity is a weak contraction.*

Let us isolate one useful fact whose proof is contained in the proof of Theorem 8.8.

Lemma 8.10. *Let T be a weak contraction of class C_0 on a separable Hilbert space $\mathfrak{H}$. The characteristic function of T coincides with an inner function $\{\mathfrak{F}, \mathfrak{F}, \Theta(\lambda)\}$ such that $\Theta(\lambda) \in I + \mathfrak{S}_1(\mathfrak{F})$ for $\lambda \in D$, and $\Theta(0) \geq O$.*

If T and Θ are as in the preceding lemma, the function $\det(\Theta(\lambda))$ is an inner function that does not depend, up to a scalar factor of absolute value one, on the particular function Θ. This inner function is called the *characteristic determinant* of T, and is denoted d_T. The characteristic determinant plays an analogous role to the characteristic polynomial of linear algebra. For instance, it was observed in the proof of Theorem 8.8 that d_T is a common inner multiple of $d_{n,\Theta}$ for $1 \leq n \leq \dim(\mathfrak{F})$.

Theorem 8.11. *Let T be a weak contraction of class C_0 on a separable Hilbert space $\mathfrak{H}$, and let $S(\Phi)$, $\Phi = \{\varphi_j : j \geq 0\}$, be its Jordan model.*

(1) *The characteristic determinant d_T is the least common inner multiple of $\{\varphi_0\varphi_1\cdots\varphi_n : n \geq 0\}$.*
(2) *The operator T is multiplicity-free if and only if $d_T = m_T$.*
(3) *We have $m_T = d_T/\mathscr{D}_1(\Theta^{\mathrm{Ad}\,1})$.*
(4) *If*

$$T = \begin{bmatrix} T_1 & * \\ O & T_2 \end{bmatrix}$$

is the triangulation associated with an invariant subspace of T, we have $d_T = d_{T_1}d_{T_2}$.

Proof. Without loss of generality, we may assume that T is invertible, its characteristic function coincides with $\{\mathfrak{F}, \mathfrak{F}, \Theta(\lambda)\}$, and $I - \Theta(\lambda) \in \mathfrak{S}_1(\mathfrak{F})$ for $\lambda \in D$. Because $\varphi_0\varphi_1\cdots\varphi_{n-1}$ is the least scalar multiple of $\Theta^{\wedge n}$, it follows that $d_T(\lambda) = \det(\Theta(\lambda))$ is a common inner multiple of $\{\varphi_0\varphi_1\cdots\varphi_n : n \geq 0\}$. Denote by u the least common inner multiple of this family, and write $d_T = uv$ and $u = u_n\varphi_0\varphi_1\cdots\varphi_{n-1}$ for some inner functions $v, u_n \in H^\infty$. There exists an inner operator-valued function

Ω_n such that $\Theta^{\wedge n}\Omega_n = \varphi_0\varphi_1\cdots\varphi_{n-1}I_{\mathfrak{F}^{\wedge n}}$. We deduce that $\Theta^{\wedge n}(\Theta^{\mathrm{Ad}\,n} - u_n v\Omega_n) = O$, and this implies $u_n v\Omega_n = \Theta^{\mathrm{Ad}\,n}$ because $\Theta^{\wedge n}$ is inner. In particular,

$$|v(0)| \geq |u_n(0)v(0)| \geq |u_n(0)v(0)||(\Omega_n(0)e,e)| = |(\Theta^{\mathrm{Ad}\,n}(0)e,e)|$$

for any unit vector $e \in \mathfrak{F}^{\wedge n}$. Let us fix an orthonormal basis $\{e_j : j \geq 1\}$ in $\mathfrak{F}$, denote by P_n the projection onto the space generated by $\{e_j : 1 \leq j \leq n\}$, and set $e(n) = e_1 \wedge e_2 \wedge \cdots \wedge e_n$. With this choice, Theorem 8.5 implies

$$|(\Theta^{\mathrm{Ad}\,n}(0)e(n),e(n))| = |\det(P_n + (I-P_n)\Theta(0)(I-P_n))|,$$

and this tends to 1 as $n \to \infty$. The preceding inequality implies $|v(0)| = 1$, and hence v is constant by the maximum modulus principle. This proves (1). Next, note that T is multiplicity free if and only if $\varphi_j = 1$ for $j \geq 1$, so that (2) follows easily from (1). To prove (3) observe that $\Omega = \Theta^{\mathrm{Ad}\,1}/\mathscr{D}_1(\Theta^{\mathrm{Ad}\,1})$ is an inner function, and $\Theta\Omega = \Omega\Theta = (d_T/\mathscr{D}_1(\Theta^{\mathrm{Ad}\,1}))I_{\mathfrak{F}}$. Because m_T is the least inner scalar multiple of Θ, this proves that m_T divides $d_T/\mathscr{D}_1(\Theta^{\mathrm{Ad}\,1})$, so that $d_T/\mathscr{D}_1(\Theta^{\mathrm{Ad}\,1}) = m_T u$ for some inner $u \in H^\infty$. The fact that m_T is a scalar multiple of Θ implies that $\Theta\Omega' = m_T I_{\mathfrak{F}}$ for some inner function Ω'. Because Θ is inner we easily obtain $u\Omega' = \Omega$: u divides all the entries of Ω. The definition of Ω implies that u is a constant of absolute value one.

It remains to prove (4). Let $\Theta = \Theta_2\Theta_1$ be a regular factorization such that the pure part of Θ_j coincides with the characteristic function of T_j (cf. Theorem VII.1.1). These functions are inner from both sides because T_1 and T_2 are of class C_0. Note that T_1 and T_2 are both invertible. Consider the polar decompositions $\Theta_2(0) = UA$, $\Theta_2(0) = BV$ with $A, B \geq O$ and U, V unitary. The fact that T_1 and T_2 are weak contractions implies that $I-A$ and $I-B$ are of trace class. Substituting $U^*\Theta V^*$ for Θ, we may assume that $U = V = I$. It is now easy to see that $I-\Theta(\lambda)$, $I-\Theta_1(\lambda)$ and $I-\Theta_2(\lambda)$ are of trace class for all $\lambda \in D$. Because $d_T = \det(\Theta)$, $d_{T_1} = \det(\Theta_1)$, and $d_{T_2} = \det(\Theta_2)$, assertion (4) follows from the multiplicative property (a) of the determinant. The theorem is proved.

4. We conclude this section with an application to operators in the class (Ω_0^+). Recall that an operator A belongs to this class if it is one-to-one, $\mathrm{Im}A \geq 0$, $\mathrm{Im}A$ has finite trace, and $\sigma(A) = \{0\}$. It was shown in Sec. VIII.4 that, given $A \in (\Omega_0^+)$ the operator $T = (A-iI)(A+iI)^{-1}$ is a weak contraction of class C_0, and its minimal function is $e_{s_A}(-\lambda)$, where the number s_A is given by (VIII.4.15). The defect indices of T are equal to the rank of $\mathrm{Im}A$. In particular, T is unitarily equivalent to a Jordan block if and only if $\mathrm{Im}A$ has rank one. It follows from Proposition VIII.4.3 that these Jordan blocks correspond to the operators A_α $(\alpha > 0)$ defined by

$$(A_\alpha f)(x) = i\alpha \int_0^x f(t)\,dt$$

for $f \in L^2(0,1)$. It is convenient to denote by A_0 the zero operator on the zero space $\{0\}$. It also follows from Sec. VIII.4 that $s_{A_\alpha} = \alpha$ for all α. The results we proved about weak contractions yield the following

Theorem 8.12. *Let A be an operator of class (Ω_0^+) on a separable Hilbert space. There exists a sequence $\alpha_0 \geq \alpha_1 \geq \cdots \geq 0$ of real numbers such that A is quasi-similar to $\bigoplus_{j=0}^{\infty} A_{\alpha_j}$ and $\sum_{j=0}^{\infty} \alpha_j = 2\mathrm{tr}\,\mathrm{Im}A$. In particular, A is unicellular if and only if $\alpha_0 = 2\mathrm{tr}\,\mathrm{Im}A$.*

Proof. Set $T = (A - iI)(A + iI)^{-1}$ and $T_\alpha = (A_\alpha - iI)(A_\alpha + iI)^{-1}$ for $\alpha \geq 0$. Theorem 8.8 implies the existence of a weak contraction of the form $\bigoplus_{j=0}^{\infty} T_{\alpha_j}$, with $\alpha_0 \geq \alpha_1 \geq \cdots \geq 0$, quasi-similar to T. The characteristic function of T_α is $e_\alpha(-\lambda) = \exp(\alpha(\lambda - 1)/(\lambda + 1))$, therefore we have $\mathrm{tr}(I - T^*T) = \sum_{j=0}^{\infty}(1 - e^{-2\alpha_j})$, and this is finite if and only if $\sum_{j=0}^{\infty} \alpha_j < \infty$. Clearly A is quasi-similar to $\bigoplus_{j=1}^{\infty} A_{\alpha_j}$. It remains to show that $\sum_{j=0}^{\infty} \alpha_j = 2\mathrm{tr}\,\mathrm{Im}A$. Denote $Q = \mathrm{Im}A$ and $\mathfrak{Q} = \overline{Q\mathfrak{H}}$. It was shown in Sec. VIII.4 that the characteristic function of T coincides with the function

$$\Theta(\lambda) = [I + 2\zeta Q^{1/2}(I + i\zeta A^*)^{-1}Q^{1/2}]|\mathfrak{Q},$$

where

$$\zeta = \frac{\lambda - 1}{\lambda + 1}.$$

Indeed, this is precisely formula (VIII.4.9) with $\zeta = -i/z$. By Theorem 8.11(1), we have $\det(\Theta(\lambda)) = \exp(c\zeta)$ with $c = \sum_{j=0}^{\infty} \alpha_j$. Apply now Corollary 8.3 to obtain

$$\left.\frac{d}{d\zeta}\det(\Theta(\lambda))\right|_{\zeta=0} = 2\mathrm{tr}A.$$

The last assertion follows from Theorem 8.11 and Corollary 4.9. This concludes the proof.

9 Notes

Jordan operators were first defined in SZ.-N.–F. [15], where the existence of quasi-similar Jordan models for operators of class $C_0(N)$ was established. The possibility of extending these results to more general operators was first indicated in SZ.-N.–F. [16]–[18], where Theorem 1.5 is proved. This result was also proved independently by HERRERO [1]. Theorems 4.2 and 4.7 are also from SZ.-N.–F. [17]. Theorem 4.10 was first stated in BERCOVICI, FOIAŞ, AND SZ.-NAGY [1], but a complete proof only appeared in BERCOVICI, KÉRCHY, FOIAŞ, AND SZ.-NAGY [1]. Theorem 5.7 was proved in BERCOVICI, FOIAŞ, AND SZ.-NAGY [1]. The beautiful idea of quasi-equivalence for finite matrices over H^∞ was introduced by NORDGREN [1]. Then MOORE AND NORDGREN [1] showed that this relation implies the quasi-similarity of the corresponding functional models, thus giving an insightful new proof of Theorem 6.6. The extension of quasi-equivalence to infinite matrices and Theorem 6.5 is due to SZ.-NAGY [15]. An abstract version of quasi-diagonalization can be found in SZŰCS [1]. The existence of Jordan models for separably acting operators of class C_0 can also be deduced by using quasi-equivalence. However, instead of the characteristic function $\Theta_T(\lambda)$ one must apply quasi-equivalence to the function $\Omega(\lambda)$ satisfying $\Theta_T(\lambda)\Omega(\lambda) = m_T(\lambda)I_{\mathfrak{D}_{T^*}}$. This approach was developed by MÜLLER [1]. Operators of class C_0 on arbitrary Hilbert spaces can also be classified by Jordan models. For nilpotent operators see APOSTOL, DOUGLAS, AND FOIAŞ [1], and for the general C_0 case BERCOVICI [2]. The material in Sections 7 and 8 is from BERCOVICI AND VOICULESCU [1]. The divisibility relations in Corollary 7.6 are part of a much larger family of relations connected with the Horn inequalities of linear algebra. For instance, with the notation of Corollary 7.6, we have $\varphi_{i+j}|\varphi_i'\varphi_j''$ for $i,j \geq 0$. For a discussion of these matters, see BERCOVICI, LI, AND SMOTZER [1,2] and LI AND MÜLLER [1]. The characteristic determinant d_T can be extended to a larger class of operators. This allows the development of an analogue of the Fredholm index in the commutant of an operator of class C_0. See BERCOVICI [1] and KÉRCHY [17] for these developments. It was shown in BERCOVICI, FOIAŞ, AND SZ.-NAGY [2] that the reflexivity of an operator of class C_0, with Jordan model $\bigoplus_{j=0}^{\infty} S(\varphi_j)$, is equivalent to the reflexivity of $S(\varphi_0/\varphi_1)$. Necessary and sufficient conditions for the reflexivity of a Jordan block $S(\varphi)$ were given by KAPUSTIN [1,2]. These conditions are that φ should not have multiple zeros, and the singular measure, defining its singular factor, should assign zero mass to any thin set in the sense of Carleson. A precursor of this result is in FOIAS [9]. Further references, as well as developments concerning invariant subspaces, reflexivity, and Fredholm theory in the context of the class C_0 can be found in BERCOVICI [3].

An operator T, acting on a separable Hilbert space, is said to be triangular if it has an upper triangular matrix relative to some orthonormal basis. A bitriangular operator is such that both T and T^* are triangular. DAVIDSON AND HERRERO [1] show that bitriangular operators have a quasisimilarity model, which is a direct sum of finite Jordan blocks. The class of bitriangular operators intersects nontrivially with the class C_0. It is not known whether a quasi-similarity theory can be constructed for a class of operators containing both the bitriangular operators and the class C_0.

ATZMON [1] characterizes contractions of class C_0 in terms of growth properties of their resolvents.

TAKAHASHI AND UCHIYAMA [1] show that an operator of class C_{00}, whose defect operators are Hilbert–Schmidt, must belong to the class C_0. The spectrum is not assumed to be a proper subset of the closed unit disk.

Bibliography

ABRAHAMSE, M. B.

[1] Toeplitz operators in multiply connected regions, *Amer. J. Math.* **96** (1974), 261–297.

[2] The Pick interpolation theorem for finitely connected domains, *Michigan Math. J.* **26** (1979), 195–203.

ABRAHAMSE, M. B. AND DOUGLAS, R. G.

[1] A class of subnormal operators related to multiply-connected domains, *Advances in Math.* **19** (1976), 106–148.

ADAMJAN, V. M. AND AROV, D. Z.

[1] On a class of scattering operators and characteristic operator-functions of contractions, *Dokl. Akad. Nauk SSSR* **160** (1965), 9–12.

[2] Unitary couplings of semi-unitary operators, *Mat. Issled.* **1**:2 (1966), 3–64.

ADAMJAN, V. M., AROV, D. Z., AND KREĬN, M.G.

[1] Bounded operators which commute with a C_{00} class contraction whose rank of nonunitarity is one, *Funkcional. Anal. i Priložen* **3**:3 (1969), 86–87.

AGLER, J.

[1] Hypercontractions and subnormality, *J. Oper. Theory* **13** (1985), 203–217.

[2] Rational dilation on an annulus, *Ann. of Math. (2)* **121** (1985), 537–563.

[3] An abstract approach to model theory, *Surveys of Some Recent Results in Oper. Theory*, Vol. II, 1–23, Pitman Res. Notes Math. Ser., 192, Longman Sci. Tech., Harlow, 1988.

AGLER, J., FRANKS, E., AND HERRERO, D. A.

[1] Spectral pictures of operators quasisimilar to the unilateral shift, *J. Reine Angew. Math.* **422** (1991), 1–20.

AGLER, J., HARLAND, J., AND RAPHAEL, B. J.

[1] Classical function theory, operator dilation theory, and machine computation on multiply-connected domains, *Mem. Amer. Math. Soc.* **191** (2008), no. 892, viii+159 pp.

AGLER, J. AND YOUNG, N. J.

[1] A commutant lifting theorem for a domain in $\mathbf{C}^2$ and spectral interpolation, *J. Funct. Anal.* **161** (1999), 452–477.

[2] The two-point spectral Nevanlinna-Pick problem, *Integral Eq. Oper. Theory* **37** (2000), 375–385.

[3] The two-by-two spectral Nevanlinna-Pick problem, *Trans. Amer. Math. Soc.* **356** (2004), 573–585.

AMBROZIE, C. AND MÜLLER, V.
[1] Invariant subspaces for polynomially bounded operators, *J. Funct. Anal.* **213** (2004), 321–345.

ANDÔ, T.
[1] On a pair of commutative contractions, *Acta Sci. Math. (Szeged)* **24** (1963), 88–90.

APOSTOL, C., DOUGLAS, R. G. AND FOIAŞ, C.
[1] Quasi-similar models for nilpotent operators, *Trans. Amer. Math. Soc.* **224** (1976), 407–415.

ARVESON, W. B.
[1] Subalgebras of C^*-algebras, *Acta. Math.* **123** (1969), 141–224.
[2] Subalgebras of C^*-algebras. II, *Acta. Math.* **128** (1972), 271–308.
[3] Subalgebras of C^*-algebras. III, *Acta. Math.* **181** (1998), 159–228.

ATZMON, A.
[1] Characterization of operators of class C_0 and a formula for their minimal function, *Acta Sci. Math. (Szeged)* **50** (1986), 191–211.
[2] Unicellular and nonunicellular dissipative operators, *Acta Sci. Math. (Szeged)* **57** (1993), 45–54.

BALAKRISHNAN, A. V.
[1] Fractional powers of closed operators and the semigroups generated by them, *Pacific J. Math.* **10** (1960), 419–437.

BALL, J. A.
[1] Models for noncontractions, *J. Math. Anal. Appl.* **52** (1975), 235–254.
[2] Factorization and model theory for contraction operators with unitary part, *Mem. Amer. Math. Soc.* **13** (1978), no. 198, iv+68 pp.
[3] Operators of class C_{00} over multiply-connected domains, *Michigan Math. J.* **25** (1978), 183–196.

BALL, J. A., FOIAS, C., HELTON, J. W. AND TANNENBAUM, A.
[1] On a local nonlinear commutant lifting theorem, *Indiana Univ. Math. J.* **36** (1987), 693–709.

BALL, J. A. AND GOHBERG, I.
[1] A commutant lifting theorem for triangular matrices with diverse applications, *Integral Eq. Oper. Theory* **8** (1985), 205–267.

BALL, J. A. AND HELTON, J. W.
[1] Shift invariant subspaces, passivity, reproducing kernels and H^∞-optimization, *Oper. Theory Adv. Appl.* **35** (1988), 265–310.
[2] Inner-outer factorization of nonlinear operators, *J. Funct. Anal.* **104** (1992), 363–413.

BALL, J. A. AND KRIETE, T. L., III
[1] Operator-valued Nevanlinna-Pick kernels and the functional models for contraction operators, *Integral Eq. Oper. Theory* **10** (1987), 17–61.

BALL, J. A., LI, W. S., TIMOTIN, D., AND TRENT, T. T.
[1] A commutant lifting theorem on the polydisc, *Indiana Univ. Math. J.* **48** (1999), 653–675.

BALL, J. A. AND LUBIN, A.
[1] On a class of contractive perturbations of restricted shifts, *Pacific J. Math.* **63** (1976), 309–323.

BALL, J. A., TRENT, T. T. AND VINNIKOV, V.
[1] Interpolation and commutant lifting for multipliers on reproducing kernel Hilbert spaces, *Oper. Theory Adv. Appl.* **122** (2001), 89–138.

BEAUZAMY, B.
[1] Spectre d'une contraction de classe C_1 et de son extension unitaire, *Publications de l'Université Paris VII. Séminaire d'Analyse fonctionelle*, 1983–84, pp. 1–8.

[2] *Introduction to Operator Theory and Invariant Subspaces*, North Holland, Amsterdam, 1988.

BEAUZAMY B. AND ROME, M.
[1] Extension unitaire et fonctions de représentation d'une contraction de classe C_1, *Arkiv för Mathematik* **23** (1985), 1–17.

BENAMARA, N.-E. AND NIKOLSKIĬ, N. K.
[1] Resolvent tests for similarity to a normal operator, *Proc. London Math. Soc. (3)* **78** (1999), 585–626.

BERBERIAN, S. K.
[1] Naĭmark's moment theorem, *Michigan Math. J.* **13** (1966), 171–184.

BERCOVICI, H.
[1] C_0-Fredholm operators, *Acta Sci. Math. (Szeged)* **42** (1980), 3–42.
[2] On the Jordan model of C_0 operators. II, *Acta Sci. Math. (Szeged)* **42** (1980), 43–56.
[3] *Operator Theory and Arithmetic in H^∞*, American Mathematical Society, Providence, Rhode Island, 1988.
[4] Factorization theorems and the structure of operators on Hilbert space, *Ann. of Math. (2)* **128** (1988), 399–413.
[5] Notes on invariant subspaces, *Bull. Amer. Math. Soc. (N.S.)* **23** (1990), 1–36.
[6] The unbounded commutant of an operator of class C_0, *Operators and Matrices*, **3** (2009), 599–605.

BERCOVICI, H., DOUGLAS, R. G., FOIAS, C., AND PEARCY, C.
[1] Confluent operator algebras and the closability property, preprint, *J. Funct. Anal.*, **258** (2010), 4122–4153.

BERCOVICI, H., FOIAŞ, C., KÉRCHY, L., AND SZ.-NAGY, B.
[1] Compléments à l'étude des opérateurs de classe C_0. IV, *Acta Sci. Math. (Szeged)* **41** (1979), 29–31.

BERCOVICI, H., FOIAS C., AND PEARCY, C. M.
[1] *Dual Algebras with Applications to Invariant Subspaces and Dilation Theory*, CBMS Regional Conference Series in Mathematics, 56, American Mathematical Society, Providence, RI, 1985.
[2] A spectral mapping theorem for functions with finite Dirichlet integral, *J. Reine Angew. Math.* **366** (1986), 1–17.

BERCOVICI, H., FOIAŞ, C., PEARCY, C. M., AND SZ.-NAGY, B.
[1] Functional models and extended spectral dominance, *Acta Sci. Math. (Szeged)* **43** (1981), 243–254.
[2] Factoring compact operator-valued functions, *Acta Sci. Math. (Szeged)* **48** (1985), 25–36.

BERCOVICI, H., FOIAŞ, C., AND SZ.-NAGY, B.
[1] Compléments à l'étude des opérateurs de classe C_0. III, *Acta Sci. Math. (Szeged)* **37** (1975), 313–332.
[2] Reflexive and hyper-reflexive operators of class C_0, *Acta Sci. Math. (Szeged)* **43** (1981), 5–13.

BERCOVICI, H., FOIAS, C., AND TANNENBAUM, A.
[1] On skew Toeplitz operators. I, *Oper. Theory Adv. Appl.* **29** (1988), 21–43.
[2] A spectral commutant lifting theorem, *Trans. Amer. Math. Soc.* **325** (1991), 741–763.
[3] On spectral tangential Nevanlinna-Pick interpolation, *J. Math. Anal. Appl.* **155** (1991), 156–176.
[4] On skew Toeplitz operators. II, *Oper. Theory Adv. Appl.* **104** (1998), 23–35.

BERCOVICI, H. AND KÉRCHY, L.
[1] Quasi-similarity and properties of the commutant of C_{11} contractions, *Acta Sci. Math. (Szeged)* **45** (1983), 67–74.
[2] On the spectra of C_{11}-contractions, *Proc. Amer. Math. Soc.* **95** (1985), 412–418.

[3] Spectral behaviour of C_{10}-contractions, *Proceedings of the 22nd Conference in Oper. Theory*, Timişoara, 2008, to appear.

BERCOVICI, H., LI, W. S., AND SMOTZER, T.
[1] A continuous version of the Littlewood-Richardson rule and its application to invariant subspaces, *Adv. Math.* **134** (1998), 278–293.
[2] Continuous versions of the Littlewood-Richardson rule, selfadjoint operators, and invariant subspaces, *J. Oper. Theory* **54** (2005), 69–92.

BERCOVICI, H. AND VOICULESCU, D.
[1] Tensor operations on characteristic functions of C_0 contractions, *Acta Sci. Math. (Szeged)* **39** (1977), 205–231.

BERGER, C. A.
[1] Normal dilations, *Doctoral dissertation*, Cornell University, 1963.
[2] A strange dilation theorem, Abstract 625–152, *Amer. Math. Soc. Notices* **12** (1965), 590.

BERGER, C. A. AND STAMPFLI, J. G.
[1] Norm relations and skew dilations, *Acta Sci. Math. (Szeged)* **28** (1967), 191–195.
[2] Mapping theorems for the numerical range, *Amer. J. Math.* **89** (1967), 1047–1055.

BEURLING, A.
[1] On two problems concerning linear transformations in Hilbert space, *Acta Math.* **81** (1948), 239–255.

BHATTACHARYYA, T., ESCHMEIER, J., AND SARKAR, J.
[1] Characteristic function of a pure commuting contractive tuple, *Integral Eq. Oper. Theory* **53** (2005), 23–32.

BISWAS, A. AND FOIAS, C.
[1] On the general intertwining lifting problem. I, *Acta Sci. Math. (Szeged)* **72** (2006), 271–298.

BISWAS, A., FOIAS, C., AND FRAZHO, A.E.
[1] Weighted commutant lifting, *Acta Sci. Math. (Szeged)* **65** (1999), 657–686.
[2] An intertwining property for positive Toeplitz operators, *J. Oper. Theory* **54** (2005), 269–290.

BOCHNER, S.
[1] Diffusion equation and stochastic processes, *Proc. Nat. Acad. Sci. U. S. A.* **35** (1949), 368–370.

DE BRANGES, L.
[1] Some Hilbert spaces of analytic functions. II, *J. Math. Anal. Appl.* **11** (1965), 44–72.
[2] *Hilbert Spaces of Entire Functions*, Prentice-Hall, Englewood Cliffs, N.J., 1968
[3] Some Hilbert spaces of entire functions, *Trans. Amer. Math. Soc.* **96** (1960), 259–295.

DE BRANGES, L. AND ROVNYAK, J.
[1] The existence of invariant subspaces, *Bull. Amer. Math. Soc.* **70** (1964), 718–721.
[2] Canonical models in quantum scattering theory, *Perturbation Theory and Its Applications in Quantum Mechanics*, ed. by C. H. Wilcox, Wiley, New York, 1966, 295–392.
[3] *Square Summable Power Series*, Holt, Rinehart and Winston, New York, 1966.

BREHMER, S.
[1] Über vetauschbare Kontraktionen des Hilbertschen Raumes, *Acta Sci. Math. (Szeged)* **22** (1961), 106–111.

BRODSKIĬ, M. S.
[1] The multiplication theorem for characteristic matrix-functions of linear operators, *Dokl. Akad. Nauk SSSR (N.S.)* **97** (1954), 761–764.
[2] Characteristic matrix functions of linear operators, *Mat. Sb. N.S.* **39 (81)** (1956), 179–200.
[3] On Jordan cells of infinite-dimensional operators, *Dokl. Akad. Nauk SSSR (N.S.)* **111** (1956), 926–929.

[4] Triangular representation of some operators with completely continuous imaginary part, *Dokl. Akad. Nauk SSSR* **133** (1960), 1271–1274; translated as *Soviet Math. Dokl.* **1** (1960) 952–955.
[5] Unicellularity criteria for Volterra operators, *Dokl. Akad. Nauk SSSR* **138** (1961), 512–514.
[6] A multiplicative representation of certain analytic operator-functions, *Dokl. Akad. Nauk SSSR* **138** (1961), 751–754.
[7] On the triangular representation of completely continuous operators with one-point spectra, *Uspehi Mat. Nauk* **16**:1 (1961), 135–141.
[8] Operators with nuclear imaginary components, *Acta Sci. Math. (Szeged)* **27** (1966), 147–155.
[9] *Triangular and Jordan Representations of Linear Operators*, Nauka, Moscow, 1966.

BRODSKIĬ, M. S., GOHBERG, I. C., KREĬN, M. G., AND MACAEV, V. I.
[1] Some new investigations in the theory of non-selfadjoint operators, *Proc. Fourth All-Union Math. Congr. (Leningrad, 1961)* **2** (1964), 261–271.

BRODSKIĬ, M. S. AND KISILEVS'KIĬ, G. È.
[1] Criterion for unicellularity of dissipative Volterra operators with nuclear imaginary components, *Izv. Akad. Nauk SSSR Ser. Mat.* **30** (1966), 1213–1228.

BRODSKIĬ, M. S. AND LIVŠIC, M. S.
[1] Spectral analysis of non-self-adjoint operators and intermediate systems, *Uspehi Mat. Nauk (N.S.)* **13:1 (79)** (1958), 3–85.

BRODSKIĬ, M. S. AND ŠMUL'JAN, JU. L.
[1] Invariant subspaces of a linear operator and divisors of its characteristic function, *Uspehi Mat. Nauk* **19:1 (115)** (1964), 143–149.

BRODSKIĬ, V. M.
[1] Multiplicative representation of the characteristic functions of contraction operators, *Dokl. Akad. Nauk SSSR* **173** (1967), 256–259.

BRODSKIĬ, V. M. AND BRODSKIĬ, M. S.
[1] The abstract triangular representation of bounded linear operators and the multiplicative expansion of their eigenfunctions, *Dokl. Akad. Nauk SSSR* **181** (1968), 511–514.
[2] Factorization of the characteristic function and invariant subspaces of a contraction operator, *Funkcional. Anal. i Priložen.* **8**:2 (1974), 63–64.

BROWN, S. W.
[1] Some invariant subspaces for subnormal operators, *Integral Eq. Oper. Theory* **1** (1978), 310–333.
[2] Contractions with spectral boundary, *Integral Eq. Oper. Theory* **11** (1988), 49–63.
[3] Full analytic subspaces for contractions with rich spectrum, *Pacific J. Math.* **132** (1988), 1–10.

BROWN, S. W. AND CHEVREAU, B.
[1] Toute contraction à calcul fonctionnel isométrique est réflexive, *C. R. Acad. Sci. Paris Sér. I Math.* **307** (1988), 185–188.

BROWN, S. W., CHEVREAU, B. AND PEARCY, C.
[1] Contractions with rich spectrum have invariant subspaces, *J. Oper. Theory* **1** (1979), 123–136.
[2] On the structure of contraction operators. II, *J. Funct. Anal.* **76** (1988), 30–55.

DE BRUIJN, N. G.
[1] On unitary equivalence of unitary dilations of contractions in Hilbert space, *Acta Sci. Math. (Szeged)* **23** (1962), 100–105.

BUNCE, J. W.
[1] Models for n-tuples of noncommuting operators, *J. Funct. Anal.* **57** (1984), 21–30.

CASSIER, G.
[1] Un exemple d'opérateur pour lequel les topologies faible et ultrafaible ne coïncident pas sur l'algèbre duale, *J. Oper. Theory* **16** (1986), 325–333.

[2] Mapping formula for functional calculus, Julia's lemma for operators and applications, *Acta Sci. Math. (Szeged)* **74** (2008), 783–805.

CASSIER, G. AND FACK, T.
[1] Contractions in von Neumann algebras, *J. Funct. Anal.* **135** (1996), 297–338.
[2] On power-bounded operators in finite von Neumann algebras, *J. Funct. Anal.* **141** (1996), 133–158.

VAN CASTEREN, J. A.
[1] A problem of Sz.-Nagy, *Acta Sci. Math. (Szeged)* **42** (1980), 189–194.
[2] Operators similar to unitary or selfadjoint ones, *Pacific J. Math.* **104** (1983), 241–255.

CHEN, K. Y., HERRERO, D. A., AND WU, P. Y.
[1] Similarity and quasisimilarity of quasinormal operators, *J. Oper. Theory* **2** (1992), 385–412.

CHEVREAU, B.
[1] Sur les contractions à calcul fonctionnel isométrique. II, *J. Oper. Theory* **20** (1988), 269–293.

CLARK, D. N.
[1] On commuting contractions, *J. Math. Anal. Appl.* **32** (1970), 590–596.
[2] One dimensional perturbations of restricted shifts, *J. Analyse Math.* **25** (1972), 169–191.
[3] Commutants that do not dilate, *Proc. Amer. Math. Soc.* **35** (1972), 483–486.

CLARY, S.
[1] Equality of spectra of quasi-similar hyponormal operators, *Proc. Amer. Math. Soc.* **53** (1975), 88–90.

COLOJOARĂ, I. AND FOIAŞ, C.
[1] *Theory of Generalized Spectral Operators*, Gordon and Breach, New York, 1968.

CONWAY, J. B.
[1] *A Course in Functional Analysis*, Springer, New York, 1990.

CONWAY, J. B. AND GILLESPIE, T. A.
[1] Is an isometry determined by its invariant subspace lattice?, *J. Oper. Theory* **22** (1989), 31–49.

COOPER, J. L. B.
[1] One-parameter semigroups of isometric operators in Hilbert space, *Ann. of Math. (2)* **48** (1947), 827–842.

CRABB, M. J. AND DAVIE, A. M.
[1] von Neumann's inequality for Hilbert space operators, *Bull. London Math. Soc.* **7** (1975), 49–50.

CURTO, R. E. AND VASILESCU, F.-H.
[1] Standard operator models in the polydisc, *Indiana Univ. Math. J.* **42** (1993), 791–810.
[2] Standard operator models in the polydisc. II, *Indiana Univ. Math. J.* **44** (1995), 727–746.

DAVIDSON, K. R. AND HERRERO, D. A.
[1] The Jordan form of a bitriangular operator, *J. Funct. Anal.* **94** (1990), 27–73.

DAVIDSON, K. R., KRIBS, D. W., AND SHPIGEL, M. E.
[1] Isometric dilations of non-commuting finite rank n-tuples, *Canad. J. Math.* **53** (2001), 506–545.

DAVIDSON, K. R. AND ZAROUF, F.
[1] Incompatibility of compact perturbations with the Sz.-Nagy–Foias functional calculus, *Proc. Amer. Math. Soc.* **121** (1994), 519–522.

DAVIS, CH.
[1] The shell of a Hilbert space operator, *Acta Sci. Math. (Szeged)* **29** (1968), 69–86.

DAVIS, CH. AND FOIAS, C.
[1] Operators with bounded characteristic function and their J-unitary dilation, *Acta Sci. Math. (Szeged)* **32** (1971), 127–139.

DEVINATZ, A.
[1] The factorization of operator valued functions, *Ann. of Math. (2)* **73** (1961), 458–495.

DIXMIER, J.
[1] *Von Neumann Algebras*, North Holland, Amsterdam, 1981.
[2] Les moyennes invariantes dans les semi-groups et leurs applications, *Acta Sci. Math. (Szeged)* **12** (1950), 213–227.

DOLPH, C. L.
[1] Positive real resolvents and linear passive Hilbert systems, *Ann. Acad. Sci. Fenn. Ser. A I No.* **336/9** (1963), 39 pp.

DOLPH, C. L. AND PENZLIN, F.
[1] On the theory of a class of non-self-adjoint operators and its applications to quantum scattering theory, *Ann. Acad. Sci. Fenn. Ser. A. I.* **263** (1959), 36 pp.

DOUGLAS, R. G.
[1] On majorization, factorization, and range inclusion of operators on Hilbert space, *Proc. Amer. Math. Soc.* **17** (1966), 413–415.
[2] On factoring positive operator functions, *J. Math. Mech.* **16** (1966), 119–126.
[3] Structure theory for operators. I, *J. Reine Angew. Math.* **232** (1968), 180–193.
[4] On extending commutative semigroups of isometries, *Bull. London Math. Soc.* **1** (1969), 157–159.
[5] On the hyperinvariant subspaces for isometries, *Math. Z.* **107** (1968), 297–300.

DOUGLAS, R. G. AND HELTON, J. W.
[1] Inner dilations of analytic matrix functions and Darlington synthesis, *Acta Sci. Math. (Szeged)* **34** (1973), 61–67.

DOUGLAS, R. G., MUHLY, P. S., AND PEARCY, C.
[1] Lifting commuting operators, *Michigan Math. J.* **15** (1968), 385–395.

DOUGLAS, R. G. AND PAULSEN, V. I.
[1] Completely bounded maps and hypo-Dirichlet algebras, *Acta Sci. Math. (Szeged)* **50** (1986), 143–157.
[2] *Hilbert Modules over Function Algebras*, Pitman Research Notes in Mathematics Series, 217. Longman Scientific & Technical, Harlow, 1989.

DOUGLAS, R. G. AND PEARCY, C.
[1] On a topology for invariant subspaces, *J. Functional Analysis* **2** (1968), 323–341.

DOUGLAS, R. G. AND YANG, R.
[1] Operator theory in the Hardy space over the bidisk. I, *Integral Eq. Oper. Theory* **38** (2000), 207–221.

DRITSCHEL, M. A. AND MCCULLOUGH, S.,
[1] The failure of rational dilation on a triply connected domain, *J. Amer. Math. Soc.* **18** (2005), 873–918.

DRURY, S. W.,
[1] A generalization of von Neumann's inequality to the complex ball, *Proc. Amer. Math. Soc.* **68** (2005), 300–304.

DUNFORD, N. AND SCHWARTZ, J. T.
[1] *Linear Operators*, Part I, Wiley, New York, 1958.

[2] *Linear Operators*, Part II, Wiley, New York, 1963.

DURSZT, E.
[1] On unitary ρ-dilations of operators, *Acta Sci. Math. (Szeged)* **27** (1966), 247–250.
[2] On the spectrum of unitary ρ-dilations, *Acta Sci. Math. (Szeged)* **28** (1967), 299–304.
[3] On the unitary part of an operator on Hilbert space, *Acta Sci. Math. (Szeged)* **31** (1970), 87–89.
[4] Factorization of operators in $\mathscr{C}_\rho$ classes, *Acta Sci. Math. (Szeged)* **37** (1975), 195–199.

DURSZT, E. AND SZ.-NAGY, B.
[1] Remark to a paper: "Models for noncommuting operators" by A. E. Frazho, *J. Funct. Anal* **52** (1983), 146–147.

ECKSTEIN, G.
[1] On the spectrum of contractions of class $C_{1\cdot}$, *Acta Sci. Math. (Szeged)* **39** (1977), 251–254.

EGERVÁRY, E.
[1] On the contractive linear transformations of n-dimensional vector space, *Acta Sci. Math. (Szeged)* **15** (1954), 178–182.

FATOU, P.
[1] Séries trigonometriques et séries de Taylor, *Acta Math.* **30** (1906), 335–400.

FIALKOW, L. A.
[1] A note on quasisimilarity of operators, *Acta Sci. Math. (Szeged)* **39** (1977), 67–85.
[2] A note on quasisimilarity. II, *Pacific J. Math.* **70** (1977), 151–162.

FISHER, S. D.
[1] *Function Theory on Planar Domains. A second course in complex analysis*, John Wiley & Sons, New York, 1983.

FOGUEL, S. R.
[1] A counterexample to a problem of Sz.-Nagy, *Proc. Amer. Math. Soc.* **15** (1964), 788–790.

FOIAŞ, C.
[1] Sur certains théorèmes de J. von Neumann concernant les ensembles spectraux, *Acta Sci. Math. (Szeged)* **18** (1957), 15–20.
[2] La mesure harmonique-spectrale et la théorie spectrale des opérateurs généraux d'un espace de Hilbert, *Bull. Soc. Math. France* **85** (1957), 263–282.
[3] On Hille's spectral theory and operational calculus for semi-groups of operators in Hilbert space, *Compositio Math.* **14** (1959), 71–73.
[4] Certaines applications des ensembles spectraux. I. Mesure harmonique-spectrale, *Stud. Cerc. Mat.* **10** (1959), 365–401.
[5] A remark on the universal model for contractions of G. C. Rota, *Com. Acad. R. P. Române* **13** (1963), 349–352.
[6] Maximality of the space H^∞ in the functional calculus, *An. Univ. Timişoara Ser. Şti. Mat.-Fiz.* **2** (1964), 77–82.
[7] Modèles fonctionnels, liaison entre les théories de la prédiction, de la fonction caractéristique et de la dilatation unitaire, *Deuxième Colloq. d'Anal. Fonct.*, Centre Belge Recherches Math., Librairie Universitaire, Louvain, 1964, 63–76.
[8] The class C_0 in the theory of decomposable operators, *Rev. Roumaine Math. Pures Appl.* **14** (1969), 1433–1440.
[9] On the scalar parts of a decomposable operator, *Rev. Roumaine Math. Pures Appl.* **17** (1972), 1181–1198.
[10] Factorisations étranges, *Acta Sci. Math. (Szeged)* **34** (1973), 85–89.

FOIAS, C. AND FRAZHO, A. E.
[1] *The Commutant Lifting Approach to Interpolation Problems*, Birkhäuser Verlag, Basel, 1990.

FOIAS, C., FRAZHO, A. E., GOHBERG, I., AND KAASHOEK, M. A.
[1] *Metric Constrained Interpolation, Commutant Lifting and Systems*, Birkhäuser Verlag, Basel, 1998.

FOIAS, C., FRAZHO, A. E., AND KAASHOEK, M. A.
[1] A weighted version of almost commutant lifting, *Oper. Theory Adv. Appl.* **129** (2001), 311–340.
[2] Contractive liftings and the commutator, *C. R. Math. Acad. Sci. Paris* **335** (2002), 431–436.
[3] Relaxation of metric constrained interpolation and a new lifting theorem, *Integral Eq. Oper. Theory* **42** (2002), 253–310.
[4] The distance to intertwining operators, contractive liftings and a related optimality result, *Integral Eq. Oper. Theory* **47** (2003), 71–89.

FOIAS, C., FRAZHO, A. E., AND LI, W. S.
[1] The exact H^2 estimate for the central H^∞ interpolant, *Oper. Theory Adv. Appl.* **64** (1983), 119–156.
[2] On H^2 minimization for the Carathéodory-Schur interpolation problem, *Integral Eq. Oper. Theory* **21** (1995), 24–32.

FOIAŞ, C. AND GEHÉR, L.
[1] Über die Weylsche Vertauschungsrelation, *Acta Sci. Math. (Szeged)* **24** (1963), 97–102.

FOIAŞ, C., GEHÉR, L., AND SZ.-NAGY, B.
[1] On the permutability condition of quantum mechanics, *Acta Sci. Math. (Szeged)* **21** (1960), 78–89.

FOIAS, C., GU, C., AND TANNENBAUM, A.
[1] Intertwining dilations, intertwining extensions and causality, *Acta Sci. Math. (Szeged)* **57** (1993), 101–123.

FOIAŞ, C. AND MLAK, W.
[1] The extended spectrum of completely non-unitary contractions and the spectral mapping theorem, *Studia Math.* **26** (1966), 239–245.

FOIAS, C., ÖZBAY, H., AND TANNENBAUM, A.
[1] *Robust Control of Infinite-Dimensional Systems*, Lecture Notes in Control and Information Sciences, **209**, Springer-Verlag, London, 1996.

FOIAS, C. AND PEARCY, C.
[1] (BCP)-operators and enrichment of invariant subspace lattices, *J. Oper. Theory* **9** (1983), 107–202.

FOIAŞ, C., PEARCY, C., AND SZ.-NAGY, B.
[1] The functional model of a contraction and the space L^1, *Acta Sci. Math. (Szeged)* **42** (1980), 201–204.
[2] Functional models and extended spectral dominance, *Acta Sci. Math. (Szeged)* **43** (1981), 243–254.

FOIAS, C. AND TANNENBAUM, A.
[1] Causality in commutant lifting theory, *J. Funct. Anal.* **118** (1993), 407–441.

FRAZHO, A. E.
[1] Models for noncommuting operators, *J. Funct. Anal.* **48** (1982), 1–11.
[2] Complements to models for commuting operators, *J. Funct. Anal.* **59** (1984), 445–461.

FRIEDRICHS, K. O.
[1] On the perturbation of continuous spectra, *Commun. Appl. Math.* **1** (1948), 361–406.

FUHRMANN, P. A.
[1] On the corona theorem and its application to spectral problems in Hilbert space, *Trans. Amer. Math. Soc.* **132** (1968), 55–66.
[2] A functional calculus in Hilbert spaces based on operator valued analytic functions, *Israel J. Math.* **6** (1968), 267–278.

FURUTA, T.
[1] A generalization of Durszt's theorem on unitary ρ-dilatations, *Proc. Japan Acad.* **43** (1967), 269–272.
[2] Relations between unitary ρ-dilatations and two norms, *Proc. Japan Acad.* **44** (1968), 16–20.

GAU, H.-L. AND WU, P. Y.
[1] Numerical range of $S(\varphi)$, *Linear and Multilinear Algebra* **45** (1998), 49–73.

GILFEATHER, F.
[1] Weighted bilateral shifts of class C_{01}, *Acta Sci. Math. (Szeged)* **32** (1971), 251–254.

GINZBURG, YU. P.
[1] On J-contractive operator functions, *Dokl. Akad. Nauk SSSR (N.S.)* **117** (1957), 171–173.
[2] The factorization of analytic matrix functions, *Dokl. Akad. Nauk SSSR* **159** (1964), 489–492.
[3] Multiplicative representations of bounded analytic operator-functions, *Dokl. Akad. Nauk SSSR* **170** (1966), 23–26.
[4] Multiplicative representations and minorants of bounded analytic operator functions, *Funkcional. Anal. i Priložen.* **1**:3 (1967), 9–23.
[5] Divisors and minorants of operator-valued functions of bounded form, *Mat. Issled.* **2**:4 (1967), 47–72.

GOHBERG, I. C. AND KREĬN, M. G.
[1] On the problem of factoring operators in a Hilbert space, *Dokl. Akad. Nauk SSSR* **147** (1962), 279–282.
[2] Factorization of operators in Hilbert space, *Acta Sci. Math. (Szeged)* **25** (1964), 90–123.
[3] On the multiplicative representation of the characteristic functions of operators close to the unitary ones, *Dokl. Akad. Nauk SSSR* **164** (1965), 732–735.
[4] *Introduction to the Theory of Linear Nonselfadjoint Operators on Hilbert Space*, Nauka, Moscow, 1965.
[5] On a description of contraction operators similar to unitary ones, *Funkcional. Anal. i Priložen.* **1** (1967), 38–60.
[6] Triangular representations of linear operators and multiplicative representations of their characteristic functions, *Dokl. Akad. Nauk SSSR* **175** (1967), 272–275.
[7] *The Theory of Volterra Operators on a Hilbert Space and Its Applications*, Nauka, Moscow, 1967.

GOLUZIN, G.M.
[1] *Geometric Theory of Functions of a Complex Variable*, GITTL, Moscow, 1952, (Russian); Transl. Math. Mono. **26**, Amer. Math. Soc., Providence, 1969, 1983.

HADWIN, D.
[1] A general view of reflexivity, *Trans. Amer. Math. Soc.* **344** (1994), 325–360.

HALMOS, P. R.
[1] Normal dilations and extensions of operators, *Summa Brasil. Math.* **2** (1950), 125–134.
[2] Shifts on Hilbert spaces, *J. Reine Angew. Math.* **208** (1961), 102–112.
[3] *Positive Definite Sequences and the Miracle of w*, A talk presented in the functional analysis seminar at the University of Michigan, July 8, 1965, 17pp.
[4] *A Hilbert Space Problem Book*, Van Nostrand, Princeton, N. J., 1967.
[5] On Foguel's answer to Nagy's question, *Proc. Amer. Math. Soc.* **15** (1964), 791–793.
[6] Ten problems in Hilbert space, *Bull. Amer. Math. Soc.* **76** (1970), 887–933.

HALPERIN, I.
[1] The unitary dilation of a contraction operator, *Duke Math. J.* **28** (1961), 563–571.
[2] Sz.-Nagy–Brehmer dilations, *Acta Sci. Math. (Szeged)* **23** (1962), 279–289.
[3] Unitary dilations which are orthogonal bilateral shift operators, *Duke Math. J.* **29** (1962), 573–580.
[4] Intrinsic description of the Sz.-Nagy–Brehmer unitary dilation, *Studia Math.* **22** (1962/1963), 211–219.
[5] Interlocking dilations, *Duke Math. J.* **30** (1963), 475–484.

HEINZ, E.
[1] Ein v. Neumannscher Satz über beschränkte Operatoren im Hilbertschen Raum, *Nachr. Akad. Wiss. Göttingen. Math.-Phys. Kl. IIa. Math.-Phys.-Chem.* **1952** (1952), 5–6.

HELSON, H.
[1] *Lectures on Invariant Subspaces*, Academic Press, New York, 1964.

HELSON, H. AND LOWDENSLAGER, D.
[1] Prediction theory and Fourier series in several variables, *Acta Math.* **99** (1958), 165–202.
[2] Prediction theory and Fourier series in several variables. II, *Acta Math.* **106** (1961), 175–213.

HELTON, J. W.
[1] The characteristic functions of operator theory and electrical network realization, *Indiana Univ. Math. J.* **22** (1972/1973), 403–414.
[2] Discrete time systems, operator models, and scattering theory, *J. Funct. Anal.* **16** (1974), 15–38.
[3] Beyond commutant lifting, *Operator Theory: Operator Algebras and Applications, Part 1, Proc. Sympos. Pure Math.* **51**, Amer. Math. Soc., Providence, 1990, 219–224.

HELTON, J. W. AND WAVRIK, J. J.
[1] Rules for computer simplification of the formulas in operator model theory and linear systems, *Oper. Theory Adv. Appl.* **73** (1994), 325–354.

HERRERO, D. A.
[1] The exceptional set of a C_0 contraction, *Trans. Amer. Math. Soc.* **173** (1972), 93–115.
[2] Quasisimilarity does not preserve the hyperlattice, *Proc. Amer. Math. Soc.* **65** (1977), 80–84.
[3] On the essential spectra of quasisimilar operators, *Canad. J. Math.* **40** (1988), 1436–1457.

HILLE, E.
[1] *Functional Analysis and Semi-Groups*, American Mathematical Society Colloquium Publications, vol. 31, New York, 1948.

HOFFMAN, K.
[1] *Banach Spaces of Analytic Functions*, Dover, New York, 1988.

HOLBROOK, J. A. R.
[1] On the power-bounded operators of Sz.-Nagy and Foiaş, *Acta Sci. Math. (Szeged)* **29** (1968), 299–310.
[2] Multiplicative properties of the numerical radius in operator theory, *J. Reine Angew. Math.* **237** (1969), 166–174.
[3] Operators similar to contractions, *Acta Sci. Math. (Szeged)* **34** (1973), 163–168.

ISTRĂŢESCU, V.
[1] A remark on a class of power-bounded operators in Hilbert space, *Acta Sci. Math. (Szeged)* **29** (1968), 311–312.

ITÔ, T.
[1] On the commutative family of subnormal operators, *J. Fac. Sci. Hokkaido Univ. Ser. I* **14** (1958), 1–15.

JULIA, G.
[1] Sur les projections des systèmes orthonormaux de l'espace Hilbertien, *C. R. Acad. Sci. Paris* **218** (1944), 892–895,
[2] Les projections des systèmes orthonormaux de l'espace Hilbertien et les opérateurs bornés, *C. R. Acad. Sci. Paris* **219** (1944), 8–11.
[3] Sur la représentation analytique des opérateurs bornés ou fermés de l'espace Hilbertien, *C. R. Acad. Sci. Paris* **219** (1944), 225–227.

KADISON, R. V. AND SINGER I. M.
[1] Three test problems in operator theory, *Pacific J. Math.* **7** (1957), 1101–1106.

KAFTAL, V., LARSON, D. AND WEISS, G.
[1] Quasitriangular subalgebras of semifinite von Neumann algebras are closed, *J. Funct. Anal.* **107** (1992), 387–401.

KALISCH, G. K.
[1] On similarity, reducing manifolds, and unitary equivalence of certain Volterra operators, *Ann. of Math. (2)* **66** (1957), 481–494.

KAPUSTIN, V. V.
[1] A criterion for the reflexivity of contractions with a defect operator of the Hilbert-Schmidt class, *Dokl. Akad. Nauk SSSR* **318** (1991), 919-922.
[2] Reflexivity of operators: General methods and a criterion for almost isometric contractions, *St. Petersburg Math. J.* **4** (1993), 319–335.

KATO, T.
[1] Fractional powers of dissipative operators, *J. Math. Soc. Japan* **13** (1961), 246–274 and **14** (1962), 242–248.
[2] Some mapping theorems for the numerical range, *Proc. Japan Acad.* **41** (1965), 652–655.

KENDALL, D. G.
[1] Unitary dilations of Markov transition operators, and the corresponding integral representations for transition-probability matrices, *Probability and statistics: The Harald Cramér Volume* (edited by Ulf Grenander), Stockholm, 1959, 139–161.
[2] Unitary dilations of one-parameter semigroups of Markov transition operators, and the corresponding integral representations for Markov processes with a countable infinity of states, *Proc. London Math. Soc. (3)* **9** (1959), 417–431.

KÉRCHY, L.
[1] Subspace lattices connected with C_{11}-contractions, *Anniversary Volume on Approximation Theory and Functional Analysis* (eds. P.L. Butzer, R.L. Stens, B. Sz.-Nagy), Birkhäuser Verlag, Basel, 1984, 89–98.
[2] Contractions being weakly similar to unitaries, *Oper. Theory Adv. Appl.* **17** (1986), 187–200.
[3] A description of invariant subspaces of C_{11}-contractions, *J. Oper. Theory* **15** (1986), 327–344.
[4] On the spectra of contractions belonging to special classes, *J. Funct. Anal.* **67** (1986), 153–166.
[5] On the residual parts of completely non-unitary contractions, *Acta Math. Hungar.* **50** (1987), 127–145.
[6] Invariant subspaces of $C_{1\cdot}$-contractions with non-reductive unitary extensions, *Bull. London Math. Soc.* **19** (1987), 161–166.
[7] Injection of shifts into contractions, *Acta Sci. Math. (Szeged)* **53** (1989), 329–338.
[8] Isometric asymptotes of power bounded operators, *Indiana Univ. Math. J.* **38** (1989), 173–188.
[9] Injection of unilateral shifts into contractions with non-vanishing unitary asymptotes, *Acta Sci. Math. (Szeged)* **61** (1995), 443–476.
[10] Operators with regular norm-sequences, *Acta Sci. Math. (Szeged)* **63** (1997), 571–605.

[11] Criteria of regularity for norm-sequences, *Integral Equations Oper. Theory* **34** (1999), 458–477.
[12] Representations with regular norm-behaviour of discrete abelian semigroups, *Acta Sci. Math. (Szeged)* **65** (1999), 701–726.
[13] Hyperinvariant subspaces of operators with non-vanishing orbits, *Proc. Amer. Math. Soc.* **127** (1999), 1363–1370.
[14] Isometries with isomorphic invariant subspace lattices, *J. Funct. Anal.* **170** (2000), 475–511.
[15] On the hyperinvariant subspace problem for asymptotically nonvanishing contractions, *Oper. Theory Adv. Appl.* **127** (2001), 399–422.
[16] Shift-type invariant subspaces of contractions, *J. Funct. Anal.* **246** (2007), 281–301.
[17] On C_0 operators with property (P), *Acta Sci. Math. (Szeged)* **42** (1980), 109–116.

KÉRCHY, L. AND LÉKA, Z.
[1] Representations with regular norm-behaviour of locally compact abelian semigroups, *Studia Math.* **183** (2007), 143–160.

KÉRCHY, L. AND MÜLLER, V.
[1] Criteria of regularity for norm-sequences. II, *Acta Sci. Math. (Szeged)* **65** (1999), 131–138.

KISILEVS'KIĬ, G. È.
[1] Conditions for unicellularity of dissipative Volterra operators with finite-dimensional imaginary component, *Dokl. Akad. Nauk SSSR* **159** (1964), 505–508.
[2] On the analogue of the Jordan theory for a certain class of infinite dimensional operators, *Internat. Congr. Math. Moscow, Abstracts of brief scientific communications*, 1966, Sect. 5, p. 54.
[3] Cyclic subspaces of dissipative operators, *Dokl. Akad. Nauk SSSR* **173** (1967), 1006–1009.
[4] A generalization of the Jordan theory to a certain class of linear operators in Hilbert space, *Dokl. Akad. Nauk SSSR* **176** (1967), 768–770.
[5] Invariant subspaces of Volterra dissipative operators with nuclear imaginary components, *Izv. Akad. Nauk SSSR Ser. Mat.* **32** (1968), 3–23.

KORÁNYI, A.
[1] On some classes of analytic functions of several variables, *Trans. Amer. Math. Soc.* **101** (1961), 520–554.

KREĬN, M. G.
[1] Analytic problems and results in the theory of linear operators on a Hilbert space, *Internat. Congr. Math. Moscow*, 1966, 189–216.

KRIETE III, T. L.
[1] Similarity of canonical models, *Bull. Amer Math. Soc.* **76** (1970), 326–330.
[2] Complete non-selfadjointness of almost selfadjoint operators, *Pacific J. Math.* **42** (1972), 413–437.
[3] Canonical models and the self-adjoint parts of dissipative operators, *J. Funct. Anal.* **23** (1976), 39–94.

KUPIN, S. AND TREIL, S.
[1] Linear resolvent growth of a weak contraction does not imply its similarity to a normal operator, *Illinois J. Math.* **45** (2001), 229–242.

LANGER, H.
[1] Ein Zerspaltungssatz für Operatoren im Hilbertraum, *Acta Math. Acad. Sci. Hungar.* **12** (1961), 441–445.
[2] Über die Wurzeln eines maximalen dissipativen operators, *Acta Math. Acad. Sci. Hungar.* **13** (1962), 415–424.

LANGER, H. AND NOLLAU, V.
[1] Einige Bemerkungen über dissipative Operatoren im Hilbertraum, *Wiss. Z. Techn. Univ. Dresden* **15** (1966), 669–673.

LAX, P. D.
[1] Translation invariant spaces, *Acta Math.* **101** (1959), 163–178.
[2] Translation invariant spaces, *Proc. Internat. Sympos. Linear Spaces* Pergamon, Oxford and Jerusalem Academic Press, Jerusalem, 1961, 299–306.

LAX, P. D. AND PHILLIPS, R. S.
[1] Scattering theory, *Bull. Amer. Math. Soc.* **70** (1964), 130–142.
[2] *Scattering theory*, Academic Press, New York, 1967.

LEBOW, A.
[1] On von Neumann's theory of spectral sets, *J. Math. Anal. Appl.* **7** (1963), 64–90.

LI, W. S. AND MÜLLER, V.
[1] Littlewood-Richardson sequences associated with C_0-operators, *Acta Sci. Math. (Szeged)* **64** (1998), 609–625.

LIVŠIC, M. S.
[1] On a class of linear operators in Hilbert space, *Math. Sb.* **19 (61)** (1946), 239–260; *Amer. Math. Soc. Transl. (2)* **13** (1960), 61–83.
[2] Isometric operators with equal deficiency indices, quasi-unitary operators, *Mat. Sbornik N.S.* **26 (68)** (1950), 247–264.
[3] On spectral decomposition of linear nonself-adjoint operators, *Mat. Sbornik N.S.* **34 (76)** (1954), 145–199.
[4] *Operators, Oscillations, Waves. Open Systems*, Nauka, Moscow, 1966.

LIVŠIC, M. S., KRAVITSKY, N., MARKUS, A. S., AND VINNIKOV, V.
[1] *Theory of Commuting Nonselfadjoint Operators*, Kluwer Academic, Dordrecht, 1995.

LIVŠIC, M. S. AND POTAPOV, V. P.
[1] A theorem on the multiplication of characteristic matrix functions, *Doklady Akad. Nauk SSSR (N.S.)* **72** (1950), 625–628.

LOWDENSLAGER, D. B.
[1] On factoring matrix valued functions, *Ann. of Math. (2)* **78** (1963), 450–454.

MACAEV, V. I. AND PALANT, JU. A.
[1] On the powers of a bounded dissipative operator, *Ukrain. Mat. Ž.* **14** (1962), 329–337.

MAKAROV, N. G. AND VASJUNIN, V. I.
[1] A model for noncontractions and stability of the continuous spectrum, *Complex analysis and spectral theory (Leningrad, 1979/1980)*, Lecture Notes in Math., 864, Springer, Berlin-New York, 1981, 365–412.

MARTIN, R. T. W.
[1] Characterization of the unbounded bicommutant of contractions, *Operators and Matrices* **3** (2009), 589–598.

MASANI, P.
[1] The prediction theory of multivariate stochastic processes. III. Unbounded spectral densities, *Acta Math.* **104** (1960), 141–162.
[2] Shift invariant spaces and prediction theory, *Acta Math.* **107** (1962), 275–290.
[3] Isometric flows on Hilbert space, *Bull. Amer. Math. Soc.* **68** (1962), 624–632.
[4] On the representation theorem of scattering, *Bull. Amer. Math. Soc.* **74** (1968), 618–624.

McCARTHY, J.
[1] Quasisimilarity of rationally cyclic subnormal operators, *J. Oper. Theory* **24** (1990), 105–116.

McKELVEY, R.
[1] Spectral measures, generalized resolvents, and functions of positive type, *J. Math. Anal. Appl.* **11** (1965), 447–477.

MLAK, W.
[1] Characterization of completely non-unitary contractions in Hilbert spaces, *Bull. Acad. Polon. Sci. Sér. Sci. Math. Astronom. Phys.* **11** (1963), 111–113.
[2] Note on the unitary dilation of a contraction operator, *Bull. Acad. Polon. Sci. Sér. Sci. Math. Astronom. Phys.* **11** (1963), 463–467.
[3] Some prediction theoretical properties of unitary dilations, *Bull. Acad. Polon. Sci. Sér. Sci. Math. Astronom. Phys.* **12** (1964), 37–42.
[4] Representations of some algebras of generalized analytic functions, *Bull. Acad. Polon. Sci. Sér. Sci. Math. Astronom. Phys.* **13** (1965), 211–214.
[5] Unitary dilations of contraction operators, *Rozprawy Mat.* **46** (1965), 1–88.
[6] Unitary dilations in case of ordered groups, *Ann. Polon. Math.* **17** (1966), 321–328,
[7] On semi-groups of contractions in Hilbert spaces, *Studia Math.* **26** (1966), 263–272.
[8] Positive definite contraction valued functions, *Bull. Acad. Polon. Sci. Sér. Sci. Math. Astronom. Phys.* **15** (1967), 509–512.
[9] Hyponormal contractions, *Colloq. Math.* **18** (1967), 137–142.
[10] Spectral properties of Q-dilations, *Bull. Acad. Polon. Sci. Sér. Sci. Math. Astronom. Phys.* **17** (1969), 397–400.

MOELLER, J. W.
[1] On the spectra of some translation invariant spaces, *J. Math. Anal. Appl.* **4** (1962), 276–296.

MOORE, B. III AND NORDGREN, E.A.
[1] On quasiequivalence and quasisimilarity, *Acta Sci. Math. (Szeged)* **34** (1973), 311–316.

MUHLY, P. S.
[1] Commutants containing a compact operator, *Bull. Amer. Math. Soc.* **75** (1969), 353–356.
[2] Some remarks on the spectra of unitary dilations, *Studia Math.* **49** (1973/74), 139–147.

MUHLY, P. S. AND SOLEL, B.
[1] Canonical models for representations of Hardy algebras, *Integral Eq. Oper. Theory* **53** (2005), 411–452.

MÜLLER, V.
[1] Jordan models and diagonalization of the characteristic function, *Acta Sci. Math. (Szeged)* **43** (1981), 321–332.

MÜLLER, V. AND TOMILOV, Y.
[1] Quasisimilarity of power bounded operators and Blum-Hanson property, *J. Funct. Anal* **246** (2007), 385–399.

NABOKO, S. N.
[1] Conditions for similarity to unitary and selfadjoint operators, *Funktsional. Anal. i Prilozhen.* **18** (1984), 16–27.

NAĬMARK, M. A.
[1] Positive definite operator functions on a commutative group, *Bull. Acad. Sci. URSS Sér. Math. [Izvestia Akad. Nauk SSSR]* **7** (1943), 237–244.
[2] Self-adjoint extensions of the second kind of a symmetric operator, *Bull. Acad. Sci. URSS. Sér. Math. [Izvestià Akad. Nauk SSSR]* **4** (1940), 53–104.

[3] On a representation of additive operator set functions, *C. R. (Doklady) Acad. Sci. URSS (N.S.)* **41** (1943), 359–361.

NAKANO, H.
[1] On unitary dilations of bounded operators, *Acta Sci. Math. (Szeged)* **22** (1961), 286–288.

VON NEUMANN, J.
[1] Allgemeine Eigenwerttheorie Hermitischer Funktionaloperatoren, *Math. Ann.* **102** (1929), 49–131.
[2] Über einen Satz von Herrn M. H. Stone, *Ann. of Math. (2)* **33** (1932), 567–573.
[3] Die Eindeutigkeit der Schrödingerschen Operatoren, *Math. Ann.* **104** (1931), 570–578.
[4] Eine Spektraltheorie für allgemeine Operatoren eines unitären Raumes, *Math. Nachr.* **4** (1951), 258–281.

NIKOLSKIĬ, N. K.
[1] Multicyclicity phenomenon. I. An introduction and maxi-formulas, *Oper. Theory Adv. Appl.* **42** (1989), 9-57.
[2] *Treatise on the Shift operator. Spectral Function Theory. With an Appendix by S. V. Hruščev and V. V. Peller*, Springer-Verlag, Berlin, 1986.
[3] *Operators, Functions, and Systems: an Easy Reading. Vol. 1. Hardy, Hankel, and Toeplitz*, American Mathematical Society, Providence, RI, 2002.
[4] *Operators, Functions, and Systems: an Easy Reading. Vol. 2. Model Operators and Systems*, American Mathematical Society, Providence, RI, 2002.

NIKOLSKIĬ, N. K. AND HRUŠČEV, S. V.
[1] A functional model and some problems of the spectral theory of functions, *Trudy Mat. Inst. Steklov.* **176** (1987), 97–210, 327.

NIKOLSKIĬ, N. K. AND TREIL, S.
[1] Linear resolvent growth of rank one perturbation of a unitary operator does not imply its similarity to a normal operator, *J. Anal. Math.* **87** (2002), 415–431.

NIKOLSKIĬ, N. K. AND VASYUNIN, V. I.
[1] A unified approach to function models, and the transcription problem, *The Gohberg Anniversary Collection, Vol. II, Oper. Theory Adv. Appl.*, **41**, Birkhäuser, Basel, 1989, 405–434.

NOLLAU, V.
[1] Über Potenzen von linearen Operatoren in Banachschen Räumen, *Acta Sci. Math. (Szeged)* **28** (1967), 107–121.
[2] Über den Logarithmus abgeschlossener Operatoren in Banachschen Räumen, *Acta Sci. Math. (Szeged)* **30** (1969), 161–174.

NORDGREN, E. A.
[1] On quasiequivalence of matrices over H^∞, *Acta Sci. Math. (Szeged)* **34** (1973), 301–310.

OKUBO, K. AND ANDO, T.
[1] Constants related to operators of class C_ρ, *Manuscripta Math.* **16** (1975), 385–394.

PALEY, R. E. A. C. AND WIENER, N.
[1] *Fourier transforms in the complex domain*, American Mathematical Society, Providence, RI, 1987.

PARROTT, S.
[1] Unitary dilations for commuting contractions, *Pacific J. Math.* **34** (1970), 481–490.

PATA, V. AND ZUCCHI, A.
[1] Hyperinvariant subspaces of C_0-operators over a multiply connected region, *Integral Eq. Oper. Theory* **36** (2000), 241–250.

PAULSEN, V.
[1] Every completely polynomially bounded operator is similar to a contraction, *J. Funct. Anal.* **55** (1984), 1–17.
[2] *Completely bounded maps and operator algebras*, Cambridge University Press, Cambridge, UK, 2002.

PEARCY, C.
[1] An elementary proof of the power inequality for the numerical radius, *Michigan Math. J.* **13** (1966), 289–291.

PHILLIPS, R. S.
[1] On the generation of semigroups of linear operators, *Pacific J. Math.* **2** (1952), 343–369.
[2] Dissipative operators and hyperbolic systems of partial differential equations, *Trans. Amer. Math. Soc.* **90** (1959), 193–254.
[3] On a theorem due to Sz.-Nagy, *Pacific J. Math.* **9** (1959), 169–173.

PISIER, G.
[1] A polynomially bounded operator on Hilbert space which is not similar to a contraction, *J. Amer. Math. Soc.* **10** (1997), 351–369.

PLESSNER, A.
[1] Zur Spektraltheorie maximaler Operatoren, *C. R. (Doklady) Acad. Sci. URSS (N. S.)* **22** (1939), 227–230.
[2] Über Funktionen eines maximalen Operators, *C. R. (Doklady) Acad. Sci. URSS (N. S.)* **23** (1939), 327–330.
[3] Über halbunitäre Operatoren, *C. R. (Doklady) Acad. Sci. URSS (N. S.)* **25** (1939), 710–712.

POLJACKIĬ, V. T.
[1] The reduction to triangular form of quasi-unitary operators, *Dokl. Akad. Nauk SSSR* **113** (1957), 756–759.
[2] The reduction to triangular form of certain non-unitary operators, *Dissertation*, Kiev, 1960.
[3] The reduction to triangular form of operators of class K, *Proc. Odessa Ped. Inst.* **24** (1959), 13–15.

POPESCU, G.
[1] Isometric dilations for infinite sequences of noncommuting operators, *Trans. Amer. Math. Soc.* **316** (1989), 523–536.
[2] Characteristic functions for infinite sequences of noncommuting operators, *J. Oper. Theory* **22** (1989), 51–71.
[3] von Neumann inequality for $(B(\mathscr{H})^n)_1$, *Math. Scand.* **68** (1991), 292–304.
[4] Multi-analytic operators on Fock spaces, *Math. Ann.* **303** (1995), 31–46.
[5] Poisson transforms on some C^*-algebras generated by isometries, *J. Funct. Anal.* **161** (1999), 27–61.
[6] Operator theory on noncommutative varieties, *Indiana Univ. Math. J.* **55** (2006), 389–442.
[7] Unitary invariants in multivariable operator theory, *Mem. Amer. Math. Soc.* **200** (2009), no. 941, vi+91pp.
[8] Operator theory on noncommutative domains, *Mem. Amer. Math. Soc.* textbf205 (2010), no. 963, vi+124pp.

POTAPOV, V. P.
[1] The multiplicative structure of J-contractive matrix functions, *Trudy Moskov. Mat. Obšč.* **4** (1955), 125–236.

PRIVALOV, I. I.
[1] *Randeigenschaften analytischer Funktionen*, VEB Deutcher Verlag, Berlin, 1956.

RÁCZ, A.

[1] Sur les transformations de classe $\mathscr{C}_\rho$ dans l'espace de Hilbert, *Acta Sci. Math. (Szeged)* **28** (1967), 305–309.

[2] Sur un théorème de W. Mlak, *Bull. Acad. Polon. Sci. Sér. Sci. Math. Astronom. Phys.* **17** (1969), 393–396.

RIESZ, F. AND SZ.-NAGY, B.

[*Func. Anal.*] *Functional Analysis*, translation of *Leçons d'Analyse Fonctionelle*, 2nd ed. (Budapest, 1953), Dover, New York, 1990.

[1] Über Kontraktionen des Hilbertschen Raumes, *Acta Sci. Math. (Szeged)* **10** (1943), 202–205.

ROSENBLUM, M. AND ROVNYAK, J.

[1] *Hardy Classes and Operator Theory*, Oxford University Press, New York, 1985.

ROSENTHAL, P.

[1] A note on unicellular operators, *Proc. Amer. Math. Soc.* **19** (1968), 505–506.

ROTA, G.-C.

[1] On models for linear operators, *Comm. Pure Appl. Math.* **13** (1960), 469–472.

ROVNYAK, J.

[1] Some Hilbert spaces of analytic functions, *Dissertation*, Yale, 1963.

SAHNOVIČ, L. A.

[1] On reduction of Volterra operators to the simplest form and on inverse problems, *Izv. Akad. Nauk SSSR. Ser. Mat.* **21** (1957), 235–262.

[2] Reduction to diagonal form of non-selfadjoint operators with continuous spectrum, *Mat. Sb. N.S.* **44 (86)** (1958), 509–548.

[3] The reduction of non-selfadjoint operators to triangular form, *Izv. Vysš. Učebn. Zaved. Matematika* **1 (8)** (1959), 180–186.

[4] A study of the "triangular form" of non-selfadjoint operators, *Izv. Vysš. Učebn. Zaved. Matematika* **4 (11)** (1959), 141–149.

[5] Dissipative operators with an absolutely continuous spectrum, *Dokl. Akad. Nauk SSSR* **167** (1966), 760–763.

[6] Nonunitary operators with absolutely continuous spectrum on the unit cirle, *Dokl. Akad. Nauk SSSR* **181** (1968), 558–561.

[7] Dissipative operators with absolutely continuous spectrum, *Trudy Moskov. Mat. Obšč.* **19** (1968), 211–270.

[8] Operators, similar to unitary operators, with absolutely continuous spectrum, *Funkcional. Anal. i Priložen.* **2** (1968), 51–63.

[9] Dissipative Volterra operators, *Mat. Sb. (N.S.)* **76 (118)** (1968), 323–343.

[10] Nonunitary operators with absolutely continuous spectrum, *Izv. Akad. Nauk SSSR Ser. Mat.* **33** (1969), 52–64.

SARASON, D.

[1] On spectral sets having connected complement, *Acta Sci. Math. (Szeged)* **26** (1965), 289–299.

[2] A remark on the Volterra operator, *J. Math. Anal. Appl.* **12** (1965), 244–246.

[3] Generalized interpolation in H^∞, *Trans. Amer. Math. Soc.* **127** (1967), 179–203.

[4] Invariant subspaces and unstarred operator algebras, *Pacific J. Math.* **17** (1966), 511–517.

[5] Unbounded Toeplitz operators, *Integral Eq. Oper. Theory* **61** (2008), 281–298.

[6] Unbounded operators commuting with restricted backward shifts, *Oper. Matrices* **2** (2008), 583–601.

SCHÄFFER, J. J.

[1] On unitary dilations of contractions, *Proc. Amer. Math. Soc.* **6** (1955), 322.

SCHREIBER, M.

[1] Unitary dilations of operators, *Duke Math. J.* **23** (1956), 579–594.

[2] A functional calculus for general operators in Hilbert space, *Trans. Amer. Math. Soc.* **87** (1958), 108–118.
[3] On the spectrum of a contraction, *Proc. Amer. Math. Soc.* **12** (1961), 709–713.
[4] Absolutely continuous operators, *Duke Math. J.* **29** (1962), 175–190.

SINAĬ, JA. G.
[1] Dynamical systems with countable Lebesgue spectrum. I, *Izv. Akad. Nauk SSSR Ser. Mat.* **25** (1961), 899–924.

ŠMUL′JAN, JU. L.
[1] Operators with degenerate characteristic functions, *Doklady Akad. Nauk SSSR (N.S.)* **93** (1953), 985–988.
[2] Some questions in the theory of operators with a finite non-Hermitian rank, *Mat. Sb. (N.S.)* **57** **(99)** (1962), 105–136.
[3] The optimal factorization of non-negative matrix functions, *Teor. Verojatnost. i Primenen* **9** (1964), 382–386.

ŠTRAUS, A. V.
[1] On a class of regular operator-functions, *Doklady Akad. Nauk SSSR (N.S.)* **70** (1950), 577–580.
[2] Spectral functions of a symmetric operator with finite defect indices, *Kuĭbyshev. Gos. Ped. Inst. Uchen. Zap.* **11** (1951), 17–66.
[3] Characteristic functions of linear operators, *Dokl. Akad. Nauk SSSR* **126** (1959), 514–516.
[4] Characteristic functions of linear operators, *Izv. Akad. Nauk SSSR Ser. Mat.* **24** (1960), 43–74.

SUCIU, I.
[1] Unitary dilations in case of a partially ordered group, *Bull. Acad. Polon. Sci. Sér. Sci. Math. Astronom. Phys.* **15** (1967), 271–275.

ŠVARCMAN, JA. S.
[1] A functional model of a completely continuous dissipative assembly, *Mat. Issled.* **3**:3 (1968), 126–138.

SZEGŐ, G.
[1] Über die Randwerte analytische Funktionen, *Math. Ann.* **84** (1921), 232–244.

SZ.-NAGY, B.
[I] Sur les contractions de l'espace de Hilbert, *Acta Sci. Math. (Szeged)* **15** (1953), 87–92.
[I bis] Transformations de l'espace de Hilbert, fonctions de type positif sur un groupe, *Acta Sci. Math. (Szeged)* **15** (1954), 104–114.
[II] Sur les contractions de l'espace de Hilbert. II, *Acta Sci. Math. (Szeged)* **18** (1957), 1–14.
[P] Extensions of linear transformations in Hilbert space which extend beyond this space (Appendix to F. Riesz and B. Sz.-Nagy, *Functional analysis*, Dover, New York, 1990). Translation of "Prolongements des transformations de l'espace de Hilbert qui sortent de cet espace", Budapest, 1955.
[1] Transformations of Hilbert space, positive definite functions on a semigroup, *Usp. Mat. Nauk* **11:6 (72)** (1956), 173–182.
[2] On Schäffer's construction of unitary dilations, *Ann. Univ. Sci. Budapest. Eötvös Sect. Math.* **3–4** (1960/1961), 343–346.
[3] Spectral sets and normal dilations of operators, *Proc. Internat. Congress Math., 1958*, Cambridge, New York, 1960, 412–422.
[4] Bemerkungen zur vorstehenden Arbeit des Herrn S. Brehmer, *Acta Sci. Math. (Szeged)* **22** (1961), 112–114.
[5] Un calcul fonctionnel pour les opérateurs linéaires de l'espace Hilbertien et certaines de ses applications, *Studia. Math.*, sér. spéc. **I**, Conférence d'analyse fonctionnelle, Varsovie, 4-10. X. 1960 (1963), 119–127.
[6] The "outer functions" and their role in functional calculus, *Proc. Internat. Congr. Mathematicians (Stockholm, 1962)*, Djursholm, 1963, 421–425.

[7] Un calcul fonctionnel pour les contractions. Sur la structure des dilatations unitaires des opérateurs de l'espace de Hilbert, *Seminari 1962/63 Anal. Alg. Geom. e Topol., Vol. 2, Ist. Naz. Alta Mat.*, 525–528.
[8] Isometric flows in Hilbert space, *Proc. Cambridge Philos. Soc.* **60** (1964), 45–49.
[9] Positive definite kernels generated by operator-valued analytic functions, *Acta Sci. Math. (Szeged)* **26** (1965), 191–192.
[10] Positiv-definite, durch Operatoren erzeugte Funktionen, *Wiss. Z. Techn. Univ. Dresden* **15** (1966), 219–222.
[11] Completely continuous operators with uniformly bounded iterates, *Magyar Tud. Akad. Mat. Kutató Int. Közl.* **4** (1959), 89–93.
[12] Products of operators of classes C_ρ, *Rev. Roumaine Math. Pures Appl.* **13** (1968), 897-899.
[13] Hilbertraum-Operatoren der Klasse C_0, *Abstract Spaces and Approximation (Proc. Conf., Oberwolfach, 1968)*, Birkhäuser, Basel, 1969, 72–81.
[14] Sous-espaces invariants d'un opérateur et factorisation de sa fonction caractéristique, *Actes du Congrès Intern. Math. Nice* (1970), 3426–3429.
[15] Diagonalization of matrices over H^∞, *Acta Sci. Math. (Szeged)* **38** (1976), 233-258.

SZ.-NAGY, B. AND FOIAŞ, C.
[III] Sur les contractions de l'espace de Hilbert. III, *Acta Sci. Math. (Szeged)* **19** (1958), 26–46.
[IV] Sur les contractions de l'espace de Hilbert. IV, *Acta Sci. Math. (Szeged)* **21** (1960), 251–259.
[V] Sur les contractions de l'espace de Hilbert. V. Translations bilatérales, *Acta Sci. Math. (Szeged)* **23** (1962), 106–129.
[VI] Sur les contractions de l'espace de Hilbert. VI. Calcul functionnel, *Acta Sci. Math. (Szeged)* **23** (1962), 130–167.
[VII] Sur les contractions de l'espace de Hilbert. VII. Triangulations canoniques. Fonctions minimum, *Acta Sci. Math. (Szeged)* **25** (1964), 12–37.
[VIII] Sur les contractions de l'espace de Hilbert. VIII. Fonctions caractéristiques. Modéles fonctionnels, *Acta Sci. Math. (Szeged)* **25** (1964), 38–71.
[IX] Sur les contractions de l'espace de Hilbert. IX. Factorisation de la fonction caractéristique. Sous-espaces invariants., *Acta Sci. Math. (Szeged)* **25** (1964), 283–316.
[IX*] Corrections et compléments aux Contractions IX, *Acta Sci. Math. (Szeged)* **26** (1965), 193–196.
[X] Sur les contractions de l'espace de Hilbert. X. Contractions similaires à des transformations unitaires, *Acta Sci. Math. (Szeged)* **26** (1965), 79–91.
[XI] Sur les contractions de l'espace de Hilbert. XI. Transformations unicellulaires, *Acta Sci. Math. (Szeged)* **26** (1965), 301–324 and **27** (1966), 265.
[XII] Sur les contractions de l'espace de Hilbert. XII. Fonction intérieures, admettant des facteurs extérieurs, *Acta Sci. Math. (Szeged)* **27** (1966), 27–33.
[1] Une relation parmi les vecteurs propres d'un opérateur de l'espace de Hilbert et de l'opérateur adjoint, *Acta Sci. Math. (Szeged)* **20** (1959), 91–96.
[2] Modèles fonctionnels des contractions de l'espace de Hilbert. La fonction caractéristique, *C. R. Acad. Sci. Paris* **256** (1963), 3236–3238.
[3] Propriétés des fonctions caractéristiques, modèles triangulaires et une classification des contractions de l'espace de Hilbert, *C. R. Acad. Sci. Paris* **256** (1963), 3413–3415.
[4] Une caractérisation des sous-espaces invariants pour une contraction de l'espace de Hilbert, *C. R. Acad. Sci. Paris* **258** (1964), 3426–3429.
[5] Quasi-similitude des opérateurs et sous-espaces invariants, *C. R. Acad. Sci. Paris* **261** (1965), 3938–3940.
[6] On certain classes of power-bounded operators in Hilbert space, *Acta Sci. Math. (Szeged)* **27** (1966), 17–25.
[7] Décomposition spectrale des contractions presque unitaires, *C. R. Acad. Sci. Paris Sér. A-B* **262** (1966), 440–442.

[8] Forme triangulaire d'une contraction et factorisation de la fonction caractéristique, *Acta Sci. Math. (Szeged)* **28** (1967), 201–212.
[9] Echelles continues de sous-espaces invariants, *Acta Sci. Math. (Szeged)* **28** (1967), 213–220.
[10] Similitude des opérateurs de class $\mathscr{C}_\rho$ à des contractions, *C. R. Acad. Sci. Paris Sér. A-B* **264** (1967), 1063–1065.
[11] Commutants de certains opérateurs, *Acta Sci. Math. (Szeged)* **29** (1968), 1–17.
[12] Dilatation des commutants d'opérateurs, *C. R. Acad. Sci. Paris Sér. A-B* **266** (1968), 493–495.
[13] Vecteurs cycliques et quasi-affinités, *Studia Math.* **31** (1968), 35–42.
[14] Opérateurs sans multiplicité, *Acta Sci. Math. (Szeged)* **30** (1969), 1–18.
[15] Modèle de Jordan pour une classe d'opérateurs de l'espace de Hilbert, *Acta Sci. Math. (Szeged)* **31** (1970), 91–115.
[16] Compléments à l'étude des opérateurs de classe C_0, *Acta Sci. Math. (Szeged)* **31** (1970), 281–296.
[17] Compléments à l'étude des opérateurs de classe C_0. II, *Acta Sci. Math. (Szeged)* **33** (1971), 113–116.
[18] Local characterization of operators of class C_0, *J. Funct. Anal.* **8** (1971), 76–81.
[19] Vecteurs cycliques et commutativité des commutants, *Acta Sci. Math. (Szeged)* **32** (1971), 177–183.
[20] The "lifting theorem" for intertwining operators and some new applications, *Indiana Univ. Math. J.* **20** (1971), 901–904.
[21] Echelles continues de sous-espaces invariants. II, *Acta Sci. Math. (Szeged)* **33** (1972), 355–356.
[22] On the structure of intertwining operators, *Acta Sci. Math. (Szeged)* **35** (1973), 225–254.
[23] Regular factorizations of contractions, *Proc. Amer. Math. Soc.* **43** (1974), 91–93.
[24] Jordan model for contractions of class $C_{\cdot 0}$, *Acta Sci. Math. (Szeged)* **36** (1974), 305–322.
[25] An application of dilation theory to hyponormal operators, *Acta Sci. Math. (Szeged)* **37** (1975), 155–159.
[26] Commutants and bicommutants of operators of class C_0, *Acta Sci. Math. (Szeged)* **38** (1976), 311–315.
[27] On contractions similar to isometries and Toeplitz operators, *Ann. Acad. Sci. Fenn. Ser. A I Math.* **2** (1976), 553–564.
[28] Vecteurs cycliques et commutativité des commutants. II, *Acta Sci. Math. (Szeged)* **39** (1977), 169–174.
[29] On injections, intertwining operators of class C_0, *Acta Sci. Math. (Szeged)* **40** (1978), 163–167.
[30] The function model of a contraction and the space L^1/H_0^1, *Acta Sci. Math. (Szeged)* **41** (1979), 403–410.
[31] Contractions without cyclic vectors, *Proc. Amer. Math. Soc.* **87** (1983), 671–674.
[32] Toeplitz type operators and hyponormality, *Oper. Theory Adv. Appl.* **11** (1983), 371–388.

SZŰCS, J.
[1] Diagonalization theorems for matrices over certain domains, *Acta Sci. Math. (Szeged)* **36** (1974), 193–201.

TAKAHASHI, K.
[1] The factorization in the commutant of a unitary operator, *Hokkaido Math. J.* **8** (1979), 253–259.
[2] $C_{1\cdot}$-contractions with Hilbert–Schmidt defect operators, *J. Oper. Theory* **12** (1984), 331–347.
[3] Contractions with the bicommutant property, *Proc. Amer. Math. Soc.* **93** (1985), 91–95.
[4] The reflexivity of contractions with non-reductive $*$-residual parts, *Michigan Math. J.* **34** (1987), 153–159.
[5] On quasisimilarity for analytic Toeplitz operators, *Canad. Math. Bull.* **31** (1988), 111–116.

[6] On contractions without disjoint invariant subspaces, *Proc. Amer. Math. Soc.* **110** (1990), 935–937.
[7] Injection of unilateral shifts into contractions, *Acta Sci. Math. (Szeged)* **57** (1993), 263–276.

TAKAHASHI, K. AND UCHIYAMA, M.
[1] Every C_{00} contraction with Hilbert–Schmidt defect operator is of class C_0, *J. Oper. Theory* **10** (1983), 331–335.

TEODORESCU, R. I.
[1] Sur les décompositions directes des contractions de l'espace de Hilbert, *J. Funct. Anal.* **18** (1975), 414–428.
[2] The direct decompositions of contractions, *Stud. Cerc. Mat.* **29** (1977), 57–84.
[3] Factorisations régulières et sousespaces hyperinvariants, *Acta Sci. Math. (Szeged)* **40** (1978), 389–396.
[4] Sur l'unicité de la décomposition des contractions en somme directe, *J. Funct. Anal.* **31** (1979), 245–254.

THORHAUER, P.
[1] Bemerkungen zu einem Satz über vertauschbare Kontraktion eines Hilbertschen Raumes, *Wiss. Z. Techn. Hochsch. Otto von Guericke Magdeburg* **5** (1961), 109–110.
[2] Schäfferartige Konstruktionen vertauschbarer Dilatationen, *Dissertation*, Magdeburg, 1962.

TREIL, S. R.
[1] Angles between co-invariant subspaces, and the operator corona problem. The Szőkefalvi-Nagy problem, *Dokl. Akad. Nauk SSSR* **302** (1988), 1063–1068.
[2] Geometric methods in spectral theory of vector-valued functions: some recent results, *Oper. Theory Adv. Appl.* **42**, Birkhäuser, Basel, 1989, 209–280.
[3] An operator Corona theorem, *Indiana Univ. Math. J.* **53** (2004), 1763–1780.
[4] Lower bounds in the matrix Corona theorem and the codimension one conjecture, *Geom. Funct. Anal.* **14** (2004), 1118–1133.

TREIL, S. AND VOLBERG, A.
[1] A fixed point approach to Nehari's problem and its applications, *Oper. Theory Adv. Appl.* **71** (1992), 165–186.

TREIL, S. R. AND WICK, B. D.
[1] Analytic projections, corona problem and geometry of holomorphic vector bundles, *J. Amer. Math. Soc.* **22** (2009), 55–76.

UCHIYAMA, M.
[1] Hyperinvariant subspaces of operators of class $C_0(N)$, *Acta Sci. Math. (Szeged)* **39** (1977), 179–184.
[2] Hyperinvariant subspaces for contractions of class C_0, *Hokkaido Math. J.* **6** (1977), 260–272.
[3] Double commutants of C_0 contractions, *Proc. Amer. Math. Soc.* **69** (1978), 283–288.
[4] Quasisimilarity of restricted C_0 contractions, *Acta Sci. Math. (Szeged)* **41** (1979), 429–433.
[5] Contractions with (σ, c) defect operators, *J. Oper. Theory* **12** (1984), 221–233.

VAROPOULOS, N. TH.
[1] On an inequality of von Neumann and an application of the metric theory of tensor products to operator theory, *J. Funct. Anal.* **16** (1974), 83–100.

VASILESCU, F.-H.
[1] An operator-valued Poisson kernel, *J. Funct. Anal.* **110** (1992), 47–72.

VASYUNIN, V. I.
[1] The construction of the B. Szőkefalvi-Nagy and C. Foiaş functional model, *Zap. Nauchn. Sem. Leningrad. Otdel. Mat. Inst. Steklov. (LOMI)* **73** (1977), 16–23, 229.

WIENER, N.
[1] On the factorization of matrices, *Comment. Math. Helv.* **29** (1955), 97–111.

WIENER, N. AND AKUTOWICZ, E. J.
[1] A factorization of positive Hermitian matrices, *J. Math. Mech.* **8** (1959), 111–120.

WIENER, N. AND MASANI, P.
[1] The prediction theory of multivariate stochastic processes. I. The regularity condition, *Acta Math.* **98** (1957), 111–150.
[2] The prediction theory of multivariate stochastic processes. II. The linear predictor, *Acta Math.* **99** (1958), 93–137.

WILLIAMS, L. R.
[1] A quasisimilarity model for algebraic operators, *Acta Sci. Math. (Szeged)* **40** (1978), 185–188.

WOLD, H.
[1] *A study in the analysis of stationary time series*, Stockholm, 1938, 2nd ed. 1954.

WU, P. Y.
[1] Commutants of $C_0(N)$ contractions, *Acta Sci. Math. (Szeged)* **38** (1976), 193–202.
[2] Jordan model for weak contractions, *Acta Sci. Math. (Szeged)* **40** (1978), 189–196.
[3] C_{11} contractions are reflexive, *Proc. Amer. Math. Soc.* **77** (1979), 68-72.
[4] On the reflexivity of $C_0(N)$ contractions, *Proc. Amer. Math. Soc.* **79** (1980), 405–409.
[5] C_{11} contractions are reflexive. II, *Proc. Amer. Math. Soc.* **82** (1981), 226-230.
[6] Unitary dilations and numerical ranges, *J. Oper. Theory* **38** (1997), 25–42.
[7] Polar decompositions of $C_0(N)$ contractions, *Integral Eq. Oper. Theory* **56** (2006), 559–569.

WU, P. Y. AND TAKAHASHI, K.
[1] Dilation to unilateral shifts, *Semigroups of operators: theory and applications*, Birkhäuser, Basel, 2000, 364–367.
[2] Singular unitary dilations, *Integral Eq. Oper. Theory* **33** (1999), 231–247.

YAKUBOVICH, D. V.
[1] A linearly similar Sz.-Nagy–Foias model in a domain, *Algebra i Analiz* **15** (2003), 190–237.
[2] Nagy–Foiaş type functional models of nondissipative operators in parabolic domains, *J. Oper. Theory* **60** (2008), 3–28.

YANG, L. M.
[1] Quasisimilarity of hyponormal and subdecomposable operators, *J. Funct. Anal.* **112** (1993), 204–217.

YANG, R.
[1] Operator theory in the Hardy space over the bidisk. II, *Integral Eq. Oper. Theory* **42** (2002), 99–124.
[2] Operator theory in the Hardy space over the bidisk. III, *J. Funct. Anal.* **186** (2001), 521–545.
[3] On two-variable Jordan blocks, *Acta Sci. Math. (Szeged)* **69** (2003), 739–754.

YOSIDA, K.
[1] Fractional powers of infinitesimal generators and the analyticity of the semi-groups generated by them, *Proc. Japan Acad.* **36** (1960), 86–89.
[2] *Functional Analysis*, Springer, Berlin, 1965.

ZASUHIN, V. N.
[1] On the theory of one dimensional stationary processes, *Dokl. Akad. Nauk SSSR* **33** (1941), 435–437.

ZUCCHI, A.
[1] Operators of class C_0 with spectra in multiply connected regions, *Mem. Amer. Math. Soc.* **127** (1997), no. 607, viii+52 pp.

Notation Index

Author Index

Subject Index